Engineering Optics 2.0

Xiangang Luo

Engineering Optics 2.0

A Revolution in Optical Theories, Materials, Devices and Systems

Xiangang Luo
State Key Laboratory of Optical
Technologies on Nano-Fabrication
and Micro-Engineering
Institute of Optics and Electronics
Chinese Academy of Sciences
Chengdu, Sichuan, China

ISBN 978-981-13-5754-1 ISBN 978-981-13-5755-8 (eBook)
https://doi.org/10.1007/978-981-13-5755-8

Library of Congress Control Number: 2019931512

This Springer imprint is published by the registered company Springer Nature Singapore Pte Ltd.
The registered company address is: 152 Beach Road, #21-01/04 Gateway East, Singapore 189721, Singapore

Foreword I

Across the whole history of modern sciences and technologies, new discoveries and innovations are always the great sources of scientific advancement. This is particularly true for optics and its engineering applications. At the turn of the century, a series of unusual optical phenomena in subwavelength patterned structures have been discovered. For instance, the extraordinary optical transmission through subwavelength hole arrays discovered by Thomas Ebbesen et al. has marked the emergence of a new area called plasmonics. Since 2003, the extraordinary Young's double-slit interference phenomena have revealed the short wavelength and localized phase-shifting mechanism of meta-surface waves, which cannot be explained by traditional geometric and physical optics. Based on these phenomena, a couple of new theories that break the limits of conventional optical laws have been established. Meanwhile, novel optical materials, devices, and optical systems have been developed with performances far beyond their traditional counterparts. Taking the super-resolution lithography as an excellent example, with plasmonic superlenses, hyperlenses, and cavity lenses, one could realize 22 nm resolution at an operating wavelength of 365 nm, which is about one magnitude smaller than that of the state-of-the-art technology for projection lithography ($\sim$230 nm resolution).

If we take a close look at the history of modern optics, the subwavelength optical engineering can be said to be the third generation of optical revolution. Practically speaking, the first revolution of modern optics is the transformation from geometric optics to physical optics, which was marked by the famous double-slit interference experiment conducted by Thomas Young in 1801. The second revolution is the quantum optics, which uses semiclassical and quantum-mechanical physics to investigate phenomena involving light and its interactions with matter at atomic levels. Coincidently, it was Young's double-slit interference of single electrons, which lies at the heart of quantum optics, as stated by Richard Feynman. Now, the subwavelength Young's interference has opened up a complete new door toward the gap between classical and quantum optics. Considering the great success of subwavelength optical engineering in recent decades, it is the right time to say that the third optical revolution, which was termed Engineering Optics 2.0, has started. The foundation of this area is formally marked by two publications entitled

Subwavelength Artificial Structures: Opening a New Era for Engineering Optics and *Engineering Optics 2.0: A Revolution in Optical Materials, Devices, and Systems*.

This book is a timely and comprehensive milestone devoted to Engineering Optics 2.0. The backgrounds, motivations, new phenomena, theories, technologies, and rudimentary applications, such as super-resolution imaging and lithography, flat optics, perfect light absorbers, and radiators, have been discussed in details. It can also serve very well as a versatile handbook or reference for researchers and students in optical engineering.

Singapore

Minghui Hong
PhD., FSEng, FOSA, FSPIE, FIAPLE & FIES
Professor, Department of Electrical
and Computer Engineering

Director, Optical Science and Engineering Centre
National University of Singapore

Chairman, Phaos Technology Pte Ltd

Foreword II

Engineering optics is an old discipline that deals with the basic properties of light and their applications in imaging, display, data storage, remote sensing, telecommunication, and many others. During the past half century, engineering optics has experienced rapid progress owing to the inventions of many revolutionary tools such as lasers and electronic computers. Especially with the significant development in fabrication and computing capacity during the recent several decades, it was found that many classic optical laws, including the diffraction limit, the Snell's law, the Fresnel's equations, could be generalized if the permittivity and permeability vary rapidly across a space smaller than the wavelength. This has led to the formation of three important research directions named photonic crystals, plasmonics, and metamaterials. In 2010, all these three ones were listed in the 23 milestones in the history of photons, along with other famous ones such as Maxwell's equations, special relativity, and lasers. It is believed that the third generation of optical revolution is coming. As a pioneer and an internationally leading authority on this topic, Prof. Xiangang Luo proposed an initiative called Engineering Optics 2.0, as part of the efforts to turn the groundbreaking theoretical advances to practical products. Like Web 2.0 and Artificial Intelligence 2.0, which transform the norm forms of Internet and artificial intelligence, Engineering Optics 2.0 has opened up a new era for industrial applications of optics.

Historically, Engineering Optics 2.0 was triggered by the observation of many extraordinary physical effects such as the photonic bandgap, negative refraction, extraordinary optical transmission (EOT), and extraordinary Young's interference (EYI). Taking the EYI as an example, the two extraordinary interference effects discovered in 2003 and 2007 (see more details in a recent review published in *Advances in Optics and Photonics* entitled *Subwavelength interference of light on structured surfaces*) are the key to realize sub-diffraction-limited imaging and flat optical elements and systems based on localized phase shift. They have played groundbreaking roles in the next-generation engineering optics, just like their counterparts in classic wave optics and quantum physics. It is well known that the Young's interference of photons directly demonstrated the wave nature of light and closed the debate on the character of light. Meanwhile, the double-slit interference of single electrons is the key to understand quantum physics. As a collective combination of photons and electrons, the surface plasmon polaritons (SPPs) lie at the heart of subwavelength optics, which will also find far-reaching implications.

This book is a systematic investigation of the history, theories, and technologies involved in Engineering Optics 2.0. Although the research area is still expanding, this positive contribution will definitely provide the readers extraordinarily enlarged views and clear guidelines.

Melbourne, Australia

Min Gu
RMIT University
Fellow of the Australian Academy of Science and the Australian Academy of Technological Sciences and Engineering
Foreign Fellow of the Chinese Academy of Engineering
AIP, OSA, SPIE, InstP, IEEE Fellow

Preface

During the past few decades, benefited from the rapid development of subwavelength structured materials, micro-/nanofabrication and characterization techniques, electronic computer and advanced algorithms, modern engineering optics has entered a new phase, termed Engineering Optics 2.0 (EO 2.0), where fundamental limitations of classic optical laws involving reflection, refraction, diffraction, absorption, and radiation could be defeated. The very first time when I was attracted by the subwavelength structures was in 2003, when I was doing Young's double-slit interference experiment with metallic nanoslits. By exciting the strongly localized plasmonic modes, I found that many commonsense optical laws are not valid in the subwavelength scale. These extraordinary phenomena opened a wealth of new research directions in the subsequent investigations. In 2017, fourteen years after the extraordinary Young's interference was discovered, I thought it might be a proper time to write a book on this topic, since the optical components, functionalities, and architectures based on subwavelength structures have been experiencing considerable revolutions. One of the most significant characteristics of EO 2.0 is its flat nature. The gradient metasurface lenses could generalize the classic laws of reflection and refraction on a flat surface, providing a promising alternative to traditional curved lenses and mirrors.

While this book by no means encompasses all the work or publications related to this topic, I hope that the theme covered here provides a fairly comprehensive overview of the main issues of EO 2.0 to offer sufficient references to senior-level undergraduates or graduate students. First of all, a brief history of traditional engineering optics (i.e., EO 1.0) based on classic optical theories and technologies is reviewed. The great challenges of EO 1.0 and the opportunities and approaches enabled by subwavelength artificial structures were then discussed in Chap. 1. The theoretical backgrounds and the fundamental principles of EO 2.0, including generalized diffraction theory, generalized laws of refraction and reflection, generalized absorption and radiation theory, are presented in Chap. 2. The material basis, modeling, and simulation methodology, as well as fabrication methods, are, respectively, discussed in Chaps. 3, 4, and 5. As important branches of EO 2.0, the sub-diffraction-limited microscopy, lithography, and telescopy are introduced in

Chaps. 6, 7, and 8. Various planar metalenses and meta-mirrors based on generalized laws of reflection and refraction and their applications in imaging, optical field generation and manipulation, structural color and meta-holography are depicted in Chaps. 9, 10, and 11. The polarization manipulation, detection, and imaging based on anisotropic and chiral artificial materials are presented in Chap. 12. Perfect light absorption and its applications in solar cells and biological sensing are discussed in Chap. 13. Thermal radiation engineering and its applications in thermophotovoltaic, daytime radiative cooling, thermal cloak, and camouflage are depicted in Chap. 14. Also, novel light-emitting diode, nanolasers, and optical phased arrays are presented in this chapter. In each chapter, the corresponding limits and shortages in EO 1.0 are firstly discussed and then more attentions are paid on the recent development in EO 2.0.

This book would never have been possible without a number of individuals who provided assistance and support. Special thanks to Prof. Xicheng Lu, Prof. Bingkun Zhou, Prof. Zuyan Xu, Prof. Jishen Li, Prof. Hequan Wu, Prof. Yueguang Lv, and Prof. Guozhen Yang for their help. I would also like to thank my students who assisted in the revision of many chapters.

Chengdu, China
2018

Xiangang Luo

Contents

Chapter 1
Introduction to Engineering Optics 2.0

Abstract In recent years, modern engineering optics has entered a new phase termed Engineering Optics 2.0 (EO 2.0) and broken the fundamental limitations of classic optical laws with respect to many aspects of optics. This change is enabled by the rapid development of micro-/nanofabrication and characterization techniques, as well as the advancement of electronic computers and numerical simulation algorithms. In this chapter, we give a detailed introduction of the background and progress of this new area.

Keywords Engineering optics · Nanofabrication · Flat optics · Diffraction limit

1.1 Definition of Engineering Optics 2.0

Over the last century, the science of light has experienced a transition from pure scientific research to an engineering-oriented subject. This subject, known as engineering optics, is founded on the classic optical theories of light established during the last four hundred years and seeks practical applications by controlling and harnessing the information and energy carried by light and other kinds of electromagnetic radiations [1, 2]. Many optical instruments such as telescopes, microscopes, interferometers, lithography machines, and fiber communication systems have emerged and greatly improved our capacity to understand and change the world. For example, optical telescopes have greatly spurred the revolution of astrophysics and cosmography, by demonstrating the fact that neither the earth nor the sun is the center of our universe. On the other hand, microscopes have brought us to a new world much smaller than that we can see with our naked eyes, and this small world is the base of the current life sciences.

In spite of these great successes, however, engineering optics has faced grand challenges in fulfilling its promise to the ever-increasing scientific pursuits [3]. For instance, according to the diffraction limit theory proposed by Ernest Abbe and Lord Rayleigh in 1873 and 1879, a higher resolution of microscope and telescope requires either the reduction in operational wavelength or the increase in aperture size. For many specific objects such as molecules and stardust, the smallest operational wave-

X. Luo, *Engineering Optics 2.0*, https://doi.org/10.1007/978-981-13-5755-8_1

length is limited by the absorption/scattering process; thus a larger aperture becomes the only possible choice [4]. However, with traditional technologies, the increase in telescope aperture requires much higher cost and longer manufacturing time for both ground-based and space-based telescopes. In microscopes and lithographic systems, the highest numerical aperture is ultimately limited by the immersion materials and hardly to be further improved beyond 1.4.

In recent decades, along with the rapid development of micro-/nanofabrication and numerical/experimental characterization techniques, modern engineering optics has entered a new phase termed **Engineering Optics 2.0** (**EO 2.0**) [2, 5], which breaks the limitation of classic optical laws associating with reflection, refraction, diffraction, absorption and radiation from the subwavelength scale. Table 1.1 shows a comparison of Engineering Optics 2.0 and 1.0 (EO 1.0), i.e., traditional Engineering optics based on classic optical theories and technologies [2].

It can be clearly seen that EO 2.0 gives promising alternatives to classical optical materials, elements, and systems in many domains of engineering optics. Multi-functional materials, planar or conformal optical elements, and light-weight yet high-performance optical systems are predicted, which are either impossible or have great challenges for EO 1.0. Before talking about EO 2.0, let us first review the history and technological challenges of EO 1.0.

Table 1.1 Comparison of Engineering Optics 1.0 and 2.0 [2]

Properties	Status of EO 1.0	Promise of EO 2.0
Refraction and reflection	• Propagation direction is determined by refractive index and surface shape • Bulky and heavy lenses and mirrors	• Tunable propagation direction without changing the geometric shape • Thin and lightweight lenses, mirrors, and other functional optical elements
Diffraction	• Resolution is limited by wavelength and aperture • High cost and long time are required for large optical systems	• Increased resolution beyond the diffraction limit • Reduced time and cost bearable for common users
Radiation	• Electromagnetic radiation is limited by the size of antenna • Thermal radiation is limited by the temperature	• Enhanced radiation for small antenna • Boosted thermal radiation at given temperature
Absorption	• Absorption takes place inside volume materials • Larger thickness is required for wider bandwidth	• Absorption takes place along the surface • Broadband absorption realized with vanishing thickness

1.2 Basics of Engineering Optics 1.0

1.2.1 Historical and Theoretical Remarks

Engineering optics deals with the engineering aspects of optics that seeks to apply optical theories to practical applications in areas ranging from imaging, micro-/nanofabrication, remote sensing, and communications to illumination, display, and energy harvest [1]. From a historical view, there are many phases in the development of optical theories and technologies. In the following, we would like to give a concise description of these nodes [2].

(1) Prior to 1600s: The geometric optics

Geometric optics is the first systematic theory of light developed by Isaac Newton et al., which treats light as particles propagating along straight lines. As early as in the ancient Greek, some basic phenomena of geometric optics have already been described, and it was well known that the reflection angle should be equal to the incidence angle. However, the research of light develops very slowly until the emergence of high transparency glasses and the explicit equation for refraction (Snell's law) in 1621. Snell's law and Fermat's principle given in 1657 (Fig. 1.1) form the basis for the design of optical lenses, which are the building blocks of optical instruments such as eyeglasses, telescopes, and microscopes.

Comparatively speaking, it is striking to study the development of optics in ancient China during this period. Although it was widely accepted that China has no real optics in those time, ancient Chinese people have indeed observed many optical phenomena and constructed numerous functional optical devices, such as the bronze mirrors and lenses. One special kind of mirror is known as Chinese magic mirror, which could reflect light to form a pattern the same as that carved in the back of the mirror [15]. As stressed by Joseph Needham, the Mohist Canons, written in the Spring and the Autumn-Warring States period (800 BC–200 BC) by Mozi and his followers, have described eight terms on optics, including the small hole imaging experiment and the formation of shadow [16]. Besides optics, Mozi may also be a precursor to Newton's first law of motion by stating that the cessation of motion is due to the opposing force … if there is no opposing force … the motion will never stop.

(2) 1600s–1860s: The emergence of wave optics

Although geometric optics could explain many natural optical phenomena, it is not satisfactory enough since it cannot interpret many others, such as why light does not collide with each other during propagation. As an alternative, Robert Hooke, Christian Huygens, and some other scientists proposed a hypothesis that light may be one kind of wave like those propagating along the water surface. However, because the authority of Newton in the eighteenth century, the undulatory theory of light was not widely accepted until the groundbreaking experiment in 1801 made by Thomas Young [17]. By directly observing the interference pattern of light passing two slits,

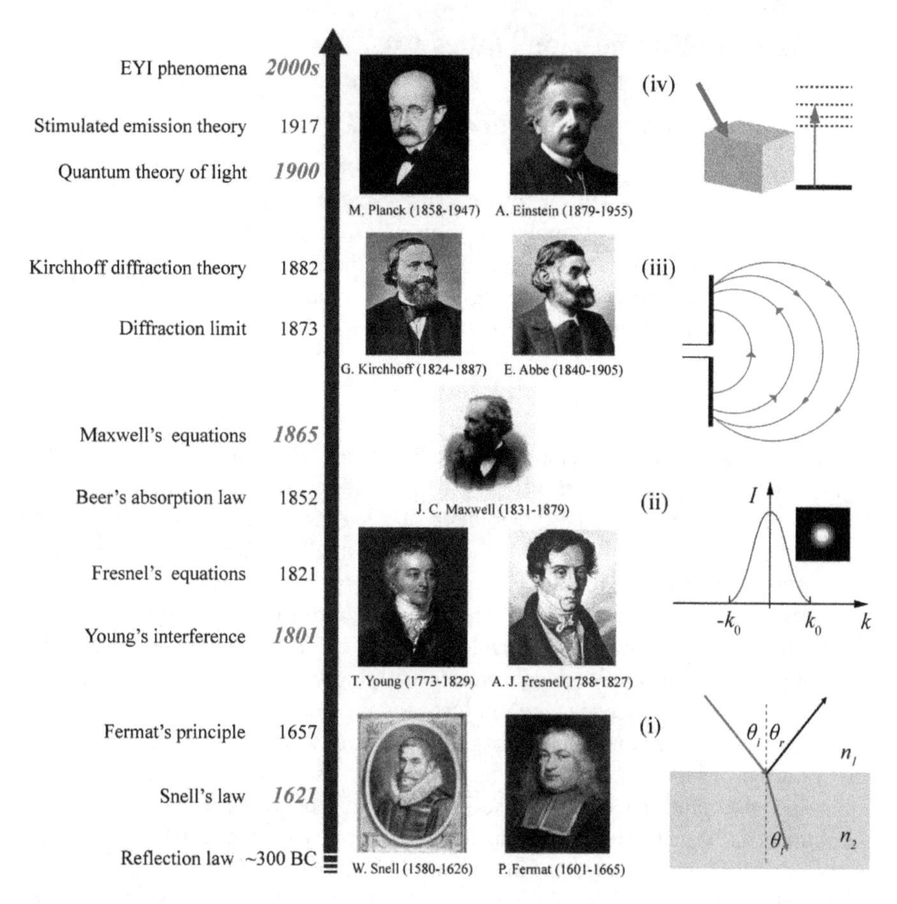

Fig. 1.1 Representative theories and discoveries in classic engineering optics. (i–iv) Schematics of the laws for reflection/refraction, diffraction, radiation and absorption, respectively. EYI means extraordinary Young's interference, which will be explained in detail in the following section. Reproduced with permission from [2]. Copyright 2018, American Chemical Society. All portraits are reproduced from Wikipedia [6–14]

this experiment sets a definitive conclusion about the debate on the corpuscular and wave theories of light. As a result, this simple experiment was listed as one of the most beautiful experiments in physics [18]. Subsequently, the diffraction theory of light was written in a more rigorous mathematic basis by Fresnel and Kirchhoff et al. and this theory is thought to be accurate enough for most traditional instrumental optics [19].

Compared with geometric optics, one big change of wave optics is the so-called diffraction limit [20, 21], which dominated the roadmap of traditional optical engineering, i.e., the increase in optical resolution must be accompanied by the increase in aperture size and/or reduction in wavelength. To some extent, this is the starting point of EO 2.0 [2].

Note that Young's interference experiment of light is a milestone in the transition from geometric optics to wave optics. More interestingly, the quantum theory shows that electrons are some kinds of matter waves, and even single electron could form interference patterns after passing from two slits. In this case, it may be concluded that a single electron can simultaneously pass two slits and interfere with itself. This experiment was thought to be the heart of quantum mechanics and listed as the most beautiful physical experiment across the whole history [18, 22].

(3) 1860s: The electromagnetic optics

In seemingly independent research into the electric and magnetic physics, James Maxwell unified the fundamental electromagnetic laws to deduce a surprising wave equation and predicted a wave with velocity close to that measured for light in the 1860s. This discovery opened a new epoch for the optical researches, and light was finally treated as electromagnetic fluctuation [19]. In Maxwell's theory, the reflection and refraction are described by matching the boundary conditions. The generation of electromagnetic waves can be interpreted using the oscillating of electrons, while the absorption is related to the damping of electrons and lattice resonances.

As shown in Fig. 1.2, the current optical researches are based on Maxwell's equations and their approximations [23]. First of all, with accurate material parameters, Maxwell's equations are the most accurate description of light–matter interaction. However, the rigorous solution of Maxwell's equations in complex materials and geometers are very difficult and time-consuming. In many practical conditions, proper approximations must be made to ensure the fast calculation. For linear optical system, the time dependence of light fields can be neglected. When the vectorial effects are further ignored, one can deal the light waves with a scalar wave equation.

It is well known that the geometric optics is an approximation of wave optics under the case when the wavelength is approaching zero. When analytical approximation is adopted to describe the geometry, higher order aberration can also be neglected. At last, with linear approximation, the geometric optics would reduce to the paraxial optics.

1.2.2 Material Basis

The invention of novel functional materials always plays critical roles in the developing history of human civilization throughout the "stone age," "iron age," and the so-called "silicon age." This is particularly true for the control of light, i.e., electromagnetic wave induced by electron oscillation or quantum transition [3]. In principle, optical materials can be divided into reflective and transmissive media. The transmissive materials include glass, crystals, plastics, glues and cements, etc. First of all, glass is the most important class of material for optical applications. The appearance of glass greatly promoted the development of modern optics. Based on optical glass, the telescopes and microscopes were invented and extremely enlarged our observing boundary. These scientific implements have made a giant contribution to the develop-

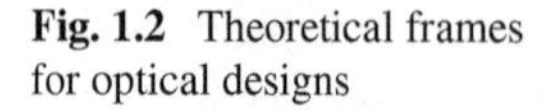

Fig. 1.2 Theoretical frames for optical designs

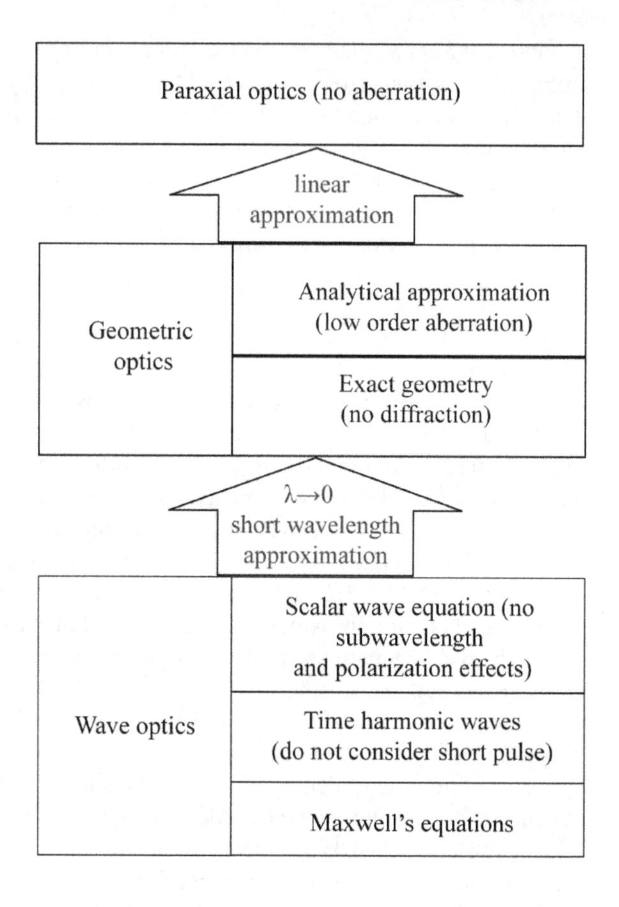

ment of astronomy and biology, and changed human's perception of the universe and microorganism. In order to satisfy the rigorous requirement in various environments, different types of glass were explored. The known suppliers of optical glasses offer about 200 different types. Recently, it has been attempted to avoid toxic adulterants like lead and cadmium as far as possible, limiting the number to about 100 types of glass. The most important commercial suppliers of optical glasses are Schott, Hoya, Ohara, Sovirel, Sumita, Hikari, and Corning [23].

The glass diagram is a good description of different glass types, where n is the refractive index and v is the Abbe number (a larger v indicates a smaller dispersion). As shown in Fig. 1.3, in the n–v plot of Schott glass, a higher refractive index, often leads to larger dispersion.

As the science and technology advance, the systems of optical materials get richer. Many kinds of crystals, such as MgF_2, GaF_2, etc. have been used to fabricate optical devices such as prism, lens, waveplate, and so on. Table 1.2 presents the material properties of some representative optical crystals.

Generally speaking, compared with optical glasses, optical plastics have the advantages of lower manufacturing cost, lower density, and better ultraviolet and

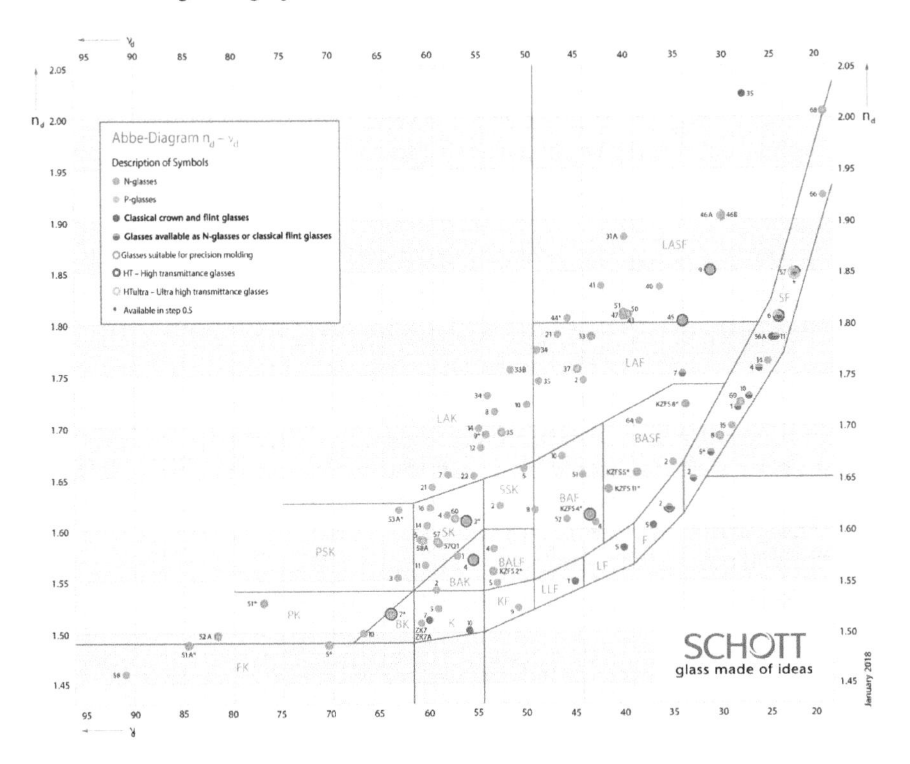

Fig. 1.3 Schott glass map 2018

Table 1.2 Properties of typical optical crystal

Crystal name	Composition formula	Transmitted wavelength (μm)	Surface reflectance (%)	Density (g/cm^3)	Crystal classification
Calcium Fluoride	CaF_2	0.13–8	3.1 (1 μm)	3.18	Isotropic
Sapphire	Al_2O_3	0.23–5	7.6 (0.9 μm)	3.99	Uniaxial
Zinc Selenide	ZnSe	0.6–20	17 (3 μm)	5.27	Isotropic
Silicon	Si	1.2–8	30 (3 μm)	2.33	Isotropic
Germanium	Ge	2–20	36 (5 μm)	5.33	Isotropic
Barium Fluoride	BaF_2	0.18–12	3.6 (1 μm)	4.89	Isotropic
Lithium Fluoride	LiF	0.12–8	2.7 (1 μm)	2.64	Isotropic
Magnesium Fluoride	MgF_2	0.12–7	2.5 (0.9 μm)	3.17	Uniaxial

infrared transmittance. The refractive index of optical plastics usually ranges from 1.46 to 1.64, and the density is generally half or even smaller than half the density of glass (2–3 g/cm^3). However, optical plastics tend to have higher dispersion than inorganic glasses with the same refractive index. Furthermore, the wear resistance and heat resistance of optical plastics are inferior to that of glass, which limits their applications in complex environments.

1.2.3 Design Methods

The optical design process is schematically shown in Fig. 1.4 [23], from which we can see that simple optical simulation is based on the ray tracing method. Since it only considers the propagation and bend of light through an object with feature size far larger than the wavelength, this technique is particularly useful in describing geometrical aspects of imaging, including longitudinal and transverse aberrations. Nevertheless, for an object with feature size comparable with the wavelength, the wave aberration should also be considered in the optical design. Therefore, more advanced modulation transfer function (MTF) simulation, field curvature, and distortion should be evaluated.

To begin with, the type of optical system should be determined, e.g. telescope, microscope, or others. Then, according to the requirement on the performance, the systematic parameters could be selected reasonably, including foci, aperture, field of view, and so on. At the same time, other parameters could be calculated from them and a feasible initial structure type may be chosen carefully from the existing designs. Also, the relevant mechanical and electric systems should be taken into consideration. After all, there is no possibility for the entire system to function well without them. Subsequently, the ray tracing method is used to calculate the trend of chief and marginal rays for an ideal optical system. After that, reasonable material and curvature of lenses are selected delicately.

Now it is the turn of optimization, which plays the crucial role in optical design process. With the help of computer software, a series of data of optical system is obtained based on the ray tracing method, including optical path length and longitudinal and transverse aberrations. These geometrical aberrations could be presented in a geometrical spot diagram, which shows the resolution to some extent. And the geometrical optical transfer function (OTF) can be calculated by taking a Fourier transform on the geometrical spot diagram. The geometrical OTF illustrates the performance of system perfectly in most cases, while it is somewhat inaccurate in optical system approximating diffraction limitation. Fortunately, there is another mean to solve this problem. The optical system aberration could also be expressed in the form of wave aberration, which is a more reliable way and could be regarded as an additional phase function for the pupil function. Through wave aberration, some other results can be calculated, including the rms value of aberration, point spread function, Strehl number, and OTF. These results provide more accurate numerical evaluation of the performance of system and ensure the designer get a good design

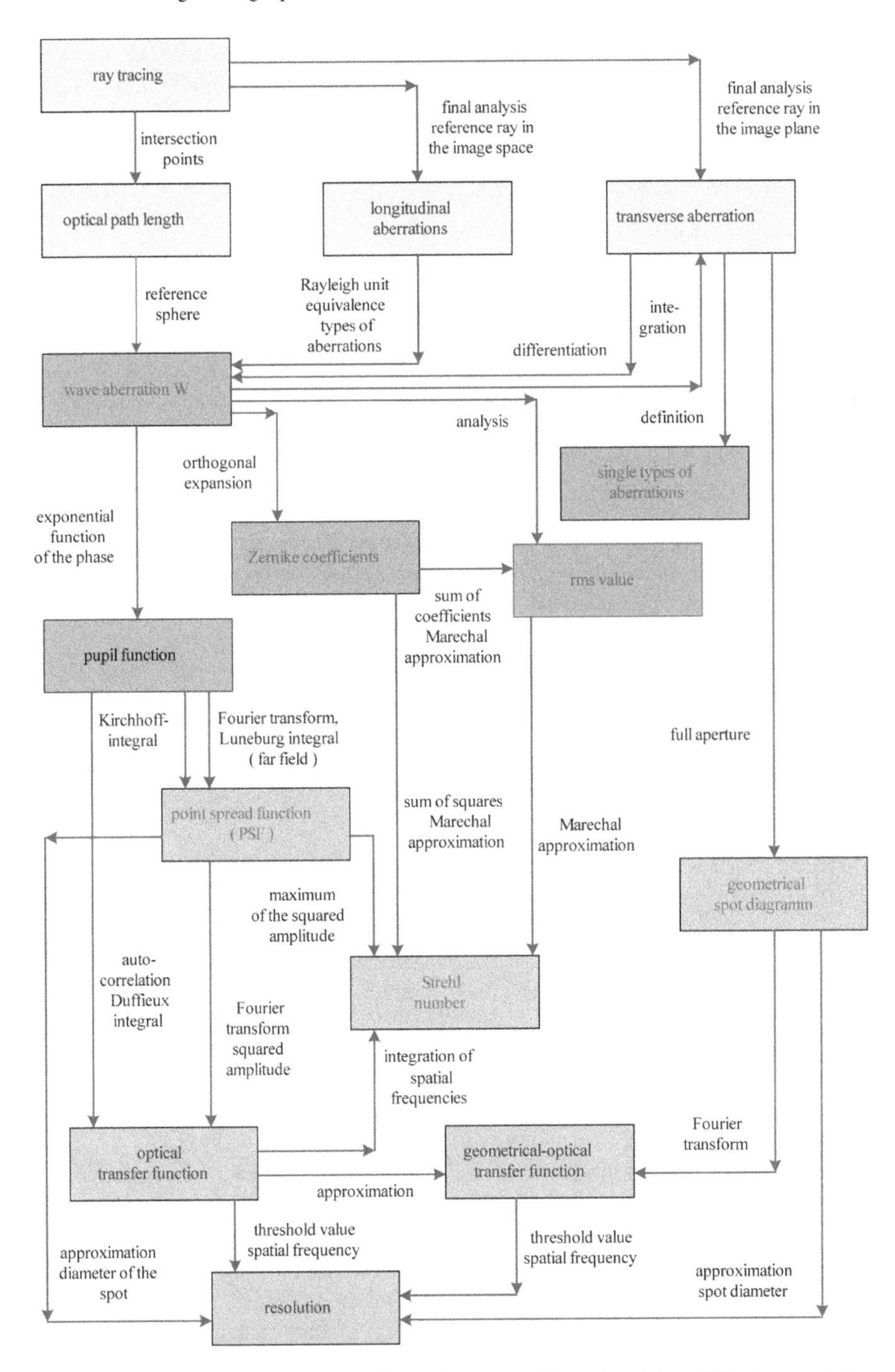

Fig. 1.4 Schematic model description of an optical system. Reproduced from [23] with permission, Copyright 2005, WILEY-VCH Verlag GmbH & Co. KGaA, Weinheim

result. Then a toleration analysis is indispensable, since it is inevitable that there are some sorts of manufacture and assembly errors. An excellent optical design result should also have a reasonable tolerance.

1.2.4 Manufacturing Technologies

Over the past decades, optical fabrication has experienced dramatically change in many aspects. Especially, many novel fabrication and measuring technologies such as computerized numerical control (CNC), stressed mirror polishing (SMP), and magnetorheological finishing (MRF) have been proposed. Without loss of generality, the milestones in optical fabrication are adapted from [24] in the following. For the details of the fabrication techniques, we recommend the readers to read other references [25, 26].

1226: Broad sheet glass was first produced in Sussex, England.
1500s: The method of making mirrors out of plate glass was invented by glassmakers in Italy. By covering the back of the glass with mercury, near-perfect and undistorted reflection was realized.
1810: The first known spherometer was invented by French optician Robert-Aglaé Cauchoix. They were widely used by opticians and astronomers to help grind lenses and curved mirrors.
1888: Machine-rolled glass was first developed, enabling patterns to be introduced.
1896: Frank Randall formed a partnership with Francis Stickney to begin manufacturing dial indicators for general industry.
1904: Ernst Abbe established the measuring instrument department at the Zeiss Works and supplied dial indicators, which have become an important tool determining the runout of an outer diameter as well as assisting with centering parts on a fixture.
1957: A computer numerical control system was developed by collaboration between the Massachusetts Institute of Technology and the Air Force Materiel Command at the Wright-Patterson Air Force Base and the Aerospace Industries Association. The invention paved the way for automated tools such as grinding, pre-polishing, and centering of optics, enabling cost-effective production for manufacturers.
1960s: Addition of coatings to surfaces and the development of the first optical test equipment—swept-tuned instruments.
1970s: Interferometers or test plates are the essential tools for determining the surface accuracy of an optic. The development of digital phase-shifting interferometry in around 1970 at Bell Labs was a crucial innovation. Without this technology, the high-precision optical surfaces used in semiconductor lithography would not be possible.
1998: The first commercially available deterministic polishing machine based on magneto-rheological finishing was introduced. This machine and technology enabled the widespread use of aspheres with highly complex geometries for high-performance optical systems.

2000s: Stressed mirror polishing has been used to apply precise forces to mirrors during their fabrication. This is very important for the latest ground-based astronomy programs, including the Thirty Meter Telescope and the European Extremely Large Telescope.
2012: The first of seven aspheric mirror segments (diameter: 8.4 m) for the 24.5-m Giant Magellan Telescope primary mirror was completed. It is the largest off-axis aspheric mirror in the world.

1.3 Great Challenges of EO 1.0

1.3.1 Diffraction Optics

The diffraction limit is one of the most notorious barriers for optical engineers since its discovery by Ernst Abbe and Lord Rayleigh in 1870s [19, 27]. Recently, it was realized that the diffraction limit is not restricted to the imaging resolution. As shown in Fig. 1.5, many other physical quantities are actually induced by the diffraction limit.

Fig. 1.5 Various limitations induced by the diffraction limit

Resolution limit	$\delta > \frac{0.61\lambda}{NA}$ $\delta\theta > \frac{1.22\lambda}{D}$
Depth of focus limit	$\Delta f \approx \frac{\lambda}{NA^2}$
Field of view limit	$\alpha \leq \frac{4\lambda f^2}{R^3}$
Number of supported OAM modes	$l \leq R\beta_0$
Diffractive intensity limit	$I_{max} \leq \left(\frac{\pi D^2}{4\lambda f}\right)^2 I_i$
Diffractive bandwidth limit	$\frac{\Delta\lambda}{\lambda} \leq \frac{\lambda}{f \cdot NA^2}$
Absorption bandwidth limit	$d > \frac{(\lambda_{max} - \lambda_{min})\Gamma_0}{172}$

δ: spatial resolution
δθ: angular resolution
NA: numerical apeture
Δf: depth of focus
R: radius of a lens or waveguide
β_0: propagation constant
f: focal length
D: diameter of apeture
α: field of view
λ: operation wavelength
d: device thickness
λmax: maximum operation wavelength
λmax: minimum operation wavelength
Γ_0: reflectance in dB
Δλ: operation bandwidth

In microscopy, the diffraction limit sets a minimal resolution about half the wavelength of light [5, 28]. For visible light with a shortest wavelength of ~400 nm, this indicates that all microscopies, no matter how to improve the lens system's quality, have a highest resolution of about 200 nm. Although a resolution reduction with a factor of $1/n$ can be obtained by increasing the refractive index (n) of the surrounding materials, the available materials with ultra-high refractive index can hardly be found. Other methods to increase the resolution include using light with a shorter wavelength (such as X-ray). However, the usage of shorter wavelength is prohibited by the fact that such light (higher energy per photon) tends to damage the samples. Similar to the microscopy, the angular resolution of telescope is also limited by the diffraction. Based on the diffraction formula given by Airy, Lord Rayleigh showed that the telescope resolution is proportional to λ/D, where λ is the light wavelength and D is the diameter of the aperture. Once again, the usage of shorter wavelength may not be favored by astronomical and other observations since some particles would absorb strongly at these wavelengths, as illustrated by a series of infrared views of the Messier 81 galaxy [4].

1.3.2 *Refractive and Reflective Optics*

As discussed above, it seems that the most feasible way to increase the resolution of telescope is to increase the aperture constantly [5]. However, growth of refractive telescope's aperture is extremely hard when the diameter becomes larger than about 1 m. The deeper physics behind is associated with the laws of reflection and refraction, which are two cornerstones of geometric optics. According to the Snell's law of refraction, a refractive lens should be constructed using curved surfaces to bend light correctly (the word lens comes from the Latin name of the lentil, which has a similar curved shape). Although large reflective telescopes have much smaller weight than lenses at the same aperture, the precise fabrication and measurement are still very challenging.

Besides the long fabrication time, one important issue for large-aperture optical system is the cost. In general, the price of a telescope is proportional to the diameter of the main telescope mirror in the form of 2.6th power [29]. If the four eight-meter very large telescope (VLT) telescopes that cost about $100 million each, and a 20-m telescope would need about $1 billion, and a 100-m telescope would be $70 billion. Although astronomers could build multiple copies of a smaller telescope to reach a desired equivalent size and reduce the cost, the equivalence in size does not mean equivalence in capability. Used as ordinary telescopes, the array would have increased sensitivity but unchanged resolving power compared with a single telescope. When used as an interferometer, the array would offer higher resolution but lower sensitivity.

Owing to the complexity of optical telescopes, a simple cost model may not apply to all the cases. As illustrated in Fig. 1.6, the NASA Advanced Mission Cost Model estimates optical telescope cost based on mission mass, difficulty level, and quantity

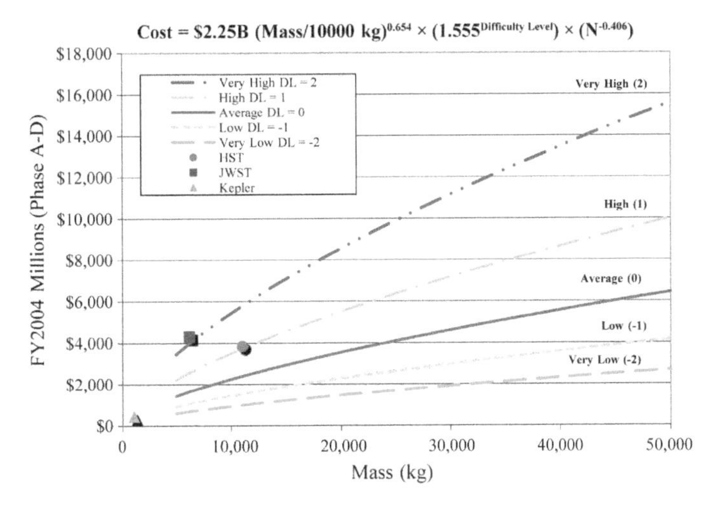

Fig. 1.6 NASA Advanced Mission Cost Model. While the parametric equation might emphasize mass as the dominant mission cost driver, difficulty level might be a larger cost driver than mass. The model reports cost in 2004 US dollars. Plotted on the model are phase A to D costs in FY09 dollars for HST, JWST, and Kepler. HST data come from the REDSTAR database. JWST cost is from NASA FY10 budget, and Kepler cost is from NASA FY09 budget. Reproduced with permission from [30]. Copyright 2010, SPIE

[30]. For different difficulty levels, the increasing rate also varies. According to this model, the cost of a space telescope is not solely dependent on the aperture size. If we could increase the aperture without increasing the weight, the cost may be greatly reduced in future.

For mass production of the optical components, the above price law can be broken, so that the cost per component decreases dramatically. Instead of the usual hyperboloidal primary mirror, which would require that each mirror segment is custom-made according to its position, a 100-m telescope could have a spherical mirror, whose segments are all identical in shape. An assembly line may create all the 3048 segments at the rate of one every two days. The trade-off is that a spherical shape introduces a distortion into the light. To compensate, the telescope would have to be equipped with a device known as a corrector, similar to the one that fixed the vision of Hubble [29].

Besides larger aperture telescopes, one of the main challenges of traditional refractive and reflective optical systems is the construction of high-precision optical system at the deep and extreme ultraviolet band. Figure 1.7 shows the photograph and schematic of the projection optical system for the ASML lithography system operating at 193 nm [31]. The height of the lens is more than 1 m and comprised of more than 20 lenses and mirrors. The obtained half-pitch resolution is about 36.5 nm, which is a bit larger than the theoretical limit calculated by $k_1\lambda/\mathrm{NA}$, where $k_1 = 0.25$, $\mathrm{NA} = 1.35$.

Fig. 1.7 Lens for DUV photolithography with the highest NA: Starlith 1900 from Carl Zeiss. The optical design and ray path are schematics and given only as an illustration. The inset depicts a resist structure of 36.5-nm half-pitch, obtained with the lens at full scanning speed. Reproduced from [31] with permission. Copyright 2007, Nature Publishing Group

1.3.3 Optical Absorption

Absorption and radiation of light are two other cornerstones of optical engineering. However, traditional absorbers and radiators are suffering from some seemingly insurmountable barriers [3]. On the one hand, Max Planck's blackbody radiation formula sets a maximal value for the thermal emittance (one body cannot radiate more energy than the blackbody at the same temperature). On the other hand, Planck also noted that a practical blackbody cannot be constructed with infinite small thickness, and a minimal thickness is required for total absorption [32]. In the early twentieth century, Planck did not give a rigorous mathematical description of the relation between the thickness and absorption efficiency. Along with the rapid development of microwave theory after the World War II, many efforts have been devoted to investigate this relationship, which was finally addressed by Rozanov in 2000 [33]. For non-magnetic absorbers, the thickness could be written as:

$$h > \frac{(\lambda_{\max} - \lambda_{\min})\Gamma}{172} = \frac{\Delta\lambda\Gamma}{172}, \tag{1.3.1}$$

Table 1.3 Typical performances of flexible foam sheet broadband microwave absorbers from ECCOSORB® AN (Emerson & Cuming Microwave Products)

	Reflectivity range (>17 dB)	Actual thickness (cm)	Minimal thickness (cm)	Weight (kg/piece)
AN-73	>7.5 GHz	1.0	0.39	0.50
AN-74	>3.5 GHz	1.9	0.85	0.70
AN-75	>2.4 GHz	2.9	1.24	0.80
AN-77	>1.2 GHz	5.7	2.47	1.50
AN-79	>600 MHz	11.4	4.94	2.95

The standard size is 61 cm × 61 cm

where Γ is the reflectance in decibel (dB). To obtain a 20 dB reflectance, the minimal thickness is larger than 0.116 $\Delta\lambda$. Considering the frequencies between 0.3 and 30 GHz, the required thickness should be at least 116 mm. Table 1.3 shows the typical performances of flexible foam sheet broadband absorbers, from which we can see the actual thicknesses of these microwave absorbers are about two times the theoretical minimal thicknesses. Therefore, how to further approach the minimal thickness limit of the absorbers is a great challenge [34].

The thickness requirements of absorbers are also embodied in the photovoltaic devices. Thin-film solar cells with low materials and processing cost provide a viable pathway toward large-scale implementation of photovoltaic technology. However, the reduced thickness causes an important challenge to maintain the high absorption efficiency. As indicated in Fig. 1.8a, a large fraction of the solar spectrum, in particular in the intense 600–1100 nm spectral range, is poorly absorbed in a 2-μm-thick crystalline Si film [35]. This is because the charge carriers generated far away from the *p*–*n* junction are not effectively collected, owing to bulk recombination (Fig. 1.8b). Obviously, solar cell design and materials synthesis considerations are strongly restricted by these opposing requirements for optical absorption thickness and carrier collection length. How to simultaneously increase the absorption yet decrease the carrier combination is a great challenge for thin-film solar cells.

1.3.4 Polarization Optics

Unlike the longitudinal waves such as acoustic waves in air, electromagnetic fields have strong polarization effects, which mean that the alignment of electric and magnetic fields perpendicular to the propagation direction may greatly change the propagation properties [36, 37]. In general, polarized light can be classified into linearly polarized, circularly polarized, elliptically polarized, and unpolarized. Owing to the orthogonality, both the linear and circular polarizations can be utilized as orthogonal states for any specific polarization state.

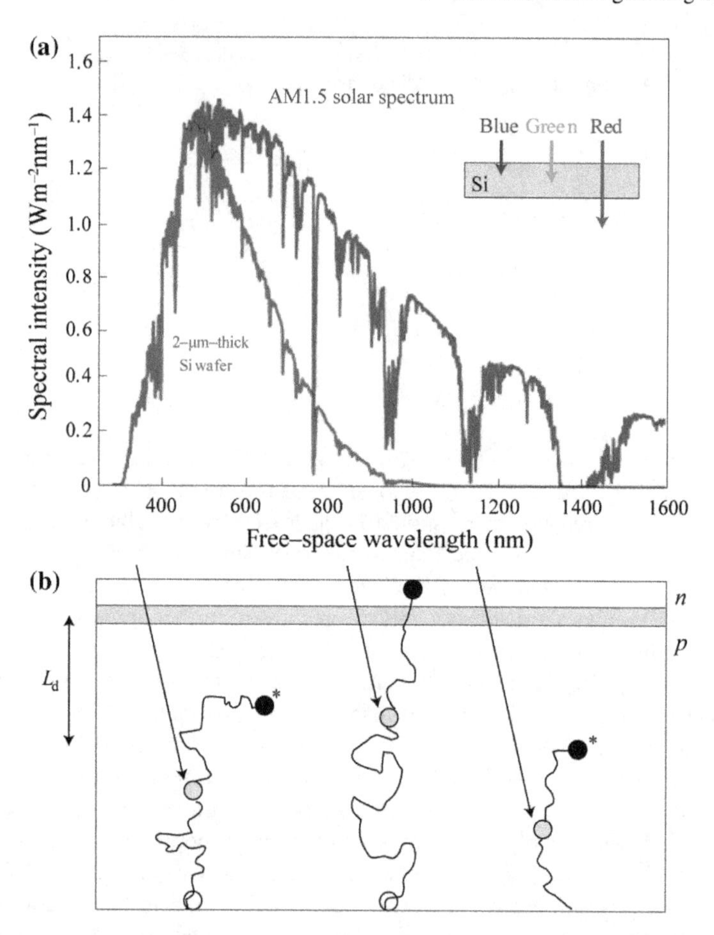

Fig. 1.8 Optical absorption and carrier diffusion requirements in a thin-film solar cell. **a** AM1.5 solar spectrum, together with a graph that indicates the solar energy absorbed in a 2-μm-thick crystalline Si film (assuming single-pass absorption and no reflection). **b** Schematic indicating carrier diffusion from the region where photocarriers are generated to the *p*–*n* junction. Reproduced with permission from [35]. Copyright 2010, Springer Nature

There are many materials and methods that can be used to manipulate the polarization of light. These materials include anisotropic and chiral ones. First of all, anisotropic polarizing devices are represented by the polarizers made of metallic gratings and waveplates. While metallic gratings selectively transmit one polarization and reflect the other in a broad wavelength band, waveplates rely on the phase difference induced by the difference in refractive index along the slow and fast axes (Fig. 1.9); thus the operational bandwidth of simple waveplate is relatively small. Traditionally, broadband waveplates are realized by stacking multilayered anisotropic films [38]. Besides anisotropic materials, chiral materials, which have different refractive indexes for opposite circular polarizations, are also a promising

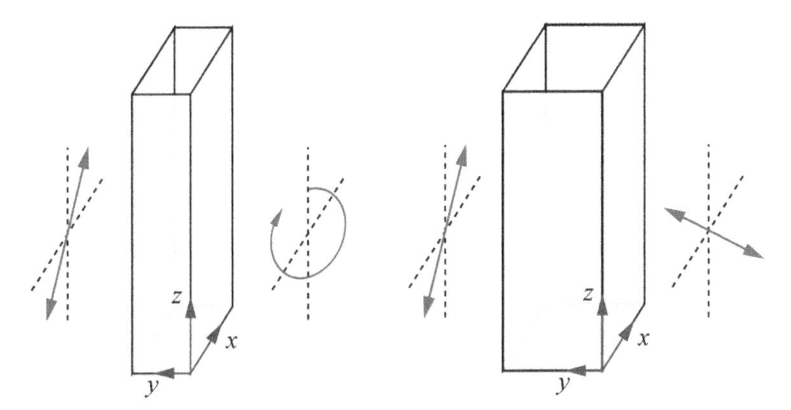

Fig. 1.9 Role of quarter wave plate (QWP) and half wave pate (HWP) on polarization conversion. The x- and y-axes represent the slow and fast axes of the waveplate

approach for polarization manipulation. For instance, such materials may be used to rotate a linear polarization or filter one particular circular polarization [39, 40].

The waveplate represents one particular example that illustrates the merits and demerits of traditional polarizing optical elements. As discussed before, there are two factors that should be considered for a waveplate. First of all, the thickness of the waveplate is determined by the difference in refractive index and the operational wavelength. Consequently, the thickness of waveplates may be very large in the long wavelength spectrum. Secondly, the bandwidth of waveplate is also limited because the difference in phase shift is calculated by the product of refractive index, wavenumber, and propagation length. When the difference in refractive index is a constant, the phase shift would be linearly proportional to the wavenumber (inversely proportional to the wavelength). The above analyses indicate that there is an intrinsic relation between the thickness and bandwidth of waveplate. Without loss of generality, we expect that the bandwidth thickness limit of polarizers can be borrowed from the Rozanov limit via the concept of artificial magnetic conductor based on the following two facts [41]:

(1) When a resistive sheet with a resistance of 377 Ω is placed above a perfect magnetic conductor, frequency-independent perfect absorption would be obtained with near-zero thickness. Consequently, the thickness limit for the perfect absorber is just the same as that for artificial magnetic conductor.
(2) The thickness-bandwidth relation for reflective waveplate can be deducted following a similar approach. If a reflector is anisotropic with one direction behaving as perfect electronic conductor and the other acting as perfect magnetic conductor, then there is always a 180° phase shift between the two orthogonal directions. Therefore, such anisotropic device would be a natural achromatic half-wave plate operating in the reflection mode. In practice, the bandwidth is determined by the artificial magnetic conductor.

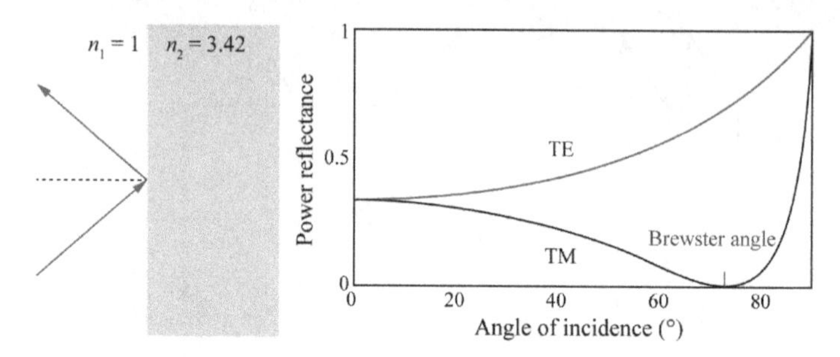

Fig. 1.10 Power reflectance of plane waves at the boundary between air ($n_1 = 1$) and silicon ($n_2 = 3.42$) for both TE and TM polarizations

Note that in a more general case the anisotropic effect also refers to the asymmetry of light–matter interaction. For instance, the Brewster effect at oblique incidence for an isotropic material has also been widely exploited to realize polarization filtering. This is because the reflection and transmission coefficients of transverse electrical (TE) and transverse magnetic (TM) waves at oblique incidence are not equal, which can be illustrated by the Fresnel's equation [42]. As shown in Fig. 1.10, when two polarized light incident from air to silicon, the power reflectance is quite different at oblique incidence. In particular, at the Brewster angle, $\theta_B = \arctan(n_2/n_1)$, the reflection of TM polarization is zero, which provides a method to create high transparency windows without anti-reflection coatings.

When the incident angle is larger than the Brewster angle, the reflection phases for TE and TM-polarization become different. This effect can also be exploited to realize polarization conversion. As illustrated in Fig. 1.11, this can be realized by means of two total internal reflections in a parallelepiped prism (Fresnel rhomb). After each reflection, an additional 45° phase shift is generated; thus two reflections are adequate to convert a linear polarization to circular polarization. Since the phase shift is only dependent on the incident angle and refractive index, such device is intrinsically broadband. It should be noted that although it has the advantage of broadband operation, Fresnel rhomb is currently not widely utilized because waveplates are more compact and do not change the beam position [42].

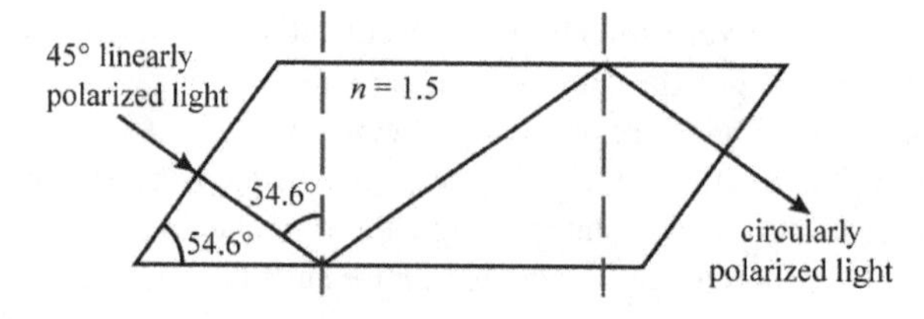

Fig. 1.11 Fresnel rhomb used to convert linearly polarized beam into circularly polarized one achromatically. The refractive index is 1.5, and the internal reflection angle is 54.6°

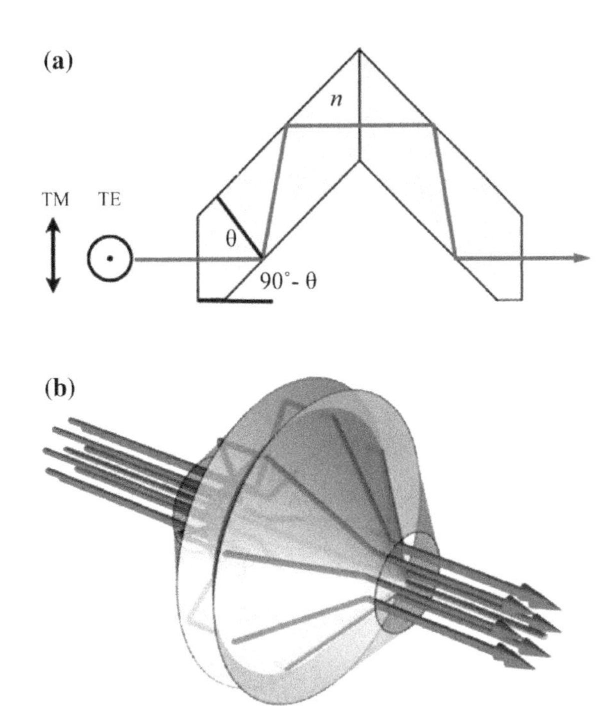

Fig. 1.12 **a** Schematic of two attached Fresnel rhombs. The relative phase change between TE and TM polarization after four appropriately designed reflections is π radians. **b** The 3D design of an achromatic OAM generator formed by rotating the design of part **a** about its optical axis. Reproduced with permission from [43]. Copyright 2014, IOP Publishing Ltd and Deutsche Physikalische Gesellschaft

The phase shift induced in the total reflection can also be applied in the generation of optical vortex beam, i.e., a beam carrying helical phase shift and orbital angular momenta (OAM). As shown in Fig. 1.12, when two attached Fresnel rhombs are rotated about the axis, an achromatic geometric phase could be introduced [43].

1.3.5 Radiation and Emission

Electromagnetic radiation coming from light sources is one hot topic of optical research. In general, radiations have many different origins, ranging from the oscillation of electrons to spontaneous and stimulated emission of atomic systems, which are described by the classic electrodynamics and quantum theories, respectively [3]. In the optical history, constructing artificial light sources has greatly changed the human society. In ancient times, the manmade light source is often blaming fires. After the invention of electricity, electric bulbs have lightened the world over the last hundred years. In the twentieth century, the emergence and development of microwave and laser reshaped the optical and electromagnetic society once again. Within the framework of Maxwell's equations, the boundary between light wave and microwave is becoming more and more blur. Anyway, the developing trends of electromagnetic sources are clear, including: (1) higher power, (2) smaller size, (3) higher efficiency, and (4) on-demand bandwidth. However, single radiators or emitters usually have only limited efficiency due to either the impedance mismatch or the limited density of state. For engineering optics, we are more concerned about quantum light sources such as thermal radiators and lasers. In the following discussion, the current status and developing trends of subwavelength optical engineering with such light sources are outlined. More detailed discussion including optical and microwave antennas can be found in the following chapters.

Perhaps blackbody radiation is the most commonly seen electromagnetic wave in the universe, the study of which opened the door of quantum physics. Theoretically, Planck's radiation law gives the spectral radiant emittance of the blackbody as a function of wavelength and temperature [44]:

$$M_{\mathrm{bb}}(\lambda, T) = \frac{c_1}{\lambda^5\left(e^{c_2/\lambda T} - 1\right)}\left(\mathrm{W/cm^2}\,\mu\mathrm{m}\right) \tag{1.3.2}$$

where c_1 is the first radiation constant (3.7415×10^{-16} W m^4), c_2 is the second radiation constant (1.4388×10^{-2} m K), T is the absolute temperature (K), and λ is the wavelength (m).

For a Lambertian source, M is related to the spectral radiance by

$$M = \int_0^{2\pi} d\phi \int_0^{\pi/2} L\cos\theta\sin\theta d\theta = \pi L \tag{1.3.3}$$

Obviously, Planck's radiation law sets a fundamental limit on the emittance of materials. Traditional approaches to increase the emittance rely on the increase in temperature. Once temperature is fixed, the maximal radiation is also limited. Figure 1.13 shows the spectral radiant emittance for blackbodies at different temperatures. The visible range is highlighted by the color stripe. For a real thermal emitter, the emittance must be smaller than the blackbody radiator, since the emissivity cannot be larger than unit.

The emissivity of a given object can be obtained from Kirchhoff's law, which states that the amount of radiation absorbed by any object is equal to the amount of radiation that is emitted by this object under thermal equilibrium. Although being transparent for visible light, glass made of SiO_2 has large emissivity in the middle- and far-infrared spectrum. In contrast, metals often have high reflectivity above 90%, and thus emissivity below 0.1. Well-polished high-conductivity metals may have emissivity smaller than 0.01.

For practical applications, thermal radiation may be required to increase or decrease in particular spectral range. For instance, narrowband infrared radiation is needed to increase the efficiency of thermal-photovoltaic applications [45, 46]. By firstly converting solar energy into thermal energy and then to a narrowband thermal emission just above the bandgap of a single-junction solar cell, the originally wasted energy in classic solar cells may be reutilized (Note that only photons with proper energy could be efficiently converted to photoelectrons). On the other hand, for infrared camouflage applications, the thermal radiation in the atmosphere win-

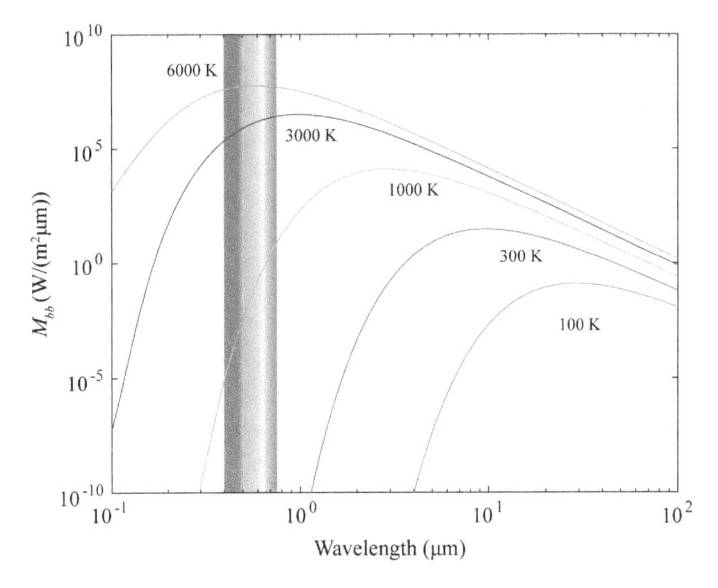

Fig. 1.13 Radiation spectrum of blackbody under different temperatures. The color bar represents the visible spectral regime

dow should be minimized or controlled to be compatible with the environment. It is worth noting that the requirements for various camouflage techniques may conflict with each other: While thermal infrared camouflage wishes the infrared radiation to be as small as possible, the threat of infrared laser detection such as CO_2 laser requires a high absorption at the laser wavelength. According to the Kirchhoff's law, a high absorption is equal to high thermal radiation; thus it seems that the two functions cannot be realized simultaneously with traditional materials [47].

Terahertz (THz) radiation is another example where classic radiation law presents great challenges for practical applications [48]. In the electromagnetic spectrum, THz lies between photonics and electronics, which is called as a "terahertz gap" (Fig. 1.14). Within the visible spectrum, red light has a longer wavelength and lower frequency than blue light. The sources for visible light are often thermal radiators and lasers, which are called photonic sources. In contrast, electromagnetic waves at much lower frequencies are usually generated using electronics and oscillators. In the terahertz frequency region between these two spectra, wave is more difficult both to generate and detect.

In the THz range, a blackbody source at 2000 K radiates only negligible power over the entire THz range [48]. From the electronic side, the power from oscillators goes down quickly with increasing frequency (the output power at 300 GHz is only several milliwatts). Figure 1.15 shows the drop in power from both the electronic and photonic sides, which demonstrates the challenges in making high-power THz region source.

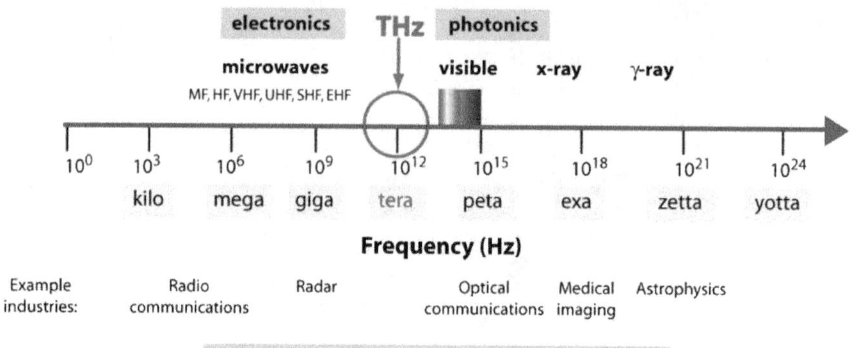

Fig. 1.14 Schematic of the electromagnetic spectrum showing that THz light lies between electronics and photonics. Reproduced from [48] with permission. Copyright 2006, IOP Publishing Ltd.

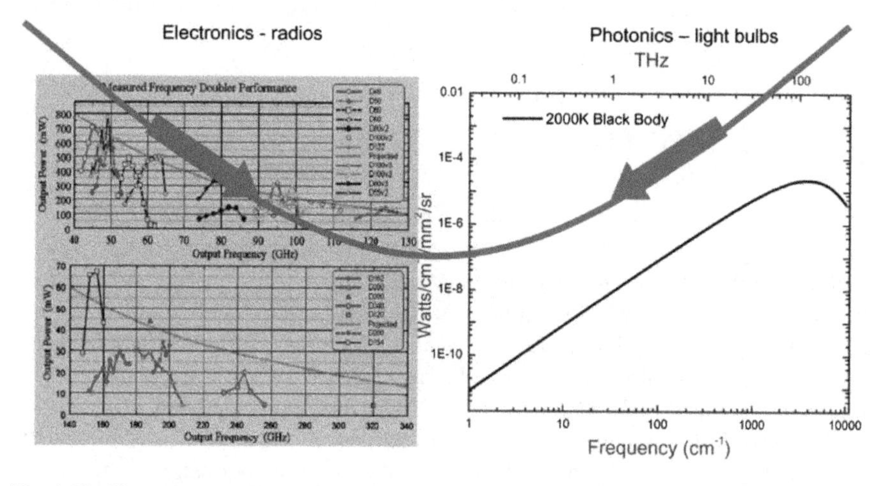

Fig. 1.15 Illustration of the real THz gap—the decline in power as electronics goes to higher frequencies, and photonics goes to lower frequencies. Reproduced from [48] with permission. Copyright 2006, IOP Publishing Ltd.

1.4 Opportunities Enabled by Subwavelength Structures

1.4.1 Emergence of Subwavelength Structures

In the last several decades, marvelous opportunities have been provided to overcome the above dilemmas by controlling light–matter interaction in the subwavelength scale, i.e., the functional structures have characteristic dimensions smaller than the wavelengths [3]. This is possible as a result of the rapid development of enabling technologies in nanophotonics [49]. First and most importantly, traditional optical engineers do not exploit light scattered by subwavelength structures. The surfaces of current lenses and mirrors need to be polished to minimize the surface roughness to the extent of one to three magnitudes smaller than the wavelength. As a result, like Feynman's statement on nanotechnology [50], *there is plenty of room at the bottom* that has not been utilized in optics. Secondly, since the 1990s when the fabrication capability of semiconductor manufacturing has become smaller than 250 nm, the experimental investigation of subwavelength light–matter interaction becomes possible. Currently, the rapid development of nanotechnologies has ensured characteristic dimension up to 10 nm, which may greatly increase the accuracy of these subwavelength structures. Thirdly, modern electronic computers and computational algorithms have facilitated the rigorous designs of complex structures with multiscale characteristic dimensions. The calculation speed of personal computer has increased up to one billion times per second, and the supercomputer has reached to more than 200,000 trillion calculations per second (200 petaflops). Thanks to these factors, many substantial discoveries that may change the entire optical sciences and technologies have been realized by exploiting the subwavelength artificial structures. In the following, we would like to discuss some details about the precision fabrication methods used to fabricate the subwavelength structures.

Optical lithography is one of the major methods used to fabricate micro- and nanopatterns. The attainable structure size in photolithography is a direct consequence of the diffraction-limited resolution of the projection optics. Over the last forty years, the resolution has been shrunk from 1 μm to sub-10 nm, as a result of the wavelength reduction from 436 to 193 nm and 13 nm [51]. According to the Moore' law proposed in 1965, the number of transistors on a silicon chip would approximately double every two years. Since 1975, the microelectronics industry has actually double the number of transistors every 1.5 years [31]. As shown in Fig. 1.16, the reduction in characteristic dimension is associated with the increasing performance of micro-processors of electronic computers. The main frequency has been increased from 33 MHz of Intel 486 to more than 5 GHz of Intel 8086 K.

Besides traditional optical projection lithography, there are many other techniques for the fabrication of nanopatterns, which typically include laser direct writing (LDW), electron-beam lithography (EBL), focused ion beam (FIB) milling, and nanoimprint. Like projection lithography, LDW is also limited by the diffraction limit, which has a minimal resolution of 300-600 nm, which is not suitable for the fabrication of nanostructured patterns (it can be used to fabricate subwavelength structures for mid-infrared regime). In contrast, particle beam lithography like EBL and FIB uses focused particles to generate a highly concentrated spot. Owing to the extremely short effective wavelength, the ultimate resolution can be easily extended to be smaller than 100 nm, although the direct writing in a scanning mode over a large area requires a tremendous amount of time.

Nanoimprint lithography (NIL) is another promising technique for nanofabrication, which uses mechanical deformation of materials to replicate nanostructures [52]. Conventional thermal NIL employs heat to cure a polymer-coated substrate, and a nanostructural master mold is first pressed against and then detached from the substrate. Another type of NIL is the so-called UV NIL that employs a liquid-phase

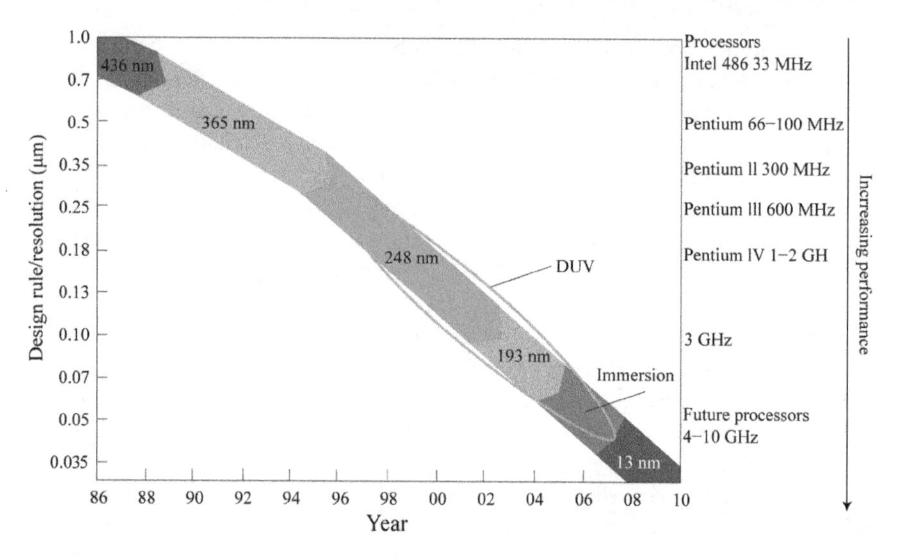

Fig. 1.16 Increasing the resolution of photolithography and the increasing of performance of processors. Reproduced from [31] with permission. Copyright 2007, Nature Publishing Group

polymer spin-coated on a substrate and uses UV radiation to generate solidified polymer. As a replica process, NIL is very cost-efficient compared to direct writing. However, the repeated mechanical contact may damage the masks.

Regardless of the mechanism of fabrication processes, the above techniques have ensured the high-precision fabrication of artificially designed subwavelength structures. Figure 1.17 illustrates a schematic diagram of the fabrication of flat metalenses using projection lithography [53]. A wafer is first deposited with a film stack comprised of the amorphous (α)-Si, photoresist (SPR700-1.0), and contrast enhancement material (CEM). The pattern of the metalens contained in the photomask is then projected on the wafer and replicated by repeatedly exposing and incrementally stepping the wafer position. In this way, throughputs as high as hundreds of wafers per hour (wph) may be achieved. Subsequently, the pattern is etched into the a-Si, forming the metalens. Finally, after residual photoresist is removed, the wafer can be diced into individual metalenses.

In another example, Hu et al. realized subwavelength structures on α-Si on a 12-in. wafer by 193 nm ArF deep UV (DUV) immersion lithography [54]. Cylindrical α-Si pillars with critical dimension below 100 nm are designed to structural colors. As shown in Fig. 1.18, the device fabrication started with a 12-in. Si wafer. A 70-nm-thick SiN layer and a 130-nm-thick α-Si layer were deposited successively on the Si substrate. The designed metasurface was then patterned by immersion lithography, followed by the inductively coupled plasma (ICP) etching. Instead of an ASML lithography machine, the Nikon immersion scanner with a resolution down to 40 nm was used in this experiment. After photoresist was removed, the polymer generated in the a-Si etch process was removed using wet clean processes.

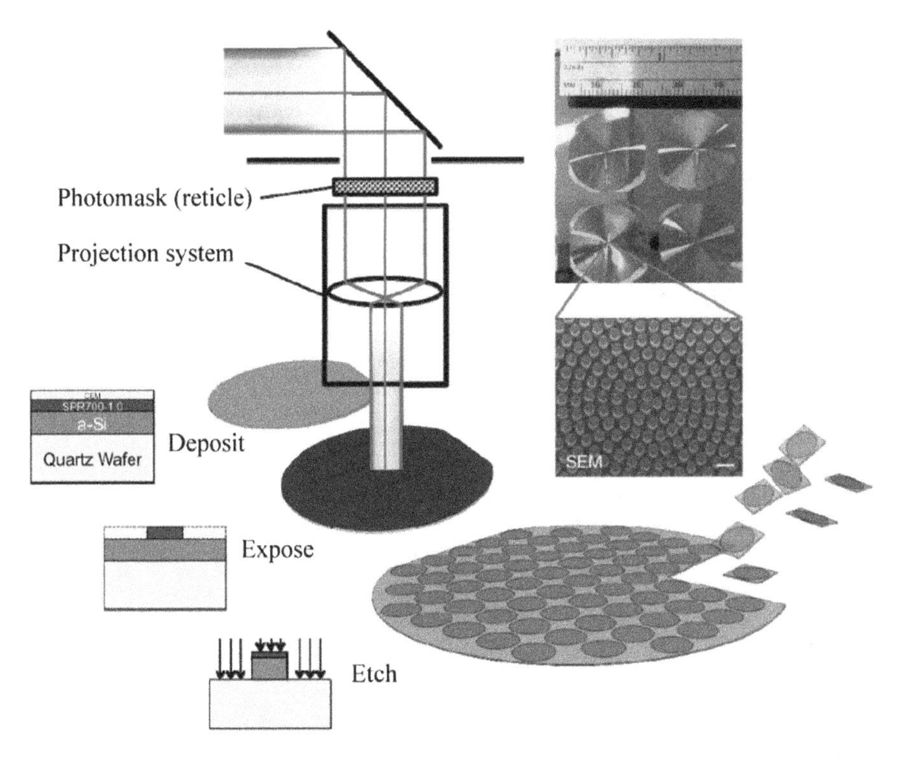

Fig. 1.17 Schematic of the production of metalenses using photolithographic stepper technology. A photograph of fabricated metalens (upper right) with a diameter of 2 cm is shown in comparison with a ruler. A scanning electron microscope (SEM) of the metalens center (center right) shows the microscopic posts comprising the metalens. Scale bar: 2 μm. Reproduced from [53] with permission. Copyright 2018, Optical Society of America

It has been demonstrated that subwavelength structures fabricated using the above technologies could meet the requirement of some practical applications. For instance, Zeitner et al. reported the design and fabrication of the grating for the Radial-Velocity-Spectrometer of the GAIA-mission of the European Space Agency (ESA) [55]. As shown in the SEM picture of Fig. 1.19a, the optimized grating pattern has a sub-structure within one grating period consisting of three 1D-bars and only one row of pillars, which was fabricated by electron-beam lithography and reactive ion etching. One of the full-size gratings is shown in Fig. 1.19b, where specially polished 9-in. sized substrates were used in order to obtain a wave-front error in the subapertures below 5 nm (rms) and polarization independent diffraction efficiency above 80%.

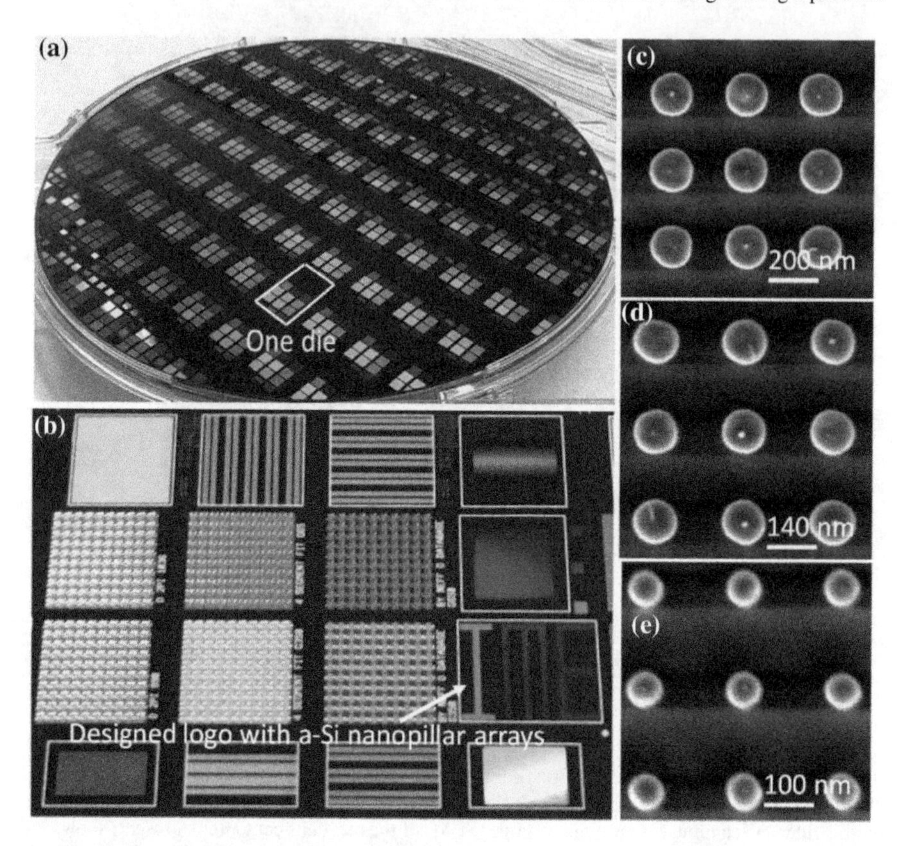

Fig. 1.18 **a** Photograph of the fabricated 12-in. wafer. **b** Close view of one die on the wafer with the colored logo "IME." **c–e** SEM images of a-Si nanopillar arrays for the three letters "I," "M," and "E." Reproduced from [54] with permission. Copyright 2018, Optical Society of America

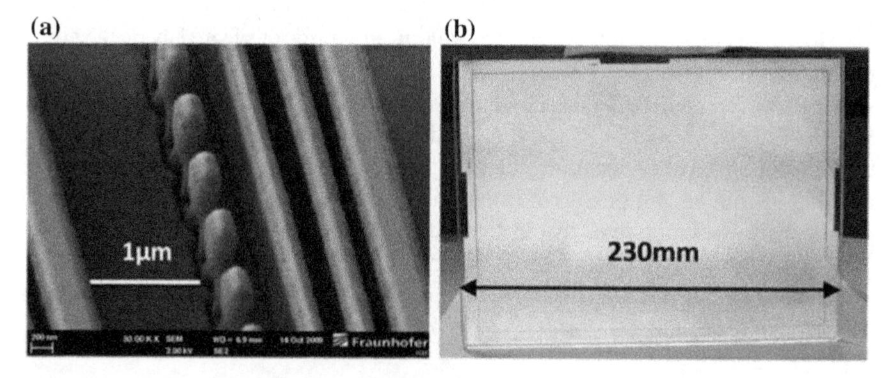

Fig. 1.19 GAIA spectrometer grating. **a** SEM image of the fused silica grating structure. **b** Photograph of the "Flight Model" of the grating. Reproduced from [55] with permission. Copyright 2012, Springer, Berlin, Heidelberg

1.4.2 Sub-diffraction Optics

As noted in previous discussions, the wave nature and diffraction effect of light have set a fundamental limitation on the resolution of optical imaging. In Fourier optics, this is related to the loss of high spatial frequency components, especially the evanescent waves which decay exponentially away from the objects [5].

One straightforward approach to beat the diffraction limit is to utilize the evanescent wave within the near field. To differentiate with traditional optics that deals with light in the far field, this new researching area is called near-field optics. One of the main tools in near-field optics is *scanning near-field optical microscopy*, known as SNOM or NSOM. As early as in 1928, encouraged by Albert Einstein, E. H. Synge proposed to scan a small aperture above a sample to achieve higher resolution [56]. Since both the size of the aperture and the distance between the aperture and object are much smaller than the wavelength, the evanescent wave generated by the object could be scattered and converted to propagating wave.

Although SNOM could break the diffraction limit, it relies on the near-field scanning and cannot be applied to direct imaging. Based on the concept of negative refractive index materials (NIM) proposed by Veselago and other precursors [57], John Pendry found that the negative perfect lens could defeat the diffraction limit in the near field [58]. As shown in Fig. 1.20, it was shown that a slab with a refractive index of -1 cannot only act as a flat lens, but also can amplify the evanescent waves thus realize sub-diffraction-limited imaging. More interestingly, for a particularly polarized wave, materials with merely negative permittivity or permeability can operate as an approximate perfect lens. In practice, this condition can be achieved using noble metal films under transverse magnetic polarized illumination.

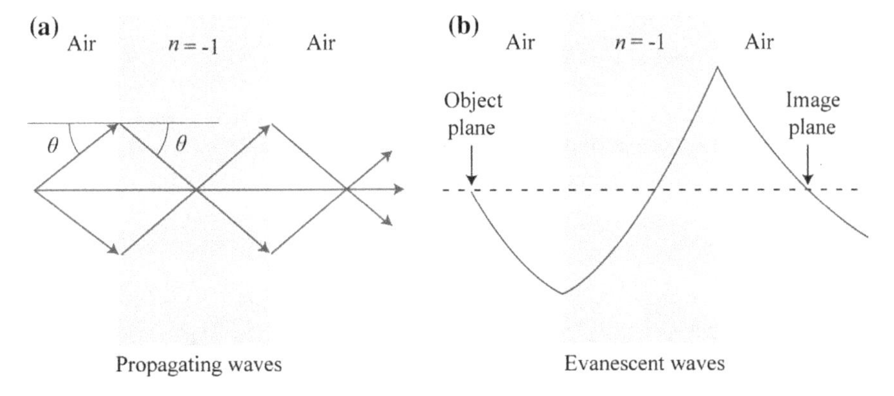

Fig. 1.20 Slab of negative refractive index medium acts as a perfect lens. **a** A negative flat lens brings all the diverging rays from an object into a focused image. **b** The lens can also enhance the evanescent waves across the lens, so the amplitude of the evanescent waves is identical at the object and the image plane. Reproduced from [57] with permission. Copyright 2008, Nature Publishing Group

To experimentally demonstrate the superlens imaging effect, both objects and detectors must be placed at the near field of the superlens, which is often impractical in the optical regime, because the wavelength is only several hundreds of nanometers. Figure 1.21 is a simple example to overcome this obstacle [59]. By perforating a thin silver film with periodic nanoslits and placing a photoresist layer below the film, the near-field optical intensity can be directly recorded. This experiment serves as two roles: First, the nanoslits with a periodicity of Λ (300 nm in the experiment) excite surface plasmon polaritons (SPPs) that interference at the top of the grating; second, these interference patterns are imaged to the bottom layer by the superlens. Note that this is in contrary to ordinary guided mode resonance, where the periodicity of grating is equal to the effective wavelength of the waveguide modes. In this experiment, the effective wavelength is about 100 nm, which is only about one-third of the periodicity. As demonstrated in a subsequent analysis, the effective wavelength of SPP could be controlled by the thickness and by introducing more metal–dielectric multilayers. It was shown that the thinner the film, the shorter the effective wavelength for both a single metallic film [60] and metal–dielectric multilayers [61].

The above experiment can also be considered as an extraordinary Young's interference (EYI) experiment [62, 63]. It is well known that Young's double slits interference is the key to demonstrate the wave characteristic of both the light and electron. The most beautiful experiment in physics is thought to be the interference of single electrons in Young's double slit [18].

Because interference is the basis of wave optics, and Young's double slits experiment is the simplest and most powerful method to demonstrate the interference effect, any approach to break the diffraction limit should generate abnormal effects in the double slits experiment, as shown by the above EYI phenomena assisted by coupled surface plasmons. Like the photons and electrons, the coupled plasmons may have far-reaching influences and applications in future physics and optics [3, 64]. Consequently, it has been listed as one milestone in the history of engineering optics (Fig. 1.1).

In the quasi-static limit, metal–dielectric multilayers can be approximated as an effective anisotropic medium with hyperbolic dispersion, which allows the propagation of high spatial components. Figure 1.22 shows the OTF of metal–dielectric multilayers with a total thickness of $\lambda/2\pi$. According to effective medium theory, the

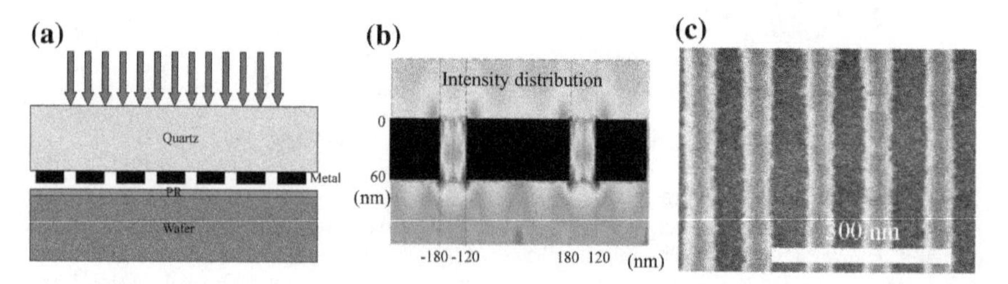

Fig. 1.21 **a** Schematic representation of surface plasmon lithography. **b** Simulation results. **c** SEM image of the lithography results. Reproduced from [59] with permission

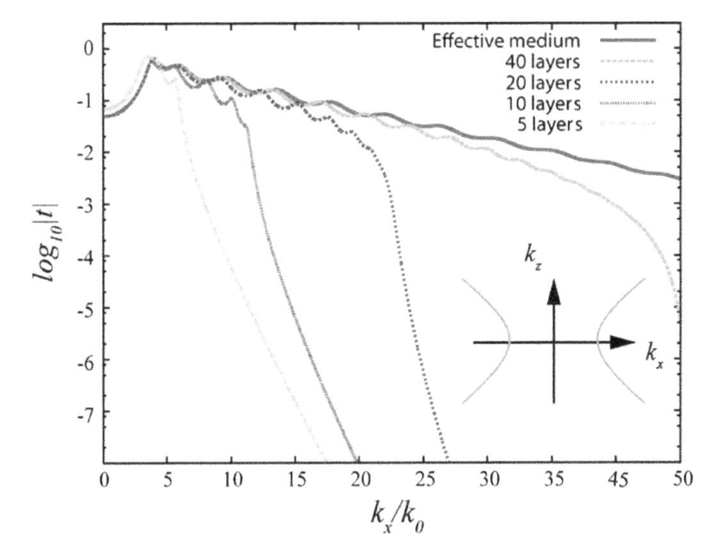

Fig. 1.22 Transmission as a function of k_x for various layer widths. The inset is the hyperbolic dispersion. The total slab width is maintained at $\lambda/2\pi$ in each case, while the number of individual layers is adjusted. For 40-layer films, the thickness of each layer is only ~$\lambda/250$. The material parameters used correspond to layers of Ag and ZnS–SiO_2, embedded in crystalline $Ge_2Sb_2Te_5$, for a light wavelength 650 nm [61]

transmission at $40k_0$ is still larger than 0.01. However, when composed of realistic structures with 5 layers, the transmission would reduce dramatically for horizontal wavevectors larger than $7k_0$. Along with the reduction in thickness, the transmission spectral approaches that of effective medium theory gradually, leading to a greatly increased imaging resolution [61].

1.4.3 Planar Optics

Classic geometric optics is based on Fermat's principle and the laws of refraction and reflection. Owing to these rigorous geometric restrictions, one could only resort to the geometric shape to control the propagation direction of light. This is why most optical elements such as lenses and mirrors used in telescopes and microscopes bear curved surfaces.

Along with the development of subwavelength optics, it was recently discovered that proper arrangement of subwavelength structures may change the reflection and refraction directions of light in a controllable way. Once again, taking the famous Young's double slits interference as an example, classic optical theory assumes that the width of slits only change the transmission intensity; thus the interference pattern would not change dramatically when the slit width is varied. However, this is not true in the subwavelength domain.

Figure 1.23 illustrates an EYI experiment with unequal slit widths [65]. When the widths are equal, a bright interference fringe would be observed at the center. However, when the widths are changed as 100 and 25 nm, respectively, the interference pattern would shift horizontally to obtain zero intensity at the center. According to Fermat's principle, this must be induced by an additional phase shift inside the slits.

Naturally, one may imagine that if many nanoslits with gradient widths are introduced, a gradient phase shift would be obtained within a macroscopically flat interface, as indicated in Fig. 1.24a. Actually, this design approach has been intensively studied to construct various flat optical devices such as lenses [66, 67], deflectors [68–70], and unidirectional surface plasmon sources [71]. The nanoslits have also been extended to two-dimensional structures with polarization-independent performance [72–74]. Note that this approach is different from that based on gradient dielectric structures (Fig. 1.24b) [75], which could be approximately treated as gradient optical materials. In fact, gradient optics has a long history more than one hundred years [76]. By utilizing the Snell's law microscopically, it can be seen that the light ray would follow a curved path [77]. In contrast, in the metallic slits light waves would propagate straightly as waveguide modes, making the coupling between adjacent neglectable. In this case, the optical performances can be solely described by a phase gradient along the surface. The reflection and refraction directions are then determined by the generalized laws of reflection and refraction [68, 78, 79].

In the ideal condition, the gradient phase shift should be introduced at a material interface with vanishing thickness (Fig. 1.24c); thus the phase is locally controlled. In the plasmonic nanoslits, the propagation constant is much larger than the vacuum wavenumber, thus deep subwavelength thickness is sufficient to realize complete

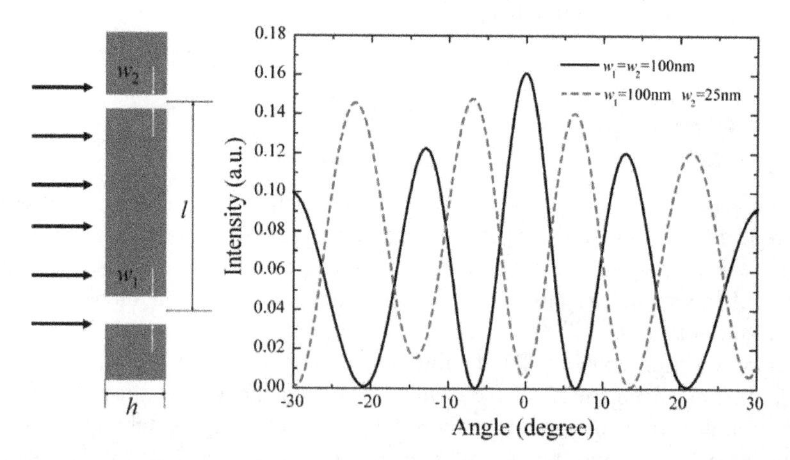

Fig. 1.23 EYI experiment with non-equal widths. The angular shift of far-field energy distribution leads to a dark fringe at the normal direction ($\theta = 0°$). Adapted from [65] with permission. Copyright 2007, Optical Society of America

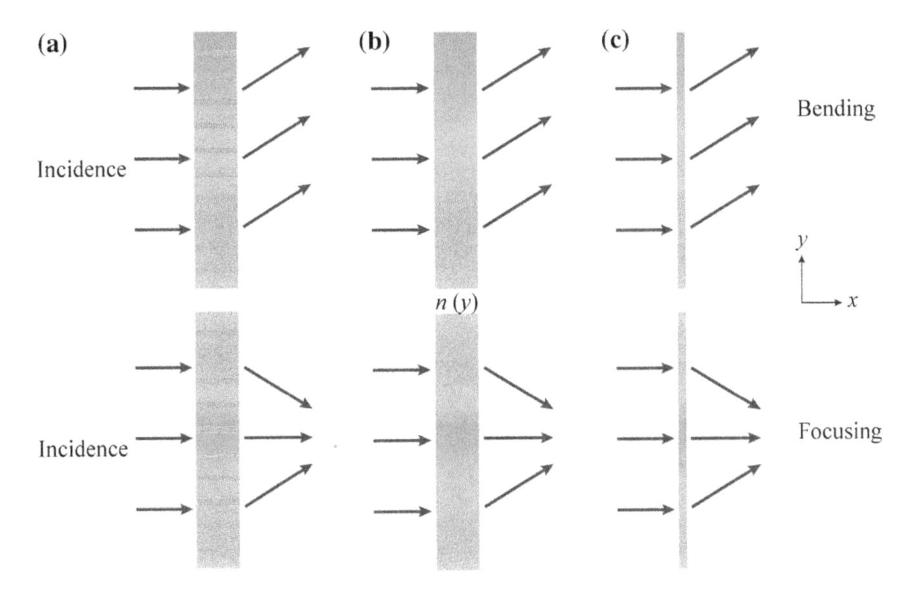

Fig. 1.24 Three configurations for the manipulation of electromagnetic waves in free space. Two phenomena—bending and focusing of light—are considered. **a** Gradient nanoslits with gradient width. To obtain a spatial gradient, the slits can be filled with different dielectrics or their width can be engineered. **b** Gradient index profiles along the y-direction; these profiles are obtained with standard techniques for metamaterials. The spatial gradient along the y-direction is represented by the color gradient. **c** Spatially varying resonant structures with subwavelength dimensions are very thin. Reproduced from [80] with permission. Copyright 2016, Macmillan Publishers Limited

phase shift in $[0, 2\pi]$. However, a larger propagation constant means a higher Fresnel reflection, leading to a smaller efficiency. To further reduce the thickness while maintain the efficiency, other phase modulation mechanisms should be exploited.

Figure 1.25 presents a novel design based on ultrathin V-shaped nanoantennas. Owing to the excitation of symmetric and anti-symmetric resonant modes, a linearly polarized incidence is converted to its cross polarization. The phase shift accompanying this conversion can be varied by tuning the geometric sizes and orientations. As shown in the inset of Fig. 1.25, four V-shaped nanoantennas are sufficient to realize an eight-level coding for arbitrary phase distribution. While the phase shifts of the elements 1–4 are realized by tuning the sizes, the elements 5–8 are obtained by simply rotating the previous four elements with an angle of 90°. This effect can be understood using the geometric phase principle [79], which has also been widely utilized to realize the generalized Snell's law [81, 82].

As a result of the fundamental role of generalized laws of reflection and reflection, this technology is thought to be disruptive [3, 84, 85]. By replacing traditional bulky lenses and mirrors with flat optical devices, one can not only make optical system much more compact, but also construct new functionalities not possible for traditional optics [86, 87].

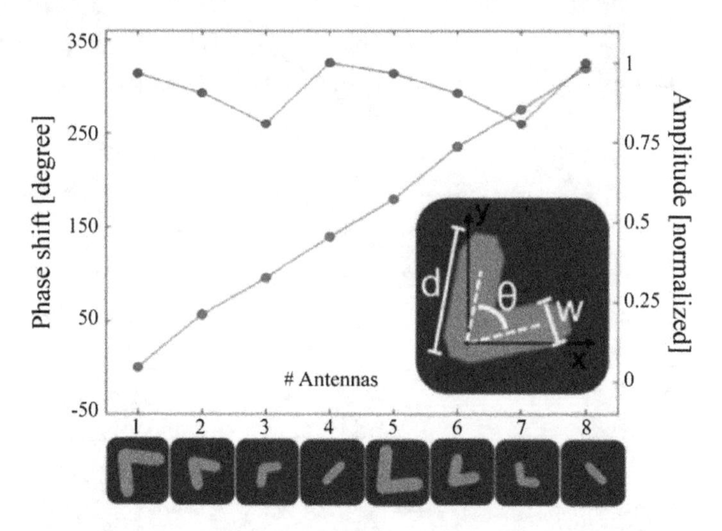

Fig. 1.25 Finite-difference time-domain simulations are used to obtain the phase shifts and scattering amplitudes in cross-polarization for the eight elements used in the metasurfaces. The parameters characterizing the elements from 1 to 4 are d = 180, 140, 130, and 85 nm, and θ = 79, 68, 104, and 175°. Elements from 5 to 8 are obtained by rotating the first set of elements by an angle of 90° counterclockwise. The width of each antenna is fixed at w = 50 nm. Reproduced from [83] with permission. Copyright 2012, American Chemical Society

1.4.4 Modulation of Polarization Properties

Polarization plays an important role in electromagnetic waves since a majority of phenomena are polarization sensitive. Manipulation of polarization has been a hot topic for quite a long time [36]. Traditional birefringent and chiral medium have the ability of polarization modulation. However, these traditional media need large thickness to accumulate enough phase shifts between the perpendicular components of the incident electric fields. Furthermore, waveplates based on a single layer anisotropic material are intrinsic narrowband, since the phase difference is generally dependent on the frequency.

With the help of subwavelength structures, one can greatly improve the performance and enable high conversion efficiency in a wide band with only an ultrathin thickness. As shown in Fig. 1.26, artificial waveplates had already been proposed in 1983. When the period of the grating is much smaller than the wavelength, the gratings may be treated as effective artificial material with very large anisotropy. Such structures can be easily fabricated via holographic or interference lithography even in the visible frequency [88].

When the refractive index and geometric parameters are chosen properly, structural resonance may take place to alter the frequency response of such grating. As shown in Fig. 1.27, the phase retardation for a subwavelength silicon grating is close to $\pi/2$ in a wide frequency range. In this case, the effective refractive index cannot be

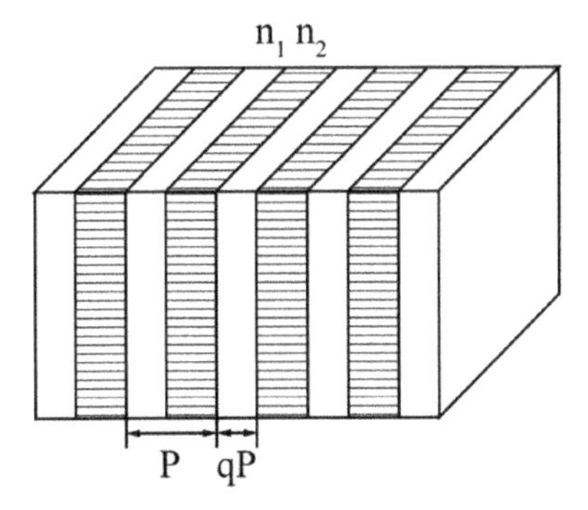

Fig. 1.26 Illustration of the geometry of a square profile grating on a substrate. Usually the index of refraction of the substrate and n_2 would be the same and $n_1 = 1$

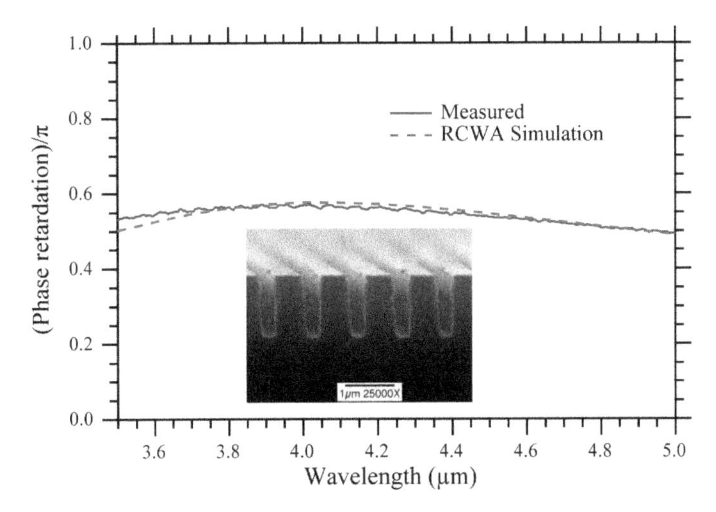

Fig. 1.27 Measured and simulated phase retardation as a function of wavelength. Inset is the SEM image of silicon grating. Reproduced from [90] with permission. Copyright 1999, Optical Society of America

simply calculated using common equivalent materials theory. Instead, a high-order approximation must be made [89].

For transmissive structures, resonance induced by artificial structures often leads to a rapid increase and reduction in the transmission coefficients. Consequently, the bandwidth of a single layer structure is still limited. To overcome this problem, reflective waveplates have been proposed by combining anisotropic surface structures, dielectric spacers, and a reflective ground plane [36, 91, 92]. Theoretical models predict that if the dispersion of the anisotropic metasurfaces is modified to mimic the ideal dispersion as shown in Fig. 1.28a, broadband waveplates with phase difference of either 90° or 180° can be easily obtained. Since the reflection amplitudes of

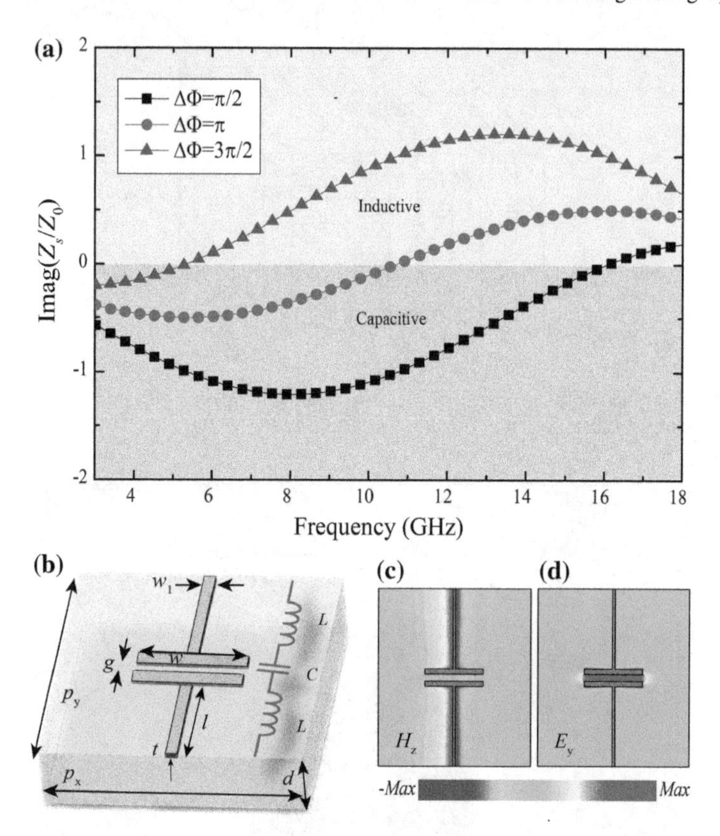

Fig. 1.28 **a** Imaginary parts of the optimal sheet impedances for $\Delta\Phi = \pi/2$, π, and $3\pi/2$. The real part is zero since no material loss is considered. The region above zero is inductive, and the region below zero is capacitive. **b** Schematic of the unit cell of meta-mirror. **c** Magnetic field along z-direction, which is related to the inductance L. **d** Electric field distribution along y-axis, which determines the capacitance C. The values in **c** and **d** are normalized with their maxima. Reproduced with permission from [36]. Copyright 2013, American Institute of Physics

both the two orthogonal components are close to unity, the polarization conversion efficiency may be optimized to be much higher than transmissive devices.

One typical anisotropic metasurface composed of I-shaped metallic elements is shown in Fig. 1.28b. Obviously, for the polarization perpendicular to the long lines, the metasurface acts as air without any phase shift. In contrary, for the other polarization, the metasurface acts as a serially connected capacitors and inductors. The capacitors are attributed to the y-polarized electric fields perpendicular to the gap formed by the two parallel metallic stripes. The inductors are associated with the magnetic energy stored in the long metallic stripe. By properly tuning the capacitance and inductance, the dispersive characteristics of the metasurfaces can be modified to mimic the ideal dispersion curves.

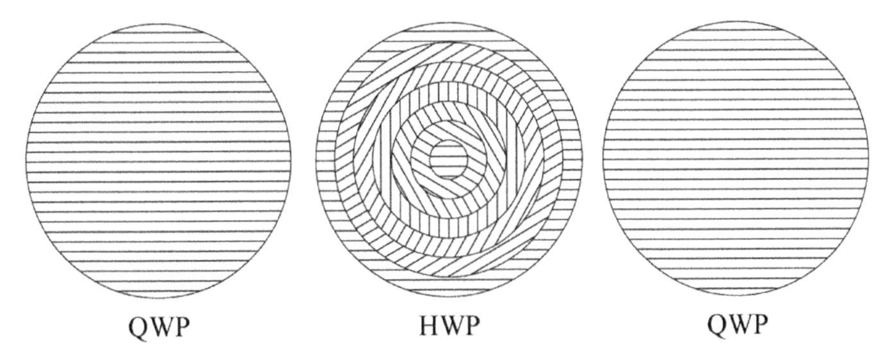

Fig. 1.29 Cross sections normal to the beam of the three elements constituting a geometric phase lens. The first and the third elements are QWPs and the middle element is an HWP whose principal axis is oriented at an angle which is a function of the radial distance from the center. Adapted with permission from [98]. Copyright 1997, Elsevier Science B.V.

We stress that the anisotropic subwavelength structures are not only useful for broadband polarization control with minimal thickness, they would also find applications in polarization-dependent wavefront control based on the photonic spin–orbit interaction and geometric phase [70, 93, 94]. As discussed in the geometric phase theory, an inhomogeneous array of subwavelength waveplates (Fig. 1.29), both in transmissive mode and reflective mode, bear a spin-dependent phase shift depending on the orientation angles of the main axes of these waveplates. This effect was originally discovered in the microwave regime [95], and then in the optical regime, known as Pancharatnam–Berry phase [96–98]. Owing to its additional degrees of freedoms, it has provided tremendous new aspects on the modulation of polarization and phase front.

1.4.5 Strategies for Dispersion Engineering

As demonstrated in many researches, the dispersion diagram can be used as a rule for the design of subwavelength structures and devices [99, 100]. Figure 1.30 shows the typical dispersion curves for waves in plasmonic and dielectric waveguides [59, 67, 79, 101, 102], Lorentz-type resonant metasurfaces [36, 103], as well as the dispersionless geometric phase [82, 98, 104].

The most conspicuous property of the plasmonic dispersion is the fact that its propagation constant could be much larger than that in the dielectric host, as a result of the collective resonance of the free electrons and photons [106]. This property is responsible for the EYI phenomena [59, 63] and has led to many practical applications such as plasmonic nanolithography, ultrathin plasmonic flat lens, and color filtering [51, 67, 101].

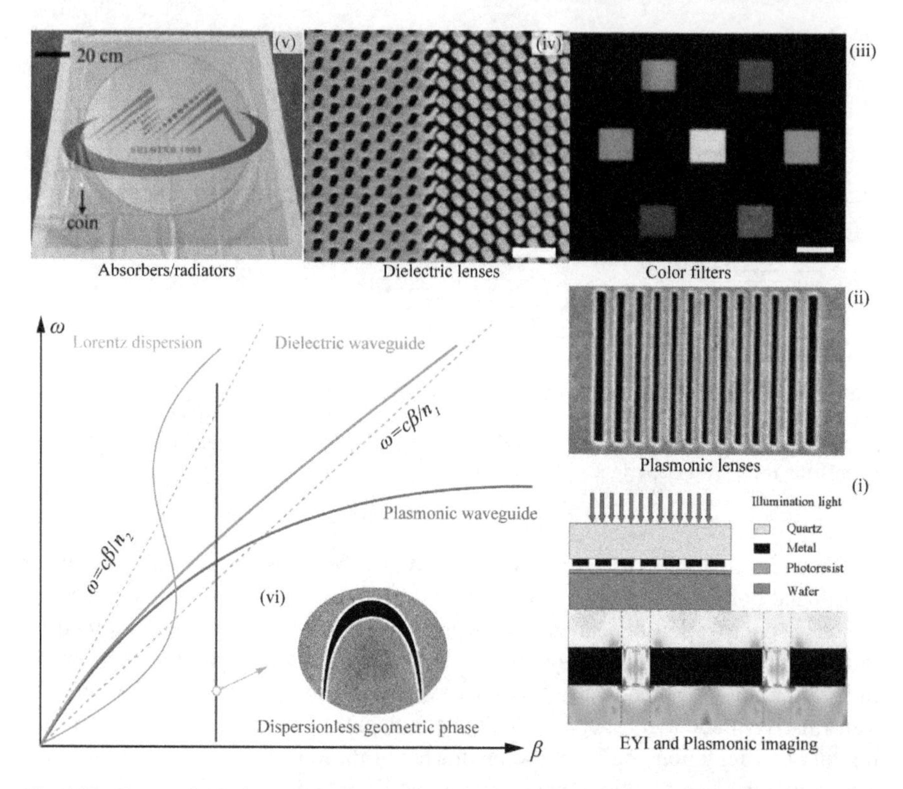

Fig. 1.30 Four typical dispersion curves in subwavelength structured optical systems and their applications. The red, green, blue, and purple lines correspond to SPPs, dielectric waveguiding modes, Lorentz-type dispersion, and dispersionless geometric phase. The figures in (i–vi) are adapted with permission from (i) [59], Copyright 2004, American Institute of Physics; (ii) [67], Copyright 2009, American Chemical Society; (iii) [101], Copyright 2010, Macmillan Publishers Limited; (iv) [102], Copyright 2018, The Authors; (v) [105], Copyright 2017, The Japan Society of Applied Physics, and (vii) [104], Copyright 2017, The Authors

As can be seen in the dispersion curve of the fundamental dielectric waveguiding modes, the propagation constant is limited in a region defined by the refractive index of the core and cladding materials. As a result, the dielectric devices often require much larger thickness to obtain a particular phase shift than their plasmonic counterparts. Nevertheless, since dielectric generally has smaller optical loss in the visible band, dielectric metasurfaces have become a hot topic in recent years [81, 107–110]. Noted that the research into subwavelength dielectric optical elements has a venerable history dating back to 1990s [85]; thus many existing technologies could be utilized in the design of novel optical devices and systems.

Besides the plasmonic and dielectric waveguiding modes, the Lorentz-type dispersion in subwavelength structured materials has also been intensively studied over the years. Both broadband absorbers and polarization convertors can be designed with Lorentz-type dispersion [36, 91, 103, 105, 111]. Since Lorentz dispersion is used as a basic building block for materials with complex dispersion obeying the Kramers–Kronig relations [112], it may be generalized to almost all kinds of metasurfaces.

The geometric phase induced by polarization conversion can also be plotted in the dispersion diagram. Different from the propagating phase defined as the product of propagation constant and distance, the dispersionless geometric phase may be written as $2\sigma\xi$ (Here $\sigma = \pm 1$ denotes the left-handed circular polarization and right-handed circular polarization, ξ is the orientation angle of the main axis), which is independent of the working frequency and does not change with the thickness d [79, 82]. If we consider the wave in such a structure has an effective propagation constant β, the dispersion relation will become $\beta = 2\sigma\xi/d$, which is also independent of the frequency. This property is helpful to realize ultrathin phase modulators and the generalized laws of reflection and refraction [78, 82, 113].

In principle, most metasurfaces can be designed with the help of the above four dispersion curves. As shown in the top and right insets of Fig. 1.30, the realized devices or functionalities include (from i to vi) EYI interference and plasmonic super-resolution imaging, plasmonic flat lenses, structural color, high-efficiency dielectric flat lenses, spectrally selective absorbers/radiators, etc. In other words, these dispersion curves form the key of dispersion engineering of waves in subwavelength structures. As will be discussed in the following chapters, subwavelength optical engineering (Engineering Optics 2.0) is more concerned about the on-demand surface waves existing at the interfaces of subwavelength structures, i.e., the meta-surface-waves (M-waves) [79, 114]. The dispersion engineering has been proven to be a universal approach to realize on-demand control of the optical and electromagnetic properties.

1.5 The Third Optical Revolution

Historically speaking, optics has been studied by more than four hundred years. It is helpful to study the important affairs that changed optical science and technologies. As indicated in Fig. 1.31, there are mainly three revolutions that transformed this discipline dramatically.

The first optical revolution begins in the sixteenth century, which finally evolved into the well-known geometric and physical optics. Geometrical optics [1], or the so-called ray optics, describes light propagation in terms of rays. Geometrical optics does not account for wave optical effects such as diffraction and interference. It is a good approximation when the wavelength is very small compared with the size of structures and is particularly useful in describing geometrical aspects of imaging, including optical aberrations. According to the laws of reflection and refraction, high-precision smooth surface has been pursued to control the direction of rays. However, the existence of diffraction, as marked by Young's interference in 1801, makes the geometric optics not valid in a scale comparable to the operating wavelength. For example, as a result of the wave nature of light, the resolution of imaging optical system cannot be infinitely increased even if the optical aberration could be completely eliminated. Instead, the diffraction limit indicates that one should increase the aperture and (or) decrease the wavelength to obtain higher imaging resolution.

The second revolution of optics is the quantum optics, which emerged from the study of blackbody radiation and photoemission effect in the early twentieth century. According to quantum theory, light may be considered not only as an electromagnetic wave but also as a "stream" of particles called photons. These particles should not be considered as classical billiard balls, but as quantum mechanical particles described by a wavefunction spread over a finite region. Each particle carries one quantum of energy of $h\nu$, where h is Planck's constant and ν is the frequency of light. Note that Young's interference is also very important in quantum physics. Richard Feynman stated that electrons' Young's double-slit experiment is in "the heart of quantum mechanics," and "*absolutely* impossible, to explain in any classical way" [22].

The third revolution of optics is the subwavelength optics and subwavelength electromagnetics, which mainly investigate the gap between the classic and quantum optics, i.e., the feature size is in the subwavelength scale where the quantum effect is relatively weak and conventional optical theories are not valid any more. As highlighted by the EYI effects assisted by surface plasmons and M-waves [59, 62, 63, 101, 115], at this scale light–matter interaction may be completely different from traditional predictions. As can be seen from the above discussion, Young's interferences of photons, electrons, and plasmons actually marked the three generations of optical revolutions. Figure 1.32 shows a schematic of the three interference experiments.

The emergence of this new field can also be understood from the fabrication side. Similar to the motivation of "Atom to product" program proposed by DARPA, traditionally there is a gap between the fabrication of nanoscale structures and macroscopic optical devices. In traditional optics, optical elements such as lenses and mirrors are fabricated using mechanic tools, making the smallest characteristic dimension much larger than the wavelength (the roughness are much smaller than wave-

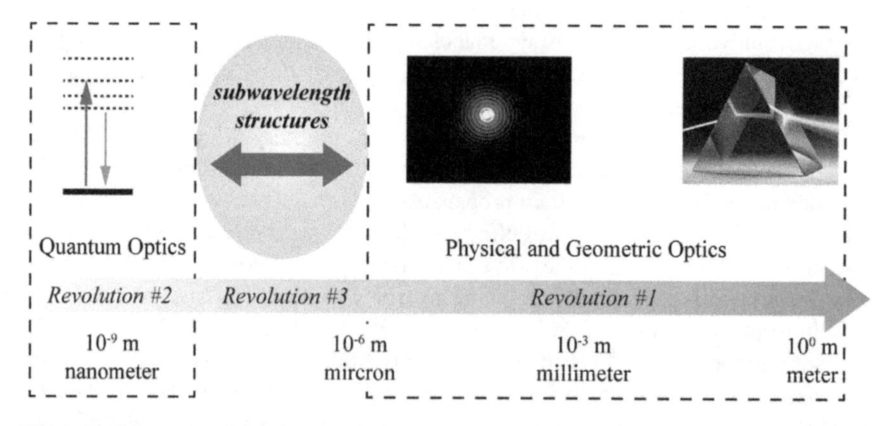

Fig. 1.31 Schematic of the three optical revolutions. The third-generation optical revolution is based on artificial subwavelength structures

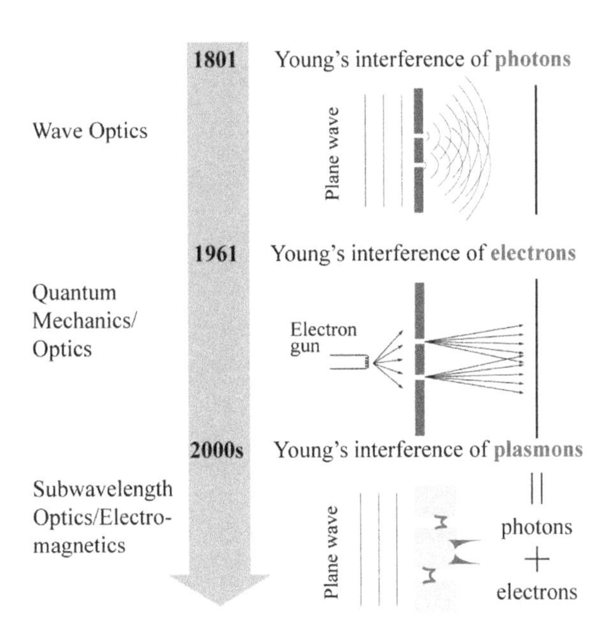

Fig. 1.32 Comparison of the three optical revolutions and three Young's interference experiments

length). Traditional quantum optical devices often utilize thin films or chemically synthesized nanostructures to control the quantum properties. Although the thickness of these films may be only several or tens of nanometers, the horizontal dimensions are still much larger than the wavelength.

Recently, the discipline of subwavelength optics and photonics has undergone tremendous progress, particularly in nanostructured engineered materials: metamaterials, plasmonic and dielectric subwavelength structures, subwavelength-integrated waveguides and photonic crystals. The novel optical properties found in these structures have opened new prospects for controlling and manipulating light at subwavelength scale in unconventional manners. To differentiate it from traditional optics and highlight its engineering applications, we term this new area as Engineering Optics 2.0 [2, 3]. As we have discussed elsewhere, the catenary optical fields and subwavelength catenary structures are playing important roles in Engineering Optics 2.0 [3, 82, 116].

The above timeline has also reflected by the milestones of photonics given by Nature [117]. As shown in Table 1.4, the third optical revolution has started as early as the 1980s and represented by the photonic crystals, plasmonics, and metamaterials.

At the last of this section, we would like to point out that besides the previous theoretical breakthroughs, there are some other important developing trends in EO 2.0:

(1) From natural materials to artificial structured materials.

The material properties are critical for the performances of optical devices and systems. With strong light–matter interaction in subwavelength structures, exotic materials parameters beyond natural materials could be obtained. As shown in Fig. 1.33,

Table 1.4 Timeline of photonics milestones [117]

Milestones	Time	Description
1	1600s–1800s	Debate on the character of light
2	1861	Maxwell's equations
3	1900	Planck's theory of black-body radiation
4	1905	Special relativity
5	1923	Compton effect
6	1947	Quantum electrodynamics
7	1948	Holograms
8	1954	Solar cells
9	1960	The laser
10	1961	Nonlinear optics
11	1963	Quantum optics
12	1964	Bell inequality
13	1966	Optical fibers
14	1970	CCD cameras
15		Semiconductor lasers
16	1981	High-resolution laser spectroscopy and frequency metrology
17	1982–1985	Quantum information
18	1987	Photonic crystals
19	1993	Blue light-emitting diodes
20	1998	Plasmonics
21	2000	Metamaterials
22	2001	Attosecond science
23	2006	Cavity optomechanics

both the permittivity and permeability could be arbitrarily tuned by controlling the electric and magnetic resonances. In contrast, natural materials often have only electric or magnetic response, and limited values of permittivity and permeability at a specific wavelength range. For instance, the available high-index dielectrics are often limited to TiO_2, Si, Ge, and Te, etc.

(2) The combination of computational optics and subwavelength optics.

In the last half century, electronic computers have changed our world dramatically. Combined with advanced algorithms and software, it has revolutionized the design of optical materials, devices, and systems. However, model-based design is still suffering from low efficiency and inadequate performance. In future, more efficient optimization algorithms should be developed based on artificial intelligence and big data base.

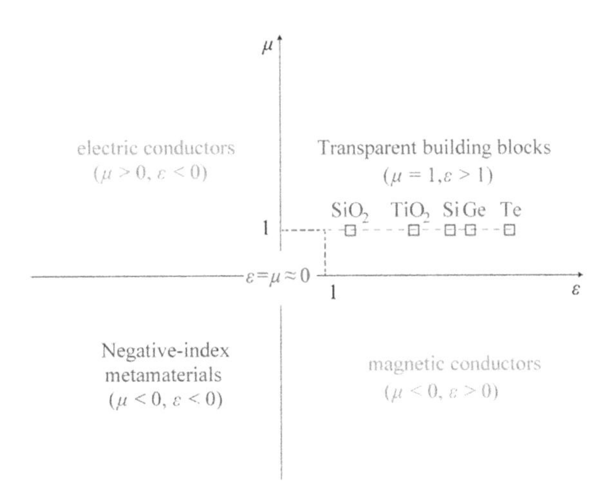

Fig. 1.33 Electric permittivity (ε) and magnetic permeability (μ) form four quadrants that represent the entire range of the isotropic electromagnetic response. All four quadrants can be covered by designing subwavelength structures. Reproduced from [110] with permission. Copyright 2016, Macmillan Publishers Limited

In imaging applications, computational optics has emerged as a new scheme for optical imaging [118, 119]. Compared with a traditional optical system which relies only on the hardware, the computational optics represents a combination of hardware and software. By coding the aperture with amplitude or phase distribution, the image obtained in the detector plane is an intermediate image, which must be digitally processed to reconstruct the real image. Through this approach, many novel applications such as seeing through scattering medium and reconstructing high-resolution images from low-resolution ones could be realized. In particular, when combined with subwavelength optics, the optical performances may be further expanded.

(3) From adaptive optics to smart optics

Adaptive optics refers to the adaptive adjusting of the optical system to realize diffraction-limited optical performance in a dynamic-changing environment. Traditionally, it has been widely used in astronomical observation to eliminate the influence of atmosphere turbulence, and in laser wavefront shaping.

In future, optical applications shall go to smart, which means the design, assembly, and adjusting may be completed in a more automatic way in response to the user's requirements [3, 120].

1.6 Overview of the Book

The organization of the book is as follows.

Chap. 2 gives the generalized optical theories in EO 2.0, including the generalized diffraction theory, generalized laws of refraction and reflection, generalized absorption, and radiation theory. These generalized optical theories, involving the light diffraction, refraction, reflection, absorption and radiation, break the fundamental limits of traditional optical theories and offer opportunities to various novel engineering applications.

Chap. 3 involves the material basis of EO 2.0, whose optical responses depend not only on the atomic/molecular properties but also the subwavelength geometries and spatial arrangements. The commonly used materials in EO 2.0 include noble metal, refractory plasmonic materials, transparent conducting oxides, III–V semiconductors, dielectric materials, phase change materials, two-dimensional materials, van der Waals materials, and some unique metamaterials, e.g., negative-index, near-zero-index, ultra-high-index, and hyperbolic metamaterials.

Chap. 4 presents the intelligent design platform of subwavelength structures, which includes the materials database and auto-modeling program that calls full-wave simulation software and intelligent optimizing without supervision.

Chap. 5 first reviews the status of manufacturing techniques for EO 1.0 and summarizes the challenges. Then, the advances in fabrication techniques for EO 2.0 are introduced. The major technological challenges need to be solved are discussed.

Chap. 6 discusses the newly developed optical sub-diffraction-limited microscopy technologies, which are based on the concepts of negative refractive lens, superlens, hyperlens, as well as intriguing phenomena of photonic nanojet and super-oscillation. Besides, sub-diffraction phase-contrast microscopy and surface microscopy are also introduced.

Chap. 7 discusses the recently emerged sub-diffraction-limited lithography technologies, including the subwavelength interference and sub-diffraction propagation of coupled SPPs, and the field enhancement of localized surface plasmon resonances. The operation principles, technical processes, illumination conditions, and equipments of sub-diffraction lithography are discussed in detail.

Chap. 8 discusses another branch of sub-diffraction-limited optics, i.e., sub-diffraction-limited telescopies, which have made great progresses based on recent investigations on pupil filters, metasurfaces, as well as super-oscillation effects.

Chap. 9 mainly focuses on the planar metalens in EO 2.0, which are theoretically based on the generalized laws of refraction and reflection. Some highly desired functional metalenses that are extremely difficult to realize in traditional optics have been introduced, including that with high numerical aperture, wide field-of-view, achromatic and multispectral imaging. The membrane imaging devices with advantages in large optical aperture, light weight, and low cost have also been discussed.

Chap. 10 mainly focuses on the generation and manipulation of various special beams through planar meta-devices. Some strange light–matter interactions (e.g., photonic spin Hall effect) and special beams (e.g., vortex beam, Bessel beam, and Airy beams) are generated through very compact devices, which may replace their traditional counterparts to realize better performance.

Chap. 11 pays special attention on the structural colors and meta-holographic display technologies. Benefiting from the powerful spectra and phase engineering ability, different types of structural colors and holograms have been introduced encompassing transmissive, reflective, polarization independent/dependent, tunable configurations.

Chap. 12 deals with the polarization manipulation in EO 2.0. Benefiting from the strong anisotropy and chirality in subwavelength structures, minimized waveplates, polarization rotators, polarization detection and imaging devices are introduced, which will significantly simplify the polarization-dependent optical systems.

Chap. 13 discusses the perfect optical absorption in EO 2.0. On the one hand, ultranarrow band absorbers are constructed by wave-impedance matching and free-bound two-wave exchanging. On the other hand, broadband absorbers are realized by modes hybrization, dispersion engineering, and optimization algorithm. The conventional thickness-bandwidth limit of the absorbers is discussed and avoided under some special conditions. Their potential applications in solar cells and biochemical sensing are also investigated.

Chap. 14 focuses on various radiation and optical phased array technologies in EO 2.0. Thermal radiation manipulation in both far field and near field are discussed, including spectrum, direction-selective radiation, super-Plankian radiation, and their intriguing applications in thermophotovoltaics, daytime radiation cooling, thermal cloak, and camouflage. High efficient light-emitting diodes, micro- and nanolasers, as well as chip-based optical phased arrays, are also discussed.

References

1. K. Iizuka, *Engineering Optics*, 3rd ed. (Springer, 2008)
2. X. Luo, Engineering optics 2.0: a revolution in optical materials, devices, and systems. ACS Photonics **5**, 4724–4738 (2018)
3. X. Luo, Subwavelength artificial structures: opening a new era for engineering optics. Adv. Mater. **25**, 1804680 (2019)
4. A. Nedelcu, V. Guériaux, A. Berurier, N. Brière de l'Isle, O. Huet, Multispectral and polarimetric imaging in the LWIR: Intersubband detectors as a versatile solution. Infrared Phys. Technol. **59**, 125–132 (2013)
5. X. Luo, Subwavelength optical engineering with metasurface waves. Adv. Opt. Mater. **6**, 1701201 (2018)
6. Willebrord Snellius, https://commons.wikimedia.org/wiki/File:Willebrord_Snellius.jpg
7. Pierre de Fermat, https://commons.wikimedia.org/wiki/File:Pierre_de_Fermat.jpg
8. Thomas Young, https://commons.wikimedia.org/wiki/File:LifeOfThomasYoung1855PeacockG.jpg
9. Augustin Fresnel, https://commons.wikimedia.org/wiki/File:Augustin_Fresnel.jpg
10. James Clerk Maxwell, https://commons.wikimedia.org/wiki/File:James_Clerk_Maxwell.png
11. Gustav Robert Kirchhoff, https://commons.wikimedia.org/wiki/File:Gustav_Robert_Kirchhoff.jpg
12. Ernst Abbe, https://commons.wikimedia.org/wiki/File:Ernst_Abbe.jpg
13. Max Planck, https://commons.wikimedia.org/wiki/File:Max_Planck_1933.jpg
14. Albert Einstein, https://commons.wikimedia.org/wiki/File:Albert_Einstein_(Nobel).png
15. D.B. Swinson, Chinese "magic" mirrors. Phys. Teach. **30**, 295–299 (1992)
16. R.K. Temple, *The Genius of China: 3,000 Years of Science, Discovery, and Invention* (Inner Traditions Rochester, VT, 2007)
17. E.S. Barr, Men and milestones in optics II. Thomas Young. Appl. Opt. **2**, 639–647 (1963)
18. R.P. Crease, The most beautiful experiment. Phys. World **15**, 19 (2002)
19. M. Born, E. Wolf, *Principles of Optics*: *Electromagnetic Theory of Propagation, Interference and Diffraction of Light*, 7th ed. (Cambridge University Press, 1999)

20. M. Pu, C. Wang, Y. Wang, X. Luo, Subwavelength electromagnetics below the diffraction limit. Acta Phys. Sin. **66**, 144101 (2017)
21. F. Qin, M. Hong, Breaking the diffraction limit in far field by planar metalens. Sci. China Phys. Mech. Astron. **60**, 044231 (2017)
22. R.P. Feynman, R.B. Leighton, M. Sands, *The Feynman Lectures on Physics* (Basic Books, 1963)
23. H. Gross, *Fundamentals of Technical Optics* (Wiley, 2005)
24. M. Freebody, Great strides in optical fabrication. Photonics Spectra **10**, 42–47 (2016)
25. R. Williamson, *Field Guide to Optical Fabrication* (SPIE, 2011)
26. F.Z. Fang, X.D. Zhang, A. Weckenmann, G.X. Zhang, C. Evans, Manufacturing and measurement of freeform optics. CIRP Ann. **62**, 823–846 (2013)
27. L. Rayleigh, XXXI. Investigations in optics, with special reference to the Spectroscope. Philos. Mag. Ser. **5**(8), 261–274 (1879)
28. L.W. Chen, Y. Zhou, M.X. Wu, M.H. Hong, Remote-mode microsphere nano-imaging: new boundaries for optical microscopes. Opto-Electron. Adv. **1**, 170001 (2018)
29. R. Gilmozzi, Giant telescopes of the future. Sci. Am. **5**, 66–71 (2006)
30. H.P. Stahl, Survey of cost models for space telescopes. Opt. Eng. **49**, 053005 (2010)
31. M. Totzeck, W. Ulrich, A. Gohnermeier, W. Kaiser, Semiconductor fabrication: pushing deep ultraviolet lithography to its limits. Nat. Photonics **1**, 629–631 (2007)
32. M. Planck, *The Theory of Heat Radiation* (P. Blakiston's Son & Co., 1914)
33. K.N. Rozanov, Ultimate thickness to bandwidth ratio of radar absorbers. IEEE Trans. Antennas Propag. **48**, 1230–1234 (2000)
34. Y. Wang, X. Ma, X. Li, M. Pu, X. Luo, Perfect electromagnetic and sound absorption via subwavelength holes array. Opto-Electron. Adv. **1**, 180013 (2018)
35. H.A. Atwater, A. Polman, Plasmonics for improved photovoltaic devices. Nat. Mater. **9**, 205–213 (2010)
36. M. Pu, P. Chen, Y. Wang, Z. Zhao, C. Huang, C. Wang, X. Ma, X. Luo, Anisotropic meta-mirror for achromatic electromagnetic polarization manipulation. Appl. Phys. Lett. **102**, 131906 (2013)
37. X. Ma, M. Pu, X. Li, Y. Guo, X. Luo, All-metallic wide-angle metasurfaces for multifunctional polarization manipulation. Opto-Electron. Adv. **2**, 180023 (2019)
38. Y.-J. Jen, A. Lakhtakia, C.-W. Yu, C.-F. Lin, M.-J. Lin, S.-H. Wang, J.-R. Lai, Biologically inspired achromatic waveplates for visible light. Nat. Commun. **2**, 363 (2011)
39. K. Robbie, M.J. Brett, A. Lakhtakia, Chiral sculptured thin films. Nature **384**, 616 (1996)
40. J.K. Gansel, M. Thiel, M.S. Rill, M. Decker, K. Bade, V. Saile, G. von Freymann, S. Linden, M. Wegener, Gold helix photonic metamaterial as broadband circular polarizer. Science **325**, 1513–1515 (2009)
41. X. Luo, M. Pu, X. Ma, X. Li, Taming the electromagnetic boundaries via metasurfaces: from theory and fabrication to functional devices. Int. J. Antennas Propag. **2015**, 204127 (2015)
42. A.I. Lvovsky, Fresnel equations, in *Encyclopedia of Optical Engineering* (Taylor & Francis, 2007), pp. 1–6
43. F. Bouchard, H. Mand, M. Mirhosseini, E. Karimi, R.W. Boyd, Achromatic orbital angular momentum generator. New J. Phys. **16**, 123006 (2014)
44. M. Vollmer, K.-P. Mollmann, *Infrared thermal imaging: fundamentals, research and applications* (Wiley-VCH Verlag GmbH & Co, KGaA, 2010)
45. M. Song, H. Yu, C. Hu, M. Pu, Z. Zhang, J. Luo, X. Luo, Conversion of broadband energy to narrowband emission through double-sided metamaterials. Opt. Express **21**, 32207–32216 (2013)
46. S. Fan, Photovoltaics: an alternative "Sun" for solar cells. Nat. Nanotechnol. **9**, 92–93 (2014)
47. X. Xie, X. Li, M. Pu, X. Ma, K. Liu, Y. Guo, X. Luo, Plasmonic metasurfaces for simultaneous thermal infrared invisibility and holographic illusion. Adv. Funct. Mater. **28**, 1706673 (2018)
48. G.P. Williams, Filling the THz gap—high power sources and applications. Rep. Prog. Phys. **69**, 301 (2006)

49. Committee on Nanophotonics Accessibility and Applicability, National Research Council, *Nanophotonics: Accessibility and Applicability* (National Academies Press, 2008).
50. R.P. Feynman, There's plenty of room at the bottom. Eng. Sci. **23**, 22–36 (1960)
51. X. Luo, Plasmonic metalens for nanofabrication. Natl. Sci. Rev. **5**, 137–138 (2018)
52. V.-C. Su, C.H. Chu, G. Sun, D.P. Tsai, Advances in optical metasurfaces: fabrication and applications [Invited]. Opt. Express **26**, 13148–13182 (2018)
53. A. She, S. Zhang, S. Shian, D.R. Clarke, F. Capasso, Large area metalenses: design, characterization, and mass manufacturing. Opt. Express **26**, 1573–1585 (2018)
54. T. Hu, C.-K. Tseng, Y.H. Fu, Z. Xu, Y. Dong, S. Wang, K.H. Lai, V. Bliznetsov, S. Zhu, Q. Lin, Y. Gu, Demonstration of color display metasurfaces via immersion lithography on a 12-inch silicon wafer. Opt. Express **26**, 19548–19554 (2018)
55. U.D. Zeitner, M. Oliva, F. Fuchs, D. Michaelis, T. Benkenstein, T. Harzendorf, E.-B. Kley, High performance diffraction gratings made by e-beam lithography. Appl. Phys. A **109**, 789–796 (2012)
56. E.H. Synge, A suggested model for extending microscopic resolution into the ultra-microscopic region. Philos. Mag. **6**, 356–362 (1928)
57. X. Zhang, Z. Liu, Superlenses to overcome the diffraction limit. Nat. Mater. **7**, 435–441 (2008)
58. J.B. Pendry, Negative refraction makes a perfect lens. Phys. Rev. Lett. **85**, 3966–3969 (2000)
59. X. Luo, T. Ishihara, Surface plasmon resonant interference nanolithography technique. Appl. Phys. Lett. **84**, 4780–4782 (2004)
60. X. Luo, T. Ishihara, Subwavelength photolithography based on surface-plasmon polariton resonance. Opt. Express **12**, 3055–3065 (2004)
61. B. Wood, J.B. Pendry, D.P. Tsai, Directed subwavelength imaging using a layered metal-dielectric system. Phys. Rev. B **74**, 115116 (2006)
62. X. Luo, D. Tsai, M. Gu, M. Hong, Subwavelength interference of light on structured surfaces. Adv. Opt. Photonics **10**, 757–842 (2018)
63. M. Pu, Y. Guo, X. Li, X. Ma, X. Luo, Revisitation of extraordinary Young's interference: from catenary optical fields to spin-orbit interaction in metasurfaces. ACS Photonics **5**, 3198–3204 (2018)
64. M. Rahmani, G. Leo, I. Brener, A. Zayats, S. Maier, C. De Angelis, H. Tan, V. F. Gili, F. Karouta, R. Oulton, Nonlinear frequency conversion in optical nanoantennas and metasurfaces: materials evolution and fabrication. Opto-Electron. Adv. **1**, 180021 (2018)
65. H. Shi, X. Luo, C. Du, Young's interference of double metallic nanoslit with different widths. Opt. Express **15**, 11321–11327 (2007)
66. T. Xu, C. Du, C. Wang, X. Luo, Subwavelength imaging by metallic slab lens with nanoslits. Appl. Phys. Lett. **91**, 201501 (2007)
67. L. Verslegers, P.B. Catrysse, Z. Yu, J.S. White, E.S. Barnard, M.L. Brongersma, S. Fan, Planar lenses based on nanoscale slit arrays in a metallic film. Nano Lett. **9**, 235–238 (2009)
68. T. Xu, C. Wang, C. Du, X. Luo, Plasmonic beam deflector. Opt. Express **16**, 4753–4759 (2008)
69. J. Yan, Y. Guo, M. Pu, X. Li, X. Ma, X. Luo, High-efficiency multi-wavelength metasurface with complete independent phase control. Chin. Opt. Lett. **16**, 050003 (2018)
70. Y. Guo, J. Yan, M. Pu, X. Li, X. Ma, Z. Zhao, X. Luo, Ultra-wideband manipulation of electromagnetic waves by bilayer scattering engineered gradient metasurface. RSC Adv. **8**, 13061–13066 (2018)
71. T. Xu, Y. Zhao, D. Gan, C. Wang, C. Du, X. Luo, Directional excitation of surface plasmons with subwavelength slits. Appl. Phys. Lett. **92**, 101501 (2008)
72. J. Sun, X. Wang, T. Xu, Z.A. Kudyshev, A.N. Cartwright, N.M. Litchinitser, Spinning light on the nanoscale. Nano Lett. **14**, 2726–2729 (2014)
73. L. Lin, X.M. Goh, L.P. McGuinness, A. Roberts, Plasmonic lenses formed by two-dimensional nanometric cross-shaped aperture arrays for Fresnel-region focusing. Nano Lett. **10**, 1936 (2010)
74. S. Ishii, V.M. Shalaev, A.V. Kildishev, Holey-metal lenses: sieving single modes with proper phases. Nano Lett. **13**, 159–163 (2013)

75. P. Lalanne, S. Astilean, P. Chavel, E. Cambril, H. Launois, Blazed binary subwavelength gratings with efficiencies larger than those of conventional échelette gratings. Opt. Lett. **23**, 1081–1083 (1998)
76. D.T. Moore, Gradient-index optics-A review. Appl. Opt. **19**, 1035–1038 (1980)
77. J. Evans, M. Rosenquist, "F = ma" optics. Am. J. Phys. **54**, 876–883 (1986)
78. N. Yu, P. Genevet, M.A. Kats, F. Aieta, J.-P. Tetienne, F. Capasso, Z. Gaburro, Light propagation with phase discontinuities: generalized laws of reflection and refraction. Science **334**, 333–337 (2011)
79. X. Luo, Principles of electromagnetic waves in metasurfaces. Sci. China-Phys. Mech. Astron. **58**, 594201 (2015)
80. Y. Xu, Y. Fu, H. Chen, Planar gradient metamaterials. Nat. Rev. Mater. **1**, 16067 (2016)
81. M. Khorasaninejad, W.T. Chen, R.C. Devlin, J. Oh, A.Y. Zhu, F. Capasso, Metalenses at visible wavelengths: diffraction-limited focusing and subwavelength resolution imaging. Science **352**, 1190–1194 (2016)
82. M. Pu, X. Li, X. Ma, Y. Wang, Z. Zhao, C. Wang, C. Hu, P. Gao, C. Huang, H. Ren, X. Li, F. Qin, J. Yang, M. Gu, M. Hong, X. Luo, Catenary optics for achromatic generation of perfect optical angular momentum. Sci. Adv. **1**, e1500396 (2015)
83. F. Aieta, P. Genevet, M.A. Kats, N. Yu, R. Blanchard, Z. Gaburro, F. Capasso, Aberration-free ultra-thin flat lenses and axicons at telecom wavelengths based on plasmonic metasurfaces. Nano Lett. **12**, 4932–4936 (2012)
84. F. Capasso, The future and promise of flat optics: a personal perspective. Nanophotonics **7**, 953 (2018)
85. P. Lalanne, P. Chavel, Metalenses at visible wavelengths: past, present, perspectives. Laser Photonics Rev. **11**, 1600295 (2017)
86. G. Cao, X. Gan, H. Lin, B. Jia, An accurate design of graphene oxide ultrathin flat lens based on Rayleigh-Sommerfeld theory. Opto-Electron. Adv. **1**, 180012 (2018)
87. S. Wang, X. Ouyang, Z. Feng, Y. Cao, M. Gu, X. Li, Diffractive photonic applications mediated by laser reduced graphene oxides. Opto-Electron. Adv. **1**, 170002 (2018)
88. D.C. Flanders, Submicrometer periodicity gratings as artificial anisotropic dielectrics. Appl. Phys. Lett. **42**, 492–494 (1983)
89. P. Lalanne, J.-P. Hugonin, High-order effective-medium theory of subwavelength gratings in classical mounting: application to volume holograms. J. Opt. Soc. Am. A **15**, 1843–1851 (1998)
90. G. Nordin, P. Deguzman, Broadband form birefringent quarter-wave plate for the mid-infrared wavelength region. Opt. Express **5**, 163–168 (1999)
91. Y. Guo, Y. Wang, M. Pu, Z. Zhao, X. Wu, X. Ma, C. Wang, L. Yan, X. Luo, Dispersion management of anisotropic metamirror for super-octave bandwidth polarization conversion. Sci. Rep. **5**, 8434 (2015)
92. Y. Guo, L. Yan, W. Pan, B. Luo, Achromatic polarization manipulation by dispersion management of anisotropic meta-mirror with dual-metasurface. Opt. Express **23**, 27566–27575 (2015)
93. M. Pu, Z. Zhao, Y. Wang, X. Li, X. Ma, C. Hu, C. Wang, C. Huang, X. Luo, Spatially and spectrally engineered spin-orbit interaction for achromatic virtual shaping. Sci. Rep. **5**, 9822 (2015)
94. Y. Guo, L. Yan, W. Pan, L. Shao, Scattering engineering in continuously shaped metasurface: an approach for electromagnetic illusion. Sci. Rep. **6**, 30154 (2016)
95. A.G. Fox, An adjustable wave-guide phase changer. Proc. IRE **35**, 1489–1498 (1947)
96. S. Pancharatnam, Generalized theory of interference, and its applications. Part I. Coherent pencils. Proc. Indian Acad. Sci. **44**, 247–262 (1956)
97. M.V. Berry, Quantal phase factors accompanying adiabatic changes. Proc. R. Soc. Lond. Math. Phys. Eng. Sci. **392**, 45–57 (1984)
98. R. Bhandari, Polarization of light and topological phases. Phys. Rep. **281**, 1–64 (1997)
99. M. Pu, X. Ma, Y. Guo, X. Li, X. Luo, Methodologies for on-demand dispersion engineering of meta-surface waves. Adv. Opt. Mater. (2019)

100. X. Luo, D. Tsai, M. Gu, M. Hong, *Extraordinary optical fields in nanostructures: from sub-diffraction-limited optics to sensing and energy conversion* (Chem. Soc, Rev, 2019)
101. T. Xu, Y.-K. Wu, X. Luo, L.J. Guo, Plasmonic nanoresonators for high-resolution colour filtering and spectral imaging. Nat. Commun. **1**, 59 (2010)
102. S. Wang, P.C. Wu, V.-C. Su, Y.-C. Lai, M.-K. Chen, H.Y. Kuo, B.H. Chen, Y.H. Chen, T.-T. Huang, J.-H. Wang, R.-M. Lin, C.-H. Kuan, T. Li, Z. Wang, S. Zhu, D.P. Tsai, A broadband achromatic metalens in the visible. Nat. Nanotechnol. **13**, 227–232 (2018)
103. Q. Feng, M. Pu, C. Hu, X. Luo, Engineering the dispersion of metamaterial surface for broadband infrared absorption. Opt. Lett. **37**, 2133–2135 (2012)
104. X. Luo, M. Pu, X. Li, X. Ma, Broadband spin Hall effect of light in single nanoapertures. Light Sci. Appl. **6**, e16276 (2017)
105. Y. Huang, M. Pu, P. Gao, Z. Zhao, X. Li, X. Ma, X. Luo, Ultra-broadband large-scale infrared perfect absorber with optical transparency. Appl. Phys. Express **10**, 112601 (2017)
106. H.A. Atwater, The promise of plasmonics. Sci. Am. **296**, 56–62 (2007)
107. D. Lin, P. Fan, E. Hasman, M.L. Brongersma, Dielectric gradient metasurface optical elements. Science **345**, 298–302 (2014)
108. A. Arbabi, Y. Horie, M. Bagheri, A. Faraon, Dielectric metasurfaces for complete control of phase and polarization with subwavelength spatial resolution and high transmission. Nat. Nanotechnol. **10**, 937–943 (2015)
109. A.I. Kuznetsov, A.E. Miroshnichenko, M.L. Brongersma, Y.S. Kivshar, B. Luk'yanchuk, Optically resonant dielectric nanostructures. Science **354**, aag2472 (2016)
110. S. Jahani, Z. Jacob, All-dielectric metamaterials. Nat. Nanotechnol. **11**, 23–36 (2016)
111. Y. Huang, J. Luo, M. Pu, Y. Guo, Z. Zhao, X. Ma, X. Li, X. Luo Catenary electromagnetics for ultrabroadband lightweight absorbers and large-scale flat antennas. Adv. Sci. 1801691 (2019)
112. C.A. Dirdal, J. Skaar, Superpositions of Lorentzians as the class of causal functions. Phys. Rev. A **88**, 033834 (2013)
113. Z. Zhao, M. Pu, Y. Wang, X. Luo, The generalized laws of refraction and reflection. Opto-Electron. Eng. **44**, 129–139 (2017)
114. M. Pu, X. Ma, Y. Guo, X. Li, X. Luo, Theory of microscopic meta-surface waves based on catenary optical fields and dispersion. Opt. Express **26**, 19555–19562 (2018)
115. H.F. Schouten, N. Kuzmin, G. Dubois, T.D. Visser, G. Gbur, P.F.A. Alkemade, H. Blok, G.W.'t Hooft, D. Lenstra, E.R. Eliel, Plasmon-assisted two-slit transmission: young's experiment revisited. Phys. Rev. Lett. **94**, 053901 (2005)
116. X. Luo, *Catenary Optics* (Springer Singapore, 2019)
117. Nature Milestones: Photons, www.nature.com/milestones/photons
118. A. Levin, R. Fergus, F. Durand, W.T. Freeman, Image and depth from a conventional camera with a coded aperture. ACM Trans. Graph. **26**, 70 (2007)
119. Y. Altmann, S. McLaughlin, M.J. Padgett, V.K. Goyal, A.O. Hero, D. Faccio, Quantum-inspired computational imaging. Science **361**, eaat2298 (2018)
120. A. Nemati, Q. Wang, M. Hong, J. Teng, Tunable and reconfigurable metasurfaces and metadevices. Opto-Electron. Adv. **1**, 180009 (2018)

Chapter 2
Theoretical Basis

Abstract This chapter first describes the theories and laws in classic optics, and highlights their drawbacks and challenges in engineering applications. Then the macroscopic and microscopic theories of meta-surface-waves are presented as a cornerstone of EO 2.0. Subsequently, the generalized theories and laws of diffraction, reflection and refraction, absorption, and radiation are discussed in detail.

Keywords Snell's law · Fresnel's equations · Diffraction limit · Generalized laws of reflection and refraction

2.1 Theories and Laws in Classic Optics

2.1.1 From Fermat's Principle to Snell's Law and Fresnel's Equations

Reflection and refraction are two most basic phenomena of light–matter interaction at a flat interface. The law of reflection, which assumes the reflected light lies in the incident plane and the reflection angle equals to the incident one, was known to ancient Greeks. The law of refraction, however, was later discovered. Although this law is named after Snell, earlier description can be found in the book written by Ibn Sahl as early as in 984. As an alternative description, the laws of reflection and refraction follow the *Principle of Least Time* proposed by Pierre de Fermat in 1662. After the invention of Maxwell's equations in 1865, it was found that the laws of reflection and refraction could be completely derived using the boundary conditions regarding the electric and magnetic fields. Furthermore, using the boundary conditions, the amplitudes of the reflection and refraction parts can be obtained, agreeing well with Fresnel's equations that have been obtained using a "wrong" theory derived from aether. Since then, the laws of reflection and refraction have been widely utilized in the design of various mirrors and lenses. By milling the shapes and polishing the surfaces, the geometric aberrations can be minimized to focus light into a tiny spot or to form clear images with telescope or microscopy systems.

X. Luo, *Engineering Optics 2.0*, https://doi.org/10.1007/978-981-13-5755-8_2

In the following, the laws of reflection and refraction are derived from both the Fermat's principle and Maxwell's equations [1, 2]. According to Fermat's principle, optical rays traveling between two points, A and B, follow a path such that the time between the two points is an extremum related to neighboring paths. Using the variation symbol δ, this is expressed as:

$$\delta \int_{A}^{B} n(\mathbf{r})\,\mathrm{d}s = 0. \tag{2.1.1}$$

Although the variation means that the optical path length is either minimized or maximized, or is a point of inflection, it is usually a minimum; thus, light rays are said to travel along the path of least time.

Consider the light ray reflected by a mirror as shown in Fig. 2.1. A ray of light from point A is reflected by the mirror surface at point P before arriving at point B, which has a horizontal distance l from point A. Supposing the vertical values of A and B are h_1 and h_2, one can calculate the length of each path and divide the length by the speed of light to determine the time required to travel between the two points:

$$t = \frac{\sqrt{x^2 + h_1^2}}{c} + \frac{\sqrt{(l - x)^2 + h_2^2}}{c}. \tag{2.1.2}$$

To minimize the time, let us set the derivative of the time with respect to x equal to zero,

$$\frac{\mathrm{d}t}{\mathrm{d}x} = \frac{x}{c\sqrt{x^2 + h_1^2}} + \frac{-(l - x)}{c\sqrt{(l - x)^2 + h_2^2}} = 0. \tag{2.1.3}$$

There is

$$\frac{x}{\sqrt{x^2 + h_1^2}} = \frac{(l - x)}{\sqrt{(l - x)^2 + h_2^2}}. \tag{2.1.4}$$

Obviously, it means that the reflection angle should be equal to the incident angle:

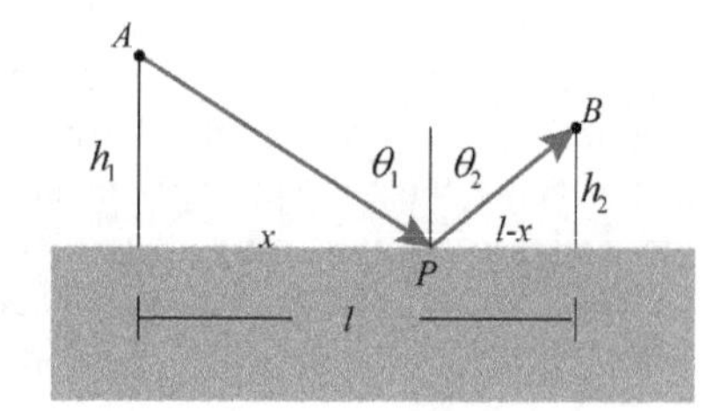

Fig. 2.1 Schematic of the shortest optical path connecting the two points A and B via the mirror

$$\theta_1 = \theta_2. \tag{2.1.5}$$

Fermat's principle can also be applied to the refraction case. To obtain the time to travel between the two points as shown in Fig. 2.2, the speed of light in the medium should be considered to obtain the time:

$$t = \frac{\sqrt{x^2 + h_1^2}}{c/n_1} + \frac{\sqrt{(l-x)^2 + h_2^2}}{c/n_2}. \tag{2.1.6}$$

Once again, the derivation of time with respect to x is:

$$\frac{\mathrm{d}t}{\mathrm{d}x} = \frac{2n_1 x}{c\sqrt{x^2 + h_1^2}} + \frac{-2n_2(l-x)}{c\sqrt{(l-x)^2 + h_2^2}} = 0. \tag{2.1.7}$$

There is

$$\frac{n_1 x}{\sqrt{x^2 + h_1^2}} = \frac{n_2(l-x)}{\sqrt{(l-x)^2 + h_2^2}}, \tag{2.1.8}$$

which can be also written as

$$n_1 \sin\theta_1 = n_2 \sin\theta_2. \tag{2.1.9}$$

This is just Snell's law.

Now let us consider the above problem in the framework of Maxwell's equations. The incident, reflected, and refracted plane waves could be written as:

$$\begin{aligned} \boldsymbol{E}_i &= \boldsymbol{E}_0 \exp(in_1 \boldsymbol{k} \cdot \boldsymbol{r} - i\omega t) \\ \boldsymbol{H}_i &= \boldsymbol{H}_0 \exp(in_1 \boldsymbol{k} \cdot \boldsymbol{r} - i\omega t) \end{aligned}, \tag{2.1.10}$$

$$\begin{aligned} \boldsymbol{E}_r &= r\boldsymbol{E}_0 \exp(in_1 \boldsymbol{k}_r \cdot \boldsymbol{r} - i\omega t) \\ \boldsymbol{H}_r &= r\boldsymbol{H}_0 \exp(in_1 \boldsymbol{k}_r \cdot \boldsymbol{r} - i\omega t) \end{aligned}, \tag{2.1.11}$$

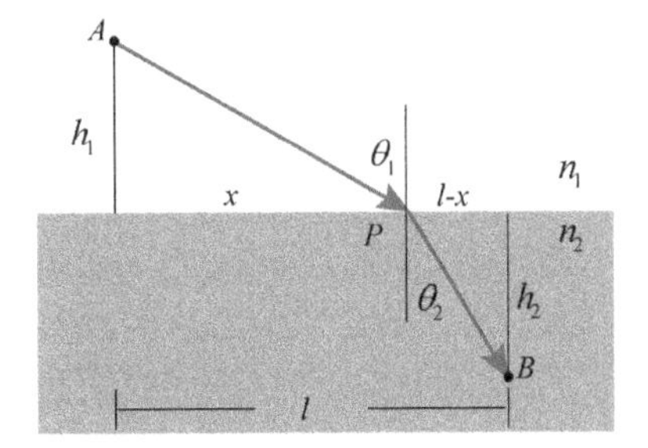

Fig. 2.2 Schematic of the refraction geometry. A and B are two points in media of refractive indices n_1 and n_2

and

$$\begin{aligned} \boldsymbol{E}_t &= t\boldsymbol{E}_t \exp(in_2\boldsymbol{k}_t \cdot \boldsymbol{r} - i\omega t) \\ \boldsymbol{H}_t &= t\boldsymbol{H}_t \exp(in_2\boldsymbol{k}_t \cdot \boldsymbol{r} - i\omega t) \end{aligned}, \tag{2.1.12}$$

where r and t are the reflection and transmission coefficients. According to the boundary conditions, i.e., the tangential electric and magnetic fields should be continuous at the interface; thus, there is:

$$k_{i\|} = k_{r\|} = k_{t\|}. \tag{2.1.13}$$

If θ_i, θ_r, and θ_t are the angles defined by the incident, reflected, and refracted rays with respect to the normal, there are:

$$n_1 k_0 \sin\theta_i = n_1 k_0 \sin\theta_r = n_2 k_0 \sin\theta_t. \tag{2.1.14}$$

This is actually the laws of reflection and refraction:

$$n_1 \sin\theta_i = n_1 \sin\theta_r = n_2 \sin\theta_t. \tag{2.1.15}$$

Fermat's principle and the laws of reflection and refraction could only determine the propagation direction of light rays. To get the reflection and transmission coefficients, Fresnel's equations should be utilized. Here we give a simple explanation of this important issue. As shown in Fig. 2.3, the boundary conditions require that the tangent components of both electric fields E and magnetic fields H are continuous across the interface between two dielectric media. For both the transverse electric (TE) and transverse magnetic (TM) polarizations, the boundary conditions can be written as

$$\begin{cases} 1 + r = t \\ Y_1(1 - r) = Y_2 t \end{cases}, \tag{2.1.16}$$

where $r = E_{yr}/E_{yi}$, $t = E_{yt}/E_{yi}$ for TE polarization and $r = E_{xr}/E_{xi}$, $t = E_{xt}/E_{xi}$ for TM polarization. The horizontal admittances for each medium are defined as $Y_i = H_{xi}/E_{yi}$ for TE polarization and $Y_i = H_{yi}/E_{xi}$ for TM polarization (here i is the index of layer). From Maxwell's equations, the following relations can be obtained:

$$\begin{cases} \nabla \times E = -\mu\mu_0 \dfrac{\partial H}{\partial t} \\ \nabla \times H = \varepsilon\varepsilon_0 \dfrac{\partial E}{\partial t} \end{cases}. \tag{2.1.17}$$

Appling Eq. (2.1.17) to Eq. (2.1.16), the horizontal admittances can be written as:

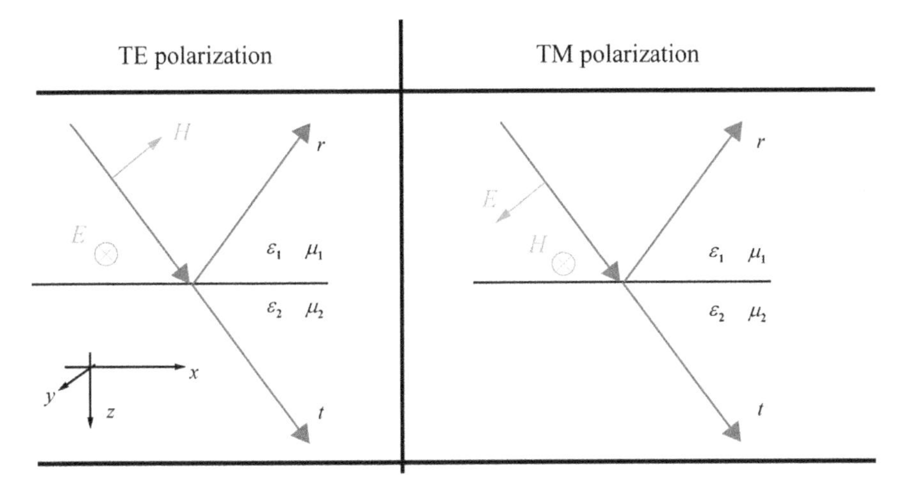

Fig. 2.3 Boundary conditions at the interface between two media

$$Y_i = \begin{cases} \dfrac{k_{zi}}{\mu_i \mu_0 \omega} = \dfrac{\sqrt{\varepsilon_i \mu_i k_0^2 - k_x^2}}{\mu_i \mu_0 \omega} & \text{for TE} \\ \dfrac{\varepsilon_i \varepsilon_0 \omega}{k_{zi}} = \dfrac{\varepsilon_i \varepsilon_0 \omega}{\sqrt{\varepsilon_i \mu_i k_0^2 - k_x^2}} & \text{for TM} \end{cases}. \tag{2.1.18}$$

Here k_x is the horizontal wavevector in this system, which is continuous for all planar multilayers. By assuming that light is illuminated from air, the incidence angle is $\theta_i = \arcsin(k_x/k_0)$. From the above equations, the reflection and transmission coefficients can be then written as:

$$\begin{cases} t = \dfrac{2Y_1}{Y_1 + Y_2} \\ r = \dfrac{Y_1 - Y_2}{Y_1 + Y_2} \end{cases}. \tag{2.1.19}$$

It should be noted that this is identical to the traditional Fresnel's equations, except that the horizontal wavevector instead of the ambiguous incidence angle is used here. For non-magnetic materials, Fresnel's equation can be also written as:

$$\begin{cases} t = \dfrac{2\sqrt{\varepsilon_1 k_0^2 - k_x^2}}{\sqrt{\varepsilon_1 k_0^2 - k_x^2} + \sqrt{\varepsilon_2 k_0^2 - k_x^2}} \\ r = \dfrac{\sqrt{\varepsilon_1 k_0^2 - k_x^2} - \sqrt{\varepsilon_2 k_0^2 - k_x^2}}{\sqrt{\varepsilon_1 k_0^2 - k_x^2} + \sqrt{\varepsilon_2 k_0^2 - k_x^2}} \end{cases} \tag{2.1.20}$$

for TE wave, and

$$\begin{cases} t = \dfrac{2\dfrac{\varepsilon_1}{\sqrt{\varepsilon_1 k_0^2 - k_x^2}}}{\dfrac{\varepsilon_1}{\sqrt{\varepsilon_1 k_0^2 - k_x^2}} + \dfrac{\varepsilon_2}{\sqrt{\varepsilon_2 k_0^2 - k_x^2}}} \\ r = \dfrac{\dfrac{\varepsilon_1}{\sqrt{\varepsilon_1 k_0^2 - k_x^2}} - \dfrac{\varepsilon_2}{\sqrt{\varepsilon_2 k_0^2 - k_x^2}}}{\dfrac{\varepsilon_1}{\sqrt{\varepsilon_1 k_0^2 - k_x^2}} + \dfrac{\varepsilon_2}{\sqrt{\varepsilon_2 k_0^2 - k_x^2}}} \end{cases} \tag{2.1.21}$$

for TM wave.

Considering the fact that $k_x = n_1 k_0 \sin\theta_1 = n_2 k_0 \sin\theta_2$, the above equations could be written as:

$$\begin{cases} t = \dfrac{2n_1 \cos\theta_1}{n_1 \cos\theta_1 + n_2 \cos\theta_2} \\ r = \dfrac{n_1 \cos\theta_1 - n_2 \cos\theta_2}{n_1 \cos\theta_1 + n_2 \cos\theta_2} \end{cases} \tag{2.1.22}$$

for TE wave and

$$\begin{cases} t = \dfrac{2n_1 \sec\theta_1}{n_1 \sec\theta_1 + n_2 \sec\theta_2} \\ r = \dfrac{n_1 \sec\theta_1 - n_2 \sec\theta_2}{n_1 \sec\theta_1 + n_2 \sec\theta_2} \end{cases} \tag{2.1.23}$$

for TM wave. Note that Eq. (2.1.23) can be also written as:

$$\begin{cases} t = \dfrac{2n_1 \cos\theta_2}{n_1 \cos\theta_2 + n_2 \cos\theta_1} \\ r = \dfrac{n_1 \cos\theta_2 - n_2 \cos\theta_1}{n_1 \cos\theta_2 + n_2 \cos\theta_1} \end{cases}. \tag{2.1.24}$$

The difference between Eq. (2.1.19) and traditional Fresnel's equations is because the coefficients are defined with respect to the tangential electric components. Obviously, this impedance description is more useful in the description of evanescent waves where the horizontal wavevector is larger than that in the background materials.

2.1.2 *From Kirchhoff's Diffraction Theory to the Diffraction Limit*

Diffraction theory of light is one of the backbones of modern optics. There are many pioneers and scientists who contributed to the wave theory of light, but the transition from corpuscular theory to wave theory takes almost two hundred years during the seventeenth and eighteenth centuries. Perhaps Robert Hooke was the first to think that light consists of rapid vibrations propagated instantaneously, while Christian Huygens proposed the celebrated principle named after him, which states that every point of the wavefront can be regarded as the center of a new disturbance propagated in the form of spherical waves, and the wavefront at any later instant may be regarded as the envelope of these wavelets. By supplementing Huygens's construction with the postulate that the secondary wavelets mutually interfere, Fresnel gave a more rigorous description of the diffraction phenomena, and this was called the Huygens–Fresnel principle. Although this principle can be adequately used to design diffractive devices such as Fresnel's zone plates, it is not accurate enough until Gustav Robert Kirchhoff put it on a sounder mathematical basis by showing that this principle can be viewed as an approximate form of a certain integral theorem. Kirchhoff's theory can be applied to the diffraction of scalar waves and is usually quite adequate for the treatment of the majority of problems in engineering optics [3].

Kirchhoff's diffraction integral is a solution of the scalar, time-independent wave equation using Green's relation [3]:

$$\oint_S \left(U\frac{\partial G}{\partial \vec{n}} - G\frac{\partial U}{\partial \vec{n}} \right) \mathrm{d}S = - \iiint_V (U\Delta G - G\Delta U)\mathrm{d}V, \tag{2.1.25}$$

where V denotes an arbitrary closed volume with surface S, $\vec{n}$ is the inner normal vector of this surface, U and G are functions that have a second continuous derivative in V, $\mathrm{d}S$ and $\mathrm{d}V$ are infinitesimal surface and volume elements, respectively.

If U and G both satisfy the free-space Helmholtz equation

$$\begin{aligned} \Delta U + k^2 U = 0 \\ \Delta G + k^2 G = 0 \end{aligned}, \tag{2.1.26}$$

the volume integral is zero and there is

$$\oint_S \left(U\frac{\partial G}{\partial \vec{n}} - G\frac{\partial U}{\partial \vec{n}} \right) dS = 0. \tag{2.1.27}$$

Now consider G as an analogue to the "probe-charge" of electrostatic theory: A spherical wave (the free-space Green's function) that emanates from a point $\vec{r}$ inside the volume V (Fig. 2.4):

$$G = \frac{\mathrm{e}^{ik\left|\vec{r}-\vec{r}'\right|}}{\left|\vec{r}-\vec{r}'\right|}. \tag{2.1.28}$$

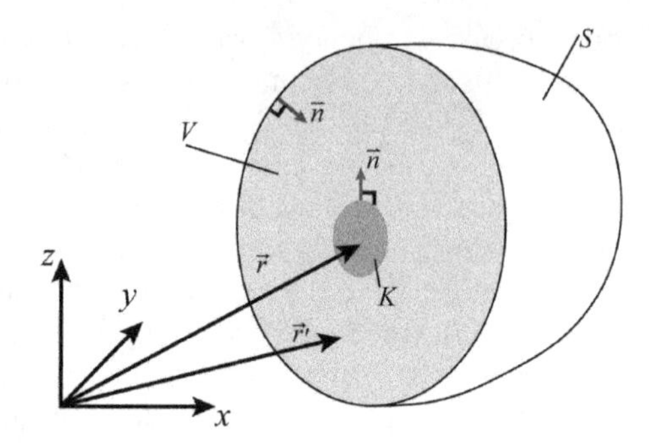

Fig. 2.4 Slice through the volume V in which Green's relation is applied. A small sphere around the point of observation is excluded from the integration

Inserting it into Green's relation, the point of origin of the spherical wave should be excluded because there it becomes singular. To do this, a small sphere is drawn around $\vec{r}$ and Green's relation is applied to the volume that remains

$$\oint_S \left(U\frac{\partial G}{\partial \vec{n}} - G\frac{\partial U}{\partial \vec{n}}\right)\mathrm{d}S + \oint_K \left(U\frac{\partial G}{\partial \vec{n}} - G\frac{\partial U}{\partial \vec{n}}\right)\mathrm{d}S = 0, \tag{2.1.29}$$

where S is the outer surface of the volume and K is the surface of the recessed sphere.

The integral over the surface of the recessed sphere in the limit of vanishing radius provides a constant contribution

$$P = \lim_{R\to 0}\oint_K \left(U\frac{\partial G}{\partial \vec{n}} - G\frac{\partial U}{\partial \vec{n}}\right)\mathrm{d}S. \tag{2.1.30}$$

Since

$$G = \frac{\mathrm{e}^{ikR}}{R}, \tag{2.1.31}$$

there is

$$\frac{\partial G}{\partial \vec{n}} = \vec{\nabla}G\cdot\vec{n} = \left(ik - \frac{1}{R}\right)\frac{\mathrm{e}^{ikR}}{R} \tag{2.1.32}$$

and

$$P = \lim_{R\to 0}\left(U\left(ik - \frac{1}{R}\right)\frac{\mathrm{e}^{ikR}}{R}4\pi R^2\right) - \lim_{R\to 0} G\oint_K \frac{\partial U}{\partial \vec{n}}\mathrm{d}S. \tag{2.1.33}$$

The second term in the above equation is zero because

$$\begin{aligned}\lim_{R\to 0} G\oint_K \frac{\partial U}{\partial \vec{n}}\mathrm{d}S &= \lim_{R\to 0}\frac{\mathrm{e}^{ikR}}{R}\oint_K \frac{\partial U}{\partial \vec{n}}R^2\mathrm{d}\Omega \\ &= \lim_{R\to 0}\mathrm{e}^{ikR}\oint_K \frac{\partial U}{\partial \vec{n}}R\mathrm{d}\Omega = 0\end{aligned}, \tag{2.1.34}$$

where Ω is the stereo angle ($\mathrm{d}S = R^2\mathrm{d}\Omega$). As a result, there is

$$P = \lim_{R\to 0}\left(U\left(ik - \frac{1}{R}\right)\frac{\mathrm{e}^{ikR}}{R}4\pi R^2\right) = -4\pi U. \tag{2.1.35}$$

Finally, there is

$$\oint_S \left(U\frac{\partial G}{\partial \vec{n}} - G\frac{\partial U}{\partial \vec{n}}\right)\mathrm{d}S = 4\pi U, \tag{2.1.36}$$

and Kirchhoff's diffraction integral is written as

$$U(\vec{r}) = \frac{1}{4\pi}\oint_S \left(U\left(\vec{r}'\right)\frac{\partial}{\partial \vec{n}}\frac{\mathrm{e}^{ik\left|\vec{r}-\vec{r}'\right|}}{\left|\vec{r}-\vec{r}'\right|} - \frac{\mathrm{e}^{ik\left|\vec{r}-\vec{r}'\right|}}{\left|\vec{r}-\vec{r}'\right|}\frac{\partial}{\partial \vec{n}}U\left(\vec{r}'\right)\right)\mathrm{d}S. \tag{2.1.37}$$

It should be noted that Kirchhoff's diffraction integral is only suitable for scalar wavefunction. In the subwavelength scale, the vectorial property is significant; thus, this simple equation is not valid. In the mesofield and far field, the diffraction can be simplified into Fresnel and Fraunhofer's diffraction. Here we would like to use the latter to derive the well-known diffraction limit in traditional optics. Considering the diffraction of light when the following equation is met,

$$z \gg \frac{\pi D^2}{\lambda}, \tag{2.1.38}$$

there is

$$U(\vec{r}) = \frac{i\mathrm{e}^{ikz}}{\lambda|z|}\mathrm{e}^{ik\frac{x^2+y^2}{2z}}\iint_A U(\vec{r})\mathrm{e}^{-ik\frac{xx'+yy'}{z}}\mathrm{d}x'\mathrm{d}y', \tag{2.1.39}$$

where D is the aperture of the screen and λ is the wavelength. Note that this is just a Fourier transform of the field and can be recast into the following form

$$U(\vec{r}) = \frac{i\mathrm{e}^{ikz}}{\lambda|z|}\mathrm{e}^{ik\frac{x^2+y^2}{2z}}F\{U(\vec{r})\}, \tag{2.1.40}$$

where the angular frequencies are $\upsilon_x = x/\lambda z,\ \upsilon_y = y/\lambda z$.

For circular aperture with a diameter of D, Fresnel's diffraction field can be integrated in the polar coordinate:

$$U(r) = \frac{i2\pi}{\lambda z}\exp\left(-\frac{i2\pi z}{\lambda}\right)\exp\left(-\frac{i\pi r^2}{\lambda z}\right)\int_0^{D/2}\exp\left(-\frac{i2\pi r_1^2}{\lambda z}\right)J_0\left(\frac{2\pi r r_1}{\lambda z}\right)r_1 \mathrm{d}r_1. \tag{2.1.41}$$

In the Fraunhofer regime, the intensity can be written as:

$$I(x,y) = I_0\left(\frac{2J_1\left(\frac{\pi D}{\lambda z}r\right)}{\frac{\pi D}{\lambda z}r}\right)^2, \tag{2.1.42}$$

where J_1 is the Bessel function of the first kind and I_0 is the peak intensity. The function $2J_1(x)/x$ is also termed the Airy function. By equaling the energy in the input surface and the output surface

$$P = \int_0^\infty I r\,\mathrm{d}r\int_0^{2\pi}\mathrm{d}\theta = 8\pi I_0\left(\frac{\lambda z}{\pi D}\right)^2\int_0^\infty\frac{(J_1(r'))^2}{r'}\mathrm{d}r' = \pi D^2 I_i/4, \tag{2.1.43}$$

the peak intensity could be written in a form of the intensity in the input plane:

$$I_0 = \left(\frac{\pi D^2}{4\lambda z}\right)^2 I_i. \tag{2.1.44}$$

Here the following equality is utilized

$$\int_0^\infty\frac{(J_1(x))^2}{x}\mathrm{d}x = \frac{1}{2}. \tag{2.1.45}$$

Now let us consider the problem when a focusing lens with a focal length of f is added after the circular aperture, the intensity distribution at the focal plane can be written as:

$$I(x,y) = I_0\left(\frac{2J_1\left(\frac{\pi D}{\lambda f}r\right)}{\frac{\pi D}{\lambda f}r}\right)^2. \tag{2.1.46}$$

The radius of the focal spot (defined by the first zero point) is then

$$r_0 = 1.22\frac{\lambda f}{D}, \tag{2.1.47}$$

which is obviously determined by the wavelength of light. The above equation is only valid for $f \gg D$. For $f < D$, this equation should be revised as:

$$r_0 = 0.61\frac{\lambda}{\sin\theta}. \quad (2.1.48)$$

This is just the resolution limit of microscope. When the background material is not vacuum, it is expressed by the numerical aperture (NA):

$$r_0 = 0.61\frac{\lambda}{n\ \sin\theta} = 0.61\frac{\lambda}{\mathrm{NA}}. \quad (2.1.49)$$

2.1.3 Absorption and Radiation Theory

In many materials, light will attenuate as they propagate owing to the absorption induced by ohmic loss in metals and relaxation of dielectric resonance. Although the absorption is not desired for cases such as plasmonic circuitry and transformation optical devices, these losses can be utilized in perfect light absorbers, either to use for energy harvest or stealth applications.

In classic absorption theory, Beer–Lambert law relates the attenuation to the properties of material through which the light is traveling. In general, Beer–Lambert law comprises two independent laws: Beer's law states that the absorbance is proportional to the concentrations of the attenuating species; Lambert's law states that absorbance is directly proportional to its thickness [4].

If the interfacial reflection is ignored (applicable when the refractive index is small), the transmittance of material is related to its optical depth d and attenuation coefficient α as

$$T = \mathrm{e}^{-2\alpha d}. \quad (2.1.50)$$

The absorbance is then

$$A = 1 - \mathrm{e}^{-2\alpha d}. \quad (2.1.51)$$

Obviously, a very large thickness is required to obtain near-perfect absorption (such as larger than 99%). If the interfacial reflectance R is not negligible, the absorption can never reach 100%

$$A = 1 - R - \mathrm{e}^{-2\alpha d}. \quad (2.1.52)$$

In the quantum theory, the absorption and radiation are described by the transition between different energy bands. When an atom or molecule absorbs one photon, its energy state would be lifted. On the contrary, when an atom or molecule at the excited state loses a photon, the energy state would get down.

In general, the absorption and radiation of a large quantity of photons could only be described statistically. While the absorption is described by the imaginary parts of the refractive index, Planck's radiation law gives the spectral radiant emittance of the blackbody as a function of wavelength and temperature [5]:

$$M_{bb}(\lambda, T) = \frac{c_1}{\lambda^5 \left(e^{c_2/\lambda T} - 1\right)}, \tag{2.1.53}$$

where c_1 is the first radiation constant (3.7415×10^{-16} W m^4), c_2 the second radiation constant (1.4388×10^{-2} m K), T the absolute temperature (K), and λ the wavelength (m). For a Lambertian source which emits or reflects a radiance that is independent of angle, M is related to the spectral radiance by

$$M = \int_0^{2\pi} d\phi \int_0^{\pi/2} L \cos\theta \sin\theta d\theta = \pi L. \tag{2.1.54}$$

Obviously, Planck's radiation law sets a fundamental limit on the emittance of materials.

2.2 Macroscopic and Microscopic Meta-surface-wave

One basic idea in EO 2.0 is utilizing structured surfaces to control the electromagnetic waves. However, it is difficult to design with so many design freedoms. To address this problem, a concept of meta-surface-wave (M-wave) has been proposed, which is defined as a special kind of on-demand electromagnetic excitation, either propagating along the surface or tunneling through the surface with abrupt phase and amplitude change [6]. The M-waves share some fundamental properties such as extremely short wavelength, abrupt phase change, and strong chromatic dispersion, which make them different from traditional bulk waves. Furthermore, the M-wave provides a simple platform for the design of various meta-devices.

2.2.1 *Theory of Surface Plasmon Polariton*

In general, surface plasmon polariton (SPP) wave is one kind of typical M-waves. Using boundary conditions shown in Fig. 2.5, the electric and magnetic fields for TM polarization can be written as [6, 7]:

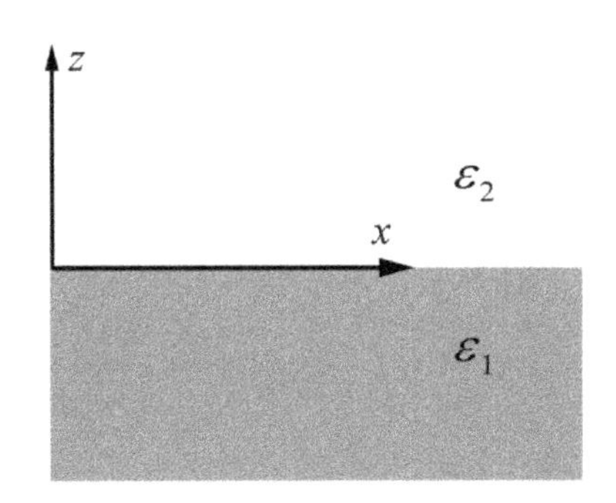

Fig. 2.5 Geometry for SPP propagation at a single interface between metal and dielectric

$$\begin{aligned} H_y(z) &= A_1 \mathrm{e}^{i\beta x}\mathrm{e}^{k_1 z} \\ E_x(z) &= -iA_1\frac{1}{\omega\varepsilon_0\varepsilon_1}k_1\mathrm{e}^{i\beta x}\mathrm{e}^{k_1 z} \\ E_z(z) &= -A_1\frac{\beta}{\omega\varepsilon_0\varepsilon_1}\mathrm{e}^{i\beta x}\mathrm{e}^{k_1 z} \end{aligned} \tag{2.2.1}$$

for $z < 0$, and

$$\begin{aligned} H_y(z) &= A_2 \mathrm{e}^{i\beta x}\mathrm{e}^{-k_2 z} \\ E_x(z) &= iA_2\frac{1}{\omega\varepsilon_0\varepsilon_2}k_2\mathrm{e}^{i\beta x}\mathrm{e}^{-k_2 z} \\ E_z(z) &= -A_2\frac{\beta}{\omega\varepsilon_0\varepsilon_2}\mathrm{e}^{i\beta x}\mathrm{e}^{-k_2 z} \end{aligned} \tag{2.2.2}$$

for $z > 0$. Taking the electric fields at $z > 0$ as an example, its x and z components have a phase shift of 90°, which may be reversed if the propagation direction is inverted.

The wavevector relations of the waves can be written as:

$$\begin{aligned} k_1^2 &= \beta^2 - k_0^2\varepsilon_1 \\ k_2^2 &= \beta^2 - k_0^2\varepsilon_2 \end{aligned}. \tag{2.2.3}$$

The continuity of E_x requires

$$\frac{k_1}{\varepsilon_1} = -\frac{k_2}{\varepsilon_2}. \tag{2.2.4}$$

Combining Eqs. (2.2.3) and (2.2.4), one can obtain the propagation constant as

$$\beta = k_0\sqrt{\frac{\varepsilon_1\varepsilon_2}{\varepsilon_1+\varepsilon_2}}. \tag{2.2.5}$$

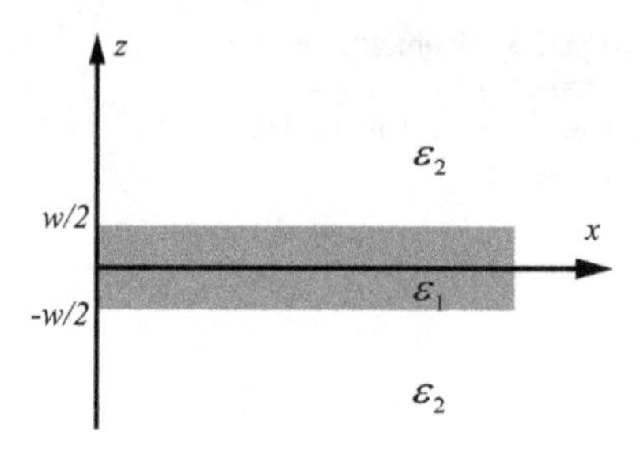

Fig. 2.6 Geometry of a three-layer system consisting of a thin layer sandwiched between two infinite half spaces

When SPPs at different interfaces couple to each other, the dispersion equations would change dramatically. For simplicity of discussion, here we consider a three-layered structure with either an insulator–metal–insulator (IMI) or metal–insulator–metal (MIM) configuration. The materials for the two cladding layers are set to be the same as shown in Fig. 2.6; thus, the modes would possess good symmetry.

For TM polarization, the electromagnetic fields can be written as:

$$\begin{aligned} H_y &= A\mathrm{e}^{i\beta x}\mathrm{e}^{-k_2 z} \\ E_x &= iA\frac{1}{\omega\varepsilon_0\varepsilon_2}k_2\mathrm{e}^{i\beta x}\mathrm{e}^{-k_2 z} \\ E_z &= -A\frac{\beta}{\omega\varepsilon_0\varepsilon_2}\mathrm{e}^{i\beta x}\mathrm{e}^{-k_2 z} \end{aligned} \tag{2.2.6}$$

for the region $z > w/2$, and

$$\begin{aligned} H_y &= Be^{i\beta x}e^{k_2 z} \\ E_x &= -iB\frac{1}{\omega\varepsilon_0\varepsilon_2}k_2 e^{i\beta x}e^{k_2 z} \\ E_z &= -B\frac{\beta}{\omega\varepsilon_0\varepsilon_2}e^{i\beta x}e^{k_2 z} \end{aligned} \tag{2.2.7}$$

for region $z < -w/2$.

In the central region, the fields are a summation or interference of two counter-propagating evanescent waves:

$$\begin{aligned} H_y &= C\mathrm{e}^{i\beta x}\mathrm{e}^{k_1 z} + D\mathrm{e}^{i\beta x}\mathrm{e}^{-k_1 z} \\ E_x &= -iC\tfrac{1}{\omega\varepsilon_0\varepsilon_1}k_1\mathrm{e}^{i\beta x}\mathrm{e}^{k_1 z} + iD\tfrac{1}{\omega\varepsilon_0\varepsilon_1}k_1\mathrm{e}^{i\beta x}\mathrm{e}^{-k_1 z}\,. \\ E_z &= C\tfrac{\beta}{\omega\varepsilon_0\varepsilon_1}\mathrm{e}^{i\beta x}\mathrm{e}^{k_1 z} + D\tfrac{\beta}{\omega\varepsilon_0\varepsilon_1}\mathrm{e}^{i\beta x}\mathrm{e}^{-k_1 z} \end{aligned} \tag{2.2.8}$$

By applying the boundary conditions, one can obtain:

$$A\mathrm{e}^{-k_2 w/2} = C\mathrm{e}^{k_1 w/2} + D\mathrm{e}^{-k_1 w/2}$$

$$\frac{A}{\varepsilon_2}k_2 e^{-k_2 w/2} = -\frac{C}{\varepsilon_1}k_1 e^{k_1 w/2} + \frac{D}{\varepsilon_1}k_1 e^{-k_1 w/2} \tag{2.2.9}$$

at $z = w/2$ and

$$\begin{aligned} Be^{-k_2 w/2} &= Ce^{-k_1 w/2} + De^{k_1 w/2} \\ -\frac{B}{\varepsilon_2}k_2 e^{-k_2 w/2} &= -\frac{C}{\varepsilon_1}k_1 e^{-k_1 w/2} + \frac{D}{\varepsilon_1}k_1 e^{k_1 w/2} \end{aligned} \tag{2.2.10}$$

at $z = -w/2$. By solving these equations with

$$\begin{aligned} k_1^2 &= \beta^2 - k_0^2\varepsilon_1 \\ k_2^2 &= \beta^2 - k_0^2\varepsilon_2 \end{aligned}, \tag{2.2.11}$$

the dispersion relations can be obtained:

$$\begin{aligned} \tanh\sqrt{\beta^2 - k_0^2\varepsilon_1}\,w/2 &= -\frac{\varepsilon_1\sqrt{\beta^2 - k_0^2\varepsilon_2}}{\varepsilon_2\sqrt{\beta^2 - k_0^2\varepsilon_1}} \\ \tanh\sqrt{\beta^2 - k_0^2\varepsilon_1}\,w/2 &= -\frac{\varepsilon_2\sqrt{\beta^2 - k_0^2\varepsilon_1}}{\varepsilon_1\sqrt{\beta^2 - k_0^2\varepsilon_2}} \end{aligned} \tag{2.2.12}$$

for odd and even modes, respectively. Note that here the symmetry is defined by the symmetry of transverse electric field E_x. Regarding to the magnetic field (H_y), the symmetry would reverse.

Since these structures are symmetric, there are $C = -D$ for the symmetric (even) mode, and $C = D$ for the anti-symmetric (odd) mode. Consequently, E_x is in proportional to $\exp(k_1 z) + \exp(-k_1 z)$, i.e., a catenary function. For the anti-symmetric mode, although E_x is in proportional to $\exp(k_1 z) - \exp(-k_1 z)$, H_y and E_z are characterized by the catenary function. More generally, if the three layers are not symmetric, the magnitudes of C and D are not equal; the field curves would be asymmetric catenaries. This intriguing property forms one basis of the catenary optics [8, 9].

2.2.2 *Spoof Surface Plasmon Polariton*

The bulk plasmon frequency of a metal material is determined by the electrons inside it. Due to the limitations of natural metal materials, the bulk plasmon frequency of ordinary metal is generally in the optical band. It was demonstrated that metallic wire array can greatly reduce the bulk plasmon frequency, thus realizing a negative permittivity in the microwave and terahertz bands [10, 11]. In 2004, Pendry et al. further proposed the use of subwavelength aperture arrays on metals to support the

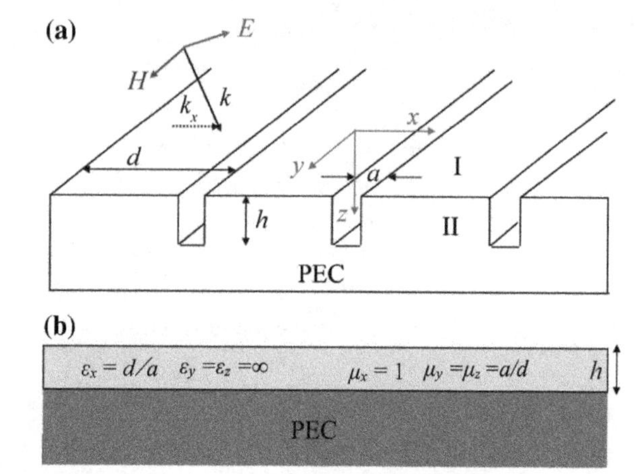

Fig. 2.7 A one-dimensional array of grooves of width a and depth h separated by a distance of d. Adapted from [13] with permission. Copyright 2005, IOP Publishing Ltd.

propagation of surface electromagnetic waves [12, 13]. The corresponding mode is defined spoof surface plasmons (SSPs).

SSPs can be excited by a one-dimensional periodic groove structure in perfect electric conductors (PEC) [13]. As shown in Fig. 2.7, the width of the groove is a, the depth is h, and the period is d. A p-polarized plane wave is incident on the surface of the perfect conductor with a wavevector parallel component of k_x. The electromagnetic fields of the incident wave are

$$\begin{aligned}\vec{E}_{\text{inc}} &= (1, 0, -k_x/k_z)\frac{1}{\sqrt{d}}e^{ik_x x}e^{ik_z z}\\ \vec{H}_{\text{inc}} &= (0, k_0/k_z, 0)\frac{1}{\sqrt{d}}e^{ik_x x}e^{ik_z z}\end{aligned}, \tag{2.2.13}$$

where $k_0 = \omega/c$ is the wave number. The reflected wave of the nth diffraction order can be expressed as

$$\begin{aligned}\vec{E}_{\text{ref},n} &= \left(1, 0, k_x^{(n)}/k_z^{(n)}\right)\frac{1}{\sqrt{d}}e^{-ik_x^{(n)}x}e^{ik_z^{(n)}z}\\ \vec{H}_{\text{ref},n} &= \left(0, -k_0^{(n)}/k_z^{(n)}, 0\right)\frac{1}{\sqrt{d}}e^{-ik_x^{(n)}x}e^{ik_z^{(n)}z}\end{aligned}, \tag{2.2.14}$$

where $k_x^{(n)} = k_x + 2n\pi/d$ ($n = -\infty, \ldots, 0, \ldots, \infty$). Assuming that the wavelength of the incident light is much larger than the width of the grooves, that is $\lambda_0 \gg a$, only the fundamental transverse electromagnetic (TEM) mode in the grooves is considered.

The electric field and magnetic field in region I can be expressed as the summation of the incident plane wave and the reflected wave.

$$\begin{aligned}\vec{E}_{\text{I}} &= \vec{E}_{\text{inc}} + \sum_{n=-\infty}^{\infty} \rho_n \vec{E}_{\text{ref},n}\\ \vec{H}_{\text{I}} &= \vec{H}_{\text{inc}} + \sum_{n=-\infty}^{\infty} \rho_n \vec{H}_{\text{ref},n}\end{aligned}, \tag{2.2.15}$$

where ρ_n is the reflection coefficient of the nth diffraction order. The electric field and the magnetic field in region II can be written as a linear superposition of the TEM modes of both forward and backward propagation

$$\begin{aligned}\vec{E}_{\text{II}} &= C^+\vec{E}_{\text{TEM},+} + C^-\vec{E}_{\text{TEM},-} \\ \vec{H}_{\text{II}} &= C^+\vec{H}_{\text{TEM},+} + C^-\vec{H}_{\text{TEM},-}\end{aligned}. \tag{2.2.16}$$

An expression of the reflection coefficient ρ_n can be easily extracted by matching the boundary conditions

$$\rho_n = -\delta_{n0} - \frac{2i\tan(k_0h)S_0S_nk_0/k_z}{1 - i\tan(k_0h)\sum_{n=-\infty}^{\infty} S_n^2k_0/k_z^{(n)}}, \tag{2.2.17}$$

where S_n represents the overlap integral between the nth-order plane wave and the TEM mode

$$S_n = \frac{1}{\sqrt{ad}}\int_{-a/2}^{a/2} e^{ik_x^{(n)}x}\mathrm{d}x = \sqrt{\frac{a}{d}}\frac{\sin\left(k_x^{(n)}a/2\right)}{k_x^{(n)}a/2}. \tag{2.2.18}$$

In principle, the surface bands can be calculated by analyzing the zeros of the denominator of Eq. (2.1.17). However, the solution to Eq. (2.1.17) is complicated. The calculation can be simpler if it is assumed that $\lambda_0 \gg d$. Then, all the diffraction orders can be neglected except the specular one and ρ_0 takes the form

$$\rho_0 = -\frac{1 + iS_0^2\tan(k_0h)k_0/k_z}{1 - iS_0^2\tan(k_0h)k_0/k_z}. \tag{2.2.19}$$

For the case $k_x > k_0$, the dispersion relation of the SSP can be obtained as

$$\frac{\sqrt{k_x^2 - k_0^2}}{k_0} = S_0^2\tan(k_0h). \tag{2.2.20}$$

This is the dispersion relation of the surface modes supported by a 1D array of grooves in the limit $\lambda_0 \gg d$ and $\lambda_0 \gg a$. It should be noted that the same dispersion relation could be obtained if the array of grooves is replaced by a single homogeneous but anisotropic layer (thickness h) on top of the surface of a perfect conductor as shown in Fig. 2.7b. The homogeneous layer would have the following parameters:

$$\varepsilon_x = d/a \quad \varepsilon_y = \varepsilon_z = \infty. \tag{2.2.21}$$

As light propagates in the grooves in the y or z directions with the velocity of light, it can be obtained that

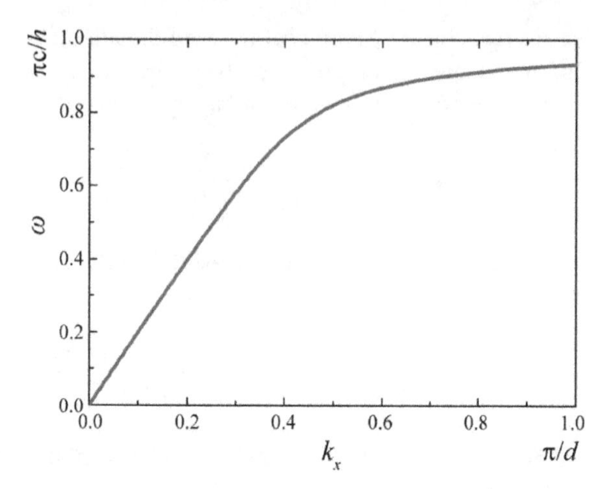

Fig. 2.8 Dispersion relation of the SSP supported by a 1D array of grooves with geometric parameters $a/d = 0.2$ and $h/d = 1$. Adapted from [13] with permission. Copyright 2005, IOP Publishing Ltd.

$$\mu_y = \mu_z = \frac{1}{\varepsilon_1} \quad \mu_x = 1. \tag{2.2.22}$$

By extending this formula to the case $k_x > k_0$, the dispersion relation of the surface modes can be calculated as:

$$\frac{\sqrt{k_x^2 - k_0^2}}{k_0} = \frac{a}{d}\tan(k_0 h). \tag{2.2.23}$$

Figure 2.8 shows the dispersion curve of SSP for a specific case when $a = 0.2d$ and $h = d$. It is worth noting the similarities between this dispersion and the one associated with the bands of SPPs supported by the surfaces of real metals in the optical band. In a SPP band, at large k_x, ω approaches $\omega_p/\sqrt{2}$, whereas in this case, ω approaches $\pi c/2h$, i.e., the frequency location of a cavity waveguide mode inside the groove.

The SSP modes in 2D hole array and groove array have also been deduced. The details can be found in Refs. [12–14].

2.2.3 Catenary Optical Fields and Catenary Dispersion

From a microscopic point of view, there are nearly infinite numbers of interfaces between the constitutive materials in metamaterials (artificial structured materials with extraordinary performances not existed in nature) and metasurfaces (the two-dimensional counterparts of metamaterials). Although these interfaces may not support surface waves when the dimension is much larger than the wavelength, strong modifications must be considered in the deep subwavelength scale. This can be understood by using the generalized Helmholtz equation [9, 15]:

$$\nabla^2 \vec{E} + \nabla\left[\vec{E} \cdot \frac{\nabla\varepsilon}{\varepsilon}\right] + k_0^2 \varepsilon \mu \vec{E} = 0, \tag{2.2.24}$$

where $\vec{E}$ is the electric field, ε and μ are the permittivity and permeability, k_0 is the vacuum wave number. At the boundaries,$\nabla\varepsilon$ approaches infinity and is responsible for the coupling between free-space waves and localized modes. To understand this equation, waves propagating through thin slits perforated in PEC are investigated [16]. As shown in Fig. 2.9, the 2D metallic slab waveguide (the dimension along the y-axis is infinite) supports TEM waves without cutoff frequency, and the propagation constant is equal to that in vacuum. When the thickness of waveguide is reduced to much smaller than the operational wavelength, strong scattering would occur at the edges, which makes the fields distribution change dramatically. Similar to the SPP fields at the slit edges [17], this new kind of field takes a form of hyperbolic cosine catenary function as a result of the evanescent coupling, which means that the localized wave has a large vertical propagation constant β along the z-direction and an imaginary horizontal component α along the x-direction.

These catenary-shaped interfacial waves may be treated as one special vertically propagating M-wave. As shown in Fig. 2.10, the electromagnetic properties of simple metallic gratings can be obtained by using impedance theory. To investigate the influence of gap width on the optical properties, the electric fields along the central line are illustrated in Fig. 2.10. Since the scattering fields are mainly evanescent waves, the amplitude distribution follows a hyperbolic cosine catenary shape. Using equivalent circuit model [18], the normalized admittance Y can be written as:

$$Y_{\text{eff}} = \frac{1}{Z_{\text{eff}}} = -i\frac{2p\left(n_1^2 + n_2^2\right)}{\lambda} \ln \csc \frac{\pi w}{2p} \tag{2.2.25}$$

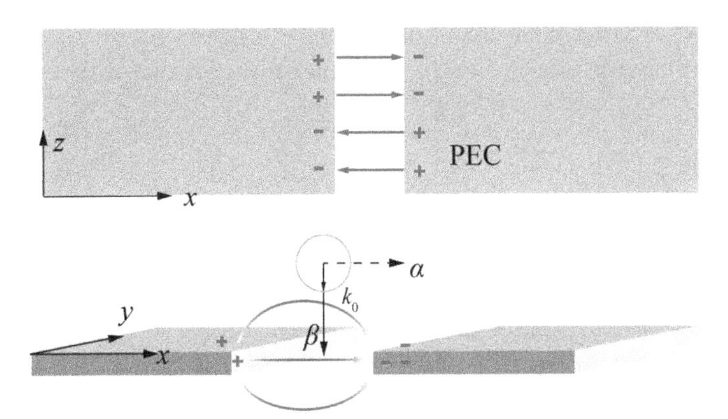

Fig. 2.9 Waves in the microscopic regime. The top panel shows the electric fields in a thick slit cut in PEC. The bottom panel shows the M-wave in a thin slit. Reproduced from [16] with permission. Copyright 2018, Optical Society of America

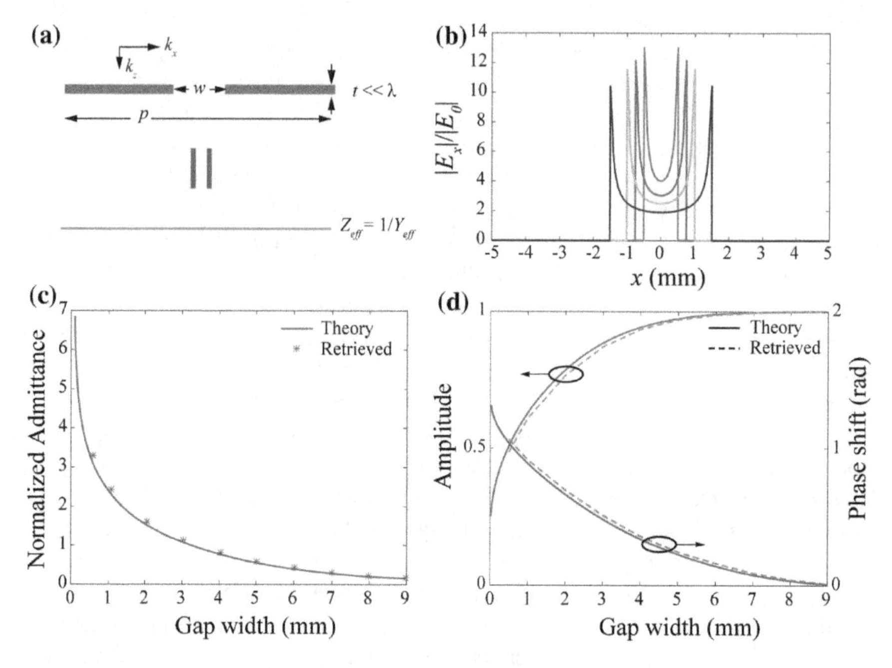

Fig. 2.10 Vertical M-waves associated with catenary optical fields. **a** Equivalent of a thin sheet perforated with subwavelength slits. The period p and thickness t are 10 and 0.1 mm. **b** E_x distribution at 10 GHz for slit widths of 1, 1.5, 2, and 3 mm. **c** Normalized admittance for different gap widths. The curve is half of a catenary curve of equal strength. **d** Transmission amplitude and phase calculated using FEM and impedance theory. Reproduced from [16] with permission. Copyright 2018, Optical Society of America

where Z_{eff} is the normalized surface impedance, p is the period of grating, w is the width of the slit, λ is the wavelength, n_1 and n_2 are the refractive indexes for the background materials. Using the equivalent impedances theory [15], the transmission and reflection coefficients can be easily obtained.

The catenary dispersion can be induced by applying the stationary electric analysis [19]. Figure 2.11a is the schematic of an array of metallic slits, which is the same as in Fig. 2.10a. A transverse electric plane wave is normally incident on this array. In order to deduce the explicit functions to describe this subwavelength structure, a modified treatment is applied as shown in Fig. 2.11b that PEC sheets are inserted in the metallic films. Owing to the symmetry of the structure, the insertion of PEC sheets parallel to y-axis and perpendicular to the plane of the slits through the central lines of the films does not disturb the electric field. Therefore, the electric field of the whole array can be divided into numerous identical fields such as (I), (II), and (III) in Fig. 2.11b. If we can obtain the models for any of the small unit, we can apply them to other units as well. Thus, to simply the case, unit (II) in Fig. 2.11b was chosen as an example that the field is constrained from $x = -p/2$ to $x = p/2$.

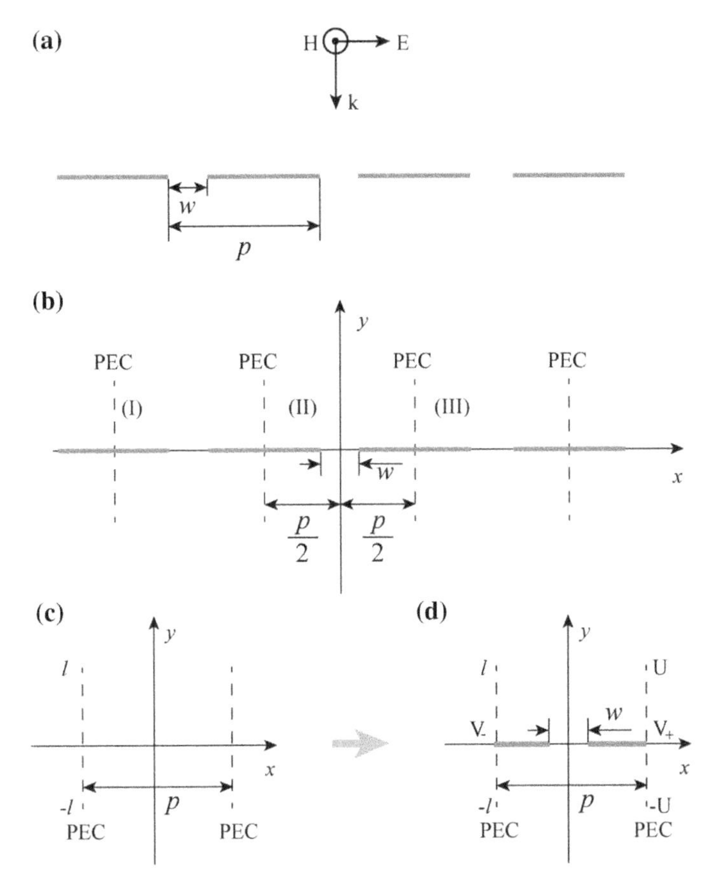

Fig. 2.11 **a** Schematic of an array of metallic slits. **b** The modified treatment to the original structure that PEC sheets are inserted in the central lines of the metallic films. **c** A single unit cell in **b** with metallic slits removed. **d** A single unit cell as shown in **b**. Reproduced from [19] with permission. Copyright 2018, The Authors

As the electric field vibrates perpendicularly to the edges of the slits, the structure behaves as a purely capacitive device. Firstly, the background capacity of the modified structure is calculated by removing the metallic slits and maintaining the PEC sheets as shown in Fig. 2.11c. In this case, the plane wave propagates in the negative direction of y-axis between two parallel PEC sheets. If we further constrain the space in y-axis direction from $y = -l$ to $y = l$, the capacity of the structure in Fig. 2.11c is

$$C'' = 2l\varepsilon_0, \tag{2.2.26}$$

where ε_0 is the vacuum permittivity. Then we return to the case as shown in Fig. 2.11d. The existence of the slits will distort the original electric field distributions, and the influence can be described by the theory of Schwarz's conformal transformations in terms of equipotentials V and lines of force U at (x, y) plane.

$$\sin(V + iU) = \csc(\pi w/2p)\sin[(x + iy)\pi/p]. \tag{2.2.27}$$

By employing this function, the field distributions in x–y plane can be calculated. As the potential difference between two identical conductors is $V = \pi$, we assume that the potentials at $x = \pm p/2$ equal to $V_{\pm} = \pm\pi/2$. Then, at the boundary of $x = p/2$, Eq. (2.2.27) can be calculated as

$$\sin\left(\frac{\pi}{2} + iU\right) = \csc\left(\frac{\pi w}{2p}\right)\sin\left(\frac{\pi}{2} + i\frac{y\pi}{p}\right), \tag{2.2.28}$$

that can be simplified as

$$\cos(iU) = \csc\left(\frac{\pi w}{2p}\right)\cos\left(i\frac{y\pi}{p}\right). \tag{2.2.29}$$

As $\cos(ia) = \cosh(a)$, U can be expressed as

$$U(y) = \pm\cosh^{-1}\left[\csc\left(\frac{\pi w}{2p}\right)\cosh\left(\frac{y\pi}{p}\right)\right]. \tag{2.2.30}$$

Next, the total charge Q on the right PEC sheet can be calculated by the characteristic capacity $\kappa = p\varepsilon_0$ times U at the limits of the space from $y = -l$ to $y = l$.

$$Q = \kappa[U(l) - U(-l)] = 2\kappa U(l). \tag{2.2.31}$$

Thus, the capacity of the structure in Fig. 2.11d can be calculated by

$$C' = \frac{Q}{V} = \frac{2\kappa}{\pi}\cosh^{-1}\left[\csc\left(\frac{\pi w}{2p}\right)\cosh\left(\frac{\pi l}{p}\right)\right]. \tag{2.2.32}$$

Therefore, the difference in capacity is due to the existence of the slits and it can be deduced by

$$C = \lim_{l\to\infty}\left(C' - C''\right) = \frac{2\kappa}{\pi}\ln\csc\frac{\pi w}{2p}. \tag{2.2.33}$$

The corresponding impedance of the structure can be expressed by

$$Z_1 = 1/i\omega C = \frac{1}{i\frac{4p}{\lambda}\ln\csc\left(\frac{\pi w}{2p}\right)} = \frac{1}{4iF}, \tag{2.2.34}$$

where $\omega = 2\pi c/\lambda$ is the angular frequency in the operation band, c is the vacuum light velocity.

2.3 Generalized Laws of Refraction and Reflection

2.3.1 Extending Snell's Law and Fresnel's Equations

(1) Beyond Snell's law

The refraction and reflection of light are the most basic concepts in geometric optics, and they are also the fundamental laws of interaction between plane waves and interfaces in physical optics. It is well known that the specific form of the law of light refraction is given by Snell in 1621, which was further generalized in many cases. In the 1960s, after the birth of nonlinear optics, Snell's law produced a corresponding nonlinear form [20]. After the metasurface appears, the "M-wave-assisted" generalized laws of refraction and reflection [6] are formed. As shown in Fig. 2.12, electromagnetic waves can be refracted or reflected in any direction by introducing subwavelength structures at the interface and generating a phase gradient.

As a cornerstone of traditional optics, the laws of reflection and refraction are responsible for the technology convention which uses curved geometry to control the propagation direction of light [21, 22]. The curved shapes have two different kinds of understanding: First, all natural lenses including homogeneous and gradient index eyes are curved, which means that curved geometry is a natural selection with optimized performance in the natural evolution. Second, as our human's ambition for higher resolution develops, the aperture size of both lenses and mirrors must be greatly increased to larger than several meters. According to the laws of reflection and refraction, the thickness or height should grow correspondingly, leading to a significant increase of weight and cost [23]. In such a condition, flat optics devices such as diffractive optical elements (DOEs) have been intensively investigated [24, 25]. Nevertheless, since the element size in traditional DOEs is much larger than the operational wavelength, the numerical aperture must be very small, leading to a very long and even unpractical optical path.

In order to replace bulky and heavy optical elements with thinner and compact ones, it was found that nanoslit array with gradient width could introduce localized gradient phase by tuning the coupling strength of catenary optical fields [26]. Similar

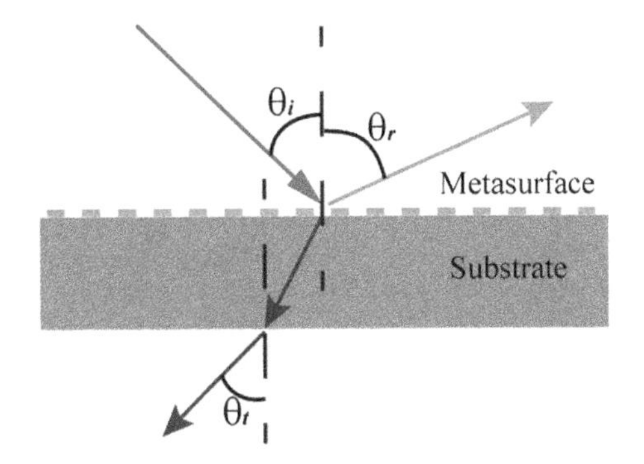

Fig. 2.12 Generalized laws of refraction and reflection

to the width-dependent coupling in a single metallic film [27], the gap plasmons in MIM slits can be tuned via the width. Then, a nanoslit lens in the visible regime was experimentally demonstrated in 2009 [28]. Recently, the above flat lensing approach, along with the gradient metasurfaces based on geometric phase [29–31], is referred to as M-wave-assisted laws of reflection and refraction (MLRR) [8], also named generalized laws of reflection and refraction [32, 33].

Supposing the metasurface has a linear phase shift, its phase gradient would be a constant and the boundary condition should be revised as:

$$\boldsymbol{k}_{i\parallel} \cdot \boldsymbol{r} + \nabla\Phi \cdot \boldsymbol{r} = \boldsymbol{k}_{r\parallel} \cdot \boldsymbol{r} = \boldsymbol{k}_{t\parallel} \cdot \boldsymbol{r}. \tag{2.3.1}$$

In the plane formed by the normal line and outgoing light, the MLRR could be written as:

$$n_1 k_0 \sin\theta_i + \nabla\Phi = n_1 k_0 \sin\theta_r = n_2 k_0 \sin\theta_t. \tag{2.3.2}$$

Considering the different phase shift in reflection and transmission, the above equation could be written as:

$$\begin{aligned} n_1 k_0 \sin\theta_i + \nabla\Phi_r &= n_1 k_0 \sin\theta_r \\ n_1 k_0 \sin\theta_i + \nabla\Phi_t &= n_2 k_0 \sin\theta_t \end{aligned}, \tag{2.3.3}$$

where $\nabla\Phi_r$ and $\nabla\Phi_t$ are the transmissive and reflective phase gradient in the metasurface plane, respectively, and can be understood as horizontal wavevectors. n_1 and n_2 are the refractive indices of media at the incident and transmit sides, θ_i, θ_r, and θ_t are the angles for incident, reflected, and refracted light. Since $\nabla\Phi$ is a vector, the reflective and refractive angles can not only be tuned in amplitudes, but also inside or outside the incident plane. With this unprecedented ability, traditional design procedure of optical system may be completely changed.

The 3D extension of the generalized laws of refraction and reflection can be realized through a rotation of the plane of incidence by an arbitrary angle with respect to the phase gradient which results in out-of-plane reflection and refraction, where the directions of the reflected and refracted wavevectors are characterized by the angles $\theta_{r(t)}$ (the angle between $\overrightarrow{k}_{r(t)}$ and its projection on the xy-plane) and $\varphi_{r(t)}$ (the angle formed by the projection of $\overrightarrow{k}_{r(t)}$ on the xy-plane and the x-axis) as defined in Fig. 2.13. The generalized law of reflection in 3D at an interface with phase gradient is then given by [34]

$$\begin{cases} \cos\theta_r \sin\varphi_r = \frac{1}{n_i k_0}\frac{\mathrm{d}\Phi}{\mathrm{d}x} \\ \sin\theta_r - \sin\theta_i = \frac{1}{n_i k_0}\frac{\mathrm{d}\Phi}{\mathrm{d}y} \end{cases}. \tag{2.3.4}$$

For refraction, the 3D Snell's law is given by

Fig. 2.13 Schematics used to describe the 3D generalized reflection at an interface

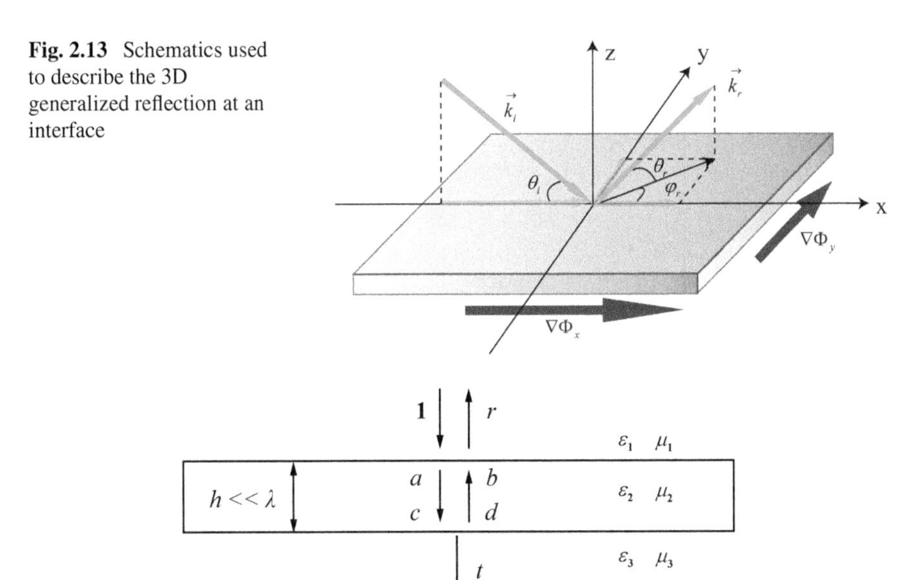

Fig. 2.14 Boundary conditions for a slab embedded between the two half spaces. Adapted from [6] with permission. Copyright 2015, Science China Press and Springer-Verlag Berlin Heidelberg

$$\begin{cases} \cos\theta_t \sin\varphi_t = \frac{1}{n_t k_0}\frac{\mathrm{d}\Phi}{\mathrm{d}x} \\ n_t \sin\theta_t - n_i \sin\theta_i = \frac{1}{k_0}\frac{\mathrm{d}\Phi}{\mathrm{d}y} \end{cases}. \tag{2.3.5}$$

(2) Generalized Fresnel's equation

The key to the generalized laws of reflection and refraction is the gradient phase shift. In order to calculate the phase shift, the boundary condition must be utilized to deduce the generalized Fresnel's equations [6]. In the following, we confine ourselves in the macroscopic description of metasurface. In such a way, the optical properties of metasurface are characterized by its permittivity ε and permeability μ. By utilizing the boundary conditions and transfer matrix, the optical response of an arbitrary form monochromatic wave can be obtained.

As shown in Fig. 2.14, for a thin slab sandwiched by two different medium, the boundary conditions can be written as

$$\begin{cases} 1 + r = a + b \\ Y_1(1 - r) = Y_2(a - b) \end{cases} \tag{2.3.6}$$

and

$$\begin{cases} c + d = t \\ Y_2(c - d) = Y_3 t \end{cases}, \tag{2.3.7}$$

where 1, 2, and 3 denote the three layers, a, b, c, and d are the coefficients for the counterpropagating waves inside the slab. By definition, there are

$$\begin{cases} a = c\exp(-ik_{z2}h) \\ b = d\exp(ik_{z2}h) \end{cases}. \tag{2.3.8}$$

For extremely thin metasurface where $|k_z h| \ll 1$, we have

$$\begin{cases} a = c(1 - ik_{z2}h) \\ b = d(1 + ik_{z2}h) \end{cases}. \tag{2.3.9}$$

From Eq. (2.3.9), one can derive that

$$\begin{cases} a + b = (c + d) + ik_{z2}h(c - d) \\ a - b = (c - d) - ik_{z2}h(c + d) \end{cases}. \tag{2.3.10}$$

Here we focus on the case when the permittivity and/or permeability of the middle layer are larger than the surrounding space. For non-magnetic layer, $Y_2 \gg Y_3$, thus $c - d \ll c + d$, so Eq. (2.3.10) approximates as

$$\begin{cases} a + b = (c + d) \\ a - b = \dfrac{Y_3}{Y_2}t - ik_{z2}ht \end{cases}. \tag{2.3.11}$$

Inserting Eq. (2.3.11) into Eq. (2.3.7), there is

$$\begin{cases} 1 + r = t \\ Y_1(1 - r) = Y_3 t - iY_2 k_z ht \end{cases}. \tag{2.3.12}$$

Comparing it with the electric impedance boundary conditions

$$\begin{cases} 1 + r = t \\ Y_1(1 - r) = Y_3 t + Y_e t \end{cases}, \tag{2.3.13}$$

one can derive the electric admittance for this slab

$$Y_e = -iY_2 k_{z2} h = \begin{cases} -i\frac{\varepsilon_2 k_0^2 - k_x^2}{\mu_0 \omega} h \approx -i\omega\varepsilon_2\varepsilon_0 h \\ -i\omega\varepsilon_2\varepsilon_0 h \end{cases}. \tag{2.3.14}$$

When the permeability μ is much larger than ε, $Y_2 \ll Y_3$, thus $c - d \gg c + d$, so Eq. (2.3.10) approximates as:

$$\begin{cases} a + b = c + d - ik_{z2}h\,(c - d) \\ a - b = c - d \end{cases}. \tag{2.3.15}$$

As a result, there is

$$\begin{cases} 1 + r = t - ik_{z2}h\dfrac{Y_3}{Y_2}t \\ Y_1(1 - r) = Y_3 t \end{cases}. \tag{2.3.16}$$

Comparing Eq. (2.3.16) with the magnetic impedance boundary conditions

$$\begin{cases} 1 + r = t + Z_{\mathrm{m}} Y_3 t \\ Y_1(1 - r) = Y_3 t \end{cases}, \tag{2.3.17}$$

the magnetic impedance can be obtained:

$$Z_{\mathrm{m}} = -i\frac{k_{z2}h}{Y_2} = \begin{cases} -i\omega\mu_2\mu_0 h \\ -i\frac{\varepsilon_2\mu_2 k_0^2 - k_x^2}{\varepsilon_2\varepsilon_0\omega}h \approx -i\omega\mu_2\mu_0 h \end{cases}. \tag{2.3.18}$$

Clearly, the above assumption is valid even for oblique incidence as long as $k_z h$ is much smaller than unit and the constitutive parameters are much larger than unit. Equations (2.3.14) and (2.3.18) are valid for homogeneous materials. For metasurfaces with subwavelength structures, the effective impedances can be obtained by applying the boundary conditions in a similar way.

In the following, we only discuss the case of thin structures without magnetic response. From Eq. (2.3.12), the reflection and transmission coefficients of an extremely thin slab are

$$\begin{cases} r = \dfrac{Y_1 - Y_3 - Y_{\mathrm{e}}}{Y_1 + Y_3 + Y_{\mathrm{e}}} \\ t = \dfrac{2Y_1}{Y_1 + Y_3 + Y_{\mathrm{e}}} \end{cases}. \tag{2.3.19}$$

Equation (2.3.19) is the M-wave-assisted Fresnel's equations, also called generalized Fresnel's equations described with sheet impedance theory [6]. The generalized Fresnel's equation has also been used to retrieve effective material parameter for thin sheets such as graphene and single-layer metamaterials. Based on this equation, the vectorial effects of electromagnetic waves, including amplitude, phase, and polarization could all be controlled almost arbitrarily.

Fresnel's equations can also be extended to include the magnetic resonance. The following general boundary conditions should be satisfied [35]:

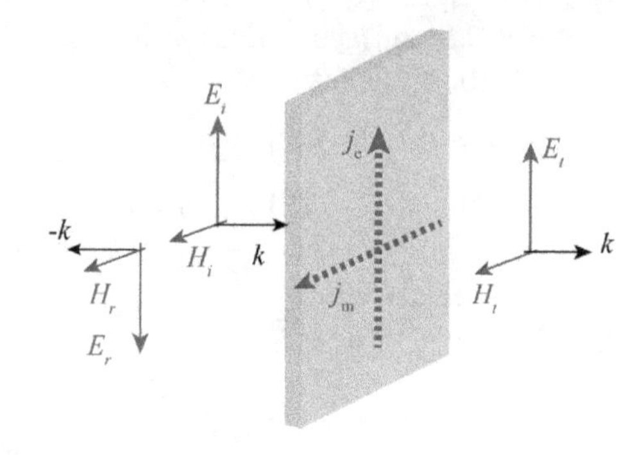

Fig. 2.15 Schematic of sheet impedance description for thin surface. The magnetic and electric responses are described by equivalent magnetic and electric sheet currents. Reproduced from [35] with permission. Copyright 2013, Optical Society of America

$$\begin{aligned}\vec{E}_i + \vec{E}_r &= \vec{E}_t + \hat{n} \times \vec{j}_{\mathrm{m}} \\ \vec{H}_i + \vec{H}_r &= \vec{H}_t + \hat{n} \times \vec{j}_{\mathrm{e}}\end{aligned}, \tag{2.3.20}$$

where j_{m} and j_{e} are the sheet magnetic and electronic current, respectively.

Figure 2.15 illustrates the schematic of the sheet impedance description for metasurface. By definition, there are $E_r = rE_i$, $H_r = rH_i$, $E_t = tE_i$ and $H_t = tH_i$, where $E_i = Z_0 H_i$ is the incident electric field, $Z_0 = 1/Y_0 = 377\ \Omega$ is the vacuum impedance, r and t are the reflection and transmission coefficients. The sheet current j_{m} and j_{e} are related to the average magnetic and electric fields through their corresponding admittance or impedance:

$$\begin{aligned}\vec{j}_{\mathrm{e}} &= \frac{1}{Z_{\mathrm{e}}}\vec{E}_{\mathrm{average}} = Y_{\mathrm{e}}\vec{E}_{\mathrm{average}} = Y_{\mathrm{e}}(1 + r + t)\vec{E}_i/2 \\ \vec{j}_{\mathrm{m}} &= Z_{\mathrm{m}}\vec{H}_{\mathrm{average}} = Z_{\mathrm{m}}(1 - r + t)\vec{H}_i/2\end{aligned} \tag{2.3.21}$$

Combining Eqs. (2.3.20) and (2.3.21), one can obtain that:

$$\begin{aligned}Y_{\mathrm{e}} &= 2Y_0\frac{1 - r - t}{1 + r + t} \\ Z_{\mathrm{m}} &= 2Z_0\frac{1 + r - t}{1 - r + t}\end{aligned} \tag{2.3.22}$$

and

$$\begin{aligned}r &= \frac{1}{2}\left(\frac{2Y_0 - Y_{\mathrm{e}}}{2Y_0 + Y_{\mathrm{e}}} + \frac{Z_{\mathrm{m}} - 2Z_0}{Z_{\mathrm{m}} + 2Z_0}\right) \\ t &= \frac{1}{2}\left(\frac{2Y_0 - Y_{\mathrm{e}}}{2Y_0 + Y_{\mathrm{e}}} - \frac{Z_{\mathrm{m}} - 2Z_0}{Z_{\mathrm{m}} + 2Z_0}\right)\end{aligned}. \tag{2.3.23}$$

2.3.2 Propagation Phase Engineering

Propagation phase refers to the phase accumulation through optical path difference in the propagation process of electromagnetic waves. According to the electromagnetic theory, when electromagnetic waves propagate in a uniform medium with refractive index of n for a distance d, the accumulated propagation phase of the electromagnetic wave can be expressed as $\varphi = nk_0 d$, where $k_0 = 2\pi/\lambda$ is the wave number in free space. Conventional phase-type optical elements always adopt a curved surface to shape the wavefront of the electromagnetic wave. For binary optics, phase engineering is implemented discretely using different steps. Both curved surface elements and binary optical elements face two problems: One is that the device is non-planar, which is not easy for integration and conformal design; the other is that most exist optical materials have a relatively low refractive index n, so in order to achieve full 2π phase modulation, a large thickness d is required. In addition to the modulation of the transmission phase accumulation by thickness d, another approach is to adjust the refractive index n. By spatially varying the refractive index n, a planar phase-type optical element design can be realized while the thickness d remains unchanged. In addition, if the refractive index n can be sufficiently large, the thickness of the device can be effectively reduced.

At present, the commonly used methods for realization of propagation phase engineering can be divided into two categories: One is based on effective index modulation of waveguide, including surface plasmon waveguide and dielectric waveguide, and the other is based on effective medium theory. The main difference between these two categories is the way for modulating the effective index.

Surface plasmon waveguide, also called plasmonic waveguide, is a special kind of slow wave waveguide which supports the propagation of SPPs along the boundary of metallic and dielectric surfaces while keeping them bounded near the surface without radiation away. For the MIM plasmonic waveguide shown in Fig. 2.16, the dispersion equation of the fundamental mode is given in Eq. (2.2.12),

$$\tanh\left(\sqrt{k_{\mathrm{sp}}^2 - k_0^2\varepsilon_{\mathrm{d}}}\,w/2\right) = \frac{-\varepsilon_{\mathrm{d}}\sqrt{k_{\mathrm{sp}}^2 - k_0^2\varepsilon_{\mathrm{m}}}}{\varepsilon_{\mathrm{m}}\sqrt{k_{\mathrm{sp}}^2 - k_0^2\varepsilon_{\mathrm{d}}}},$$

where ε_{m} and ε_{d} are the permittivities of metal and dielectric, respectively, w is the width of the dielectric slit. From this equation, we obtain that the propagation constant of the MIM waveguide is directly related to the width w. Therefore, the slit

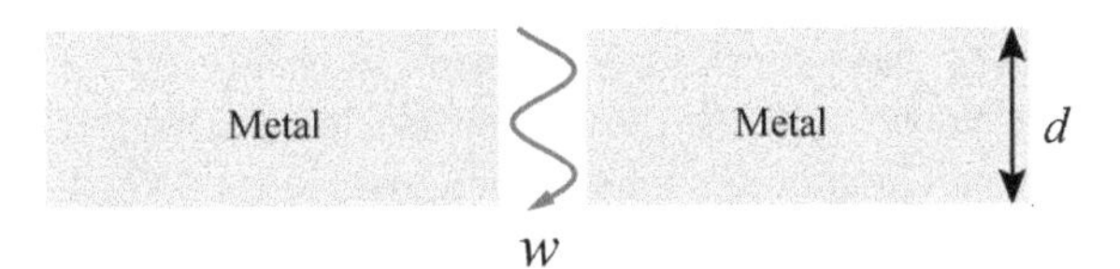

Fig. 2.16 Schematic of a MIM waveguide

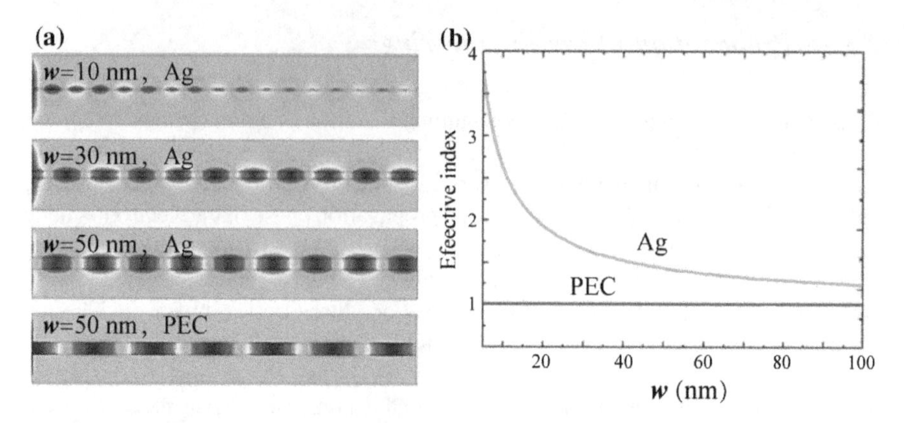

Fig. 2.17 **a** Simulated magnetic field distribution of the modes with different slits width for TM illumination, $\lambda = 633$ nm. **b** The relationship between the waveguide effective index and the slit width

width can be used to modulate the propagation constant of the surface plasmon in the slit, thereby controlling the optical propagation phase with an unchanged thickness d. Figure 2.17a shows the magnetic field distribution of TM electromagnetic waves propagating in nanoslits decorated in silver films. The lengths of the slits are 2 μm, and the widths are 10, 30, and 50 nm, respectively. The medium in the slit is air (refractive index $n = 1$). It can be seen that the surface plasmon modes have different effective wavelengths in the nanoslits with different widths, that is, they are corresponding to different complex propagation constants. The real and imaginary parts of the complex propagation constants determine the phase velocity and loss of the plasmonic modes in the slits, respectively. The smaller the width of the slit, the larger the propagation constant.

By solving the dispersion equation, the propagation constant of the plasmonic mode k_{sp} can be calculated as a function of the width of the nanoslit. We define the effective index of the plasmonic mode as real part of the ratio of the propagation constant in SPP waveguide to the corresponding wave number in free space, i.e., $\mathrm{Re}(k_{sp}/k_0)$. The effective index of the plasmonic mode is plotted in Fig. 2.17b. It can be seen that, as the slit width becomes smaller, the effective index of the plasmonic mode becomes larger. When the slit width is reduced to less than 20 nm, the propagation constant of the plasmonic mode increases dramatically. Another point worth noting is in the case when the metal material is the perfect conductor, it does not support the plasmonic mode, but a TEM mode when the slit is of subwavelength width. Compared with the plasmonic mode in the silver film, the propagation constant of the electromagnetic wave in the PEC waveguide is smaller. The effective index keeps to be 1 when the width of the nanoslit changes.

This novel phase modulation mechanism has been verified by an extraordinary Young's double-slit interference experiment [36]. When light passes through two metallic slits of the same thickness but different widths (Fig. 2.18a), the intensity at the center of the interference fringes is minimum as shown in Fig. 2.18b, which

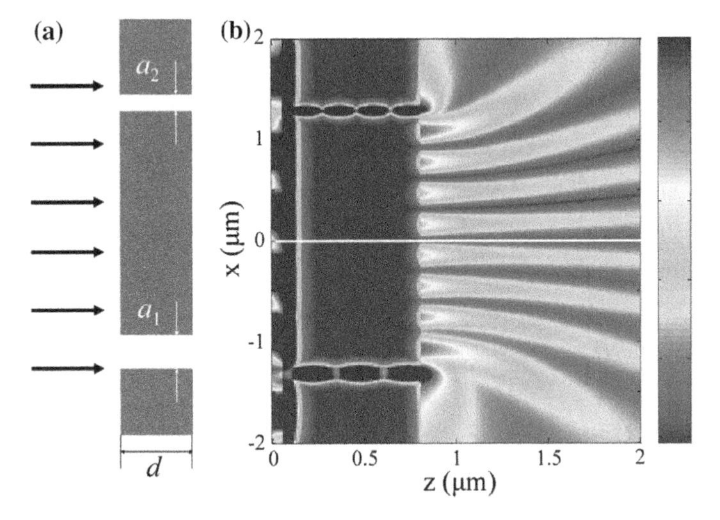

Fig. 2.18 **a** Sketch of metallic double nanoslits with different widths. **b** Near-field $|H_y|$ distribution pattern, $\lambda = 633$ nm. Reproduced from [36] with permission. Copyright 2007, Optical Society of America

seems contrary to our common sense. It is well known in classical Young's double-slit experiment that a bright fringe should be generated after the interference of the diffraction waves from two slits. The mechanism of this extraordinary Young's double-slit interference is the phase engineering of the plasmonic modes in nanoslits. In this case, the phase retardation difference between the two slits can be calculated as:

$$\Delta\varphi = \frac{2\pi d(n_2 - n_1)}{\lambda}, \tag{2.3.24}$$

where n_1 and n_2 are the effective index of slit 1 and slit 2, respectively, $d = 700$ nm is the thickness of the slits. The values of n_1 and n_2 can be calculated to be 1.72 and 1.24, which leads to $\Delta\varphi \approx \pi$ and thus the dark fringe appears in the center.

Dielectric waveguide with high-refractive index nanoridge or nanorod is another scheme of waveguide-based propagation phase engineering [37]. Figure 2.19a is the schematic of a dielectric ridge waveguide (DRW) with an amorphous-silicon nanoridge on a glass substrate, and the DRW has a deep subwavelength width. Figure 2.19b plots the electric field (real $[E_y]$) distribution in xz-plane for the DRW with different widths, which shows an obvious phase difference between the two DRWs. It should be noted that, contrary to the case of SPP waveguide, the propagation constant of DRWs increases with the width of DRW. An effective mode index can be modulated from $n_{\text{eff}} \approx 1$ (when the light is mostly in air) to $n_{\text{eff}} \approx n_{\text{a-Si}}$ (when the light is mostly in a-Si) as shown in Fig. 2.19c.

Another type of propagation phase engineering is based on the effective medium theory. Metamaterials are typical equivalent mediums, which are constructed by

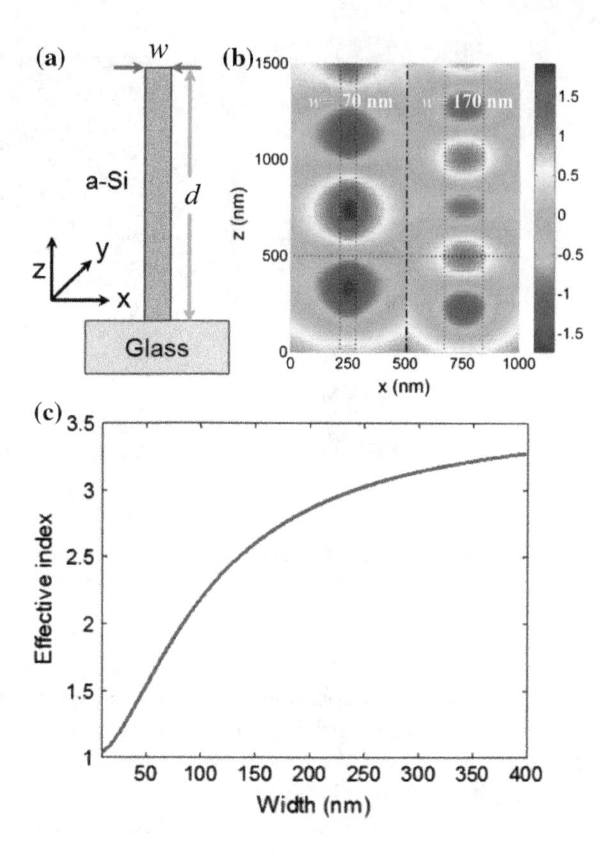

Fig. 2.19 **a** Schematic diagram of DRW. **b** Electric field (real $[E_y]$) distribution in xz-plane showing the emergence of a phase difference between two DRWs with different widths. $\lambda = 1425$ nm. In the simulation, the DRW is infinitely extended along y. **c** Effective index of DRW as a function of its width for parallel polarization (E_y). Reproduced from [37] with permission. Copyright 2015, American Chemical Society

artificial structures with unique electromagnetic properties that are not available in nature. The effective refractive index of metamaterials can be manipulated by design of the inclusions, geometries, and the scales of the structures. Figure 2.20a shows a meta-hologram consisting of three types of metamaterials with unit cells of circular patch, rectangular patch, and I-shaped patch [38]. The metamaterials consist of three layers of gold elements in a SiO_2 matrix over a Ge substrate. An arbitrary effective refractive index value between 2.1 and 5.4, for an overall index contrast of 3.3, can be obtained by interpolating the physical dimensions of the elements as shown in Fig. 2.20b. This effective index modulation leads to the phase engineering in a planar way.

2.3.3 *Circuit-Type Phase Engineering*

In the following, the reflection phase and the transmission phase are deduced separately for the characteristics of the reflective and transmissive metasurfaces [39].

(1) Transmissive metasurfaces

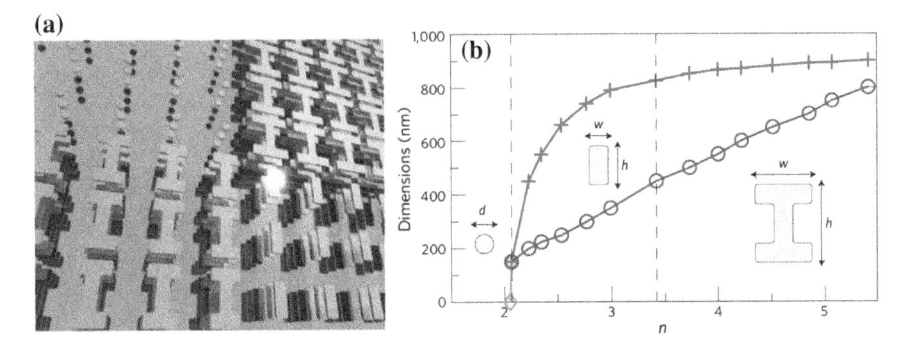

Fig. 2.20 **a** Artistic rendering of a section of metamaterial hologram. **b** Relationship between the dimensions of the metamaterial elements and the real part of their effective refractive index. The refractive index is presented for a wavelength of 10.6 μm and for an electric field polarized along the long axis of the rectangular patches, or the axis of the I-beams. The refractive index is adjusted by controlling the diameter *d* (green diamonds) of circular patches, or the width *w* (blue circles) and height *h* (red crosses) of rectangular patches or I-beams. The refractive index regions covered by the various metamaterial elements are separated by vertical dashed lines. Reproduced from [38] with permission. Copyright 2012, Springer Nature

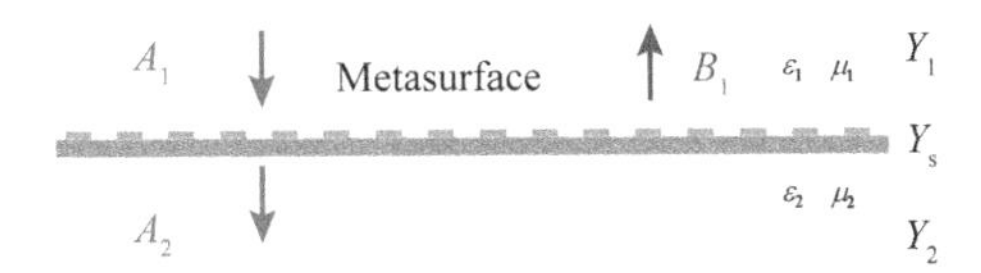

Fig. 2.21 Circuit-type phase engineering for transmissive metasurface. Adapted from [39] with permission. Copyright 2017, Institute of Optics and Electronics, Chinese Academy of Sciences

Figure 2.21 is a schematic diagram of a single-layered metasurface. The electromagnetic wave illuminates from medium 1 to medium 2 through the metasurface. The permittivity and permeability of medium 1 and medium 2 are ε_1, μ_1 and ε_2, μ_2 respectively. Similarly, the equivalent admittances of medium 1 and medium 2 are Y_1 and Y_2 respectively, and the metasurface is treated as a two-dimensional sheet with effective admittance Y_s.

For a single-layered transmissive metasurface, the boundary conditions are

$$\begin{array}{l} A_1 + B_1 = A_2 \\ Y_1(A_1 - B_1) = Y_2 A_2 + Y_s A_2 \end{array}. \tag{2.3.25}$$

After transformation, we have:

$$\begin{array}{l} 2A_1 Y_1 = A_2(Y_1 + Y_2 + Y_s) \\ 2B_1 Y_1 = A_2(Y_1 - Y_2 - Y_s) \end{array}. \tag{2.3.26}$$

According to the definition of reflection coefficient and transmission coefficient, we obtain,

$$\begin{aligned} r &= \frac{B_1}{A_1} = \frac{(Y_1-Y_2-Y_S)}{(Y_1+Y_2+Y_S)} \\ t &= \frac{A_2}{A_1} = \frac{2Y_1}{(Y_1+Y_2+Y_S)} \end{aligned}. \tag{2.3.27}$$

When electromagnetic wave illuminates from free space into the free space through the metasurface, there is $Y_1 = Y_2 = Y_0$; thus, Eq. (2.3.27) is simplified to

$$\begin{aligned} r &= \frac{-Y_S/Y_0}{(2+Y_S/Y_0)} \\ t &= \frac{2}{(2+Y_S/Y_0)} \end{aligned}. \tag{2.3.28}$$

The reflection phase and the transmission phase are

$$\begin{aligned} \Phi_r &= \arg\left(\frac{-Y_S/Y_0}{(2+Y_S/Y_0)}\right) \\ \Phi_t &= \arg\left(\frac{2}{(2+Y_S/Y_0)}\right) \end{aligned}. \tag{2.3.29}$$

(2) Reflective metasurfaces

The metasurfaces working in reflective scheme are also treated as two-dimensional sheets with effective admittance Y_s, as shown in Fig. 2.22.

The electromagnetic wave illuminates from medium 1 to medium 2 through the metasurface. The permittivity and permeability of medium 1 and medium 2 are ε_1, μ_1 and ε_2, μ_2, respectively. The tangential electric field in the transmission and reflection direction is A_i and B_i. Here, i represents the corresponding medium. The effective admittance of medium 1 and medium 2 is Y_1 and Y_2, respectively. A ground plane is added, so that all the electromagnetic energy would be reflected. The thickness of the medial dielectric layer is d. The boundary conditions can be written as

$$\begin{aligned} A_1 + B_1 &= A_2 + B_2 \\ Y_1(A_1 - B_1) &= Y_2(A_2 - B_2) + Y_s(A_2 + B_2) \end{aligned}. \tag{2.3.30}$$

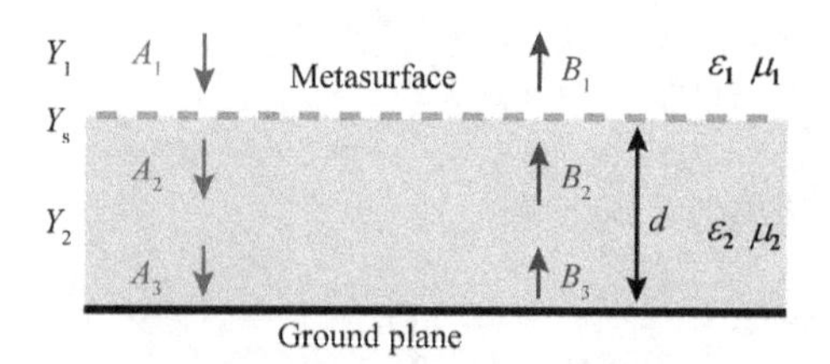

Fig. 2.22 Circuit-type phase engineering for reflective metasurface. Adapted from [39] with permission. Copyright 2017, Institute of Optics and Electronics, Chinese Academy of Sciences

Considering the propagation of electromagnetic wave in the dielectric, we have

$$\begin{pmatrix} A_2 \\ B_2 \end{pmatrix} = \begin{pmatrix} exp(-ikd) & 0 \\ 0 & exp(ikd) \end{pmatrix} \begin{pmatrix} A_3 \\ B_3 \end{pmatrix}, \quad (2.3.31)$$

where k is the wave number in the dielectric. Considering the reflectance coefficient of the ground plane is -1, we have

$$\begin{pmatrix} A_1 \\ B_1 \end{pmatrix} = \frac{1}{2Y_1} \begin{pmatrix} (Y_1 + Y_s + Y_2)\exp(-ikd) - (Y_1 + Y_s - Y_2)\exp(ikd) \\ (Y_1 - Y_s - Y_2)\exp(-ikd) - (Y_1 - Y_s + Y_2)\exp(ikd) \end{pmatrix}. \quad (2.3.32)$$

Assuming no magnetic materials are used in the design, and the illumination space is free space, we can obtain that $Y_1 = Y_0 = 1/377$. So, the reflection coefficient for the reflective mode can be then written as:

$$r = \frac{B_1}{A_1} = \frac{\left(1 - \sqrt{\varepsilon_2} - Y_s/Y_0\right) - \left(1 + \sqrt{\varepsilon_2} - Y_s/Y_0\right)\exp(2ikd)}{\left(1 + \sqrt{\varepsilon_2} + Y_s/Y_0\right) - \left(1 - \sqrt{\varepsilon_2} + Y_s/Y_0\right)\exp(2ikd)}. \quad (2.3.33)$$

The phase shift of the reflection is:

$$\Phi_r = \arg\left(\frac{\left(1 - \sqrt{\varepsilon_2} - Y_s/Y_0\right) - \left(1 + \sqrt{\varepsilon_2} - Y_s/Y_0\right)\exp(2ikd)}{\left(1 + \sqrt{\varepsilon_2} + Y_s/Y_0\right) - \left(1 - \sqrt{\varepsilon_2} + Y_s/Y_0\right)\exp(2ikd)}\right). \quad (2.3.34)$$

From Eqs. (2.3.33) and (2.3.34), we obtain both the amplitude and phase response after the electromagnetic waves interact with the reflective metasurfaces. This can be an important guidance in the design of the metasurfaces to obtain a broadband response.

2.3.4 Geometric Phase Engineering

The concept of geometric phase was first described generally by Michael Berry in 1984. He found when the parameters of a quantum mechanical wavefunction are slowly cycled around a circuit, the phase of the wavefunction need not return to its original value. That means there is a phase, whose value is independent of the transmitted optical path or dynamic process, and only related to the geometric path of the system's evolution. Berry's research found that when an adiabatic physical system evolves from an initial state along a path for a period and returns to the initial state, its final state is not equivalent to the initial state, and an additional phase is generated [40]. Since this phase factor is only related to the geometric path of the evolution of the system, this type of phase was subsequently named as geometric phase or Berry's phase. Geometric phase widely exists in a variety of physical systems, including quantum systems, optical systems, and even classical mechanical systems. Compared with the optical devices based on conventional propagation phase, the geometric phase-based devices can be infinitely thin in principle and the loss is usually lower.

Although the geometric phase is well known by Berry's research, the discovery of this phase is much earlier in design of phase changer in microwave band [41, 42]. In 1956, Pancharatnam [43] from the Raman Research Institute discovered the anomalous interference fringes that depend on the polarization state during his study of the light absorption of biaxial crystals and found an additional phase factor in the polarization transformation of electromagnetic waves. So the geometric phase is also called Pancharatnam phase, or more usually Pancharatnam–Berry phase (PB phase). Pancharatnam provided an accessible derivation of the geometric phase entirely using basic geometry and gave the expression of the geometric phase of the light wave based on Poincare's sphere. Poincare's sphere is a common way of describing the polarization state of electromagnetic waves. As shown in Fig. 2.23, on a spherical surface with a unit radius, each point on the surface corresponds to a polarization state. The North and South Poles correspond to right-handed and left-handed circularly polarized states, respectively. The equator corresponds to a polarization state in which the left and right circularly polarized components are equal, that is, linear polarization. The different longitudes on the equator correspond to different linear polarization angles. The points between the equator and the pole are the elliptical polarization states.

As shown in Fig. 2.23, when the polarization of the light wave returns from the North Pole through the equator and the South Pole back to the North Pole, the phase of the light wave changes to half the solid angle of the closed path. This geometric phase can also be understood in another equivalent way: When the polarization state of the light moves from the North Pole to the South Pole, the paths of different longitudes will produce a phase difference equal to the geometric phase described above.

The transmission of electromagnetic waves in the anisotropic structures can be described by Jones matrix. The basic expression is [39]:

$$\begin{bmatrix} E_{xout} \\ E_{yout} \end{bmatrix} = J_{\zeta} \begin{bmatrix} E_{xin} \\ E_{yin} \end{bmatrix}, \tag{2.3.35}$$

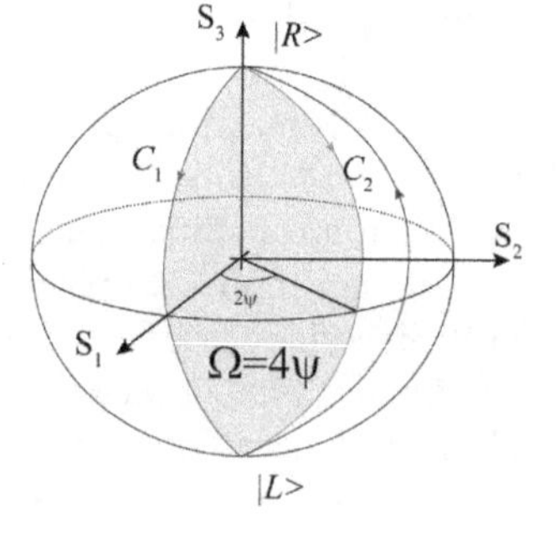

Fig. 2.23 Poincare's sphere representation of polarization states. Adapted from [39] with permission. Copyright 2017, Institute of Optics and Electronics, Chinese Academy of Sciences

where E_{xout} and E_{yout} are the x-polarized and y-polarized component of the output electromagnetic wave, respectively, E_{xin} and E_{yin} are the x-polarized component and the y-polarized component of the input electromagnetic wave, and J_ζ represents a 2 × 2 Jones matrix:

$$J_\zeta = \begin{bmatrix} J_{11} & J_{12} \\ J_{21} & J_{22} \end{bmatrix}. \tag{2.3.36}$$

For anisotropic structures, the principal axes of the local coordinate are assumed to be u and v, respectively, and the angle between the u-axis and the x-axis is ζ, as shown in Fig. 2.24.

Assuming that the complex amplitudes of the transmission in the two principal directions are t_u and t_v, respectively, the Jones matrix of the anisotropic structures can be expressed as:

$$J_\zeta = \begin{bmatrix} \cos\zeta & \sin\zeta \\ -\sin\zeta & \cos\zeta \end{bmatrix} J_g \begin{bmatrix} \cos\zeta & -\sin\zeta \\ \sin\zeta & \cos\zeta \end{bmatrix}, \tag{2.3.37}$$

where J_g is the transmission matrix of the complex amplitude, which can be expressed as:

$$J_g = \begin{bmatrix} t_u & 0 \\ 0 & t_v \end{bmatrix}. \tag{2.3.38}$$

Substituting Eq. (2.3.38) into Eq. (2.3.37), we can get:

$$J_\zeta = \begin{bmatrix} t_u\cos^2\zeta + t_v\sin^2\zeta & (-t_u + t_v)\sin\zeta\cos\zeta \\ (-t_u + t_v)\sin\zeta\cos\zeta & t_u\sin^2\zeta + t_v\cos^2\zeta \end{bmatrix}. \tag{2.3.39}$$

If the incident electromagnetic wave is linearly polarized, the polarization states of the x-polarized and y-polarized incident electromagnetic waves can be, respectively, described by Jones vectors,

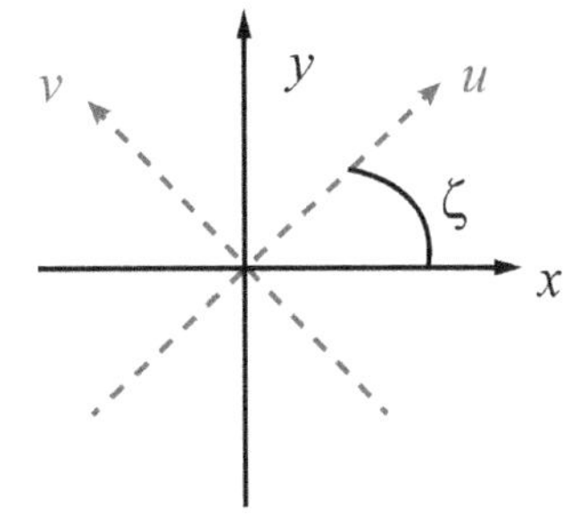

Fig. 2.24 Coordinate for the anisotropic structures

$$\begin{bmatrix} E_{xin} \\ E_{yin} \end{bmatrix} = \begin{bmatrix} 1 \\ 0 \end{bmatrix}, \tag{2.3.40}$$

and

$$\begin{bmatrix} E_{xin} \\ E_{yin} \end{bmatrix} = \begin{bmatrix} 0 \\ 1 \end{bmatrix}. \tag{2.3.41}$$

The corresponding output electric fields are

$$\begin{bmatrix} E_{xout} \\ E_{yout} \end{bmatrix} = J_{\zeta} \begin{bmatrix} 1 \\ 0 \end{bmatrix}, \tag{2.3.42}$$

and

$$\begin{bmatrix} E_{xout} \\ E_{yout} \end{bmatrix} = J_{\zeta} \begin{bmatrix} 0 \\ 1 \end{bmatrix}. \tag{2.3.43}$$

Substituting Eq. (2.3.39) into Eqs. (2.3.42) and (2.3.43), we obtain the expression of the output electric fields under the condition of linear polarization incidence:

$$\begin{aligned} &\begin{bmatrix} E_{xout} \\ E_{yout} \end{bmatrix} \\ &= \begin{bmatrix} t_u \cos^2 \zeta + t_v \sin^2 \zeta & (-t_u + t_v) \sin \zeta \cos \zeta \\ (-t_u + t_v) \sin \zeta \cos \zeta & t_u \sin^2 \zeta + t_v \cos^2 \zeta \end{bmatrix} \begin{bmatrix} 1 \\ 0 \end{bmatrix} \\ &= \begin{bmatrix} t_u \cos^2 \zeta + t_v \sin^2 \zeta \\ (-t_u + t_v) \sin \zeta \cos \zeta \end{bmatrix} \end{aligned} \tag{2.3.44}$$

and

$$\begin{aligned} &\begin{bmatrix} E_{xout} \\ E_{yout} \end{bmatrix} \\ &= \begin{bmatrix} t_u \cos^2 \zeta + t_v \sin^2 \zeta & (-t_u + t_v) \sin \zeta \cos \zeta \\ (-t_u + t_v) \sin \zeta \cos \zeta & t_u \sin^2 \zeta + t_v \cos^2 \zeta \end{bmatrix} \begin{bmatrix} 0 \\ 1 \end{bmatrix}. \\ &= \begin{bmatrix} (-t_u + t_v) \sin \zeta \cos \zeta \\ t_u \sin^2 \zeta + t_v \cos^2 \zeta \end{bmatrix} \end{aligned} \tag{2.3.45}$$

Equations (2.3.44) and (2.3.45) are the expressions of output electric fields for x and y linear polarization incidence. Obviously, after the interaction of the electromagnetic wave with the anisotropic structures, the orthogonally polarized electromagnetic component with an amplitude of $(-t_u + t_v) \sin\zeta \cos\zeta$ is generated, apart from the electromagnetic component with the original polarization state.

The following is an analysis of the case where the anisotropic structure is rotated by $\pm 90°$ around the z-axis. According to the Eqs. (2.3.44) and (2.3.45), the angle ζ between the u-axis and the x-axis becomes $\zeta \pm \pi/2$, and the output electric field corresponding to the x-polarization and y-polarization incidence is obtained as:

$$\begin{bmatrix} E_{xout} \\ E_{yout} \end{bmatrix} = \begin{bmatrix} t_u \sin^2(\zeta) + t_v \cos^2(\zeta) \\ (t_u - t_v)\sin(\zeta)\cos(\zeta) \end{bmatrix}, \tag{2.3.46}$$

and

$$\begin{bmatrix} E_{xout} \\ E_{yout} \end{bmatrix} = \begin{bmatrix} (t_u - t_v)\sin\zeta\cos\zeta \\ t_u \cos^2\zeta + t_v \sin^2\zeta \end{bmatrix}. \tag{2.3.47}$$

Obviously, the orthogonally polarized electromagnetic components are equal in magnitude to that in Eqs. (2.3.44) and (2.3.45), but have a phases difference of π.

For circularly polarized incident electromagnetic waves, the polarization state can be expressed as a Jones vector

$$\begin{bmatrix} E_{xin} \\ E_{yin} \end{bmatrix} = \begin{bmatrix} 1 \\ i\sigma \end{bmatrix}, \tag{2.3.48}$$

where $\sigma = \pm 1$ corresponds to the right-handed and left-handed circular polarization state, respectively. The polarization state corresponding to the output electromagnetic wave is:

$$\begin{bmatrix} E_{xout} \\ E_{yout} \end{bmatrix} = \frac{J_\zeta}{\sqrt{2}} \begin{bmatrix} 1 \\ i\sigma \end{bmatrix}. \tag{2.3.49}$$

Substituting Eq. (2.3.39) into Eq. (2.3.49), we obtain:

$$\begin{aligned} \begin{bmatrix} E_{xout} \\ E_{yout} \end{bmatrix} &= \frac{1}{\sqrt{2}} \begin{bmatrix} t_u \cos^2\zeta + t_v \sin^2\zeta & (-t_u + t_v)\sin\zeta\cos\zeta \\ (-t_u + t_v)\sin\zeta\cos\zeta & t_u \sin^2\zeta + t_v \cos^2\zeta \end{bmatrix} \begin{bmatrix} 1 \\ i\sigma \end{bmatrix}. \\ &= \frac{1}{2\sqrt{2}} \left((t_u + t_v) \begin{bmatrix} 1 \\ i\sigma \end{bmatrix} + (t_u - t_v) e^{-2i\sigma\zeta} \begin{bmatrix} 1 \\ -i\sigma \end{bmatrix} \right) \end{aligned} \tag{2.3.50}$$

It can be seen that after the interactions between incident left (right) circularly polarized electromagnetic wave with the anisotropic structure, the output electromagnetic field generates a right (left) circularly polarized electromagnetic component (orthogonally polarized with the incidence) with complex amplitude of $(t_u - t_v)e^{-2i\sigma\zeta}/2\sqrt{2}$, in addition to the original polarized electromagnetic wave with

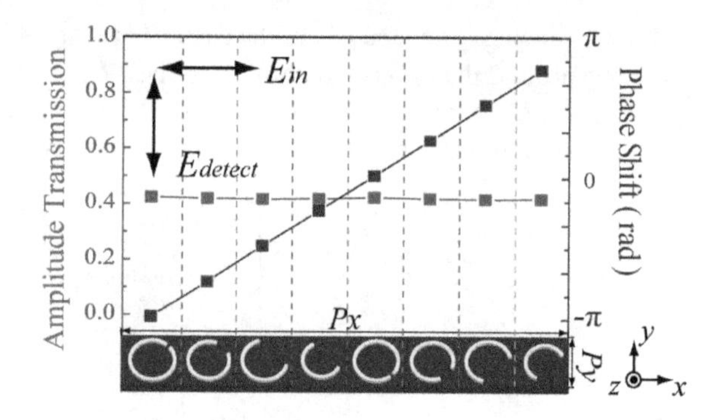

Fig. 2.25 Amplitude transmission and phase shift (the y component of the E field) of the C-shaped antennas at the bottom under x-polarized incidence. Reproduced from [44] with permission. Copyright 2013, Wiley-VCH Verlag GmbH & Co. KGaA, Weinheim

complex amplitude of $(t_u + t_v)/2\sqrt{2}$. And the orthogonally polarized electromagnetic waves carry an additional geometric phase of $-2\sigma\zeta$.

Similarly, for anisotropic reflective materials, one only needs to replace the complex amplitude of the transmission coefficient t_u and t_v in Eq. (2.3.50) with the complex amplitude of the reflection coefficient r_u and r_v. Then we obtain the corresponding expression of the output electric field:

$$\begin{bmatrix} E_{xout} \\ E_{yout} \end{bmatrix} = \frac{1}{2\sqrt{2}}\left((r_u + r_v)\begin{bmatrix} 1 \\ i\sigma \end{bmatrix} + (r_u - r_v)\mathrm{e}^{-2i\sigma\zeta}\begin{bmatrix} 1 \\ -i\sigma \end{bmatrix}\right). \tag{2.3.51}$$

Recalling the expression of the output E-fields under linear polarization incidence from Eqs. (2.3.44) and (2.3.45), the orthogonally polarized electric component generated by the anisotropic structure is $(-t_u + t_v)\sin\zeta\cos\zeta$ for both x- and y-polarization illumination. That means when the anisotropic structure is rotated by 90°, the output orthogonally polarized electromagnetic wave has a phase shift of π and same amplitude with that from the initial structure. This property is widely applied in design of metasurfaces including V-shaped and C-shaped antennas. Figure 2.25 shows the unit cells of eight C-shaped antennas [44]. The simulations demonstrated that the amplitudes of the cross-polarized radiation scattered by the eight antennas are nearly equal, with phases in $\pi/4$ increments. The right four antennas are created from the left four antennas by simply rotating 90° clockwise with a phase shift of π to the corresponding elements.

2.4 Generalized Theory of Diffraction

The generalized diffraction can overcome the principal fundamental hurdle of diffraction limit that is faced by traditional diffraction theory in EO 1.0. In traditional optics, the spatial resolution of microscopes was ultimately limited by the diffraction of light wave. Since the wavelength is the key of diffraction limit, Abbe suggested using light with shorter wavelength to obtain higher resolution. In 2014, the Nobel Prize in

Chemistry was awarded to Stefan Hell et al. for the development of super-resolved fluorescence microscopy. However, this super-resolution technique is limited to fluorescence samples. As Hell said, such systems do not break the diffraction limit of the lens systems [45].

2.4.1 *Sub-diffraction-Limited Optics with Evanescent Waves*

Based on the Fourier optics theory, making use of evanescent fields or recovery of evanescent fields is extremely important to overcome the obstacle of diffraction limit. Different schemes, such as scanning near-field optical microscopy (SNOM) [46, 47], superlenses [48], and various forms of field concentrators [49], have been exploited. The proper far-field optical superlens requires bulk negative-index materials that are still to be developed. Other designs, though offering substantial advances, are united by a common severe limitation that the object being imaged or stimulated must be in the immediate proximity of the superlens or field concentrator.

The emergence of M-wave permits a low-cost way to break the diffraction barrier by all-optical methods [6]. Since the short wavelength and directional propagation properties associated with hyperbolic dispersion are two fundamental ingredients in plasmonic super-resolution imaging, they form the foundation stones for super-resolution imaging beyond diffraction limit. As shown in Fig. 2.26, the sub-diffraction-limited imaging model of a planar plasmonic lens can be written in the form of [6]

$$\begin{bmatrix} E_x(x, y, z) \\ E_y(x, y, z) \\ E_z(x, y, z) \end{bmatrix} = \int_{-\infty}^{\infty}\int_{-\infty}^{\infty} \begin{bmatrix} A_x(k_x, k_y) \\ A_y(k_x, k_y) \\ \frac{k_x A_x + k_y A_y}{-k_z} \end{bmatrix} \times \exp(ik_x x + ik_y y)\,\mathrm{d}k_x \mathrm{d}k_y, \quad (2.4.1)$$

and

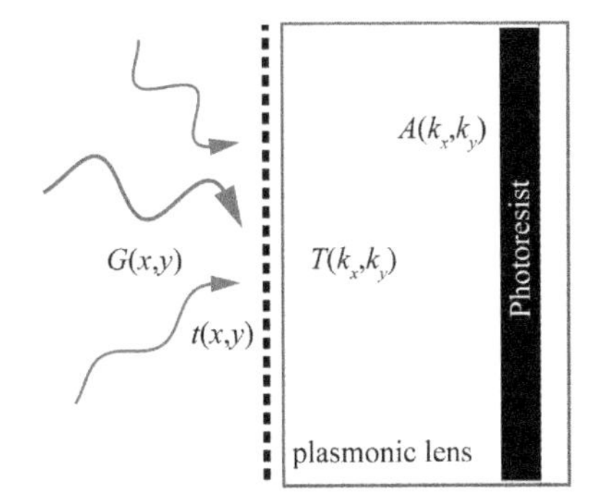

Fig. 2.26 A simple model of plasmonic imaging lithography. Adapted from [6] with permission. Copyright 2015, Science China Press and Springer-Verlag Berlin Heidelberg

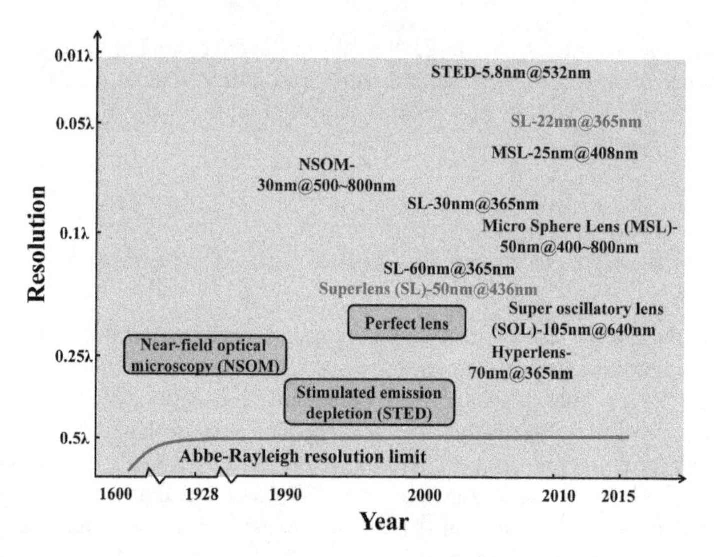

Fig. 2.27 Development of sub-diffraction-limited technologies. Adapted from [6] with permission. Copyright 2015, Science China Press and Springer-Verlag Berlin Heidelberg

$$\begin{bmatrix} A_x(k_x, k_y) \\ A_y(k_x, k_y) \end{bmatrix} = \int_{-\infty}^{\infty}\int_{-\infty}^{\infty} \begin{bmatrix} G_x(x, y)t_x(x, y)T_x(k_x, k_y) \\ G_y(x, y)t_y(x, y)T_y(k_x, k_y) \end{bmatrix} \times \exp\left(-ik_x x - ik_y y\right)\mathrm{d}x\mathrm{d}y, \tag{2.4.2}$$

where $G(x, y)$, $t(x, y)$, and $T(k_x, k_y)$ are the illumination function, transmission function of object and optical transfer function (OTF) of plasmonic lens, respectively. The OTF can be obtained through transfer matrix. If there is reflection plane in the plasmonic lens, the OTF should be determined by using the fields in the photoresist layer, implying that the images can also be modulated by the reflecting layer [50].

By utilizing the imaging model, 50 nm pitch interference pattern was experimentally observed with a 436 nm light source in 2004 [51]. This interference experiment also demonstrated the superlens effect, since the interference fringes at the two sides of the silver film are nearly identical. This conclusion can also be drawn from the fact that the wavevector of surface plasmon is broaden when the wavevector approaches infinity. In 2005, Fang et al. [52] demonstrated this idea with 60 nm resolution achieved at a wavelength of 365 nm. More recently, the short wavelength and the superlens effect were exploited in the experimental demonstration of half-pitch 32 and 22 nm lithography with high aspect ratio [53], achieving the highest resolution of imaging lithography as far as we know (Fig. 2.27). It has also been shown that the resolution capability can be pushed down to 16 nm for gratings and 9 nm for isolated patterns through the self-aligned multiple patterning technique [54].

It should be noted that there is a trade-off between the resolution and the working distance at a given wavelength for near-field imaging, which is the so-called near-

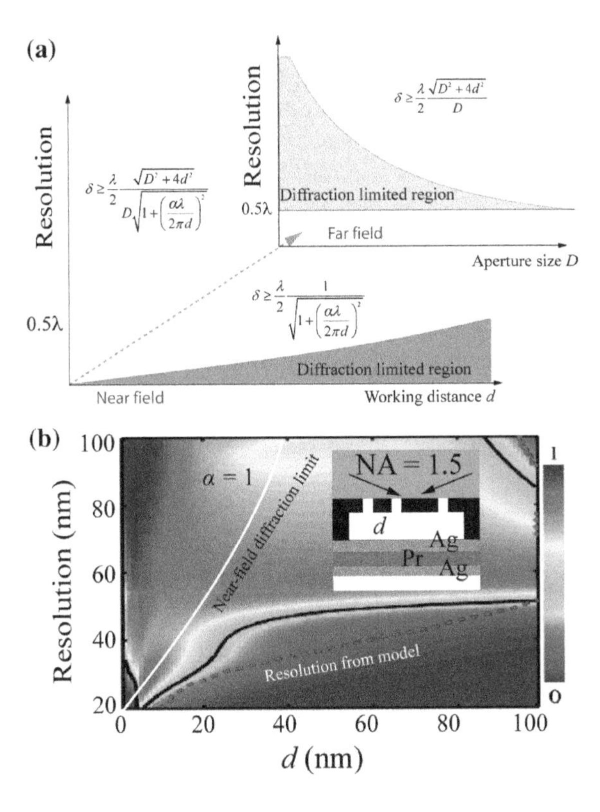

Fig. 2.28 **a** Schematic of the far-field and near-field diffraction limit ($\alpha = 1$). **b** Imaging contrast for plasmonic lens with different working distances and resolutions. Inset shows the schematic of the plasmonic lens. Adapted from [6] with permission. Copyright 2015, Science China Press and Springer-Verlag Berlin Heidelberg

field diffraction limit. As illustrated in Fig. 2.28, the near-field diffraction limit was defined according to the working distance at which the highest spatial component decays to its $1/(\alpha e)$ in amplitude:

$$\delta \geq \frac{\lambda}{2} \frac{1}{\sqrt{1 + \left(\frac{\alpha\lambda}{2\pi d}\right)^2}}, \tag{2.4.3}$$

where δ is the resolution at a given working distance d between the image and lens. Obviously, this equation could be combined with the far-field diffraction limit to give a universal definition:

$$\delta \geq \frac{\lambda}{2} \frac{\sqrt{D^2 + 4d^2}}{D\sqrt{1 + \left(\frac{\alpha\lambda}{2\pi d}\right)^2}}, \tag{2.4.4}$$

where D is the aperture diameter of the lens and d is the working distance between the image and object. When the working distance is much larger than the wavelength, Eq. (2.4.4) is reduced to $\delta \geq 0.5\lambda/\text{NA}$, i.e., the classic diffraction limit proposed by Abbe.

The use of plasmonic lens and reflective slab provides a route to break the near-field diffraction limit. To compare the resolution of plasmonic lenses with the near-field diffraction limit, a plasmonic cavity lens composed of Ag-photoresist-Ag incorporating high spatial frequency spectrum off-axis illumination was proposed [55]. As illustrated in Fig. 2.28b (α is chosen to be 1), the imaging contrast for a plasmonic lens shows that the near-field diffraction limit is surpassed. This approach remarkably enhances the object's subwavelength information and damps negative contribution from the longitudinal electric field component in the imaging region. Experimental images of well-resolved 60 nm half-pitch patterns under 365 nm ultraviolet light were demonstrated at an air distance of 80 nm between the mask patterns and plasmonic cavity lens, approximately fourfold longer than that of the superlens scheme.

2.4.2 Sub-diffraction-Limited Optics Without Evanescent Waves

So far, the main direction of research aiming to break the traditional diffraction limit seeks to exploit the evanescent components containing fine detail of the electromagnetic field distribution. In contrast, super-oscillation sub-diffraction-limited imaging is considered as a novel approach to realize super-resolution without evanescent waves. The concept of super-oscillation was proposed in microwave regime and optics to improve the resolution of imaging in the far-field as early as 1950s [56]. Later, this abnormal phenomenon was understood in terms of super-oscillation, which occurs in a region where the band-limited functions are able to oscillate faster than their highest Fourier components [57]. Seemingly counterintuitive, this anomalous effect may be understood by investigating the difference between complex amplitude and intensity [8]. As depicted in Fig. 2.29, if one constructs an intensity function as $|\sin(2\pi x) + 0.99|^2$, a small peak can be obtained with very small width. Though this example is extremely simple, it reveals some important aspects of super-oscillation interference. First, although the local intensity may oscillate more rapidly than the highest Fourier components, the amplitude often does not have this property. As a result, it is the conversion from complex amplitude to intensity that results in this effect. Notably, if one could record and reconstruct complex optical fields, the time-reversal process could be used to realize super-resolution focusing and imaging [58]; second, the super-oscillation is weak and accompanied with strong side lobes. In general, the narrower the super-oscillatory lobe, the higher the side lobe will be, which poses a great challenge for the practical applications [59].

A super-oscillatory function is a band-limited function $f(x)$ oscillating faster than its fastest Fourier component, which is taken to be the initial state of a freely evolving quantum wavefunction ψ [60]. Where super-oscillations occur, the functions are exponentially weak, because the different Fourier components exhibit almost-perfect destructive interference. This weakness is the mechanism by which super-oscillations

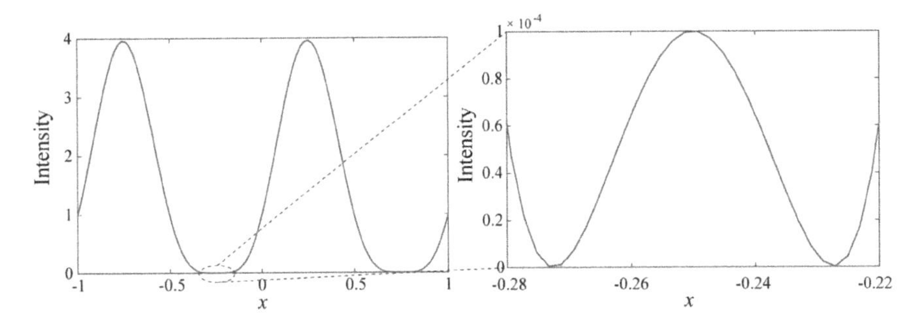

Fig. 2.29 A 1D super-oscillatory function (solid blue line) and its magnified view. Reproduced from [8] with permission. Copyright 2018, Optical Society of America

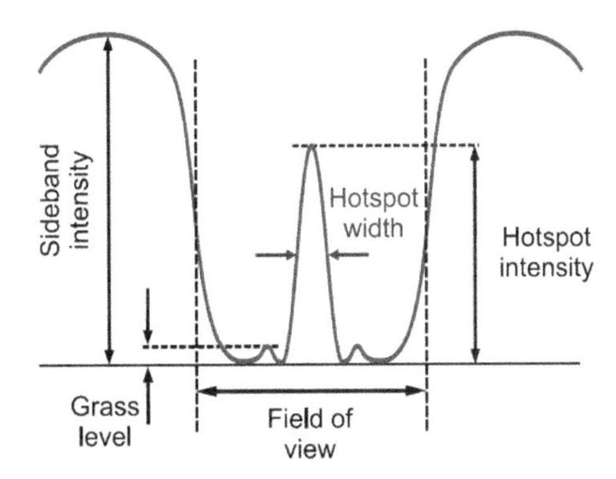

Fig. 2.30 Schematic of optical super-oscillation. An arbitrary field distribution with an arbitrarily small hot spot within a limited area $[-D/2, D/2]$

evade the uncertainty principle, because the principle is a relation between variances, and variances are insensitive to extremely small values.

A commonly used design algorithm to construct super-oscillation function comprises the following steps [61]: initially, the desired subwavelength hot spot is presented as a series of prolate spheroidal wavefunctions, which can be truncated when a satisfactory level of approximation is achieved; at the second step, this series of prolate spheroidal wavefunctions is presented as a series of plane waves, and using the scalar angular spectrum description of light propagating from the mask to the super-oscillating feature, the required complex mask transmission function $t(x)$ can be readily derived. Figure 2.30 shows the field distribution of a typical optical super-oscillation. Within a limited field of view region $[-D/2, D/2]$, a sub-diffraction-limited field can be superimposed by a series of orthogonal prolate spheroidal wavefunctions $\psi_n(c, x)$ that are band-limited to the frequency domain $[-k_0, k_0]$, which is given as [61]:

$$h_N(x) = \sum_{n=0}^{n=N} a_n(c)\psi_n(c, x). \tag{2.4.5}$$

Here a_n(c) is the modulation factor of the wavefunction, which depends on a constant $c = \pi D/\lambda$, and the Fourier transform function of $h_N(x)$ is given by:

$$H_N(u) = \sum_{n=0}^{n=N} \frac{\pi a_n \psi_n\left(c, \frac{uD}{2k_0}\right)}{i^n R_{0n}^{(1)}(c, 1)}, \tag{2.4.6}$$

where $R_{0n}^{(1)}(c, 1)$ is a radial prolate spheroidal wavefunction of the first kind, and

$$h_N(x) = \int_{-k_0}^{k_0} H_N(u) \mathrm{e}^{iux} \mathrm{d}u. \tag{2.4.7}$$

Assuming a mask with a complex transmission function $t(x)$ is illuminated at normal incidence with a plane monochromatic wave $E(x, z = 0) = 1$ at a wavelength $\lambda = 2\pi/k_0$, a prescribed field distribution $f(x)$ within a limited region $[-D/2, D/2]$ at a distance z will be generated. In the scalar angular spectrum description of light propagation, the field at a point (x, z) is

$$E(x, z) = \int_{-k_0}^{k_0} T(u) \exp(iux) \exp\left(iz\sqrt{k_0^2 - u^2}\right) du, \tag{2.4.8}$$

where $T(u)$ is the Fourier transform of $t(x)$. We now approximate $h(x) = E(x, z)$, and the required transmission function $t(x)$ of the mask can be obtained:

$$t(x) = \sum_{n=0}^{n=N} \int_{-k_0}^{k_0} \frac{\pi a_n \psi_n\left(c, \frac{uD}{2k_0}\right)}{i^n R_{0n}^{(1)}(c, 1)} \exp\left(iux - iz\sqrt{k_0^2 - u^2}\right) du. \tag{2.4.9}$$

2.5 Generalized Theory of Absorption and Radiation

2.5.1 Generalized Absorption Theory

(1) Two-wave exchange

In many cases, the M-waves propagating along the surface will be recoupled to propagating waves by the scatters themselves. This is because these evanescent waves are indeed leaky wave and not a rigorous bound wave. Taking the prism excitation of surface plasmon as an example, the evanescent wave tends to rescattered into the prism. At some specific condition, when the directly reflected wave destructively interferes with the rescattered wave, there will be zero reflection at the output channel.

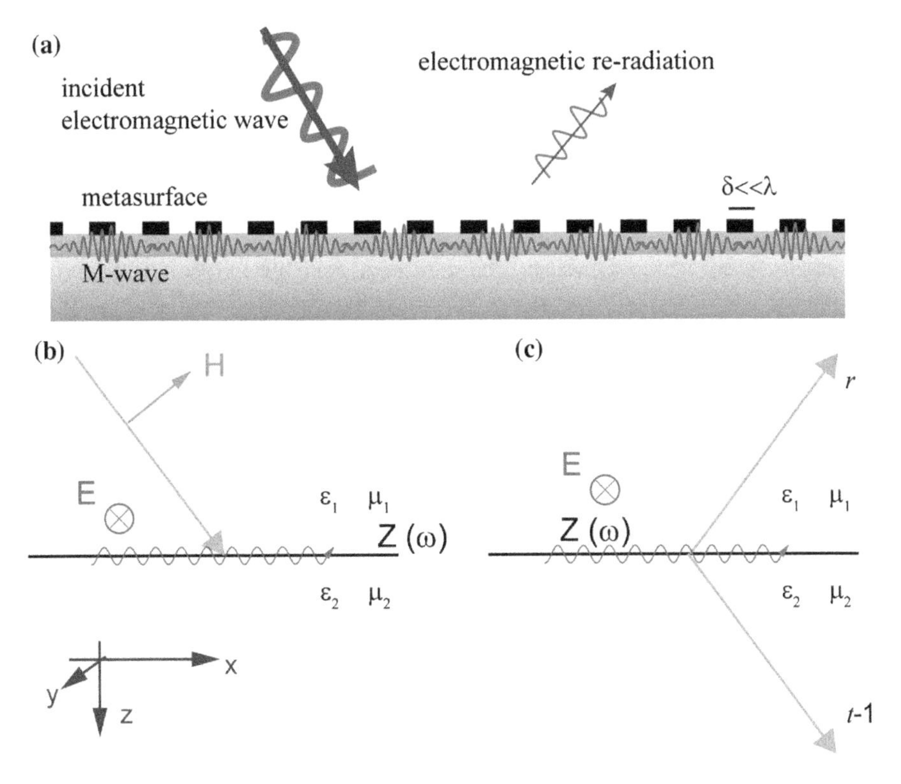

Fig. 2.31 Exchange between the propagating waves and M-waves. **a** Schematic of the principle. **b** Excitation of M-wave by propagating wave. **c** Conversion of the M-wave to propagating wave. Adapted from [6] with permission. Copyright 2015, Science China Press and Springer-Verlag Berlin Heidelberg

This condition is termed critical coupling [7]. For local optical excitation by needle, microsphere, or single scatter, the critical coupling condition is difficult to meet. Nevertheless, such condition has been demonstrated to be feasible for area excitation such as prism excitation and grating excitation. Indeed, the critical coupling of light into a metallic grating has been observed as early as 1902 when Robert Wood was investigating the diffraction spectrum of a metal grating [62].

Figure 2.31 illustrates the principle of wave exchange in subwavelength structures and a hybrid numerical method used to simulate the electromagnetic behavior [6]. First, the near-field electromagnetic responses of subwavelength structures must be solved using the most generalized Maxwell's wave equation, which is:

$$\nabla \times \varepsilon^{-1} \nabla \times \mathbf{H} = k_0^2 \mu \mathbf{H}, \tag{2.5.1}$$

for the magnetic fields $\mathbf{H}$. Here ε, μ, and k_0 are the relative permittivity, permeability, and free-space wave number. With some mathematical manipulation, Eq. (2.5.1) can be rewritten as:

$$\nabla \times \nabla \times \mathbf{H} + \varepsilon \nabla \varepsilon^{-1} \times \nabla \times \mathbf{H} = k_0^2 \varepsilon \mu \mathbf{H}. \tag{2.5.2}$$

Obviously, the second term of Eq. (2.5.2) illustrates the role of subwavelength inclusions, which makes the waves behave different from that in free space. After the reradiation waves leave the subwavelength structures, vectorial diffraction theory is used to calculate the far-field diffractions (either reflection or transmission), which has been shown to be very powerful in terms of computational capability [29, 63].

Owing to the unusual properties such as strong field localization and enhancement, the bound waves have been demonstrated to be useful in sub-diffraction-limited waveguiding, imaging, and lithography. Meanwhile, there are also diverse applications of the re-emitted waves since their phases, amplitudes, and polarizations have been thoroughly modulated by the subwavelength structures [64].

As shown in Fig. 2.31, the reflection and transmission problem can be rewritten as a scattering and reradiation problem of the bounded wave. The propagating wave and bounded wave (M-wave) could exchange with each other with the help of metasurface, which we defined as a process of two-wave exchange. In the far-fields, we only care about propagating wave. The total electromagnetic fields can be written as:

$$E_{\text{total}} = E_{\text{in}} + E_{\text{sc}} = E_{\text{in}} + S(E_{\text{in}}). \tag{2.5.3}$$

Here E_{sc} is the scattered electric fields, which is determined by the incident electric field E_{in} via the structure function S.

As depicted in Fig. 2.31c, the scattered fields are identical to the radiation fields of the M-waves. For homogeneous metasurfaces, the forward and backward scattering coefficients are $t - 1$ and r, respectively. From Eqs. (2.3.13) and (2.3.15), we find that the forward and backward waves are in phase and out of phase with each other for purely electric and magnetic metasurfaces. In a normal non-magnetic metasurface, there is

$$S = \frac{Y_1 - Y_3 - Y_{\text{e}}}{Y_1 + Y_3 + Y_{\text{e}}} E_i. \tag{2.5.4}$$

In general, the incident electric fields can be expanded using plane wave spectrum. In this case, the S function can be rewritten as:

$$S(k_x, k_y) = \frac{Y_1(k_x, k_y) - Y_3(k_x, k_y) - Y_{\text{e}}(k_x, k_y)}{Y_1(k_x, k_y) + Y_3(k_x, k_y) + Y_{\text{e}}(k_x, k_y)} E_i(k_x, k_y). \tag{2.5.5}$$

When S equals to 0, the metasurface does not scatter incident electromagnetic wave, and all incident wave is transmitted. On the other hand, it should be noted

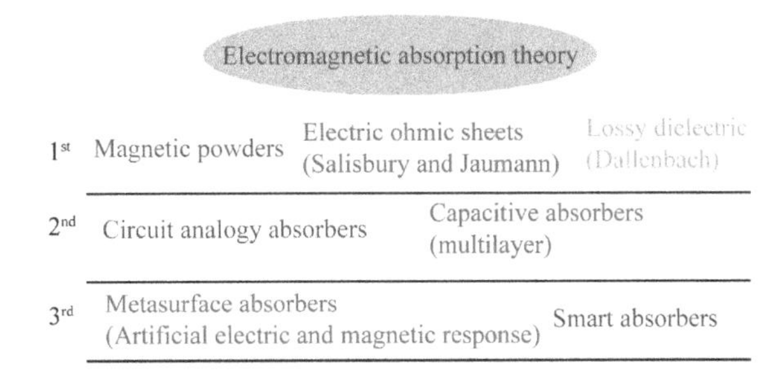

Fig. 2.32 Road map for electromagnetic absorbers. Reproduced from [6] with permission. Copyright 2015, Science China Press and Springer-Verlag Berlin Heidelberg

that the reflection and transmission cannot be simultaneously reduced to zero when a single metasurface is illuminated by a plane wave. Nevertheless, it was demonstrated in 1934 that the absorption coefficient can be enhanced up to 50% in a wide frequency band [65].

Figure 2.32 shows that the development of electromagnetic absorbers could be categorized into three stages [6, 66]. The first stage is using magnetic material and multilayered conductive films; the second is circuit analogy absorber (CAA); the third is metasurface-based absorber. To achieve perfect absorption with metasurface, additional reflective plane is often employed to suppress the reradiation of M-wave [67–70] and realize perfect conversion from propagating waves to bounded waves. Since the absorption does not depend on the magnetic loss, the metasurface absorber can be successfully utilized in environment with extreme conditions such as high temperatures. Interestingly, the absorption can be well described by the transfer-matrix and catenary models for the impedance dispersion [19, 71].

(2) Coherent absorption

In traditional thin-film interference theory, materials are often lossless and the refractive index is a real number. In this section, we show that the large imaginary part of metallic thin film may lead to many unusual interference effects. Since the imaginary part is associated with absorption of light, we will focus on the applications of interference in electromagnetic absorbers.

Without loss of generality, the absorption of a thin film with homogeneous electromagnetic parameters will be discussed first. The reflection (r) and transmission (t) coefficients of light normally incident upon a dielectric slab in air can be calculated using the Fresnel–Airy formulae as [72]:

$$r = \frac{\left(n^2 - 1\right)\left(-1 + \mathrm{e}^{i2nkd}\right)}{(n+1)^2 - (n-1)^2\mathrm{e}^{i2nkd}}, \tag{2.5.6}$$

$$t = \frac{4n\mathrm{e}^{inkd}}{(n+1)^2 - (n-1)^2\mathrm{e}^{i2nkd}}, \tag{2.5.7}$$

where n is the refractive index and d is the thickness of the slab. As early as 1934, it was discovered that a thin film of metal can absorb up to 50% of the incident light

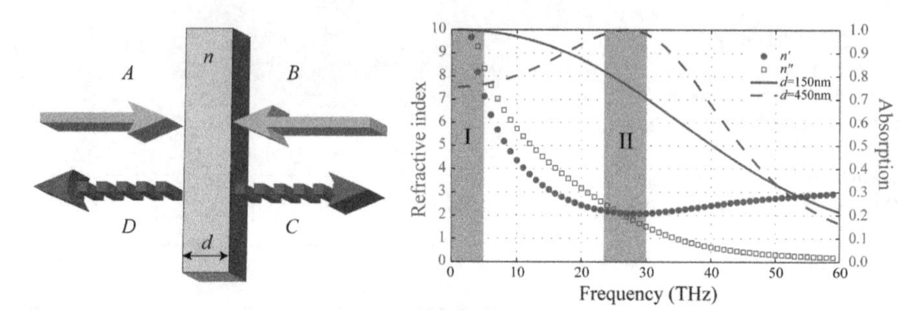

Fig. 2.33 Schematic of the thin-film CPA. The right panel shows the refractive index and absorption spectrum for doped silicon film. The doping concentration is 4e^{19} cm^{-3}. Adapted from [72] with permission. Copyright 2012, Optical society of America

[65]. Meanwhile, the transmission and reflection intensities are equal to 25%. Based on the anti-lasing concept [73], it was recently demonstrated that the interference of two oppositely propagating coherent beam in a heavily doped silicon film could lead to broadband perfect absorption with absorbance larger than 99.99% [72], which is called thin-film coherent perfect absorber (CPA). On the basis of transfer-matrix theory, the perfect absorption condition for the thin-film CPA can be written as:

$$n' \approx n'' \approx \frac{1}{\sqrt{kd}} = \sqrt{\frac{c}{\omega d}}. \tag{2.5.8}$$

Obviously, material with specific dispersion characteristics should be used to obtain a broadband CPA since the required complex refractive index is frequency dependent. Fortunately, it has been shown that metal and metal-like materials such as doped semiconductor are natural candidates for such application. Figure 2.33 illustrates the absorption curves of doped silicon film for two characteristic thickness, named Woltersdorff thickness (150 nm) and Plasmon thickness (450 nm).

For a structured film exhibiting simultaneous electric and magnetic responses, the general CPA condition is:

$$\exp(inkd) = \pm\frac{1 - Z}{1 + Z}, \tag{2.5.9}$$

where k is the wavevector in free space and d is the total thickness of the effective slab. The $\pm$ signs are corresponding to the symmetrical and anti-symmetric inputs, respectively. Such mechanism can be realized by utilizing coherent plasmon hybridization in a metamaterial film comprised of MIM structure [74].

The concept of thin-film CPA was experimentally demonstrated in the microwave range with both resistive sheet and graphene layer [75, 76]. As realized more recently, a thin layer of resistive sheet can be used as a near-perfect absorber in the radio frequency [77], where the thickness–wavelength ratio is as small as 8×10^{-5}, implying that the broadband CPA may bypass the Planck–Rozanov limit for absorbers [78–80].

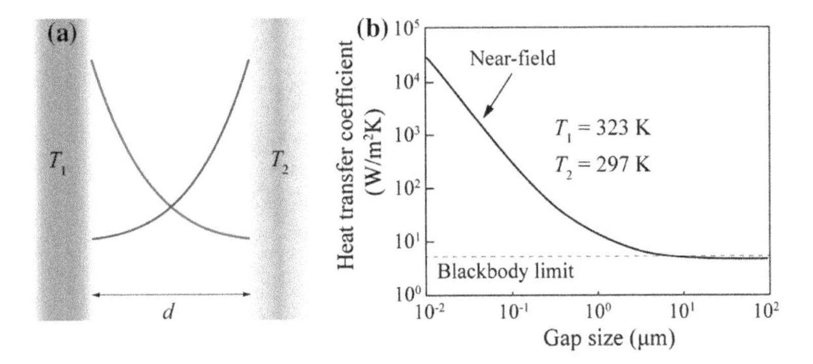

Fig. 2.34 **a** Schematic of the near-field coupling of two objects with temperatures of T_1 and T_2. The gap distance d is much smaller than the wavelength. Adapted from [85] with permission. Copyright 1999, IOP Publishing Ltd. **b** Calculated heat transfer coefficient for variant gap size. Reproduced from [84] with permission. Copyright 2008, American Institute of Physics

2.5.2 Generalized Radiation Theory

In principle, electromagnetic radiation has many different sources, ranging from the oscillation of electrons to spontaneous and stimulated emission of atomic systems, which are described by the classic electrodynamics and quantum theories, respectively [21, 81]. In either case, single radiators or emitters usually have only limited efficiency due to either the impedance mismatch or the limited density of state. For engineering optics, we are more concerned about the quantum light sources such as thermal radiators and lasers.

First of all, although Planck's formula presents a universal description of the thermal radiation of ideal blackbody, recent results have shown that the thermal emission of structured materials can exceed this limit [82, 83]. There are two physical mechanisms [21]: On the one hand, the near-field radiation is not included in Planck's original formula. Owing to the evanescent coupling described by a catenary function (Fig. 2.34a), the radiative heat transfer coefficient can be larger than the far-field value by several orders of magnitude (Fig. 2.34b) [84].

On the other hand, since Planck's law is defined upon infinite large absorber, it is possible to break this limit with finite-size absorbers. For instance, it is well known that the scattering and absorption cross section of a small object can be much larger than the geometric cross section in Mie's scattering theory, which means that such object could absorb much more light than its geometric area can collect. According to Kirchhoff's theory, at thermal equilibrium condition the emissivity of a material equals its absorptivity; therefore, the amount of thermal radiation can be much larger than that defined with the object's geometric size. This effect can also be illustrated by the Poynting vector in a resonant absorber comprised of a metallic patch, dielectric spacer, and reflective layer. Figure 2.35a shows the schematic illustration of the unit cell of a structured absorber comprised of a metallic patch, SiO_2 spacer, and a gold mirror. From the power flow in the absorber as shown in Fig. 2.35b, almost all

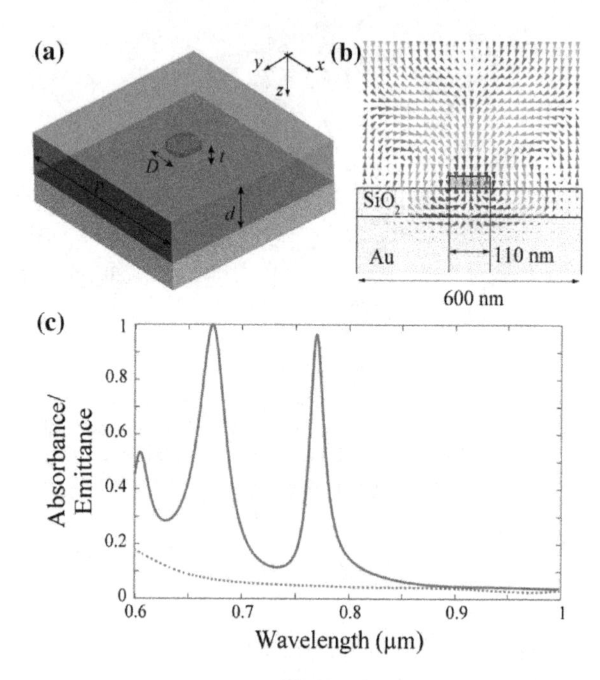

Fig. 2.35 **a** Schematic of the unit cell of a structured absorber. **b** Illustration of the power flow for a structured absorber. The diameter, thickness, and period of the patch are 110, 30, and 600 nm. **c** Absorbance/emittance for structures with (solid red curve) and without (dashed blue curve) metallic patches. Reproduced from [81] with permission. Copyright 2008, American Institute of Physics

the incident light energy at $\lambda = 673$ nm can be concentrated into the cavity formed between the patch and reflective layer, although the patch is very small compared with the unit cell. When the effective absorption/emission area is defined by the patch size, the Planck's law can be violated in the far field. When normalized to ideal blackbody with area equal to that of patches, the maximal enhancement factor could be as large as 38 at $\lambda = 673$ nm as shown in Fig. 2.35c. Of course, once the size of radiators goes to much larger than the wavelength, the diffraction effect would be much smaller, thus there would be negligible enhancement over the classic law. Moreover, it should be noted that the integral power emitted at all wavelengths remains sub-Planckian for any body formed by passive and causal components [83]. Based on above considerations, in many practical applications such as radiative cooling, the focus of interest is moved from breaking the Planck's limit to the optimization of spectral selectivity with judicious selection of constitutive materials and subwavelength structures. Several recent designs with metal–dielectric multilayers, silica photonic crystals, and glass–polymer hybrid structures have proved that it is possible to maximize the mid-infrared radiation while minimize the visible and near-infrared absorption [86–89]. Such materials may be of significant importance in the thermal control of satellites, buildings, and solar cell panels.

Besides thermal emission, artificial subwavelength structures have also been intensively investigated to enhance the quantum emissions from radiators such as fluorescent molecules and quantum dots. One basic mechanism of such enhancement is the well-known Purcell effect developed in 1946 [90], which states that the spontaneous emission rate is deeply dependent on the environment where the emitter is localized. The direct way to increase this rate is to increase the quality factor of the resonant cavity while reduce the mode volume. Two promising approaches that have been found so far are dielectric photonic crystals and metal nanoparticles [91]. While photonic crystals have much higher quality factor than the lossy metallic structures, the plasmonic modes confined at metal surface ensure much smaller mode volume. Consequently, the two approaches have their own merits and demerits, which should be compromised depending on the particular applications.

Directional radiation with miniature light source is another dream of many researchers, which is, however, strictly restricted by the diffraction nature of light. As early as the early twentieth century, Albert Einstein has considered a needle stick antenna, which could send arbitrarily large fraction of the emitted energy into an arbitrarily small solid angle [92]. After 2002, the optical beaming effect and many similar phenomena have been intensively studied to achieve directional emission of laser, thermal emission, fluorescent emission, as well as microwave directional radiation [93–96]. In principle, the effective radiation aperture has been increased over the geometric aperture, so that the diffraction limit could be surpassed.

As a concluding remark of this section, we note that structured materials may play an important role in nanolaser systems. Starting from the plasmonic nanolasers and spasers [97–99], the nanoscale coherent light sources have promised more integrated optical circuits. Nevertheless, owing to the large loss at the nanoscale and the difficulty in minimization of electrodes, there are still great challenges to transform these devices to practical applications.

References

1. Fermat's Principle and the Laws of Reflection and Refraction, http://scipp.ucsc.edu/~haber/ph5B/fermat09.pdf
2. M. Born, E. Wolf, *Principle of Optics*, 7th edn. (Pergamon, Oxford, UK, 2007)
3. W. Singer, M. Totzek, H. Gross, *Physical Image Formation* (Wiley, 2005)
4. Beer-Lambert law, https://en.wikipedia.org/Beer-Lambert_law
5. M. Vollmer, K.-P. Mollmann, *Infrared Thermal Imaging: Fundamentals, Research and Applications* (Wiley-VCH Verlag GmbH & Co. KGaA, Germany, 2010)
6. X. Luo, Principles of electromagnetic waves in metasurfaces. Sci. China Phys. Mech. Astron. **58**, 594201 (2015)
7. S.A. Maier, *Plasmonis: Fundamentals and Applications* (Springer, 2007)
8. X. Luo, D. Tsai, M. Gu, M. Hong, Subwavelength interference of light on structured surfaces. Adv. Opt. Photon. **10**, 757–842 (2018)
9. M. Pu, Y. Guo, X. Li, X. Ma, X. Luo, Revisitation of extraordinary Young's interference: from catenary optical fields to spin-orbit interaction in metasurfaces. ACS Photonics **5**, 3198–3204 (2018)
10. J.B. Pendry, A.J. Holden, W.J. Stewart, I. Youngs, Extremely low frequency plasmons in metallic mesostructures. Phys. Rev. Lett. **76**, 4773–4776 (1996)
11. W. Rotman, Plasma simulation by artificial dielectrics and parallel-plate media. IRE Trans. Antennas Propag. **10**, 82–95 (1962)
12. J.B. Pendry, L. Martín-Moreno, F.J. Garcia-Vidal, Mimicking surface plasmons with structured surfaces. Science **305**, 847–848 (2004)

13. F.J. Garcia-Vidal, L. Martín-Moreno, J.B. Pendry, Surfaces with holes in them: new plasmonic metamaterials. J. Opt. A Pure Appl. Opt. **7**, S97 (2005)
14. S.A. Maier, S.R. Andrews, L. Martín-Moreno, F.J. García-Vidal, Terahertz surface plasmon-polariton propagation and focusing on periodically corrugated metal wires. Phys. Rev. Lett. **97**, 176805 (2006)
15. X. Luo, Subwavelength optical engineering with metasurface waves. Adv. Opt. Mater. **6**, 1701201 (2018)
16. M. Pu, X. Ma, Y. Guo, X. Li, X. Luo, Theory of microscopic meta-surface waves based on catenary optical fields and dispersion. Opt. Express **26**, 19555–19562 (2018)
17. X. Luo, I. Teruya, Sub 100 nm lithography based on plasmon polariton resonance, in *Digest of Papers* (IEEE, 2003), pp. 138–139
18. L.B. Whitbourn, R.C. Compton, Equivalent-circuit formulas for metal grid reflectors at a dielectric boundary. Appl. Opt. **24**, 217–220 (1985)
19. Y. Huang, J. Luo, M. Pu, Y. Guo, Z. Zhao, X. Ma, X. Li, X. Luo, Catenary electromagnetics for ultrabroadband lightweight absorbers and large-scale flat antennas. Adv. Sci. 1801691 (2019)
20. J. Ducuing, N. Bloembergen, Observation of reflected light harmonics at the boundary of piezoelectric crystals. Phys. Rev. Lett. **10**, 474–476 (1963)
21. X. Luo, Subwavelength artificial structures: opening a new era for engineering optics. Adv. Mater. **31**, 1804680 (2019)
22. X. Luo, Engineering optics 2.0: a revolution in optical materials, devices, and systems. ACS Photonics **5**, 4724–4728 (2018)
23. H.P. Stahl, Survey of cost models for space telescopes. Opt. Eng. **49**, 053005 (2010)
24. R.A. Hyde, Eyeglass. 1. Very large aperture diffractive telescopes. Appl. Opt. **38**, 4198–4212 (1999)
25. P.D. Atcheson, C. Stewart, J. Domber, K. Whiteaker, J. Cole, P. Spuhler, A. Seltzer, J.A. Britten, S.N. Dixit, B. Farmer, L. Smith, MOIRE: initial demonstration of a transmissive diffractive membrane optic for large lightweight optical telescopes, in *SPIE Astronomical Telescopes + Instrumentation* (SPIE, 2012), p. 14
26. T. Xu, C. Wang, C. Du, X. Luo, Plasmonic beam deflector. Opt. Express **16**, 4753–4759 (2008)
27. X.G. Luo, T. Ishihara, Subwavelength photolithography based on surface-plasmon polariton resonance. Opt. Express **12**, 3055–3065 (2004)
28. L. Verslegers, P.B. Catrysse, Z. Yu, J.S. White, E.S. Barnard, M.L. Brongersma, S. Fan, Planar lenses based on nanoscale slit arrays in a metallic film. Nano Lett. **9**, 235–238 (2008)
29. M. Pu, X. Li, X. Ma, Y. Wang, Z. Zhao, C. Wang, C. Hu, P. Gao, C. Huang, H. Ren, X. Li, F. Qin, J. Yang, M. Gu, M. Hong, X. Luo, Catenary optics for achromatic generation of perfect optical angular momentum. Sci. Adv. **1**, e1500396 (2015)
30. X. Li, M. Pu, Y. Wang, X. Ma, Y. Li, H. Gao, Z. Zhao, P. Gao, C. Wang, X. Luo, Dynamic control of the extraordinary optical scattering in semicontinuous 2d metamaterials. Adv. Opt. Mater. **4**, 659–663 (2016)
31. X. Li, M. Pu, Z. Zhao, X. Ma, J. Jin, Y. Wang, P. Gao, X. Luo, Catenary nanostructures as compact Bessel beam generators. Sci. Rep. **6**, 20524 (2016)
32. N. Yu, P. Genevet, M.A. Kats, F. Aieta, J.-P. Tetienne, F. Capasso, Z. Gaburro, Light propagation with phase discontinuities: generalized laws of reflection and refraction. Science **334**, 333–337 (2011)
33. N. Yu, F. Capasso, Flat optics with designer metasurfaces. Nat. Mater. **13**, 139–150 (2014)
34. F. Aieta, P. Genevet, N. Yu, M.A. Kats, Z. Gaburro, F. Capasso, Out-of-plane reflection and refraction of light by anisotropic optical antenna metasurfaces with phase discontinuities. Nano Lett. **12**, 1702–1706 (2012)
35. M. Pu, C. Hu, C. Huang, C. Wang, Z. Zhao, Y. Wang, X. Luo, Investigation of Fano resonance in planar metamaterial with perturbed periodicity. Opt. Express **21**, 992 (2013)
36. H. Shi, X. Luo, C. Du, Young's interference of double metallic nanoslit with different widths. Opt. Express **15**, 11321–11327 (2007)
37. M. Khorasaninejad, F. Capasso, Broadband multifunctional efficient meta-gratings based on dielectric waveguide phase shifters. Nano Lett. **15**, 6709–6715 (2015)

38. S. Larouche, Y.-J. Tsai, T. Tyler, N.M. Jokerst, D.R. Smith, Infrared metamaterial phase holograms. Nat. Mater. **11**, 450–454 (2012)
39. X. Li, X. Ma, X. Luo, Principles and applications of metasurfaces with phase modulation. Opto-Electron. Eng. **44**, 255–275 (2017)
40. M.V. Berry, Quantal phase factors accompanying adiabatic changes. Proc. R. Soc. Lond. A **392**, 45–57 (1984)
41. A.G. Fox, An adjustable wave-guide phase changer. Proc. IRE **35**, 1489–1498 (1947)
42. W. Sichak, D.J. Levine, Microwave high-speed continuous phase shifter. Proc. IRE **43**, 1661–1663 (1955)
43. S. Pancharatnam, Generalized theory of interference, and its applications. Proc. Indian Acad. Sci. **44**, 247–262 (1956)
44. X. Zhang, Z. Tian, W. Yue, J. Gu, S. Zhang, J. Han, W. Zhang, Broadband terahertz wave deflection based on c-shape complex metamaterials with phase discontinuities. Adv. Mater. **25**, 4567–4572 (2013)
45. Kavli Foundation, http://www.kavlifoundation.org
46. D.W. Pohl, W. Denk, M. Lanz, Optical stethoscopy: image recording with resolution λ/20. Appl. Phys. Lett. **44**, 651–653 (1984)
47. E.A. Ash, G. Nicholls, Super-resolution aperture scanning microscope. Nature **237**, 510 (1972)
48. J.B. Pendry, Negative refraction makes a perfect lens. Phys. Rev. Lett. **85**, 3966–3969 (2000)
49. W. Wang, L. Lin, J. Ma, C. Wang, J. Cui, C. Du, X. Luo, Electromagnetic concentrators with reduced material parameters based on coordinate transformation. Opt. Express **16**, 11431–11437 (2008)
50. C. Wang, P. Gao, Z. Zhao, N. Yao, Y. Wang, L. Liu, K. Liu, X. Luo, Deep sub-wavelength imaging lithography by a reflective plasmonic slab. Opt. Express **21**, 20683–20691 (2013)
51. X. Luo, T. Ishihara, Surface plasmon resonant interference nanolithography technique. Appl. Phys. Lett. **84**, 4780–4782 (2004)
52. N. Fang, H. Lee, C. Sun, X. Zhang, Sub-diffraction-limited optical imaging with a silver superlens. Science **308**, 534–537 (2005)
53. P. Gao, N. Yao, C. Wang, Z. Zhao, Y. Luo, Y. Wang, G. Gao, K. Liu, C. Zhao, X. Luo, Enhancing aspect profile of half-pitch 32 nm and 22 nm lithography with plasmonic cavity lens. Appl. Phys. Lett. **106**, 093110 (2015)
54. X. Luo, Plasmonic metalens for nanofabrication. Natl. Sci. Rev. **5**, 137–138 (2018)
55. Z. Zhao, Y. Luo, W. Zhang, C. Wang, P. Gao, Y. Wang, M. Pu, N. Yao, C. Zhao, X. Luo, Going far beyond the near-field diffraction limit via plasmonic cavity lens with high spatial frequency spectrum off-axis illumination. Sci. Rep. **5**, 15320 (2015)
56. G.T. Di Francia, Super-gain antennas and optical resolving power. Il Nuovo Cimento **9**, 426–438 (1952)
57. E.T.F. Rogers, N.I. Zheludev, Optical super-oscillations: sub-wavelength light focusing and super-resolution imaging. J. Opt. **15**, 094008 (2013)
58. G. Lerosey, J. de Rosny, A. Tourin, M. Fink, Focusing beyond the diffraction limit with far-field time reversal. Science **315**, 1120–1122 (2007)
59. C. Wang, D. Tang, Y. Wang, Z. Zhao, J. Wang, M. Pu, Y. Zhang, W. Yan, P. Gao, X. Luo, Super-resolution optical telescopes with local light diffraction shrinkage. Sci. Rep. **5**, 18485 (2015)
60. M.V. Berry, S. Popescu, Evolution of quantum superoscillations and optical superresolution without evanescent waves. J. Phys. A Math. Gen. **39**, 6965 (2006)
61. F.M. Huang, N.I. Zheludev, Super-resolution without evanescent waves. Nano Lett. **9**, 1249–1254 (2009)
62. R.W. Wood, On a remarkable case of uneven distribution of light in a diffraction grating spectrum. Proc. Phys. Soc. London **18**, 269 (1902)
63. A. Ciattoni, B. Crosignani, P. Di Porto, Vectorial free-space optical propagation: a simple approach for generating all-order nonparaxial corrections. Opt. Commun. **177**, 9–13 (2000)
64. M. Pu, X. Ma, X. Li, Y. Guo, X. Luo, Merging plasmonics and metamaterials by two-dimensional subwavelength structures. J. Mater. Chem. C **5**, 4361–4278 (2017)

65. W. Woltersdorff, Über die optischen Konstanten dünner Metallschichten im langwelligen Ultrarot. Zeitschrift für Physik A Hadrons and Nuclei **91**, 230–252 (1934)
66. E.F. Knott, J.F. Shaeffer, M.T. Tuley, *Radar Cross Section*, 2nd edn. (SciTech Publishing, USA, 2004)
67. W.W. Salisbury, *Absorbent Body for Electromagnetic Waves* (1952)
68. N.I. Landy, S. Sajuyigbe, J.J. Mock, D.R. Smith, W.J. Padilla, Perfect metamaterial absorber. Phys. Rev. Lett. **100**, 207402 (2008)
69. C. Hu, Z. Zhao, X. Chen, X. Luo, Realizing near-perfect absorption at visible frequencies. Opt. Express **17**, 11039–11044 (2009)
70. M. Pu, M. Wang, C. Hu, C. Huang, Z. Zhao, Y. Wang, X. Luo, Engineering heavily doped silicon for broadband absorber in the terahertz regime. Opt. Express **20**, 25513–25519 (2012)
71. M. Pu, C. Hu, M. Wang, C. Huang, Z. Zhao, C. Wang, Q. Feng, X. Luo, Design principles for infrared wide-angle perfect absorber based on plasmonic structure. Opt. Express **19**, 17413–17420 (2011)
72. M. Pu, Q. Feng, M. Wang, C. Hu, C. Huang, X. Ma, Z. Zhao, C. Wang, X. Luo, Ultrathin broadband nearly perfect absorber with symmetrical coherent illumination. Opt. Express **20**, 2246–2254 (2012)
73. W. Wan, Y. Chong, L. Ge, H. Noh, A.D. Stone, H. Cao, Time-reversed lasing and interferometric control of absorption. Science **331**, 889–892 (2011)
74. M. Pu, Q. Feng, C. Hu, X. Luo, Perfect absorption of light by coherently induced plasmon hybridization in ultrathin metamaterial film. Plasmonics **7**, 733–738 (2012)
75. S. Li, J. Luo, S. Anwar, S. Li, W. Lu, Z.H. Hang, Y. Lai, B. Hou, M. Shen, C. Wang, Broadband perfect absorption of ultrathin conductive films with coherent illumination: superabsorption of microwave radiation. Phys. Rev. B **91**, 220301 (2015)
76. S. Li, Q. Duan, S. Li, Q. Yin, W. Lu, L. Li, B. Gu, B. Hou, W. Wen, Perfect electromagnetic absorption at one-atom-thick scale. Appl. Phys. Lett. **107**, 181112 (2015)
77. C. Yan, M. Pu, J. Luo, Y. Huang, X. Li, X. Ma, X. Luo, Coherent perfect absorption of electromagnetic wave in subwavelength structures. Opt. Laser Technol. **101**, 499–506 (2018)
78. M. Hong, Metasurface wave in planar nano-photonics. Sci. Bull. **61**, 112–113 (2016)
79. K.N. Rozanov, Ultimate thickness to bandwidth ratio of radar absorbers. IEEE Trans. Antennas Propag. **48**, 1230–1234 (2000)
80. D. Wang, Q. Huang, C. Qiu, M. Hong, Selective excitation of resonances in gammadion metamaterials for terahertz wave manipulation. Sci. China Phys. Mech. Astron. **58**, 08420 (2015)
81. X. Luo, M. Pu, X. Ma, X. Li, Taming the electromagnetic boundaries via metasurfaces: from theory and fabrication to functional devices. Int. J. Antenn. Propag. **2015**, 204127 (2015)
82. Y. Guo, C.L. Cortes, S. Molesky, Z. Jacob, Broadband super-Planckian thermal emission from hyperbolic metamaterials. Appl. Phys. Lett. **101**, 131106 (2012)
83. I.M. Stanislav, R.S. Constantin, A.T. Sergei, Overcoming black body radiation limit in free space: metamaterial superemitter. New J. Phys. **18**, 013034 (2016)
84. L. Hu, A. Narayanaswamy, X. Chen, G. Chen, Near-field thermal radiation between two closely spaced glass plates exceeding Planck's blackbody radiation law. Appl. Phys. Lett. **92**, 133106 (2008)
85. J.B. Pendry, Radiative exchange of heat between nanostructures. J. Phys. Condens. Matter **11**, 6621 (1999)
86. H. Yijia, P. Mingbo, G. Ping, Z. Zeyu, L. Xiong, M. Xiaoliang, L. Xiangang, Ultra-broadband large-scale infrared perfect absorber with optical transparency. Appl. Phys. Express **10**, 112601 (2017)
87. A.P. Raman, M.A. Anoma, L. Zhu, E. Rephaeli, S. Fan, Passive radiative cooling below ambient air temperature under direct sunlight. Nature **515**, 540 (2014)
88. L. Zhu, A.P. Raman, S. Fan, Radiative cooling of solar absorbers using a visibly transparent photonic crystal thermal blackbody. PNAS **112**, 12282–12287 (2015)
89. Y. Zhai, Y. Ma, S.N. David, D. Zhao, R. Lou, G. Tan, R. Yang, X. Yin, Scalable-manufactured randomized glass-polymer hybrid metamaterial for daytime radiative cooling. Sci. **339**, 1045–1047 (2017)

90. E.M. Purcell, Spontaneous emission probabilities at radio frequencies. Phys. Rev. Appl. **69**, 681 (1946)
91. M. Pelton, Modified spontaneous emission in nanophotonic structures. Nat. Photon. **9**, 427 (2015)
92. N.I. Zheludev, What diffraction limit? Nat. Mater. **7**, 420–422 (2008)
93. J.-J. Greffet, R. Carminati, K. Joulain, J.-P. Mulet, S. Mainguy, Y. Chen, Coherent emission of light by thermal sources. Nature **416**, 61 (2002)
94. H.J. Lezec, A. Degiron, E. Devaux, R.A. Linke, L. Martin-Moreno, F.J. Garcia-Vidal, T.W. Ebbesen, Beaming light from a subwavelength aperture. Science **297**, 820–822 (2002)
95. H. Aouani, O. Mahboub, N. Bonod, E. Devaux, E. Popov, H. Rigneault, T.W. Ebbesen, J. Wenger, Bright unidirectional fluorescence emission of molecules in a nanoaperture with plasmonic corrugations. Nano Lett. **11**, 637–644 (2011)
96. H. Caglayan, I. Bulu, E. Ozbay, Beaming of electromagnetic waves emitted through a subwavelength annular aperture. J. Opt. Soc. Am. B **23**, 419–422 (2006)
97. R.F. Oulton, V.J. Sorger, T. Zentgraf, R.M. Ma, C. Gladden, L. Dai, G. Bartal, X. Zhang, Plasmon lasers at deep subwavelength scale. Nature **461**, 629–632 (2009)
98. N.I. Zheludev, S.L. Prosvirnin, N. Papasimakis, V.A. Fedotov, Lasing spaser. Nat. Photon. **2**, 351–354 (2008)
99. E. Plum, V.A. Fedotov, P. Kuo, D.P. Tsai, N.I. Zheludev, Towards the lasing spaser: controlling metamaterial optical response with semiconductor quantum dots. Opt. Express **17**, 8548–8551 (2009)

Chapter 3
Material Basis

Abstract The electromagnetic properties of materials determine the way of light—matter interaction. Traditional engineering optics (i.e., EO 1.0) relies on the natural occurring materials whose electromagnetic properties are greatly restricted by the molecules or atoms. When combined with traditional laws of reflection and refraction, the optical systems are often complex and bulky to perform a special function. Distinct from EO 1.0, the material basis of EO 2.0 not only includes the natural occurring materials but also recently emerging artificial materials, whose physical properties are engineered by assembling microscopic and nanoscopic structures in unusual combinations. In this chapter, the commonly used natural materials and some unique metamaterials, e.g., negative-index metamaterials, near-zero index metamaterials, ultra-high index metamaterials, and hyperbolic metamaterials are introduced.

Keywords Optical materials · Plasmonic materials · Phase-change materials · Two-dimensional materials · Metamaterials

3.1 Introduction

The invention of novel functional materials always played critical roles in the developing history of human civilization throughout the "stone age", "iron age", and the so-called silicon age [1]. This is particularly true for the control of light, i.e., the electromagnetic wave induced by electron oscillation or quantum transition. Since light is ubiquitous in our world and its applications exist nearly everywhere from information technology to green energy, it is widely regarded as one of the most vital elements of the modern society. Traditionally, the basic materials for engineering optical applications are mainly transparent glasses, reflective metals, and optical thin films, which selectively refract and reflect light according to Fermat's principle developed in 1657.

The emergence of new optical materials and nanotechnologies in the last three decades has provided marvelous opportunities to change the traditional optics. For example, the use of plasmonic materials with negative dielectric permittivity is one of the most feasible ways to circumvent the diffraction limit and achieve localization

X. Luo, *Engineering Optics 2.0*, https://doi.org/10.1007/978-981-13-5755-8_3

of electromagnetic energy (at optical frequencies) into nanoscale regions as small as a few nanometers [2]. Furthermore, different from natural materials whose properties are primarily determined by the chemical constituents and bonds, metamaterials offer a significantly broader range of material properties by engineering the geometries and arrangements of the subwavelength building blocks (meta-atoms) [3].

The abundance material properties of metamaterials propel the rapid development of transformation optics (TO) [4, 5]. TO is a concept that enables extreme control over the flow of light, which suggests a general approach to optical design in which the required optical path and functionality are achieved by spatially varying the optical properties of materials, in contrast to the traditional methods of shaping the surface curvature of lenses to refract light. This control often requires a certain distribution of permittivity and permeability within a given space. Since natural materials cannot provide such arbitrary distributions in material's optical properties, metamaterials must be utilized.

Although metamaterials have achieved great success in the entire electromagnetic spectrum, great challenges still exist in the large-scale fabrication and the design of devices with broadband response, especially in the visible frequency range [6]. As alternatives, metasurfaces were proposed to tune the behavior of electromagnetic wave across one interface. Since metasurfaces are extremely thin and much easier to fabricate than metamaterials, they have become one of the most promising researching areas in subwavelength electromagnetics and optics [7, 8]. The building blocks of metasurfaces can be either metallic or dielectric structured nanoparticles or subwavelength apertures in a metallic or dielectric thin film. So far, numerous material platforms have been proposed to realize metamaterials and metasurfaces [9], as indicated in Fig. 3.1. In the following sections, the basic materials are discussed in detail.

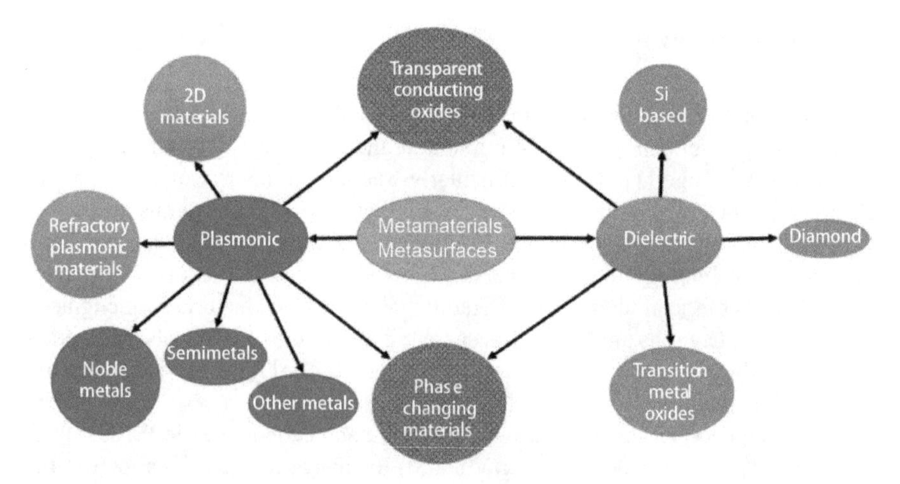

Fig. 3.1 Material platforms for optical metamaterials and metasurfaces used in EO 2.0. Reproduced from [9] with permission. Copyright 2018, The Authors

3.2 Natural Materials

3.2.1 Metals

Over a wide frequency range, the optical properties of metals can be explained by a plasma model [10], which can be derived from the motion equation of a free electron in an external electric field **E**:

$$-e\mathbf{E} = m^* \frac{d\mathbf{x}^2}{dt^2} + m^* \gamma \frac{d\mathbf{x}}{dt}. \tag{3.2.1}$$

If a harmonic time dependence $\mathbf{E}(t) = E_0 e^{-i\omega t}$ of the driving field is assumed, a particular solution of this equation describing the oscillation of the electron is $\mathbf{x}(t) = \mathbf{x}_0 e^{-i\omega t}$. The complex amplitude $\mathbf{x}_0$ can be written as:

$$\mathbf{x}(t) = \frac{e}{m^*(\omega^2 + i\gamma\omega)} \mathbf{E}(t), \tag{3.2.2}$$

where e is the electron charge, m^* is the effective mass of an electron, $\gamma = 1/\tau$ is the collision frequency, and τ is the relaxation time. The displaced electrons contribute to the macroscopic polarization $\mathbf{P} = -Ne\mathbf{x}$, given by:

$$\mathbf{P} = \frac{-Ne^2}{m^*(\omega^2 + i\gamma\omega)} \mathbf{E} \tag{3.2.3}$$

where N is the electrons density. Inserting Eq. (3.2.3) into $\mathbf{D} = \varepsilon_0 \mathbf{E} + \mathbf{P}$ yields:

$$\mathbf{D} = \varepsilon_0 \mathbf{E} + \mathbf{P} = \varepsilon_0 (1 - \frac{\omega_p^2}{\omega^2 + i\gamma\omega}) \mathbf{E} \tag{3.2.4}$$

where $\omega_p^2 = \frac{Ne^2}{\varepsilon_0 m^*}$ is the plasmon frequency of the free electron gas. Then, the dielectric function of the free electron gas can be written as:

$$\varepsilon(\omega) = 1 - \frac{\omega_p^2}{\omega^2 + i\gamma\omega} \tag{3.2.5}$$

The real and imaginary components of this complex dielectric function $\varepsilon(\omega) = \varepsilon_1(\omega) + i\varepsilon_2(\omega)$ are given by:

$$\varepsilon_1(\omega) = 1 - \frac{\omega_p^2 \tau^2}{1 + \omega^2 \tau^2}$$

$$\varepsilon_2(\omega) = \frac{\omega_p^2 \tau}{\omega(1 + \omega^2 \tau^2)} \tag{3.2.6}$$

Table 3.1 Drude model parameters for metals. ω_{int} is the frequency of onset for interband transitions. Drude parameters tabulated are not valid beyond this frequency [11]

Material	ε_∞	ω_p (eV)	γ (eV)	ω_{int} (eV)
Silver	3.7	9.2	0.02	3.9
Gold	11.70	60	0.390	0.0495
Copper	15.10	50	0.24	0.0965
Aluminum	10.91	1000	0.068	0.017

For many noble metals, the original Drude model in Eq. (3.2.5) should be refined to include a dielectric constant ε_∞ [10]:

$$\varepsilon(\omega) = \varepsilon_\infty - \frac{\omega_p^2}{\omega^2 + i\gamma\omega} \tag{3.2.7}$$

Furthermore, owing to the interband transition, more Lorentz oscillators must be included in the dielectric function at higher frequencies ($\omega > \omega_{int}$). The typical values of the parameters for some metals are shown in Table 3.1.

3.2.2 *Refractory Plasmonic Materials*

Although plasmonic materials offer several advantages, such as high electrical conductivity and low optical loss, the lack of thermal stability is particularly detrimental for applications at high temperatures, e.g., solar thermophotovoltaics and heat-assisted magnetic recording techniques. During the past few years, refractory plasmonic materials have been proposed for high-temperature applications of advanced plasmonic and metamaterial devices [12, 13]. One example is transition metal nitrides, such as titanium nitride (TiN) and zirconium nitride (ZrN). Figure 3.2 shows that TiN exhibit plasmonic properties comparable to those of gold in the visible and near-infrared (NIR) spectrum because of large free carrier concentrations ($\approx 10^{22}$ cm^{-3}), smaller interband losses, and a small negative real permittivity. Furthermore, the melting point of TiN (2930 °C) is much larger than that of bulk Au (1063 °C) and Ag (961 °C) and thus can boost the performance of many heat-assisted plasmonic devices [14].

3.2.3 *Semiconductors*

Semiconductors are conventionally regarded as dielectric materials for frequencies above several hundred THz. However, if the carrier concentration in semiconductors is increased to an extent, semiconductors can actually exhibit metal-like optical properties, i.e., negative real permittivity and thus as potential materials for plasmonics.

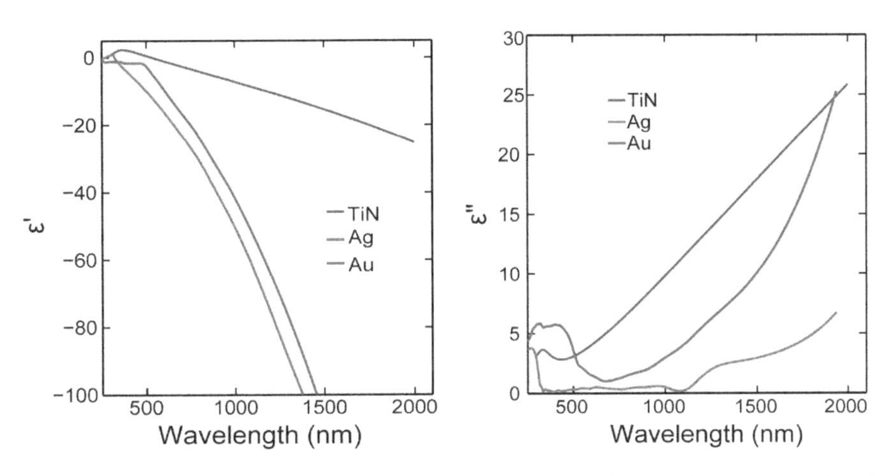

Fig. 3.2 Dielectric function of TiN in comparison with conventional plasmonic materials: gold and silver. Adapted from [14] with permission. Copyright 2012, Optical Society of America

Based on the Drude model in Eq. (3.2.7), one can estimate the required minimum carrier concentration N [15]:

$$\begin{aligned} \omega_p^2 &> \varepsilon_\infty\left(\omega_c^2+\gamma^2\right) \\ N &> \frac{\varepsilon_0 m^*}{e^2}\varepsilon_\infty\left(\omega_c^2+\gamma^2\right) \end{aligned} \tag{3.2.8}$$

where ω_c is the transition frequency of permittivity changes from positive to negative. Table 3.2 [15] shows the estimates of carrier concentration that would be required to obtain $\mathrm{Re}(\varepsilon) = -1$ at the telecommunication wavelength of 1.55 μm. It can be found that high doping with carrier concentration about 10^{21} cm^{-3} is needed for silicon to obtain the metal-like properties. Another important property of semiconductors listed in Table 3.2 is their optical bandgap (corresponds to the onset of interband transitions), which cause additional optical losses. Hence, the optical bandgap needs to be larger than the frequency spectrum of interest. Clearly, there are many semiconductors that can be useful for applications in the NIR and longer wavelengths when they are doped heavily [15].

(1) Transparent conducting oxides

Transparent conductive oxides (TCO), e.g., indium tin oxide (ITO) and Al-doped ZnO, are materials optically transparent and electrically conductive. These materials have been widely used in photovoltaics, organic light-emitting diodes, displays, and electro-optics devices in general. The advantages of TCOs include complementary metal–oxide–semiconductor (CMOS) compatibility, low losses, and high melting points compared to noble metals [16]. Moreover, the permittivity of TCOs can be flexibly engineered resulting in the epsilon-near-zero (ENZ) behavior in a tunable spectral range that significantly enhances nonlinear optical generation processes.

Table 3.2 Comparisons of different heavily doped semiconductors as potential alternative plasmonic materials [15]

Material	Background permittivity (ε_∞)	Carrier mobility when heavily doped ($cm^2\ V^{-1}\ s^{-1}$)	Effective mass (m)	Relaxation rate (eV)	Carrier concentration required to achieve Re(ε) = −1 at 1.55 μm ($\times 10^{20}\ cm^{-3}$)	Im(ε) or losses at 1.55 μm
n-Si	11.70	80	0.270	0.0536	16.0	0.8508
p-Si	11.70	60	0.390	0.0495	23.1	0.7853
n-SiGe	15.10	50	0.24	0.0965	18.2	1.9414
n-GaAs	10.91	1000	0.068	0.017	3.76	0.2534
p-GaAs	10.91	60	0.44	0.0438	24.4	0.6528
n-InP	9.55	700	0.078	0.0212	3.82	0.2796
n-GaN	5.04	50	0.24	0.0965	6.83	0.7283
p-GaN	5.24	5	1.4	0.1654	42.3	1.290
Al:ZnO	3.80	47.6	0.38	0.064	8.52	0.384
Ga:ZnO	3.80	30.96	0.38	0.0984	8.59	0.5904
ITO	3.80	36	0.38	0.0846	8.56	0.5077

Plasmonic TCOs have also recently been utilized to achieve enhanced electro-optical modulation [17], switching, enhanced light–matter interaction, and negative refraction.

ITO is an existing material with index ($n \sim 2$) less than that of silicon used for metasurface. Due to its two key properties (transparency and conductive), ITO has been widely used in photonic devices. The conductive property of ITO is already being exploited to realize reconfigurable metasurfaces, using field-effect modulation of the refractive index to control the modes and resonances within adjacent scatterers. This index modulation enabled dynamic beam deflection by applying a spatially varying voltage to a gold grating. The change in the ITO's refractive index would induce a change in the scattering properties of the gold structures [18].

The properties of ITO film, i.e., the conductivity and carrier concentration, could be controlled during the film deposition, which in turn will change the dielectric constant and the tunability of the refractive index of ITO film. The ENZ wavelength (λ_{ENZ}) can be tuned in the visible or near-infrared spectrum with a low loss, which is given by [19]:

$$\lambda_{\text{ENZ}} = \frac{2\pi c}{\omega_{\text{ENZ}}} = \frac{2\pi c}{\sqrt{\frac{\omega_p^2}{\varepsilon_\infty} - \Gamma^2}} = \frac{2\pi c}{\sqrt{\frac{Ne^2}{\varepsilon_0 \varepsilon_\infty m^*} - \left(\frac{e}{\mu m^*}\right)^2}} \tag{3.2.9}$$

From the above equation, one can find that changing plasma frequency will change the ENZ wavelength. If the electron density N increases, the plasma frequency increases and the ENZ wavelength decreases.

A commonly used structure to tune the electron density in the ITO layer is shown in Fig. 3.3a [17]. When the DC voltage increased from 0 to 2.5 V, the electron density can be increased from $N = 1 \times 10^{21}$ cm^{-3} to $N = 2.8 \times 10^{22}$ cm^{-3} in an electron accumulation layer of 5 nm ITO, which results in a refractive index change as shown in Fig. 3.3b. One can see that the refractive index (n) decreases and the extinction coefficient (k) increases as the bias voltage increases. Furthermore, the change in refractive index due to bias voltage is much larger at longer wavelengths as it is getting closer to ENZ [17].

Heavily doped zinc oxide, as another kind of TCOs, is a highly conducting material that has been used in applications such as liquid-crystal displays. Zinc oxide can be heavily doped ($\sim 10^{21}$ cm^{-3}) by trivalent dopants such as aluminum (i.e., Al:ZnO,

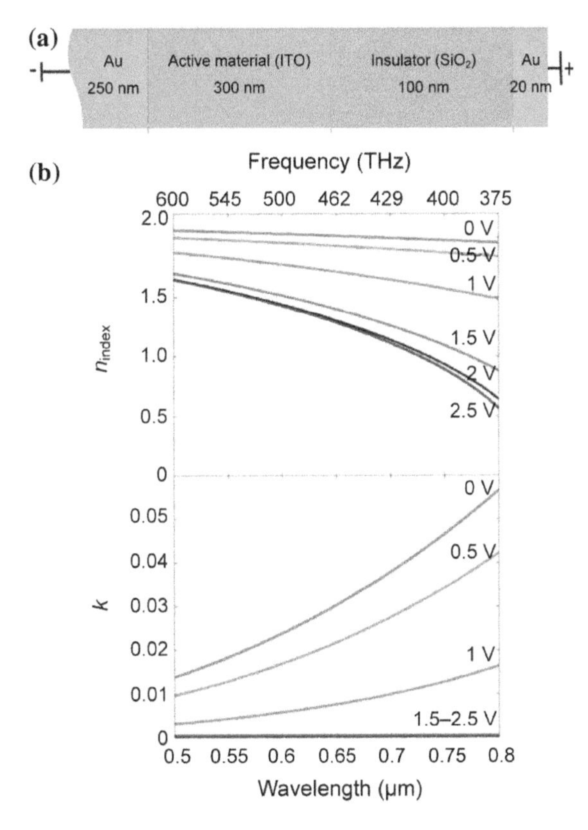

Fig. 3.3 **a** The metal-oxide-semiconductor heterostructure with ITO as the active material. **b** The refractive index (n) and extinction coefficient (k) in a 5 nm accumulation layer, extracted data from the ellipsometry measurements. Adapted from [17] with permission. Copyright 2010, American Chemical Society

herein referred to as AZO) and gallium (i.e., Ga:ZnO, herein referred to as GZO). Ultra-high doping results in a large carrier concentration and enables Drude metal like optical properties in the near-IR. AZO exhibits the low losses, about five times lower than the loss in silver in the near-IR, owing to its small Drude damping coefficient [20, 21].

(2) III–V Semiconductors

In recent decades, III–V semiconductors have provided the materials platform for many technologies such as high-speed switching, power electronics, and optoelectronics. These materials exhibit a wide tunability in the optical bandgap that can be controlled by varying the composition of their ternary and quaternary compounds.

In the optical regime, gallium nitride (GaN) possesses a relatively high-refractive index ($n > 2.4$) [22]. Benefiting from its high transparency through the whole visible spectrum, it has been widely used an active material for blue-, cyan-, and green-emitting LEDs and laser for general lighting, and backlighting. Moreover, GaN is a good material choice for high-index metasurface used in photonic applications. GaN is wide bandgap semiconductor with a direct bandgap of about 3.3 eV. With slightly higher doping, GaN could be turned to be plasmonic material at the telecommunication wavelength, as shown in Fig. 3.4 [15]. Clearly, GaN holds some promise for being a low-loss alternative plasmonic material in the NIR.

Gallium arsenide (GaAs) is a widely used III–V semiconductor with a relatively high-refractive index of ~3.5. Compared with silicon, GaAs has a direct bandgap structure with a bandgap width of 1.42 eV (at 300 K) [15], which is the range of the optimum bandgap required for solar cells. Thus, under the same absorption coefficient, the thickness of GaAs is much smaller than that of silicon. Pikhtin and

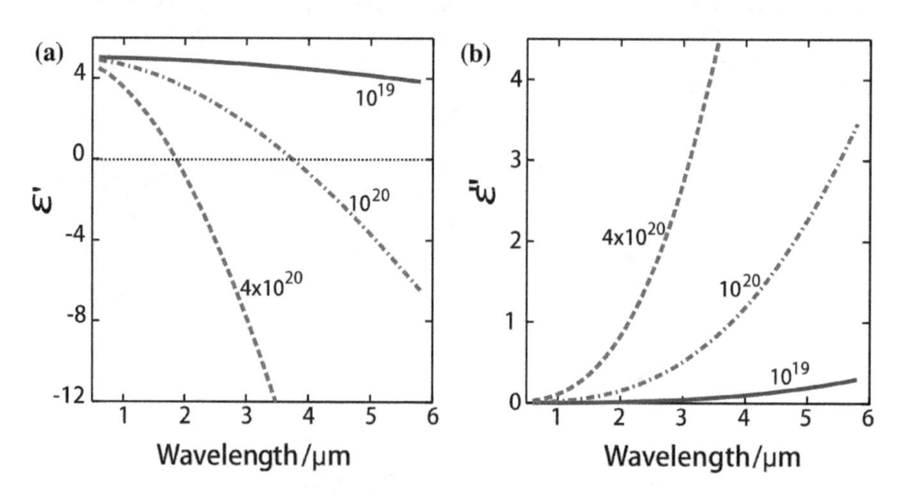

Fig. 3.4 Optical properties of heavily doped GaN as expected from Drude model. The plot shows **a** real and **b** imaginary parts of the dielectric function for doping concentrations mentioned in the units of cm^{-3}. Reproduced from [15] with permission. Copyright 2013, WILEY-VCH Verlag GmbH & Co. KGaA, Weinheim

Yas'kov [23] have measured the index of refraction in the near-IR. They fitted their data to an oscillator formula of the form:

$$n^2 = 1 + \frac{A}{\pi}\ln\frac{E_1^2 - (\hbar\omega)^2}{E_0^2 - (\hbar\omega)^2} + \frac{G_1}{E_1^2 - (\hbar\omega)^2} + \frac{G_2}{E_2^2 - (\hbar\omega)^2} + \frac{G_3}{E_3^2 - (\hbar\omega)^2} \quad (3.2.10)$$

with the parameters $E_0 = 1.428$ eV, $E_1 = 3.0$ eV, $E_2 = 5.1$ eV, $E_3 = 0.0333$ eV, $G_1 = 39.194$ eV^2, $G_2 = 136.08$ eV^2, $G_3 = 0.00218$ eV^2, and $A = 0.7/E_0^{1/2} = 0.5858$.

3.2.4 Dielectric Materials

The rapid progress of all-dielectric metamaterials has been fueled by the introduction of fabrication approaches beyond the conventional electron beam lithography and photolithography techniques. Then, all-dielectric metamaterials play a more and more important role in many fields. Table 3.3 summarizes the deposition technique, refractive index, and transparency region of typical dielectric materials.

(1) High-index materials

One important application of high-index materials is the emerging field of flat photonics. The fundamental drawback of low transmission efficiency can be circumvented using low-loss and high-index all-dielectric metasurfaces, where the high-refractive index ensures strong confinement of the light that ultimately allows full control of the phase, amplitude, and polarization of light [24]. Silicon has emerged as the most widely employed material platform to date for dielectric nanostructures, which is

Table 3.3 Deposition techniques, refractive index, and transparent region of typically dielectric materials

Materials	Deposition techniques	Refractive index	Transparent range
Germanium	Magnetron sputtering	$n \sim 4$ in IR	IR
Silicon	Magnetron sputtering	$n \sim 3.4$ in MIR	2.7–25 μm
Silicon carbide	Magnetron sputtering	$n \sim 2.7$ in visible	Visible and NIR
Titanium dioxide	E-beam evaporation; ALD	$n \sim 2.5$ in visible	Visible
Gallium nitride	Molecular beam epitaxy; MOCVD	$n \sim 2.4$ in visible	Visible
Silicon nitride	Magnetron sputtering; PECVD	$n \sim 2$ in visible	Visible
Aluminum Oxide	ALD; Magnetron sputtering; E-beam evaporation	$n \sim 1.7$ in visible and IR	0.4–7.7 μm
Silicon dioxide	E-beam evaporation; Magnetron sputtering	$n \sim 1.5$ in visible	Visible

not only due to its fine optical properties, but also attributed to the CMOS compatibility and low cost. At photon energies below its fundamental electronic bandgap at 1.1 μm wavelength [25], intrinsic crystalline silicon exhibits a high-refractive index and negligible absorption losses. Nevertheless, the absorption losses become notable for shorter wavelengths, thus affecting the visible spectral range.

Germanium (Ge) is another standard dielectric material that is commonly used along with the silicon platform for electronic devices [15]. Ge is attractive for its higher electron mobility and smaller optical bandgap than silicon, which can allow the fabrication of photodetectors at the telecommunication frequency. Compared with silicon, Ge possesses a higher refractive index ($n \sim 4$) and lower material loss in infrared spectral range. Furthermore, Ge has broader middle infrared (MIR) transparency range than Si. Due to these extraordinary features, germanium-based material platforms are particularly interesting for the realization of MIR photonic devices.

Titanium dioxide (TiO_2) has gained popularity as a metasurface material because of its relatively high index ($n \sim 2.6$) while being transparent across the visible regime. The optical properties of TiO_2 make it a great choice for metasurface building blocks [26]. Figure 3.5 shows the measured optical properties of a TiO_2 film from UV (243 nm) to near-infrared (1000 nm) wavelengths [27]. Over the visible spectrum, the refractive index ranges from 2.63 to 2.34 and remains relatively flat between $\lambda =$ 500 nm and $\lambda = 750$ nm ($\Delta n = 0.09$). Below the wavelength of 500 nm, the index of refraction increases rapidly. For wavelengths shorter than 360 nm, the imaginary part of the refractive index, k, begins to take on nonzero values, a result of interband absorption [27].

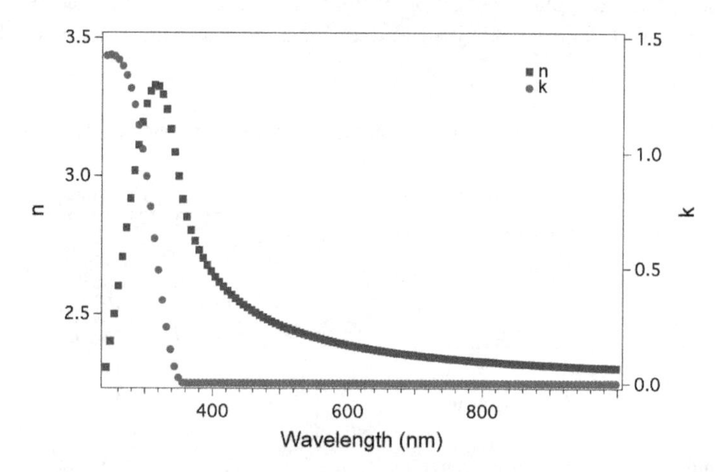

Fig. 3.5 Measured real (blue squares) and imaginary (red circles) part of the refractive index (n and k) of amorphous TiO_2 as a function of wavelength. Reproduced from [27] with permission. Copyright 2016, The Authors

(2) Low-index materials

Compared with high-index materials, low-index materials ($n < 2$) are also desirable for many applications such as broadband anti-reflection coatings. Silicon nitride (SiN) ($n \sim 2$) possesses a transparency window extending from the infrared down to the near-ultraviolet regime and thus is potential as a versatile metasurface material at different wavelengths across the transparent region. Furthermore, SiN is a CMOS-compatible material, enabling more streamlined integration with foundry infrastructure already in use for semiconductor micro-fabrication [22]. Al_2O_3 ($n \sim$ 1.7) and SiO_2 ($n \sim 1.5$) are known as transparency substrate in metalens and are widely used as dielectric spacer in plasmonic nanostructure.

3.2.5 Phase Transition and Phase-Change Materials

Phase transition and phase-change materials (PTMs and PCMs) are promising and emerging photonic materials since their molecular structures and optical properties can be drastically modified by external stimuli such as thermal heating, electrical pulse, laser pulse, or electric field [9]. The major difference between PTMs and PCMs is whether the process reverses naturally: PTMs (e.g., vanadium dioxide (VO_2)) are reversible (volatile) and will return to their initial state after removing the external stimulus; in contrast, PCMs (e.g., germanium-antimony-tellurium (GST)) are irreversible (nonvolatile) and will remain in a fixed state of matter unless an input excitation "resets" the PCM back to its original state [9].

(1) Vanadium oxide

Vanadium oxide (VO_2) is a well-known PTM which exhibits an insulator-to-metal phase transition at around 68 °C [28]. Unlike intrinsic crystalline vanadium dioxide, thin VO_2 films consist typically of many nanocrystal domains or nanograins. It was observed that the transition in such films at the nanoscale occurs gradually, with different grains undergoing phase transition at different temperatures, thus forming a mixture of the two phases at intermediate temperatures.

For a mixture of two linear dielectrics, using the Clausius–Mossotti relation, the effective dielectric constant can be written as:

$$\frac{\varepsilon_{\text{mix}} - 1}{\varepsilon_{\text{mix}} + 2} = \eta \frac{\varepsilon_i - 1}{\varepsilon_m + 2} + (1 - \eta) \frac{\varepsilon_i - 1}{\varepsilon_m + 2} \tag{3.2.11}$$

where ε_i and ε_m are the permittivies in semiconductor state and metal state, η is the volume fraction of the material with permittivity of ε_i. Figures 3.6 and 3.7 depict the measured real and imaging parts of permittivity for various mixtures of dielectric phase and metallic phase VO_2 during the heating and cooling process. Obviously, VO_2 exhibits reversible property after a recycle of heating and cooling process.

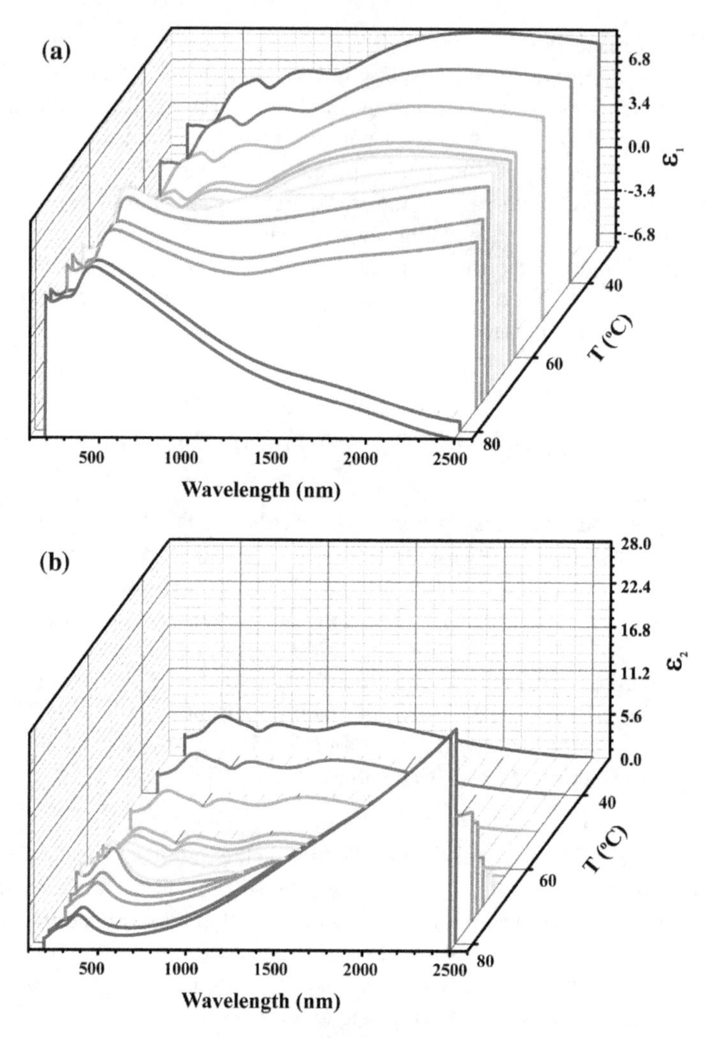

Fig. 3.6 Dielectric constant of VO_2 in the heating process

(2) Germanium-antimony-tellurium

Germanium-antimony-tellurium ($Ge_xSb_yTe_z$, GST) is one of the most representative PCMs, which is widely exploited in rewritable optical disk storage technology and nonvolatile electronic memories due to its good thermal stability, high switching speed, and large number of achievable rewriting cycles [29]. When annealed to a temperature between the glass transition and the melting point, GST transforms from an amorphous state into a metastable cubic crystalline state, and a short high-density laser pulse melts and quickly quenches the material back to its amorphous phase, with a pronounced contrast of dielectric properties observed between the two phases, as indicated in Fig. 3.8 [30].

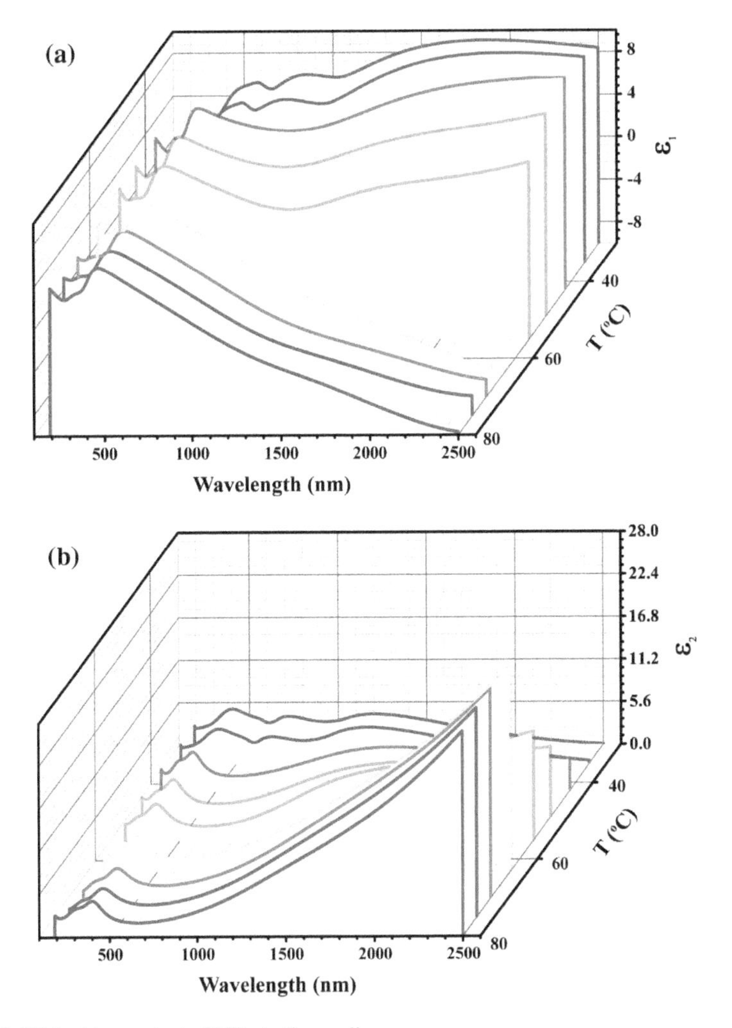

Fig. 3.7 Dielectric constant of VO_2 in the cooling process

Various compositions of GST have been tested. Figure 3.9 shows the real (ε_1) and imaginary (ε_2) parts of the measured dielectric function of these GSTs under amorphous and crystalline states. Obviously, the dielectric functions of the crystalline and amorphous phases differ widely for each kind of GST. Below the bandgap without interband switching excitation, the refractive index of the crystalline phase can increase by 50% because of the resonant bonding [30, 31]. Owing to GST's scalability, adaptability with CMOS technology, and tunable optical properties, it has been commercialized for a variety of optoelectrical applications including optical disk storage.

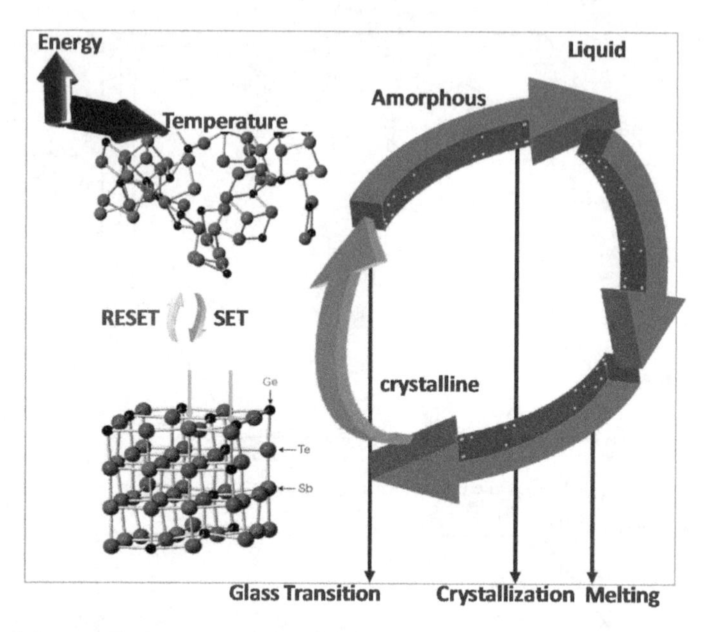

Fig. 3.8 Schematic illustration of phase transition of GST between amorphous and crystalline phases. Reproduced from [30] with permission. Copyright 2017, The Authors

3.2.6 Flexible Substrate Materials

Flexible substrates provide ideal platforms for exploring some of the unique characteristics that arise in metamaterials via mechanical deformation [32]. The most commonly used flexible substrates for metamaterials are polydimethylsiloxane (PDMS) and polyimide, due to their widespread use in flexible electronics. Other flexible substrates utilized for metamaterial devices include metaflex, polyethylene naphthalene (PEN), polyethylene terephthalate (PET), polymethylmethacrylate (PMMA), and polystyrene. One unique property of PDMS lies on the fact that it could be stretched [33].

Table 3.4 lists the important electromagnetic, electrical, and mechanical properties of some popular polymer substrates for metamaterials. Based on the application and regime of operational frequency, a flexible substrate with low absorption loss and the desired mechanical properties can be chosen [32].

3.2.7 Two-Dimensional Materials and van der Waals Materials

Since the exfoliation of graphene in 2004 [34], two-dimensional (2D) materials have received great attention because of qualitative changes in their physical and chemical

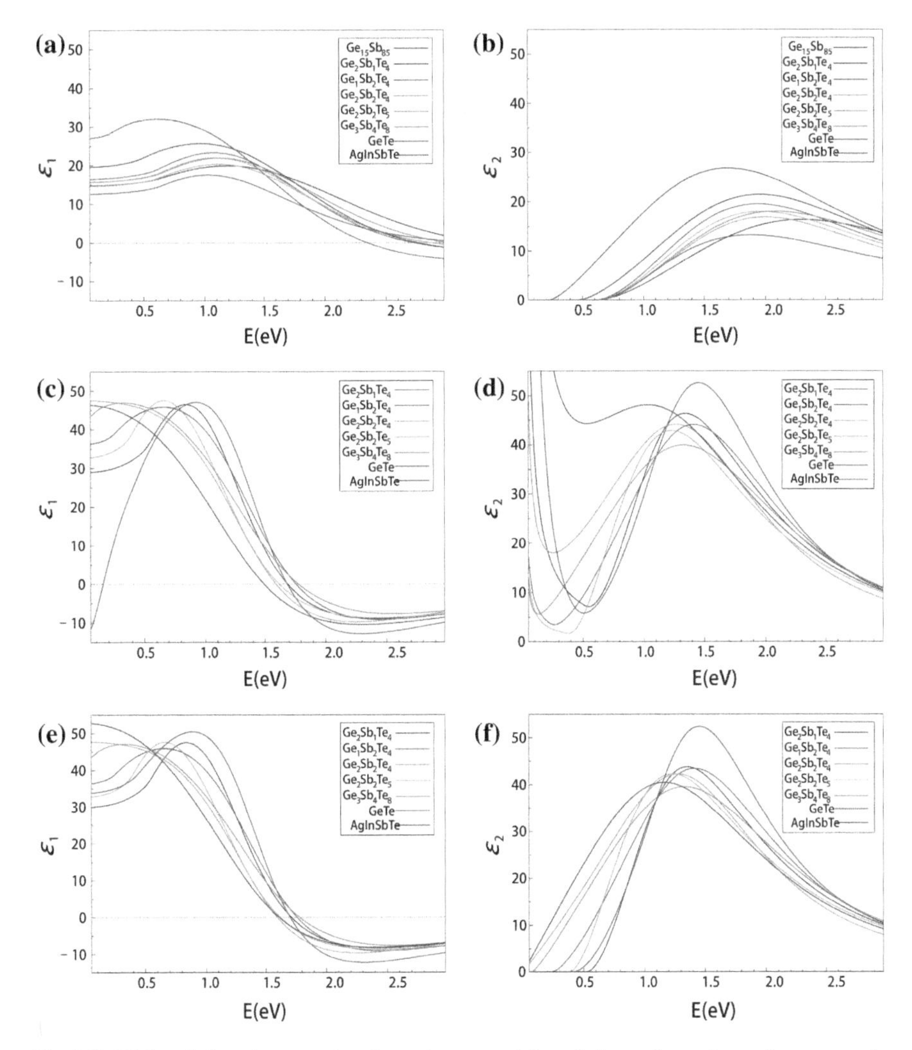

Fig. 3.9 Dielectric function ε_1 and ε_2 for various materials. **a, b** Amorphous phase-change sample. **c, d** Crystalline sample including Drude-type contribution. **e, f** Crystalline sample with Drude contribution subtracted. Reproduced from [31] with permission. Copyright 2008, Macmillan Publishers Limited

properties due to quantum size effect. For general 2D materials, the transports of charge carriers, heat, and photon will be strongly confined in a plane, leading to remarkable variations in the electronic and optical properties [35]. A sketch of the 2D material library can be found in Fig. 3.10 [36].

(1) Graphene

As a 2D material with unprecedented properties, graphene has attracted tremendous attention for future electronic and photonic applications. Graphene has a low carrier

Table 3.4 Electromagnetic, electrical, and mechanical properties of polymers commonly used as flexible substrates [32]

Material	Dielectric permittivity (0.2–2.5 THz)	Loss tangent	Absorption coefficient cm^{-1} (@1THz)	Young's modulus *E* (GPa)	Operating temperature (°C)	Planarization
PDMS	2.35	0.020–0.060	13	7.5×10^{-4}	−45 ~ 200	Moderate
Polyimide	3.24	0.031	12	2.5	−269 ~ 400	Good
PET	2.86	0.053–0.072	25	4.0	−80 ~ 180	Excellent
PEN	2.56	0.003	1	5.2	–	Excellent
BCB	2.65	0.001–0.009	3	2.9	–	Excellent
PMMA	2.22	0.042–0.070	22	3.1	–	Excellent
Polypropylene	2.25	0.008	2	2.0	0 ~ 135	Moderate
Parylene	3.00	0.120	–	–	80	Excellent
SU8	2.89	0.140	25	–	200	Excellent
Polystyrene	2.53	–	11	3.1	62	Poor

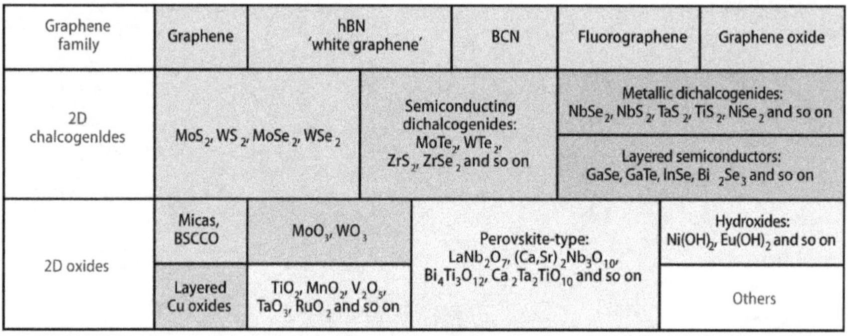

Fig. 3.10 A sketch of 2D materials library. Reproduced from [36] with permission. Copyright 2013, Macmillan Publishers Limited

density of states near the Dirac point; hence, its Fermi level can be tuned with small bias gate voltage [37, 38]. Graphene exhibits very good electric, mechanical, and optical properties: room temperature electron mobility of 2.5×10^5 cm^2 V^{-1} s^{-1} [39]; very high thermal conductivity (above 3000 W mK^{-1}); optical absorption of exactly $\pi\alpha < 2.3\%$ (α is the fine structure constant) [40]. Such properties lead to novel ways of manipulating the interaction between light and matter.

The Kubo formula is often adopted to model the surface conductivity of a graphene monolayer [41, 42]:

$$\sigma_s(\omega, \mu_c, \tau, T) = -\frac{ie^2(\omega + i\tau^{-1})}{\pi\hbar^2} \times \left[\int_{-\infty}^{+\infty} \frac{|E|}{(\omega + i\tau^{-1})^2} \frac{\partial f_d(E)}{\partial E} \mathrm{d}E - \int_0^{+\infty} \frac{\partial f_d(-E) - \partial f_d(E)}{(\omega + i\tau^{-1})^2 - 4(E/\hbar)^2} \mathrm{d}E \right] \tag{3.2.12}$$

where $f_d = 1/(1 + \exp[(E - \mu_c)/(k_B T)])$ is the Fermi–Dirac distribution, E is the energy, μ_c is the chemical potential, T is the temperature, e is the electron charge, $\hbar$ is the reduced Planck's constant, and τ is the momentum relaxation time (inverse of the electron-phonon scattering rate), due to the carrier intraband scattering. The first term in Eq. (3.2.12) corresponds to the intraband electron photon scattering process, which can be evaluated as [43]

$$\sigma_{\text{intra}} = i\frac{e^2 k_B T}{\pi\hbar^2(\omega + i\tau^{-1})}\left[\frac{\mu_c}{k_B T} + 2\ln\left(\exp\left(-\frac{\mu_c}{k_B T}\right) + 1\right)\right] \tag{3.2.13}$$

where the real part of σ_{intra}, associated with τ, contributes to energy absorption or dissipation. The second term corresponds to the direct interband electron transition and, for $\hbar\omega$, $|\mu_c| >> k_B T$ and it can be approximated as:

$$\sigma_{\text{inter}} = i\frac{e^2}{4\pi\hbar}\ln\left[\frac{2|\mu_c| - \hbar(\omega + i\tau^{-1})}{2|\mu_c| + \hbar(\omega + i\tau^{-1})}\right] \tag{3.2.14}$$

From above equations, it is found that, in the THz and far-infrared region, the intraband contribution (Eq. 3.2.13) dominates the surface conductivity.

(2) Transition metal dichalcogenide

Since the successful exfoliation of monolayer graphene, the discovery of other 2D materials has surged. Transition metal dichalcogenide (TMDC) crystals (e.g., MoS_2, $MoSe_2$, WS_2, and WSe_2) have emerged as a new class of 2D materials that display distinctive properties [35]. Different from graphene, they exhibit a transition to direct bandgap semiconductors at monolayer thickness, offering access to the valley degree of freedom by optical helicity. The materials thus have attracted much interest for applications in optoelectronics as light emitters, detectors, and photovoltaic devices.

Figure 3.11 shows the complex in-plane dielectric function of monolayers of four transition metal dichalcogenides for photon energies from 1.5 to 3 eV [44], which was obtained from reflection spectra using a Kramers–Kronig constrained variational analysis. The complex dielectric functions of the samples can be modeled as a function of photon energy E using a superposition of Lorentzian oscillators.

(3) Black phosphorus

Similar to graphene, it is possible to exfoliate few layers and thin-film black phosphorus (BP) from its bulk crystal, owing to the weak van der Waals interaction among the layer. The crystal structure is shown in Fig. 3.12 [45]. In its bulk form, black phosphorus has a bandgap of ≈0.3 eV, which makes BP interact strongly with

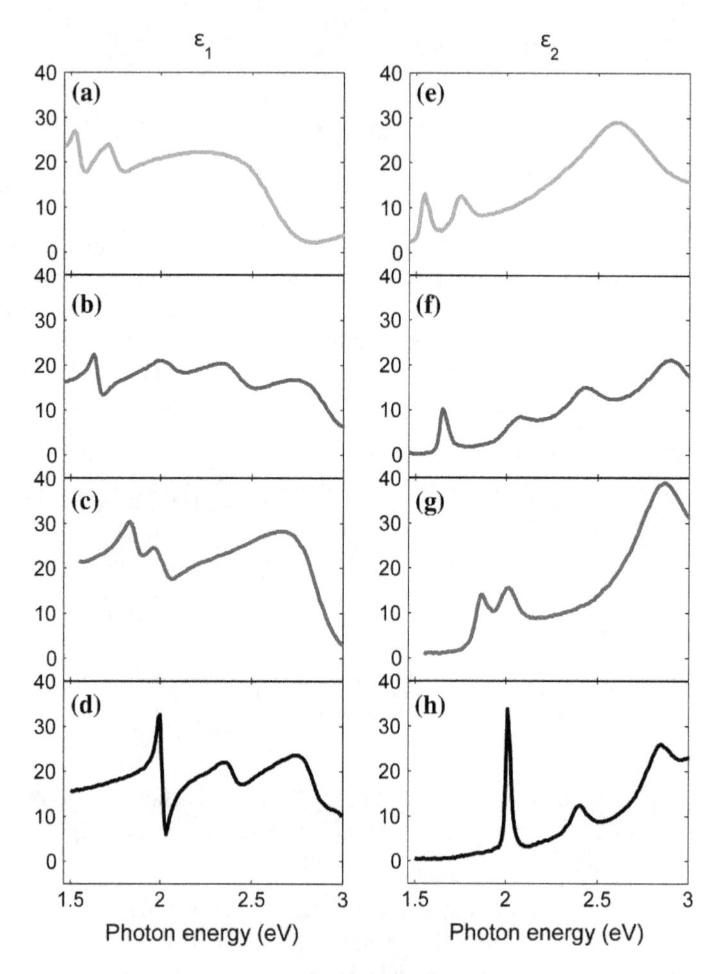

Fig. 3.11 Optical response of monolayers of $MoSe_2$, WSe_2, MoS_2, and WS_2 exfoliated on fused silica: **a–d** Real part of the dielectric function, ε_1. **e–h** Imaginary part of the dielectric function, ε_2. Reproduced from [44] with permission. Copyright 2014, American Physical Society

mid-infrared photons and those with even higher energy. The bandgap increases as the thickness of BP thin film goes down, and finally reaches $\approx$2 eV for the monolayer. Moreover, the bandgap of thin-film BP can be efficiently tuned simply by an electrical gating approach. Another distinct feature of black phosphorus is the strong in-plane anisotropy [46], which originates from the puckered arrangement of phosphorus atoms.

(4) Hexagonal boron nitride

Hexagonal boron nitride (*h*-BN) is a polar dielectric material that is particularly well suited for nanophotonic components. *h*-BN exhibits an extremely large crystalline anisotropy resulting from strong, in-plane covalent bonding of boron and nitrogen

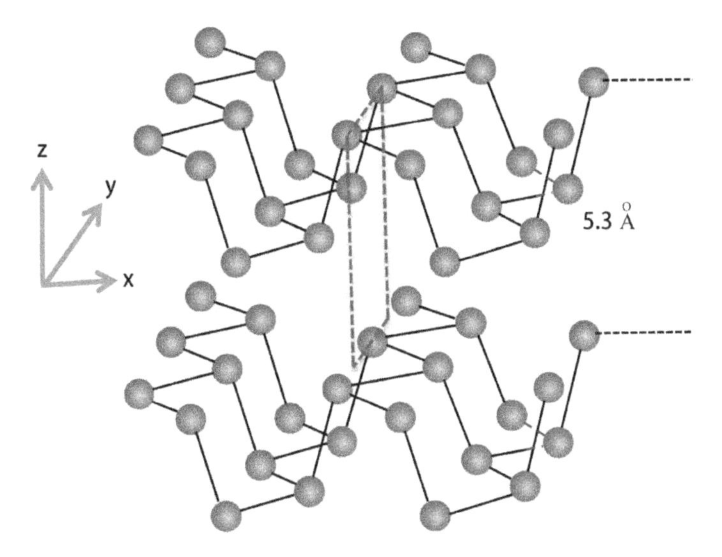

Fig. 3.12 Layered crystal structure of black phosphorus, showing the relative positioning of two adjacent puckered sheets with linked phosphorus atoms. The BP layer-to-layer spacing is around 0.53 nm. Reproduced with [47] from permission. Copyright 2014, Macmillan Publishers Limited

atoms and weak, out-of-plane, interlayer van der Waals bonding [48]. This gives rise to two spectrally distinct bands where phonon polaritons (PhPs) can be supported, designated as the lower (LR, ~760–820 cm^{-1}), and upper (UR, ~1365–1610 cm^{-1}) Reststrahlen bands where the dielectric permittivities along orthogonal crystal axes are opposite in sign. Such materials are referred to as hyperbolic material. Specifically, the in-plane permittivity is isotropic and is defined with a single value, ε_t, which is negative (positive) in the UR (LR), whereas the out-of-plane component ε_z is positive (negative), as shown in Fig. 3.13 [48].

The uniaxial material h-BN is characterized by a relative permittivity of $\varepsilon_{r,h\mathrm{BN}} = \left[\varepsilon_{\parallel,h\mathrm{BN}}, \varepsilon_{\parallel,h\mathrm{BN}}, \varepsilon_{\perp,h\mathrm{BN}}\right]$, where $\varepsilon_x = \varepsilon_y = \varepsilon_{\parallel,h\mathrm{BN}}$ and $\varepsilon_z = \varepsilon_{\perp,h\mathrm{BN}}$. The anisotropic permittivity of h-BN is modeled as [50]:

$$\varepsilon_u = \varepsilon_{\infty,u}\left[1 + \frac{\omega_{\mathrm{LO},u}^2 - \omega_{\mathrm{TO},u}^2}{\omega_{\mathrm{LO},u}^2 - \omega^2 - i\omega\gamma_u}\right], \tag{3.2.15}$$

where $u = x, y$ represents the transverse (crystal plane) and $u = z$ represents the z-axis. ε_∞ and γ represent the high-frequency dielectric permittivity and the damping constant, respectively, $\varepsilon_{\infty,x} = \varepsilon_{\infty,y} = 4.87$, $\gamma_x = \gamma_y = 5\ \mathrm{cm}^{-1}$, $\omega_{\mathrm{TO},x} = \omega_{\mathrm{TO},y} = 1370\,\mathrm{cm}^{-1}$, $\omega_{\mathrm{LO},x} = \omega_{\mathrm{LO},y} = 1610\,\mathrm{cm}^{-1}$, $\varepsilon_{\infty,y} = 2.95$, $\gamma_z = 4\,\mathrm{cm}^{-1}$, $\omega_{\mathrm{TO},z} = 780\,\mathrm{cm}^{-1}$, and $\omega_{\mathrm{LO},z} = 830\ \mathrm{cm}^{-1}$.

(5) α-MoO$_3$

Recently, α-MoO_3, as another in-plane anisotropic and ultra-low-loss polaritons

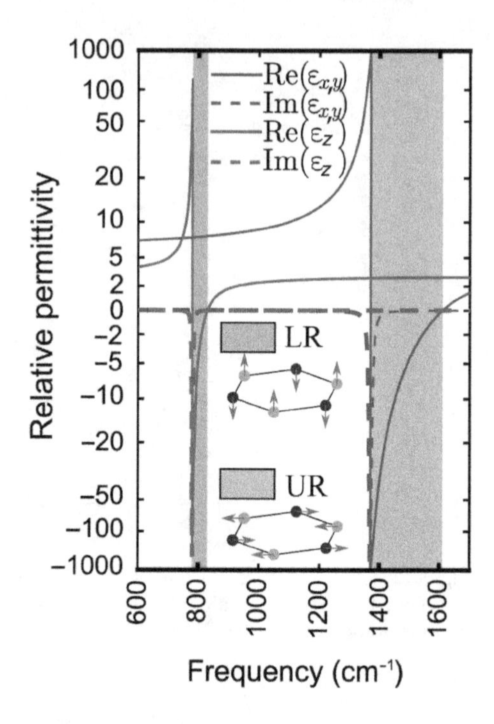

Fig. 3.13 In-plane and out-of-plane relative permittivity of *h*-BN. The two colored bands are the LR band (where $\varepsilon_z < 0$) and the UR band (where $\varepsilon_{x,y} < 0$). Reproduced from [49] with permission. Copyright 2018, The Authors

in a natural van der Waals crystal has been reported [51]. Figure 3.14 shows the orthorhombic crystal structure of α-MoO_3, in which layers formed by distorted MoO_6 octahedra are weakly bound by van der Waals forces and all three lattice constants (*a*, *b* and *c*) are different. α-MoO_3 has strong in-plane structural anisotropy, caused by the interlayer spacing of the [100] facet differing from that of the [001] facet by as much as 7.2%, which leads to the highly anisotropic response. Indeed, the different directional vibrations of the α-MoO_3 crystal structure yield two infrared Reststrahlen bands between about 820 cm^{-1} and 1010 cm^{-1} [51].

3.2.8 Perovskite Materials

Perovskites (Fig. 3.15) with the general formula ABX_3 (A = organic ammonium cation, Cs^+; B = Pb^{2+}, Sn^{2+}; X = Cl^-, Br^-, I^-) have been widely studied as a new family of semiconductor materials for a variety of applications due to their remarkable features, including long charge carrier lifetime and diffusion length, intense photoluminescence (PL), high light absorption coefficients, easily tunable properties, rich phase diagram, and low-temperature solution process ability [52].

For perovskite solar cells, the mostly studied materials include $CH_3NH_3PbI_3$, $CH_3NH_3PbI_{3-x}Cl_x$, $CH_3NH_3PbBr_3$, $CH_3NH_3Pb(I_{1-x}Br_x)_3$, $HC(NH_2)_2PbI_3$, $HC(NH_2)_2Pb(I_{1-x}Br_x)_3$, and $CH_3NH_3SnI_3$. Table 3.5 summarizes the material,

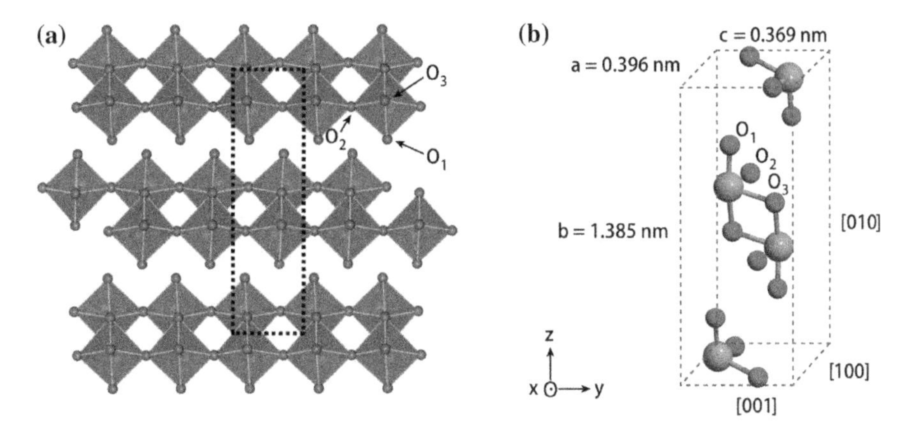

Fig. 3.14 **a** The orthorhombic lattice structure of layered α-MoO_3 (red spheres, oxygen atoms). **b** Schematic of the unit cell of α-MoO_3. Blue spheres, molybdenum atoms. Reproduced from [51] with permission. Copyright 2018, Springer Nature Limited

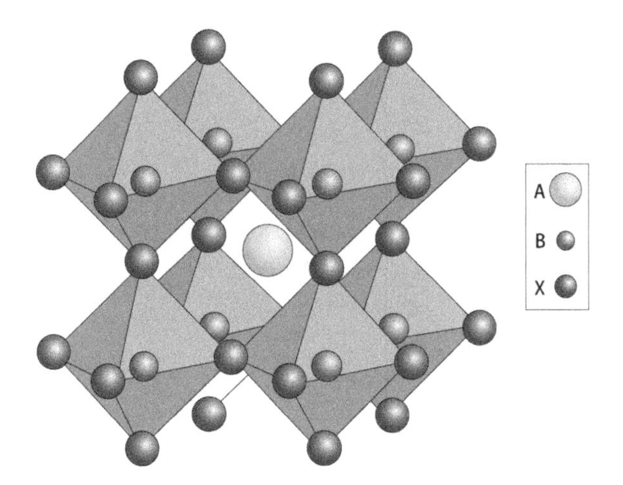

Fig. 3.15 Cubic perovskite crystal structure. Reproduced from [53] with permission. Copyright 2014, Macmillan Publishers Limited

cell configuration, and reported performances of perovskite solar cells [54], where MA = $CH_3NH_3^+$ (methylammonium lead halide perovskite) and PCE represents the power conversion efficiency. It is noted that hole transport material (HTM) materials (e.g., spiro-MeOTAD, thiophene derivative P3HT and tryarylamine-based PTAA) are important for high-efficiency perovskite solar cells [54].

3.3 Artificially Structured Materials

Artificially structured materials have become the basis of EO 2.0 [55]. Distinct with natural occurring materials that the optical constants are determined by the inherent

Table 3.5 Photovoltaic performance of pervoskite solar cells depending on materials and cell configurations

Materials	Cell configuration	J_{sc} (mA/cm^2)	V_{oc} (V)	FF	PCE (%)
$MAPbI_3$	mp-TiO_2/$MAPbI_3$/spiro-MeOTAD	17.6	0.888	0.62	9.7
	$MAPbI_3$/PCBM	10.32	0.60	0.63	3.9
	ZnO nanorod/$MAPbI_3$/spiro-MeOTAD	16.98	1.02	0.51	8.9
	mp-TiO_2/$MAPbI_3$/P3HT-MWNT	14.8	0.76	0.57	6.45
	rutile TiO_2 nanorod/$MAPbI_3$/spiro-MeOTAD	15.6	0.9555	0.63	9.4
	mp-TiO_2/$MAPbI_3$/PTAA	16.5	0.997	0.727	12
	mp-TiO_2/$MAPbI_3$(2-step)/spiro-MeOTAD	20	0.933	0.73	15
	mp-TiO_2/$MAPbI_3$/spiro-MeOTAD	17.3	1.07	0.59	10.8
	mp-TiO_2/$MAPbI_3$(Toluene)/PTAA	19.58	1.105	0.76	16.46
	mp-TiO_2/$MAPbI_3$ cuboid/spiro-MeOTAD	21.64	1.056	0.741	17.01
	ZnO/$MAPbI_3$/spiro-MeOTAD	20.08	0.991	0.56	11.13
	NiO/$MAPbI_3$/PCBM	13.24	1.04	0.69	9.51
	polyTPD/$MAPbI_3$/PCBM	16.12	1.05	0.67	12.04
	ZnO (25 nm)/$MAPbI_3$/spiro-MeOTAD	20.4	1.03	0.749	15.7
	NiO (45 nm)/$MAPbI_3$/PCBM	16.27	0.882	0.635	9.11
	mp-TiO_2/$MAPbI_3$/CuSCN	19.7	1.016	0.62	12.4
	mp-TiO_2/$MAPbI_3$/Cul	17.8	0.55	0.62	6.0
$MAPbI_{3-x}Cl_x$	mp-Al_2O_3/$MAPbI_{3-x}Cl_x$/spiro-MeOTAD	17.8	0.98	0.63	10.9
	$MAPbI_{3-x}Cl_x$/P3HT	20.8	0.921	0.542	10.4
	(Al_2O_3 + $MAPbI_{3-x}Cl_x$)/spiro-MeOTAD at $T < 110$ °C	12.78	0.925	0.61	7.16
	$MAPbI_{3-x}Cl_x$(evap.)/spiro-MeOTAD	21.5	1.07	0.67	15.4
	bl-Y:TiO_2/$MAPbI_{3-x}Cl_x$(evap.)/spiro-MeOTAD	22.75	1.13	0.75	19.3
$MAPbBr_3$	mp-TiO_2/$MAPbBr_3$/PCBTDPP	4.47	1.16	0.59	3.04
	mp-TiO_2/$MAPbBr_3$/PDI	1.08	1.30	0.4	0.56
	mp-TiO_2/$MAPbBr_3$/PIF8-TAA	6.1	1.40	0.79	6.7
	mp-Al_2O_3/$MAPbBr_{3-x}Cl_x$/CBP	4.0	1.50	0.46	2.7

(continued)

Table 3.5 (continued)

Materials	Cell configuration	J_{sc} (mA/cm^2)	V_{oc} (V)	FF	PCE (%)
$MAPbBr_3$	mp-TiO_2/$MAPb(I_{0.8}Br_{0.2})_3$/PTAA	19.3	0.91	0.702	12.3
$FAPbI_3$	mp-TiO_2/$FAPbI_3$/PTAA/spiro-MeOTAD	19.3	0.91	0.702	12.3
	$FAPbI_3$/PTAA/spiro-MeOTAD	23.3	0.94	0.65	14.2
	mp-TiO_2/$FAPbI_3$/$MAPbI_3$(ETL)/spiro-MeOTAD	20.97	1.032	0.74	16.01
	mp-TiO_2/$(MA_{0.6}FA_{0.4})PbI_3$/spiro-MeOTAD	21.2	1.003	0.7	14.9
$MASnX_3$	mp-TiO_2/$MASnI_3$/spiro-MeOTAD	16.8	0.88	0.42	6.4
	mp-TiO_2/$MASnI_3$/spiro-MeOTAD	16.3	0.68	0.48	5.23
	mp-TiO_2/$MASnIBr_2$/spiro-MeOTAD	12.30	0.82	0.57	5.73
$MASn_{0.5}Pb_{0.5}I_3$	mp-TiO_2/$MASn_xPb_{(1-x)}I_3$/P3HT	20.04	0.42	0.50	4.18

Jsc, Voc, and FF stand for short-circuit current density, open-circuit voltage, and fill factor (FF), respectively. mp, bl, and ETL represent mesoporous, blocking layer and extremely thin layer, respectively [54]

molecules and atoms, the electromagnetic properties of artificially structured materials are mainly determined by the geometries and arrangements of building blocks, which offer unprecedented freedoms to tailor the effective materials parameters. So far, numerous artificially structured materials have been developed, including negative-index media [56], zero-index media [57], and chiral materials [58], etc.

As the basic geometric structures, metallic and dielectric holes, slits, and stripes are commonly used in optical diffraction experiments. A serial of extraordinary optical phenomena including extraordinary optical transmission (EOT) [59] and extraordinary Young's interference (EYI) [60–63] have been discussed based on such structures. At present, the artificial combination of a slits array and stripes array has been developed into metasurfaces with optical properties that go beyond those of natural materials and their interfaces, as indicated in Fig. 3.16.

Although offering superb performance in subwavelength concentration of light, metallic nanostructures suffer from intrinsic absorption losses at optical frequencies, severely limiting their efficiency [25]. Recent researches suggest that designed dielectric building blocks provide similar opportunities to their metallic counterparts regarding the design of resonant nanostructures with an optical response tailored by geometry [25]. The types of silicon-based nanostructures are summarized, including individual silicon nanoresonators, silicon nanoparticle clusters, silicon metasurfaces, and silicon metamaterials, which are illustrated in Fig. 3.17 [25].

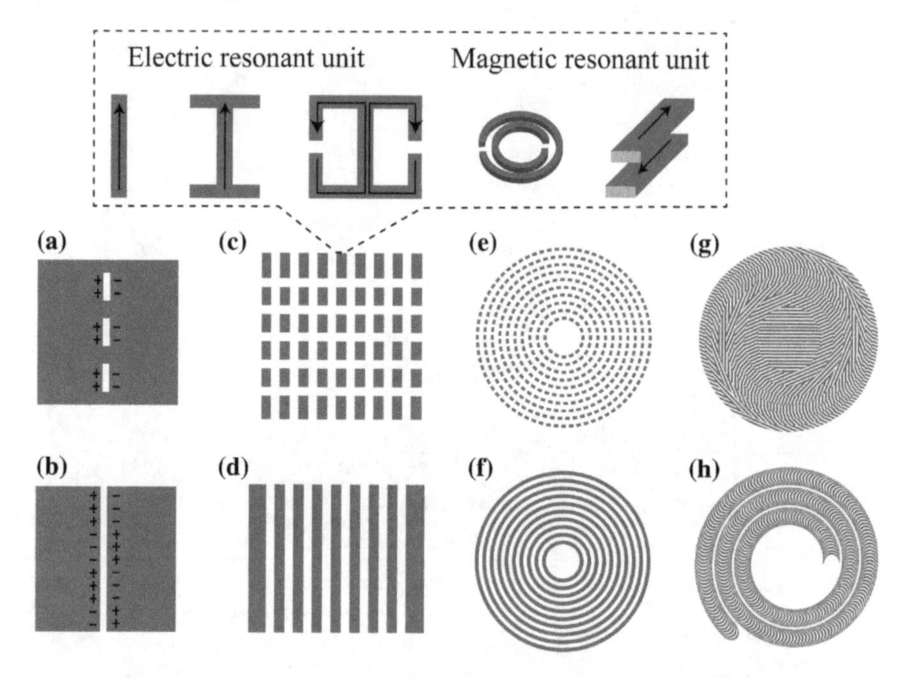

Fig. 3.16 From building blocks to functional metasurfaces. Reproduced from [64] with permission. Copyright 2018, IOP Publishing Ltd

3.3.1 Effective Medium Theory

There are two approaches to describe the behavior of ensembles of the artificially structured materials: the macroscopic and the microscopic. In the macroscopic approach, the element size is much smaller than the wavelength at the frequency of operation and thus the detail in the material can be neglected. In this case, the artificially structured materials can be described as a homogeneous, effective medium whose electromagnetic response is defined by an effective permittivity and permeability that may not be available in natural materials. In the microscopic picture, the ensemble is considered as an array of coupled resonators, each with a resonant frequency and coupled to neighboring elements (or indeed to all other elements). In the following, we mainly discuss the former approach.

From the viewpoint of metamaterials, the alternative multilayers shown in Fig. 3.18 can be taken as an anisotropic media, whose effective parameters can be obtained using effective medium theory (EMT). For the two electric fields polarized parallel and perpendicular to the multilayers, the effective permittivity can be written as [65–68]:

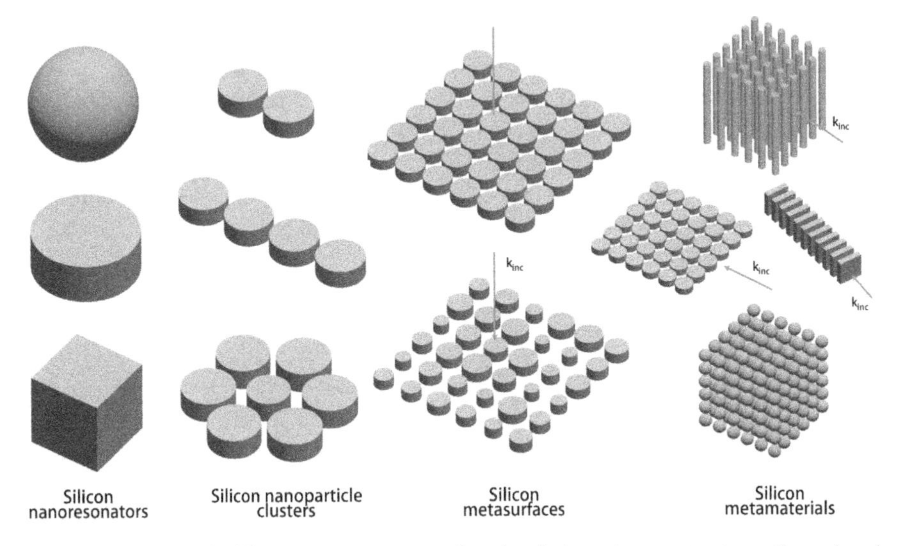

Fig. 3.17 From single silicon nanoresonators to functional photonic nanostructures. Reproduced from [25] with permission. Copyright 2017, Macmillan Publishers Limited, part of Springer Nature

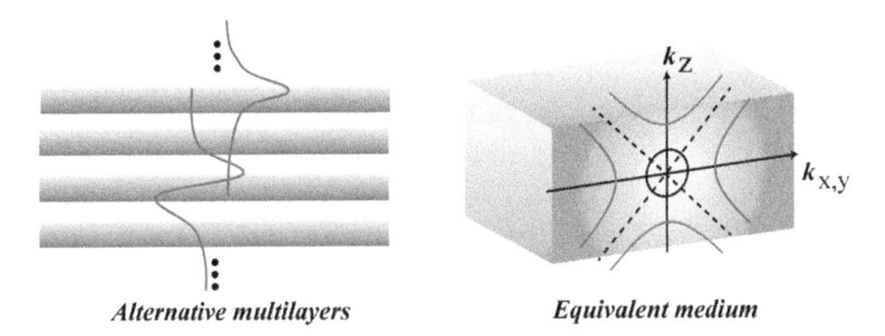

Fig. 3.18 Dispersion relation in alternative multilayers and equivalent anisotropic materials with different topologies of iso-frequency contours. Reproduced from [68] with permission. Copyright 2017, Institute of Optics and Electronics, Chinese Academy of Sciences

$$\begin{cases} \varepsilon_x = \varepsilon_y = \varepsilon_{\parallel} = \dfrac{\varepsilon_1 d_1 + \varepsilon_2 d_2}{d_1 + d_2} \\ \dfrac{1}{\varepsilon_z} = \dfrac{1}{\varepsilon_{\perp}} = \dfrac{d_1/\varepsilon_1 + d_2/\varepsilon_2}{d_1 + d_2} \end{cases} \tag{3.3.1}$$

where d_1 and d_2 are the thickness of the alternating metallic and dielectric films. Clearly, the value of parallel component is between ε_1 and ε_2. When it satisfies the condition

$$\frac{d_1}{\varepsilon_1} = -\frac{d_2}{\varepsilon_2} \tag{3.3.2}$$

The perpendicular component can be much larger than 1, resulting in extremely short wavelength. This can only be achieved by metal–dielectric multilayers. Compared with the permittivity match condition of single metal slab, metal–dielectric multilayers offer more design flexibility in material and operating frequency selection for sub-diffraction imaging [69].

For a homogeneous anisotropic structure under the illumination of TM polarized incidences, the propagation of light satisfies the following dispersion relationship

$$\frac{k_x^2 + k_y^2}{\varepsilon_\perp} + \frac{k_z^2}{\varepsilon_\parallel} = k_0^2 \tag{3.3.3}$$

The topology of the frequency contour will change from a closed elliptical dispersion to an open extreme anisotropic one when the materials and the fill ratio of alternating layer vary [70]. When $\varepsilon_\parallel \varepsilon_\perp < 0$, the dispersion curve of this effective medium is hyperbolic, thus enabling the directive transmission, spatial frequency filtering, and negative refraction [71–76]. The short wavelength and hyperbolic dispersion of SPP could be used to achieve far-field imaging of subwavelength objects. By using curved multilayers, the subwavelength object can be magnified and the near-field information could be transferred to the far-field, and vice versa [65, 77–79]

It has been proposed that electromagnetic metamaterials—composite structured materials formed by either periodic or random arrays of scattering elements—should respond to electromagnetic radiation as homogeneous materials, at least in the long-wavelength limit [56, 80–82]. The common method of characterizing the electromagnetic scattering properties of a homogeneous material is to identify its impedance (z) and refractive index (n). For example, photonic bandgap materials are characterized by dispersion curves from which an effective index can be extracted, even for bands well above the first bandgap. Alternatively, one can choose electric permittivity $\varepsilon = n/z$ and magnetic permeability $\mu = nz$ for more direct physical meanings.

Smith et al. [83] presented a method to determine the effective permittivity and permeability from the reflection and transmission coefficients by exploring transfer matrix simulations on finite lengths of electromagnetic metamaterials. For a 1D slab of homogeneous material (in vacuum) shown in Fig. 3.19, there is [83]:

$$\cos(nkd) = \frac{1}{2t}\left[1 - \left(r^2 - t^2\right)\right] \tag{3.3.4}$$

$$z = \pm\sqrt{\frac{(1+r)^2 - t^2}{(1-r)^2 - t^2}}, \tag{3.3.5}$$

where r and t are the reflection and transmission coefficients. Note that while the expressions for n and z are relatively uncomplicated, they are complex functions with multiple branches, the interpretation of which can lead to ambiguities in determining the final expressions for ε and μ. Additional knowledge about the material is helpful to resolve these ambiguities. For example, if the material is passive, the requirement

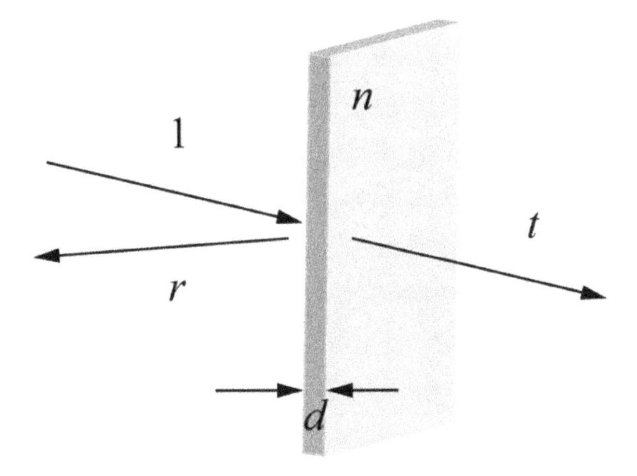

Fig. 3.19 Schematic of S-parameter retrieve

that $\mathrm{Re}(z) > 0$ fixes the choice of sign in Eq. (3.3.5). Likewise, $\mathrm{Im}(n) > 0$ leads to an unambiguous result for $\mathrm{Im}(n)$ [83]:

$$\mathrm{Im}(n) = \pm \mathrm{Im}\left(\frac{\cos^{-1}\left(\frac{1}{2t}\left[1 - \left(r^2 - t^2\right)\right]\right)}{kd} \right) \tag{3.3.6}$$

$$\mathrm{Re}(n) = \pm \mathrm{Re}\left(\frac{\cos^{-1}\left(\frac{1}{2t}\left[1 - \left(r^2 - t^2\right)\right]\right)}{kd} \right) + \frac{2\pi m}{kd} \tag{3.3.7}$$

where m is an integer. Once the refractive index n and the impedance z are determined, the permittivity ε and permeability μ can also be determined.

3.3.2 Negative-Index Materials

The negative-index materials (NIMs), firstly considered by Veselago in 1967 [84] simultaneously possess a negative dielectric permittivity ($\varepsilon < 0$) and a negative magnetic permeability ($\mu < 0$). Veselago's idea remained obscure since there is no such negative refractive materials were known to exist at any frequency. Fortunately, by making use of artificially structured metamaterials, scientists can circumvent this limitation. Resonance is the key to achieving this kind of negative response and it can be introduced artificially by building small circuits to mimic the magnetic or electrical response of a material [80, 81].

In 2001, the first metamaterials exhibiting a negative index of refraction at microwave frequencies were experimentally reported by combining metallic wire arrays with negative electric permittivity and split-ring-resonators (SRRs) array with negative magnetic permeability [56]. Subsequently, NIMs at telecommunication wavelengths were proposed [85]. The structures in Fig. 3.20 consist of double-plate

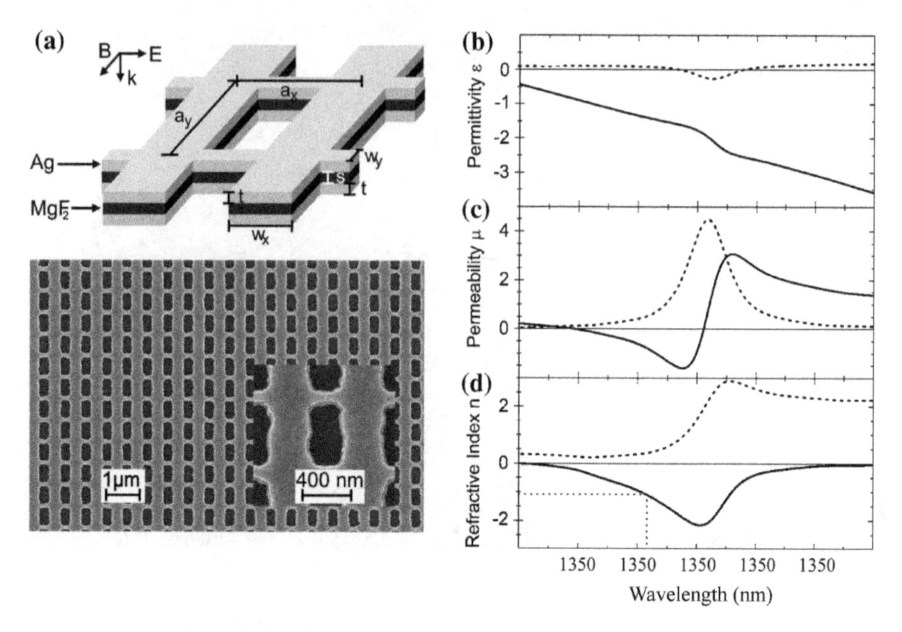

Fig. 3.20 **a** Scheme and SEM of the negative-index metamaterial design. Retrieved permittivity ε (**b**), permeability μ (**c**), and refractive index n (**d**) Reproduced from [85] with permission. Copyright 2006, Optical Society of America

(or double-wire) pairs as "magnetic atoms" and long wires as "electric atoms" (just a diluted Drude metal). The key to optimizing the figure of merit (FOM, defined as $-\text{Re}(n)/\text{Im}(n)$) of this structure lays in tuning the combination of wire widths, metal thickness, and spacer thickness. Recently, a 3D negative refractive index optical metamaterial with a high FOM of 3.5 was proposed [86].

Figure 3.21a and b show a diagram and SEM image of 3D fishnet metamaterial, which is fabricated on a multilayer metal–dielectric stack by using focused ion-beam (FIB) milling. The measured refractive index of the 3D fishnet metamaterial at various wavelengths is shown in Fig. 3.21c, from which one can see the refractive index varies from $n = 0.63 \pm 0.05$ at 1200 nm to $n = -1.23 \pm 0.34$ at 1775 nm [86].

To acquire a clear understanding of optical response of the 3D metamaterial, the composites of 3D array of metal wires are investigated by RCWA. The first constituent is a 3D array of metal wires aligned with the polarization direction of the incident electric field (Fig. 3.22a) [86]. This array behaves as an effective medium with lowered volumetric plasma frequency (220 THz), below which wave propagation is not allowed because of negative effective permittivity [86]. The second constituent is a 3D array of metal strips along the direction of the magnetic field (Fig. 3.22b), in which induced antisymmetric conductive currents across the dielectric layers give rise to a magnetic bandgap between 135 and 210 THz [86]. Finally, these two structures are merged to form the 3D fishnet metamaterial, for which the dispersion relation is shown in Fig. 3.23c. A propagation band with negative slope

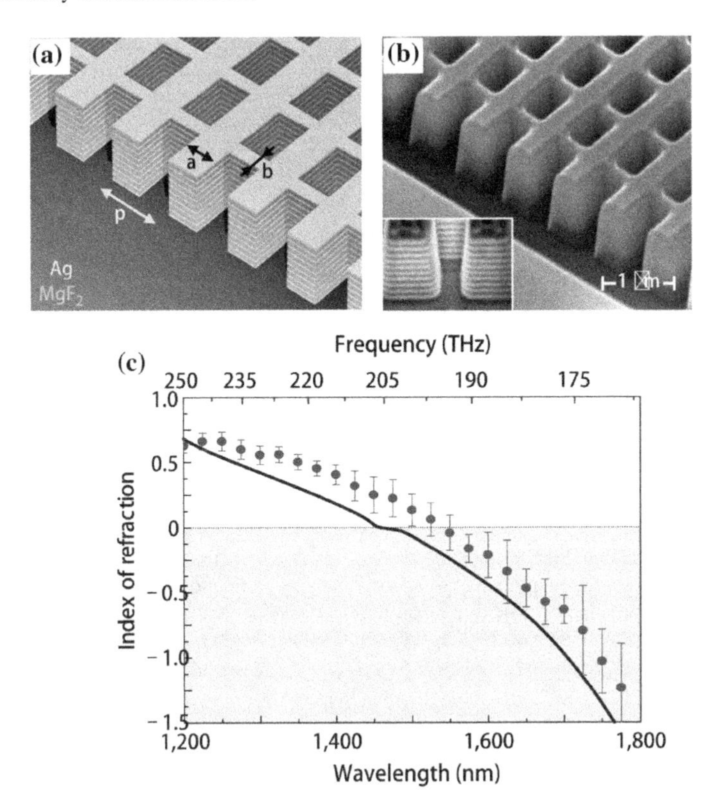

Fig. 3.21 **a**, **b** Diagram and SEM image of fabricated fishnet structure. **c** Measurements and simulation of the fishnet refractive index. The measurement agrees closely with the simulated refractive index using the coupled-wave analysis method (black line). Reproduced from [86] with permission. Copyright 2008, Macmillan Publishers Limited

appears in the overlapped region of the forbidden gaps of both electric and magnetic media, demonstrating that the negative-index behavior in the 3D cascaded fishnet does indeed result from the fact that both the electric permittivity and the magnetic permeability are negative [86].

3.3.3 Near-Zero Index Materials

Structures with near-zero parameters at a given frequency can be classified as ENZ, $\varepsilon \approx 0$, mu-near-zero (MNZ), $\mu \approx 0$, and epsilon-and-mu-near-zero (EMNZ), $\mu \approx 0$ and $\varepsilon \approx 0$ media [57]. All aforementioned classes exhibit a near-zero index of refraction $n = \sqrt{\varepsilon\mu} \approx 0$ at the frequency of interest and can be jointly addressed as zero-index media. Among them, the metallic mesh of thin wires is the simplest

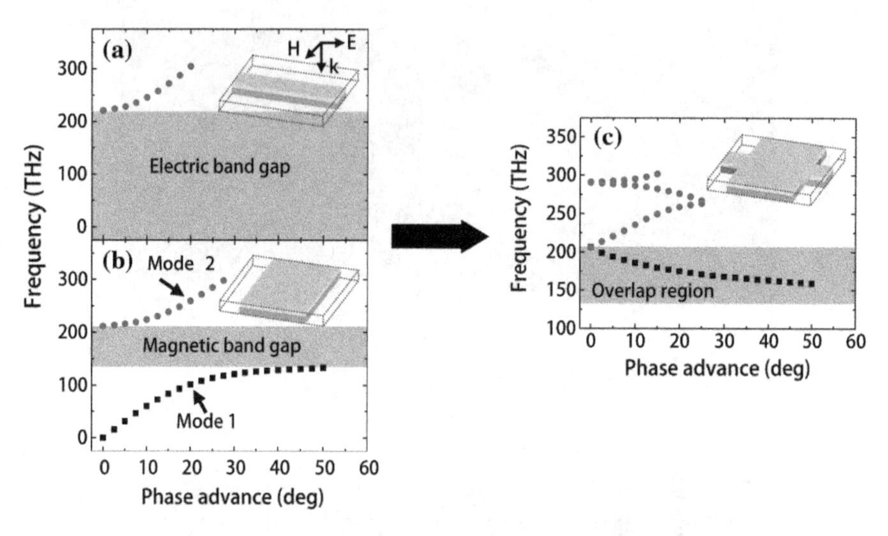

Fig. 3.22 **a** Dispersion relation for a 3D array of metal wires aligned along the electric field E, where k denotes the incident propagation vector. **b** Dispersion relation of a 3D array of metal strips along the magnetic field H. **c** The dispersion for the 3D fishnet structure. A dispersion curve with negative slope appears within the overlapped region of the electric bandgap and magnetic bandgap if both structures are combined. Produced from [86] with permission. Copyright 2008, Macmillan Publishers Limited

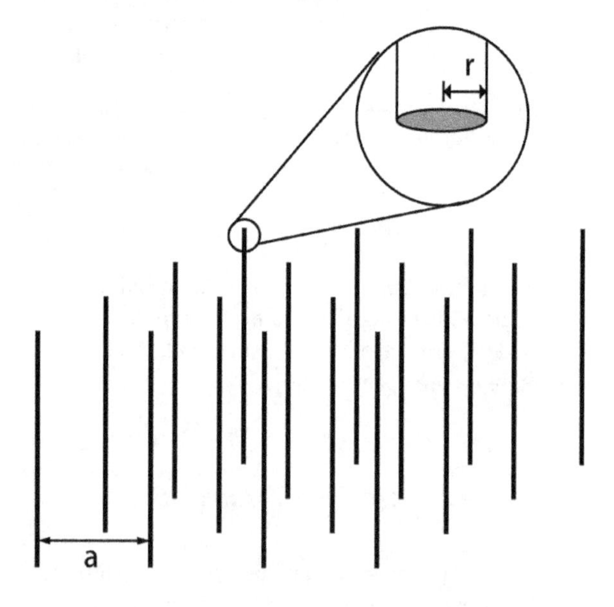

Fig. 3.23 Periodic structure composed of infinite metallic wires arranged in two arrays. Adapted from [80] with permission. Copyright 1996, American Physical Society

type of metamaterials potentially used to form ENZ medium, since the Drude-like dispersion behavior can be tuned by adjusting the radius and period of the wires.

For 2D infinite metallic wires shown in Fig. 3.23, the density of these active electrons in the structure as a whole is given by the fraction of space occupied by the wire (N is the electron density in the metallic wire) [80]:

$$n_{\text{eff}} = N\frac{\pi r^2}{a^2}, \tag{3.3.8}$$

and the effective mass of the electrons can be given by:

$$m_{\text{eff}} = \frac{\mu_0 \pi r^2 e^2 N}{2\pi} \ln(a/r). \tag{3.3.9}$$

Therefore, the plasma frequency can be written as [80]:

$$\omega_p^2 = \frac{n_{\text{eff}} e^2}{\varepsilon_0 m_{\text{eff}}} = \frac{2\pi c_0^2}{a^2 \ln(a/r)}. \tag{3.3.10}$$

Although this model revels that extreme low plasma frequency can be realized by engineering the geometries of metallic wires array, this model requires that the wires be very thin and the cross sections of the wires be circles, which greatly restricts the freedom of design and fabrication. This model may fail in some instance, such as when the shape of wire is not circular or when the wire dimensions exceed the skin depth of the electromagnetic field.

In order to overcome the above problems, a refined plasma frequency model for ENZ materials consisting of arranged metallic wires with arbitrary cross section is proposed, as shown in Fig. 3.24 [87]. This model, taking into account of the skin effect, breaks the limitation on the shape of the cross section of wires. Numerical simulations results verify that the plasma frequency of arranged metallic wires is determined by the lattice spacing and the outer perimeter of the cross section [87]. Supposing the density of the electrons in these wires is n_0, the number of active electrons per unit length of the wire is expressed approximately as $N = l \times \delta \times n_0$ when $\delta << l$. Therefore, the effective electron density n_{eff} in the whole cell can be written as [87]:

$$n_{\text{eff}} = \frac{l \cdot \delta \cdot n_0}{a^2} \tag{3.3.11}$$

We can see that n_{eff} depends on the lattice constant a, the perimeter l, and the skin depth δ, where the skin depth δ is determined by both the wires material and the frequency of the incident electromagnetic wave.

The effective electron mass in the skin of the wire is determined as:

$$m_{\text{eff}} = \frac{\mu_0 l \delta n_0 e^2}{2\pi}\left[\ln\left(\frac{2\sqrt{\pi}a}{l}\right) + \frac{l^2}{8\pi a^2} - \frac{1}{2}\right] \tag{3.3.12}$$

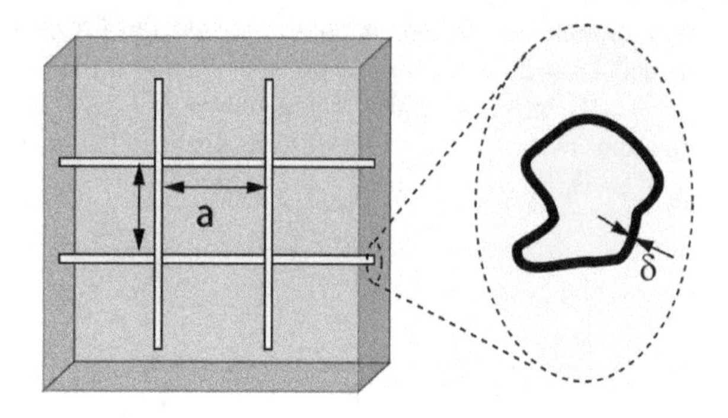

Fig. 3.24 A schematic view of a 2D periodic infinite copper mesh. The lattice spacing is a, the arbitrary cross section of the wires whose perimeter of the outer bounding is l, and the skin depth is δ. Adapted from [87] with permission. Copyright 2009, Springer-Verlag

and the plasma frequency of metallic wires with arbitrary cross section can be expressed as [87]:

$$\omega_p^2 = \frac{2 \cdot \pi \cdot c_0^2}{a^2\left[\ln(\frac{l}{2 \cdot \sqrt{\pi} \cdot a}) - \frac{l^2}{8 \cdot \pi \cdot a^2} + \frac{1}{2}\right]} \tag{3.3.13}$$

Comparisons between the refined model and the simulation results for wires with rectangular and ring cross section are shown in Fig. 3.25, from which one can see that they are matched well, verifying the feasibility of refined model [87].

3.3.4 Ultra-High Index Materials

The key to creating the desired high-index behavior lies in the existence of subwavelength propagating modes. For a metal film with 1D periodic cut-through slits (Fig. 3.26a), regardless of how small the width is, there exists a propagating TEM mode, with the electric field pointing in the x-direction, which permits perfect transmission of light through subwavelength slit arrays [88]. The properties of the metal film for the TM polarization asymptotically approach those of a dielectric slab with a uniquely defined refractive index n and a width L, as depicted in Fig. 3.26b [88]. In order to verify the above theory, the transmission spectra and magnetic field distributions of the both structures are presented and compared in Fig. 3.26c and d, from which one can see that the theoretical model operates well [88].

Although the above method provides a theoretical implementation of high-index artificial materials, it requires a high aspect ratio etching and thus poses a great challenge to fabrication. Subsequently, terahertz metamaterials composed of strongly coupled unit cells were demonstrated to exhibit an extremely high index of refraction [89]. By drastically increasing the effective permittivity through strong capacitive coupling and decreasing the diamagnetic response with a thin metallic structure in the unit cell, a peak refractive index of 38.6 along with a low-frequency quasi-static

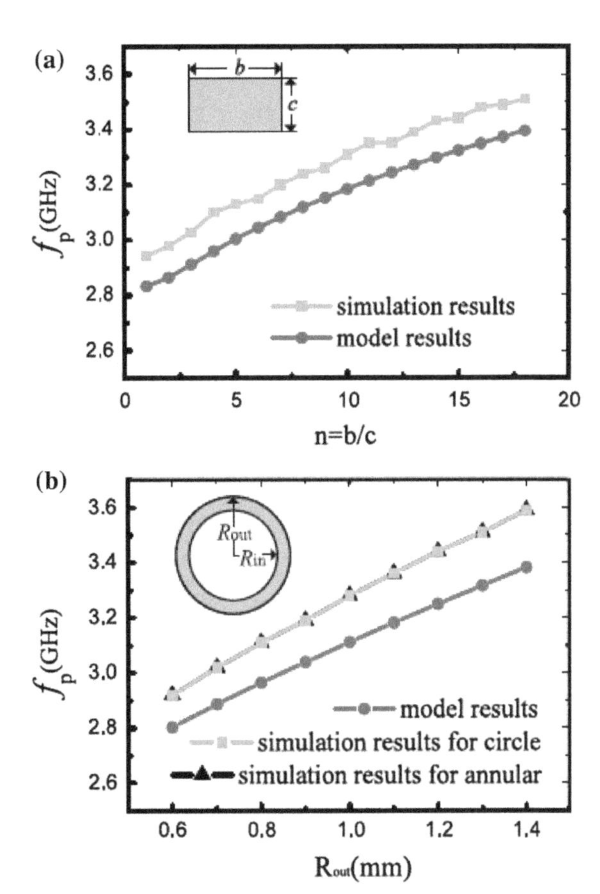

Fig. 3.25 Results derived by the combination of FEM and standard *S*-parameters retrieval method and corresponding results calculated by the arbitrary cross-sectional model (**a**) for rectangle cross section, red square marks are the simulated ω_p, and green circle marks are the calculated results. **b** For annual and circle cross section, green square marks and black triangle marks denote the simulated result of annual and circular, while the red circle marks denote the calculated results. The insets are illustrations of various shapes of the wire cross section. Adapted from [87] with permission. Copyright 2009, Springer-Verlag

value of over 20 was experimentally realized for a single-layer terahertz metamaterial, while maintaining low losses.

The basic building block (single-layer unit cell) of the proposed high-refractive-index terahertz metamaterial is shown in Fig. 3.27a, where strongly coupled thin "I"-shaped metallic paths are embedded symmetrically in the substrate [89]. The gap width plays a crucial role to tailor the value of the refractive index owing to the fact that the scaling of the charge accumulation and effective permittivity shows different asymptotic behaviors with the gap width. In the strongly coupled regime, a large amount of surface charge is accumulated on each arm of the metallic patch capacitor. This leads to an extremely large dipole moment in the unit cell, as the accumulated charge is inversely proportional to the gap width in the strongly coupled regime. This extreme charge accumulation contributes to a huge dipole moment inside the unit cell (or a large polarization density) and ultimately leads to a large effective permittivity. Besides enhance the effective permittivity, thin "I"-shaped metallic structure can effectively reduce the diamagnetic effect since it has a small area subtended by the current loop. Consequently, a high-refractive index can be generated [89].

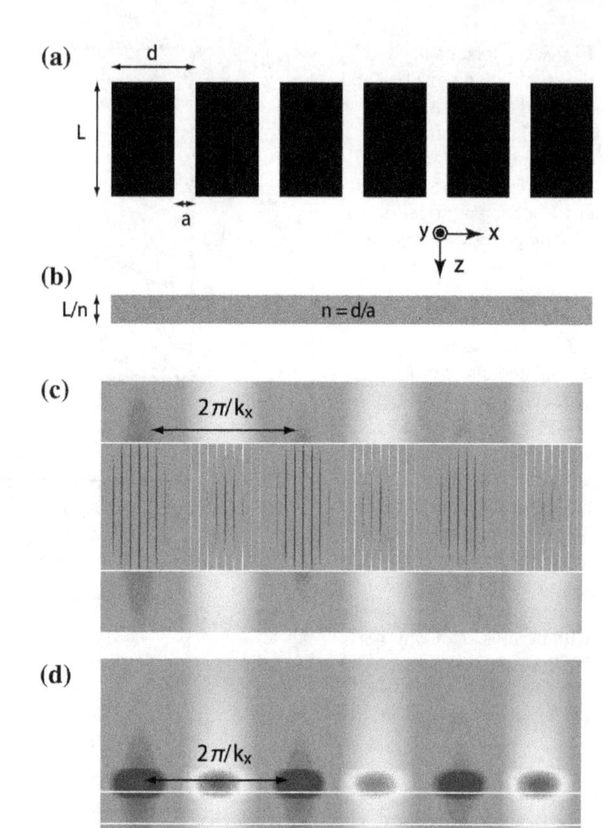

Fig. 3.26 **a** Schematic of the metal film with periodic slits. **b** The equivalent effective dielectric slab corresponding to (**a**). The effective refractive index is $n \equiv d/a$, and the thickness is L/n. **c** Snapshots of the *Hy* field distributions of the fundamental waveguide modes of the metal film and **d** the corresponding effective dielectric slab. The normalized excitation frequency is d/λ =0.0516. Reproduced from [88] with permission. Copyright 2005, American Physical Society

Figure 3.27c shows that a peak relative permittivity of 583 appears at a frequency of 0.504 THz and of 122 at the quasi-static limit. The magnetic permeability, however, remains near unity over the whole frequency domain, except near the frequency of the electric resonance [89]. Benefiting from this extreme enhancement of permittivity along with the suppression of diamagnetism, a peak index of refraction of $n = 27.25$ at 0.516 THz is achieved, along with a value of $n = 11.1$ at the quasi-static limit [89]. Note that, one can further improve the equivalent index by increasing the layers of the composites (insets in Fig. 3.27b). For example, a maximum bulk refractive index of 33.2 along with a value of around 8 at the quasi-static limit is realized by quasi-three-dimension [89].

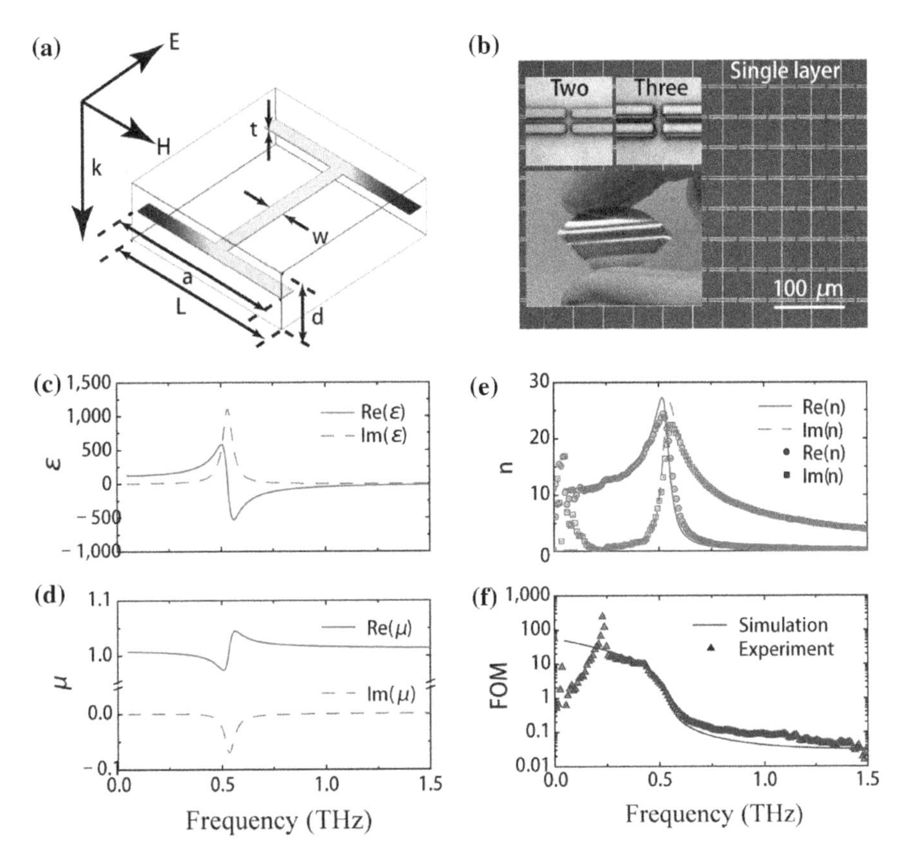

Fig. 3.27 **a** Unit cell structure of the high-index metamaterial. **b** Optical micrograph of fabricated single-layer metamaterial. **c** Effective permittivity and permeability. **d** Extracted from the characterization of a single-layer metamaterial unit cell. **e** Experimentally obtained complex refractive index, along with the corresponding numerically extracted values from the *S*-parameter retrieval method. **f** The experimental and simulated figure of merit. Reproduced from [89] with permission. Copyright 2011, Macmillan Publishers Limited

3.3.5 Hyperbolic Metamaterials

Among the varieties of metamaterials proposed in recent years, hyperbolic metamaterials (HMMs) have rapidly gained much attention in nanophotonics, thanks to their unparalleled ability to access and manipulate the near-field of a light emitter or a light scatterer. The term HMM refers to a medium for which the permittivity and permeability tensor elements (along principal axes) are of opposite signs, whose equifrequency contour (EFC) is hyperbolic [67].

The relative signs of permittivities determine the type of EFC, e.g., an ellipse or a hyperbola. When $\varepsilon_x > 0$ and $\varepsilon_z > 0$, the EFC is an ellipse. When $\varepsilon_x > 0$ and $\varepsilon_z < 0$, the corresponding EFC is shown in Fig. 3.28a, referred to as *Type I* hyperbolic

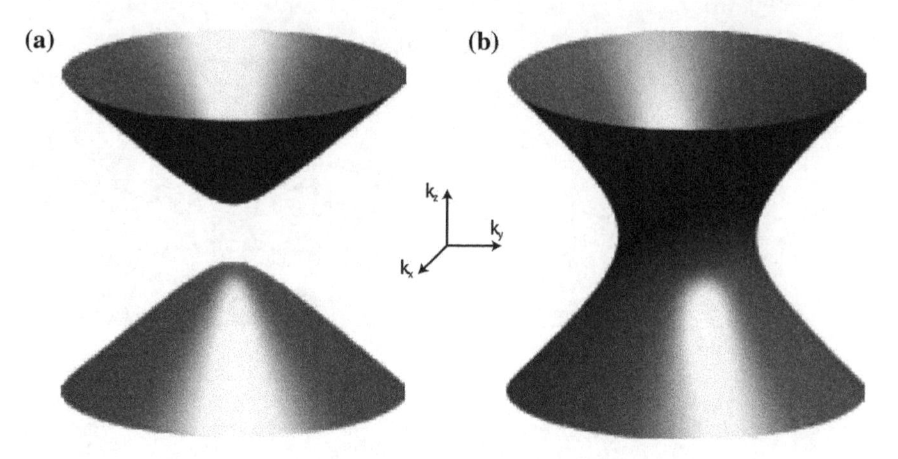

Fig. 3.28 Equifrequency surfaces of extraordinary waves in hyperbolic metamaterials. **a**, **b** Equifrequency surfaces given by $\omega(k) = \text{constant}$ for $\varepsilon_{zz} < 0$, $\varepsilon_{xx} = \varepsilon_{yy} > 0$ (**a**) and $\varepsilon_{zz} > 0$, $\varepsilon_{xx} = \varepsilon_{yy} < 0$ (**b**). Reproduced from [67] with permission. Copyright 2013, Macmillan Publishers Limited

metamaterials. When the signs of ε_x and ε_z are flipped, i.e., $\varepsilon_x < 0$ and $\varepsilon_z > 0$, the corresponding EFC is shown in Fig. 3.28b, referred to as *Type II* hyperbolic metamaterials [67].

Many novel and unique properties result from this hyperbolic EFC [90]. First, hyperbolic EFC gives rise to an omnidirectional negative refraction for a certain polarization of the electromagnetic wave. Second, hyperbolic metamaterials can support the transmission of evanescent wave components and transform them into propagating wave. This property forms the basis for the so-called hyperlens [65, 78, 91], enabling subwavelength imaging in the far-field. Next, hyperboloid EFC enables nanometer-scale 3D cavities with a very large *Q*-value [92]. Finally, the hyperbolic EFC is shown to greatly enhance the photonic density of states in a broad range of frequencies, paving the way for boosting the efficiency of solar cells by controlling the emission of the dye [93].

Currently, several different structures have been shown to implement the hyperbolic metamaterials, including layered metal–dielectric structures in planar profile (Fig. 3.29a), curved (Fig. 3.29b) and fishnet profile (Fig. 3.29c), as well as a lattice of metallic nanowires embedded in a dielectric matrix, termed nanowire array (Fig. 3.29d) [67]. In the limit of infinitesimal layer thickness, the optical response of these discontinuous media can be homogenized via EMT in Eq. (3.3.1), resulting in a hyperbolic effective permittivity tensor [67].

Table 3.6 collects the experimentally demonstrated material combinations that lead to layered metal–dielectric structure with hyperbolic properties [67]. By tuning these parameters such that $\varepsilon_{\parallel}\varepsilon_{\perp} < 0$, one can attain the hyperbolic regime. For cylinder shaped metal–dielectric multilayers shown in Fig. 3.29b, the parallel ($\varepsilon_{\parallel}$) and perpendicular ($\varepsilon_{\perp}$) should be replaced by ε_{θ} and ε_r. Furthermore, a magnetic

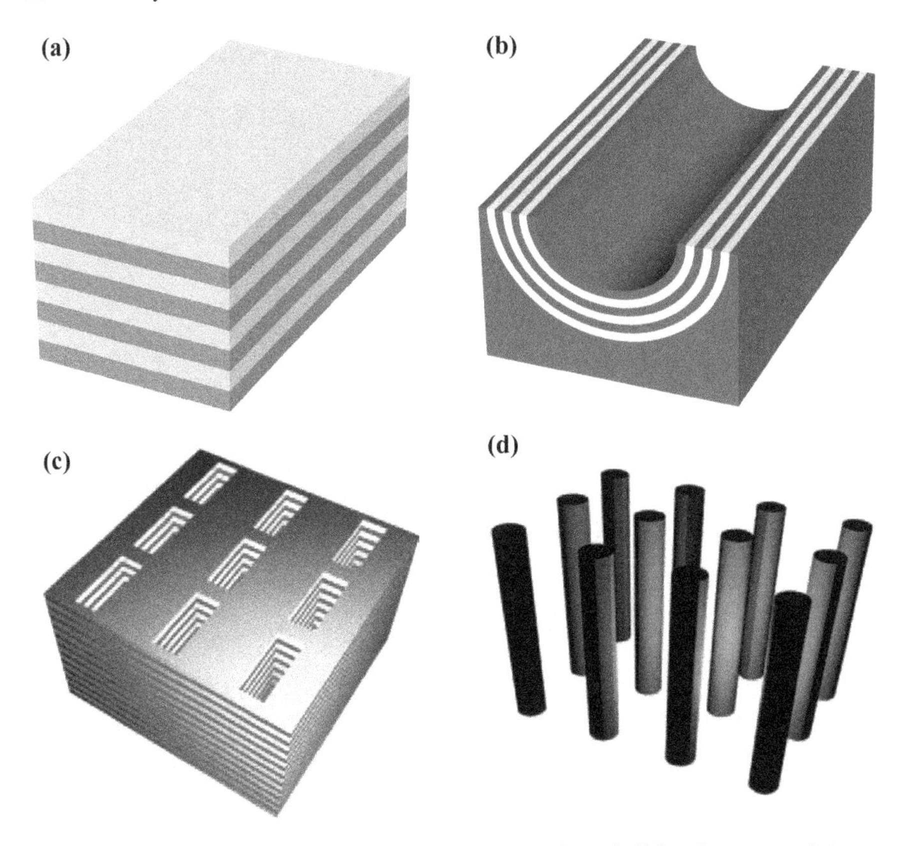

Fig. 3.29 Examples of hyperbolic metamaterials. **a** Layered metal–dielectric structure; **b** hyperlens; **c** multilayer fishnet; **d** nanorod arrays. Reproduced from [67] with permission. Copyright 2013, Macmillan Publishers Limited

hyperbolic medium with $\mu_{\|}\mu_{\perp} < 0$ in fishnet-like structure (Fig. 3.29c) has also been predicted [67].

For metallic wires or nanorods arrays oriented along the z-axis, the effective permittivity tensors take the form [67]:

$$\varepsilon_{xx} = \varepsilon_{yy} = \frac{[(1+p)\varepsilon_m + (1-p)\varepsilon_d]\varepsilon_d}{(1-p)\varepsilon_m + (1+p)\varepsilon_d}$$
$$\varepsilon_{zz} = p\varepsilon_m + (1-p)\varepsilon_d \tag{3.3.14}$$

where p is the percentual area occupied by nanorods in an xy section of the medium.

Table 3.6 Material combinations for the multilayer hyperbolic metamaterials [94]

Range	Materials	Period (nm)	p	No. of periods
UV	Ag/Al_2O_3	70	0.5	8
Visible	Au/Al_2O_3	38	0.5	8
	Au/Al_2O_3	40	0.5	8
	Au/Al_2O_3	43	0.35	10
	Au/TiO_2	48	0.33	4
	Ag/PMMA	55	0.45	10
	Ag/LiF	75	0.4	8
	Ag/MgF_2	50	0.4	8
	Ag/MgF_2	60	0.42	7
	Ag/TiO_2	31	0.29	5
	Ag/Ti_3O_5	60	0.5	9
	Ag/SiO_2	30	0.5	3
	Ag/Si	20	0.5	15
IR	AlInAs/InGaAs	160	0.5	50
	Al:ZnO/ZnO	120	0.5	16
	Ag/Ge	50	0.4	3

References

1. X. Luo, Subwavelength artificial structures: opening a new era for engineering optics. Adv. Mater. **0**, 1804680 (2018)
2. D.K. Gramotnev, S.I. Bozhevolnyi, Plasmonics beyond the diffraction limit. Nat. Photonics **4**, 83–91 (2010)
3. V.G. Veselago, E.E. Narimanov, The left hand of brightness: past, present and future of negative index materials. Nat. Mater. **5**, 759–762 (2006)
4. H. Chen, C.T. Chan, P. Sheng, Transformation optics and metamaterials. Nat. Mater. **9**, 387–396 (2010)
5. H. Ma, T. Cui, Three-dimensional broadband and broad-angle transformation-optics lens. Nat. Commun. **1**, 124 (2010)
6. N. Meinzer, W.L. Barnes, I.R. Hooper, Plasmonic meta-atoms and metasurfaces. Nat. Photonics **8**, 889–898 (2014)
7. X. Luo, Subwavelength electromagnetics. Front. Optoelectron. **9**, 138–150 (2016)
8. M. Pu, C. Wang, Y. Wang, X. Luo, Subwavelength electromagnetics below the diffraction limit. Acta Phys. Sin. **66**, 144101 (2017)
9. S.M. Choudhury, D. Wang, K. Chaudhuri, C. DeVault, A.V. Kildishev, A. Boltasseva, V.M. Shalaev, Material platforms for optical metasurfaces. Nanophotonics **7**, 959 (2018)
10. S.A. Maier in *Plasmonics: Fundamentals and Applications* (Springer Science & Business Media, 2007)
11. P.R. West, S. Ishii, G.V. Naik, N.K. Emani, V.M. Shalaev, A. Boltasseva, Searching for better plasmonic materials. Laser Photonics Rev. **4**, 795–808 (2010)
12. U. Guler, A. Boltasseva, V.M. Shalaev, Refractory plasmonics. Science **344**, 263–264 (2014)
13. X. Luo, Principles of electromagnetic waves in metasurfaces. Sci. China-Phys. Mech. Astron. **58**, 594201 (2015)

14. G.V. Naik, J.L. Schroeder, X. Ni, A.V. Kildishev, T.D. Sands, A. Boltasseva, Titanium nitride as a plasmonic material for visible and near-infrared wavelengths. Opt. Mater. Express **2**, 478–489 (2012)
15. G.V. Naik, V.M. Shalaev, A. Boltasseva, Alternative plasmonic materials: beyond gold and silver. Adv. Mater. **25**, 3264–3294 (2013)
16. Y. Wang, A.C. Overvig, S. Shrestha, R. Zhang, R. Wang, N. Yu, L. Dal Negro, Tunability of indium tin oxide materials for mid-infrared plasmonics applications. Opt. Mater. Express **7**, 2727–2739 (2017)
17. E. Feigenbaum, K. Diest, H.A. Atwater, Unity-order index change in transparent conducting oxides at visible frequencies. Nano Lett. **10**, 2111–2116 (2010)
18. Y.-W. Huang, H.W.H. Lee, R. Sokhoyan, R.A. Pala, K. Thyagarajan, S. Han, D.P. Tsai, H.A. Atwater, Gate-tunable conducting oxide metasurfaces. Nano Lett. **16**, 5319–5325 (2016)
19. A. Nemati, Q. Wang, M. Hong, J. Teng, Tunable and reconfigurable metasurfaces and metadevices. Opto-Electron. Adv. **1**, 180009 (2018)
20. G.V. Naik, J. Liu, A.V. Kildishev, V.M. Shalaev, A. Boltasseva, Demonstration of Al:ZnO as a plasmonic component for near-infrared metamaterials. Proc. Natl. Acad. Sci. **109**, 8834–8838 (2012)
21. G.V. Naik, J. Kim, A. Boltasseva, Oxides and nitrides as alternative plasmonic materials in the optical range [Invited]. Opt. Mater. Express **1**, 1090–1099 (2011)
22. S. Colburn, A. Zhan, E. Bayati, J. Whitehead, A. Ryou, L. Huang, A. Majumdar, Broadband transparent and CMOS-compatible flat optics with silicon nitride metasurfaces. Opt. Mater. Express **8**, 2330–2344 (2018)
23. A.N. Pikhtin, A.D. Yas'kov, Dispersion of the refractive index in semiconductors with diamond and zinc-blend structures. Sov. Phys. Semicond. **12**, 622–626 (1978)
24. A.I. Kuznetsov, A.E. Miroshnichenko, M.L. Brongersma, Y.S. Kivshar, B. Luk'yanchuk, Optically resonant dielectric nanostructures. Science **354**, 2472 (2016)
25. I. Staude, J. Schilling, Metamaterial-inspired silicon nanophotonics. Nat. Photonics **11**, 274–284 (2017)
26. M. Khorasaninejad, W.T. Chen, R.C. Devlin, J. Oh, A.Y. Zhu, F. Capasso, Metalenses at visible wavelengths: diffraction-limited focusing and subwavelength resolution imaging. Science **352**, 1190 (2016)
27. R.C. Devlin, M. Khorasaninejad, W.T. Chen, J. Oh, F. Capasso, Broadband high-efficiency dielectric metasurfaces for the visible spectrum. Proc. Natl. Acad. Sci. (2016)
28. C.N. Berglund, H.J. Guggenheim, Electronic properties of VO2 near the semiconductor-metal transition. Phys. Rev. **185**, 1022–1033 (1969)
29. Q. Wang, E.T.F. Rogers, B. Gholipour, C.-M. Wang, G. Yuan, J. Teng, N.I. Zheludev, Optically reconfigurable metasurfaces and photonic devices based on phase change materials. Nat. Photonics **10**, 60–65 (2016)
30. N. Raeis-Hosseini, J. Rho, metasurfaces based on phase-change material as a reconfigurable platform for multifunctional devices. Materials **10**, 1046 (2017)
31. K. Shportko, S. Kremers, M. Woda, D. Lencer, J. Robertson, M. Wuttig, Resonant bonding in crystalline phase-change materials. Nat. Mater. **7**, 653–658 (2008)
32. S. Walia, C.M. Shah, P. Gutruf, H. Nili, D.R. Chowdhury, W. Withayachumnankul, M. Bhaskaran, S. Sriram, Flexible metasurfaces and metamaterials: a review of materials and fabrication processes at micro- and nano-scales. Appl. Phys. Rev. **2**, 011303 (2015)
33. S. Song, X. Ma, M. Pu, X. Li, K. Liu, P. Gao, Z. Zhao, Y. Wang, C. Wang, X. Luo, Actively tunable structural color rendering with tensile substrate. Adv. Opt. Mater. **5**, 1600829 (2017)
34. K.S. Novoselov, A.K. Geim, S.V. Morozov, D. Jiang, Y. Zhang, S.V. Dubonos, I.V. Grigorieva, A.A. Firsov, Electric field effect in atomically thin carbon films. Science **306**, 666 (2004)
35. M. Zeng, Y. Xiao, J. Liu, K. Yang, L. Fu, Exploring two-dimensional materials toward the next-generation circuits: from monomer design to assembly control. Chem. Rev. **118**, 6236–6296 (2018)
36. A.K. Geim, I.V. Grigorieva, Van der Waals heterostructures. Nature **499**, 419 (2013)

37. K.S. Novoselov, A.K. Geim, S.V. Morozov, D. Jiang, M.I.K.I. Grigorieva, S.V. Dubonos, A.A. Firsov, Two-dimensional gas of massless Dirac fermions in graphene. Nature **438**, 197–200 (2005)
38. F. Wang, Y. Zhang, C. Tian, C. Girit, A. Zettl, M. Crommie, Y.R. Shen, Gate-variable optical transitions in graphene. Science **320**, 206–209 (2008)
39. K.S. Novoselov, V.I. Fal′ko, L. Colombo, P.R. Gellert, M.G. Schwab, K. Kim, A roadmap for graphene. Nature **490**, 192 (2012)
40. R.R. Nair, P. Blake, A.N. Grigorenko, K.S. Novoselov, T.J. Booth, T. Stauber, N.M.R. Peres, A.K. Geim, Fine structure constant defines visual transparency of graphene. Science **320**, 1308 (2008)
41. G.W. Hanson, Dyadic Green's functions and guided surface waves for a surface conductivity model of graphene. J. Appl. Phys. **103**, 064302 (2008)
42. G.W. Hanson, Quasi-transverse electromagnetic modes supported by a graphene parallel-plate waveguide. J. Appl. Phys. **104**, 084314 (2008)
43. P.-Y. Chen, A. Alù, Atomically thin surface cloak using graphene monolayers. ACS Nano **5**, 5855–5863 (2011)
44. Y. Li, A. Chernikov, X. Zhang, A. Rigosi, H.M. Hill, A.M. van der Zande, D.A. Chenet, E.-M. Shih, J. Hone, T.F. Heinz, Measurement of the optical dielectric function of monolayer transition-metal dichalcogenides: MoS2, MoSe2, WS2, and WSe2. Phys Rev B **90**, 205422 (2014)
45. B. Deng, R. Frisenda, C. Li, X. Chen, A. Castellanos-Gomez, F. Xia, Progress on black phosphorus photonics. Adv. Opt. Mater. **6**, 1800365 (2018)
46. X. Wang, A.M. Jones, K.L. Seyler, V. Tran, Y. Jia, H. Zhao, H. Wang, L. Yang, X. Xu, F. Xia, Highly anisotropic and robust excitons in monolayer black phosphorus. Nat. Nanotechnol. **10**, 517 (2015)
47. F. Xia, H. Wang, Y. Jia, Rediscovering black phosphorus as an anisotropic layered material for optoelectronics and electronics. Nat. Commun. **5**, 4458 (2014)
48. A.J. Giles, S. Dai, I. Vurgaftman, T. Hoffman, S. Liu, L. Lindsay, C.T. Ellis, N. Assefa, I. Chatzakis, T.L. Reinecke, J.G. Tischler, M.M. Fogler, J.H. Edgar, D.N. Basov, J.D. Caldwell, Ultralow-loss polaritons in isotopically pure boron nitride. Nat. Mater. **17**, 134 (2017)
49. M. Tamagnone, A. Ambrosio, K. Chaudhary, L.A. Jauregui, P. Kim, W.L. Wilson, F. Capasso, Ultra-confined mid-infrared resonant phonon polaritons in van der Waals nanostructures. Sci. Adv. **4** (2018)
50. X. Lin, Y. Shen, I. Kaminer, H. Chen, M. Soljačić, Transverse-electric Brewster effect enabled by nonmagnetic two-dimensional materials. Phys. Rev. A **94**, 023836 (2016)
51. W. Ma, P. Alonso-González, S. Li, A.Y. Nikitin, J. Yuan, J. Martín-Sánchez, J. Taboada-Gutiérrez, I. Amenabar, P. Li, S. Vélez, C. Tollan, Z. Dai, Y. Zhang, S. Sriram, K. Kalantar-Zadeh, S.-T. Lee, R. Hillenbrand, Q. Bao, In-plane anisotropic and ultra-low-loss polaritons in a natural van der Waals crystal. Nature **562**, 557–562 (2018)
52. L. Dou, Emerging two-dimensional halide perovskite nanomaterials. J. Mater. Chem. C **5**, 11165–11173 (2017)
53. M.A. Green, A. Ho-Baillie, H.J. Snaith, The emergence of perovskite solar cells. Nat. Photonics **8**, 506 (2014)
54. H.S. Jung, N.-G. Park, Perovskite solar cells: from materials to devices. Small **11**, 10–25 (2014)
55. X. Luo, Subwavelength artificial structures: opening a new era for engineering optics. Adv. Mater. 1804680 (2018)
56. R. Shelby, D. Smith, S. Schultz, Experimental verification of a negative index of refraction. Science **292**, 77–79 (2001)
57. I. Liberal, N. Engheta, Near-zero refractive index photonics. Nat. Photonics **11**, 149 (2017)
58. B. Bai, Y. Svirko, J. Turunen, T. Vallius, Optical activity in planar chiral metamaterials: theoretical study. Phys. Rev. A **76**, 023811 (2007)
59. T.W. Ebbesen, H.J. Lezec, H.F. Ghaemi, T. Thio, P.A. Wolff, Extraordinary optical transmission through sub-wavelength hole arrays. Nature **391**, 667–669 (1998)

60. X. Luo, T. Ishihara, Surface plasmon resonant interference nanolithography technique. Appl. Phys. Lett. **84**, 4780–4782 (2004)
61. H. Shi, X. Luo, C. Du, Young's interference of double metallic nanoslit with different widths. Opt. Express **15**, 11321–11327 (2007)
62. M. Pu, Y. Guo, X. Li, X. Ma, X. Luo, Revisitation of extraordinary Young's interference: from catenary optical fields to spin-orbit interaction in metasurfaces. ACS Photonics **5**, 3198–3204 (2018)
63. X. Luo, D. Tsai, M. Gu, M. Hong, Subwavelength interference of light on structured surfaces. Adv. Opt. Photonics **10**, 757–842 (2018)
64. Y. Guo, M. Pu, X. Li, X. Ma, P. Gao, Y. Wang, X. Luo, Functional metasurfaces based on metallic and dielectric subwavelength slits and stripes array. J. Phys. Condens. Matter **30**, 144003 (2018)
65. Z. Jacob, L.V. Alekseyev, E. Narimanov, Optical hyperlens: far-field imaging beyond the diffraction limit. Opt. Express **14**, 8247–8256 (2006)
66. A.V. Kildishev, E.E. Narimanov, Impedance-matched hyperlens. Opt. Lett. **32**, 3432–3434 (2007)
67. A. Poddubny, I. Iorsh, P. Belov, Y. Kivshar, Hyperbolic metamaterials. Nat. Photonics **7**, 948–957 (2013)
68. Y. Guo, M. Pu, X. Ma, X. Li, X. Luo, Advances of dispersion-engineered metamaterials. Opto-Electron. Eng. **44**, 3–22 (2017)
69. Y. Xiong, Z. Liu, C. Sun, X. Zhang, Two-dimensional Imaging by far-field superlens at visible wavelengths. Nano Lett. **7**, 3360–3365 (2007)
70. H.N.S. Krishnamoorthy, Z. Jacob, E. Narimanov, I. Kretzschmar, V.M. Menon, Topological transitions in metamaterials. Science **336**, 205–209 (2012)
71. S.A. Ramakrishna, J.B. Pendry, M.C.K. Wiltshire, W.J. Stewart, Imaging the near field. J. Mod. Opt. **50**, 1419–1430 (2003)
72. B. Wood, J.B. Pendry, D.P. Tsai, Directed subwavelength imaging using a layered metal-dielectric system. Phys. Rev. B **74**, 115116 (2006)
73. C. Wang, P. Gao, X. Tao, Z. Zhao, M. Pu, P. Chen, X. Luo, Far field observation and theoretical analyses of light directional imaging in metamaterial with stacked metal-dielectric films. Appl. Phys. Lett. **103**, 031911 (2013)
74. Z. Guo, Z.Y. Zhao, L.S. Yan, P. Gao, C.T. Wang, N. Yao, K.P. Liu, B. Jiang, X. Luo, Moiré fringes characterization of surface plasmon transmission and filtering in multi metal-dielectric films. Appl. Phys. Lett. **105**, 141107 (2014)
75. T. Xu, A. Agrawal, M. Abashin, K.J. Chau, H.J. Lezec, All-angle negative refraction and active flat lensing of ultraviolet light. Nature **497**, 470–474 (2013)
76. R. Maas, E. Verhagen, J. Parsons, A. Polman, Negative refractive index and higher-order harmonics in layered metallodielectric optical metamaterials. ACS Photonics **1**, 670–676 (2014)
77. G. Ren, C. Wang, G. Yi, X. Tao, X. Luo, Subwavelength demagnification imaging and lithography using hyperlens with a plasmonic reflector layer. Plasmonics **8**, 1065–1072 (2013)
78. L. Liu, K. Liu, Z. Zhao, C. Wang, P. Gao, X. Luo, Sub-diffraction demagnification imaging lithography by hyperlens with plasmonic reflector layer. RSC Adv. **6**, 95973–95978 (2016)
79. J. Sun, T. Xu, N.M. Litchinitser, Experimental demonstration of demagnifying hyperlens. Nano Lett. **16**, 7905–7909 (2016)
80. J. Pendry, A. Holden, W. Stewart, I. Youngs, Extremely low frequency plasmons in metallic mesostructures. Phys. Rev. Lett. **76**, 4773–4776 (1996)
81. J.B. Pendry, A.J. Holden, D.J. Robbins, W.J. Stewart, Magnetism from conductors and enhanced nonlinear phenomena. IEEE Trans. Microw. Theory Tech. **47**, 2075–2084 (1999)
82. D. Smith, W. Padilla, D. Vier, S. Nemat-Nasser, S. Schultz, Composite medium with simultaneously negative permeability and permittivity. Phys. Rev. Lett. **84**, 4184–4187 (2000)
83. D.R. Smith, S. Schultz, P. Markoš, C.M. Soukoulis, Determination of effective permittivity and permeability of metamaterials from reflection and transmission coefficients. Phys. Rev. B **65**, 195104 (2002)

84. V.G. Veselago, The electrodynamics of substances with simultaneously negative values of ε and μ. Sov. Phys. USPEKHI **10**, 509–514 (1968)
85. G. Dolling, C. Enkrich, M. Wegener, C. Soukoulis, S. Linden, Low-loss negative-index metamaterial at telecommunication wavelengths. Opt. Lett. **31**, 1800–1802 (2006)
86. J. Valentine, S. Zhang, T. Zentgraf, E. Ulin-Avila, D.A. Genov, G. Bartal, X. Zhang, Three-dimensional optical metamaterial with a negative refractive index. Nature **455**, 376 (2008)
87. L. Liu, H. Shi, X. Luo, X. Wei, C. Du, A plasma frequency modulation model for constructing structure material with arbitrary cross-section thin metallic wires. Appl. Phys. A **95**, 563–566 (2009)
88. J.T. Shen, P.B. Catrysse, S. Fan, Mechanism for designing metallic metamaterials with a high index of refraction. Phys. Rev. Lett. **94**, 197401 (2005)
89. M. Choi, S.H. Lee, Y. Kim, S.B. Kang, J. Shin, M.H. Kwak, K.-Y. Kang, Y.-H. Lee, N. Park, B. Min, A terahertz metamaterial with unnaturally high refractive index. Nature **470**, 369–373 (2011)
90. J. Sun, N.M. Litchinitser, J. Zhou, Indefinite by nature: from ultraviolet to terahertz. ACS Photonics **1**, 293–303 (2014)
91. Z. Liu, H. Lee, Y. Xiong, C. Sun, X. Zhang, Far-field optical hyperlens magnifying sub-diffraction-limited objects. Science **315**, 1686 (2007)
92. X. Yang, J. Yao, J. Rho, X. Yin, X. Zhang, Experimental realization of three-dimensional indefinite cavities at the nanoscale with anomalous scaling laws. Nat. Photonics **6**, 450 (2012)
93. T.U. Tumkur, L. Gu, J.K. Kitur, E.E. Narimanov, M.A. Noginov, Control of absorption with hyperbolic metamaterials. Appl. Phys. Lett. **100**, 161103 (2012)
94. L. Ferrari, C. Wu, D. Lepage, X. Zhang, Z. Liu, Hyperbolic metamaterials and their applications. Prog. Quantum Electron. **40**, 1–40 (2015)

Chapter 4
Numerical Modeling and Intelligent Designs

Abstract The material basis of EO 2.0 is subwavelength structured materials, which possess many intriguing electromagnetic properties that do not exist in nature. The complicated physical mechanisms that describe these light–matter interactions at the nanoscale often cannot be explained by conventional macroscopic theories. Therefore, new analysis and numerical simulation methods must be exploited to give an accurate prediction of the optical performance. Furthermore, since there are so many design freedoms in subwavelength structured materials, traditional trial-and-error method suffers from low efficiency and locally optimized solution. Thanks to the rapid developments in big data and artificial intelligence, some scientific problems that classically require human perception or intricate mechanisms have recently been solved. The concept of design databases, e.g., materials database and inverse design database, has also been proposed. In this chapter, we present some frequently used modeling and optimizing methods of subwavelength structures and show the concept of artificial intelligent materials design platform.

Keywords Artificial intelligence · Optimizing algorithm · Transfer matrix method · Gerchberg–Saxton algorithm

4.1 Introduction

For simple periodic subwavelength structures, e.g., subwavelength gratings, alternative metal–dielectric multilayers, and wire arrays, numerical analysis methods including transfer matrix method (TMM) and rigorous coupled-wave analysis (RCWA) are often adopted to analyze the electromagnetic scattering and transmission properties. For more complex structures, full-wave simulations are required to obtain the precise electromagnetic responses. Generally, subwavelength structured materials are simulated using softwares based on finite element methods (FEM) or finite integration techniques (FIT). These methods enable the designers to study the scattering parameters (S-parameters) of the metamaterial structure, and use S-parameter retrieval method to obtain the effective material parameters. It is clear that the material characteristics can be engineered through optimizing the structural parameters such as the

X. Luo, *Engineering Optics 2.0*, https://doi.org/10.1007/978-981-13-5755-8_4

thickness of dielectric, size, and shape of the patterns in the metamaterial structure [1]. Nevertheless, full-dimensional parameter sweeping is not a good choice since it is inefficient and time-consuming. Soft computing plays an important role in the design and optimization of various problems in engineering field [1]. Soft computing is an optimization technique to find the solution to a problem which is very difficult to solve or with less mathematical formulations regarding the problem domain.

Various evolutionary soft computing techniques such as genetic algorithm (GA), genetic programming (GP), simulated annealing (SA), and particle swarm optimization (PSO) have been utilized in the optimization of subwavelength structured materials. Most of the soft computing techniques are inspired from biological phenomena and the social behavior of biological populations. Soft computing methods do not require extensive mathematical formulation of the problem. Thus, the necessity of exclusive domain-specific knowledge can be reduced. Also, it can handle multiple variables and multiple objective functions simultaneously. These make the soft computing techniques quite useful in electromagnetic applications to provide a cost-effective solution to the user in less computational time [1].

In many occasions, it is desired not only to know the electromagnetic responses of subwavelength structured materials, but also to give the concrete design that can generate special transmission spectra, i.e., reversed design [2]. In order to obtain the optimized reversed design, artificial neural network (ANN) and the concept of deep learning have been introduced into EO 2.0 during the past few years and achieved a great success. Here, we present the architecture of the intelligent design platform and database working on an operating system, as displayed in Fig. 4.1.

The workflow of the intelligent design platform is shown in Fig. 4.2. The user can add or import the expected data in the design platform, whose characters can be compared with those in databases. The initial design can be obtained after calling the databases of design platform, and then, a serial of modeling methods is utilized for performance analysis. During the optimizing process, appropriate optimizing

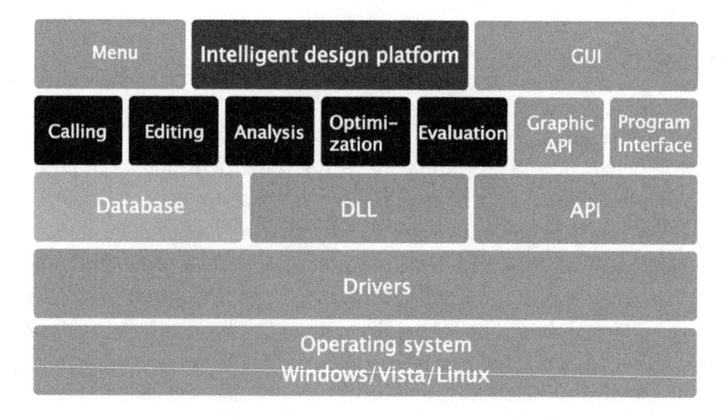

Fig. 4.1 Architecture of the intelligent design platform and database

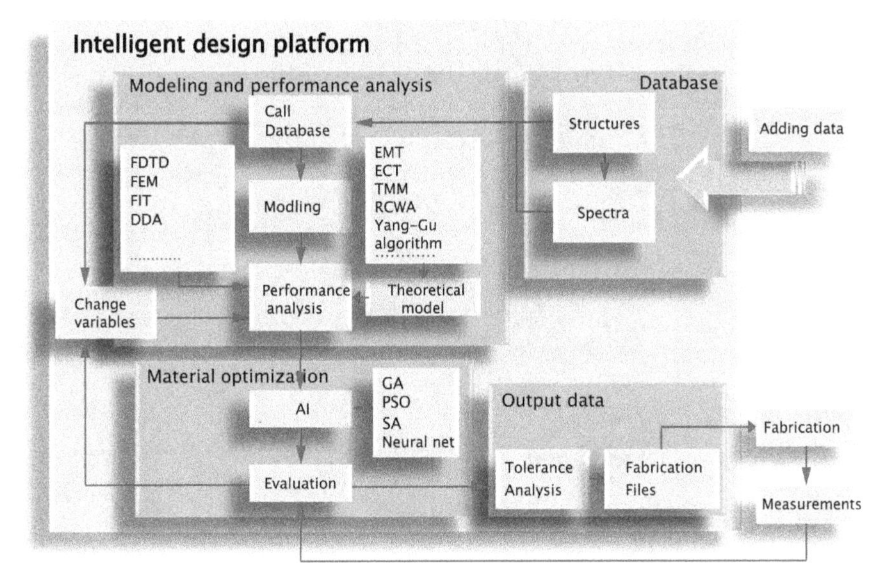

Fig. 4.2 Workflow of the intelligent design platform

algorithm or intelligent network is adopted to obtain the optimal design. After the tolerance analysis, the final fabrication files can be given.

One of the cores of the intelligent design platform is establishing the database, which includes the available structures and materials in EO 2.0. One potential organization of the database is illustrated in Fig. 4.3. It should be noted that the database can be updated by the users by importing the existing structures and their optical responses through the interfaces of the platform.

4.2 Design Methods for Multilayers and Gratings

Multilayers and gratings are simple but important structures in EO 2.0. On the one hand, one can exploit the multilayers to tune dispersion topology of the equivalent materials from elliptical to hyperbolic, for sub-diffraction-limited imaging and lithography. On the other hand, the multilayers can behave as filters readily to be used in absorption and radiative devices. Gratings are often utilized as polarizers [3], artificial waveplates [4–6], and an effective method to realize the conversion between surface wave and free-space waves [7].

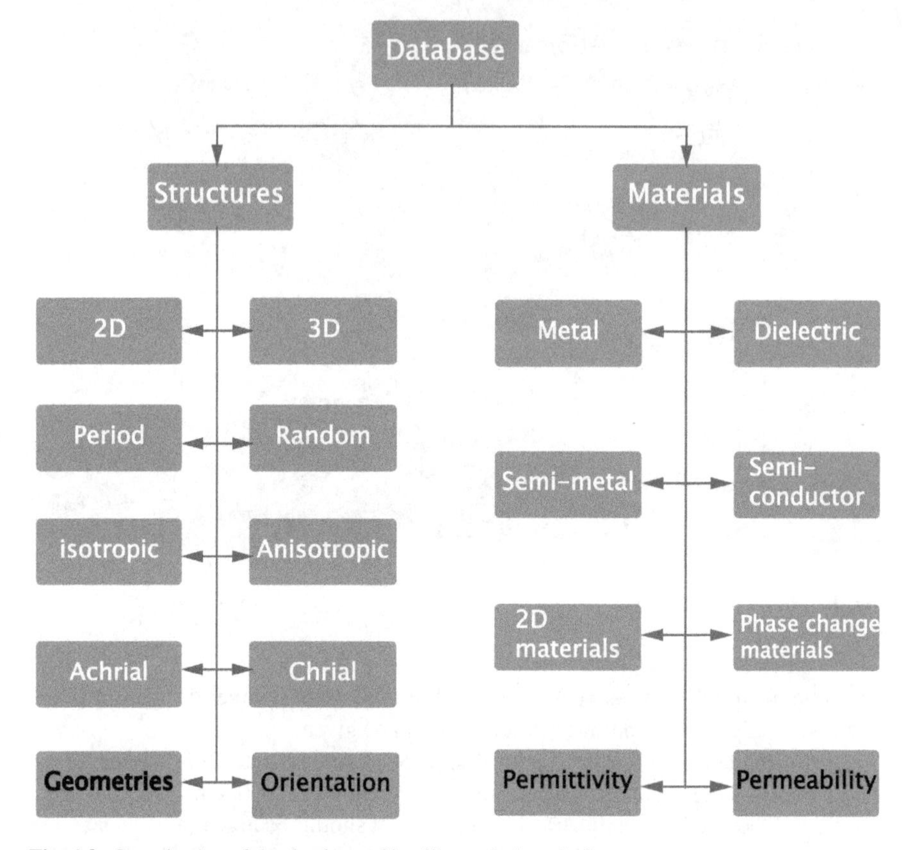

Fig. 4.3 Organization of the database of intelligent design platform

4.2.1 Transfer Matrix Method

(1) Planar multilayers

Transfer matrix method (TMM) is a powerful tool in the analysis of light propagation through layered dielectric media. The central idea lies in the fact that electric or magnetic fields in one layer can be related to those in other layers through a transfer matrix.

For multilayers shown in Fig. 4.4, considering non-magnetic material ($\mu = 1$) and transverse magnetically (TM) polarized wave illumination, the following continuity conditions (the tangential continuity of the electric field E and the magnetic field H) should be satisfied:

$$\begin{aligned} H_y^j = H_y^{j+1} &\Rightarrow a_j e^{ik_j d_j} + b_j e^{-ik_j d_j} = a_{j+1} + b_{j+1} \\ E_z^j = E_z^{j+1} &\Rightarrow \frac{k_j}{\varepsilon_j}\left(a_j e^{ik_j d_j} - b_j e^{-ik_j d_j}\right) = \frac{k_{j+1}}{\varepsilon_{j+1}}\left(a_{j+1} - b_{j+1}\right) \end{aligned} \quad (4.2.1)$$

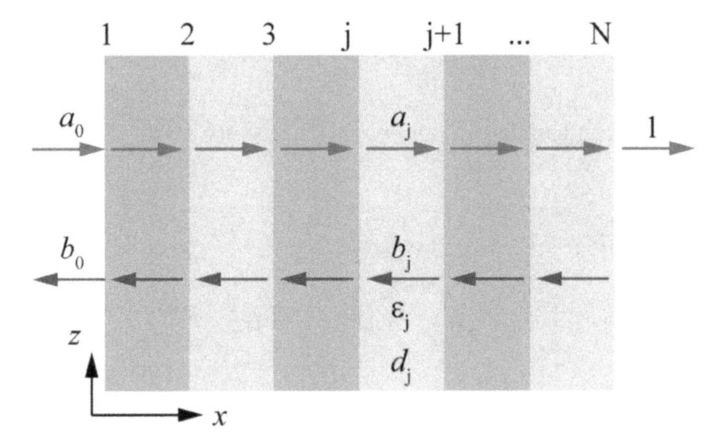

Fig. 4.4 TMM model of multiple films

where a and b represent the coefficients along different propagation directions, k is the wavevector in multilayers, and the subscripts j and $j+1$ denote the sequence number of the layers. The above equation can be written in a matrix form:

$$\begin{bmatrix} a_j \\ b_j \end{bmatrix} = T_{j,j+1} \begin{bmatrix} a_{j+1} \\ b_{j+1} \end{bmatrix} \tag{4.2.2}$$

where

$$T_{j,j+1} = \frac{1}{2} \begin{bmatrix} (1+K)e^{ik_j d_j} & (1+K)e^{ik_j d_j} \\ (1-K)e^{-ik_j d_j} & (1+K)e^{-ik_j x_j} \end{bmatrix} \tag{4.2.3}$$

and $K = \frac{k_{j+1}\varepsilon_j}{k_j \varepsilon_{j+1}}$ for TM polarization. Note that, the expression of K will be changed to be $K = \frac{k_{j+1}}{k_j}$ for transverse electric (TE) polarization. Then, the whole transfer matrix of multilayers can be expressed as:

$$T = T_{0,1} \times T_{1,2} \times T_{2,3} \times \cdots \times T_{N-1,N} \times T_{N,N+1} \tag{4.2.4}$$

The transfer matrix in Eq. (4.2.4) can be conveniently used for the calculation of multilayer films. For transmissive configuration, we have:

$$\begin{bmatrix} a_0 \\ b_0 \end{bmatrix} = T \begin{bmatrix} 1 \\ 0 \end{bmatrix} \tag{4.2.5}$$

Consequently, the transmission and reflection coefficient can be expressed as

$$t = \frac{1}{a_0} \tag{4.2.6}$$

$$r = \frac{b_0}{a_0} \tag{4.2.7}$$

For reflective configuration, assuming the reflectivity of the last film is -1, we have:

$$\begin{bmatrix} a_0 \\ b_0 \end{bmatrix} = T_{0,1} \times T_{1,2} \times T_{2,3} \times \cdots \times T_{N-1,N} \times T_{N,N+1} \begin{bmatrix} 1 \\ -1 \end{bmatrix} \tag{4.2.8}$$

In this case, the transmission is equal to zero (i.e., $T = 0$) and the reflective coefficient is

$$r = \frac{b_0}{a_0} \tag{4.2.9}$$

Then, the absorption A of the multilayers can be written as:

$$A = 1 - T - R = 1 - \left(\frac{b_0}{a_0}\right)^2 \tag{4.2.10}$$

For two-dimensional subwavelength structured materials, the boundary conditions are different from the general dielectric boundary conditions, and the above transmission matrix needs to be changed accordingly [8, 9]. It has been demonstrated that the transfer matrix method could greatly improve the calculation efficiency while simultaneously keeps the calculation accuracy [10, 11]. For a typical unit cell shown there [11], the calculation time using CST MICROWAVE STUDIO is more than 5 min, but the calculation time using transfer matrix method is less than 2 s [12]. This means that there is an enhancement of efficiency by at least two orders of magnitude.

(2) Curved multilayers

For special curved multilayer film systems, using different eigenmodes, the TMM is still applicable. Taking a cylindrical multilayer film as an example (as shown in Fig. 4.5), Richmond proposed a solution in 1975 [13]: using the cylindrical Bessel function to obtain the transmission coefficient between adjacent layers.

For TE waves (the electric field is along the axis of the cylinder), the total electric field outside the cylinder can be expressed as:

$$E_z = \sum_{n=0}^{\infty} [e_n j^{-n} J_n(k_0 r) + C_n H_n^{(2)}(k_0 r)] \cos n\phi \tag{4.2.11}$$

where k_0 is the wavevector in free space, $j = (-1)^{1/2}$, J_n is first-order Bessel function, and $H_n^{(2)}$ is Hankel function. The first item represents the serial expansion of incident electromagnetic waves, and the second item represents the serial expansion

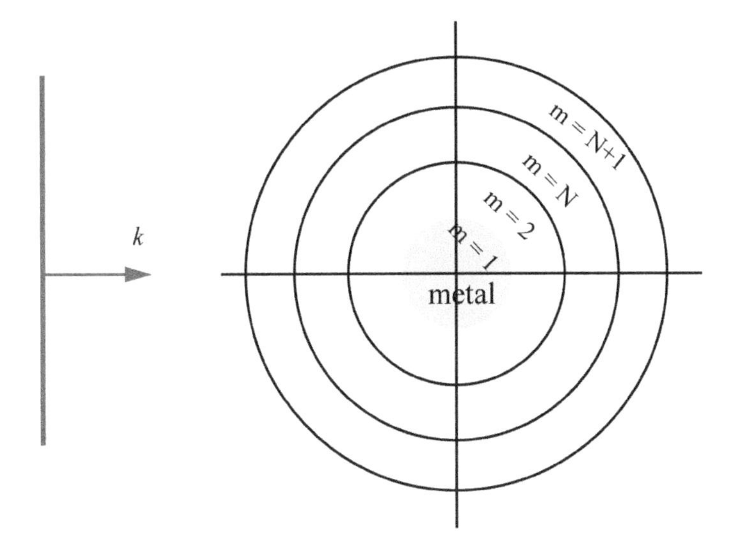

Fig. 4.5 Schematic of TMM in curved multilayers

of scattering electromagnetic waves. When $n = 0$, $e_n = 1$, and when $n > 0$, $e_n = 2$. The coefficient C_n is generally a complex coefficient.

In the mth layer (1, 2, 3… from the inside to the outside), the electric field can be expressed as:

$$E_{z,m} = \sum_{n=0}^{\infty} (A_{mn} J_n(k_m r) + B_{mn} Y_n(k_m r)) \cos n\phi \tag{4.2.12}$$

where $k_m = \omega(\varepsilon_m \mu_m)^{1/2}$ is the wavevector in mth layer, ε_m and μ_m are the dielectric constant and magnetic permeability of the layer medium, respectively, and Y_n is Bessel function of the second kind. Since there is only inwardly propagating electromagnetic waves in the first layer, unnormalized parameters are available for all orders:

$$A'_{1,n} = 1, \quad B'_{1,n} = 0. \tag{4.2.13}$$

The coefficients in each layer of medium are calculated sequentially from the inside to the outside using a recursive function and finally normalized by the formula (4.2.11). The key to parameter recursion is the matching of the electric field component in the z-direction and the magnetic field component in the azimuthal direction:

$$H_\phi = (k_m / j\omega\mu_m) \sum_{n=0}^{\infty} (A_{mn} J'_n(k_m r) + B_{mn} Y'_n(k_m r)) \cos n\phi \tag{4.2.14}$$

where the superscript indicates the derivative of the corresponding parameter. According to the above conditions, the relationship of the transfer coefficients in each layer can be obtained:

$$\begin{pmatrix} A'_{m+1,n} \\ B'_{m+1,n} \end{pmatrix} = \begin{pmatrix} U_{mn} & W_{mn} \\ V_{mn} & X_{mn} \end{pmatrix} \begin{pmatrix} A'_{m,n} \\ B'_{m,n} \end{pmatrix}, \tag{4.2.15}$$

where

$$\begin{aligned} U_{mn} &= \mu_m k_{m+1} J_n(k_m r_m) Y'_n(k_{m+1} r_m) - \mu_{m+1} k_m J'_n(k_m r_m) Y_n(k_{m+1} r_m) \\ V_{mn} &= \mu_{m+1} k_m J_n(k_{m+1} r_m) J'_n(k_m r_m) - \mu_m k_{m+1} J'_n(k_{m+1} r_m) J_n(k_m r_m) \\ W_{mn} &= \mu_m k_{m+1} Y_n(k_m r_m) Y'_n(k_{m+1} r_m) - \mu_{m+1} k_m Y'_n(k_m r_m) Y_n(k_{m+1} r_m) \\ X_{mn} &= \mu_{m+1} k_m Y'_n(k_m r_m) J_n(k_{m+1} r_m) - \mu_m k_{m+1} Y_n(k_m r_m) J'_n(k_{m+1} r_m). \end{aligned} \tag{4.2.16}$$

After a series of calculations, the final scattering coefficient is:

$$C_n = -j^{-n} e_n \left[B'_{M+1,n} / (B'_{M+1,n} + j A'_{M+1,n}) \right]. \tag{4.2.17}$$

The scattering field can be expressed as:

$$\begin{aligned} E^s &= (2j/\pi k_0 r)^{1/2} e^{-jk_0 r} \sum_{n=0}^{\infty} e_n D_n \cos n\phi \\ D_n &= j^n C_n / e_n = -B'/(B' + jA'). \end{aligned} \tag{4.2.18}$$

The total forward scatter coefficient is

$$T(0) = \sum_{n=0}^{\infty} e_n D_n, \tag{4.2.19}$$

where $T(0)$ is the case where $\phi = 0$ in the formula (4.2.18). The total backscattering power cross section is:

$$P(\pi) = (4/k_0)|T(\pi)|^2 \tag{4.2.20}$$

where $T(\pi)$ is the case of $\phi = \pi$ in the formula (4.2.18).

In 1980, Knott [14] extended the above-mentioned cylindrical transfer matrix method to a multilayer structure containing a resistive film using a boundary condition of a resistive film layer (such as a cylindrical Jaumann absorbing material).

By considering the surface current J, the boundary condition of the resistive layer is:

$$GE^+ = GE^- = J$$

$$H^+ - H^- = J. \tag{4.2.21}$$

For TE waves, the transfer matrix is changed to:

$$\begin{aligned} A_n^+ &= \frac{\pi}{2}k^+a\left[S(k^-a)\{Y_n'(k^+a) - jGZ_r^+Y_n(k^+a)\} - T(k^-a)\frac{Z_r^+}{Z_r^-}Y_n(k^+a)\right] \\ B_n^+ &= -\frac{\pi}{2}k^+a\left[S(k^-a)\{J_n'(k^+a) - jGZ_r^+J_n(k^+a)\} - T(k^-a)\frac{Z_r^+}{Z_r^-}J_n(k^+a)\right] \end{aligned} \tag{4.2.22}$$

For TM waves, the transfer matrix is:

$$\begin{aligned} A_n^+ &= \frac{\pi}{2}k^+a\left[S(k^-a)Y_n'(k^+a) - T(k^-a)\frac{Z_r^-}{Z_r^+}\{Y_n(k^+a) + jGZ_r^+Y_n'(k^+a)\}\right] \\ B_n^+ &= -\frac{\pi}{2}k^+a\left[S(k^-a)J_n'(k^+a) - T(k^-a)\frac{Z_r^-}{Z_r^+}\{J_n(k^+a) + jGZ_r^+J_n'(k^+a)\}\right] \end{aligned} \tag{4.2.23}$$

where a is the radial position of the impedance layer, and there are

$$\begin{aligned} S(k^-a) &= A_n^-J_n(k^-a) + B_n^-Y_n(k^-a) \\ S(k^-a) &= A_n^-J_n'(k^-a) + B_n^-Y_n'(k^-a). \end{aligned} \tag{4.2.24}$$

4.2.2 *Rigorous Coupled-Wave Analysis*

Rigorous coupled-wave analysis (RCWA) is a semi-analytical method in computational electromagnetics that is most typically applied to solve scattering from periodic dielectric structures. In 1981, Moharam and Gaylord [15] first proposed the RCWA method for vectorial analysis of electromagnetic wave diffraction problems of gratings. The RCWA mainly consists of three steps: (1) Obtain the expression of the electromagnetic field in the backward-diffraction region and the forward-diffraction region. (2) The electromagnetic field in the grating region is expanded in the Fourier series, and then, the coupled-wave equations are derived from Maxwell's equations. (3) The boundary continuity conditions are utilized to deduce the amplitude and diffraction efficiency of the diffracted waves.

As shown in Fig. 4.6, a monochromatic plane wave obliquely illuminates the subwavelength grating with an incident angle θ. The refractive indices of the grating reflection region and the transmission region are n_1 and n_2, respectively. The refractive indices of the grating ridge and the grating groove are n_{rd} and n_{gr}, respectively. The permittivity of the periodic grating can be expanded by Fourier series:

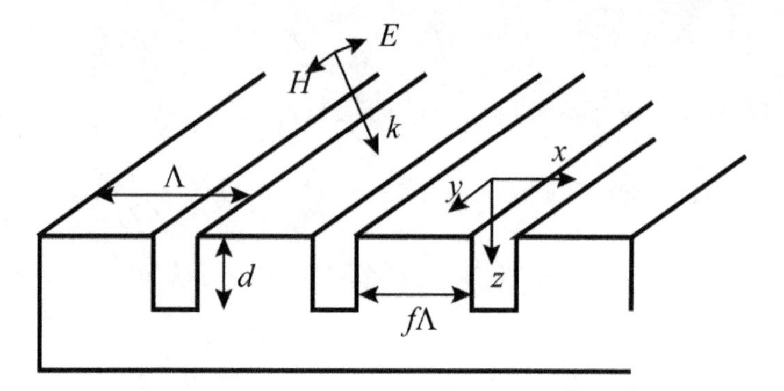

Fig. 4.6 Schematic of 1D grating

$$\varepsilon(x) = \sum_n \varepsilon_n \exp\left(i\frac{2\pi n}{\Lambda}x\right), \tag{4.2.25}$$

where Λ is the grating period, and ε_n is the nth order Fourier expansion coefficient of the permittivity. For the one-dimensional rectangular grating shown in Fig. 4.6, the expansion coefficients of each level can be expressed as:

$$\begin{aligned} \varepsilon_n &= n_{\mathrm{rd}}^2 f + n_{\mathrm{gr}}^2(1-f) \quad (n=0) \\ \varepsilon_n &= (n_{\mathrm{rd}}^2 - n_{\mathrm{gr}}^2)\frac{\sin(n\pi f)}{n\pi} \quad (n \neq 0), \end{aligned} \tag{4.2.26}$$

where f is the duty cycle of the grating ridge.

For TE polarization (also known as s-polarization) illumination, the electric fields of the backward- and frontward-diffraction regions are, respectively, expressed as:

$$E_1 = \exp(-ik_0 n_1(x\sin(\theta) + z\cos(\theta))) + \sum_m R_m \exp(-i(k_{xm}x - k_{1,zm}z)) \tag{4.2.27}$$

and

$$E_2 = \sum_m T_m \exp(-i(k_{xm}x + k_{2,zm}(z-d))), \tag{4.2.28}$$

where R_m and T_m are reflection and transmission coefficients of the mth order diffraction, k_0 and d are wavevector in free space and groove depth, and k_{xm}, $k_{1,zm}$, and $k_{2,zm}$, respectively, are:

$$k_{xm} = k_0[n_1\sin(\theta) + m\lambda/\Lambda] \tag{4.2.29}$$

and

$$k_{l,zm} = \sqrt{k_0^2 n_l^2 - k_{xm}^2} \quad l = 1, 2. \tag{4.2.30}$$

The magnetic fields of the reflection and transmission regions can be derived from Maxwell's equations:

$$\mathbf{H} = \left(\frac{i}{\omega\mu}\right)\nabla \times \mathbf{E}. \tag{4.2.31}$$

The electric field in the y-direction and the magnetic field in the x-direction in the grating region ($0 < z < d$) are expanded by the Fourier series:

$$\begin{aligned} E_{gy} &= \sum_m S_{ym}(z)\exp(-ik_{xm}x) \\ H_{gx} &= -i\left(\frac{\varepsilon_0}{\mu_0}\right)^{1/2} \sum_m U_{xm}(z)\exp(-ik_{xm}x), \end{aligned} \tag{4.2.32}$$

where $S_{ym}(z)$ and $U_{xm}(z)$ are transmission amplitude of the electric field and magnetic field of the mth order. The fields (E_{gy} and H_{gx}) in the grating region should satisfy Maxwell's equation:

$$\begin{aligned} \frac{\partial E_{gy}}{\partial z} &= i\omega\mu_0 H_{gx} \\ \frac{\partial H_{gx}}{\partial z} &= i\omega\varepsilon_0\varepsilon(x)E_{gy} + \frac{\partial H_{gz}}{\partial x}. \end{aligned} \tag{4.2.33}$$

Substituting Eq. (4.2.32) into Eq. (4.2.33) and eliminating H_{gz}, the coupled-wave equation can be obtained:

$$\begin{aligned} \frac{\partial S_{ym}}{\partial z} &= k_0 U_{xm} \\ \frac{\partial U_{xm}}{\partial z} &= \left(\frac{k_{xm}^2}{k_0}\right)S_{ym} - k_0 \sum_p \varepsilon_{(m-p)} S_{yp}. \end{aligned} \tag{4.2.34}$$

The coupled-wave equation in the form of a matrix is expressed as:

$$\begin{bmatrix} \partial S_y/\partial(z') \\ \partial U_x/\partial(z') \end{bmatrix} = \begin{bmatrix} 0 & \mathrm{I} \\ \mathrm{A} & 0 \end{bmatrix} \begin{bmatrix} \mathrm{S}_y \\ \mathrm{U}_x \end{bmatrix}, \tag{4.2.35}$$

where $z' = k_0 z$, $A = K_x^2 - E$, and I is the unit matrix. The electric and magnetic fields in the grating region can be obtained by solving the eigenvalues and eigenvectors of the matrix:

$$S_{ym}(z) = \sum_{h=1}^{n} w_{m,h}\left\{c_h^+ e^{-k_0 q_h z} + c_h^- e^{k_0 q_h (z-d)}\right\}$$

$$U_{xm}(z) = \sum_{h=1}^{n} g_{m,h}\left\{-c_h^+ e^{-k_0 q_h z} + c_h^- e^{k_0 q_h (z-d)}\right\}, \tag{4.2.36}$$

where $w_{m,h}$ is the element of eigenmatrix W, and q_h is the positive solution of the square root of the eigenvalues of matrix A. $v_{m,h} = q_h w_{m,h}$ is the element of the matrix $V = WQ$, where Q is the diagonalization matrix with matrix elements being q_h. The coefficients c_h^+ and c_h^- are determined by the boundary conditions.

According to the continuous boundary conditions of transverse electric and magnetic field components at the incident interface ($z = 0$), we have

$$\delta_{m0} + R_m = \sum_{h=1}^{n} w_{m,h}[c_h^+ + c_h^- e^{-k_0 q_h d}]$$

$$i[n_1 \delta_{m0} \cos(\theta) - (k_{1,zm}/k_0) R_m] = \sum_{h=1}^{n} g_{m,h}[c_h^+ - c_h^- e^{-k_0 q_h d}]. \tag{4.2.37}$$

Equation 4.2.37 can be rewritten in a matrix form:

$$\begin{bmatrix} \delta_{m0} \\ i n_1 \delta_{m0} \cos(\theta) \end{bmatrix} + \begin{bmatrix} I \\ -iY_1 \end{bmatrix} [R] = \begin{bmatrix} W & WX \\ V & -VX \end{bmatrix} \begin{bmatrix} C^+ \\ C^- \end{bmatrix}. \tag{4.2.38}$$

Likewise, the transverse electric and magnetic field components at the output interface ($z = d$) should satisfy:

$$\sum_{h=1}^{n} w_{m,h}\left[c_h^- + c_h^+ e^{-k_0 q_h d}\right] = T_m$$

$$\sum_{h=1}^{n} g_{m,h}\left[-c_h^- + c_h^+ e^{-k_0 q_h d}\right] = i(k_{2,zm}/k_0) T_m. \tag{4.2.39}$$

Equation (4.2.39) can also be written as:

$$\begin{bmatrix} WX & W \\ VX & -V \end{bmatrix} \begin{bmatrix} C^+ \\ C^- \end{bmatrix} = \begin{bmatrix} I \\ iY_2 \end{bmatrix} [T], \tag{4.2.40}$$

where $\delta_{m0} = 1$ for $m = 0$, and $\delta_{m0} = 0$ for $m \neq 0$; X, Y_1, and Y_2 are diagonal matrices, and the diagonal elements are $e^{-k_0 q_h d}$, $k_{1,zh}/k_0$, and $k_{2,zh}/k_0$; C^+ and C^- are the column matrixes that consist of c_h^+ and c_h^-. T and R can be deduced by combining (4.2.38) and (4.2.40). Similarly, the method can be utilized to analyze the diffraction for TM polarization (also known as p-polarization).

4.3 Full-Wave Simulation Methods

During the 1960s, both the finite element method (FEM) and the finite-difference time-domain method (FDTD) were reported for the first time in the field of computational electromagnetics. Currently, they have become two of the main methods for analyzing complex optical metamaterials.

4.3.1 FDTD

FDTD method is a fully vectorial method that naturally gives both time domain and frequency domain information and offers unique insight for the analysis of electromagnetic and photonic problems. Since this method is carried out in the time domain, the transient data from one simulation can be transformed to the frequency domain to obtain a wideband response [16].

For non-magnetic materials, FDTD solves Maxwell's equations in the following way:

$$\frac{\partial \vec{E}}{\partial t} = \frac{1}{\varepsilon(\vec{r})}\nabla \times \vec{H} - \frac{\sigma(\vec{r})}{\varepsilon(\vec{r})}\vec{E} \tag{4.3.1}$$

$$\frac{\partial \vec{H}}{\partial t} = -\frac{1}{\mu(\vec{r})}\nabla \times \vec{E} \tag{4.3.2}$$

where E and H, respectively, are the electric and magnetic fields, and $\varepsilon(r)$, $\mu(r)$, and $\sigma(r)$ are position-dependent permittivity, permeability, and conductivity, respectively. In three dimensions, Maxwell equations have six electromagnetic field components: E_x, E_y, E_z, H_x, H_y, and H_z.

By exploiting the Yee grid shown in Fig. 4.7 [17], the following six equations are obtained [16]:

$$H_x\Big|_{i,j,k}^{n+1/2} = H_x\Big|_{i,j,k}^{n-1/2} - \frac{\Delta t}{\mu_{i,j,k}}\left(\frac{E_z\Big|_{i,j+1,k}^{n} - E_z\Big|_{i,j,k}^{n}}{\Delta y} - \frac{E_y\Big|_{i,j,k+1}^{n} - E_y\Big|_{i,j,k}^{n}}{\Delta z}\right) \tag{4.3.3}$$

$$H_y\Big|_{i,j,k}^{n+1/2} = H_y\Big|_{i,j,k}^{n-1/2} - \frac{\Delta t}{\mu_{i,j,k}}\left(\frac{E_x\Big|_{i,j,k+1}^{n} - E_x\Big|_{i,j,k}^{n}}{\Delta z} - \frac{E_z\Big|_{i+1,j,k}^{n} - E_z\Big|_{i,j,k}^{n}}{\Delta x}\right) \tag{4.3.4}$$

Fig. 4.7 Schematic of Yee grid

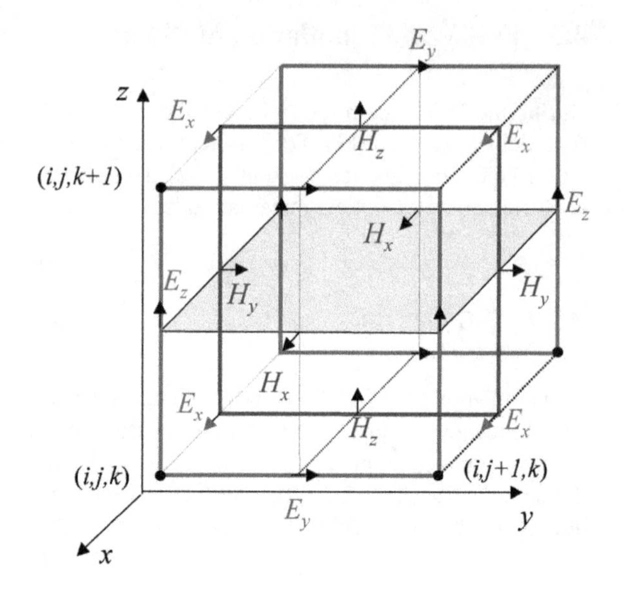

$$H_z\Big|_{i,j,k}^{n+1/2} = H_z\Big|_{i,j,k}^{n-1/2} - \frac{\Delta t}{\mu_{i,j,k}}\left(\frac{-E_y\Big|_{i+1,j,k}^{n} + E_y\Big|_{i,j,k}^{n}}{\Delta x} - \frac{E_x\Big|_{i,j+1,k}^{n} - E_x\Big|_{i,j,k}^{n}}{\Delta y}\right) \tag{4.3.5}$$

$$\begin{aligned} E_x\Big|_{i,j,k}^{n+1} &= \frac{\varepsilon_{i,j,k} - \sigma_{i,j,k}\Delta t/2}{\varepsilon_{i,j,k} + \sigma_{i,j,k}\Delta t/2} E_x\Big|_{i,j,k}^{n} \\ &- \frac{\Delta t}{\varepsilon_{i,j,k} + \sigma_{i,j,k}\Delta t/2}\left(\frac{H_z\Big|_{i,j,k}^{n+1/2} - H_z\Big|_{i,j-1,k}^{n+1/2}}{\Delta y} - \frac{H_y\Big|_{i,j,k}^{n+1/2} - H_y\Big|_{i,j,k-1}^{n+1/2}}{\Delta z}\right) \end{aligned} \tag{4.3.6}$$

$$\begin{aligned} E_y\Big|_{i,j,k}^{n+1} &= \frac{\varepsilon_{i,j,k} - \sigma_{i,j,k}\Delta t/2}{\varepsilon_{i,j,k} + \sigma_{i,j,k}\Delta t/2} E_y\Big|_{i,j,k}^{n} \\ &- \frac{\Delta t}{\varepsilon_{i,j,k} + \sigma_{i,j,k}\Delta t/2}\left(\frac{H_x\Big|_{i,j,k}^{n+1/2} - H_x\Big|_{i,j,k-1}^{n+1/2}}{\Delta z} - \frac{H_z\Big|_{i,j,k}^{n+1/2} - H_z\Big|_{i-1,j,k}^{n+1/2}}{\Delta x}\right) \end{aligned} \tag{4.3.7}$$

$$\begin{aligned} E_z\Big|_{i,j,k}^{n+1} &= \frac{\varepsilon_{i,j,k} - \sigma_{i,j,k}\Delta t/2}{\varepsilon_{i,j,k} + \sigma_{i,j,k}\Delta t/2} E_z\Big|_{i,j,k}^{n} \\ &- \frac{\Delta t}{\varepsilon_{i,j,k} + \sigma_{i,j,k}\Delta t/2}\left(\frac{H_y\Big|_{i,j,k}^{n+1/2} - H_y\Big|_{i-1,j,k}^{n+1/2}}{\Delta x} - \frac{H_x\Big|_{i,j,k}^{n+1/2} - H_x\Big|_{i,j-1,k}^{n+1/2}}{\Delta y}\right) \end{aligned} \tag{4.3.8}$$

where superscript n represents the numbers of time step, i, j, and k, respectively, are the lattices positions along the x- , y- , and z-directions, Δt is the time evolution step, and Δx, Δy, and Δz are the distance steps along the x-, y-, and z-directions. The calculations are interleaved in space and time; i.e., the new value of E_x is calculated from the previous value of E_x and the most recent values of H_y. Finally, a Fourier transform of these results yields the field magnitudes and phases at every point and every frequency.

Since the practical optical materials are dispersive, the dispersion phenomenon should be considered when the electromagnetic wave propagation is simulated by the computer. In order to ensure the calculation stability, the time step should satisfy [16]:

$$\Delta t \leq \frac{1}{c\sqrt{(\Delta x)^{-2} + (\Delta y)^{-2} + (\Delta z)^{-2}}} \tag{4.3.9}$$

For open problems such as radiation and scattering, the mesh space required is infinite, but in practical calculations, it is necessary to truncate it in a suitable place so that it can be computed in finite space. In 1994, Berenger [18] proposed a concept of a perfectly matched layer (PML). This artificially designed PML is composed of lossy conductive and magnetic conductive media and can absorb wave at any angle of incidence regardless of the frequency and polarization.

In addition, there is also a very important boundary condition in the simulation of photonic structures, where the electromagnetic field exhibits obvious periodicity. In this case, the electromagnetic field distribution of the whole space can be deduced from one unit cell. Therefore, using periodic boundary conditions will greatly shorten the simulation time and help to improve the accuracy of simulation.

Symmetric boundary conditions are used in the case that both the structure and source must be symmetric. In FDTD Solutions [19], symmetric boundaries are mirrors for the electric field, and anti-mirrors for the magnetic field. Symmetric boundary conditions are used when a problem exhibits one or more planes of symmetry. Asymmetric boundaries are anti-mirrors for the electric field, and mirrors for the magnetic field. A visual explanation of a symmetric boundary condition is shown in Fig. 4.8.

4.3.2 FEM

Compared with the FDTD method, the FEM is an inherently more complex and universal method. FEM is a numerical procedure to find stable solutions to boundary-value partial differential equations (PDE) [20]. It was not until 1969 that the method was introduced to the field of electromagnetic engineering [21].

FEM is a powerful simulation technique used to solve boundary-value problems in a variety of engineering circumstances, which has been widely used for analysis of electromagnetic fields in antennas, radar scattering, radio frequency (RF) and microwave engineering, high-speed/high-frequency circuits, wireless commu-

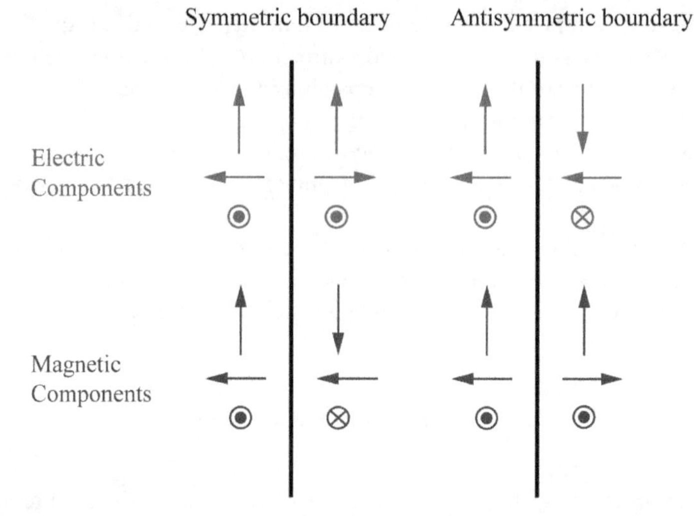

Fig. 4.8 Illustration of symmetric boundary and anti-symmetric boundary in FDTD Solutions

nication, electromagnetic compatibility, photonics, remote sensing, biomedical engineering, and space exploration [20].

In the FEM, the field of the continuous function represented by partial differential equations is divided into a finite number of small regions, each of which is replaced by a selected approximate function. As a result, the functions in the whole field are discretized, so a set of approximate algebraic equations can be obtained, and the approximate values of the functions in the field domain can be obtained by simultaneous solution [20].

One of the benefits of using the FEM is that it offers great freedom in the selection of discretization, both in the elements that may be used to discretize space and the basis functions. For linear functions in 2D and 3D, the most common elements are illustrated in Fig. 4.9. The basis functions are expressed as functions of the positions of the nodes (x and y in 2D and x, y, and z in 3D) [22].

4.4 Optimizing Algorithms

4.4.1 Holographic Algorithms

(1) Gerchberg–Saxton algorithm

The Gerchberg–Saxton (GS) algorithm [23] was proposed in 1972 by Gerchberg and Saxton to solve the problem of phase determination from image and diffraction plane

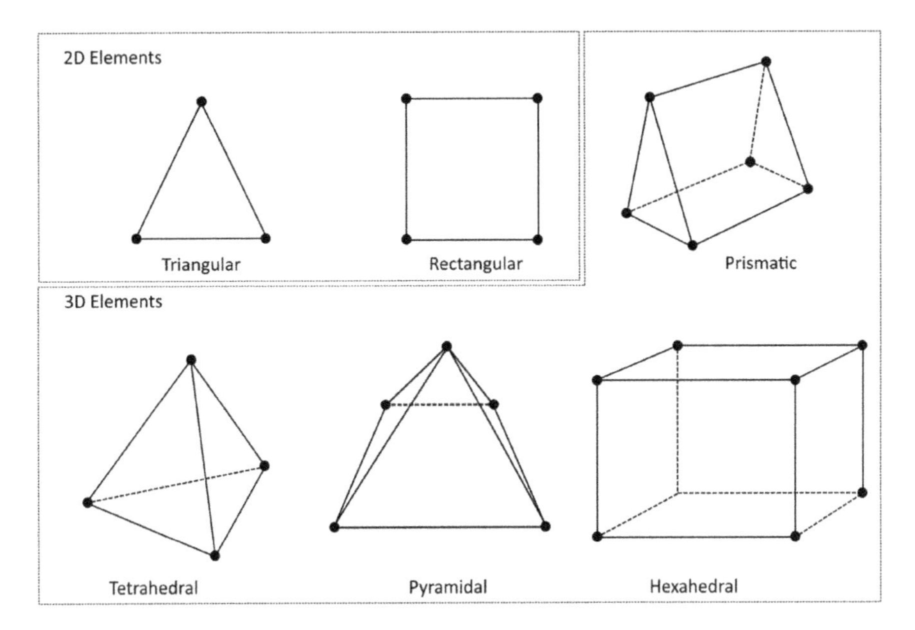

Fig. 4.9 Node placement and geometry for 2D and 3D linear elements

pictures. So far, it has also been extensively used to calculate phase-only diffractive optical elements (DOE), for which the diffraction pattern is known.

The algorithm consists of the following four steps [24], as indicated in Fig. 4.10.

- Take the Fourier transform of E_1 to propagate it to the second plane.
- Replace the amplitude of this second field, E_2, with the amplitude measured (or predefined) in the second plane. Keep the calculated phase map.

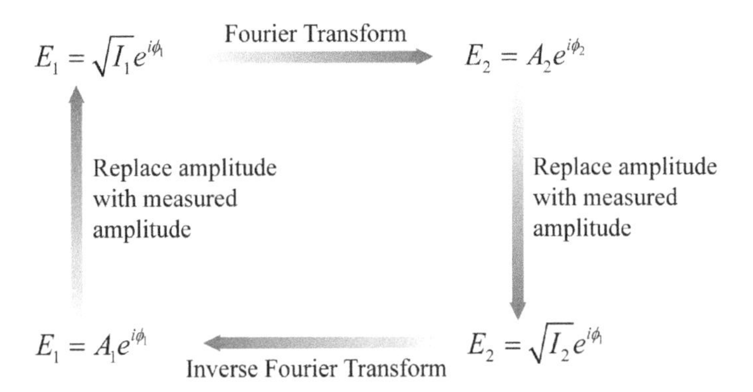

Fig. 4.10 Block diagram of the GS algorithm. I_1 and I_2 are the measured intensity. The initial estimate of ϕ_1 can be set to any phase map. With each iteration around the loop, the error between the estimated and measured amplitudes, A_2 and $\sqrt{I_2}$ or A_1 and $\sqrt{I_1}$, decreases

- Take the inverse Fourier transform of this estimate of E_2 to propagate it back to the first plane.
- Create the next estimate of E_1 by replacing the amplitude with the measured (or predefined) amplitude from the first intensity measurement. Keep the calculated phase map, and return to step 1.

After a number of iterations, this process will converge to an estimate of the actual phase map at both the first and second planes.

(2) Yang–Gu algorithm

In 1981, Yang and Gu [25, 26] put forward a general description of the amplitude-phase-retrieval problem for a unitary transform system. A set of equations for determining amplitude and phase were obtained based on a rigorous mathematical derivation. By using these equations and an iterative algorithm (Fig. 4.11), a variety of reconstruction problems can be handled.

The YG algorithm is the generalization of the original GS algorithm for including a non-unitary optical system [28]. When the transform kernel G is a unitary operator, the two algorithms are the same. An obvious difference between them can be found by comparing Fig. 4.10 with Fig. 4.11. The original GS algorithm was derived for the image reconstruction problem in systems involving a unitary transform G, whereas the YG algorithm may handle the phase-amplitude retrieval problem in a general system involving a non-unitary transform G as well.

(3) Point-source algorithm

Point-source algorithm takes the hologram as a summation of the spherical light waves of all the object points on the hologram plane. As indicated in Fig. 4.12, the spatial location of each object point is indexed by its discrete horizontal position "*m* " vertical position "*n*" and axial distance (depth) "*z*" from the hologram plane.

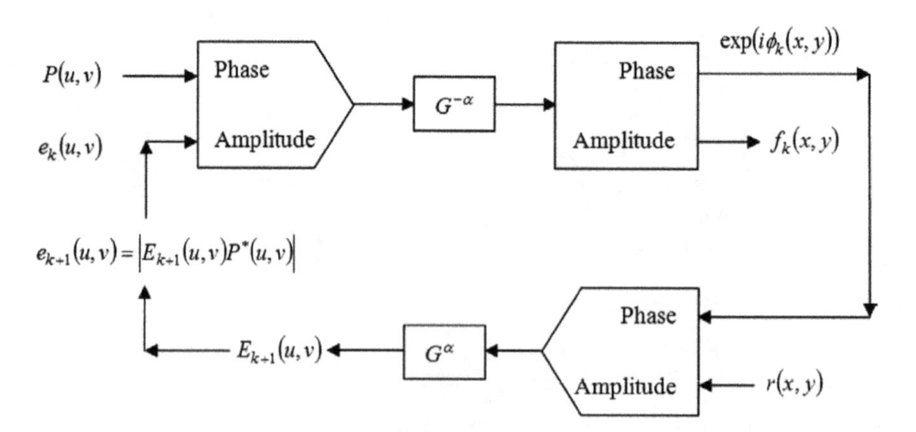

Fig. 4.11 Diagram of iterative process of Yang–Gu algorithm in gyrator domain. Reproduced from [27] with permission. Copyright 2015, Elsevier Ltd

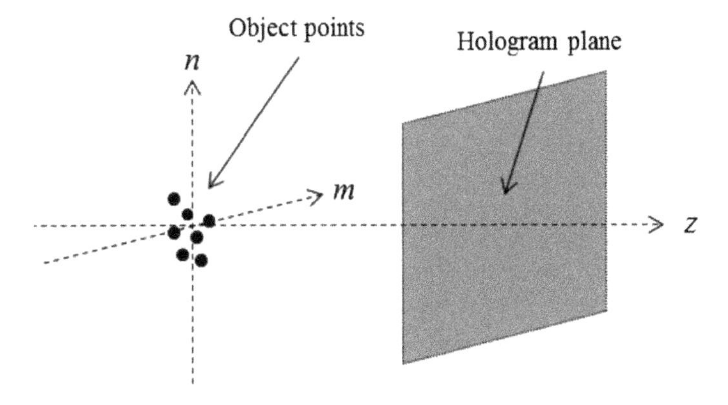

Fig. 4.12 Spatial relation between object points and the hologram plane. Reproduced from [29] with permission. Copyright 2018, Chinese Laser Press

Consequently, the optical wave on the hologram plane at an axial distance $z = z_0$ is given by the spatial impulse response of propagation [29]:

$$G(m, n; z_0) = \exp\left(\frac{i2\pi\sqrt{(m\delta)^2 + (n\delta)^2 + z_0^2}}{\lambda}\right) = \exp\left(i\omega_n\sqrt{m_x^2 + n_y^2 + z_0^2}\right) \tag{4.4.1}$$

where λ is the wavelength of light, δ is the sampling distance between adjacent pixels on the hologram, $m_x = m\delta$, and $n_y = n\delta$. Note that, the constant amplitude factor $1/i\lambda z_0$ in front of the exponential function is neglected. For an object with P object points, each locating at an axial distance z_p from the hologram, the diffracted wave on the hologram plane can be computed by the following equation,

$$H(m, n) = \sum_{p=0}^{P} A_p G(m - u_p, n - v_p, z_p) \tag{4.4.2}$$

where A_p and (u_p, v_p) are intensity and location of the pth object point.

4.4.2 Nature-Inspired Optimization Methods

For complex structures that possess a couple of design freedoms, it is not easy to obtain the optimal design. Some nature-inspired optimization methods have been developed and introduced to the electromagnetics community. In the following, we mainly focus on two optimization techniques.

(1) Genetic Algorithms

Genetic algorithm (GA) is an optimization program based on genetic inheritance and natural selection. It is inspired by Darwin's theory of natural evolution: Suitable organism survives in nature, and unsuitable organism dies out. The GA was originally introduced by Holland in 1975 [30] and later applied by Goldberg [31] to many practical problems. In the field of electromagnetics, GA has been used to solve a wide variety of problems, from antenna element design and phased array synthesis to scattering control of frequency selective surfaces and absorbers.

Figure 4.13 shows the basic workflow of the GA [32]. The evolution usually begins with a population of randomly generated individuals and is an iterative process in which the population in each iteration is called a generation. In each generation, fitness of each individual is evaluated; the fitness is usually the value of the objective function in the optimization problem to be solved. More suitable individuals are randomly selected from the current population, and the genome of each individual is modified (recombined and possibly stochastically mutated) to form a new generation. This new generation of candidate solutions is then used in the next iteration of the algorithm. Typically, the algorithm terminates when the maximum number of generations has been created or a satisfactory fitness level has been achieved.

In the GA implementation, the genes can be represented by either binary bits or by real numbers and are separately mapped to design parameters [32]. The cost is a measure of the design performance, where low cost suggests a high fitness. Fitness evaluation function is either complicated or simple and depends on optimization problem. For a particular chromosome, the fitness function returns a single value which denotes the merit of corresponding solution to the given problem. Since the algorithm maintains static population size, the chromosomes with low fitness value are eliminated, providing space for offspring with better fitness.

As indicated in Fig. 4.14, crossover is a genetic operator or a process that uses multiple parent solutions to generate an offspring solution [1, 32, 33]. After the crossover, the parents' genes are exchanged to create new offspring. A mutation is a genetic operator that selects the random position of a random chromosome and replaces the corresponding gene orbit with some other information. The probability of a mutation is determined before starting the algorithm, which determines the rate at which new gene values are propagated to the next generation. By using mutations and crossover operations, premature convergence including solutions falling into local optimums is prevented.

(2) Particle Swarm Optimization

The particle swarm optimization (PSO) algorithm was introduced in 1995 by Kennedy and Eberhart [34]. Subsequently, Poli [35] conducted an extensive survey of PSO applications. Recently, Bonyadi and Michalewicz [36] published a comprehensive review of PSO theory and experimental work.

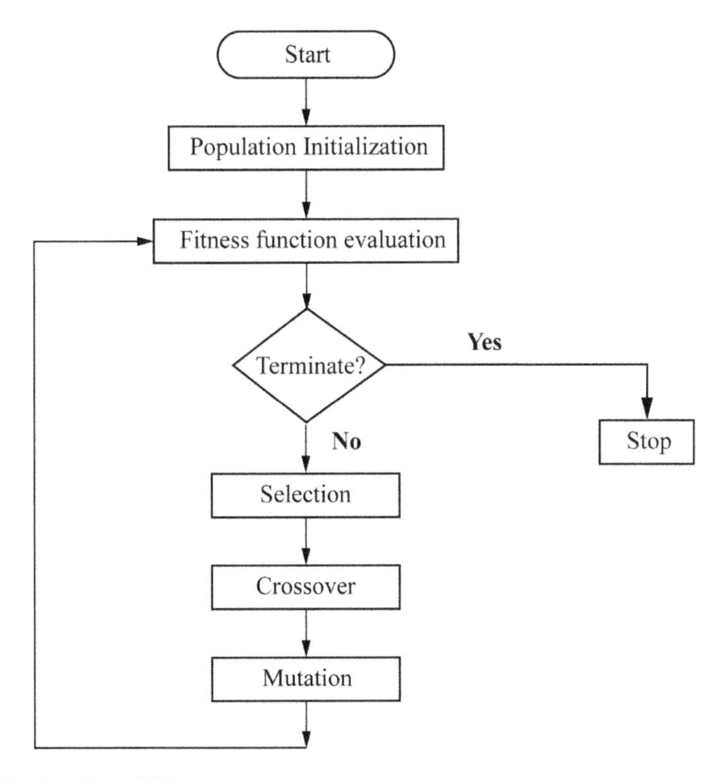

Fig. 4.13 Flowchart of GA

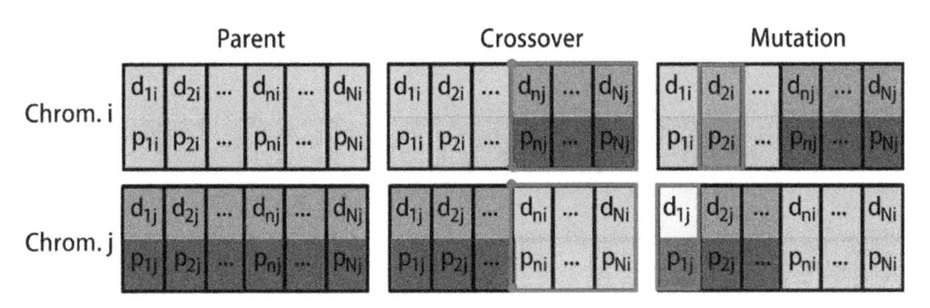

Fig. 4.14 Sketch of crossover and mutation operation between two parent chromosomes. Reproduced from [33] with permission. Copyright 2015, Macmillan Publishers Limited

A basic variant of the PSO algorithm works by a group (called a swarm) of candidate solutions (called particles) (Fig. 4.15). These particles move through the search space according to a few simple formulas. The motion of the particles is guided by their own best-known positions in the search space and the best-known position of the entire swarm. These will guide the movement of the group when an improved position is found. This process is repeated, and by doing so, it is hoped but not guaranteed that a satisfactory solution will eventually be found.

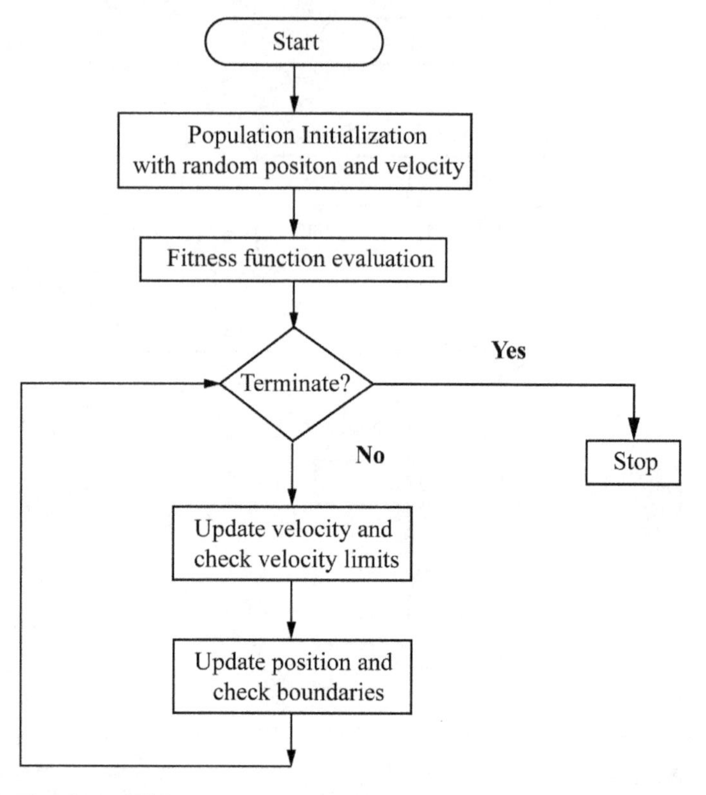

Fig. 4.15 Flowchart of PSO

4.4.3 Other Optimizing Algorithms

(1) Hill climbing

In numerical analysis, hill climbing [37] is a mathematical optimization technique that belongs to the family of local search. It is an iterative algorithm. It starts with any solution to the problem and then tries to find a better solution by gradually changing the solution. If the change produces a better solution, another incremental change is made to the new solution until no further improvements can be found. Hill climbing is a ready-to-use algorithm: Even if it is interrupted at any time before the end, it can return a valid solution.

Hill climbing can find the optimal solutions to the convex problems. For other problems, it can only find local optimums (solutions that cannot be improved by any adjacent configurations). These solutions are not necessarily the best solution (global optimal solution) of all possible solutions. In order to avoid falling into local optimum, one could use restarts (i.e., repeated local search), or more complex schemes based on iterations (like iterated local search), or on memory-less stochastic modifications (like simulated annealing).

(2) Simulated annealing

Simulated annealing (SA) [38] is a probabilistic technique used to approximate the global optimum of a given function. Particularly, it is a meta-heuristic approach that approximates global optimization in a large search space. Simulated annealing may be superior to alternatives such as gradient descent for problems where finding an approximate global optimal value is more important than finding an accurate local optimum in a fixed time.

Typically, the simulated annealing algorithm works as follows. At each time step, the algorithm randomly selects a solution close to the current solution and measures its quality. Then, it decides to move to it or maintain the current solution based on one of the two probabilities it chooses. In fact, it chooses between the two probabilities based on the fact that the new solution is better or worse than the existing solution. During the search, the temperature gradually decreases from the initial positive value to zero and affects two probabilities. At each step, the probability of moving to a better new solution either remains at 1 or changes to a positive value between 0 and 1; instead, the probability of moving to a worse new solution gradually becomes 0. These probabilities ultimately lead to a system turning to a lower energy state. Typically, this step is repeated until the system reaches a state sufficient to satisfy the application, or until a given computational budget is exhausted.

4.5 Intelligent Design and All-Optical Implementation

4.5.1 Intelligent Design Models

Except for very particular cases, most complex structures cannot be modeled analytically, so the analysis of a single design variation entails intensive numerical electromagnetic simulations. Moreover, the best performance will typically be achieved by combining several of the above approaches, resulting in a very large optimization space. Consequently, today, the design process is left to trial-and-error optimization procedures that are based on heavy electromagnetic simulations. Such procedures are not only time-consuming, but are also bound to yield sub-optimal performance or unnecessarily complex structures [32].

During the past few years, there is an increasing need to develop efficient methods that expedite the discovery and design of novel metasurface structures with custom-defined functionality. An effective method to arrive this goal is resorting deep neural networks. It should be noted that although certain progresses have been made, conventional neural network schema, like back-propagation, are impotent in the inverse design of metasurfaces, due to the enormous degrees of freedom in typical metasurface patterns. To mitigate these challenges, many new concepts have been proposed.

For instance, a generative adversarial network (GAN) is proposed [39]. The network architecture, illustrated in Fig. 4.16, includes three parts: a simulator (S), a generator (G), and a critic (D).

In the proof-of-concept design, the primary goal is to train an overfitted generator, which produces metasurface patterns in response to given input spectra T and noise z such that the Euclidean distance between the spectra of the generated pattern T' and the input spectra T is minimized [39]. All three networks are convolutional neural networks with delicate differences in detailed structures. The simulator is a pretrained model with fixed weights, taking the generated patterns as input and approximating their transmittances spectra without the use of electromagnetic simulations. The critic of this GAN accepts both the user-defined geometric data and the patterns generated from the generator and then yields a value l which is essential to compute the distance between the distributions of the two sets of data. By minimizing this distance, the critic network guides the generator to produce patterns that share common features with the input geometric data. During the training process, the weights in the generator are updated by back-propagation from the losses defined by the simulator and the critic. Whenever the losses of the simulator and critic are sufficiently small, valid patterns produced by the generator are documented. The generated patterns are finally smoothed to binary images as candidates of the metasurface design.

Furthermore, a deep-learning-based model to automatically design and optimize three-dimensional chiral metamaterials with strong chiroptical responses at predesigned wavelengths was reported [40]. This model comprises two bidirectional neural networks assembled by a partial stacking strategy, as schematically depicted in Fig. 4.17. The dataflow in the deep neural network is denoted by the yellow arrows, where all the internal nodes (design parameters, reflection spectra, and CD spectra,

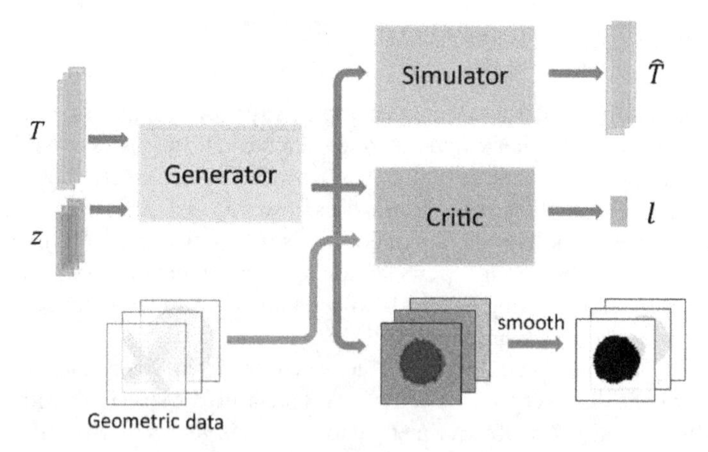

Fig. 4.16 Architecture of the network for AI-based optical design. Three networks, the generator (G), the simulator (S), and the critic (D) constitute the complete architecture. Adapted from [39] with permission. Copyright 2018, American Chemical Society

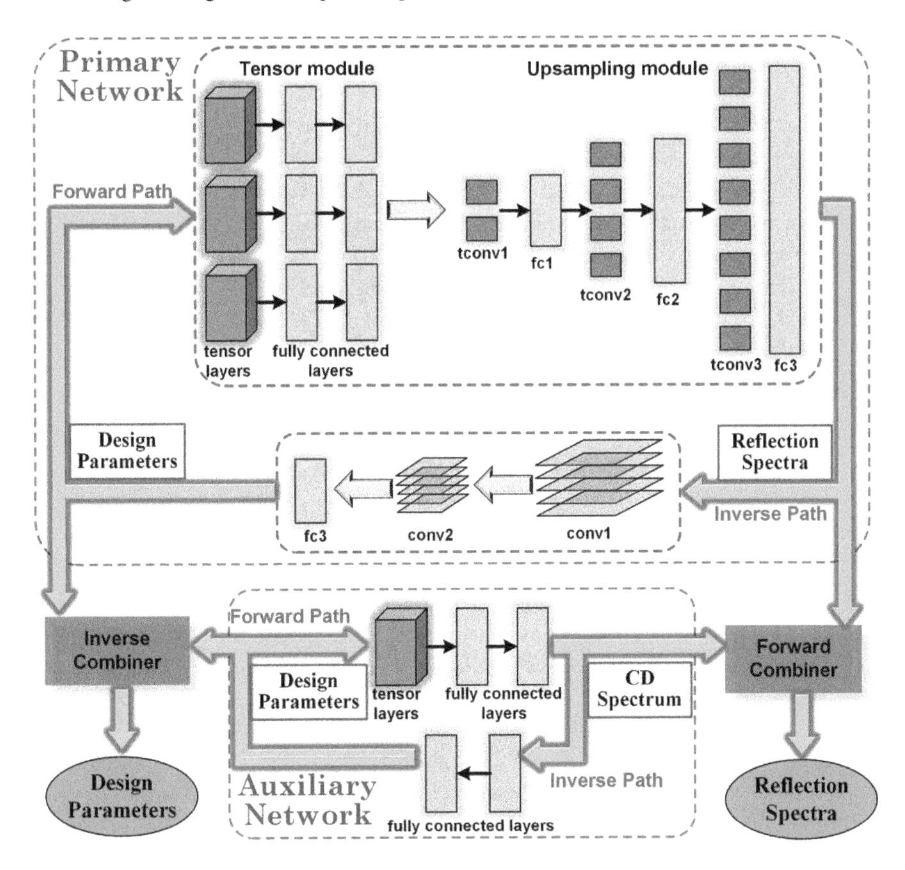

Fig. 4.17 Structure of the deep-learning model for designing chiral metamaterials. Reflection spectra, CD spectra, and design parameters are interconnected in the model (yellow arrows) and can be treated as either input or output at specific ports (fc, fully connected layer; conv, convolutional layer; tconv, transposed convolutional layer). Reproduced from [40] with permission. Copyright 2018, American Chemical Society

highlighted by the purple box) can be treated as either input or output nodes due to the nature of bidirectional mapping. The output of the tensor layer is given by:

$$\text{output}_{\text{tensor}} = f\left(D^T W_k D + V_k D + B\right) \tag{4.5.1}$$

where f is the rectified linear unit activation function, D is the row vector of five design parameters, k is the output vector dimension, N is the number of variables, W_k is a $k \times N \times N$ tensor, V_k is a $k \times N$ weight matrix, and B is a $k \times 1$ bias vector.

4.5.2 All-Optical Implementation

Benefiting from the improvements in training algorithms, network architectures, and computational powers, more complex problems can be solved by deep-learning algorithms. However, today's computing hardware is inefficient at implementing neural networks, in large part because much of it was designed for von Neumann computing schemes. New hardware platforms including graphical processing units (GPUs), application-specific integrated circuits (ASICs), and field-programmable gate arrays (FPGAs) have been utilized to improve the energy and processing efficiency. Also, all-optical implementation is a promising direction.

(1) Nanophotonic circuit-based deep neural networks

Recently, a fully optical neural network (ONN) has been proposed [41], which in principle could offer an enhancement in computational speed and power efficiency over state-of-the-art electronics for conventional inference tasks. The ONN architecture is depicted in Fig. 4.18a. Each layer of the ONN is composed of an optical interference unit (OIU) that implements optical matrix multiplication and an optical non-linearity unit (ONU) that implements the non-linear activation. In principle, the ONN can implement artificial neural networks (ANN) of arbitrary depth and dimensions fully in the optical domain. The ONU can be implemented using common optical non-linearities such as saturable absorption and bistability, which have all been demonstrated previously in photonic circuits. A schematic diagram of the proposed fully optical neural network is shown in Fig. 4.18b. The OIU was implemented using a programmable nanophotonic processor (PNP), which was composed of 56 programmable Mach–Zehnder interferometers (MZIs), each of which comprises a thermo-optic phase shifter (θ) between two 50% evanescent directional couplers, followed by another phase shifter (ϕ). The proposed architecture could be applied to other ANN algorithms where matrix multiplications and non-linear activations are heavily used, including convolutional neural networks (CNNs) and recurrent neural networks (RNNs).

(2) Diffractive deep neural networks

Another all-optical deep-learning framework is physically formed by multiple layers of diffractive surfaces that work in collaboration to optically perform an arbitrary function, which is termed as a diffractive deep neural network (D^2NN) [42]. Each point on a given layer either transmits or reflects the incoming wave, representing an artificial neuron that is connected to other neurons of the following layers through optical diffraction.

In accordance with the Huygens–Fresnel principle, each point on a given layer acts as a secondary source of a wave, the amplitude and phase of which are determined by the product of the input wave and the complex-valued transmission or reflection coefficient at that point. For coherent transmissive networks with phase-only modulation, each layer can be approximated as a thin optical element [43]. Through deep learning, the phase values of the neurons of each layer of the diffractive network are iteratively adjusted (trained) to perform a specific function by feeding training data at

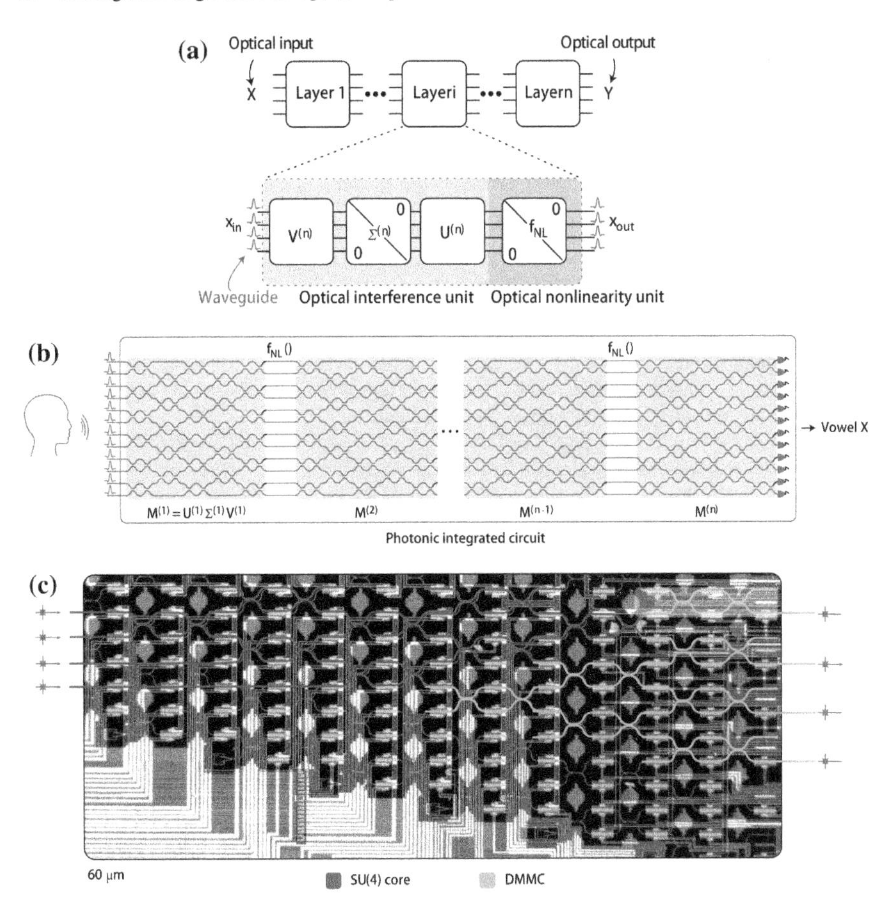

Fig. 4.18 **a** Upper: decomposition of the general neural network into individual layers. Below: optical interference and non-linearity units that compose each layer of the artificial neural network. **b** Proposal for an all-optical, fully integrated neural network. **c** The transmission curve for tuning the internal phase shifter. DMMC, diagonal matrix multiplication core. Adapted from [41] with permission. Copyright 2017, Macmillan Publishers Limited, part of Springer Nature. All rights reserved

the input layer and then computing the network's output through optical diffraction. On the basis of the calculated error with respect to the target output and determined by the desired function, the network structure and its neuron phase values are optimized via an error back-propagation algorithm, which is based on the stochastic gradient descent approach used in conventional deep learning.

References

1. B. Choudhury, *Metamaterial Inspired Electromagnetic Applications* (Springer, NY, 2017)
2. S. Molesky, Z. Lin, A.Y. Piggott, W. Jin, J. Vucković, A.W. Rodriguez, Inverse design in nanophotonics. Nat. Photon. **12**, 659–670 (2018)
3. Y. Guo, M. Pu, Z. Zhao, Y. Wang, J. Jin, P. Gao, X. Li, X. Ma, X. Luo, Merging geometric phase and plasmon retardation phase in continuously shaped metasurfaces for arbitrary orbital angular momentum generation. ACS Photon. **3**, 2022–2029 (2016)
4. X. Xie, X. Li, M. Pu, X. Ma, K. Liu, Y. Guo, X. Luo, Plasmonic metasurfaces for simultaneous thermal infrared invisibility and holographic illusion. Adv. Funct. Mater. **28**, 1706673 (2018)
5. Y. Guo, L. Yan, W. Pan, B. Luo, Generation and manipulation of orbital angular momentum by all-dielectric metasurfaces. Plasmonics **11**, 337–344 (2016)
6. Y. Wang, M. Pu, C. Hu, Z. Zhao, C. Wang, X. Luo, Dynamic manipulation of polarization states using anisotropic meta-surface. Opt. Commun. **319**, 14–16 (2014)
7. M. Pu, X. Ma, X. Li, Y. Guo, X. Luo, Merging plasmonics and metamaterials by two-dimensional subwavelength structures. J. Mater. Chem. C **5**, 4361–4378 (2017)
8. M. Pu, C. Hu, M. Wang, C. Huang, Z. Zhao, C. Wang, Q. Feng, X. Luo, Design principles for infrared wide-angle perfect absorber based on plasmonic structure. Opt. Express **19**, 17413–17420 (2011)
9. M. Pu, C. Hu, C. Huang, C. Wang, Z. Zhao, Y. Wang, X. Luo, Investigation of Fano resonance in planar metamaterial with perturbed periodicity. Opt. Express **21**, 992–1001 (2013)
10. Q. Feng, M. Pu, C. Hu, X. Luo, Engineering the dispersion of metamaterial surface for broadband infrared absorption. Opt. Lett. **37**, 2133–2135 (2012)
11. Y. Huang, J. Luo, M. Pu, Y. Guo, Z. Zhao, X. Ma, X. Li, X. Luo, Catenary electromagnetics for ultrabroadband lightweight absorbers and large-scale flat antennas. Adv. Sci. 1801691 (2018)
12. X. Luo, D. Tsai, M. Gu, M. Hong, Extraordinary optical fields in nanostructures: from sub-diffraction-limited optics to sensing and energy conversion. Chem. Soc. Rev. (2019)
13. H.E. Bussey, J.H. Richmond, Scattering by a lossy dielectric circular cylindrical multilayer, numerical values. IEEE Trans. Antennas Propag. **23**, 723–725 (1975)
14. E.F. Knott, K. Langseth, Performance degradation of Jaumann absorbers due to curvature. IEEE Trans. Antennas Propag. **28**, 137–139 (1980)
15. M.G. Moharam, T.K. Gaylord, Rigorous coupled-wave analysis of planar-grating diffraction. J. Opt. Soc. Am. A **71**, 811–818 (1981)
16. F. Capolino, *Theory and Phenomena of Metamaterials* (CRC Press, Boca Raton, 2009)
17. K.S. Yee, Numerical solution of initial boundary value problems involving Maxwell's equations in isotropic media. IEEE Trans. Antennas Propag. **14**, 302–307 (1966)
18. J.-P. Berenger, A perfectly matched layer for the absorption of electromagnetic waves. J. Comput. Phys. **114**, 185–200 (1994)
19. FDTD Solutions. https://www.lumerical.com/
20. https://en.wikipedia.org/wiki/Finite_element_method
21. P. Silvester, High-order polynomial triangular finite elements for potential problems. Int. J. Eng. Sci. **7**, 849–861 (1969)
22. Comsol_FEM. http://uk.comsol.com/multiphysics/finite-element-method
23. R.W. Gerchberg, W.O. Saxton, A practical algorithm for the determination of phase from image and diffraction plane pictures. Optik **35**, 237–250 (1972)
24. J. Kay, N.J. Kasdin, R. Belikov, Wavefront correction in a shaped-pupil coronagraph using a Gerchberg-Saxton-based estimation scheme. Proc. SPIE 6691, Astronomical Adaptive Optics Systems and Applications III, 66910D (2007)
25. B. Gu, G. Yang, On the phase retrieval problem in optical and electronic microscopy. Acta. Opt. Sin. **1**, 517–552 (1981)
26. G. Yang, B. Gu, On the amplitude-phase retrieval problem in optical systems. Acta. Phys. Sin. **30**, 410–413 (1981)

27. L. Sui, B. Liu, Q. Wang, Y. Li, J. Liang, Color image encryption by using Yang-Gu mixture amplitude-phase retrieval algorithm in gyrator transform domain and two-dimensional sine logistic modulation map. Opt. Lasers Eng. **75**, 17–26 (2015)
28. G. Yang, B. Dong, B. Gu, J. Zhuang, O.K. Ersoy, Gerchberg-Saxton and Yang-Gu algorithms for phase retrieval in a nonunitary transform system: a comparison. Appl. Opt. **33**, 209–218 (1994)
29. P.W.M. Tsang, T.-C. Poon, Y.M. Wu, Review of fast methods for point-based computer-generated holography. Photon. Res. **6**, 837–846 (2018)
30. J.H. Holland, On quantifying agricultural and water management practices from low spatial resolution RS data using genetic algorithms: a numerical study for mixed pixel environment. Adv. Water Resour. **28**, 856–870 (1975)
31. D.S. Weile, E. Michielssen, D.E. Goldberg, Genetic algorithm design of Pareto optimal broad-band microwave absorbers. IEEE Trans. Electromagn. Compat. **38**, 518–525 (1996)
32. K. Diest, *Numerical Methods for Metamaterial Design* (Springer, NY, 2013)
33. K. Huang, H. Liu, F.J. Garcia-Vidal, M. Hong, B. Luk'yanchuk, J. Teng, C.-W. Qiu, Ultrahigh-capacity non-periodic photon sieves operating in visible light. Nat. Commun. **6**, 7059 (2015)
34. J. Kennedy, R. Eberhart, *Particle Swarm Optimization,* in *Proceedings of IEEE International Conference on Neural Networks (ICNN'95)*, (1995)
35. R. Poli, J. Kennedy, T. Blackwell, Particle swarm optimization. Swarm Intell. **1**, 33–57 (2007)
36. M.R. Bonyadi, Z. Michalewicz, Particle swarm optimization for single objective continuous space problems: a review. Evol. Comput. **25**, 1–54 (2016)
37. Hill climbing. https://en.wikipedia.org/wiki/Hill_climbing
38. Simulated annealing. https://en.wikipedia.org/wiki/Simulated_annealing
39. Z. Liu, D. Zhu, S.P. Rodrigues, K.-T. Lee, W. Cai, A generative model for the inverse design of metasurfaces. Nano Lett. **18**, 6570–6576 (2018)
40. W. Ma, F. Cheng, Y. Liu, Deep-learning-enabled on-demand design of chiral metamaterials. ACS Nano **12**, 6326–6334 (2018)
41. Y. Shen, N.C. Harris, S. Skirlo, M. Prabhu, T. Baehr-Jones, M. Hochberg, X. Sun, S. Zhao, H. Larochelle, D. Englund, M. Soljačić, Deep learning with coherent nanophotonic circuits. Nat. Photonics **11**, 441–446 (2017)
42. X. Lin, Y. Rivenson, N.T. Yardimci, M. Veli, Y. Luo, M. Jarrahi, A. Ozcan, All-optical machine learning using diffractive deep neural networks. Science **361**, 1004–1008 (2018)
43. S. Wang, X. Ouyang, Z. Feng, Y. Cao, M. Gu, and X. Li, Diffractive photonic applications mediated by laser reduced graphene oxides. Opto-Electron. Adv. **1**, 170002 (2018)

Chapter 5
Fabrication Techniques

Abstract Different from well-established and highly refined fabrication processes in EO 1.0, the fabrication techniques in EO 2.0 are still imperfect, which need to be carefully investigated to form systematic processing methods. In this chapter, we first introduce the status of manufacturing techniques in EO 1.0, including the fabrication of refractive, reflective, and diffractive optical elements. The challenges of the manufacturing techniques for EO 1.0 are also summarized. Then, we will introduce the progresses of fabrication techniques in EO 2.0, such as the layered fabrication techniques, direct-writing techniques, and subwavelength structures fabrication techniques. The principles and implementations of these methods will be stated in detail. Some technological challenges in EO 2.0 are also discussed, including large-aperture manufacturing, conformal flexible manufacturing, and super-molecular and super-atom manufacturing.

Keywords Micro-/nanofabrication · Optical fabrication

5.1 Status and Challenges of Manufacturing Techniques for EO 1.0

EO 1.0 involves traditional reflective optics, refractive optics, and diffractive optics. Over the past decades, manufacturing in EO 1.0 is mainly based on conventional optical manufacturing and micro-optical fabrication. Conventional optical manufacturing can process planar, spherical, aspheric, freeform optical components. Micro-optical fabrication, which evolves from semiconductor fabrication technology, is the basic fabrication method for diffractive elements. In this section, we will introduce these two categories of methods in detail.

X. Luo, *Engineering Optics 2.0*, https://doi.org/10.1007/978-981-13-5755-8_5

5.1.1 Manufacture of Refractive and Reflective Optical Elements

The basic processes of the conventional optical manufacturing contain glass blank manufacturing, generation, blocking and grinding, polishing, and surface finishing, as shown in Fig. 5.1. The glass blank is the basic material to form the final lens, which can be acquired by batch mixing, pouring and casting, annealing, physical characterization and measurements of refraction index.

The goal of generation process is to generate the shape of the lens from the blank and get close to its final size and curvature. Loose abrasive grinding and diamond cutting are two commonly used methods for the generation process. Loose abrasive grinding is an archaic process, but it is still used today [1]. It results in a rough surface because the removal rate of loose abrasive grinding is fast. Different from loose abrasive grinding, diamond cutting is a latter method which can program the removal function on demand.

The following step is blocking, in which the lenses are mounted onto a convex or concave surface with pitch or wax [1]. And then by using inverse spheres with the same radii ground together for grinding, the hills and valleys on the lens surface will wear away.

After that, the lens may go through several stages of polishing depending on how precise the generation is [1]. The polishing process combines the mechanical and chemical effects to remove subsurface damage and reduce surface roughness. However, it is difficult or time-consuming to reach the specified surface accuracy because of tool wear, edge roll-off effect, force loading of workpiece, etc. The most

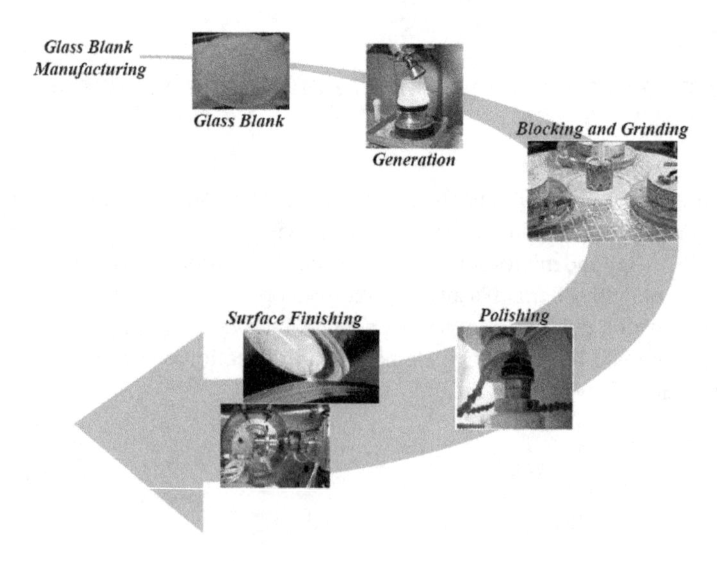

Fig. 5.1 A typical process flow of conventional optical manufacturing

conventional polishing method is pitch polishing. The shape deviation (PV) and surface roughness (RMS) can be controlled to 300 nm and 0.2–0.5 nm, respectively. But the correction of local surface deviation is limited. For higher production parts, high-speed computer numerical control (CNC) machines can also be used.

For high-quality surface, especially in precision optics, the requirement of surface finishing becomes more and more stringent. The surface finishing process is a necessarily additional step to deterministically correct the surface contour to its desired accuracy. In this process, the typical machining methods include magnetorheological finishing (MRF) and ion-beam figuring (IBF), and so on [2].

MRF is a precise surface finishing technology that is commonly used in the production of high-quality optical lenses. MRF process can make significant improvements in surface roughness, and has been adopted by major manufacturers of precision optics. In the aspect of form accuracy and micro-roughness, the MRF process has the ability to produce optical surfaces to tight tolerances. The planar and aspheric surfaces can be made with materials ranging from glass (including fused silica, ultra-low thermal expansion (ULE) quartz glass) to single-crystalline materials (including silicone and calcium fluoride) or polycrystalline materials (including SiC). The PV and RMS of the lens can be controlled to 10 nm and 0.3 nm, respectively.

IBF is a non-abrasive technique for finely correcting the contour of precision optics. This figuring process utilizes a Kaufman-type ion source where plasma is generated in a discharge chamber. In IBF process, a non-varied ion-beam energy maintains a constant sputtering rate (or material removal rate) and profile which is very important for optical deterministic figuring method to gain a constant beam removal function. Then, this temporally and spatially stable ion beam is held perpendicular to the optical surface at a fixed distance with ion source controlled by a 5-axis CNC system. So, the IBF has the best machining precision with the PV and RMS being 5 nm and 0.2 nm, respectively. But the main limitation of IBF is its low removal rate.

5.1.2 Manufacture of Diffractive Optical Elements

Typical diffractive optical elements (DOEs) have multi-level micro-reliefs or continuous micro-reliefs, with features ranging from submicron to millimeter dimensions and relief amplitudes of a few microns. Novel structures can be realized, complementing and exceeding the functionalities of traditional lenses, prisms, and mirrors.

5.1.2.1 VLSI Technology

Compared with traditional optical elements, the curved surface of the micro-lenses is more difficult to define as the dimension shrink. To solve this problem, the shape of a continuous-relief structure may be approximated by a stepped profile known as binary optical elements, which can be fabricated by discrete optical lithography steps. Figure 5.2 gives an example of continuous Fresnel-like lens (Fig. 5.2a) that can be approximated using a multi-level concentric structure (Fig. 5.2b) [3].

Ideally, the efficiency tends to 100% when the profile is close to continuous. But the number of phase levels is fundamentally limited by the resolution of the lithography. The maximum number of phase levels, P_{max}, can be expressed as [3]:

$$P_{max} = \frac{\lambda}{\sin\theta \cdot l_{min}} = 2^N \tag{5.1.1}$$

where λ is the wavelength, θ is the first-order diffraction angle, l_{min} is the minimum feature size that can be fabricated with the available technology, and N is the number of required lithography steps. For a given number of phase levels, the maximum diffraction efficiency profile as a function of the first-order diffraction angle, $\eta(\theta)$, is given by [3]:

$$\eta(\theta) = \left(\frac{\sin[(\pi l_{min} \sin\theta)/\lambda]}{(\pi l_{min} \sin\theta)/\lambda} \right)^2 \tag{5.1.2}$$

Then by integrating $\eta(\theta)$ over the angle, the maximum efficiency of a diffractive lens can be estimated.

Except for the efficiency, the feature height of the micro-optical structures and lateral position of zone are critical parameters. In order to generate a phase shift of 2π, the depth of the micro-optical structures with refractive index n_h has the maximum value of [3]:

$$h_{max} = \frac{\lambda}{2\pi(n_h - 1)} \tag{5.1.3}$$

where λ is the vacuum wavelength. h_{max} defines the zone depth of a Fresnel-like lens, such that, for P_{max} phase levels, the depth of each phase step is given by [3]:

$$h_{step} = \frac{h_{max}}{P_{max}} = \frac{\lambda}{2\pi P_{max}(n_h - 1)} \tag{5.1.4}$$

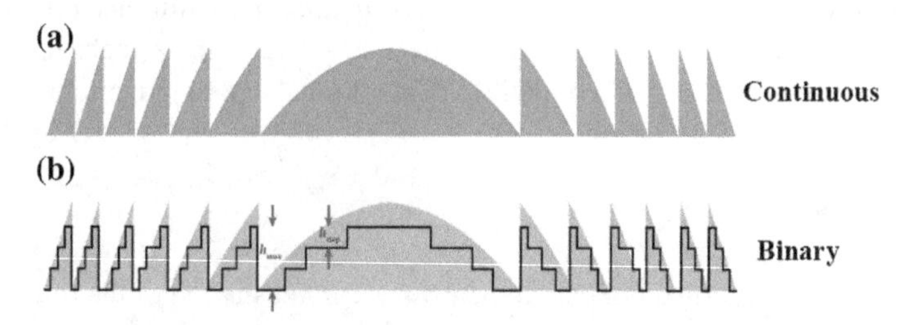

Fig. 5.2 Binary approximation of a continuous Fresnel lens profile (**a**) using a multi-level center structure (**b**)

For a Fresnel-like lens, the lateral position of zone m is given by [3]:

$$x_m = \sqrt{\frac{2\pi\lambda f}{n_h}} \tag{5.1.5}$$

where f is the focal length. So that the space between two adjacent zones is given by [3]:

$$\Delta x = x_{m+1} - x_m = \sqrt{\frac{2\lambda f}{n_h}}(\sqrt{m+1} - \sqrt{m}) \tag{5.1.6}$$

where Δx defines the smallest required mask dimension, which decreases as m increases. Thus, lateral tolerances provide a limitation of the size and number of zones.

Figure 5.3 shows a typical binary optics process to generate an eight-phase level Fresnel-like lens by using the very large-scale integration (VLSI) technology. Three masks are used in this case. Each mask has a pattern roughly twice as large as the previous one, so that half of the features generated are transferred to a lower level by the subsequent etch step.

VLSI technology is one of the basic process in binary optics, which includes substrate preparing, deposition, photoresist coating, soft bake/prebake, alignment, exposure, post-exposure bake, etching, and post-processing. Typical processes may consist of many sequential steps [4].

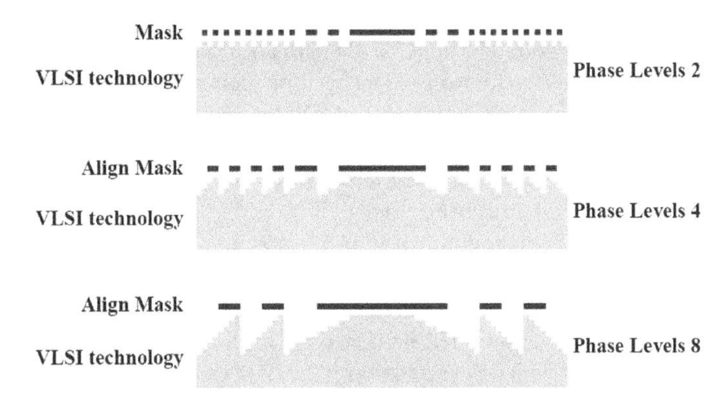

Fig. 5.3 Typical processes to generate an eight-phase level Fresnel-like lens using the VLSI techniques of binary optics. Three steps, using three masks (black) are shown

- Substrates Preparing

The standard starting substrate for VLSI process usually is the silicon wafer, with diameters ranging from 1 to 12 in. Fused silica (SiO_2), polymer, and other materials are also employed in micro-optical fabrication.

Many fabrication failures or low yields may be attributed to inadequate cleaning. So, cleanliness is the utmost important step in substrates preparing. Process is usually performed in an isolated environment of clean rooms, and the substrates are repeatedly cleaned using chemical methods. Except normal chemical cleaning process to remove the stains and particles from substrate surface, the substrate has to be thoroughly dried so that photoresist or film can well adhere to the surface.

- Photoresist Coating

There are many types of photoresists, and each photoresist is specially designed to suit a specific application. Generally, photoresist contains the following four components: resin, solvent, photoactive compound (PAC), and additives. By the way of promoting the adhesion between photoresist and substrate surface, a thin hexamethyldisilazane (HMDS) layer is coated on it. During the spin coating process, a precise amount of liquid photoresist is dropped at the center of the substrate. By high-speed spinning, the photoresist drops spread to the whole surface uniformly. With the solvent evaporation, the film thickness reaches the desired value. Commonly, the thickness of photoresist can be controlled by the spin speed. Additionally, the flow and rheological properties of the resist do affect the coating process and need to be considered for optimal results.

- Soft Bake/Prebake

After spin coating the resist on the substrate, the next processing step is soft bake, which is also called prebake. The purpose of baking is to densify the film and remove residual solvent, so that the photoresist layer becomes completely dry and solid. Other consequences of soft baking include reduced free volume and polymer slack, which have been considered useful for improving resist processing performance. Soft bake also improves the adhesion of the resist to the substrate, promotes resist uniformity on the substrate, and facilitates better line width control during etching.

A typical soft bake temperature on a hot plate is 90–120 °C for 60 s, and then a cooling step is performed on the cooling plate to achieve uniform substrate temperature control. Temperature control is very important for the soft baking process, especially for chemically amplified photoresists. In order to achieve the stringent temperature uniformity which is necessary for the tight critical dimension (CD) control for advanced lithography, careful consideration of the vacuum hotplate design including uniform substrate-to-hot-plate contact, airflow, thermal management, and contamination control, is required.

- Alignment

Once the resist-coated substrate is soft baked and cooled, the substrate is then sent into the wafer stage of the exposure tool. The substrate is raised or lowered within the focal length of the optical system, and then aligned with the pattern on the reticle/mask to ensure that the pattern can be transferred to the appropriate location on the surface of the resist-coated substrate. It is absolutely necessary to ensure each new pattern is placed in the correct position at the top of the previous layer. Specifically, alignment can be defined as the process of determining the position, orientation, and distortion of select patterns already on the wafer and then placing them in correct relation to the projected image from the reticle/mask. Alignment should be fast, repeatable, accurate, and precise. The outcome of the alignment process, or how accurately each successive pattern is matched to the previous layer, is known as overlay, which is a critical factor in determining if the final device will function properly or not.

- Exposure

Exposure is the following step after precise alignment. Exposure radiation illuminates the mask and projects a transmitted or reflected image of the mask onto the photoresist. The images will cause a spatial distribution in the photoresist where the photochemistry happens.

Ordinarily, mask aligners and i-line stepper are the main tools for micro-optical fabrication. Mask aligners are low-cost systems that can accommodate substrates having non-standard thicknesses, diameters, and geometries encountered in micro-optical fabrication. The overlay and resolution of mask aligner can achieve $\leq \pm 0.5$ μm and ≤ 0.8 μm, respectively, with vacuum contact. However, particulate contamination can compromise mask life and introduce local distortion. For some applications where the resolution is not critical, proximity exposure is acceptable, with limited resolution of ≥ 2 μm which depends on the gap between the substrate and the mask. I-line steppers are high-resolution and high-throughput optical lithography machines with 2× to 5× reduction ratio. So, the positioning platform and alignment algorithms of i-line stepper are more advanced than mask aligners, which can achieve overlay accuracy with less than 20 nm across the wafer. Choosing an appropriate exposure dose is of a great consequence to exposure.

- Post-exposure Bake

The baking of the exposed but undeveloped resist film is referred to as post-exposure bake (PEB). For diazonaphthoquinone (DNQ) or novolak photoresist, the role of PEB is to smooth the standing wave by thermally induced diffusion of photoacids, which is caused by the interference of the incident and reflected waves. Such a standing wave produces a uniform distribution of the radiation intensity in the resist film, which can be partially eliminated by PEB. Ordinarily, the temperature of PEB is lower than soft bake.

- Development

After the PEB for conventional or chemically amplified photoresists, the photoresist is developed in a liquid chemical developer to dissolve the short chains of the photoresist. The soluble region of the positive resist is the exposed region, and for the negative resist, it is the unexposed region. Three kinds of methods are used for development, for instance, immersion, spray, and puddle. Puddle is a composite method, which combines both the immersion and spray methods.

- Descum

Usually, there is a very thin layer remaining left on the substrate after development which is called scum. This phenomenon is especially noticeable for deep and narrow pattern, because the developer is difficult to reach the bottom. But this very thin layer of scum may affect subsequent pattern transfer process. So, after photoresist development, a descumming process is sometime required. Oxygen plasma etching with a short time is generally used for the descumming process.

- Hard Bake

Hard bake is an optional process which can enhance the adhesion of photoresist pattern to the underlayer and enhance the etch resistance of photoresist patterns. But it should be aware that hard bake could make photoresist patterns more difficult to remove.

- Pattern Transfer

Ordinarily, photoresist acts as a masking layer, not a functional pattern, which should be transferred onto a functional material using wet (chemical) or dry (plasma) etching processes. Depending on the material intended to be etched, with possible requirements for selectivity over underlayers, a wide scope of wet chemical etchants may be employed. Besides crystallographic, most wet etching is isotropic, i.e., material removed equally in all directions. Dry etching is an anisotropic process that remove materials by a combination of physical sputtering and/or chemical reaction mechanisms, like reactive ion etching (RIE), ion-beam etching (IBE), et al. Extrinsic parameters, such as the gas chemistry, chamber pressure, ion energy, et al. determine the profile anisotropy, linewidth acuity, etch rate, selectivity, and uniformity.

- Post-processing

After the pattern transfer process, the photoresist needs to be removed. Wet and dry process can be used to dissolve the photoresist. The wet process is to remove the photoresist with various acidic or alkaline or organic solvents. While, the dry process is to remove the photoresist by oxygen plasma etching.

In VLSI process, mask linewidth errors, alignment errors, and etch depth errors may lead to reductions in efficiency. Therefore, how to minimize these tolerances is the key of micro-optical fabrication.

5.1.2.2 Direct-Write Lithography

Standard DOEs, kinoforms, refractive micro-lens, etc. have in common that the optical surface is defined by a continuous relief. Although VLSI technology can be used to define an approximately continuous surface profiles, these processes are typically optimized for digital response. Techniques have been developed to generate intermediate responses using either laser-beam or electron-beam direct-write lithography. This process directly exposes the resist spin-coated on the surface of the substrate with a variable dose of laser or electron beam, and the exposure depth is controlled by the exposure dose.

- Laser-Beam Direct-Write

The use of lasers for direct-write applications began with the development of laser. Laser direct-write is a process that takes advantage of the small feature sizes that can be easily achieved by focusing a laser beam onto a very small area. For laser direct-write purposes, the spot size, which corresponds to the system's resolution, the depth of focus, and the laser-beam intensity are three of the most important parameters. For Gaussian laser beam of circular cross section, the focused minimum radius, w_0 is normally determined by the diffraction limit of the imaging system, so the theoretical resolution at the focal point is given by:

$$w_0 = \frac{2\lambda f}{\pi d} \tag{5.1.7}$$

where λ is the wavelength of the laser radiation, d is the diameter of the limiting aperture before the focusing lens, and f is the focal length of the lens. Thus, the principal way of increasing the resolution is by reducing either the wavelength or the ratio of f/d. Furthermore, the rules of diffraction dictate that a focused laser spot cannot remain so as it propagates in free space. The depth of focus (DOF) is given by:

$$\text{DOF} = \pm\frac{\pi w_0^2}{\lambda} \tag{5.1.8}$$

For laser direct-write applications, the DOF indicates how much variation can be tolerated in the distance between the sample and the focal point. According to Eq. (5.1.8), there is an inverse relationship between the need for high resolution and a practical depth of field.

In the raster scan mode (Fig. 5.4), the total area to be exposed is split up into rows or columns, which are scanned one after the other at constant speed with a small spot of light. For an unrestricted focused laser beam, the laser-beam intensity distribution I can be well approximated by a Gaussian function:

$$I(r, \sigma) = \frac{I_0}{\sigma\sqrt{2\pi}} \exp\left(-\left(\frac{r}{\sigma\sqrt{2}}\right)^2\right) \tag{5.1.9}$$

where I_0 is the dose constant, r is the radial coordinated and the beam center is located at $r = 0$, σ is the standard deviation of the Gaussian distribution. During exposing, since the beam moves in a pixel-by-pixel motion, the superimposed or

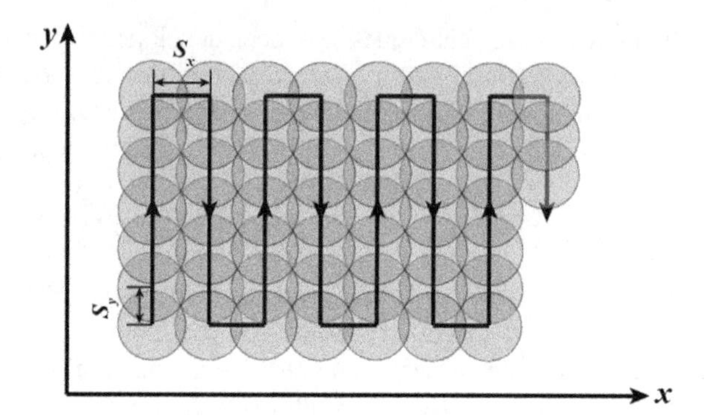

Fig. 5.4 Schematic diagram of a typical raster scan exposure

overlaid laser-beam intensity over the target can be conveniently expressed by the Cartesian coordinates (x, y) as:

$$I(x, y, \sigma) = \frac{I_0}{\sigma\sqrt{2\pi}} \sum_{m=0}^{M} \exp\left(-\left(\frac{x - mS_x}{\sigma\sqrt{2}}\right)^2\right) \times \sum_{n=0}^{N} \exp\left(-\left(\frac{y - nS_y}{\sigma\sqrt{2}}\right)^2\right) \tag{5.1.10}$$

where M is the total number of pixels involved in the x-direction, N is the total number of scanning pixels in the y-direction, and S_x and S_y are the pixels spacing along and across the scanning lines, respectively.

A typical laser direct-writing system with Cartesian coordinates is shown schematically in Fig. 5.5 [5]. In this system, a HeCd laser ($\lambda = 442$ nm) is used for exposing. The laser-beam intensity is modulated using an acousto-optical modulator allowing

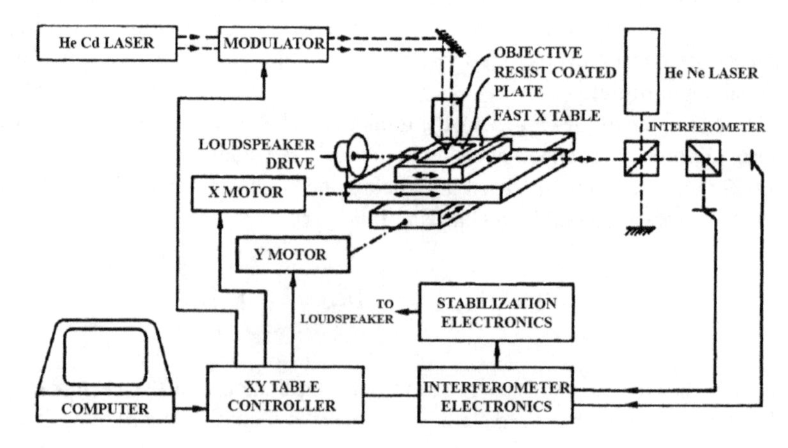

Fig. 5.5 Schematic diagram of a typical laser direct-writing system with Cartesian coordinates. Reprinted from [5] with permission. Copyright 1983, SPIE

up to hundred gray levels, and is focused by a microscope objective onto the surface of a photoresist-coated substrate. Focused spot with diameters down to about 1 μm is readily achieved by suitable choice of the objective and beam diameter. The photoresist-coated substrate is mounted on a motorized *XY* positioning stage interfaced through a controller unit to a computer, which generates the codes required to move the substrate in a raster scan movement. The controller executes these codes in sequence and also relays data to the modulator to set the laser-beam intensity to the required value at the beginning of each line scan.

Positioning accuracy and optical spot size limit the minimum feature size of laser direct-written optical structures, and the DOF of the optics limits the usable photoresist height to about the same level. Optimized exposure and development processes can yield very smooth optical surfaces.

- Electron-Beam Direct-Write Lithography

Electron-beam direct-write lithography is a new type of micro- and nanofabrication technology which was originally developed from the application of focused electron beams in the 1960s, like scanning electron microscope technology. It is a processing method for forming a pattern by using some polymers which are sensitive to electrons, and the polymer used is called an electron-beam resist.

The diameter of the focused spot is related to the wavelength of the electron beams. A simple formula for the wavelength of a moving particle is defined by de Broglie, which is given by:

$$\lambda = \frac{h}{p} = \frac{h}{m \cdot v} \tag{5.1.11}$$

where h is Planck's constant (4.135×10^{-15} eV·s), p is the momentum, m is the mass of the electron (9.11×10^{-31} kg), and v is its speed. An electron's velocity can be determined from its kinetic energy, by the formula:

$$E_{\text{in}} = \frac{1}{2} m v^2 \tag{5.1.12}$$

Thus, the de Broglie wavelength for electrons with kinetic energy E_{in} is:

$$\lambda = \frac{h}{\sqrt{2mE_{\text{in}}}} = \frac{1.226}{\sqrt{E_{\text{in}}}} [\text{nm}] \tag{5.1.13}$$

where E_{in} is expressed in eV. For example, if an electron was accelerated to have the energy of 10 keV, it would have $\lambda = 0.12$ Å. But due to the size of the cathode, the aberration in the electron lens, etc., the focused spot is expanded. In addition, when electrons interact with the resist, it will produce forward and backward scatterings, which are the reason for broadening of exposure area in resist.

The electron-beam direct-writing system (Fig. 5.6a) is mainly divided into an electron optical system, a vacuum system, a workpiece stage, and a high-speed pattern generator [6]. The electron optical system is used to generate an electron

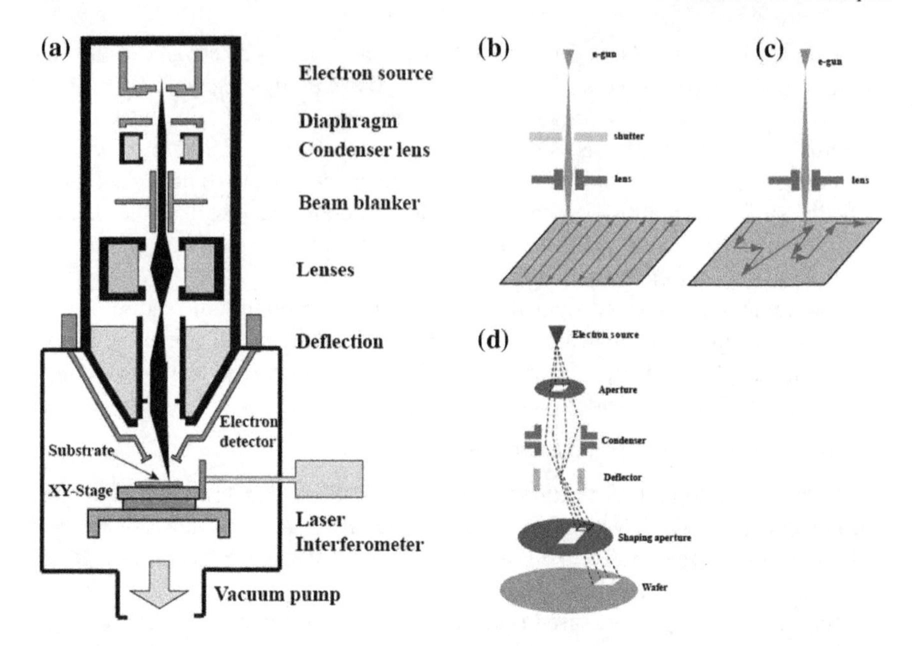

Fig. 5.6 **a** Typical schematic of an electron-beam direct-writing system. **b** Raster scan. **c** Vector scan. **d** Variable shape beam. **a** Reprinted from [6] with permission. Copyright 1997, Taylor & Francis Ltd. **b–c** Reprinted from [7] with permission. Copyright 2012, Springer Science+Business Media B.V.

beam and focus it to an ultra-small spot, and at the same time realizes the switching and deflection of the spot according to the instructions. The electron optics mainly includes three parts: electron gun, electron lens, and electronic deflection system. Furthermore, vacuum systems are needed and used to reduce energy dispersion and scattering of electrons. The workpiece stage is usually fed back by a laser interferometer to achieve precise positioning of the exposure point. The high-speed pattern generator is used to convert design data into an electrical signal that control the deflector and blanker.

The electron-beam direct-writing system can be divided into raster scan (Fig. 5.6b) and vector scan (Fig. 5.6c) according to the exposure mode [7]. In raster scan, the electron beam scans the whole field continually within a predetermined deflection area, and switches on the beam blanker only where there are patterns. But in the vector scan, electron beam only scans the exposure patterns. Generally, raster scan has lower scan speed compared with the vector scan.

According to the shape of the electron beam, the electron-beam direct-writing system also can be divided into Gaussian beam and shaped beam (Fig. 5.6d). When the electron beam passes through the aperture, its beam spot shape is adjusted by the diaphragm. When the diaphragm is circular, the electrons will be distributed in a Gaussian shape, which has a higher resolution than shaped beam system, e.g., triangular or rectangular, due to its ultra-small spot. Currently, the minimum spot of a Gaussian beam can reach 0.5 nm. The shaped beam system has high efficiency and is generally applied in the fabrication of reticle.

5.1.2.3 Photoresist Reflow Lenses

The increasing applications for micro-lenses have motivated researchers to develop a simple process for continuous profile lenslets. One of the most established means is by photoresist reflow, developed in 1980s. The key of this process is to use a controlled melting of cylindrical photoresist posts, where surface tension results in a hemispherical liquid surface. In the following, we present a mathematical model to analyze the photoresist reflow process (Fig. 5.7a).

With the help of photolithography, cylindrical photoresist posts with diameter D are patterned on a substrate. The volume of the photoresist cylinder, V_{cylinder} is given by [3]:

$$V_{\text{cylinder}} = \pi (D/2)^2 t_{\text{PR}} \tag{5.1.14}$$

where t_{PR} is the initial thickness of photoresist. And the volume of a hemispherical lens, $V_{\text{hemispherical}}$, is given by [3]:

$$V_{\text{hemispherical}} = \frac{1}{3}\pi h^2(3R - h) \tag{5.1.15}$$

where R is the radius of curvature. h is the sag height of the micro-lens, which is given by:

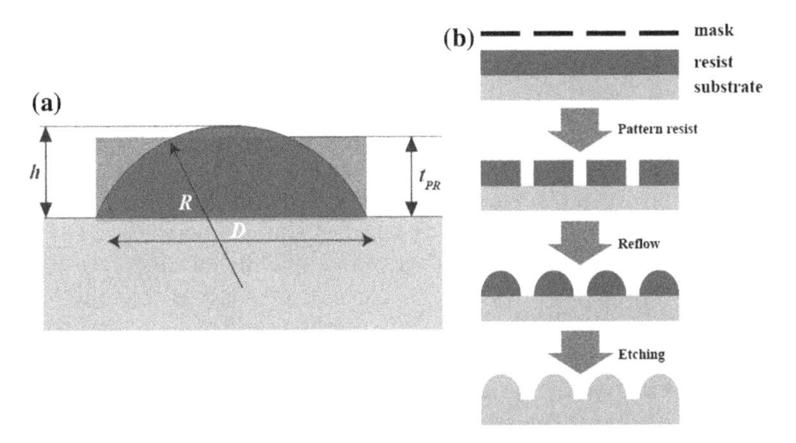

Fig. 5.7 **a** Schematic of the micro-lenses fabrication process flow using photoresist reflow. **b** The high-temperature reflow of a cylindrical post of photoresist causes the cylinder to assume a hemispherical shape and thus a segment of a spherical lens

$$h = R - \sqrt{R^2 - (D/2)^2} \tag{5.1.16}$$

Since photoresist volume is conserved during reflow, $V_{\text{cylinder}} = V_{\text{hemispherical}}$, which leads to the required initial thickness of the photoresist [3]:

$$t_{\text{PR}} = \frac{h}{6}[3 + 4(\frac{h}{D})^2] \tag{5.1.17}$$

Equation (5.1.17) indicates that the required initial thickness is only a function of sag height and diameter.

The process is schematically illustrated in Fig. 5.7b. Firstly, photoresist cylinders with diameter D and thickness t_{PR} on a substrate are defined photolithographically. Following is the reflow step, which carried out in an oven or on a hotplate, where the photolithographically defined patterns are the subjected to a high temperature, usually in the range 100–140 °C. The geometry of the micro-lens is mainly determined by surface tension and contact angle. The surface tension usually decreases with increasing temperature, and the contact angle is a function of photoresist viscosity and temperature. For complete melting, the original shape of the cylinder is irrelevant: surface tension always yields a hemisphere. After cooling, a solid continuous profile resist micro-lenses are formed. To achieve controlled and reproducible reflow, the heating and cooling ramps generally need to be precisely controlled. Although photoresist is also rather delicate, the properties of photoresist do not meet the application requirements. Then by utilizing RIE or IBE and accurately control the selectivity, continuous profile resist micro-lenses can be transferred to the substrate conformally. Typical reflow micro-lenses are limited by the initial photoresist thicknesses, and thus it is not possible to fabricate radii of curvature for lager aperture lenses. This process works well for lens with diameters less than 1 mm.

5.1.2.4 Moving Mask Method

The principle of the moving mask exposure is shown in Fig. 5.8a [8]. Different from conventional static exposure, the mask is continuously and linearly moved along with the mask holder. If the binary black and white mask remains static during the exposure, the mask pattern is copied to the resist after development. The exposure dose on the resist is either 0 or $I_0{\cdot}t$, where I_0 is the illuminating intensity of the lithographic system and t is the exposure time. If the mask is driven to move for one pitch or more pitches in the x-direction during exposure, the exposure dose on the resist is modulated continuously by the shape of binary mask. In Fig. 5.8b, an observing point on the mask moves from p to p' with the mask moving one pitch. The corresponding point p' at the resist surface goes through two intervals of transparent mask area h_2 and h_3. Defining v as the moving velocity of the system, the accumulated exposure dose at point p' of the resist is $E_p = I_0(h_2/v + h_3/v)$.

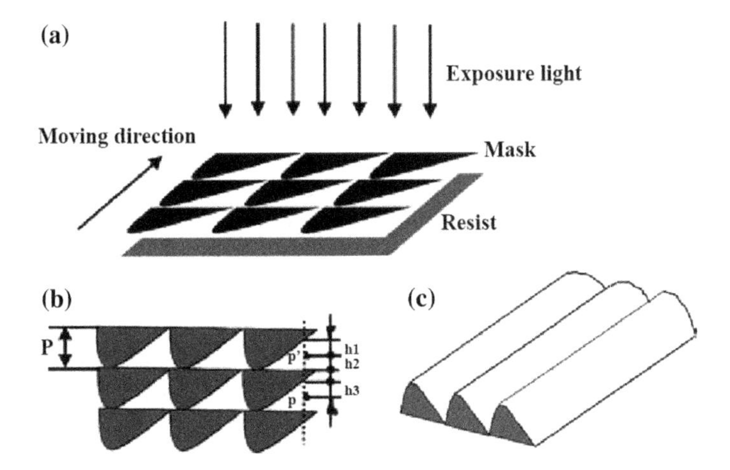

Fig. 5.8 Microprofile formation procedure using the mask-moving method. Reprinted from [8] with permission. Copyright 2004, SPIE

In Fig. 5.8b, there is a relationship of $h_1 = h_3$ and $h_2 + h_3 = h(x)$, where $h(x)$ is called the mask function. The exposure dose function toward the resist can be expressed as [8]:

$$E(x, y) = I_0 \frac{h(x)}{v} \tag{5.1.18}$$

We know, from Eq. (5.1.18), that a continuous exposure dose can be realized by a mask-moving exposure and a continuous profile is formed in the resist as shown in Fig. 5.8c.

Equation (5.1.18) shows that the exposure dose of the photoresist is proportional to the shape of the mask in one period, regardless of the coordinate position. The above equation shows that after a single exposure, the exposure dose of the resist is only a function of the coordinate x, so a single movement can only form a two-dimensional cylindrical microstructure. The preparation of three-dimensional periodic structures can be achieved by two cross-moving exposures. Finally, the exposure distribution obtained on the surface of the resist can be expressed as:

$$E(x, y) = E_x(x, y) + E_y(x, y) = I_0 \frac{h(x)}{v} + I_0 \frac{h(y)}{v} \tag{5.1.19}$$

The method is suitable to fabricate micro-lens arrays with a rotary symmetric profile. Figure 5.9 shows a schematic diagram of the rotary exposure.

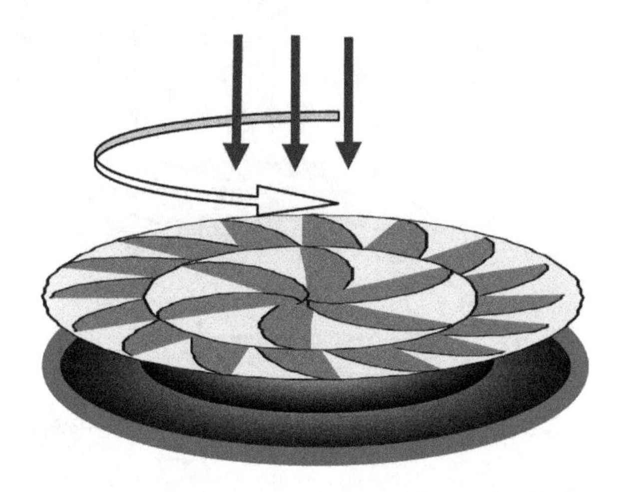

Fig. 5.9 Principle of rotary movement

5.1.2.5 Replication

The means mentioned above are direct fabrication methods, therefore the efficiencies are low. Replication is one of the most attractive methods for mass fabrication that the optical components can be replicated by using a precisely defined mater. And it allows the fabrication of micro-optical devices in very large volumes at a relatively low cost per device. The replication method mainly includes injection molding and hot embossing.

- Hot Embossing

Hot embossing is the direct transfer of the embossed pattern onto the polymeric substrate. The template is first heated, and the temperature is higher than the glass transition temperature of the polymeric material. The mold is then pressed against the heat-softened polymeric material. Then, start cooling before the pressure is removed. When the temperature drops below the glass transition temperature, the template is separated, and a relief structure is formed on the surface of the polymer material. The main disadvantage of hot embossing is the long processing time, which is attributed to the long time required to heat and cool the mold.

This process is carried out in a vacuum environment in order to prevent the forming bubbles in the molding material. Demolding is the most important part for hot embossing process. Especially, when the molded structure has a large aspect ratio, the demolding process is very likely to damage the structure formed by the molding. In addition to the pretreatment of the surface of the mold to prevent it from sticking to the molded material, the uniformity of the demolding pressure and the cooperation of the demolding pressure and temperature must be strictly controlled.

- Injection Molding

Due to the small size of the micro-optical components, the production of tiny plastic injection molds is not a simple matter of reducing the mold size, and a specially designed injection molding machine is required. Besides that, the manufacture of micro-molds is also the key to micro-injection molding. The micro-injected mold must be a metal material to withstand the mechanical shock vibration of the injection molding machine. The difference between micro-injection molding and hot embossing is that hot embossing only constructs surface-relief structures, while micro-injection molding structures are usually complex three-dimensional structures.

- UV Casting

UV casting is an industrially replication process for large-scale micro-optical components. The materials which used in this process usually contain two component resin: epoxies and a class of synthetic thermosetting polymers. These materials are firstly deposited into a mold. After sealed by a substrate, it is cured by UV flood-exposure. This technique can also be applied for double-sided micro-optical components.

5.1.2.6 Single-Point Diamond Turning

Single-point diamond turning (SPDT) is a class of ultra-precision machining processes that holds a unique place within the domain of single-point mechanical material removal processes such as turning, milling, and drilling. The SPDT machine uses a diamond-tipped bit for its machining, and it is able to machine surface finishes of a few nanometers and features that are down to about 1 μm in size. The surface roughness that produced by SPDT is typically of the order of a few nanometers. Compared with the standard VLSI techniques, SPDT can produce continuous relief DOEs on non-planar surfaces.

5.1.3 Challenges for Traditional Optical Manufacturing

As mentioned above, the main fabrication techniques in EO 1.0 are conventional optical manufacturing and micro-optical fabrication. Conventional optical manufacturing is only suitable for refractive or reflective lens. Although it can achieve atomic-level accuracy for surface processing, the curved shape has become a barrier for both the integrated optical functionalities and large-aperture optical systems. Conventional micro-optical fabrication is usually utilized to manufacture diffractive optical lenses. Compared with refractive/reflective devices, the thickness of the overall device can be reduced by application of diffractive elements. A lot of methods can be used to manufacture diffractive elements, like VLSI technology, photoresist reflow, direct-write, replication, and SPDT. VLSI is a precision processing technology for binary optics, but the requirement of typical lateral tolerances provides a limit on the size and number of zones in a binary optical approximation. Photoresist reflow usually is not able to make high-precision or large-aperture lenses due to the uncertainty and limited initial photoresist thickness. Although direct-writing technology has high resolution and processing flexibility, its throughput is low. Most replication processes, particularly those relevant for micro-optics, involve molding of thermoplastic materials, usually polymers. SPDT is an efficient way to produce micro-optics, whereas it can only process axisymmetric structures and not suitable for nanofabrication. In summary, these methods mentioned above cannot meet the need of EO 2.0 for large-area, high-resolution, and conformal fabrication. Therefore, we need to explore new fabrication methods to meet the above requirements.

5.2 Fabrication of Layered Structures

5.2.1 Ultra-Smooth and Single-Crystalline Metal Films

Metal is one of the basic materials in EO 2.0, which is used for surface plasmon polaritons (SPPs) excitation and metamaterial. In many applications, reducing intrinsic loss of the metal film is the crux for improving the performances. As the metal films produced by traditional vapor deposition consist of randomly oriented crystal grains, and usually have a rough surface morphology, the precision of the fabricated structures is limited by the size of the grains. It would be preferable to have a simple approach to produce continuous, smooth, and inexpensive single-crystalline metallic films, which could then be used as the starting material for a variety of devices.

5.2.1.1 Chemical Synthesis

Chemical synthesis is a facile route to synthesize micrometer-sized edge, single-crystal, polygonal gold nanoplates. A typical growth of large gold flakes procedure is: 50 ml of Ethylene glycol (EG) solution which contains 0.036 mmol $HAuCl_4 \cdot 4H_2O$ is heated to 95 °C by a water bath for 20 min. After that, 0.1 M aniline solution in EG was added to this solution with mild stirring. This procedure can also be modified to obtain thin flakes with large surface area (>100 μm^2) by reducing the reaction temperature to 60 °C [9]. Figure 5.10a and b shows the scanning electron microscopy (SEM) and transmission electron microscopy (TEM) image of the single-crystalline gold flakes chemically synthesized. The thickness of the synthesized flakes usually varies between 40 and 80 nm [9]. Although chemical synthesis is a simple way to synthesize single-crystal gold nanoplates, this process is inhomogeneous in size and shape, and inconvenient to place and manipulate for structural processing.

5.2.1.2 Deposition with Nucleation Layer

To address the issue of roughness in silver films, the widely employed approach to date is to use germanium (Ge) as a nucleation layer before Ag deposition [10]. The

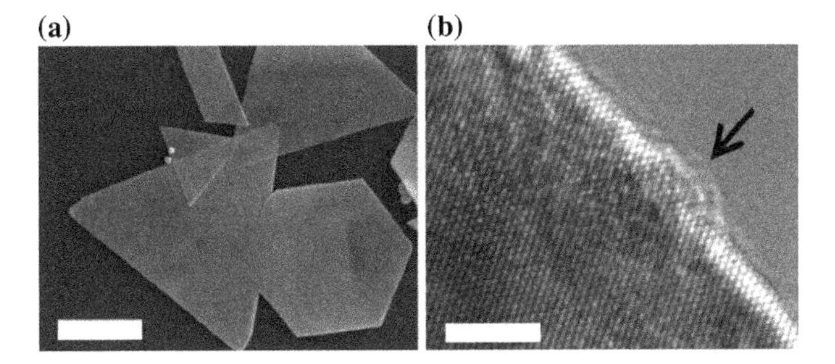

Fig. 5.10 **a** SEM image of self-assembled single-crystalline gold flakes. Scale bar: 5 μm. **b** High-resolution TEM image of a single-crystalline gold flake. Scale bar: 2 nm. Reprinted from [9] with permission. Copyright 2010, Macmillan Publishers Limited

immanent mechanism of the smooth Ag film is that the presence of thin layer of Ge leads to change of the growth kinetics (nucleation and evolution) of the electron-beam-evaporated Ag. The typical procedure of this process is: 1–2-nm Ge layer and Ag film are deposited sequentially on the substrate in an E-beam evaporator at a base pressure and at ambient temperature. The typical deposition rate is 0.01 nm/s for Ge and 0.1 nm/s for Ag [10]. The surface roughness of Ag film shows an order of magnitude improvement.

Figure 5.11a–d shows the atomic force microscopy (AFM) images and height histograms of the $Ag/SiO_2/Si(100)$ and $Ag/Ge/SiO_2/Si(100)$, respectively. In Fig. 5.11a and b, the AFM scan areas are both 1 μm × 1 μm. It can be seen that the roughness of Ag film is much smaller with a germanium nucleation layer. The peak-to-valley height of the Ag film without a germanium nucleation layer was measured to be about 34 nm, and the average has also reached 20 nm. With the help of a germanium nucleation layer, the RMS roughness of the Ag film can be reduced by at least a factor of ~10. Figure 5.11e shows the average RMS as a function of the Ge thickness for a constant Ag thickness of 15 nm. The RMS surface roughness of Ag was improved from ~6 nm to ~1 nm with 0.5-nm Ge deposited. But further increasing the thickness of Ge, the RMS roughness tends to be stable, at about 0.6 nm. However, Ge is highly lossy in the visible range and therefore the transmittance of this bilayer film is reduced.

5.2.1.3 Magnetron Sputtering Under Controlled Conditions

At controlled conditions of magnetron sputtering, smooth single-crystalline Ag films can be epitaxially grew on a well lattice-matched material [11]. The choice of underlying substrate is very important because it will influence the crystalline structure and orientation of the growing films. Some researches have demonstrated that the deposition rate and substrate temperature are the key factors for surface morphology and crystalline structure. To avoid surface-diffusion-enabled agglomeration, i.e., to

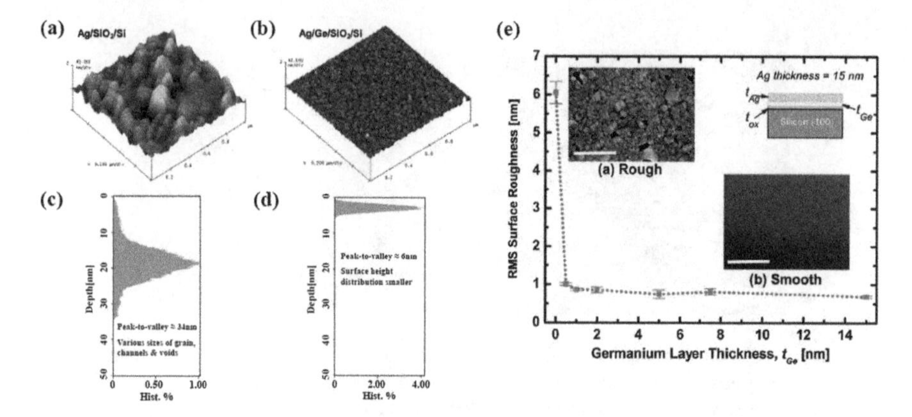

Fig. 5.11 AFM topographs of **a** 15-nm Ag film on SiO_2/Si(100), **b** 15-nm Ag film with a 2-nm Ge overlayer on SiO_2/Si(100). **c** and **d** Histograms of the 2D surface-height values from the respective topographs. **e** Plot showing the average rms surface roughness as a function of the Ge thickness for a constant Ag thickness of 15 nm (line drawn for clarity). Reprinted from [10] with permission. Copyright 2009, American Chemical Society

Fig. 5.12 **a** AFM image of single-crystalline Ag films with RMS roughness of 0.82. **b** AFM image of and polycrystalline Ag films with RMS roughness of 1.30 nm. Reprinted from [11] with permission. Copyright 2012, WILEY-VCH Verlag GmbH & Co. KGaA, Weinheim

make the surface continuous, a high deposition rate is needed. A smooth single-crystalline with RMS roughness of 0.82 ± 0.05 nm has been obtained at a substrate temperature of 350 °C and a high deposition rate of 1.65 nm/s [11]. Although the RMS surface roughness of single-crystalline (Fig. 5.12a) and polycrystalline Ag films (Fig. 5.12b) are nearly the same, the AFM images are quite different. Compared with single-crystalline films obtained via chemical synthesis process, epitaxial metallic films can provide many advantages: a flat surface over a large area, an accurately controlled thickness, and high crystallinity.

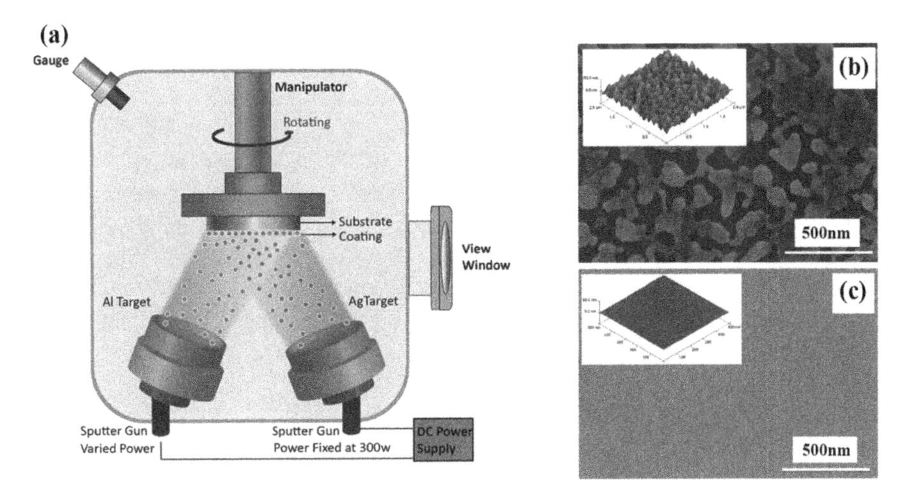

Fig. 5.13 **a** Schematic of co-deposition of Ag and Al. **b** SEM and AFM (insert) images of 9-nm pure Ag film. **c** SEM and AFM (insert) images of 9-nm Al-doped Ag film. Reprinted from [12] with permission. Copyright 2014, WILEY-VCH Verlag GmbH & Co. KGaA, Weinheim

5.2.1.4 Co-sputtering

Co-sputtering uses two or more targets to simultaneously sputter, and by changing the sputtering discharge current on different cathode targets to change the composition of the film, it has been demonstrated to be an effective method for preparing alloy films. Some researches have demonstrated ultra-thin and smooth Ag film also can be obtained by co-sputtering (Fig. 5.13a) [12]. Normally, the deposited Ag atoms initially form isolated islands due to its Volmer–Weber growth, whereas the incorporation of a small amount of Al suppresses the island growth of Ag and facilitates ultra-thin film formation. A RMS surface roughness value of 0.86-nm Al-doped Ag films can be obtained with Ag target power of 300 W and Al target power of 200 W, which translates into deposition rates of 0.9 and 0.06 nm/s, respectively [12]. The SEM characterization shows drastically different film morphologies between the 9-nm pure Ag film (Fig. 5.13b) and 9-nm doped Ag film (Fig. 5.13c). The insets in Fig. 5.13b and c shows the AFM images of 9-nm pure Ag and Al-doped Ag films, respectively. The RMS roughness of 9-nm pure Ag film is 10.8 nm. However, the roughness of Al-doped Ag films is over one order of magnitude lower, with an RMS value of 0.86 nm.

5.2.1.5 Molecular Beam Epitaxy Under Controlled Conditions

Molecular beam epitaxy (MBE) is a typical crystal epitaxial growth technique. Although it was applied to the preparation of semiconductor thin films in the early days, it has now been expanded to various materials such as metals and insulators. This method also offers an effective way to prepare ultra-smooth and single-crystalline Ag films [13], which contains two steps: low-temperature deposition followed by room temperature annealing. Firstly, a 2.5-nm-thick Ag is evaporated onto a liquid-nitrogen-cooled Si(111) substrate (~90 K) with a ultra-low deposition rate of ~1 Å/min. Secondly, it is naturally annealed to room temperature for about 1 h. In order to grow film up to a designed thickness, subsequent deposition is carried out using a similarly two-step growth process: deposition of film at 90 K, and annealing at room temperature for about 3 h. The most critical point in this process is to suppress de-wetting in an ambient environment.

As shown in Fig. 5.14a, the roughness of the oxide capped epitaxially grown single-crystalline Ag film with 45 nm thickness is only 0.36 nm. For comparison, the roughness of a 50-nm thermally deposited film is 3.27 nm, nearly an order of magnitude larger (Fig. 5.14b).

5.2.2 Layered Metal–Dielectric Hyperbolic Metamaterials

Layered metal–dielectric structures with hyperbolic dispersion have been experimentally realized across the optical spectrum. Various interesting effects have been demonstrated, including subwavelength imaging, focusing, and fluorescence lifetime engineering.

In the hyperbolic regime, the components of the dielectric tensor are negative in only one or two spatial directions. Hence, the most common realization of hyperbolic metamaterials is layered metal–dielectric structures [14].

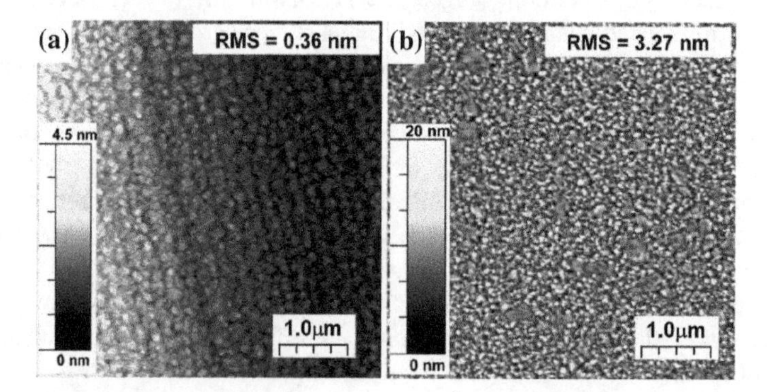

Fig. 5.14 **a** AFM image of a 45-nm epitaxial Ag film with 2-nm Al_2O_3/1.5 nm MgO capped. **b** AFM image of a 50-nm thermally deposited Ag film with 0.03 nm/s deposition rate. Reprinted from [13] with permission. Copyright 2014, WILEY-VCH Verlag GmbH & Co. KGaA, Weinheim

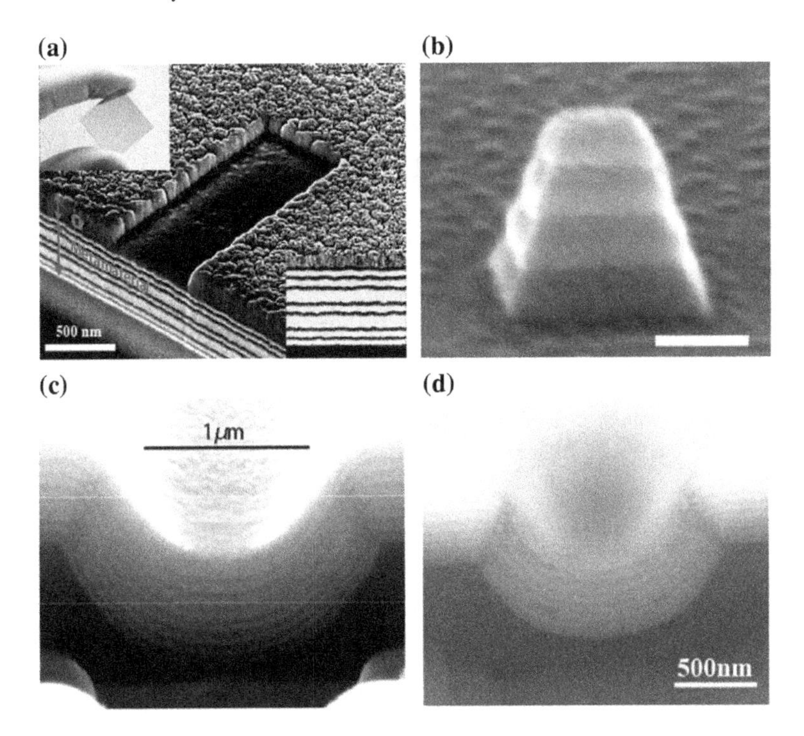

Fig. 5.15 **a** Cross-section SEM view of the negative refraction flat lens; left inset, glass slide uniformly coated with 450-nm-thick metamaterial; right inset, magnified cross section of the metamaterial layers. **b** SEM images of four pairs of silver/germanium multilayers. **c** SEM image of the cross section of a hyperlens made by depositing 16 periodic silver (35 nm) and Al_2O_3 (35 nm) layers on a cylindrical cavity in quartz substrate. **d** A scanning electron microscope image of the cross-sectional view of spherical hyperlens. **a** Reprinted from [15] with permission. Copyright 2013, Macmillan Publishers Limited. **b** Reprinted from [16] with permission. Copyright 2012, Macmillan Publishers Limited. **c** Reprinted from [17] with permission. Copyright 2007, Optical Society of America. **d** Reprinted from [18] with permission. Copyright 2010, Macmillan Publishers Limited

5.2.2.1 Direct Deposition

The most common methods that used for metal–dielectric metamaterials fabrication are sputtering and physical vapor deposition. Numerous hyperbolic metamaterials have been demonstrated based on different metal–dielectric pairs, including Ag/TiO_2, Ag/Ge, Ag/Al_2O_3, Ag/Ti_3O_5, and Ag/MgF_2. Figure 5.15a shows the cross section of Ag/TiO_2 negative refraction flat lens fabricated by radio-frequency sputtering with deposition rates: $R_{Ag} < 3.6$ Å/s, $R_{TiO_2} < 0.35$ Å/s [15]. Flat metal–dielectric multilayers can be structured by lift-off process, as shown in Fig. 5.15b. Four pairs of Ag/Ge multilayers are realized with ultra-high optical indices [16].

Metal–dielectric multilayers can not be merely deposited on flat surface, but also on non-planar surface, like half-cylindrical and hemisphere. Figure 5.15c shows eight pairs of Ag (35 nm)/Al_2O_3 (35 nm) thin layers onto half-cylindrical quartz mold,

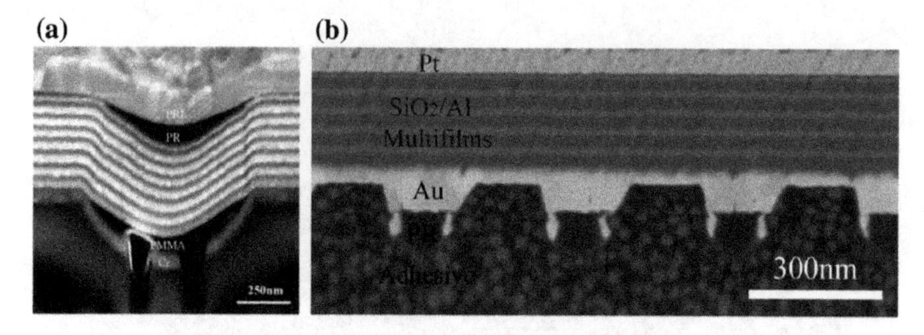

Fig. 5.16 **a** SEM picture of the cross section of hyperlens with Ag/SiO_2 multilayers. **b** SEM cross-section picture of SiO_2/Al multi-films. **a** Reprinted from [19] with permission. Copyright 2016, The Royal Society of Chemistry. **b** Reprinted from [20] with permission. Copyright 2015, WILEY-VCH Verlag GmbH & Co. KGaA, Weinheim

yielding hyperlens with inner cavity about 950 nm wide [17]. The film growth rate for Ag and Al_2O_3 was 2 nm/s, 0.25 nm/s, respectively. The surface roughness of the inner surface of hyperlens is 1.7 nm. Figure 5.15d is the SEM of the cross-sectional view of spherical hyperlens, which consists of nine pairs of Ag and Ti_3O_5 [18]. Titanium oxide layers can also be deposited by evaporating Titanium wire in the presence of oxygen.

To reduce the scattering from the surface, some methods are applied to improve the roughness of metals, like wetting and alloy method. Figure 5.16a shows the cross-sectional view of 15 layers of 20 nm Ag and 20 nm SiO_2 films that alternately deposited on the half-cylindrical quartz mold, and 1 ~ 2-nm-thick Ge film was pre-deposited beneath every Ag film as wetting material to improve surface smoothness [19]. Figure 5.16b shows cross-sectional SEM picture for SiO_2/Al multi-films. The thicknesses of SiO_2 and Al are 30 nm and 15 nm, respectively [20]. The Al target alloyed with 3% Cu was used to reduce the roughness of Al films.

5.2.2.2 Precise Control the Composition of Multilayers

Clearly, to realize high-performance hyperbolic metamaterials, the possibility of growing both metal and dielectric material components as a whole epitaxial system is very helpful.

To demonstrate negative refraction and hyperbolic dispersion in the mid-infrared region, a structure consisting of interleaved 80 nm layers of $In_{0.53}Ga_{0.47}As$ and $Al_{0.48}In_{0.52}As$ was realized (Fig. 5.17a) [21]. The layers, with a total thickness about 8.1 μm, were grown by MBE on lattice-matched InP substrates. The InGaAs layers were uniformly doped, at different densities for each sample to provide a plasma resonance of free carriers.

Reactive sputtering is a method of preparing films by mixing a proportion of active gas in an insert gas, reacting with a target atom to form a specific compound during

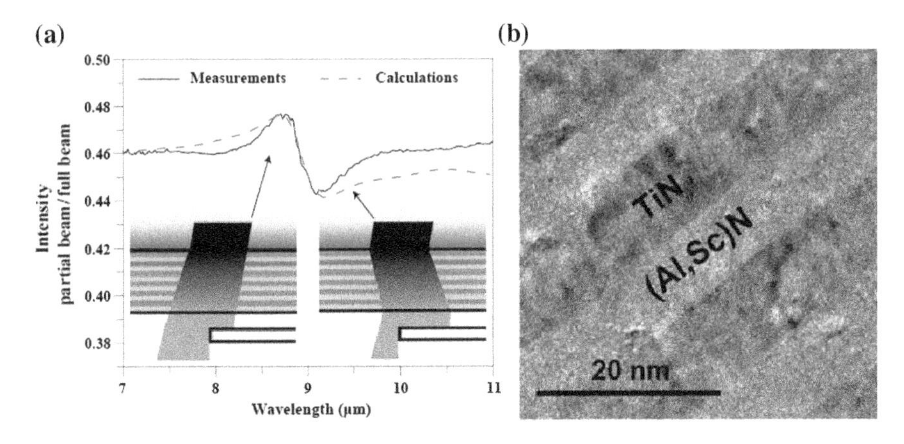

Fig. 5.17 **a** Experimental results showing negative refraction at hyperbolic metamaterial boundary. **b** Cross-sectional TEM image of TiN-$Al_{0.7}Sc_{0.3}N$ super-lattice. **a** Reprinted from [21] with permission. Copyright 2007, Nature Publishing Group. **b** Reprinted from [22] with permission. Copyright 2014, WILEY-VCH Verlag GmbH & Co. KGaA, Weinheim

sputtering, and depositing it on a substrate. This method can control the composition and properties of the deposited materials by controlling the pressure of the reactive gas, and thus can also be used to prepare a compound film [22]. Figure 5.17b shows a cross-sectional TEM image of a TiN/$Al_{0.7}Sc_{0.3}N$ super-lattice epitaxially grown on a 0.5-mm-thick, (001)-oriented magnesium oxide (MgO) substrate using reactive DC magnetron sputtering. The metamaterial was fabricated as an epitaxial stack of ten pairs of layers each consisting of an 8.5-nm-thick film of TiN and a 6.3-nm-thick film of $Al_{0.7}Sc_{0.3}N$. All depositions were performed with an Ar_2–N_2 mixture with the flow rates of Ar_2 and N_2 being 4 and 6 standard cubic cm per min, respectively, and a deposition gas pressure of 10 mTorr [22]. Whereas the Ti target was fixed at 200 W, the Al and Sc target powers were varied to achieve the desired stoichiometry of (Al, Sc)N alloy layers, i.e., $Al_{0.7}Sc_{0.3}N$.

5.3 Direct Writing of Subwavelength Patterns

5.3.1 Laser Direct Writing

The earliest laser direct writing exposes the patterns with a single focused laser beam that moved over the surface of the substrate, which is limited in both the resolution and efficiency. To improve the resolution, exposure efficient, and the capability of three-dimensional nanofabrication, some new direct-writing technologies (like zone-plate-array lithography, deep ultra-violet (DUV) laser pattern generator based on spatial-light modulator, two-photon direct laser writing, and stimulated-emission-depletion-inspired system) have been introduced.

5.3.1.1 Zone-Plate-Array Lithography

Zone-plate-array lithography (ZPAL) uses an array of Fresnel zone diffractive lenses with high NA that produce a corresponding array of tightly focused spots to direct-write patterns on a photoresist-coated substrate [23]. It allows the flexible, inexpensive, high-throughput, and scalable nanofabrication.

The ZPAL system mainly consists of four components: UV laser illumination, spatial-light modulator, Fresnel zone plates, and precision stages. The UV laser radiation is modulated by spatial-light modulator and focused to a spot whose size is approximately equal to the outer diameter of a zone. By using the spatial-light modulator as a dynamic mask to multiplex the radiation to the zone plates, almost arbitrary patterns can be generated by scanning the substrate.

A variety of micro- and nanoscale devices have been patterned, demonstrating the feasibility of this technology [24–26]. Examples of periodic patterns (dense lines and spaces) are presented in Fig. 5.18. The minimum lines have a width of 135 nm, which corresponds to a k_1 of 0.287. Figure 5.19a shows SEM images of patterns that were written using a linear array of zone plates. The patterns were written simultaneously with different zone plates, as shown in Fig. 5.19b, thereby demonstrating the parallel patterning capability of ZPAL [27]. Figure 5.19c–f shows SEM images of patterns exposed at $\lambda = 400$ nm representing: (Fig. 5.19c) a portion of a mask layout, (Fig. 5.19d) a ring resonator optical filter, (Fig. 5.19e) a photonic bandgap device, and (Fig. 5.19f) an array of elliptical rings for magnetic studies. These patterns demonstrate the versatility of the ZPAL technology and its potential for prototyping of novel designs. This technique also shows good performance regarding the mean stitching error and overlay accuracy, which is less than 30 nm and 50 nm, respectively.

5.3.1.2 DUV Laser Pattern Generator Based on Spatial-Light Modulator

As feature sizes continue to shrink, laser pattern generation is moving to DUV laser wavelengths. The principle of a DUV laser pattern generator based on spatial-light modulator (SLM) can be seen in Fig. 5.20 [28–31]. The light source used in this system is a pulsed excimer laser with a wavelength of 248 or 193 nm and high repetition rate. The pattern modulation is generated by programming the SLM, which has a flat surface composed of a million mechanical micro-mirrors. By applying a voltage to the micro-mirrors, the laser can be slightly deflected. Then the light is focused on to the photoresist-coated substrate through a high NA lens. Aperture is used to block or attenuate the deflected laser.

The stage with photoresist-coated substrate moves continuously and the interferometer commands the laser to pulse when it reaches the position for the next field.

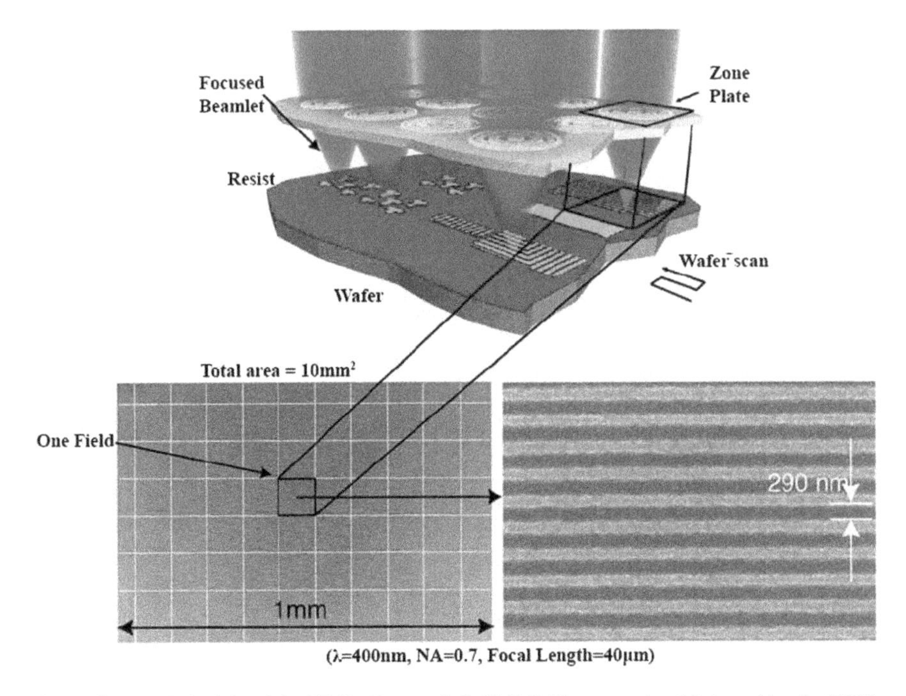

Fig. 5.18 Top: Principle of the ZPAL. Bottom-left: Full-field pattern that fabricated by the ZPAL. The dense pattern is 290 nm lines and spaces. The total patterned area contained 1000 fields, spanning a 10 mm^2 area. Bottom-right: High-magnification SEM of the center of the full-field grating. Reprinted from [23] with permission. Copyright 2004, Elsevier B.V.

Because of the short pulse time of laser, the movement of the stage is frozen and a sharp contrast image of the SLM is created in the photoresist. The SLM is reloaded with a new pattern in time for the next pulse. The pattern is stitched together by overlapping fields.

One of the targets for this technique is to manufacture photomasks at the 90 and 65 nm technology nodes [32]. This process has been developed by using a single-layer chemically amplified resists on standard AR8 chrome. Resist footing and standing waves in the resist are small (Fig. 5.21a), which allow for good plasma etch properties (Fig. 5.21b). Consequently, the etch bias is smaller than 30 nm [32]. A Bossung plot (Fig. 5.21c) shows that the iso-focal dose is slightly below nominal dose. This allows writing close to iso-focal dose with only small data sizing needed to meet the nominal linewidth. The dose sensitivity on CD can also be derived from the Bossung measurement. Close to iso-focal dose, the sensitivity is about 1.5 nm/%. The minimum resolvable dense lines are about 150 nm [32].

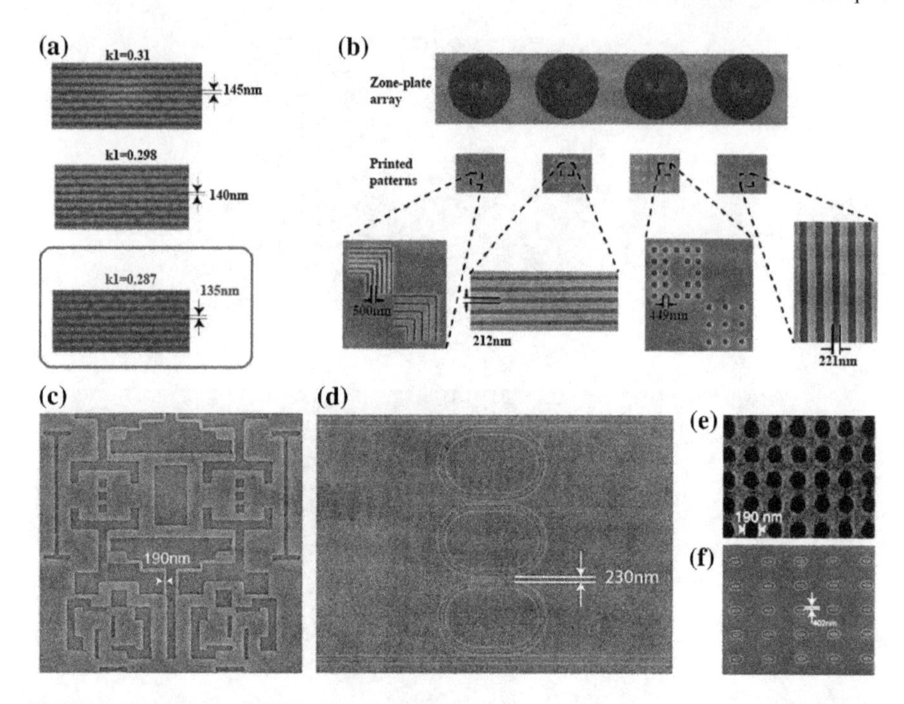

Fig. 5.19 **a** Resolution of ZPAL, SEM images of periodic patterns. **b** Parallel patterning with ZPAL system, SEM images of a variety of patterns that were exposed simultaneously. SEM images of a mask layout pattern (**c**), an optical add/drop filter (**d**), a photonic-bandgap device (**e**), an array of elliptical rings for magnetic studies (**f**). These patterns were printed using ZPAL at $\lambda = 400$ nm. **a** Reprinted from [27] with permission. Copyright 2005, SPIE. **b**, **e** and **f** Reprinted from [25] with permission. Copyright 2004, American Vacuum Society. **c** and **d** Reprinted from [24] with permission. Copyright 2004, American Vacuum Society

5.3.1.3 Two-Photon Direct Laser Writing and Stimulated-Emission-Depletion-Inspired System

Two-photon direct laser writing (TP-DLW) is based on the multi-photon polymerization and the nonlinear optical effect, which selectively cause local polymerization in suitable photoresists to allow the fabrication of arbitrary three-dimensional nanostructures.

Under the conditions of ultra-high photon density, electrons in the ground state may simultaneously or sequentially absorb two photons in an ultra-short time to cause electronic transition. This phenomenon is called a two-photon absorption. Two-photon absorption is a typical third-order nonlinear optical effect, and the formula of absorption probability is: $P = \sigma^{(2)} I^2/hv$, where $\sigma^{(2)}$ is the absorption coefficient of two-photon, I is the intensity of incident light, h is the Planck constant, v is the excitation frequency. It can be seen from the formula that the probability of two-photon absorption is proportional to the square of the incident light intensity. Since the two-photon absorption cross section of most materials is very low, two-

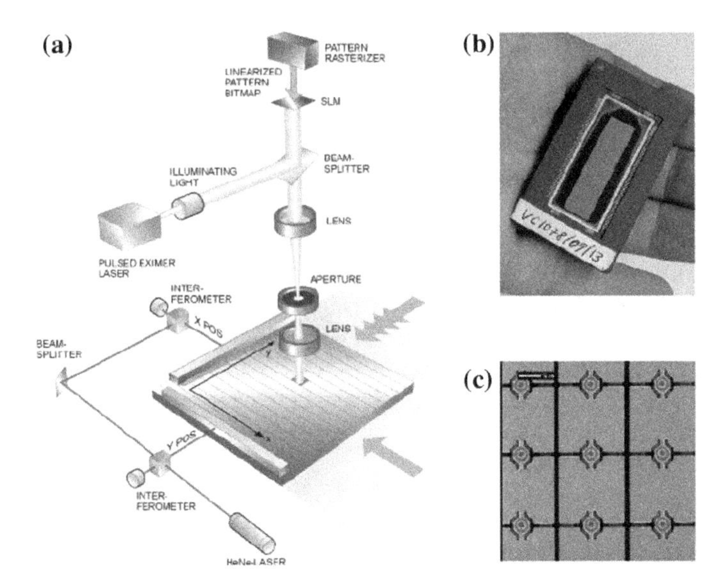

Fig. 5.20 **a** Principle of DUV direct writing based on SLM. **b** Photography of the micro-mirror array with 512 × 2048 torsional suspended mirrors. **c** The SEM micrograph details the mirror surface. Reprinted from [28] with permission. Copyright 2005, SPIE

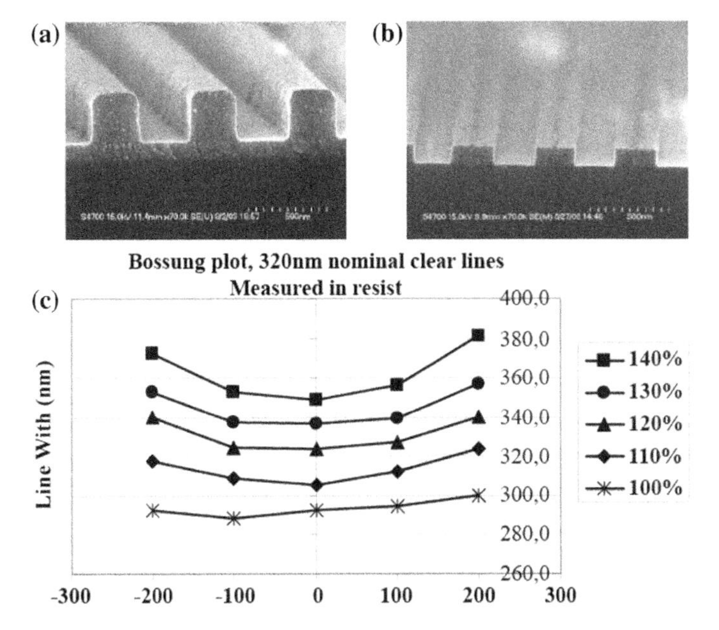

Fig. 5.21 **a** and **b** SEM pictures of dense lines and spaces before and after dry-etch, **a** 320 nm lines with 400 nm thickness resist, **b** 260 nm chrome lines after dry-etch. **c** Bossung plot for isolated clear lines with 320 nm nominal linewidth. The measurement was made in resist before etching. Reprinted from [32] with permission. Copyright 2003, SPIE

photon absorption can only occur with very high intensity illumination, generally with femtosecond laser. The two-photon absorption is localized to a small-volume pixel ("voxel"), which is typically characterized with an ellipsoidal shape. By moving the substrate relative to the fixed focal position, arbitrary paths can be written into the material.

Figure 5.22a gives a schematic of TP-DLW system, which consists of femtosecond laser and controller, optics, inverted microscope, piezoelectric 3D scanning stage, CCD, and computer. Three-dimensional structures are first designed and programmed in a specialized general writing language. Then, the system translates the designed structures to machine control signals, automatically approaches the sample, and adjusts the focus to the interface. Under the control of instructions, the sample is exposed. Finally, the exposed sample is developed and post-processed. Figure 5.22a and b give the SEM images of a chiral and a woodpile photonic crystal that fabricated by the TP-DLW. There are many advantages of TP-DLW, including true three-dimension, high spatial resolution, and small heat affect. However, only electroplate [33] and atomic layer deposition [34] are suitable for three-dimensional

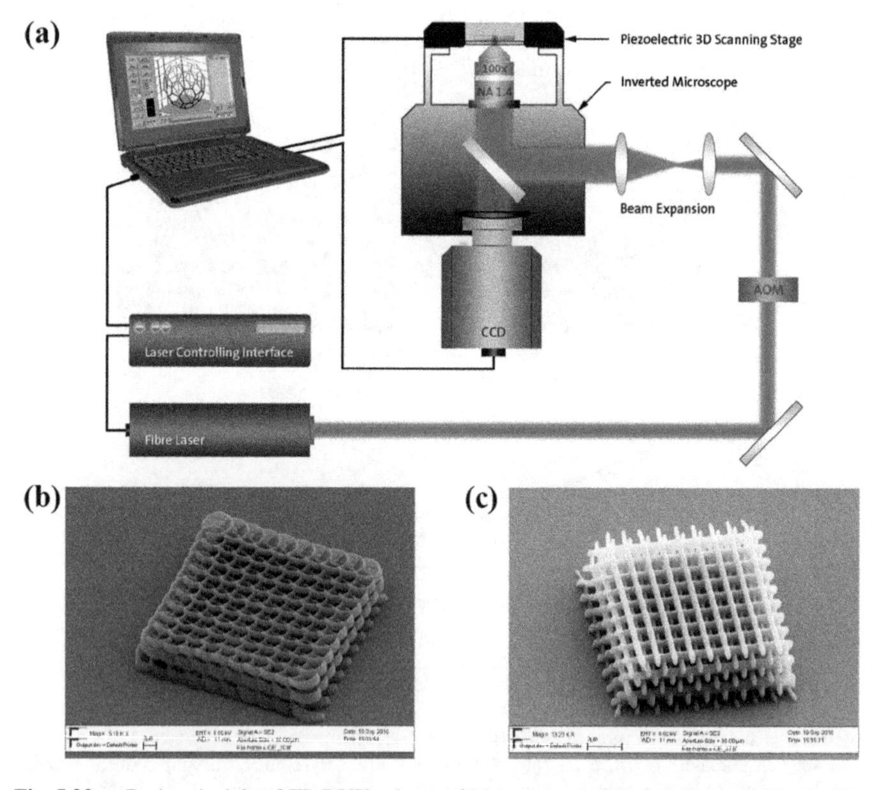

Fig. 5.22 **a** Basic principle of TP-DLW, where AOM is acousto-optical modulator. (Nanoscribe, GmbH). **b** SEM image of a chiral photonic crystal. **c** SEM image of a woodpile photonic crystal

transfer processing. Typically, the substrates should be transparent and conductive for electroplate.

Although TP-DLW can create 3D structures with almost arbitrary shapes, the resolution in 3D TP-DLW is still limited by the diffraction of light. With the development of stimulated-emission-depletion (STED) microscopy, the resolution far below the diffraction limit has been demonstrated. If DLW is combined with STED, the resolution will be greatly improved. However, this combination is not simple at all and particularly requires to develop unique photoresist material. A typical setup of STED-DLW is shown in Fig. 5.23a [35]. For multiphoton excitation, a femtosecond laser with 800 nm center wavelength combined with a pulse picker was used. And a pulsed laser with 532 nm wavelength emitting 100 ps pulses serving as the depletion laser is triggered and delayed electronically. Then, a high-NA-oil-immersion objective lens is used to coalign and focus the both lasers [35].

Figure 5.23c gives a fabrication process of gold triple-helix. In first step, a hollow polymer template is written on a transparent conductive substrate using a STED-DLW system [36], forming a polymer shell of the desired helical structure. Additionally, to avoid gold deposition between the helices, a polymer floor is written in these areas. After developing, the voids of the polymer template are filled with gold by

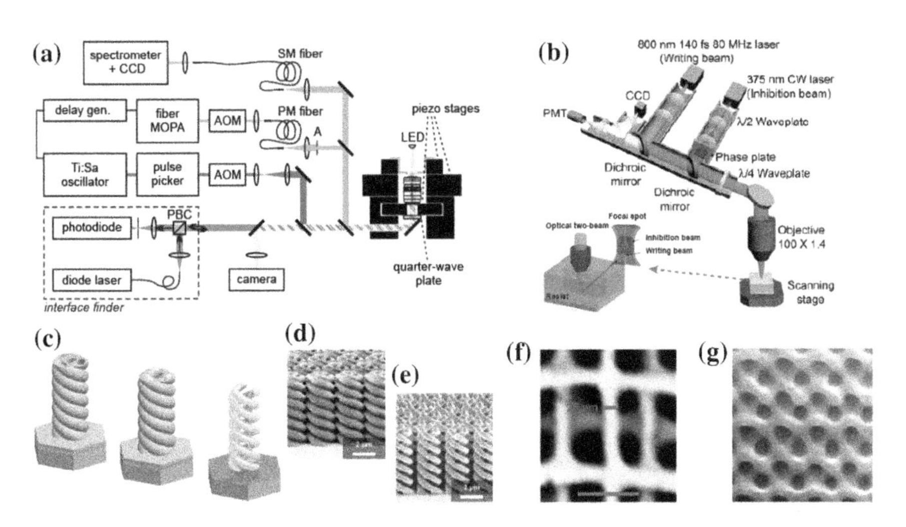

Fig. 5.23 **a** Experimental setup for STED-DLW with variable repetition rate. **b** Schematic setup for the 3D two-beam OBL. **c** Fabrication process of bi-chiral photonic crystals. **d** SEM image of polymer templates fabricated via STED-DLW before electrochemical deposition. **e** SEM image of the final gold structures after the polymer template has been removed via air-plasma etching. **f** SEM image of four-layer stacked nanowires and **g** 3D gyroid PCs with lattice constants of 300 nm fabricated by two-beam OBL. **a** Reprinted from [35] with permission. Copyright 2014, WILEY-VCH Verlag GmbH & Co. KGaA, Weinheim. **b** and **f** Reprinted from [37] with permission. Copyright 2013, Macmillan Publishers Limited. **c–e** Reprinted from [36] with permission. Copyright 2015, Optical Society of America. **g** Reprinted from [38] with permission. Copyright 2017, the Authors

electroplate, which is shown in Fig. 5.23c . Then, the polymer template is removed by oxygen-plasma etching, yielding the final gold structures as shown in Fig. 5.23e.

In order to further improve the resolution and flexibility in STED-DLW, two-beam optical-beam lithography (OBL) with high mechanical strength resin has been proposed, which utilizes a doughnut-shaped inhibition beam to inhibit the photopolymerization triggered by the writing beam at the doughnut ring (Fig. 5.23b) [37]. The minimum feature size of 9-nm isolate line and 52-nm two-line can be obtained by this method, which are 1/42 and 1/7 of the inhibition wavelength, respectively. OBL also shows an ultra-high-resolution in 3D nanofabrication. As shown in Fig. 5.23f, the minimum feature size of 22 nm can be realized. Figure 5.23g [38] shows 3D gyroid PCs with a lattice constant of 300 nm.

5.3.2 *Focused Ion-/Electron-Beam-Based Methods*

5.3.2.1 Focused Ion-/Electron-Beam-Based Reduction Processes

Focused ion-/electron-beam-based reduction processes fall into two classes, i.e., the focused ion-beam (FIB) milling and focused electron beam (FEB) induced etching. Focused ion beam milling is one of the most widely used micro-nanofabrication tools in failure analysis, circuit restructuring, TEM sample preparation and nano-optics. It can remove any type of material and does not require chemical reactions. Ion source is a core of focused ion beam, which is usually made by liquid metal. The metals that can form the ion source are Li, Be, B, Al, Si, P, Fe, Cu, Zn, Ga, Ge, As, Pd, In, Sn, Cs, Au, Pb, Bi, U. And low mass ions produce relatively low sputter yield, but penetrate deep into the sample [39]. The most commonly used liquid metal ion source (LMIS) in commercial FIB equipment is Ga.

Figure 5.24a shows a schematic of ion–solid interaction [40]. When a high-energy ion beam interacts with a solid surface, a part of the ions collides with the solid surface elastically or inelastically, and the change of motion direction produce backscattering, which is called backscattered ions. Another part of the ions can penetrate the solid surface, then it will cause a series of cascade collisions with solid atoms, and the energy is gradually transferred to the surrounding lattice and finally exhausted and stays in the crystal lattices. This phenomenon is called ion implantation. During the cascade collision of atoms, the number of atoms that are collided increases rapidly. If the momentum direction is away from the surface after the collision, and the energy reaches a certain threshold, it will cause the particles escape from the surface, which is called sputtering. The sputtered particles may be atoms, molecules, or atomic group, and the particles may be neutral or charged. During the ion–solid interaction, secondary electrons and X-ray are also emitted.

Different from FIB milling, focused electron beam (FEB) induced etching remove the materials selectively with etchant compound by electron beam irradiation, as shown in Fig. 5.24b. The etchant compound is delivered and adsorbed on the solid surface by gas injection system. When the electron beam irradiates the adsorbed

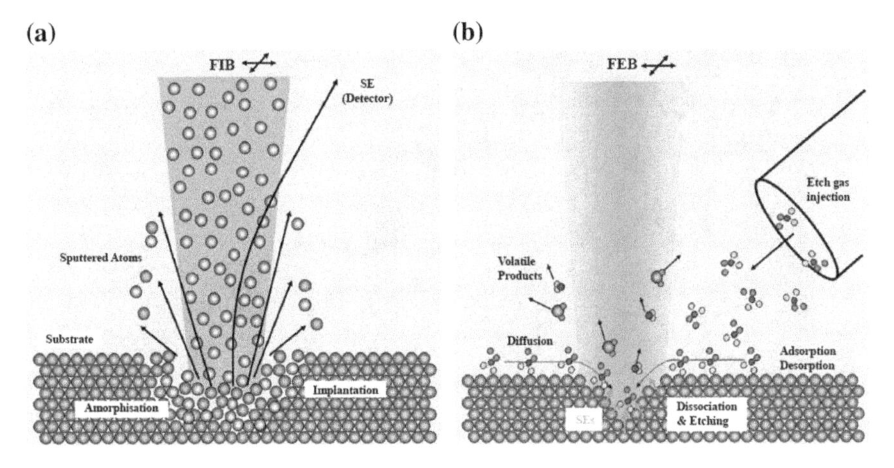

Fig. 5.24 **a** Principle of FIB milling (or sputtering). **b** FEB-induced etching: the surface adsorbed molecules dissociate under electron impact into reactive species and react to volatile compounds with the substrate material. Reprinted from [40] with permission. Copyright 2008, American Vacuum Society

compound, the etch reaction occurs. XeF_2, O_2, H_2, Cl_2, WF_2 are often used as the etchant compound [40].

The most widely used application of FIB is to prepare specimens for subsequent analysis by other analytical techniques, such as TEM. With the development of nanotechnology, new applications of FIB are constantly being developed for meta-optics. The micro-/nanopatterns can be directly formed on the materials surface by well-designed structures, and FIB is capable of cutting nanometer-sized features with a high aspect ratio. Figure 5.25a shows the two-dimensional catenary metasurface fabricated directly by FIB milling on a single layer of Au [41]. Figure 5.25b shows the SEM image of the proposed 3D fishnet pattern, which was milled on 21 alternating films of silver and magnesium fluoride [42].

Most of the 3D metamaterials require supporting substrates, which may introduce undesirable effects. The application of FIB folding for direct fabrication of substrate-free 3D plasmonic nanostructures have been demonstrated for unusual applications such as Fano resonances. The fabrication process employs FIB nanopatterning combined with in situ irradiation-induced folding of metallic thin-film structures [43], as shown in Fig. 5.26a and b. On this basis, a one-step and on-site nanokirigami method with nanoscale accuracy has been introduced by in situ cutting and buckling a suspended gold film [44]. By using the topography-guided stress equilibrium during global ion-beam irradiation, versatile buckling, rotation, and twisting of nanostructures are simultaneously or selectively achieved, as shown in Fig. 5.26c and d.

Conventional FIB system is not feasible for step and repeat processing over large areas. For this reason, ion beam lithography (IBL) system that is flexible for specific applications with multi-ion was developed. Figure 5.27a shows a photograph of an installed IBL system [45], which includes an ion beam column optics (Fig. 5.27b)

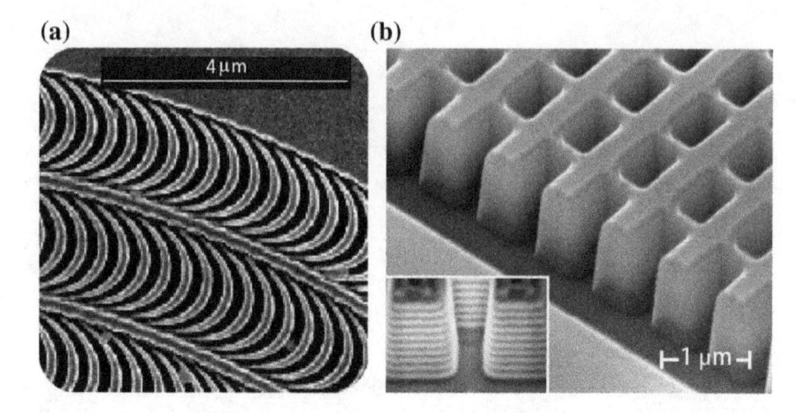

Fig. 5.25 **a** Scaled view of the catenary array fabricated directly by FIB milling. **b** SEM image of the 21-layer fishnet structure with the side etched, showing the cross section. The structure consists of alternating layers of 30 nm silver (Ag) and 50 nm magnesium fluoride (MgF_2). **a** Reprinted from [41] with permission. Copyright 2015, the Authors. **b** Reprinted from [42] with permission. Copyright 2008, Macmillan Publishers Limited

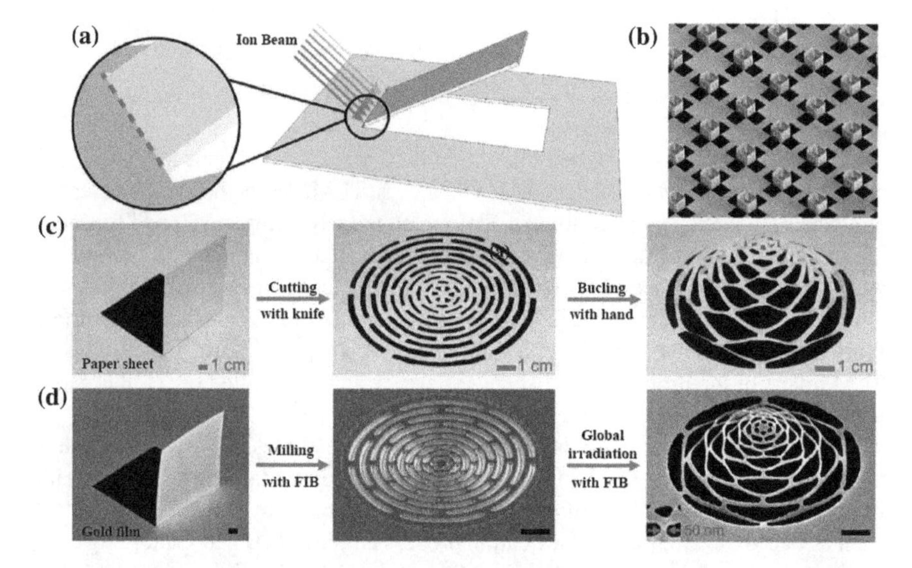

Fig. 5.26 **a** Ion-beam irradiation-induced folding of in-plane Au nanostructures, illustration of the line-scanning strategy. **b** Boxes with U-shaped holes on the walls. Scale bars: 1 mm. SEM, scanning electron microscope. **c** Camera images of the paper kirigami process of an expandable dome (corresponding to a traditional Chinese kirigami named "pulling flower"). **d** SEM images of an 80-nm-thick gold film, a 2D concentric arc pattern, and a 3D micro-dome. The high-dose FIB milling corresponds to the "cutting" process, and the global low-dose FIB irradiation of the sample area (enclosed by the dashed ellipse) corresponds to the "buckling" process in nanokirigami. The buckling direction is downward along the FIB incident direction. **a** and **b** Reprinted from [43] with permission. Copyright 2015, CIOMP. **c** and **d** Reprinted from [44] with permission. Copyright 2018, the Authors

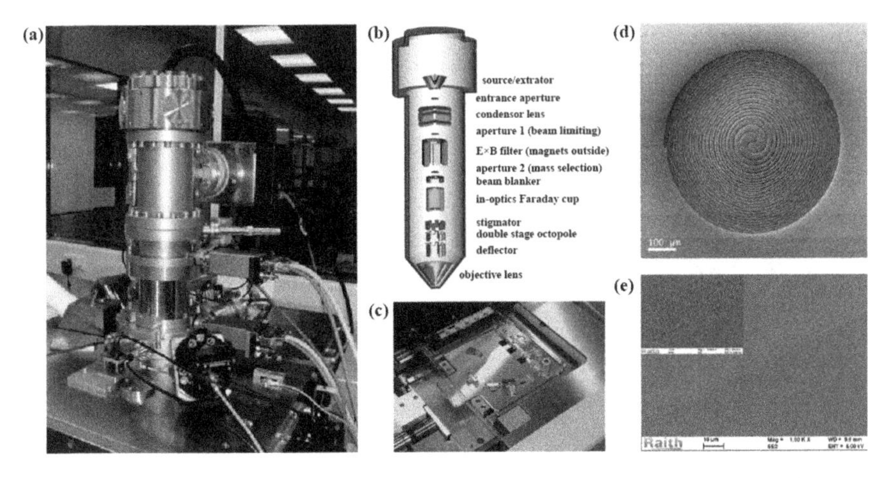

Fig. 5.27 **a** Photograph of an installed IBL system. **b** Schematic representation of the ion-beam column optics. **c** IBL sample holder positioned with a laser interferometer stage. **d** Spiral demonstration pattern written in the fixed-beam moving stage (FBMS) mode with the IBL. **e** SEM images of a photonic crystal light guiding devices fabricated by IBL milled into a Si sample about 2 mm long with a magnification of 1000 and 13000 (insert). **a** and **c**–**e** Reprinted from [45] with permission. Copyright 2011, Materials Research Society. **b** Reprinted from [46] with permission. Copyright 2011, American Institute of Physics

[46] and a sample holder positioned with a laser interferometer stage (Fig. 5.27c). The pattern in Fig. 5.27d was written with 30 keV Au in a GaAs crystal by starting at the outer diameter, spiraling into the center and then spiraling back out to the starting point parallel to the first write. The line width and length are 30 nm, and ~100 mm, respectively. A photonic crystal waveguide device is shown in Fig. 5.27e, which is significantly larger than one field of view. Here, repetitive stitching with dose control at the stitching border has been carried out to increase the fabrication area.

5.3.2.2 Focused Ion-/Electron-Beam-Based Additive Process

Focused ion-/electron-beam-based additive process uses high-energy beams to dissociate precursor molecules adsorbed on a surface of substrate into volatile and nonvolatile components, and the nonvolatile component deposits on the surface and direct form two- or three-dimensional nanostructures, as shown in Fig. 5.28a. Since the dissociation is highly localized, which only occurs within the beam spot, this method has an ultra-high spatial resolution. Moreover, the dissociation rate of precursor molecules is proportional to the beam energy and surface density of molecules.

Three-dimensional structure fabrication process by FEB-/FIB-induced deposition is illustrated in Fig. 5.28b [47]. First, a pillar is formed on position 1 at the substrate. Next, the beam is moved to position 2 and then fixed until the deposited terrace thickness exceeds an ion range which is a few tens of nanometer. Then, repeat this

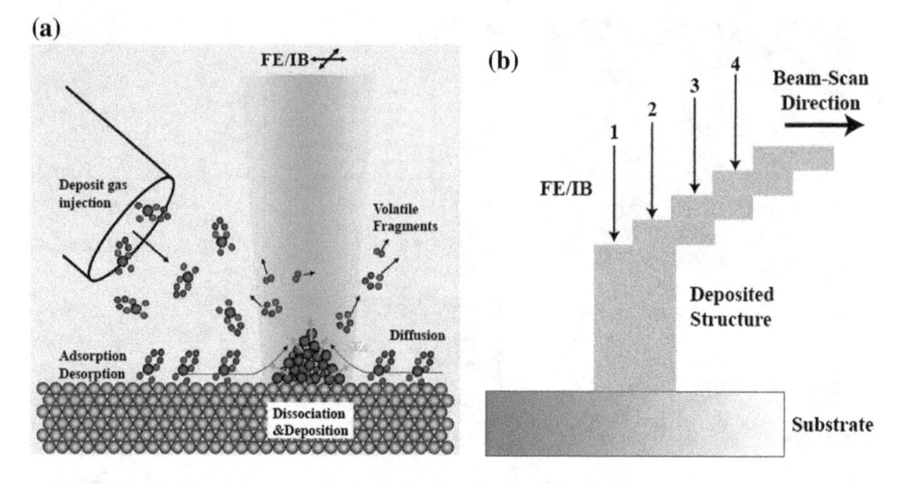

Fig. 5.28 **a** FEB-induced deposition: the nonvolatile dissociation products form the deposit growing coaxially into the beam. Volatile fragments are pumped away. **b** Fabrication process for three-dimensional nanostructure by FE/IB-induced deposition. **a** Reprinted from [40] with permission. Copyright 2008, American Vacuum Society

process to create three-dimensional structures. The sticking point to make three-dimensional structures fabrication is to adjust a beam-scan speed as the remaining ion beam within the deposited terrace.

This simple dose-correction protocol was used to fabricate an array consisting of 4×1 3D right-handed nanostructures along the x-axis with five loops, a reduced vertical pitch (VP) of 198 nm, and total height of around 1 μm, as shown in Fig. 5.29a [48]. Figure 5.29b shows the application of this concept in Pt [49]. In this case, the chiral dipoles are spatially located within the growth plane, always with nanometer-scale geometric features (external diameter (ED) = 300 nm, wire diameter (WD) = 100 nm, pitch number (N) = 1, VP = 700 nm).

The SEM image in Fig. 5.29c shows glass helices, fabricated by electron-beam-induced deposition with a step size of 5 nm [50]. The glass helices can be then conformally coated the cores with a gold shell via sputter coating. The false color image in Fig. 5.29d reveals the resulting gold coverage. Here, the gold is shown in blue and the glass core in black.

5.3.3 *Electron-Beam Direct Writing*

Electron-beam direct writing is the most flexible, and widely used nanolithography tool by far. The development of nanotechnology has placed higher demands on the resolution of electron-beam lithography. Although commercial electron-beam exposure machines can easily achieve nanostructure fabrication with the feature size below 100 nm, it is still not easy to realize resolution below 50 nm. The resolu-

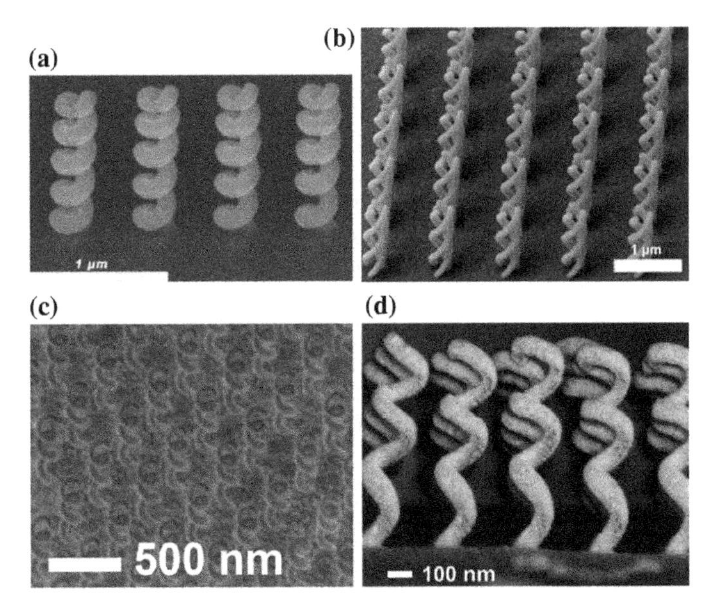

Fig. 5.29 **a** SEM image of a 4 × 1 array of chiral nanospirals grown on Si. **b** SEM of chiral dipoles intertwined by the tomographic rotatory growth in Pt. **c** Nanoscale glass helices fabricated with FEB-induced deposition. **d** Core–shell (glass-gold) nanohelices after sputter coating. **a** Reprinted from [48] with permission. Copyright 2013, WILEY-VCH Verlag GmbH & Co. KGaA, Weinheim. **b** Reprinted from [49] with permission. Copyright 2016, American Chemical Society. **c** and **d** Reprinted from [50] with permission. Copyright 2017, American Chemical Society

tion of electron-beam exposure is related to various factors. To achieve the highest resolution, generally the following aspects should be considered:

(1) High electron energy. As mentioned in Sect. 5.1.2, the wavelength of electron beam is mainly determined by the energy of electrons. The high-energy electrons have ultra-short wavelength, and result in an ultra-high resolution. Furthermore, the increase in the energy of electrons has some additional advantages. For instance, the electrons forward scattering spread can be reduced and backscattering spread is enlarged, which can avoid local concentration of backscattering electron energy deposition and reduce the proximity effect during exposure. In addition, a high-energy electron beam has a low chromatic aberration and space charge effect, which is advantageous for obtaining a smaller electron-beam spot. The high-energy electron beam also facilitates exposure of thicker resist and forms a high-aspect-ratio pattern. Therefore, high-precision electron-beam lithography systems can generally operate at an acceleration voltage of more than 100 kV.

(2) Low beam current. At a particular electron-beam acceleration voltage, the beam spot size and beam current are determined by the aperture size. Changing the aperture size actually changes the convergence angle of the electron beam. The spherical aberration of the electron-beam astigmatism is proportional to the cube of the convergence angle, and thus narrowing the convergence angle can significantly improve the resolution of the electron beam. However, the reduction in beam current will increase the exposure time and greatly reduce the imaging brightness (or signal intensity) of the detection focus or alignment mark, making focusing and alignment very difficult.

(3) High resolution, low sensitivity resist. The resolution of the electron beam is not only affected by the size of the electron-beam spot, but also by the resist itself [51]. For a positive resist, a larger molecular weight can lead lower sensitivity and higher resolution of electron-beam lithography; but for a negative resist, it is opposite. For the same resist, a thicker resist will lead to larger forward scattering of the electron beam therein and a lower resolution [52]. Reducing the thickness of the resist can reduce the effects of electron beam forward scattering and the proximity effect. However, it is hard for pattern transfer as the thickness reduction. Therefore, how to select a resist is an important factor in obtaining an ideal pattern in electron-beam exposure.

(4) Scan field size and minimum step size. In order to reduce astigmatism and distortion caused by electron-beam deflection, it is necessary to divide the whole pattern into a plurality of sub-fields during electron-beam exposure. In the sub-field, the electron beam is subjected to an exposure operation by deflection scanning. But the deflection astigmatism and distortion increase as the area of sub-field increases. Another parameter related to the scan field size is the minimum step size, which is the distance between adjacent two electron-beam scanning points. To ensure the uniformity of the exposure pattern, the minimum step size is generally required to be smaller than half of the beam spot size.

(5) Exposure dose. The exposure dose is determined by the beam current, dwell time, and minimum step size. The required exposure dose is different as the acceleration voltage of electron beam, resist, and exposure pattern are changed [53]. For a particular design pattern, an experiment is needed to determine the optimal exposure dose. In addition, when the shape and density of the pattern change, the exposure dose also needs to be changed. Therefore, the dose test before each exposure is necessary.

(6) The concentration and temperature of the developer. After the electron-beam exposure is completed, the development process also has a great influence on the final result of the pattern. Generally, the higher the developer concentration, the faster the development speed, and the shorter the development time required. On the contrary, the longer develop time is needed as the reduction of developer concentration, due to the milder reaction. The development temperature also affects the development speed. The higher the temperature, the more intense the development reaction and the faster the development speed. In addition, studies have shown that the development temperature also affects the contrast [54] and resolution [55] of the exposed pattern. The low-temperature development facilitates the fabrication of high contrast and high-resolution patterns [56].

(7) Reduce the pattern density. If the adjacent two exposure patterns are in close proximity, the exposure energy distribution due to electron scattering will extend into the adjacent pattern area, causing the exposure pattern to be distorted. Even within the same pattern, the scattering causes the energy at the edge of the pattern to be lower than the energy in the middle. It is also difficult to perform proximity effect correction for dense patterns of periodic distribution. In this case, the electron-beam proximity effect can be reduced only by thinning the resist layer.

(8) Choose low density, highly conductive substrate. High-density substrate materials have many backscattered electrons and strong proximity effects. Poor conductivity of the substrate can cause charge accumulation, affecting the accuracy of the pattern.

(9) A stable working environment. The stability of the electron-beam lithography system itself is an important condition for achieving high-resolution exposure. The stable working environment includes low vibration, low electromagnetic interference, and stable ambient temperature.

Ultra-high-resolution electron-beam direct writing is a powerful means for large-area ultra-high-precision nanostructure fabrication. Figure 5.30a gives a standard process flow of electron-beam direct writing and metal lift-off. To prevent charging effect during E-beam direct writing, transparent conductive film and ultra-thin metal film are used. The difference is that transparent conductive film usually deposit on substrate before E-beam resist coating, while ultra-thin metal film is directly deposited on E-beam resist. Bilayer E-beam photoresist with different thickness and sensitivity can be used for better lift-off. A representative sample with the densest packing of antennas with a lateral periodicity of 11 μm is shown in Fig. 5.30b [57, 58]. The length and width of the gold V-antennas are ~650 and ~130 nm, respectively. Figure 5.30c shows an asymmetric U-shape nanostructure with a period of 500 nm, which is also fabricated by this process [59]. But in lift-off process, it usually requires a ratio up to 3:1 for E-beam resist thickness and metal film thickness, so it is hard to achieve high aspect ratio. Figure 5.30d shows another standard process, which combines electron-beam direct writing with etching process to improve the aspect ratio of nanostructures. Here, the choice of etching mask is very important. According to the aspect ratio requirement of the final design, there are two paths to get an etching mask. One is using E-beam resist pattern as an etching mask directly, and another one is using lift-off process to form a hard mask. Figure 5.30e shows a result of a high aspect ratio amorphous silicon (α-Si) nanoposts metasurface etching with an aluminum oxide hard mask formed by lift-off process [60]. The metasurface consists of 715-nm-tall amorphous silicon posts with diameters ranging from 65 to 455 nm, which are arranged on a hexagonal lattice with a lattice constant of 650 nm. An example of directly use E-beam resist pattern as an etching mask is shown in Fig. 5.30f [61]. The silicon nanoblocks have a lattice constant of 800 nm, height of 270 nm.

There are two limits for the processes described above. One is that these processes are only suitable for single-layer fabrication, and the other one is that it is difficult to fabricate high-aspect-ratio nanostructures with low etch selectivity materials. Figure 5.31a schematically shows a multilayer structure [62]. The non-planar surface of the metamaterial layers prevents stacking by simple serial exposure, development and metal evaporation. Therefore, the surfaces of the layers were flattened by applying a planarization procedure with dielectric spacers. The four-layer SRR structure and the 90° twisted gold SRR dimer metamaterials fabricated by this process are shown in Fig. 5.31b and c [63], respectively.

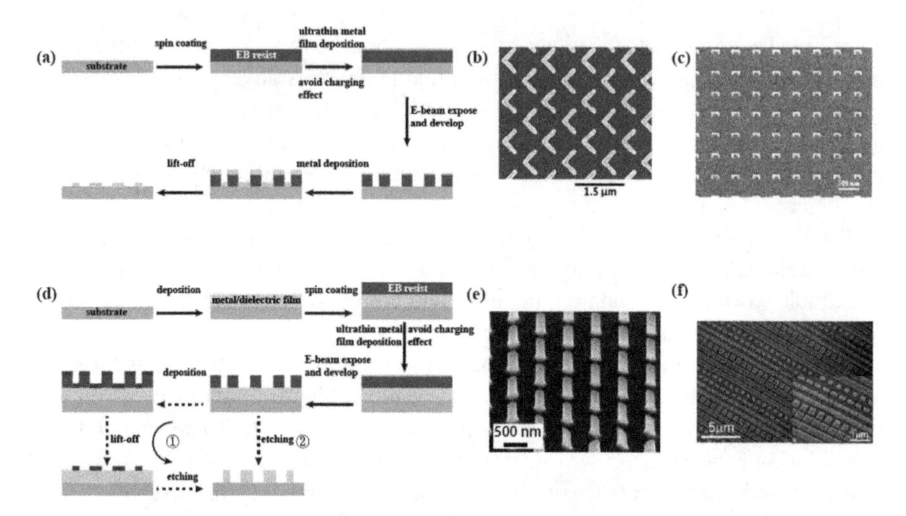

Fig. 5.30 **a** Standard electron-beam direct writing and metal lift-off process. SEM image of a representative antenna array (**b**) and asymmetric U-shape nanostructures (**c**) fabricated by process flow in **a**. **d** Standard electron-beam direct writing and etching process. **e** SEM of the tilted view of a fabricated dielectric metasurface with process flow ① in **d**. **f** SEM image of metasurfaces with nanoblocks fabricated by process flow ② in **d**. **b** Reprinted from [58] with permission. Copyright 2012, the Authors. **c** Reprinted from [59] with permission. Copyright 2015, Macmillan Publishers Limited. **e** Reprinted from [60] with permission. Copyright 2015, Macmillan Publishers Limited. **f** Reprinted from [61] with permission. Copyright 2015, American Chemical Society

To achieve high-aspect-ratio nanopatterns with low etch selectivity materials, a new process that shown in Fig. 5.31d has been proposed based on electron-beam direct writing, atomic layer deposition (ALD) planarization and etching steps [64]. The control of E-beam resist thickness is important because it determines the height of the final nanostructures. After electron-beam lithography and subsequent development, resist pattern can be formed, but this pattern is the inverse of the final nanostructures. Then transfer the exposed sample to an ALD chamber set to low temperature (blow the glass transition temperature of resist). Since the conformal ALD process fills the gaps from both sides the total ALD film thickness required is $t_{\text{film}} \geq w/2$, where w is the maximum width of all gaps. Next, residual ALD film that coats the top surface of the E-beam resist is removed by etching process and then exposes the underlying resist and the top of the nanostructures. Finally, the remaining resist is removed and only the nanostructures are remained. Figure 5.31e and f shows tilted and top SEM view of fabricated TiO_2 nanofin metalenses [65, 66]. Nanofins have a width of 95 nm, a length of 250 nm, and a height of 600 nm. Figure 5.31g and h shows tilted and top SEM view of the fabricated TiO_2 nanopillar metalens with diameters ranging from 80 to 320 nm [67].

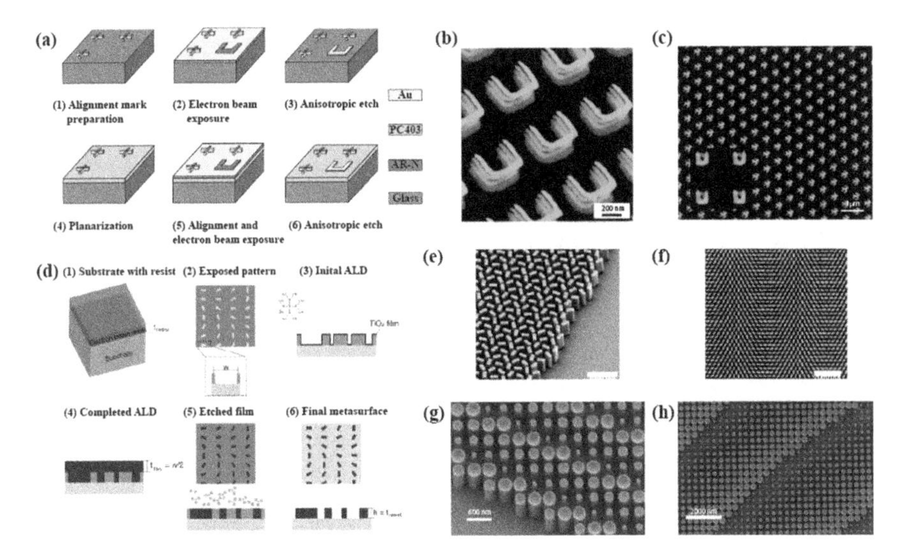

Fig. 5.31 **a** Processing scheme of layer-by-layer technique with alignment. **b** Enlarged oblique view of four-layer SRR structure. **c** Oblique views of the 90° twisted gold SRR dimer metamaterials. **d** Fabrication process for TiO_2 metasurface. SEM micrographs of fabricated TiO_2 nanofin metalenses, **e** tilted view, **f** top view. Side-view (**g**) and top view (**h**) SEM image of the edge of the TiO_2 nanopillar metalens. **a** and **b** Reprinted from [62] with permission. Copyright 2008, Nature Publishing Group. **c** Reprinted from [63] with permission. Copyright 2009, Springer Nature. **d** Reprinted from [64] with permission. Copyright 2016, the Authors. **e** and **f** Reprinted from [66] with permission. Copyright 2017, American Chemical Society. **g** and **h** Reprinted from [67] with permission. Copyright 2016, American Chemical Society

5.3.4 Data Compression for Direct Writing

In large-scale metasurfaces, there are millions or billions of individual microscopic meta-elements described macroscopically, which makes the data file describing the elements to be very large. The unmanageably large total file sizes limit the fabrication of metalenses to sizes no larger than $10^5\lambda$. For example, a $10^5\lambda$ diameter device usually consist of over 10 billion meta-elements (necessitated by the subwavelength size criterion), and each instance of which must be described by nanometer precision definitions of position, leading to a size >300 gigabytes (GB). This size makes design files hard to undergo data conversion for use with mask-writing equipment. We implement a compression algorithm that allows the file size to be reduced by many orders of magnitude. Similar techniques have also been reported [68].

Figure 5.32a shows the schematic illustration of the algorithm. To put it simply, the algorithm uses many hierarchical levels ($L_1, L_2, L_3, \ldots$) in the layout file (e.g., GDSII) to copy modular centrosymmetric subgroups ($g_1, g_2, g_3, \ldots$). In this case, only a few definitions of the rotation angle can replace a large number of unique meta-element definitions, resulting in significant reduction in the file size. The sector angles (a_1, a_2, a_3, …) of subgroups can be described as:

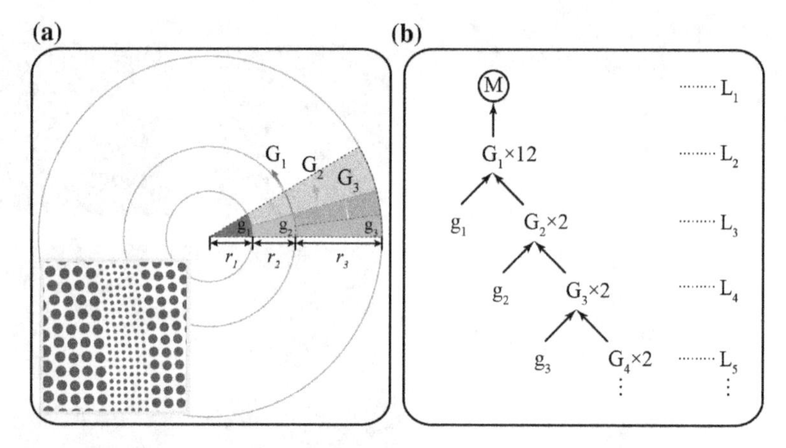

Fig. 5.32 **a** Schematic illustrations of the compression algorithm. **b** The cascade relationship of the compression algorithm

$$a_1 = 2a_2 = 4a_3 = \cdots = 2^{n-1}a_n \tag{5.3.1}$$

where $a_1 = 2\pi/m$, m and n are positive integers. As an example, m is equal to 12. In fact, the bigger m is, the smaller file size is. However, too small m will aggravate the distortion of the structures in the central area. The radial length ($r_1, r_2, r_3, \ldots$) of each subgroup depends on its arc length at both ends. Here, the backend one is two times the frontend one, so the relationship between the radial lengths can be expressed as:

$$r_{n+1} = \sum_{1}^{n} r_n \tag{5.3.2}$$

where r_1 depends on the frontend arc length (l) of the subgroup, that is, $r_1 = 12l/\pi$. The smaller the length l, the less unique meta-element definitions. However, too small one will introduce more errors to the period (p) of the meta-element. The highlighted inset in Fig. 5.32a shows the arrangement of meta-elements along the azimuthal direction. In this case, no stitching marks appear on the metalenses that are formed by a series of rotated subgroups. Considering the error and files size, we specified that $l = 10\lambda$, so the maximum error is only 2.5% with a condition of $p \leq \lambda/2$. This small error hardly affects the optical properties of the meta-element.

Figure 5.32b shows the cascade relationship among subgroups, main groups (G_1, G_2, G_3, ...), and levels. The L_1 is composed of 12 rotated G_1. G_1 consists of one g_1 and two rotated G_2, and G_n consists of one g_n and two rotated G_{n+1}. As a result, the file size depends only on the unique definitions of the meta-elements in the subgroups, because a few definitions of rotation angles hardly take up storage space. Thus, the compression ratio is the ratio of the total area of the metalens to the total area of subgroups, as given by:

$$c_n = \frac{2\pi R_n^2}{\sum_2^n a_{n-1}\left(r_{n-1}^2 - r_{n-2}^2\right)} \tag{5.3.3}$$

with

$$R_n = \sum_1^n r_{n-1} = 2^{n-2} r_1 = \frac{30 \times 2^n \lambda}{\pi} \tag{5.3.4}$$

where $r_0 = 0$ and $n \geq 2$ indicates the order of level. R_n indicates the radius of the metalens for the level n. This compression algorithm needs two levels at least. Figure 5.33a shows the compression ratio as a function the radius of the metalens. The compression ratio is nearly proportional to the radius, with a slope of approximate $0.4191/\lambda$, as depicted in the form of dash line in Fig. 5.33a. As a result, the file size is also near linearly proportional to the radius.

In some case, certain conversion software or mask-writing machine imposes an upper limit on the number of levels that any design may contain. Here, the algorithm is limited to 16 levels. The maximum radius is about $6.26 \times 10^5\lambda$. However, larger radius can be supported by the improved algorithm shown in Fig. 5.33b. The difference is that there are more subgroups in L_1 so as to reduce the original number of level. Furthermore, improved algorithm does not improve too many storage (less than 1 megabytes (MB)). In this case, the maximum radius reaches over $10^7\lambda$ (e.g., 5 m for $\lambda = 500$ nm). For example, a $10^7\lambda$ radius device may be comprised of approximately $\pi \times 10^{12}$ meta-elements with a condition of $p = \lambda/2$. The unique definitions of position result in a large size of about 92.1 terabytes (TB) without compression. However, after compression, the file size is reduced to only about 218.9 MB. For

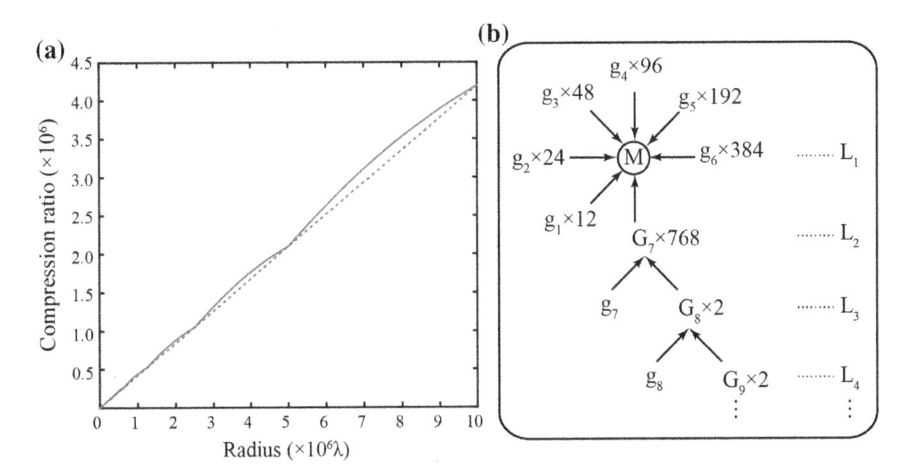

Fig. 5.33 **a** Compression ratio as a function the radius of the metalens. The solid curve and dash line depict the calculated compression ratio and linear function, respectively. **b** The cascade relationship of the improved algorithm

other radius smaller than $10^7\lambda$, corresponding file size is smaller than 218.94 MB. Furthermore, adding more subgroups into L_1 can further increase the maximum radius. We believe that this algorithm will facilitate the development of large-area metasurfaces by preventing debilitating large file sizes in the realm of terabytes and petabytes as feature densities increase and device areas continue to grow toward the meter scale and beyond.

5.4 Batch Fabrication of Subwavelength Structures

5.4.1 Laser Interference Lithography

Laser interference lithography is based on the coherent characteristics of either a continuous laser or a pulsed laser. Combining laser beams in different incident directions to control the intensity distribution in the interference field, and using photosensitive materials to record the interference field could result in periodic patterns such as gratings and lattices. Laser interference lithography does not require expensive projection optics, so it is an inexpensive lithography technique, which has a resolution of $\lambda/2 \sim \lambda/4$. And laser interference lithography has the advantages of simple system, no mask, infinite depth of focus, large exposure area, etc., making it especially suitable for processing periodic micro-/nanostructural pattern. According to the number of interfering laser beams, interference lithography can be divided into dual-beam interference lithography, three-beam interference lithography, four-beam interference lithography, and so on. In this section we will introduce dual-beam interference lithography as a typical representative. The principle of multi-beam interference lithography such as three-beam interference lithography and four-beam interference lithography is similar to that of the dual beam.

5.4.1.1 Principle of Dual-Beam Laser Interference Lithography

The principle of dual-beam laser interference lithography is shown in Fig. 5.34. When two plane waves with wavelength λ are incident on the photoresist at angles of $+\theta$ and $-\theta$, respectively, a periodic light intensity distribution along the x-axis is generated in the photoresist.

$$I(x) = 2I_0[1 + \cos(2kxn \sin\theta)] \tag{5.4.1}$$

where n is the refractive index and $k = 2\pi/\lambda$ is the propagating wave vector. Equation (5.4.1) is independent of the z-direction, indicating that the ideal interferometric lithography depth is infinite; and the actual depth of focus is determined by the overlap of the two beams and the coherence length of the incident laser beam. The

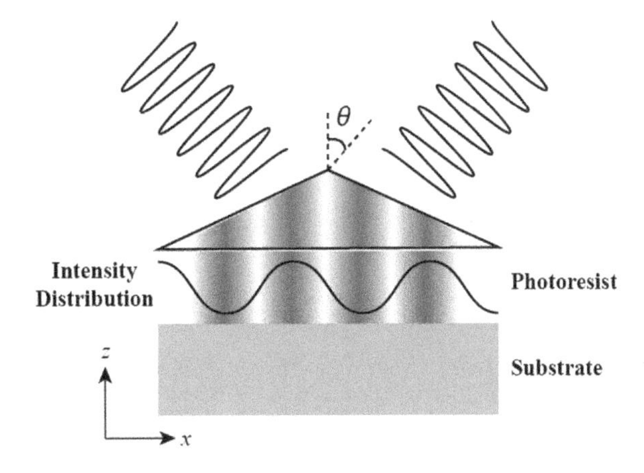

Fig. 5.34 The principle of dual-beam laser interference lithography

interfering light field changes in a sinusoidal period along the x-axis, and the period is determined by $d = \lambda/(2n \cdot \sin\theta)$.

5.4.1.2 Realization of Dual-Beam Laser Interference Lithography

The coherence and other properties of the laser determine the geometry of the interferometer for interference lithography. For single-mode (TEM_{00}) laser sources, the traditional geometry is the Lloyd's mirror arrangement, as shown in Fig. 5.35a [69]. The bottom half of the laser beam directly incident on the wafer with photoresist coated on it. The top half of the laser beam is reflected by the Lloyd mirror and is incident on the wafer. The angle θ can be adjusted by rotating the Lloyd's mirror. A similar structure is the prism geometry, as shown in Fig. 5.35b [69]. The angle is adjusted by changing the incident angle of the laser on the prism or by changing the apex angle of the prism. The prism interference geometry can also be adapted to immersion interference lithography, thereby improving the resolution of the lithography to a certain extent.

For multi-transverse-mode laser (such as 193 nm ArF laser), interferometer can be achieved by the Mach–Zehnder interference arrangement, as shown in Fig. 5.36a [69]. It ensures matching the transverse parts of the laser beam on the wafer, as shown by the solid and the dotted lines in Fig. 5.36a, thereby reducing the influence of the impure mode of the laser, which causes the disturbance of interference field. Figure 5.36b and c [70] are the interference geometries based on the diffraction grating. The diffraction grating is used to realize beam splitting and frequency selection. The period of the interfering field is $d = g/(2nm)$, where g is the period of diffraction grating, m is the diffraction order, n is the refractive index.

Figure 5.37a shows a half-pitch of 32 nm obtained by immersion interference lithography with a refractive index of 1.644 at 193 nm wavelength [71]. Figure 5.37b shows the EUV interference lithography system with a resolution of 8 nm [72].

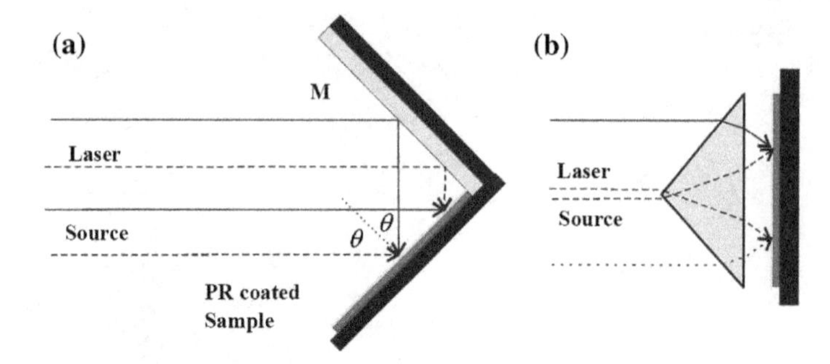

Fig. 5.35 **a** Lloyd's mirror interferometer geometries. **b** Prism interferometer geometries. Reprinted from [69] with permission. Copyright 2011, WILEY-VCH Verlag GmbH & Co. KGaA, Weinheim

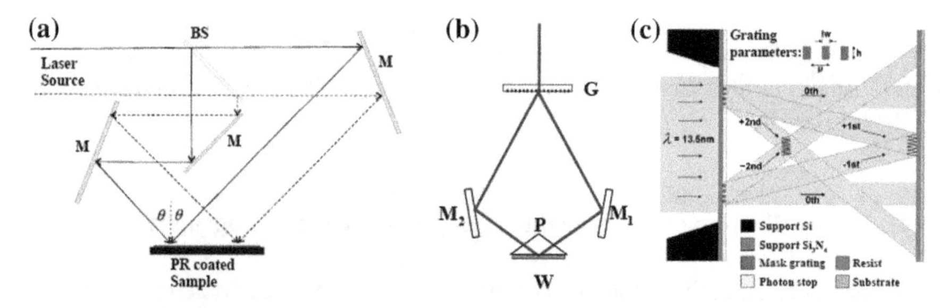

Fig. 5.36 **a** A standard Mach–Zehnder configuration. **b** and **c** The interference geometries based on the diffraction grating. **a** Reprinted from [69] with permission. Copyright 2011, WILEY-VCH Verlag GmbH & Co. KGaA, Weinheim. **c** Reprinted from [70] with permission. Copyright 2013, American Vacuum Society

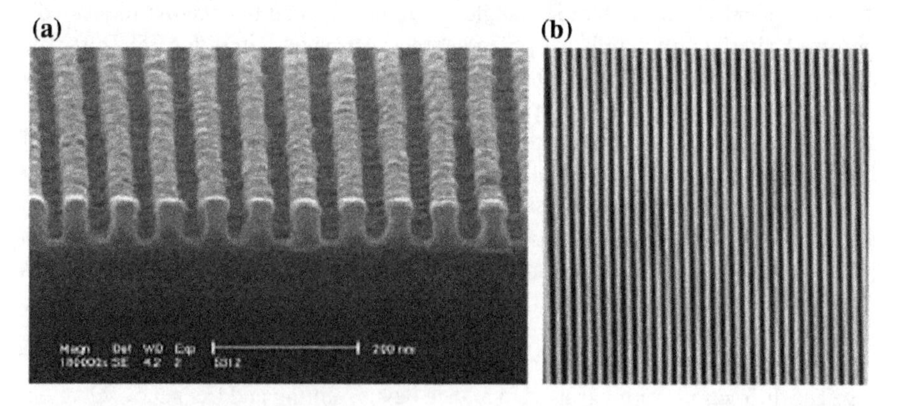

Fig. 5.37 **a** A half-pitch of 32 nm obtained by immersion interference lithography with a refractive index of 1.644 at 193 nm wavelength. **b** EUV interference lithography system with resolutions up to 8 nm. **a** Reprinted from [71] with permission. Copyright 2005, SPIE. **b** Reprinted from [72] with permission. Copyright 2013, SPIE

5.4.1.3 Influencing Factors of Laser Interference Lithography

One of the main parameters of interference lithography is the contrast of the interference fringes. The main factors that affect the contrast of the interference pattern are: non-monochromaticity of the light source, unequal intensity of the coherent beam, polarization state difference of each beam, size of the laser source, wavefront distortion of the interference beam, uniform amplitude of the interference beam, standing wave, etc.

Although traditional laser interference lithography can only be used for the fabrication of periodic structures, such structures have a large number of practical applications, such as: spectral spectroscopic elements, two- or three-dimensional photonic crystals, negative refractive materials, nanomaterial growth templates, and so on.

5.4.2 *Projection Optical Lithography*

Projection lithography is based on the principle of optical projection imaging that transfers the pattern on the mask to the photoresist-coated wafer. The schematic diagram is shown in Fig. 5.38 [73]. Projection lithography is the most widely used lithography technology in today's VLSI manufacturing. The optical projection exposure system is the basis of projection lithography. It mainly includes illumination system, projection objective system, workpiece stage, mask/reticle table, leveling and focusing system, mask/reticle and wafer alignment system, wafer transfer, pre-alignment system, and environmental control system. As the core of the lithography system, the projection objective covers almost all the core high-end technologies such as applied optics, optical manufacture, optical testing and optical film, and it is also the most difficult part of manufacturing.

Spanning a period of over 60 years and up to now comprising photons with wavelengths from the visible (436 nm) to mid-UV (365 nm), DUV (248 and 193 nm), and EUV (13.5 nm) regions, optical lithography has demonstrated remarkable longevity and is expected to continue into the foreseeable future. Resolution and depth of focus (DOF) are two important indicators for judging the exposure quality of projection lithography systems. Considering the different working manners of projection lithography (divided into dry lithography and immersion lithography), the resolution and DOF in dry and immersion lithographies can be given, respectively [74]. For dry lithography, according to the non-paraxial scaling equations, the resolution (R_{air}) is given by [74]:

$$R_{\text{air}} = k_1 \frac{\lambda}{\sin\theta} = k_1 \frac{\lambda_0/n_{\text{air}}}{\sin\theta_{\text{air}}} = k_1 \frac{\lambda_0}{\text{NA}_{\text{air}}} \tag{5.4.2}$$

where $\lambda = \lambda_0/n_{\text{air}}$ and λ is the wavelength in the imaging medium (air), θ_{air} is the refracted angle in air, k_1 is a process-dependent constant and is a measure of the

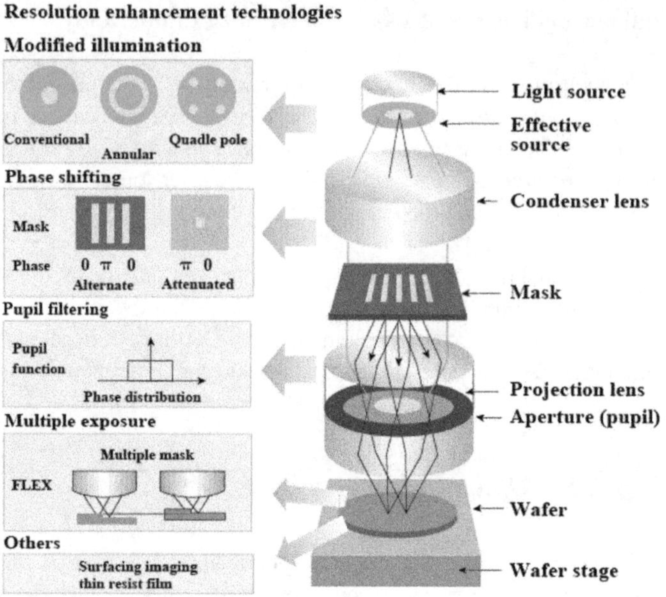

Fig. 5.38 Schematic view of a typical exposure optical system. Resolution enhancement techniques may be applied at various points in the optical path, as illustrated to the left of the figure. Reprinted from [73] with permission. Copyright 2000, Macmillan Publishers Limited

difficulty of the process, and θ_{air} is the refractive index of air (which is nearly equal to 1.0). The DOF of dry lithography is [74]:

$$\mathrm{DOF}_{\text{dry}} = k_3 \frac{\lambda_{\text{air}}}{\sin^2(\theta_{\text{air}}/2)} = k_3 \frac{n_{\text{air}}\lambda_0}{\mathrm{NHA}_{\text{air}}^2} \tag{5.4.3}$$

where $\mathrm{NHA}_{\text{air}} = n_{\text{air}}^2 \sin^2(\theta_{\text{air}}/2)$ is the numerical half-aperture and k_3 is a process-dependent constant.

Similarly, the non-paraxial scaling equations for resolution and focus depth in immersion lithographies are given by [74]:

$$R_{\text{immersion}} = k_1 \frac{\lambda_{\text{water}}}{\sin\theta_{\text{water}}} = k_1 \frac{\lambda_0/n_{\text{water}}}{\sin\theta_{\text{water}}} = k_1 \frac{\lambda_0}{\mathrm{NA}_{\text{water}}} \tag{5.4.4}$$

and

$$\mathrm{DOF}_{\text{immersion}} = k_3 \frac{\lambda_{\text{water}}}{\sin^2(\theta_{\text{water}}/2)} = k_3 \frac{n_{\text{water}}\lambda_0}{\mathrm{NHA}_{\text{water}}^2} \tag{5.4.5}$$

where $\mathrm{NHA_{water}} = n_{\mathrm{water}}^2 \sin^2(\theta_{\mathrm{water}}/2)$ is the numerical half-aperture, $\lambda_{\mathrm{water}} = \lambda_0/n_{\mathrm{water}}$ is the wavelength in the imaging medium (water), θ_{water} is the refracted angle in water.

As can be seen from the above equations, immersion lithography has many advantages compared to dry lithography. Firstly, the equivalent numerical aperture is larger than dry lithography because higher refractive index can couple higher spatial frequencies into the photoresist. Secondly, for a given diffraction order of light from the mask, the angle of the light inside water is less than that in dry lithography. These smaller angles result in smaller optical path differences between the various diffraction orders when they are out of focus; the result is a smaller degradation of the image for a given amount of defocus. In this way, immersion lithography provides a larger focus depth than that of dry lithography for a given NA.

Figure 5.39 gives an example of a metalens designed at 1550 nm, which was fabricated on a 4-inch fused silica (SiO_2) wafer substrate by a i-line stepper [68]. First, the a-Si layer with thickness of 0.6 μm was deposited on a substrate using plasma-enhanced chemical vapor deposition (PECVD). Then, photoresist with thickness of 1.1 μm and contrast enhancement material (CEM) with thickness of 0.4 μm were spun-coated on its surface [68]. The wafer was then exposed using an i-line stepper with 5× reduction. After exposure, the wafer was developed, followed by a water rinse and post-exposure bake. Then, the pattern was transferred into the a-Si by RIE.

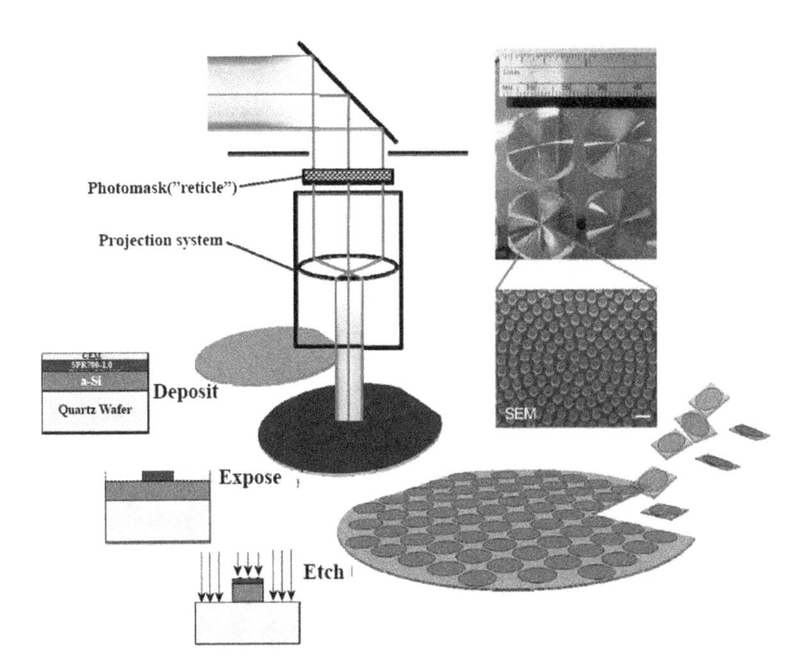

Fig. 5.39 A schematic diagram showing the production of metalenses with i-line stepper. A SEM of the metalens center (center right) shows the microscopic posts comprising the metalens. Scale bar: 2 μm. Reprinted from [68] with permission. Copyright 2018, Optical Society of America

Finally, after removing the residual photoresist, the wafer can be diced into separate individual metalens devices

Utilizing immersion scanner, the resolution can be improved to down to 40 nm. Figure 5.40a shows a color display metasurface fabrication on a 12-inch Si wafer [75]. A 70 nm-thick SiN layer and a 130 nm-thick a-Si layer were deposited on the Si substrate using PECVD. The designed metasurface was then patterned by immersion lithography, followed by the inductively coupled plasma (ICP) etching. The Nikon immersion scanner with resolution down to 40 nm was used [75]. After photoresist being removed, wet clean processes were used to remove the polymer generated in the a-Si etching process. The pillar diameters in Fig. 5.40b–d are around 166, 120, and 65 nm, respectively.

5.4.3 *NanoImprint Lithography*

Nanoimprint lithography (NIL) is derived from ancient casting techniques. The earliest research on microlithography based on imprinting or mold dates back to the 1970s. However, it does not attract widespread attention until the report of thermal imprint lithography (T-NIL) by Stephen Y. Chou [76]. As an inexpensive and high-resolution micro- and nanopattern replication technology, nanoimprinting mainly utilizes the mechanical deformation of photoresist materials to form pattern with low cost and high resolution, avoiding the resolution encountered in conventional optical lithography. Nanoimprint technology has been selected in the international semiconductor technology roadmap as a candidate for next generation lithography (NGL).

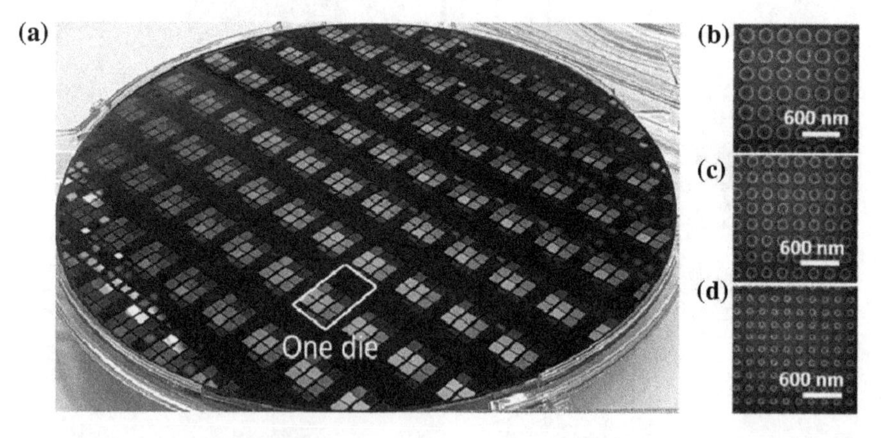

Fig. 5.40 **a** Photograph of the fabricated 12-inch wafer. SEM images of a-Si nanopillar arrays for 166 nm (**b**), 120 nm (**c**), and 65 nm (**d**), respectively. Reprinted from [75] with permission. Copyright 2018, Optical Society of America

The nanoimprint technology process mainly includes T-NIL, UV-NIL, step and flash imprint lithography and roll-type nanoimprint.

5.4.3.1 Thermal Nanoimprint

Usually, the material used for thermal nanoimprint is a thermoplastic or thermosetting polymer. Before imprinting, the polymer needs to be heated to the glass transition temperature (T_g), and then pressurize the imprint template and the polymer. After demolding, the polymer patterns are transferred into the substrate by dry etching. Generally, it is necessary to introduce O_2 plasma etch to remove the residual polymer. The thermal nanoimprint process is shown in Fig. 5.41a, which includes the following steps:

- Preparation and pretreatment of stamps. Nanoimprint stamps should have high mechanical strength and can be prepared in a variety of ways, such as the aforementioned EBL, FIB, laser interference lithography et al. Then, an anti-adhesive layer should be coated on the surface of stamp. Otherwise, the conglutination of the stamp and the polymer would affect the quality of imprint patterns.
- Heating and pressurizing. Under the pressure, heating is also required to increase the polymer temperature beyond T_g, thereby reducing the viscosity of the polymer and increasing the fluidity. Mostly, the pressure is as high as 50–100 bar.
- Pressurize and cool down, decompression and demolding process. After thermal imprint, it is necessary to cool down the system while maintains the pressure, and then release pressure and demold.

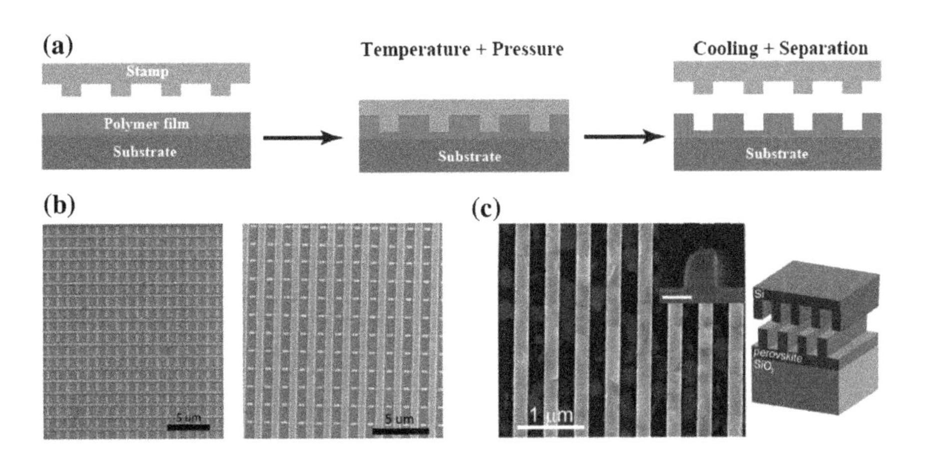

Fig. 5.41 **a** Schematics of a standard thermal nanoimprint lithography process. **b** An SEM image showing a larger area of the nanoimprinted, colloidal Au-nanocrystal-based nanoantennas fabricated on a glass substrate. **c** SEM imaging of nanoimprinted nanostripe perovskite metasurfaces. **b** Reprinted from [77] with permission. Copyright 2015, American Chemical Society. **c** Reprinted from [78] with permission. Copyright 2017, American Chemical Society

The replication performance of thermal nanoimprint is affected by the following factors: viscosity, glass transition temperature, shrinkage, hardness, and etch resistance of the polymer material; contact surface between the polymer and the stamp; and so on.

Thermal nanoimprint technology has the advantages such as good flexibility, low cost, easy fabrication for high-aspect-ratio and high-resolution structures. However, thermal nanoimprint needs to undergo a long heating and cooling process, which lowers the process efficiency and causes thermal mismatch due to the different thermal expansion coefficients of the stamp material and the polymer material. The alignment ability is also affected by heating and cooling process.

Figure 5.41b shows large-area nanoantennas fabricated using solution-based nanoimprinting and ligand exchange of colloidal Au nanocrystal [77]. Au nanocrystals were deposited by spin-coating, then lifting off the imprint resist in acetone, and immersing the patterned NC thin film in an acetone solution of thiocyanate. The dimensions of the template are designed to be larger than those of the desired final structures to compensate for shrinkage of the NC-based nanostructures upon ligand exchange. Not only polymer film, thermal nanoimprint can also be applied to perovskite films. By alloying the organic cation part of perovskites, designing a composition of a triple alloy of a mixed cation exhibiting record photovoltaic performance and increased stability. Figure 5.41c shows the SEM image of a nanoimprinted nanostripe perovskite metasurfaces [78]. The results suggest a cost-effective approach based on nanoimprint lithography. Combined with simple chemical reactions for creating a new generation of functional metasurfaces, which may pave a way toward high-efficient planar optoelectronic metadevices.

5.4.3.2 UV Nanoimprint

UV nanoimprint can be done at room temperature while requires a pressure below one atmosphere. The basic process flow is shown in Fig. 5.42a. It can be seen that there are two main differences between UV nanoimprint and thermal nanoimprint: firstly, the stamp structures need to be UV transparent; secondly, the polymer used is usually a composite photoresist. UV nanoimprint is carried out at room temperature, thus saving the time for heating and cooling; and eliminating the thermal mismatch of the substrate. In addition, the polymer used in UV nanoimprint generally has low viscosity, which improves the lifetime of the stamp. UV curable polymeric material is the key factor for the ultimate performance of UV nanoimprint, which should have low adhesion, fast cure, high resolution, low shrinkage, and good etching resistance.

Figure 5.42b and c gives an example of a plasmonic metasurface fabricated by UV nanoimprint [79]. The sample was fabricated on a 0.5-mm-thick, 4-inch Borofloat glass wafer, upon which a thin film of Ormocomp mixed with ma-T 1,050 thinner (25% w/w) was deposited. A silicon stamp with an anti-stiction coating was employed for replicating the pillar structure through room temperature nanoimprinting into the Ormocomp layer. The Ormocomp film was cured by exposure to UV light and

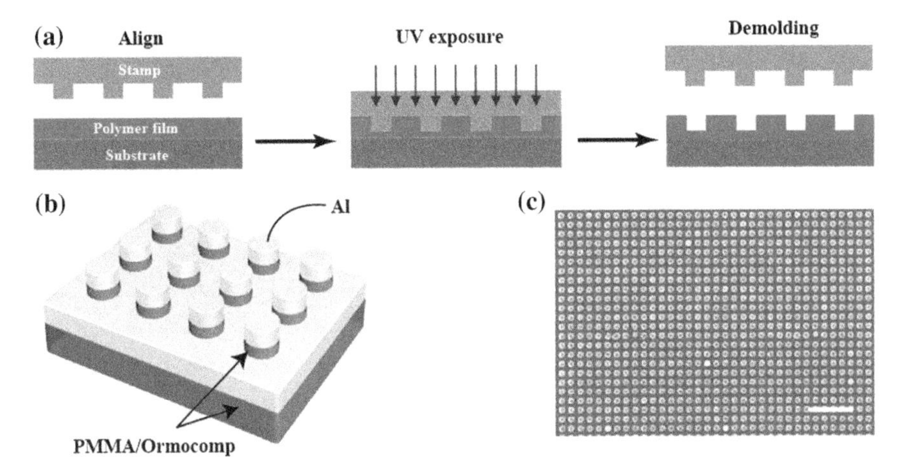

Fig. 5.42 **a** Schematic of UV-NIL: imprint resist is coated on the substrate; the stamp is imprinted into the resist, and UV curing is applied; separation of the stamp and the substrate. **b** Schematic illustrations of the plasmonic metasurface, where the periodicity, $P = 200$ nm, the thickness of Al is $T = 20$ nm, the height of the pillars is $H = 30$ nm, and the diameter of the disks is D. **c** A top-view SEM image of a plasmonic metasurface. Scale bar: 1 μm. **b** and **c** Reprinted from [79] with permission. Copyright 2016, Macmillan Publishers Limited

separated from the silicon master, and subsequently peeled from the Borofloat glass substrate.

5.4.3.3 Step and Flash Imprint Lithography

Step and flash imprint lithography (SFIL) mainly relies on chemical and low pressure mechanical processes to replicate patterns. The SFIL process is shown schematically in Fig. 5.43a [80]. The process starts with a template that the patterns are written using EBL, and then is etched into the glass blank. The imprint resist is formed from an array of picoliter-sized drops that are correlated with the pattern density on the template. The resist spreads across the imprint field as the template is lowered onto the drop array. When the surface tension of the liquid phase imprint resist has been broken, capillary action draws the resist into the template features. Once filling is completed, UV light is illuminated on the back side of the glass template, thereby driving a cross-linking action that converts the resist to a solid. The template can then be withdrawn and the process can be repeated on the next field. The sectional view of the SFIL imprint column is shown in Fig. 5.43b [80].

5.4.3.4 Roll-Type Nanoimprint

Compared with thermal nanoimprint and UV nanoimprint, roll-type nanoimprint can fabricate patterns with large area because it works in a continuous manner, which greatly improve the efficiency and cut the costs. Figure 5.44a illustrates the

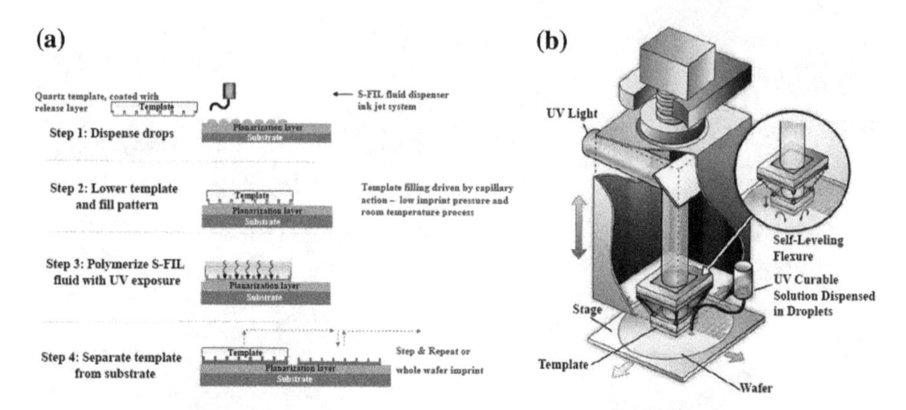

Fig. 5.43 **a** Schematic illustrations of SFIL process. **b** Sectional view of a SFIL imprint column. Reprinted from [80] with permission. Copyright 2011, SPIE

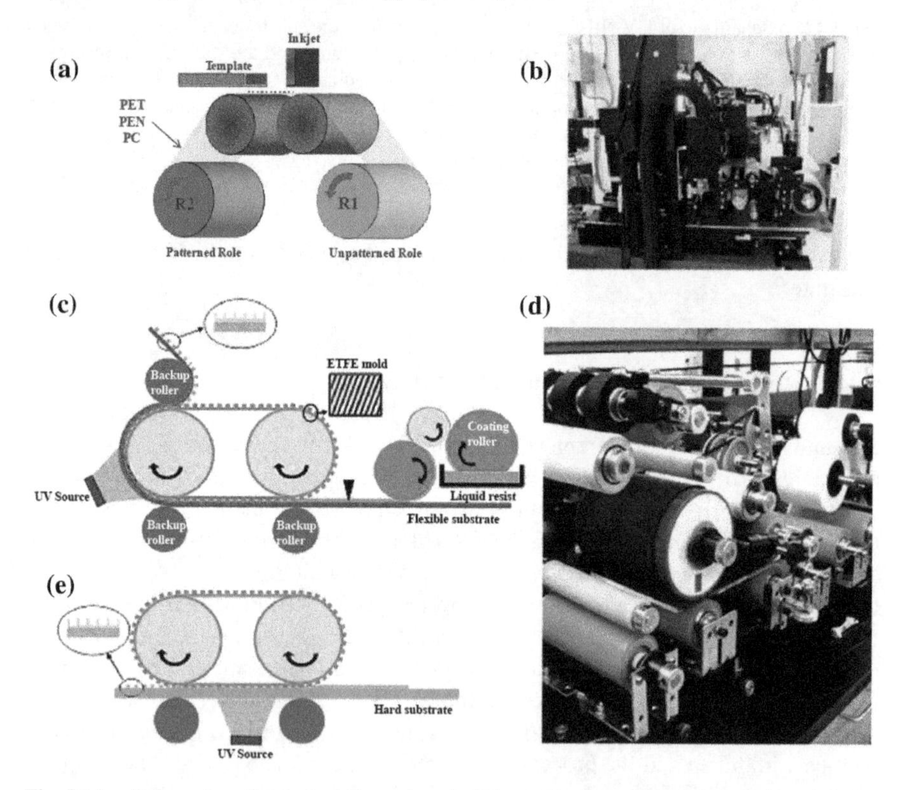

Fig. 5.44 **a** Schematics of plate-to-roll nanoimprint lithography using jet and flash. **b** A partial view of plate-to-roll imprint system. Schematics of **c** roll-to-roll nanoimprint lithography (R2RNIL) and **d** roll-to-plate nanoimprint lithography (R2PNIL) process. **e** Photograph of 6 inch capable R2R/R2PNIL apparatus. **a** Reprinted from [81] with permission. Copyright 2012, SPIE. **c–e** Reprinted from [82] with permission. Copyright 2009, American Chemical Society

basic imprint tool using a template and a roll module where the flexible film can be patterned [81]. Imprinting is performed by moving the roll module and the template is only translated up or down. The process sequence is as follows: First, fluid dispensing with picoliter volume drops is performed by moving a linear stage onto which the roller module is mounted. Drop patterns are preprogrammed based on the template pattern geometry. Once the fluid drops are dispersed on the film, the roll module is moved to the opposite side of the template, while the dispersed portion of the film is rolled backwards by the counterclockwise roller motion. Imprinting is performed by the synchronized motion of the roll module and bottom linear stage so that the dispersed portion of the film is brought into contact with the template similar to a laminating process. It is followed by a UV curing step, where a broadband UV spectrum is used and a separation step, where synchronized motions of the roll module and linear stage induce a peeling separation from the template starting at one side and ending at the other side of the template.

Figure 5.44c and d shows the overall configuration of a continuous (c) R2RNIL and (d) R2PNIL nanomanufacturing process [82], which consists of a plurality of rollers. The processes of coating, embossing, UV curing and demolding can be realized by controlling the rotation of the rollers and the pressure between the rollers and the belt. To guarantee uniform coating thickness regardless of web speed, the coating system is synchronized with the main imprinting roller. The pressure between the imprint roller and the back-up roller is controlled by a clamping device and a force sensor. The process uses a high-power UV light source to cure the embossing resist.

5.4.4 Bottom-up Fabrication

Bottom-up nanofabrication approaches are related to the construction of multi-functional nanostructural materials and devices by the self-assembly of atoms, molecules or particles. A true "bottom-up" fabrication process based on self- and directed assembly will begin to impact the transfer of nanotechnology into volume manufacturing at a level unthinkable with extensions of today's "top-down" micro-/nanofabrication. Although it is still far to realize, self-organizing functional devices are the ultimate aim of bottom-up fabrication.

5.4.4.1 Block Copolymers Self-assembly

Block copolymers (BCPs) are polymers that consist of two or more distinct polymer chains (blocks) joined by a covalent bond. The BCPs self-assembly is inseparable from the unique micro-phase separation of immiscible polymer blocks, offering a route to creating macroscopically large samples with well-ordered periodic nanostructures. Various and complex intrinsic geometric shapes can be obtained by BCPs self-assembly, including hexagonal cylinder arrays, parallel line arrays, double gyroids, and other three-dimensional structures.

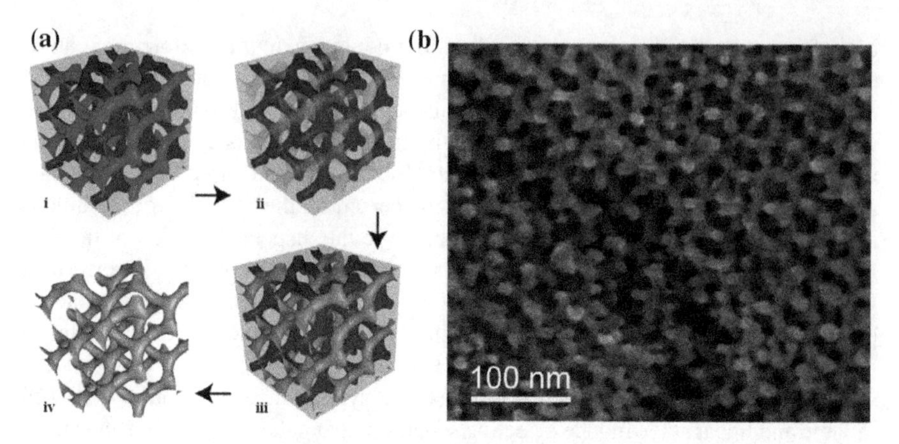

Fig. 5.45 **a** Schematic representation of sample fabrication. **b** SEM image of a metal gyroid by filling a porous polymer network. Reprinted from [83] with permission. Copyright 2012, WILEY-VCH Verlag GmbH & Co. KGaA, Weinheim

Figure 5.45 shows a 3D gold metamaterial created by BCPs self-assembly [83]. Figure 5.45a is a schematic representation of the fabrication process. Firstly, an isoprene-*block*-styrene-*block*-ethylene oxide (ISO) BCP forms two chemically distinct and interpenetrating gyroid networks (I (blue), O (red)) of opposite chirality in a matrix of the third block (gray) [83]. Secondly, the I gyroid network is then removed by selective UV and chemical etching. Thirdly, back-filled with gold (yellow) by electrodeposition. Finally, the remaining polymer was removed by plasma etching, revealing a 3D continuous gold network in air, as shown in Fig. 5.45b.

Figure 5.46 presents a tunable reflective index visible-light metasurface by BCPs self-assembly and shrinkage, which form an Au nanoparticle ensemble with sub-10 nm interparticle distance [84]. BCPs self-assembly is utilized to form a nanotemplate. A cylindrical polystyrene-block-poly(methyl methacrylate) (PS-b-PMMA) with 85-nm-thick was spin-coated on a Si substrate and thermally annealed to self-assemble into vertical hexagonal cylinder nanodomains. After Au deposition and lift-off, an Au nanoparticle array was generated. Then, a transfer mediator (PMMA) was spin-casted upon the Au nanoparticle array. Subsequently, the sample was immersed in a water bath to separate the PMMA film embedded with Au nanoparticle array from the substrate, floating at the water surface. It was carefully transferred to the shrinkage film. After complete drying, the PMMA layer was removed by an acetic acid solution and thoroughly washed. Subsequent heat treatment was performed at 180 °C. The initial and final interparticle separations were shown in Fig. 5.46b and c, respectively.

5.4.4.2 Self-assembly Monolayer Colloidal Crystal

Self-assembly monolayer colloidal crystal refers to assembly of colloidal crystal with micro- or nanoscale to form one-, two-, or three-dimensional structures. The self-assembly process mainly relies on external forces, such as electric field, Van der

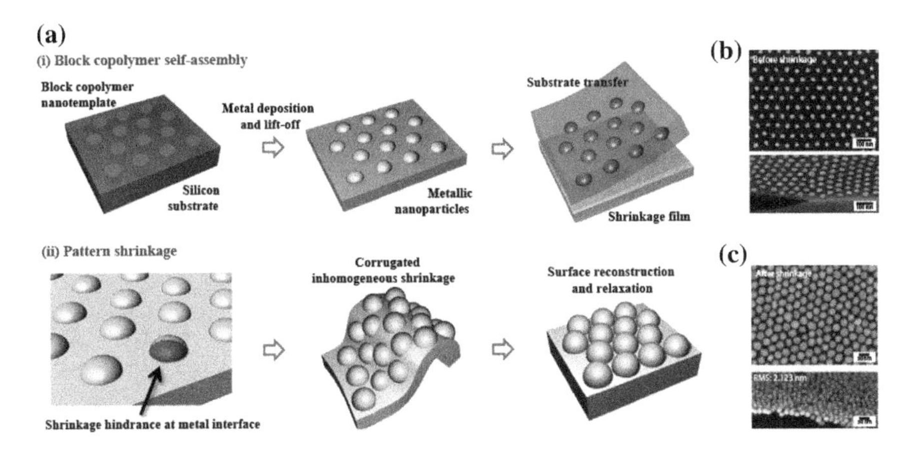

Fig. 5.46 **a** Schematic for metal nanoparticle ensemble preparation by (i) BCP self-assembly and substrate transfer and (ii) pattern shrinkage. **b** SEM images of hexagonal Au nanoparticle arrays as-prepared from BCP self-assembly (upside: plane view, downside: 60° tilted view). **c** SEM images of Au nanoparticle ensembles after complete pattern shrinkage (upside: plane view, downside: 60° tilted view). Reprinted from [84] with permission. Copyright 2016, the Author(s)

Waals force, magnetic field, and capillary force. Self-assembly needs to meet certain conditions: the diameter of colloidal crystal should be uniform and small enough that can move freely. Figure 5.47 shows a process to fabricate optical metasurface with self-assembly microsphere and shadow deposition [85]. Polystyrene spheres with 1 μm diameter were first self-assembled at an air/water interface, and then deposited on bare silicon wafers. An isotropic etching with oxygen plasma was performed to reduce the diameter of the spheres, thereby opening the gaps to the design specifications. Finally, the samples were mounted onto a custom-built, dual-axis rotation stage [85]. The relative angle between the sample and source of deposition is adjusted, as specified by the designs. Figure 5.47a and b is the schematic for single- and multi-angled deposition. Figure 5.47c–h shows the designed and experimental results of interconnected lines, asymmetric and symmetric bars with six angles of projection. This method offers a new way for development of metasurfaces with simple design and rapid fabrication. But it also presents several challenges.

5.5 Challenges of Fabrication Techniques for EO 2.0

In the above sections, we have introduced the research progress of the fabrication methods including layered structures fabrication, direct writing, laser interference lithography, projection optical lithography, nanoimprint lithography, and bottom-up fabrication, from which we can see many new principles and methods have emerged. We believe that this is just a starting point. The fabrication technology for EO 2.0 still faces many unknowns and challenges, and it is an important motivation to promote

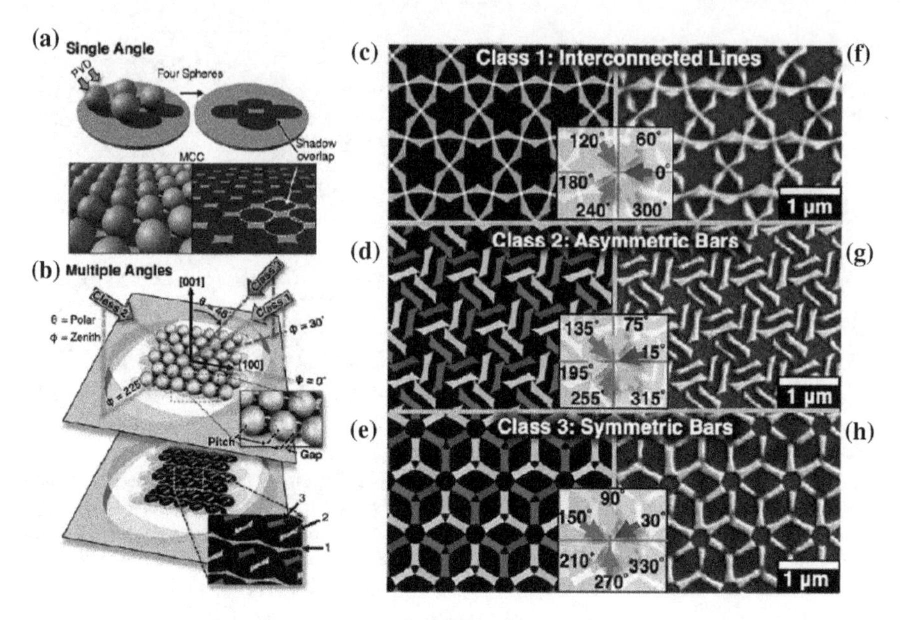

Fig. 5.47 Schematic for single- and multi-angled deposition. **a** Generation of a single bar or an array of bars by a single angle of deposition. **b** Definition of free parameters relative to the crystal axis. Here we show an example composed of three different types of features: (1) an interconnected line, (2) an asymmetric bar, and (3) a symmetric bar. **c**–**e** Composition of six angles of projection demonstrating the duplication of feature types 1–3 at intervals of $\phi = 60°$ due to the C_6 symmetry of the lattice. **f**–**h** SEM images of the fabricated patterns (20 nm of Ag on Si). Inset: The six angles by which the features are reproduced. Reprinted from [85] with permission. Copyright 2014, American Chemical Society

its leaping development. In general, the potential fabrication challenges faced by EO 2.0 mainly include:

(1) Large-aperture high-resolution direct writing

As a maskless patterning technology, direct writing is highly flexible in processing and therefore may be widely used in EO 2.0. As mentioned in the previous section, the direct writing that currently used in EO 2.0 can be divided into optical-beam direct writing, ion-beam direct writing and electron-beam direct writing. Zone-plate-array lithography and DUV laser pattern generator based on spatial-light modulator in optical-beam direct writing need to be further improved in resolution, large-aperture patterning efficiency and stability. The spatial resolution of OBL is comparable to electron-beam direct writing, but the technology is still in the researching stage, so the system is immature. Besides, the stability of large-area direct writing is unknown, and the photosensitive material still has incompatibility. Although ion-beam direct writing can directly form a high-resolution structure on the surface of the material by sputtering, the processing efficiency is low and the patterning area is only 4 inch. Electron-beam direct writing has a very high resolution, and the maximum diameter of the patterning can be up to 300 mm. Theoretically, large-aperture direct writing can be realized by introducing a large travel range stage, but it brings many problems such as an increase in the vacuum chamber, laser interference measurement in a large travel range, and patterning efficiency. Therefore, high-efficiency, high-resolution, and large-aperture direct writing is one of the problems in EO 2.0 fabrication technology which needs to be broken through.

(2) Large-aperture high-resolution pattern replication technology

Although direct-writing technology can meet the needs of research and small-volume production, the application of EO 2.0 in the future will face the efficiency, productivity and cost requirements, so large-aperture with high-resolution pattern replication technology should be developed. At present, the pattern replication technology is mainly divided into optical lithography and nanoimprint. Projection optical lithography is a commonly used high-resolution optical lithography technique. The resolution can be improved to below 40 nm by shortening the exposure wavelength and increasing the numerical aperture, but it is limited by the dimension of projection field (typically 26 mm × 33 mm). So, direct imaging of full aperture patterns is not possible. Nanoimprint lithography, a high-throughput, low-cost, and non-physical-limit lithography, is discommodious to avert the demolding and defects over large aperture.

(3) Large-aperture high-fidelity pattern transfer technology

The core of the pattern transfer technology is to transfer the replicated structure to the functional materials. The fidelity of the transfer process determines the overall performance of the final device. Lift-off and etching are two main pattern transfer methods. The height of the lift-off pattern is determined by the height of the initial structure after replication (generally, the initial structural height after replication is three times of the height of the lift-off pattern), so it is hard to transfer pattern with high aspect ratio. The wet etching is an isotropic etching method, so it cannot meet the transfer requirements with high-fidelity and high aspect ratio. Dry etching (or plasma etching) is superior to wet etching in terms of anisotropy and transfer fidelity, but most of the existing dry etching systems can only achieve pattern transfer with a diameter below 300 mm. So, in order to increase the transfer area, technical innovations are required in the new design of the vacuum chamber, antenna, and gas in.

(4) Conformal flexible fabrication technology with large-aperture

In the context of conformal and freeform optics, optical components with non-standard surfaces were developed for integration of optics into various platforms with specific geometric shapes. The conventional solution is to stack several bulky optical elements with non-standard surface profiles underneath the outermost surface of the object. The approach of conformal optics with subwavelength structures can provide a solution for decoupling the geometric shape and optical characteristics of arbitrary objects [86]. Several efforts have been made to transfer subwavelength structures to flexible substrates with the aim of tuning their frequency response using substrate deformation. But flexible subwavelength structures that can be conformed to a large-aperture non-planar arbitrarily shaped object have not been realized.

(5) Super-molecular and super-atom manufacturing methods

At present, the processing demand of EO 2.0 is mainly concentrated on the nanometer-to-micrometer. But from the perspective of its development trend, EO 2.0 will be expanded toward the direction of dynamic, nonlinear, and quantum regimes, thus ultimately requiring the molecular or atomic-level resolution. Manufacturing methods for manipulating materials at molecular or atomic scales, which greatly exceeds the current manufacturing level, require the development of molecular or even atomic-level structural constructive methods.

References

1. K. Schwertz, An introduction to the optics manufacturing process. Optomech. Rep. (2008)
2. A.Y.C. Nee, *Handbook of Manufacturing Engineering and Technology* (Springer, 2015)
3. H. Zappe, *Fundamentals of Micro-Optics* (Cambridge University Press, 2010)
4. Z. Cui, *Nanofabrication: Principles, Capabilities and Limits* (Springer, 2017)
5. M.T. Gale, K. Knop, The fabrication of fine lens arrays by laser beam writing, in *Proceedings of SPIE 0398, Industrial Applications of Laser Technology*, vol. 0398 (1983), pp. 0398–7
6. H.P. Herzig, *Micro-Optics: Elements, Systems and Applications* (Taylor & Francis Ltd, 1997)
7. B. Bharat, *Encyclopedia of Nanotechnology* (Springer Science+Business Media B.V., 2012)
8. C. Du, X. Dong, C. Qiu, Q. Deng, C. Zhou, Profile control technology for high-performance microlens array. Opt. Eng. **43**, 2595–2602 (2004)
9. J.-S. Huang, V. Callegari, P. Geisler, C. Brüning, J. Kern, J.C. Prangsma, X. Wu, T. Feichtner, J. Ziegler, P. Weinmann, M. Kamp, A. Forchel, P. Biagioni, U. Sennhauser, B. Hecht, Atomically flat single-crystalline gold nanostructures for plasmonic nanocircuitry. Nat. Commun. **1**, 150 (2010)
10. V.J. Logeeswaran, N.P. Kobayashi, M.S. Islam, W. Wu, P. Chaturvedi, N.X. Fang, S.Y. Wang, R.S. Williams, Ultrasmooth silver thin films deposited with a germanium nucleation layer. Nano Lett. **9**, 178–182 (2009)
11. J.H. Park, P. Ambwani, M. Manno, N.C. Lindquist, P. Nagpal, S.-H. Oh, C. Leighton, D.J. Norris, Single-crystalline silver films for plasmonics. Adv. Mater. **24**, 3988–3992 (2012)
12. C. Zhang, D. Zhao, D. Gu, H. Kim, T. Ling, Y.-K.R. Wu, L.J. Guo, An ultrathin, smooth, and low-loss Al-doped Ag film and its application as a transparent electrode in organic photovoltaics. Adv. Mater. **26**, 5696–5701 (2014)
13. Y. Wu, C. Zhang, N.M. Estakhri, Y. Zhao, J. Kim, M. Zhang, X.-X. Liu, G.K. Pribil, A. Alù, C.-K. Shih, X. Li, Intrinsic optical properties and enhanced plasmonic response of epitaxial silver. Adv. Mater. **26**, 6106–6110 (2014)
14. A. Poddubny, I. Iorsh, P. Belov, Y. Kivshar, Hyperbolic metamaterials. Nat. Photonics **7**, 948 (2013)
15. T. Xu, A. Agrawal, M. Abashin, K.J. Chau, H.J. Lezec, All-angle negative refraction and active flat lensing of ultraviolet light. Nature **497**, 470 (2013)
16. X. Yang, J. Yao, J. Rho, X. Yin, X. Zhang, Experimental realization of three-dimensional indefinite cavities at the nanoscale with anomalous scaling laws. Nat. Photonics **6**, 450 (2012)
17. H. Lee, Z. Liu, Y. Xiong, C. Sun, X. Zhang, Development of optical hyperlens for imaging below the diffraction limit. Opt. Express **15**, 15886–15891 (2007)
18. J. Rho, Z. Ye, Y. Xiong, X. Yin, Z. Liu, H. Choi, G. Bartal, X. Zhang, Spherical hyperlens for two-dimensional sub-diffractional imaging at visible frequencies. Nat. Commun. **1**, 143 (2010)
19. L. Liu, K. Liu, Z. Zhao, C. Wang, P. Gao, X. Luo, Sub-diffraction demagnification imaging lithography by hyperlens with plasmonic reflector layer. RSC Adv. **6**, 95973–95978 (2016)

20. G. Liang, C. Wang, Z. Zhao, Y. Wang, N. Yao, P. Gao, Y. Luo, G. Gao, Q. Zhao, X. Luo, Squeezing bulk plasmon polaritons through hyperbolic metamaterials for large area deep subwavelength interference lithography. Adv. Opt. Mater. **3**, 1248–1256 (2015)
21. A.J. Hoffman, L. Alekseyev, S.S. Howard, K.J. Franz, D. Wasserman, V.A. Podolskiy, E.E. Narimanov, D.L. Sivco, C. Gmachl, Negative refraction in semiconductor metamaterials. Nat. Mater. **6**, 946 (2007)
22. M.Y. Shalaginov, V.V. Vorobyov, J. Liu, M. Ferrera, A.V. Akimov, A. Lagutchev, A.N. Smolyaninov, V.V. Klimov, J. Irudayaraj, A.V. Kildishev, A. Boltasseva, V.M. Shalaev, Enhancement of single-photon emission from nitrogen-vacancy centers with TiN/(Al,Sc)N hyperbolic metamaterial. Laser Photonics Rev. **9**, 120–127 (2014)
23. D. Gil, R. Menon, H.I. Smith, The promise of diffractive optics in maskless lithography. Micro Nano Eng. **2003**(73–74), 35–41 (2004)
24. R. Menon, E.E. Moon, M.K. Mondol, F.J. Castaño, H.I. Smith, Scanning-spatial-phase alignment for zone-plate-array lithography. J. Vac. Sci. Technol. B Microelectron. Nanometer Struct. Process. Meas. Phenom. **22**, 3382–3385 (2004)
25. R. Menon, A. Patel, E.E. Moon, H.I. Smith, Alpha-prototype system for zone-plate-array lithography. J. Vac. Sci. Technol. B Microelectron. Nanometer Struct. Process. Meas. Phenom. **22**, 3032–3037 (2004)
26. H.I. Smith, R. Menon, A. Patel, D. Chao, M. Walsh, G. Barbastathis, Zone-plate-array lithography: a low-cost complement or competitor to scanning-electron-beam lithography. Microelectron. Eng. MNE **2005**(83), 956–961 (2006)
27. R. Menon, A. Patel, D. Chao, M. Walsh, H.I. Smith, Zone-plate-array lithography (ZPAL): optical maskless lithography for cost-effective patterning, in *Proceedings of SPIE 5751, Emerging Lithographic Technologies IX*, vol. 5751 (2005), pp. 5751–10
28. P. Björnängen, M. Ekberg, T. Öström, H.A. Fosshaug, J. Karlsson, C. Björnberg, F.K. Nikolajeff, M. Karlsson, DOE manufacture with the DUV SLM-based Sigma7300 laser pattern generator, in *Proceedings of SPIE 5377, Optical Microlithography XVII*, vol. 5377 (2004), pp. 5377–10
29. U.B. Ljungblad, P. Askebjer, T. Karlin, T. Sandstrom, H. Sjoeberg, A high-end mask writer using a spatial light modulator, in *Proceedings of SPIE 5721, MOEMS Display and Imaging Systems III*, vol. 5721 (2005), pp. 5721–10
30. H.K. Lakner, P. Duerr, U. Dauderstaedt, W. Doleschal, J. Amelung, Design and fabrication of micromirror arrays for UV lithography, in *Proceedings of SPIE 4561, MOEMS and Miniaturized Systems II*, vol. 4561 (2001), pp. 4561–10
31. H. Martinsson, T. Sandstrom, A.J. Bleeker, J.D. Hintersteiner, Current status of optical maskless lithography. J. MicroNanolithograhy MEMS MOEMS **4**, 011003-4–15 (2005)
32. J. Aman, H.A. Fosshaug, T. Hedqvist, J. Harkesjo, P. Hogfeldt, M. Jacobsson, A. Karawajczyk, J. Karlsson, M. Rosling, H.J. Sjoberg, Properties of a 248-nm DUV laser mask pattern generator for the 90-nm and 65-nm technology nodes, in *Proceedings of SPIE 5256, 23rd Annual BACUS Symposium on Photomask Technology*, vol. 5256 (2003), pp. 5256–11
33. J.K. Gansel, M. Thiel, M.S. Rill, M. Decker, K. Bade, V. Saile, G. von Freymann, S. Linden, M. Wegener, Gold helix photonic metamaterial as broadband circular polarizer. Science **325**, 1513 (2009)
34. A. Frölich, J. Fischer, T. Zebrowski, K. Busch, M. Wegener, Titania woodpiles with complete three-dimensional photonic bandgaps in the visible. Adv. Mater. **25**, 3588–3592 (2013)
35. J. Fischer, J.B. Mueller, A.S. Quick, J. Kaschke, C. Barner-Kowollik, M. Wegener, Exploring the mechanisms in STED-enhanced direct laser writing. Adv. Opt. Mater. **3**, 221–232 (2014)
36. J. Kaschke, M. Wegener, Gold triple-helix mid-infrared metamaterial by STED-inspired laser lithography. Opt. Lett. **40**, 3986–3989 (2015)
37. Z. Gan, Y. Cao, R.A. Evans, M. Gu, Three-dimensional deep sub-diffraction optical beam lithography with 9 nm feature size. Nat. Commun. **4**, 2061 (2013)
38. Z. Gan, M.D. Turner, M. Gu, Biomimetic gyroid nanostructures exceeding their natural origins. Sci. Adv. **2**, e1600084 (2016)

39. S. Tan, R. Livengood, D. Shima, J. Notte, S. McVey, Gas field ion source and liquid metal ion source charged particle material interaction study for semiconductor nanomachining applications. J. Vac. Sci. Technol. B **28**, C6F15–C6F21 (2010)
40. I. Utke, P. Hoffmann, J. Melngailis, Gas-assisted focused electron beam and ion beam processing and fabrication. J. Vac. Sci. Technol. B Microelectron. Nanometer Struct. Process. Meas. Phenom. **26**, 1197–1276 (2008)
41. M. Pu, X. Li, X. Ma, Y. Wang, Z. Zhao, C. Wang, C. Hu, P. Gao, C. Huang, H. Ren, X. Li, F. Qin, J. Yang, M. Gu, M. Hong, X. Luo, Catenary optics for achromatic generation of perfect optical angular momentum. Sci. Adv. **1**, e1500396 (2015)
42. J. Valentine, S. Zhang, T. Zentgraf, E. Ulin-Avila, D.A. Genov, G. Bartal, X. Zhang, Three-dimensional optical metamaterial with a negative refractive index. Nature **455**, 376 (2008)
43. A. Cui, Z. Liu, J. Li, T.H. Shen, X. Xia, Z. Li, Z. Gong, H. Li, B. Wang, J. Li, H. Yang, W. Li, C. Gu, Directly patterned substrate-free plasmonic "nanograter" structures with unusual Fano resonances. Light Sci. Appl. **4**, e308 (2015)
44. Z. Liu, H. Du, J. Li, L. Lu, Z.-Y. Li, N.X. Fang, Nano-kirigami with giant optical chirality. Sci. Adv. **4**, eaat4436 (2018)
45. B.R. Appleton, S. Tongay, M. Lemaitre, B. Gila, D. Hays, A. Scheuermann, J. Fridmann, Multi-ion beam lithography and processing studies. MRS Proc. **1354**, mrss11-1354-ii03-05 (2011)
46. B. Gila, B.R. Appleton, J. Fridmann, P. Mazarov, J.E. Sanabia, S. Bauerdick, L. Bruchhaus, R. Mimura, R. Jede, First results from a multi-ion beam lithography and processing system at the University Of Florida. AIP Conf. Proc. **1336**, 243–247 (2011)
47. S. Matsui, T. Kaito, J. Fujita, M. Komuro, K. Kanda, Y. Haruyama, Three-dimensional nanostructure fabrication by focused-ion-beam chemical vapor deposition. J. Vac. Sci. Technol. B Microelectron. Nanometer Struct. Process. Meas. Phenom. **18**, 3181–3184 (2000)
48. M. Esposito, V. Tasco, F. Todisco, A. Benedetti, D. Sanvitto, A. Passaseo, Three dimensional chiral metamaterial nanospirals in the visible range by vertically compensated focused ion beam induced-deposition. Adv. Opt. Mater. **2**, 154–161 (2013)
49. M. Esposito, V. Tasco, F. Todisco, M. Cuscunà, A. Benedetti, M. Scuderi, G. Nicotra, A. Passaseo, Programmable extreme chirality in the visible by helix-shaped metamaterial platform. Nano Lett. **16**, 5823–5828 (2016)
50. D. Kosters, A. de Hoogh, H. Zeijlemaker, H. Acar, N. Rotenberg, L. Kuipers, Core–shell plasmonic nanohelices. ACS Photonics **4**, 1858–1863 (2017)
51. M. Yan, S. Choi, K.R.V. Subramanian, I. Adesida, The effects of molecular weight on the exposure characteristics of poly(methylmethacrylate) developed at low temperatures. J. Vac. Sci. Technol. B Microelectron. Nanometer Struct. Process. Meas. Phenom. **26**, 2306–2310 (2008)
52. B. Cord, J. Yang, H. Duan, D.C. Joy, J. Klingfus, K.K. Berggren, Limiting factors in sub-10 nm scanning-electron-beam lithography. J. Vac. Sci. Technol. B Microelectron. Nanometer Struct. Process. Meas. Phenom. **27**, 2616–2621 (2009)
53. L.E. Ocola, A. Stein, Effect of cold development on improvement in electron-beam nanopatterning resolution and line roughness. J. Vac. Sci. Technol. B Microelectron. Nanometer Struct. Process. Meas. Phenom. **24**, 3061–3065 (2006)
54. J. Reinspach, M. Lindblom, O. von Hofsten, M. Bertilson, H.M. Hertz, A. Holmberg, Cold-developed electron-beam-patterned ZEP 7000 for fabrication of 13 nm nickel zone plates. J. Vac. Sci. Technol. B Microelectron. Nanometer Struct. Process. Meas. Phenom. **27**, 2593–2596 (2009)
55. T. Okada, J. Fujimori, M. Aida, M. Fujimura, T. Yoshizawa, M. Katsumura, T. Iida, Enhanced resolution and groove-width simulation in cold development of ZEP520A. J. Vac. Sci. Technol. B **29**, 021604 (2011)
56. J.K.W. Yang, K.K. Berggren, Using high-contrast salty development of hydrogen silsesquioxane for sub-10-nm half-pitch lithography. J. Vac. Sci. Technol. B Microelectron. Nanometer Struct. Process. Meas. Phenom. **25**, 2025–2029 (2007)

57. N. Yu, P. Genevet, M.A. Kats, F. Aieta, J.-P. Tetienne, F. Capasso, Z. Gaburro, Light propagation with phase discontinuities: generalized laws of reflection and refraction. Science **334**, 333–337 (2011)
58. M.A. Kats, P. Genevet, G. Aoust, N. Yu, R. Blanchard, F. Aieta, Z. Gaburro, F. Capasso, Giant birefringence in optical antenna arrays with widely tailorable optical anisotropy. Proc. Natl. Acad. Sci. **109**, 12364 (2012)
59. K. O'Brien, H. Suchowski, J. Rho, A. Salandrino, B. Kante, X. Yin, X. Zhang, Predicting nonlinear properties of metamaterials from the linear response. Nat. Mater. **14**, 379 (2015)
60. A. Arbabi, Y. Horie, M. Bagheri, A. Faraon, Dielectric metasurfaces for complete control of phase and polarization with subwavelength spatial resolution and high transmission. Nat. Nanotechnol. **10**, 937 (2015)
61. M.I. Shalaev, J. Sun, A. Tsukernik, A. Pandey, K. Nikolskiy, N.M. Litchinitser, High-efficiency all-dielectric metasurfaces for ultracompact beam manipulation in transmission mode. Nano Lett. **15**, 6261–6266 (2015)
62. N. Liu, H. Guo, L. Fu, S. Kaiser, H. Schweizer, H. Giessen, Three-dimensional photonic metamaterials at optical frequencies. Nat. Mater. **7**, 31 (2007)
63. N. Liu, H. Liu, S. Zhu, H. Giessen, Stereometamaterials. Nat. Photonics **3**, 157 (2009)
64. R.C. Devlin, M. Khorasaninejad, W.T. Chen, J. Oh, F. Capasso, Broadband high-efficiency dielectric metasurfaces for the visible spectrum. Proc. Natl. Acad. Sci. **113**, 10473–10478 (2016)
65. M. Khorasaninejad, W.T. Chen, R.C. Devlin, J. Oh, A.Y. Zhu, F. Capasso, Metalenses at visible wavelengths: diffraction-limited focusing and subwavelength resolution imaging. Science **352**, 1190 (2016)
66. B. Groever, W.T. Chen, F. Capasso, Meta-lens doublet in the visible region. Nano Lett. **17**, 4902–4907 (2017)
67. M. Khorasaninejad, A.Y. Zhu, C. Roques-Carmes, W.T. Chen, J. Oh, I. Mishra, R.C. Devlin, F. Capasso, Polarization-insensitive metalenses at visible wavelengths. Nano Lett. **16**, 7229–7234 (2016)
68. A. She, S. Zhang, S. Shian, D.R. Clarke, F. Capasso, Large area metalenses: design, characterization, and mass manufacturing. Opt. Express **26**, 1573–1585 (2018)
69. D. Xia, Z. Ku, S.C. Lee, S.R.J. Brueck, Nanostructures and functional materials fabricated by interferometric lithography. Adv. Mater. **23**, 147–179 (2010)
70. L. Wang, D. Fan, V.A. Guzenko, Y. Ekinci, Facile fabrication of high-resolution extreme ultraviolet interference lithography grating masks using footing strategy during electron beam writing. J. Vac. Sci. Technol. B **31**, 06F602 (2013)
71. R.H. French, H. Sewell, M.K. Yang, S. Peng, D.C. McCafferty, W. Qiu, R.C. Wheland, M.F. Lemon, L. Markoya, M.K. Crawford, Imaging of 32-nm 1:1 lines and spaces using 193-nm immersion interference lithography with second-generation immersion fluids to achieve a numerical aperture of 1.5 and ak 1 of 0.25. J. MicroNanolithograhy MEMS MOEMS **4**, 031103-4–14 (2005)
72. Y. Ekinci, M. Vockenhuber, M. Hojeij, L. Wang, N. Mojarad, Evaluation of EUV resist performance with interference lithography towards 11 nm half-pitch and beyond, in *Proceedings of SPIE 8679, Extreme Ultraviolet (EUV) Lithography IV*, vol. 8679 (2013), pp. 867910-8679–11
73. T. Ito, S. Okazaki, Pushing the limits of lithography. Nature **406**, 1027 (2000)
74. U. Okoroanyanwu, *Chemistry and Lithography* (SPIE Press, 2010)
75. T. Hu, C.-K. Tseng, Y.H. Fu, Z. Xu, Y. Dong, S. Wang, K.H. Lai, V. Bliznetsov, S. Zhu, Q. Lin, Y. Gu, Demonstration of color display metasurfaces via immersion lithography on a 12-inch silicon wafer. Opt. Express **26**, 19548–19554 (2018)
76. S. Y. Chou, P.R. Krauss, P.J. Renstrom, Imprint lithography with 25-nanometer resolution. Sci. **272**, 85 (1996)
77. W. Chen, M. Tymchenko, P. Gopalan, X. Ye, Y. Wu, M. Zhang, C.B. Murray, A. Alu, C.R. Kagan, Large-area nanoimprinted colloidal Au nanocrystal-based nanoantennas for ultrathin polarizing plasmonic metasurfaces. Nano Lett. **15**, 5254–5260 (2015)

78. S.V. Makarov, V. Milichko, E.V. Ushakova, M. Omelyanovich, A. Cerdan Pasaran, R. Haroldson, B. Balachandran, H. Wang, W. Hu, Y.S. Kivshar, A.A. Zakhidov, Multifold emission enhancement in nanoimprinted hybrid perovskite metasurfaces. ACS Photonics **4**, 728–735 (2017)
79. X. Zhu, C. Vannahme, E. Højlund-Nielsen, N.A. Mortensen, A. Kristensen, Plasmonic colour laser printing. Nat. Nanotechnol. **11**, 325 (2015)
80. T. Higashiki, T. Nakasugi, I. Yoneda, Nanoimprint lithography and future patterning for semiconductor devices. J. MicroNanolithograhy MEMS MOEMS **10**, 043008-10–8 (2011)
81. S. Ahn, M. Ganapathisubramanian, M. Miller, J. Yang, J. Choi, F. Xu, D.J. Resnick, S.V. Sreenivasan, Roll-to-roll nanopatterning using jet and flash imprint lithography, in *Proceedings of SPIE 8323, Alternative Lithographic Technologies IV*, vol. 8323 (2012), p. 83231L–8323–7
82. S.H. Ahn, L.J. Guo, Large-area roll-to-roll and roll-to-plate nanoimprint lithography: a step toward high-throughput application of continuous nanoimprinting. ACS Nano **3**, 2304–2310 (2009)
83. S. Vignolini, N.A. Yufa, P.S. Cunha, S. Guldin, I. Rushkin, M. Stefik, K. Hur, U. Wiesner, J.J. Baumberg, U. Steiner, A 3D optical metamaterial made by self-assembly. Adv. Mater. **24**, OP23–OP27 (2011)
84. J.Y. Kim, H. Kim, B.H. Kim, T. Chang, J. Lim, H.M. Jin, J.H. Mun, Y.J. Choi, K. Chung, J. Shin, S. Fan, S.O. Kim, Highly tunable refractive index visible-light metasurface from block copolymer self-assembly. Nat. Commun. **7**, 12911 (2016)
85. A. Nemiroski, M. Gonidec, J.M. Fox, P. Jean-Remy, E. Turnage, G.M. Whitesides, Engineering shadows to fabricate optical metasurfaces. ACS Nano **8**, 11061–11070 (2014)
86. M. Pu, Z. Zhao, Y. Wang, X. Li, X. Ma, C. Hu, C. Wang, C. Huang, X. Luo. Spatially and spectrally engineered spin-orbit interaction for achromatic virtual shaping. Sci. Rep. **5**, 9822 (2015)

Chapter 6
Super-resolution Microscopy

Abstract Optical microscopy is one of the most important scientific instruments in the history of mankind. It has revolutionized the field of life sciences and remains indispensable in many areas of scientific research. However, the resolution of the optical microscopy could not be enhanced infinitely through improving the amplification factor and eliminating the aberration due to the optical diffraction from a limited aperture in optical imaging system, and there exists a theoretical limit, which is named as diffraction limit. Essentially, this is attributed to the loss of high spatial frequencies that contain the details of an object. Although spatial or temporal manipulation of fluorescence microscopy has been demonstrated as an avenue of super-resolution microscopy, they require special labeling of the samples. With the development of subwavelength structured materials, superlens- and hyperlens-based super-resolution microscopies have been proposed for both intensity- and phase-contrast imaging. Furthermore, inspired by the dielectric microsphere-based photonic nanojets and far-field super-oscillation phenomena, new super-resolution microscopies have also been proposed, forming one important research direction of EO 2.0.

Keywords Microscopy · Super-resolution · Superlens · Hyperlens

6.1 Introduction

In 1873, Abbe [1] discovered a fundamental "diffraction limit" in optics: whenever an object is imaged by an optical system, such as the lens of a camera, the resolution Δ is limited by the wavelength of light λ and the numerical aperture (NA) of the objective lens, in a simple form $\Delta = 0.61\lambda/\text{NA}$. The physical origin of Abbe's diffraction limit can be illustrated by a picture of electromagnetic scattering by an object. All sources of electromagnetic waves produce two distinct types of fields: the far-field and the near-field components. The far-field component is the part that radiates far from an object and can be captured by a lens to form an image. Unfortunately, it contains only a broad-brush picture of the object, limiting the resolution to the size of the

X. Luo, *Engineering Optics 2.0*, https://doi.org/10.1007/978-981-13-5755-8_6

wavelength λ. The near field, on the other hand, contains all the finest details of an object, but its intensity drops off rapidly with distance away from the object.

To achieve super-resolution microscopy, Synge conceived and developed the idea of detecting the optical signal in the near field of the object in 1928, which ultimately results in the near-field scanning optical microscopy (NSOM) in 1980s [2–5]. As one of the most popular microscopy techniques so far, NSOM forms its image with high spatial resolution by scanning an ultra-sharp tip in the vicinity of the object and collecting signal point by point [6].

Instead of a physical tip, another family of method based on a virtual light probe has also been developed. Stimulated emission depletion (STED) fluorescence microscopy [7] utilizes a tiny light spot or line to scan the samples. STED overcomes the diffraction limit by accompanying a focused excitation beam with a spatially patterned "depletion" beam, typically in a donut shape.

Sample illumination is an important aspect of optical microscopy [8]. For example, the introduction of an opaque stop reduced the background light in dark-field microscopy [9]. Illuminating the sample at an angle enhanced the details of the patterns in the transparent specimens in oblique illumination microscopy [10]. Introducing a pinhole in a plane conjugate to the plane of observation in the sample or focusing collimated laser light onto the sample can considerably reduce out-of-focus light. This technique is known as confocal microscopy, also termed as confocal laser scanning microscopy (CLSM) [11]. By aggregating a series of position information, a super-resolution image can also be assembled (i.e., synthetic aperture imaging). However, these scanning sampling-based microscopy techniques achieve super-resolving power by sacrificing imaging speed and fluorescence label, making them uncompetitive for dynamic imaging.

The method of structured illumination microscopy (SIM) is a fast and wide-field microscopy technique [8], where the sample is excited by a series of standing waves with different orientations or phases to increase the spatial frequency detectable by the microscope. SIM works for both fluorescence imaging and non-fluorescent imaging, and the setup is relatively simple compared with other methods. Nevertheless, the linear form of SIM only extends the diffraction limit by a factor of 2, because the standing wave pattern is itself limited by diffraction.

Besides the spatial domain strategies above, time-domain methods like stochastic optical reconstruction microscopy (STORM) [12] and (fluorescence) photoactivated localization microscopy [(F)PALM] [13, 14] overcome this limit by switching on only a stochastic subset of fluorescent molecules within a field of view at any given time such that their images do not substantially overlap, allowing their positions to be localized with high precision. These molecules are then switched off (or bleached), and a stochastically different subset of molecules is switched on and localized. Iterating this process allows a super-resolution image to be constructed from numerous molecular localizations accumulated over time [15]. In essence, the spatial super-resolution is obtained at a cost of time-consuming process.

Lens-based projection imaging remains a good option for high-speed microscopy, and making a perfect lens that produces flawless images has been a dream of lens makers for centuries [16]. During the past few years, superlens- and hyperlens-based super-resolution microscopies have been proposed for both intensity and phase-contrast sub-diffraction-limited imaging.

6.2 Negative Refractive Lens Microscopy

Negative refraction is a long-sought material in optics and electromagnetics. Early in 1967, Veselago [17] proposed a material that simultaneously possesses a negative electric permittivity and a negative magnetic permeability, termed as left-handed material or negative-index material (NIM), which has an ability to realize perfect imaging. In a NIM, the refracted ray undergoes "negative refraction" and both beams stay on the same side of the normal direction. Therefore, NIMs lens allows for light rays to focus once within the lens and once outside the lens, ensuring a flat scheme for optical imaging (Fig. 6.1a). Moreover, if a lens made of NIM is placed close to an object, the near-field evanescent waves can be strongly enhanced across the lens (Fig. 6.1b) [18]. After passing the NIM lens, the evanescent waves decay again until their amplitudes reach their original level at the image plane [19]. On the other hand, the propagating waves pass through the NIM lens with both negative refraction and a reversed phase front, leading to zero phase change at the image plane. By completely recovering both propagating and evanescent waves in phase and amplitude, a perfect image may be created (Fig. 6.1c).

Although such NIM does not exist in nature, the emergence of metamaterials during the past two decades offers new opportunities to realize NIMs by building small circuits (e.g., split-ring resonator arrays and parallel metallic wire arrays) to mimic the magnetic and electrical response of a material [20, 21]. After the experimental demonstration in the microwave band [22], tremendous effort has been devoted to extend the negative refraction effect to higher frequencies. Metamaterials based on plasmonic waveguides consisting of alternating planar metal and dielectric layers offer an attractive avenue to achieve three-dimensional (3D) left-handed response at high frequency [23]. This left-handed response is attributed to the backwards electromagnetic mode naturally sustained by the waveguide in a frequency range between the bulk plasmon resonance frequency ω_P of the metal and the surface plasmon resonance frequency ω_{SP} of the metal–dielectric interface.

In 2013, Xu et al. [24] exploited the broad-angle negative refractive index achieved with left-handed metamaterial to demonstrate Veselago flat lensing in the ultraviolet, extending the experimental realization of this effect across the electromagnetic spectrum and into a frequency range with important applications such as fluorescence microscopy and photolithography, as displayed in Fig. 6.2 [24]. The ability of the flat lens to image two-dimensional objects with arbitrary shapes is demonstrated using cross (Fig. 6.2c) and ring (Fig. 6.2d) slit apertures. The aperture width is set to a constant value of 180 nm ($w_s < \lambda_0/2$) to suppress transverse electric-polarized light [24]. To maintain imaging of the full specified shapes of the two-dimensional aperture objects, circularly polarized illumination is used to probe evenly all radial directions of each object. The resulting images of the objects formed by the flat lens are shown

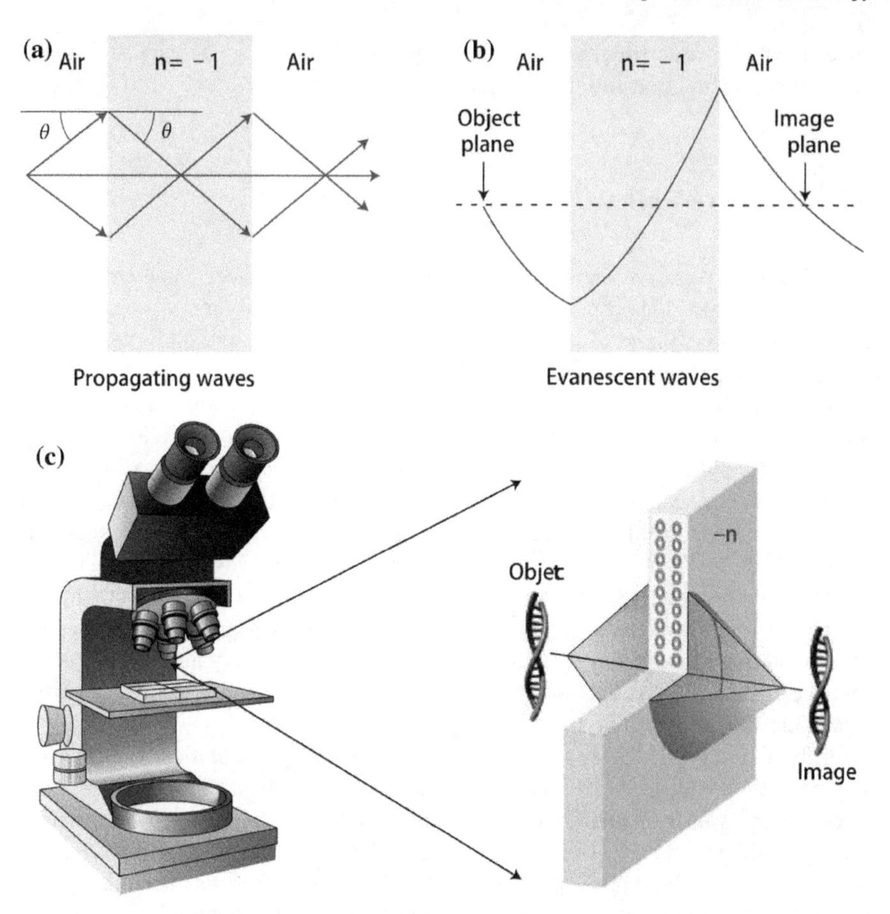

Fig. 6.1 **a** A NIM flat lens brings all the diverging rays from an object into a focused image. **b** The NIM can enhance the evanescent waves across the lens, so the amplitudes of the evanescent waves are identical at the object and the image plane. **c** A microscope based on an ideal NIM lens should focus both propagating and evanescent waves into an image with arbitrarily high resolution. Reproduced from [18] with permission. Copyright 2008, Nature Publishing Group

in Fig. 6.2d (cross) and Fig. 6.2f (ring), as recorded by the optical microscope focused at $z_f = 360$ nm [24]. The salient characteristics of the images, including point-to-point replication of features of the two-dimensional objects, are consistent with those expected from a Veselago flat lens imaging beyond the near field.

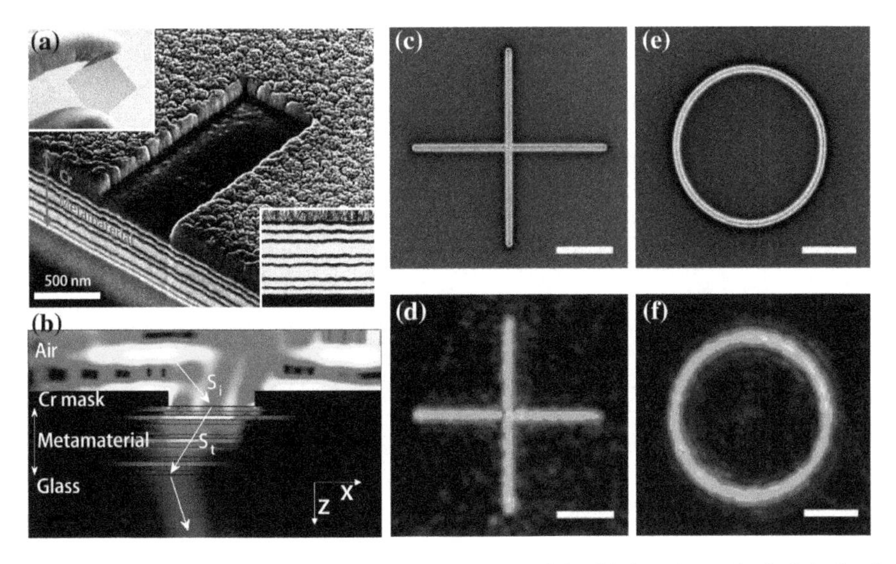

Fig. 6.2 **a** Scanning electron microscope (SEM) image of the fabricated sample. **b** Calculated profile of the time-averaged Poynting vector across the whole structures. **c**, **e** SEM of the two-dimensional imaging object. **d**, **f** Corresponding images produced by the flat lens under illumination by circularly polarized ultraviolet light. Reproduced from [24] with permission. Copyright 2013, Macmillan Publishers Limited

6.3 Superlens Microscopy

6.3.1 Operation Principle

Distinct from the Veselago negative lens that strictly depends on the simultaneous negative permittivity and permeability, Pendry proposed that a superlens based on single negative metallic slab can also realize super-resolution imaging. Such intriguing behavior relies on the fact that the metallic slab with negative permittivity can support SPPs at the metallic–dielectric interface and effectively enhance the exponentially decaying evanescent fields [19]. The above physical mechanism can be illustrated through the optical transfer function (OTF) of superlens. An asymmetric superlens is shown in Fig. 6.3, which is constructed by a thin slab with negative refractive medium of thickness d, dielectric constant ε_2, and magnetic permeability μ_2 sandwiched between medium 1 (ε_1 and μ_1) and medium 3 (ε_3 and μ_3) [25]. The black curve shows the decay of evanescent wave in positive refractive index materials such as dielectrics. The blue curve shows the enhancement of evanescent waves by a superlens, which decays again on the other side restoring the original amplitude at the image plane. The black dotted curve represents a decaying field without a superlens. The transmission coefficient for the TM-polarized wave across the slab in the electrostatic approximation can be expressed as [26]:

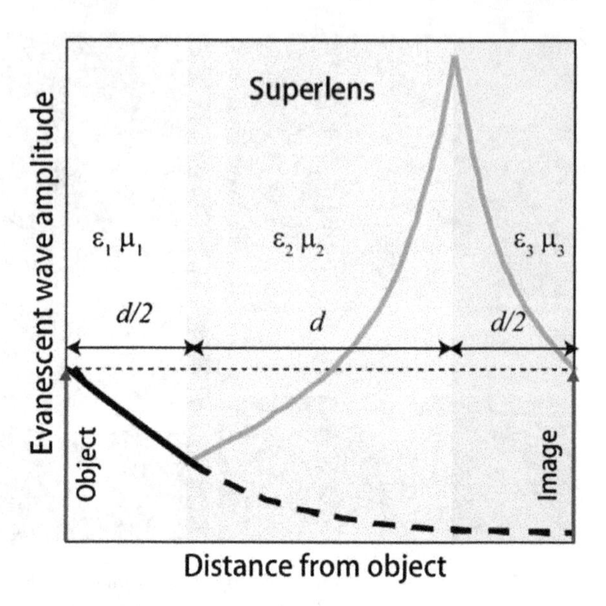

Fig. 6.3 Illustration of the evanescent field enhancement by superlens. Reproduced from [25] with permission. Copyright 2005, IOP Publishing Ltd and Deutsche Physikalische Gesellschaft

$$T_p(k_x) = \frac{4\varepsilon_2\varepsilon_3 \exp(-k_x d)}{(\varepsilon_1 + \varepsilon_2)(\varepsilon_2 + \varepsilon_3) - (\varepsilon_2 - \varepsilon_1)(\varepsilon_2 - \varepsilon_3)\exp(-2k_x d)} \tag{6.3.1}$$

The enhancement of the evanescent wave transmission depends on the condition of the dielectric constant. When $\varepsilon_2 = -\varepsilon_1$ or $\varepsilon_2 = -\varepsilon_3$, $T_p(k_x)$ is proportional to $\exp(k_x d)$ and thus shows exponential growth of the evanescent field.

6.3.2 *Near-Field Superlens Microscopy*

Pendry showed that in the near-field regime, negative permittivity alone is sufficient to realize a superlens to recover details on a subwavelength scale, which initiated a number of related experiments [19, 27]. A thin planar silver [28] sheet has been demonstrated to be a promising candidate for superlens-based ultraviolet (UV) nanomicroscopy and nanolithography. Subsequently, SiC with negative permittivity in infrared band was utilized to realize near-field superlens microscopy at the wavelength of 11 μm [29]. Experiment results demonstrated that such superlens can resolve λ/20-sized objects 880 nm away from the near-field probe.

It should be noted that, the accessible wavelengths of near-field superlens microscopy are restricted by the limited availability of suitable materials. An approach to realize a spectrally adjustable plasmonic superlens in the mid- and far-infrared is to use Si-doped GaAs, taking advantage of precisely controllable charge carrier concentration by standard semiconductor fabrication techniques [30].

Figure 6.4 shows a superlens structure consists of a highly n-doped (Si) GaAs film of thickness d sandwiched between two undoped layers of thickness $d/2$ [30]. Here, the electron density determines the plasma frequency of a conductive layer which in turn determines the superlensing wavelength. At the operation wavelength of 20 μm, not only the permittivity match condition is satisfied but also the imaginary part $\mathrm{Im}(\varepsilon)$ of both intrinsic and doped GaAs is comparably small corresponding to small dissipation [30].

Two superlens samples were prepared with a total thickness of 400 and 800 nm, respectively. The target object is a periodic arrangement of gold stripes of 2 μm width at a spacing of 10 μm. The imaging results are shown in Fig. 6.5 [30]. Additionally, a reference sample consisting of an intrinsic 400-nm-thick GaAs layer was investigated. One can see that the stripes are barely visible in the image obtained from the reference sample (Fig. 6.5b) [30]. While for a 400-nm-thickness doped GaAs superlens, the 2-μm gold stripes are imaged with an apparent width of about 3 μm, clearly breaking the diffraction limit. Nevertheless, the image obtained via the 400-nm-thick superlens is not purely induced by the superlensing effect, as can be seen by the non-vanishing near field in case of the reference structure. In order to suppress this background noise, the imaging property of 800-nm-thickness doped GaAs superlens was further investigated, as shown in Fig. 6.5c [30]. The image is sharpest at $\lambda = 22$ μm, while it becomes blurred at shorter and longer wavelengths. A FWHM of recording strip width is 3.4 μm, corresponding to a subwavelength resolution of at least $\lambda/6$.

6.3.3 Far-Field Superlens Microscopy

Although the superlensing mechanism provides a new avenue for nanoscale optical imaging, it was pointed out that the superlens could only produce sub-diffraction-limited images in the near field. This is a major drawback compared with conventional lenses which are widely used in far-field microscopy. One has to use either photoresist exposure method or near-field scanning method to record this image which considerably limits the potential applications of this exciting imaging scheme [31]. The far-field superlensing concept, directly interfaced with conventional optics and image recording instruments, enables real-time high-resolution optical imaging possibility.

In order to overcome this problem, far-field superlens (FSL) was proposed to project a sub-diffraction-resolution image to the far field [31]. The FSL used a silver slab to enhance the evanescent waves and an attached subwavelength gratings to convert evanescent waves into propagating waves by shifting their incident field wave numbers k_{in} into the various diffraction orders; i.e., $k = k_{\text{in}} + mk_{\Lambda}$, where m is the diffraction order, and Λ is the grating period, as shown in Fig. 6.6. The FSL geometric parameters, shown in Fig. 6.6a, were optimized using rigorous coupled-wave analysis (RCWA) [31]. The calculated OTF in Fig. 6.6b shows that the FSL

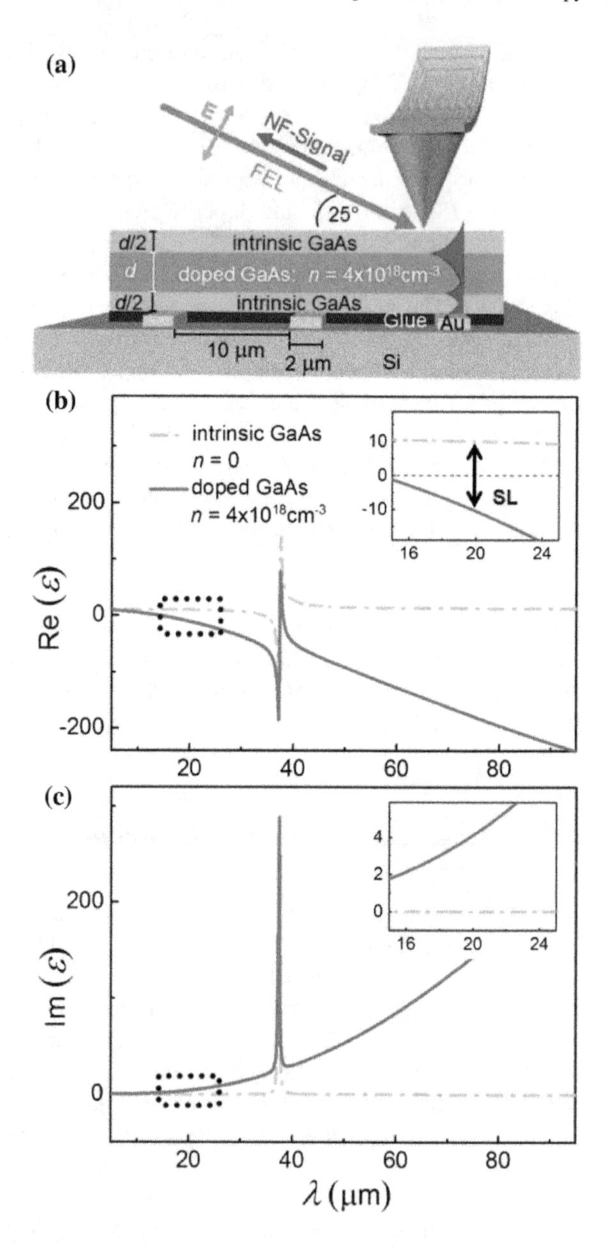

Fig. 6.4 **a** Gold stripes (width 2 μm, spacing 10 μm) are imaged through the three-layered GaAs superlens system. **b** Real part of the dielectric function Re(ε) of intrinsic and doped GaAs, fulfilling the superlensing condition $\text{Re}(\varepsilon_{\text{n-GaAs}}) = -\text{Re}(\varepsilon_{\text{GaAs}})$ at $\lambda \sim 20$ μm (see inset). **c** Imaginary part of the dielectric function Im(ε). SL: superlensing wavelength. Reproduced from [30] with permission. Copyright 2015, American Chemical Society

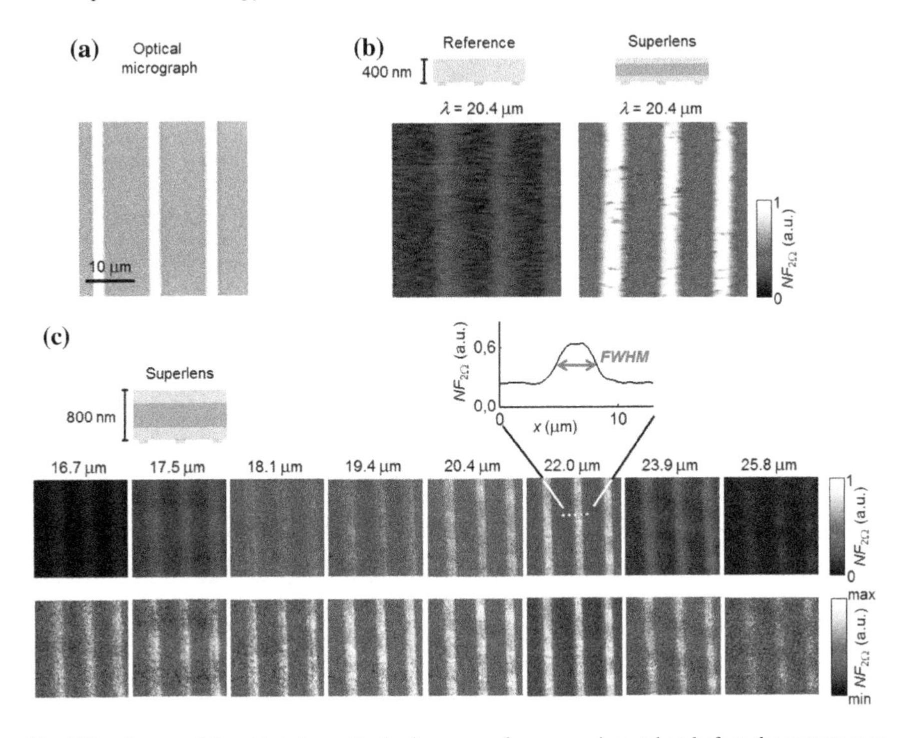

Fig. 6.5 **a** Image of the stripes by optical microscopy for comparison taken before the pattern was covered by the superlens. **b** Near-field images of gold stripes below 400-nm-intrinsic GaAs as a reference (left) and below a superlens with the same total thickness, including 200-nm doped GaAs (right). **c** Upper row: near-field images of gold stripes below an 800-nm-thick superlens. Bottom row: same measurements but with a full-range color scale applied to each image. Reproduced from [30] with permission. Copyright 2015, American Chemical Society

can indeed significantly enhance evanescent waves and effectively couple them into the far field while suppressing the zero-order diffraction.

As a simple example, a pair of line objects of 50 nm width with a 70 nm gap inscribed by focused ion beam on a 40-nm-thick Cr film on the quartz substrate was imaged (Fig. 6.7a) [31]. Due to the diffraction limit, optical imaging through a conventional optical microscope cannot resolve the line pairs (Fig. 6.7b). For FSL imaging, *s*-polarization does not excite the surface plasmons at FSL, resulting in a diffraction-limited image similar to that from a conventional optical microscope image (Fig. 6.7c) [31]. In contrast, for *p*-polarization, the evanescent waves from the object gain significant enhancement by the excitation of surface plasmon in the silver superlens. By a combination of the evanescent components from the *p*-polarization with the propagating components from *s*-polarization, the pair of lines of 50 nm width can be clearly imaged (Fig. 6.7d) [31]. The FSL image cross-sectional profiles are compared with those obtained from a regular optical microscope and control experiment using *s*-polarization (Fig. 6.7e).

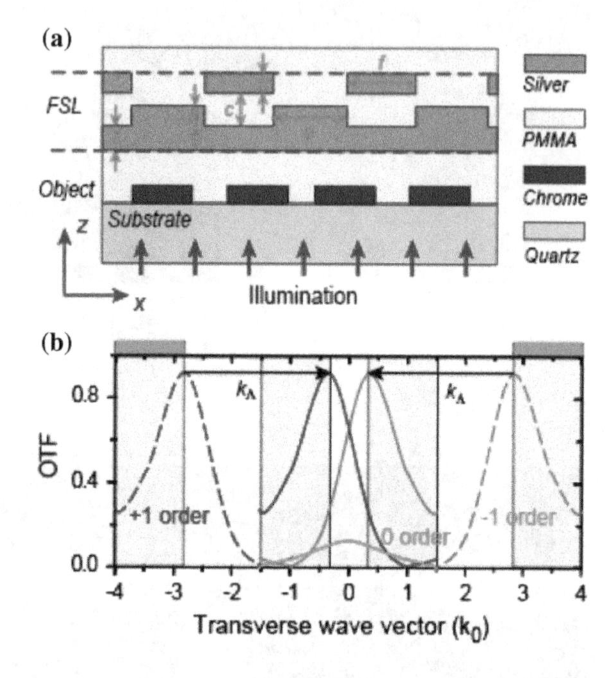

Fig. 6.6 **a** Schematic of a silver FSL and a chromium object fabricated on a quartz substrate. **b** Calculated OTF of the optimized FSL under p-polarized incident light. Reproduced from [31] with permission. Copyright 2007, American Chemical Society

By replacing the original silver slab with a silver–dielectric multilayer structure, a far-field superlens that operates at visible frequency was proposed (Fig. 6.8a) [32]. The mechanism of evanescent wave enhancement for a broad range of wavevectors comes from a well-known surface plasmon mode splitting (which will be discussed in the following section), so that a multilayer-based FSL can be designed to work at any frequency below the surface plasmon frequency [32]. Numerical calculations showed that a two-dimensional arbitrary particle of 40 nm radius that cannot be resolved by traditional microscope can be imaged with sub-diffraction-limited resolution by the multilayer FSL, as indicated in Fig. 6.8b, c [32].

6.4 Hyperlens Magnifying Microscopy

Although the far-field superlensing microscopy can overcome the near-field limit of traditional superlens, there are still some problems should be solved. First, the strict permittivity match condition should be satisfied for the sake of a broadband enhancement of evanescent waves; otherwise, the resolution and imaging quality of superlens would deteriorate rapidly. This condition limits the operating frequency and results in a narrow working bandwidth of superlens microscopy. Second, the superlens microscopy, regardless of near-field or far-field, can only realize equal-scale microscopy. Recently, there has been a growing interest in developing structures that are able to magnify subwavelength field distributions. This means that the details of the source pattern are retained while transferring the image over a certain distance, and at the same time, the pattern is linearly magnified or enlarged. Before we turn

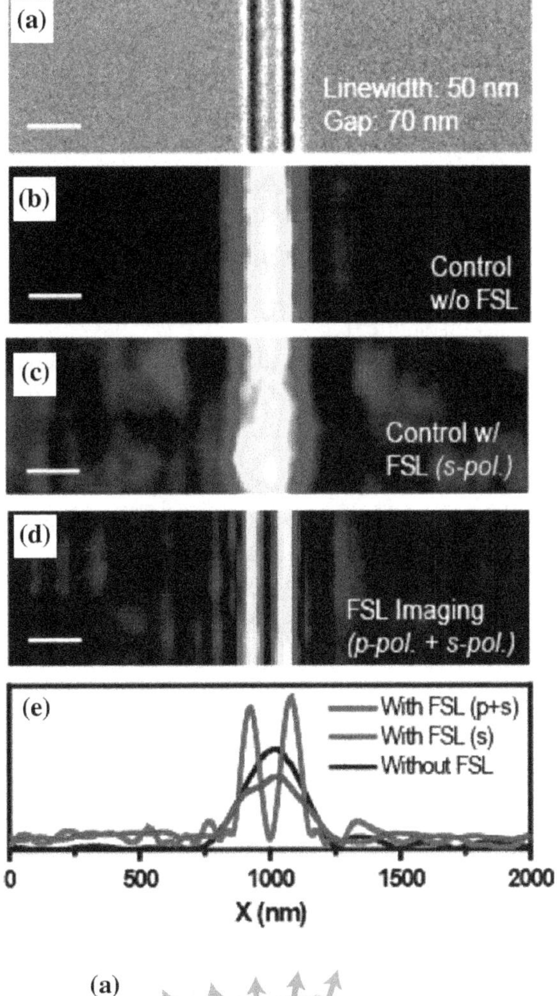

Fig. 6.7 **a** SEM image of an object. **b** Diffraction-limited image from a conventional optical microscope. **c** Reconstructed FSL images using *s*-polarization. **d** FSL image combining both *s*- and *p*-polarizations. The scale bars are 200 nm. **e** The averaged cross-sectional image profiles from **b**, **c**, and **d**, respectively. Reproduced from [31] with permission. Copyright 2007, American Chemical Society

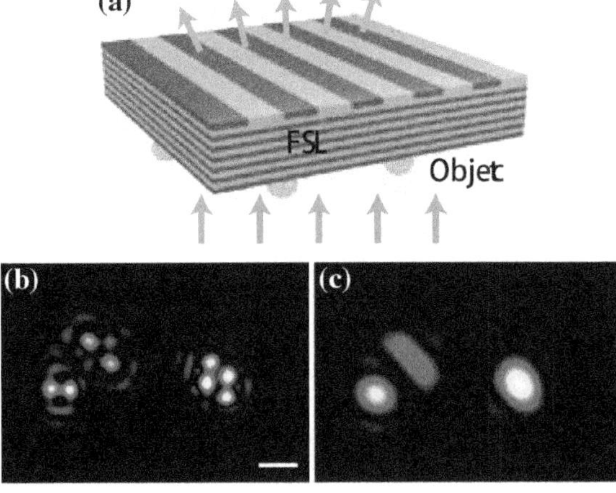

Fig. 6.8 **a** Modified visible FSL based on a silver–dielectric multilayer structure. **b** Two-dimensional sub-diffraction-limited imaging at 405 nm wavelength using the multilayer FSL. **c** Simulated real space image obtained by a conventional far-field optical system with NA = 1.5. Reproduced from [32] with permission. Copyright 2007, American Chemical Society

to the hyperlens microscopy, we would like to give some discussions about two important properties of hyperlens.

6.4.1 Broadband Evanescent Waves Enhancement in Hyperlens

The first limitation can be alleviated by replacing single metallic slab with metal–dielectric multilayers. The reason can be explained by a surface plasmon mode interaction picture (Fig. 6.9a) [32]. It is well known that surface plasmon modes split on metal thin film due to the interaction of modes on two metal surfaces. As the number of the metal thin films increases, the number of the split surface plasmon modes increases accordingly. With proper design, the split surface plasmon modes can be highly compact. Then, the transmission of the evanescent waves through the multilayer is large over a continuously broad range of wave numbers (Fig. 6.9b) [32]. Because the enhancement of the evanescent waves by metal–dielectric multilayer is due to the splitting of surface plasmon mode instead of the surface plasmon resonance excitation at the condition $|\varepsilon_m| = \varepsilon_d$, the metal–dielectric multiplayer superlens can work over a broad range of wavelengths, which is absolutely favorable in microscopy applications [32].

From the viewpoint of metamaterials, it is easy to calculate the effective anisotropic permittivity of the deep subwavelength multilayers as:

$$\begin{aligned} \varepsilon_x = \varepsilon_y &= \frac{\varepsilon_m t_m + \varepsilon_d t_d}{t_m + t_d}, \\ \frac{1}{\varepsilon_z} &= \frac{t_m/\varepsilon_m + t_d/\varepsilon_d}{t_m + t_d}, \end{aligned} \tag{6.4.1}$$

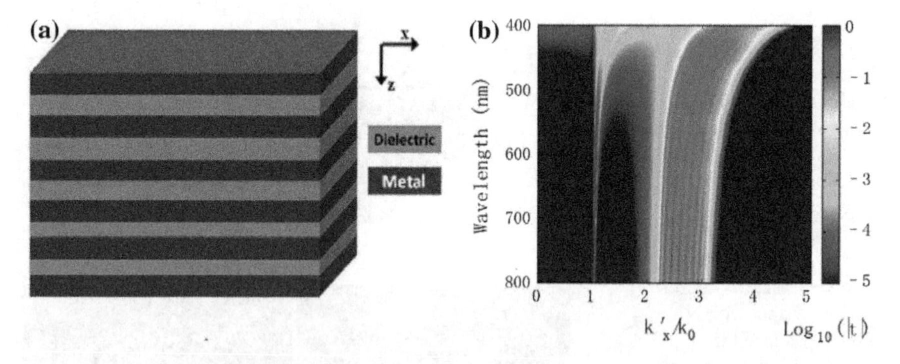

Fig. 6.9 **a** Alternatively stacked metal and dielectric layers. **b** Calculated magnitude (logarithmic scale) of the electromagnetic field transmission coefficient for a free-standing metamaterial stack composed of ten bilayers of Ag (30 nm) and SiO_2 (25 nm), for TM-polarized light. Reproduced from [33] with permission. Copyright 2014, Macmillan Publishers Limited

and the dispersion equation is:

$$\frac{k_x^2 + k_y^2}{\varepsilon_z} + \frac{k_z^2}{\varepsilon_x} = \left(\frac{\omega}{c}\right)^2. \tag{6.4.2}$$

Clearly, this dispersion curve is hyperbolic if $\varepsilon_z \varepsilon_x < 0$, which can be realized by tuning the permittivities of metal and dielectric (ε_{m} and ε_{d}), as well as the thicknesses (t_{m} and t_{d}).

As indicated in Fig. 6.10a, for the TM waves, the equi-frequency contour (EFC) surface in effective medium theory approximation shows a cylindrical hyperbolic profile, enabling light propagation with infinite large transverse wave number k_x and k_y. The OTF calculated by RCWA shows a filter window ranging from 1.6 k_0 to 3.6 k_0 in Fig. 6.10b. The peak denoted by the red dashed curve could be explained by two surface modes confined at the interface between the HMM and dielectric medium [34].

6.4.2 *Directional Propagation of Light in Hyperlens*

Besides the spatial filtering property, the hyperbolic metamaterials also possess extraordinary directional propagation ability, which is determined by the group velocity. Evanescent waves with high spatial frequency k_x tend to propagate perpendicularly to the asymptote line of hyperbolic curve [35]:

$$k_z = \sqrt{-\frac{\varepsilon_x}{\varepsilon_z}} |k_x| \tag{6.4.1}$$

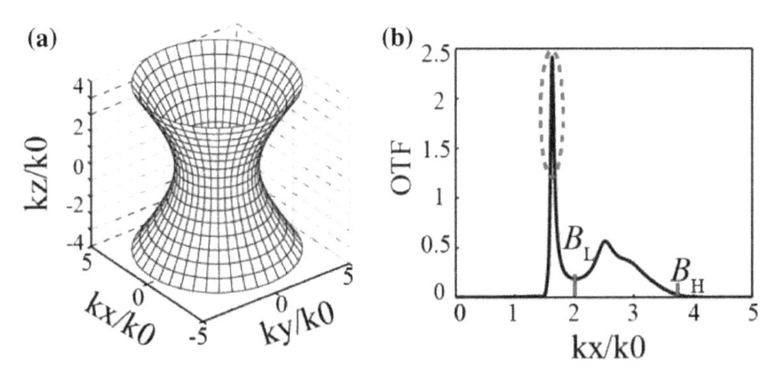

Fig. 6.10 **a** 3D plots of EFC surface in EMT approximation for metamaterial structure. **b** The OTF window of hyperbolic metamaterials calculated by RCWA. Reproduced from [34] with permission. Copyright 2018, Optical Society of America

The propagation direction of evanescent waves can be calculated by the mathematical formula $\theta = \arctan\left(\sqrt{-\varepsilon_x/\varepsilon_z}\right)$.

The simulated and experimental observation of directional propagation was realized by far-field imaging of a silicon mask slit through Ag/SiO_2 multilayers [36] (Fig. 6.11). Evanescent waves of directional imaging are transferred to far field by roughening the top Ag layer and observed with a microscope objective. Considering its no cutoff property, it is not surprising that the directed propagating light has the ability to restore the subwavelength features of line object.

6.4.3 Far-Field Magnified Microscopy

Based on the directional propagation of evanescent waves in hyperbolic metamaterial crystals, Engheta et al. theoretically proposed far-field scanless optical microscopy with a sub-diffraction-limited resolution in 2006, also termed as magnifying superlens [37]. They exploited the special dispersion characteristics of metamaterial crystals that is obliquely cut at its output plane or has a curved output surface to map the input field distribution onto the crystal's output surface with a compressed angular spectrum, resulting in a "magnified" image. The magnification at the output surface

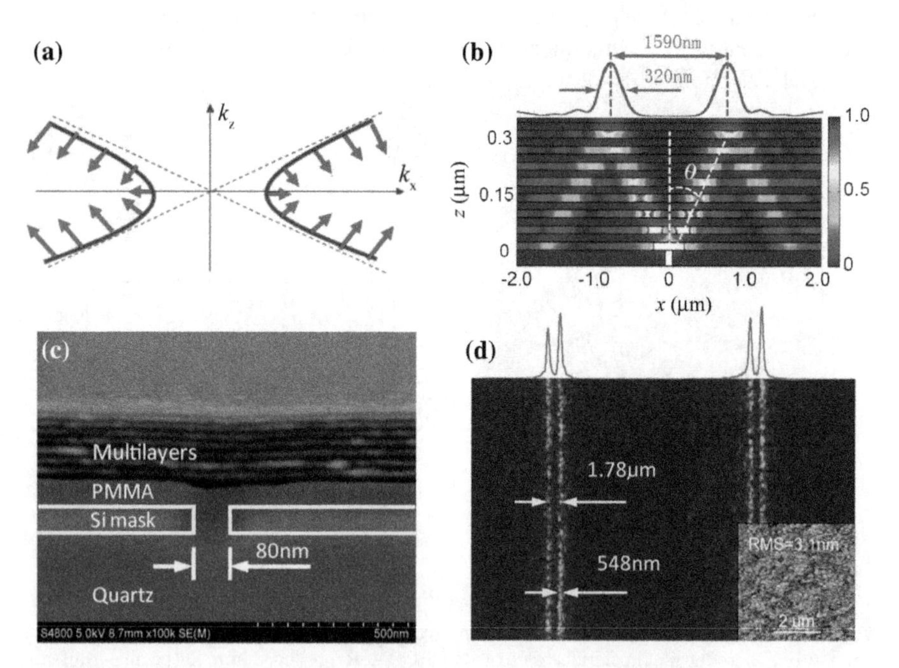

Fig. 6.11 **a** Schematic illustration of directional propagation of evanescent wave in hyperbolic metamaterials. **b** Simulated field distribution in the hyperbolic metamaterials. **c** SEM image of the cross section of a device to characterize the directional propagation property. **d** Imaging results through the multilayers obtained by CCD. Reproduced from [36] with permission. Copyright 2013, American Institute of Physics

is given simply by the ratio of the radii at the two boundaries. Magnifying superlens in visible frequency range was realized by using two-dimensional SPP confined in a concentric polymer grating placed on a gold surface [38], which were scattered by the surface roughness and generated a 3× magnification and an imaging resolution of 70 nm (~$\lambda/7$) in the far field.

Based on the same operation scheme, Jacob proposed the concept of "hyperlens" in the same year [39]. Possible implementations are shown in Fig. 6.12a, b. Owing to the conservation of angular momentum, a magnified image carried by low wave vectors will ultimately be formed at the outer boundary of the hyperlens before propagating into the far field. Then, conventional microscopy can be utilized to capture the output of hyperlens to achieve far-field super-resolution imaging [39]. Figure 6.12c–f show that two sources separated with a subwavelength distance can be clearly resolved after propagating through the hyperlens. The magnification factor is determined by the ratio between the outer and inner radius [39].

Subsequently, an experimental demonstration of the hyperlens for magnified microscopy in ultraviolet frequencies is constructed by exploiting cylindrical Ag–Al_2O_3 multilayer [40], through which a sub-diffraction-limited object (with a 130 nm center-to-center distance) was observed directly in the far field. Furthermore, a spherical hyperlens with two-dimensional super-resolution capability was experimentally demonstrated in the visible spectral region [41]. Recently, Lee et al. [42] proposed a new device consisting of a 4-inch wafer-scale spherical hyperlens array that allows high-throughput and easy-to-handle real-time biomolecular imaging. The hyperlens array is fabricated by nano-imprinting [42]. The fabricated 4-inch wafer-scale device and a SEM image of its corresponding hyperlens arrays are shown in Fig. 6.13a, b, respectively. Transmission electron microscopy (TEM) images of the fabricated hyperlens are shown in Fig. 6.13c–e [42]. A conformal and uniform multilayer is important as, otherwise, images will be distorted due to spherical aberration, scattering from defects or loss of the hyperbolic dispersion. Figure 6.13d, e show that the device has good conformality and uniformity without agglomeration or any empty area [42].

The performance of the hyperlens array device is experimentally verified through the imaging of hippocampal neurons with subwavelength dimensions. This system leads to a reliable positioning of the biological samples on the wafer-scale hyperlens array as shown in Fig. 6.14a [42]. The hyperlens array device is applied to a conventional microscope (Fig. 6.14b) and used to image a neuron. The neuron image is captured (Fig. 6.14c), and the obtained image of a bunch of legs (box in Fig. 6.14d) is selected to show the resolution of the hyperlens array device. The image magnified by the hyperlens and the objective lens indicate a separation distance of 334 nm between two legs [42]. The actual separation distance of 151 nm is obtained considering a magnification factor of 2.2 of the hyperlens. Two legs separated by 151 nm, which are sub-diffraction-limited, are clearly distinguished in this hyperlens system [42].

It should be mentioned that the curved geometry of a (de)magnifying hyperlens above causes difficulty in the practical implementation and applications. A simple way for planar imaging profile is to cut and polish hyperlens on one or both sides [43]. But this operation seems challenging for fabrication and usually delivers the

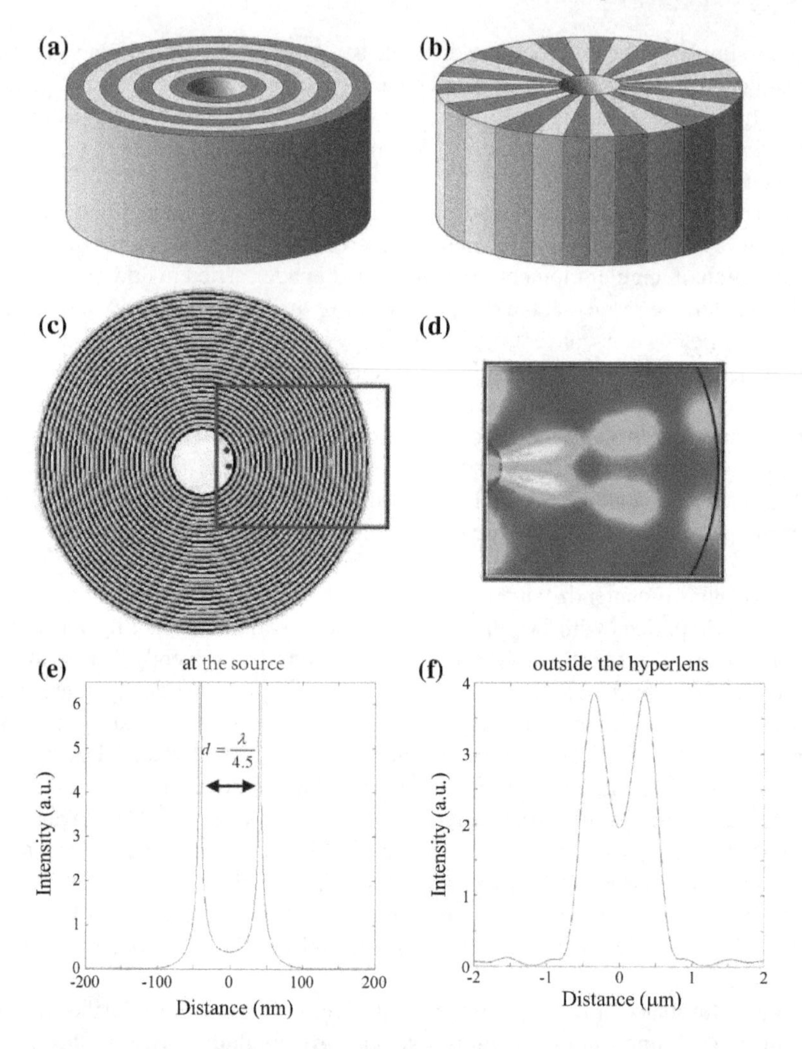

Fig. 6.12 Concentric metallic layers alternate with dielectric layers (**a**) or radially symmetric "slices" alternate in composition between metallic and dielectric (**b**) to produce ($\varepsilon_\theta > 0$, $\varepsilon_r < 0$) anisotropy. **c** Schematics of imaging by the hyperlens. **d** False color plot of intensity. **e**, **f** Demonstration of subwavelength resolution in the composite hyperlens containing two sources placed a distance $\lambda/4.5$ apart inside the core. **e** Field at the source. **f** Field outside the hyperlens. Reproduced from [39] with permission. Copyright 2006, Optical Society of America

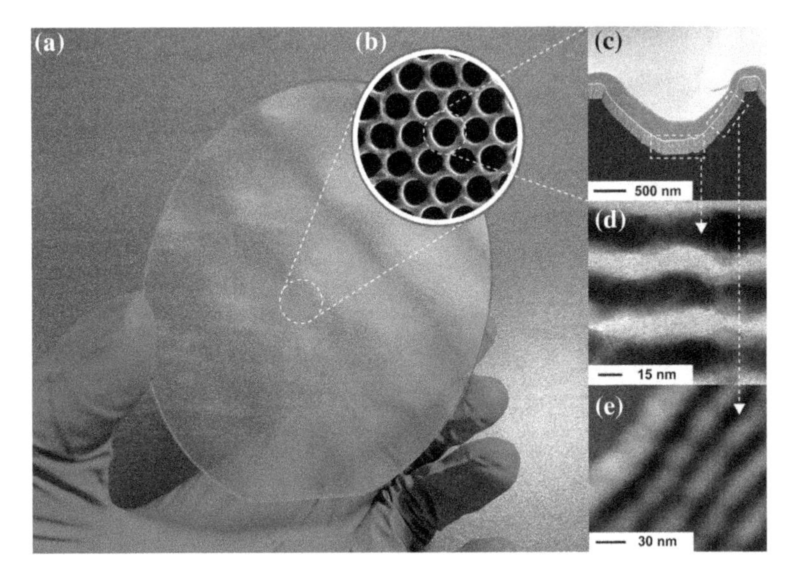

Fig. 6.13 **a** Fabricated hyperlens array device on a 4-inch wafer. **b** Top-view SEM image. **c**–**e** TEM images of a single hyperlens with different magnification. Bright and dark layers represent Ag and TiO_2 layers; 15-nm-thick layers in the center area (**d**) and sidewall (**e**) are observed. Reproduced from [42] with permission. Copyright 2018, American Chemical Society

deformation of images. As an alternative, hybrid lens combining hyperlens and planar superlens was proposed [44–46], as illustrated in Fig. 6.15. It was shown that the resolution ability of the proposed hybrid lens is high up to about $\lambda/6$.

Considering the magnification ratio is not uniform on the entire input plane, a conformal transformation theory with specially designed and complicated mathematic functions was proposed to ensure uniform magnification [47], as shown in Fig. 6.16. It is found that specifically designed trajectory route from the object plane to the image plane helps to yield imaging devices with uniform magnification ratio and improve image quality.

6.5 Super-resolution Phase-Contrast Microscopy

Besides the intensity-contrast imaging above, plasmonic cavity lens can also be utilized for sub-diffraction-limited phase-contrast imaging for transparent nano-objects. Phase-contrast microscopy is widely used in the visualization of transparent specimens that exhibit a small refractive index difference with respect to the surrounding environment. Conventional phase-contrast imaging system suffers from the diffraction-limited resolution.

According to Zernike's theory, the merit of phase-contrast imaging lies in the manipulation of the amplitude and phase of the transmitted illumination light and

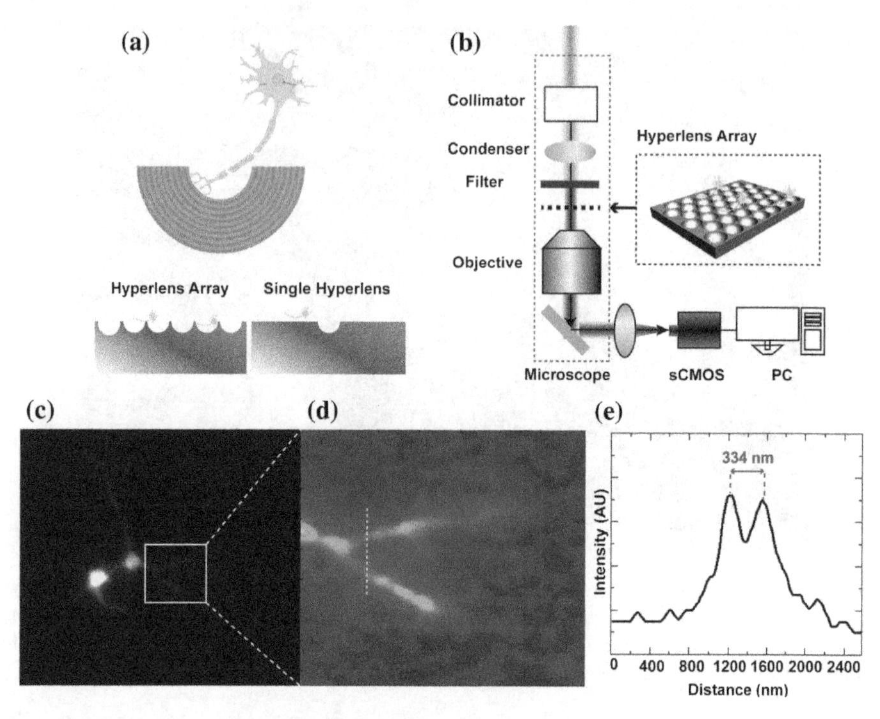

Fig. 6.14 **a** Schematic of simplified positioning of biomolecular objects on the hyperlens array. **b** Schematic of integration of the hyperlens with a conventional wide-field microscopy system for sub-diffraction-limited imaging under unpolarized white light illumination. **c** Captured neuron image passing through the hyperlens array device with a magnification factor of 2.2. **d** Zoomed-in image of neuron legs. **e** Normalized intensity profile and measured distance along the dashed line in **d**. Reproduced from [42] with permission. Copyright 2018, American Chemical Society

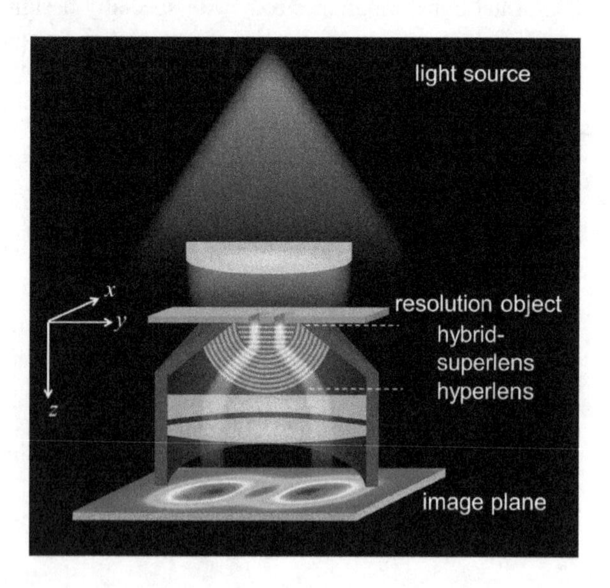

Fig. 6.15 Schematic of the hybrid-super-hyperlens. Reproduced from [46] with permission. Copyright 2013, Optical Society of America

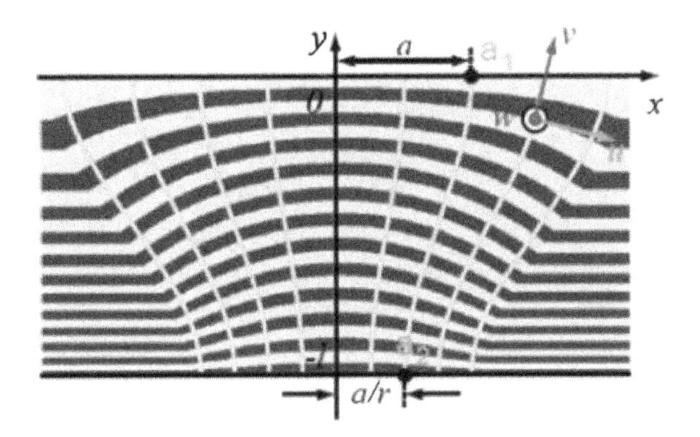

Fig. 6.16 Schematic of planar hyperlens for uniform magnification image. Reproduced from [47] with permission. Copyright 2014, Springer-Verlag Berlin Heidelberg

the scattered light of phase objects [48]. Although Zernike's method was originally proposed for imaging of transparent object features larger than a wavelength, due to the diffraction-limited resolution of conventional microscopy, it is reasonable to apply this idea for nanophase objects. The challenge is to manipulate the scattered evanescent features from nano-objects. Fortunately, the excitation of SPPs in super cavity helps to enhance the scattered light from transparent nano-objects and simultaneously suppress the transmitted light [49].

In order to illustrate it, a sub-diffraction-limited phase-contrast imaging model of nano-objects was presented (Fig. 6.17a), which is composed of a PMMA layer sandwiched between two 40-nm-thick Ag films. The transparent nanophase objects are embedded inside the PMMA layer [49]. The lower Ag film incorporated with the upper Ag film can enhance the scattered light of the nano-objects, while the upper Ag film transfers them to the near field and achieves sub-diffraction-limited imaging. The transmission amplitude ratio written as [49]:

$$T(k_\Lambda, h) = \left| \frac{t_{\pm 1}(k_\Lambda, h)}{t_0(0, h)} \right| \tag{6.5.1}$$

was calculated to obtain the optimal geometries of plasmonic cavity lens. In order to quantify the imaging performance of the plasmonic cavity lens, the imaging contrast ratio is shown in Fig. 6.17b from which one can see that the plasmonic cavity lens exhibits highest contrast and widest operation bandwidth [49]. Such property makes it be the best choice to achieve sub-diffraction-limited imaging for nano-objects. As a proof-of-concept experiment, a spatial resolution of about 64 nm and a minimum distinguishable refractive index difference down to 0.1 has been numerically demonstrated [49], as displayed in Fig. 6.17c–f [49].

In order to demonstrate the two-dimensional super-resolution phase-contrast microscopy, irregularly positioned nanocylinders with $\varepsilon_{obj} = 2$ and 2.6 were used as objects [49]. As shown in Fig. 6.18a, with the plasmonic cavity lens, the spatial profiles and permittivities of the nanocylinders can be clearly seen as bright and dark circular light spots. In contrast, superlens using a 40-nm-thick Ag film shows the

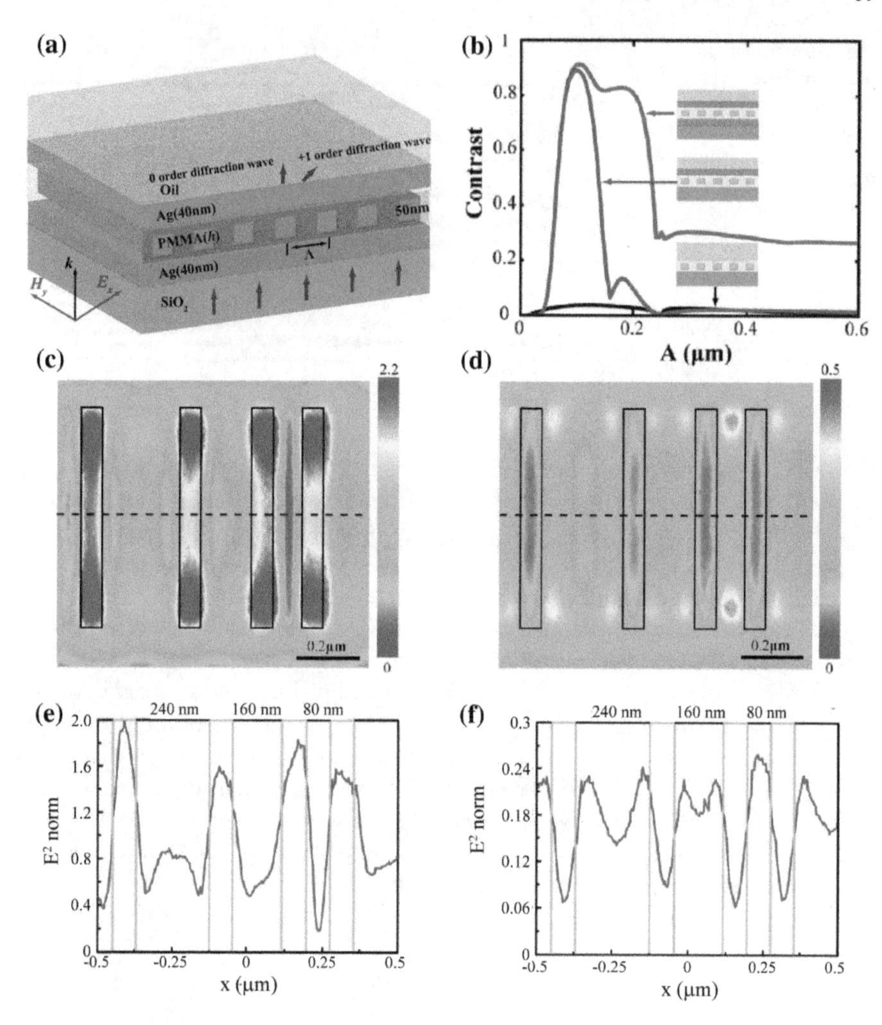

Fig. 6.17 **a** Schematic for sub-diffraction-limited phase-contrast imaging of nano-objects. **b** Calculated imaging contrast of superlens, plasmonic reflective lens, and plasmonic cavity lens. Sub-diffraction-limited imaging of randomly placed nanowires with different spacing and refractive index difference between the nanowires and PMMA layer. **c** −0.1, **d** 0.1. **e**, **f** Retrieved electric intensity along the dash lines in **c**, **d**. Reproduced from [49] with permission. Copyright 2013, IOP Publishing Ltd.

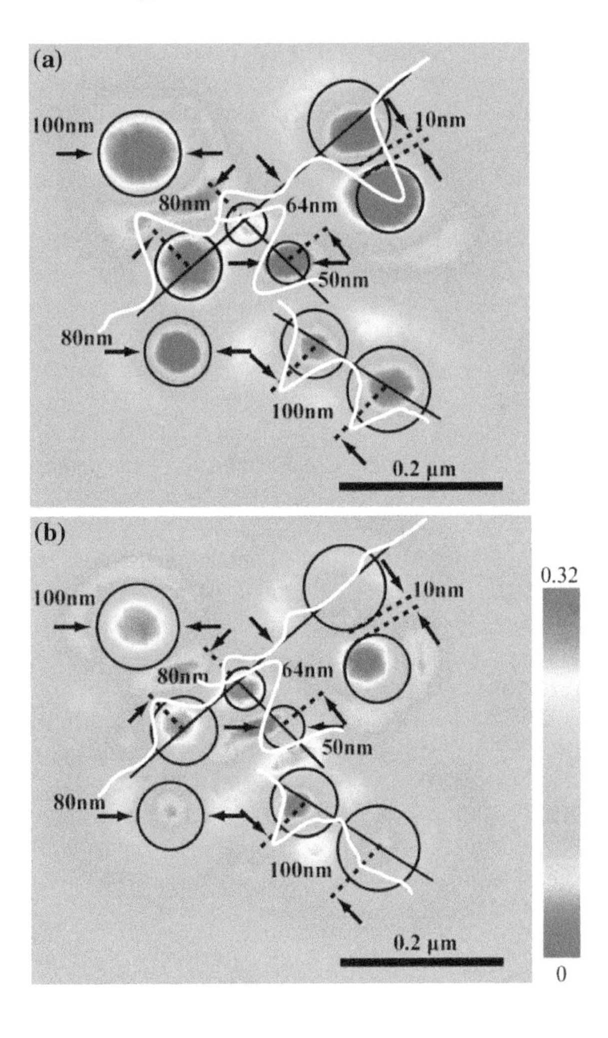

Fig. 6.18 Light distribution when imaging irregularly positioned nanocylinders with different sizes and permittivities for **a** MIM structure and **b** superlens. The solid circles indicate the positions of the nanocylinders. The white curves in **a** and **b** are the normalized electric field intensity distribution along the black lines. Reproduced from [49] with permission. Copyright 2013, IOP Publishing Ltd.

images to be greatly disturbed by the background illumination light noise, especially for those objects with dark visual imaging effects (Fig. 6.18b) [49].

6.6 Surface Imaging Microscopy

Interfacial and near-surface events, such as cell plasma membrane fusion of synaptic vesicles and the movement of single molecules during signal transduction, are key components in physical and biological systems. These events occur in the plasma membrane orientation on a distance of tens of nanometers and even below. In traditional fluorescence microscopy, the entire samples are flooded with excitation light, and the images of surface zone are easily deteriorated by the fluorescence excited outside the layer of interest. In order to reduce the unwanted light from internal samples and achieve the surface images of samples with high contrast, only the surface zone needs to be illuminated. Total internal reflection fluorescence microscopy (TIRFM) [50–52], also termed as evanescent wave fluorescence microscopy, selectively excites the fluorescently labeled entities located near the surface.

The surface imaging microscopy could also be obtained by employing surface plasmon polariton (SPP) illumination [53–55]. When the light is coupled into the metal film by using a prism [56], an optically matched microscopy objective lens [57], or diffraction gratings [58], the SPPs near the surface of metal film would be generated and the evanescent field is employed for illuminating samples with small penetration depth. In order to obtain small illumination depth and uniform illumination, single evanescent mode with high spatial frequency is highly desired. Recently, Kong et al. [59] proposed and demonstrated that deep subwavelength surface illumination imaging could be realized by launching bulk plasmon polariton (BPP) mode in hyperbolic metamaterials (HMMs) composed of multiple nanometal–dielectric films.

The schematic of BPPs illumination structure based on HMM is presented in Fig. 6.19a [59]. A monochromatic plane wave in TM-polarization impinges from the substrate side upon a 1D grating close to HMM, which is composed of alternatively stacked metal–dielectric films. The BPPs modes are formed by the mutual coupling of plasmonic polaritons fields between adjacent metal–dielectric films. For accurately describing the BPPs modes inside multi-films, RCWA is employed for calculating the OTF of HMM, as shown in Fig. 6.19b (blue line) [59]. Compared with that of a single Ag film (green line), the HMM OTF shows a window filtering effect, which greatly inhibits those light modes outside the window and generates a high-purity evanescent mode. This feature plays the key role in BPPs illumination with ultra-short illumination depth, which is defined as the distance where light intensity drops to $1/e$ of its value at the surface [59]:

$$L_p = \frac{\lambda}{4\pi\sqrt{(k_{\mathrm{BPPs}}/k_0)^2 - \varepsilon}} = \frac{\lambda}{4\pi\sqrt{(n\sin\theta + q\lambda/d)^2 - \varepsilon}} \tag{6.6.1}$$

where λ and θ are the wavelength and incidence angle of excitation light, respectively, n is the refractive index of substrate, ε is the permittivity of illuminated sample, q is the diffraction order, and d is the period of metal–dielectric films. According to Eq. (6.6.1), the illumination depth mainly depends on the incidence angle, wavelength as well as the permittivity of samples. Moreover, by controlling the incidence angle of light, the illumination depth could be continuously tuned, as shown in Fig. 6.19c [59]. At the incidence angles of 5°, 25°, 40°, and 50°, the corresponding illumination depths are 19, 28, 40, and 58 nm, respectively. In order to demonstrate the superiority of the surface microscopy based on BPP, Fig. 6.20 shows a comparison of illumination depth between the BPPs method and TIRFM. Obviously, BPPs show dramatically shrinkage of illumination depth [59].

In order to demonstrate the surface imaging capability of BPPs illumination structure, two crossed fluorescence nanoparticles with the radius of 100 nm are illuminated

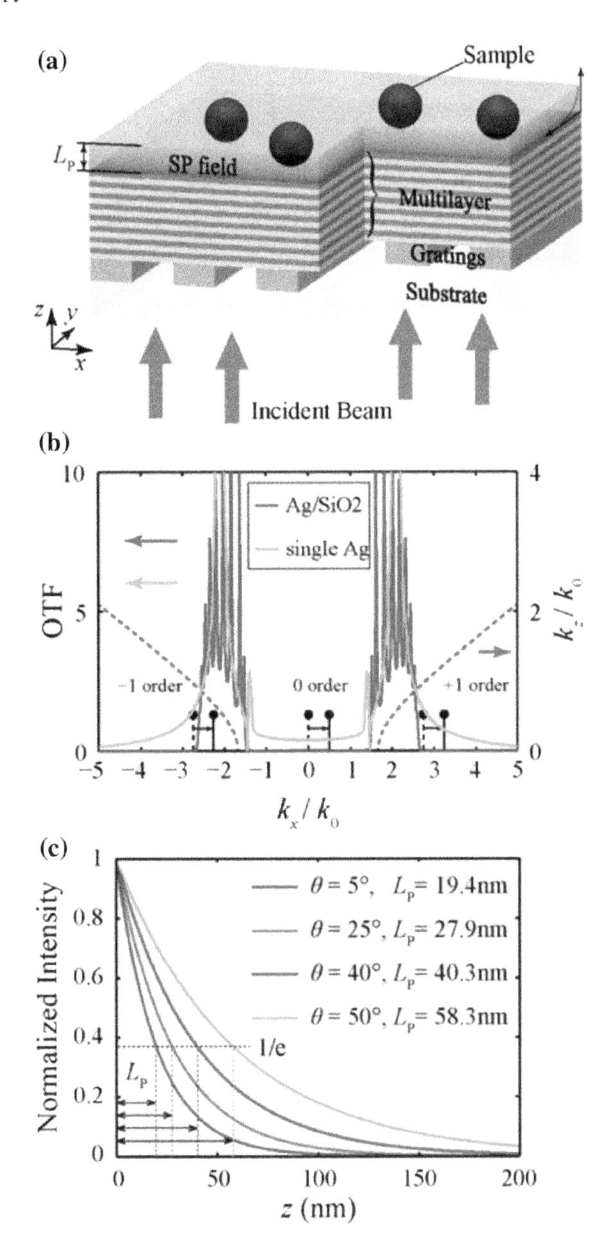

Fig. 6.19 **a** Schematic of BPPs illumination structure. **b** The OTF for the single Ag layer (green line) and Ag–SiO_2 multi-films (blue line). The red line represents the dispersion relation of Ag–SiO_2 multi-films in EMT approximation. **c** The corresponding intensity decay curves away from illumination surface when incidence angles are 5°, 25°, 40°, and 50°, respectively. Reproduced from [59] with permission. Copyright 2016, The Royal Society of Chemistry

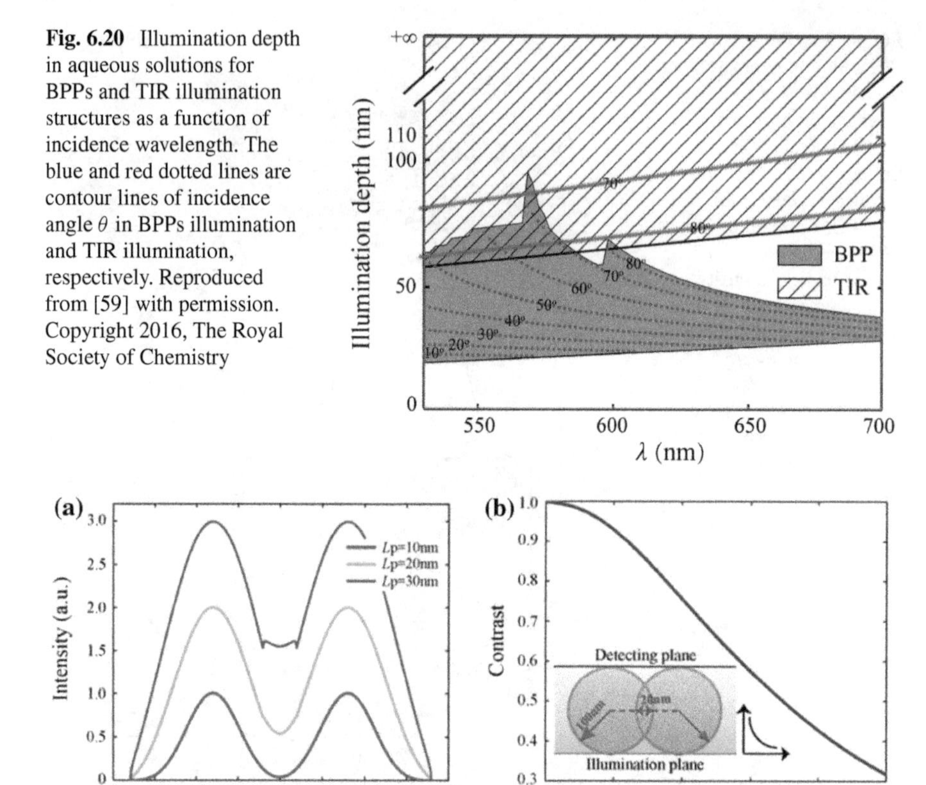

Fig. 6.20 Illumination depth in aqueous solutions for BPPs and TIR illumination structures as a function of incidence wavelength. The blue and red dotted lines are contour lines of incidence angle θ in BPPs illumination and TIR illumination, respectively. Reproduced from [59] with permission. Copyright 2016, The Royal Society of Chemistry

Fig. 6.21 **a** Images of the crossed fluorescence nanoparticles obtained in detecting plane (inset in **b**) under the illumination depth of 10, 20, and 30 nm, respectively. **b** The contrast of nanoparticles' image as the function of illumination depth. The inset shows the schematic of nanoparticles imaging. Reproduced from [59] with permission. Copyright 2016, The Royal Society of Chemistry

under the different illumination depth (depicted in the inset of Fig. 6.21b), and the images obtained in the detecting plane are shown in Fig. 6.21a [59]. Herein, the contrast is defined as $C = (I_{\max} - I_{\text{central}})/(I_{\max} + I_{\text{central}})$, where $I_{\max}$ and I_{central} are the maximum intensity and the intensity at the central position between nanoparticles along the detecting plane, respectively. Figure 6.21b shows the relationship between contrast and illumination depth. It is obvious that the contrast decreases near linearly with increasing the illumination depth [59].

A custom-built microscopy system was constructed and displayed in Fig. 6.22. A semiconductor laser with the wavelength of 532 nm was employed as the excitation source [60]. The excitation laser was impinged on the HMM from the half-cylinder sapphire prism with different incidence angles. The launched BPP mode illuminates the dyed specimen, and the radiated fluorescence is collected by the 40× objective along with a tube lens (0.65X). Meanwhile, a notch filter was mounted

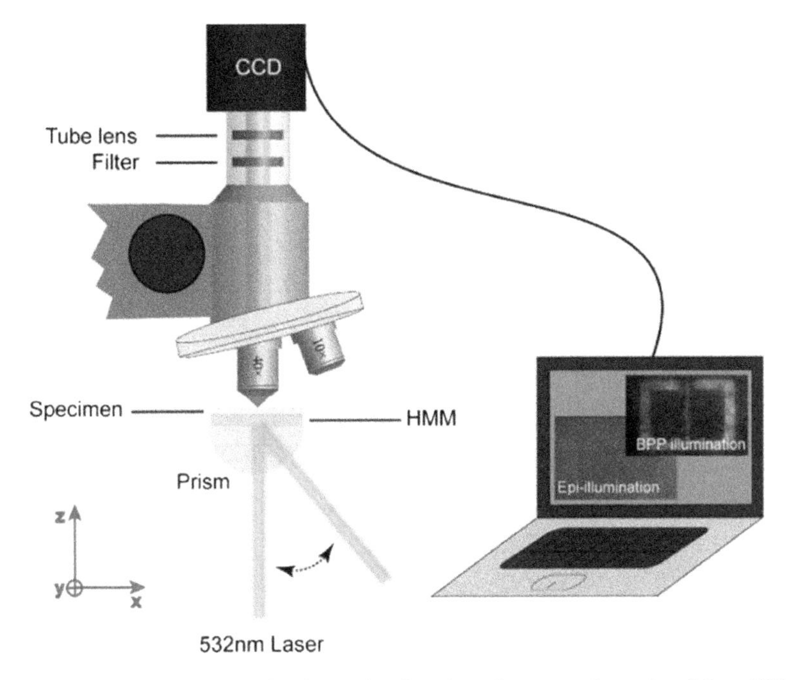

Fig. 6.22 Setup of the BPPs illuminating surface imaging microscopy. Reproduced from [60] with permission. Copyright 2018, The Royal Society of Chemistry

between the objective and the tube lens for eliminating possible excitation laser from the emission fluorescent. Ultimately, the images of the specimen were captured by intensity-sensitive CCD detector [60].

When focusing the fluorescent nanoparticle, the images under different incidence angles from the substrate side of HMM were captured and displayed in the insets of Fig. 6.23 [60]. The dots are the cross-sectional intensities of a single nanoparticle extracted from the detected images (along the lines in the insets). One can see that the real illumination depths agree well with the predicted ones and changes with the illumination angles [60].

When evaluating the surface imaging capability of the BPP illumination structure with tunable ultra-short illumination depth, the fluorescence nanoparticle was replaced by the fluorescence pattern with 27 nm rhodamine 6G/photoresist film. The imaging results are shown in Fig. 6.24 [60]. Obviously, under epi-illumination, the contribution from Mega 520 background dominates and the contrast of the pattern image was drastically reduced. On the contrary, under the surface illumination, the background fluorescence was suppressed to some extent and thus the distinct pattern image could be observed. Furthermore, by increasing the incidence angle from 20° to 40°, the illumination depth was improved from 25 to 38 nm correspondingly [60]. Because longer penetration depth means more background fluorescence, the pattern image contrast decreases slightly with raising the illumination depth. The intensity

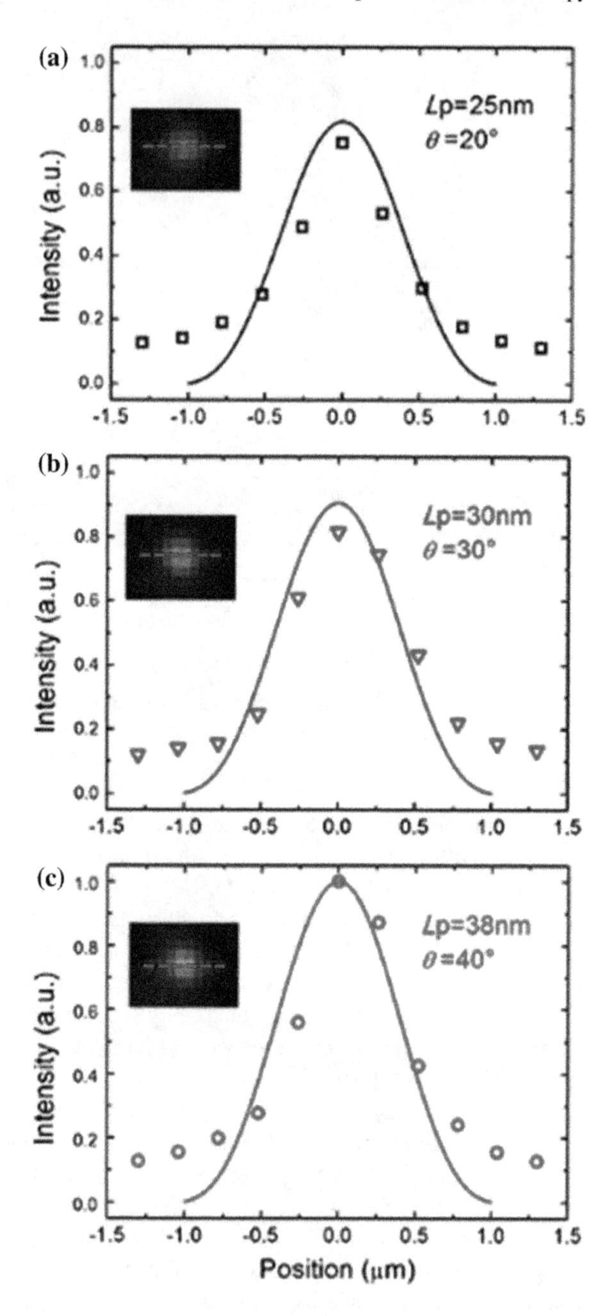

Fig. 6.23 Experimental and simulated cross-sectional intensities of a single fluorescence nanoparticle with a radius of 20 nm. The penetration depth is **a** 25 nm, **b** 30 nm, **c** 38 nm, respectively. The corresponding incidence angle is **a** 20°, **b** 30°, and **c** 40°, respectively. The insets are the corresponding images of the illuminating nanoparticle. Reproduced from [60] with permission. Copyright 2018, The Royal Society of Chemistry

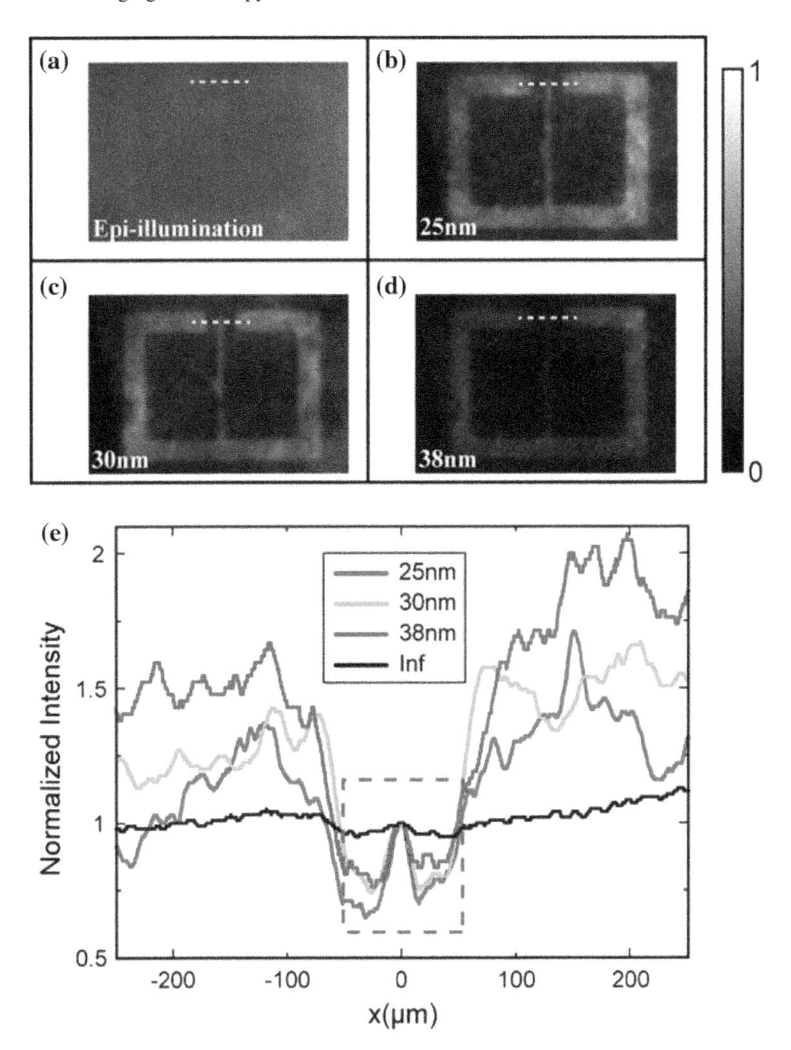

Fig. 6.24 Images of the fluorescence pattern. **a** Epi-illumination. BPPs illumination with the penetration depth of **b** 25 nm, **c** 30 nm, and **d** 38 nm. **e** The normalized intensities along the dotted lines in **a**–**d**. Reproduced from [60] with permission. Copyright 2018, The Royal Society of Chemistry

lines across the dotted lines were plotted in Fig. 6.24e. One can observe that the lowest valley is achieved under the surface illumination with penetration depth of 25 nm (incidence angle of 20°). When increasing the illumination depth, the valley rises and the pattern image contrast is weakened. For the epi-illumination, the valley almost disappears and the distinct surface image could not be obtained [60].

6.7 Microsphere and Micro-cylinder Microscopy

Different from the high-performance and sophisticated super-resolution techniques based on superlens and hyperlens mentioned above, a dielectric lens of spherical/cylindrical shape with appropriate refractive index can also image objects with a precision well beyond the classical diffraction limit. The super-resolution capability of such spheres was suggested to stem from their extraordinary sharp focusing properties, so-called photonic nanojets (PNJs). These specific properties cannot be achieved by classical Gaussian beams through high numerical aperture objectives. Chen et al. first reported it in 2004 through finite-difference time-domain (FDTD) modeling of cylindrical structures under plane wave illumination [61]. Since then, many research groups have reported a broad range of microsphere diameters from 2λ to more than 50λ theoretically and experimentally [62–65].

6.7.1 Photonic Nanojets Generated by Engineered Microspheres

Many approaches to modify the optical properties of the PNJs have been proposed, including changing the refractive index and diameter of the microspheres, varying the illumination conditions and the shapes of the microspheres. For example, in order to improve the propagation distance or decrease the FWHM of PNJs, original proposal of microsphere has been replaced by concentric microsphere-micro-cylinder [66], two-layer dielectric microsphere [67], gradient index microsphere [68], and hemisphere shaped shell [69]. Another effective way to tune the photonic nanojet is fabricating microstructure on the spherical surface and modifying the contribution of different field components to the total field [70]. As shown in Fig. 6.25a–f, by decorating the microspheres with concentric rings, the beam waist of the focal spot can be decreased. For an engineered microsphere with four uniformly distributed rings etched at a depth of 1.2 μm and width of 0.25 μm, it has been demonstrated to generate PNJ with a FWHM of 0.485λ ($\lambda = 400$ nm). Compared to the case without the rings, a shrinkage of beam size up to 28.0% was achieved. Besides, this approach introduces freedom in designing the surface structure of the microsphere and manipulating the interaction of the microsphere with incident beam. Also, one can employ a center-covered engineered microsphere to shrink the focal spot size since the total electric field intensity pattern at the cross-sectional perpendicular to the polarization direction is determined by the transverse field component [71]. As illuminated in Fig. 6.25g–l, by tuning the cover ratio of the engineered microsphere, the beam waist can be medicated. At a wavelength of 633 nm, a focal spot of 245 nm (0.387λ) was achieved experimentally under the plane wave illumination.

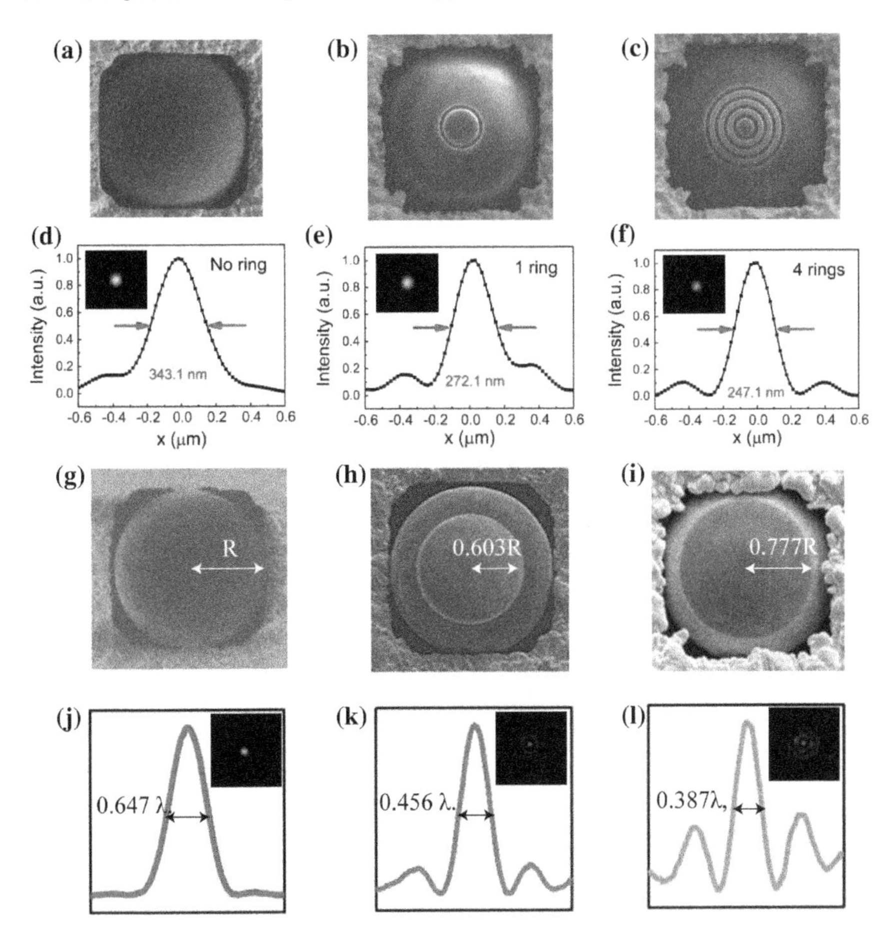

Fig. 6.25 SEM of **a** microsphere, **b** 1-concentric-ring microsphere (1-CRMS), **c** 4-CRMS, and **d–f** corresponding intensity profile. Insets are the normalized intensity distribution of the cross section of the photonic nanojet along the optical axis captured by the CMOS camera. SEM of the microsphere covered by Pt mask with different cover ratios **g** 0, **h** 0.603, **i** 0.777, and **j–l** corresponding intensity profile. Insets are the normalized intensity distribution of the cross section of the photonic nanojet along the optical axis captured by the CMOS camera. Reproduced from **a–f** [70] and **g–l** [71] with permission. **a–f** Copyright 2015, Optical Society of America. **g–f** Copyright 2016, The Authors

6.7.2 *Nanoscopy Based on Dielectric Microsphere and Cylinder*

In 2011, a white light optical microscope based on fused silica microspheres was first demonstrated [72] with the imaging resolution around ~50 nm (λ/8) and magnification up to 8× in air. Figure 6.26 illustrates the schematic of a transmission-mode white light microsphere nanoscope. The microspheres are placed on the top of the object surface by self-assembly. A halogen lamp with a peak wavelength of 600 nm is used as the white light illumination source. The microsphere superlenses collect the underlying near-field object information, magnify it (forming virtual images which keep the same orientation as the objects in the far field), and pick it up by a conventional 80× objective lens [72].

In the experiments, gratings consisting of 360-nm-wide lines and spaced 130 nm apart, were imaged using 4.74 μm diameter microspheres. As can be seen from Fig. 6.27a, b, only those lines with particles on top of them have been resolved. As a control experiment, the lines without particles on top mix together and form a bright spot which cannot be directly resolved by the optical microscope because of the diffraction limit [72]. The magnified image corresponds to a 4.17× magnification factor. Figure 6.27c, d show a fishnet gold-coated anodic aluminum oxide (AAO) membrane imaged with 4.74 μm diameter microspheres [72]. The pores are 50 nm in diameter and spaced 50 nm apart. As it can be seen, the microsphere nanoscope resolves these tiny pores that are well beyond the diffraction limit, giving a resolution between $\lambda/8$ ($\lambda = 400$ nm) and $\lambda/14$ ($\lambda = 750$ nm) in the visible spectrum range. It is important to note that the magnification in this case is around 8×. Further experiments have confirmed that the proposal can also be operated in reflection manner (Fig. 6.28), and both the transmission and reflection modes can achieve 50 nm resolution [72].

The mechanism of super-resolution imaging was explained as PNJ-enhanced sub-diffraction-limited illumination and conversion of the near-field evanescent waves to magnified far-field propagating waves by the microspheres. Compared to traditional

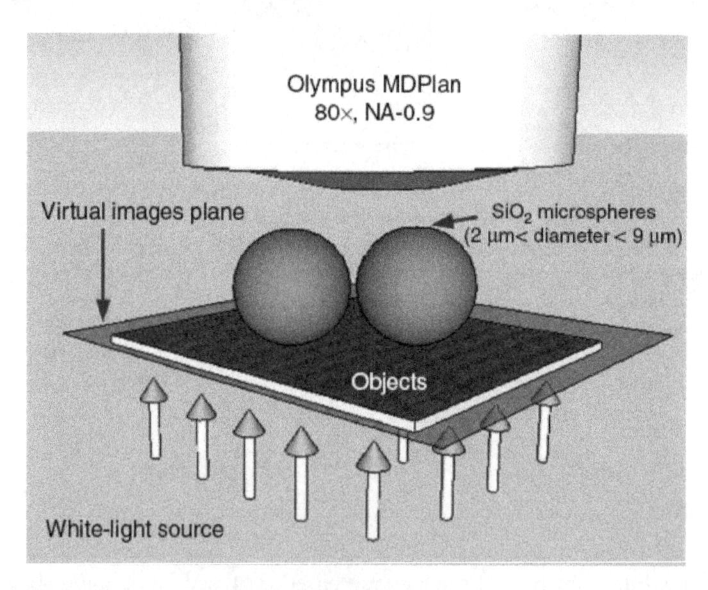

Fig. 6.26 Schematic of the transmission-mode microsphere superlens integrated with a classical optical microscope. The spheres collect the near-field object information and form virtual images that can be captured by the conventional lens. Reproduced from [72] with permission. Copyright 2011, Macmillan Publishers Limited

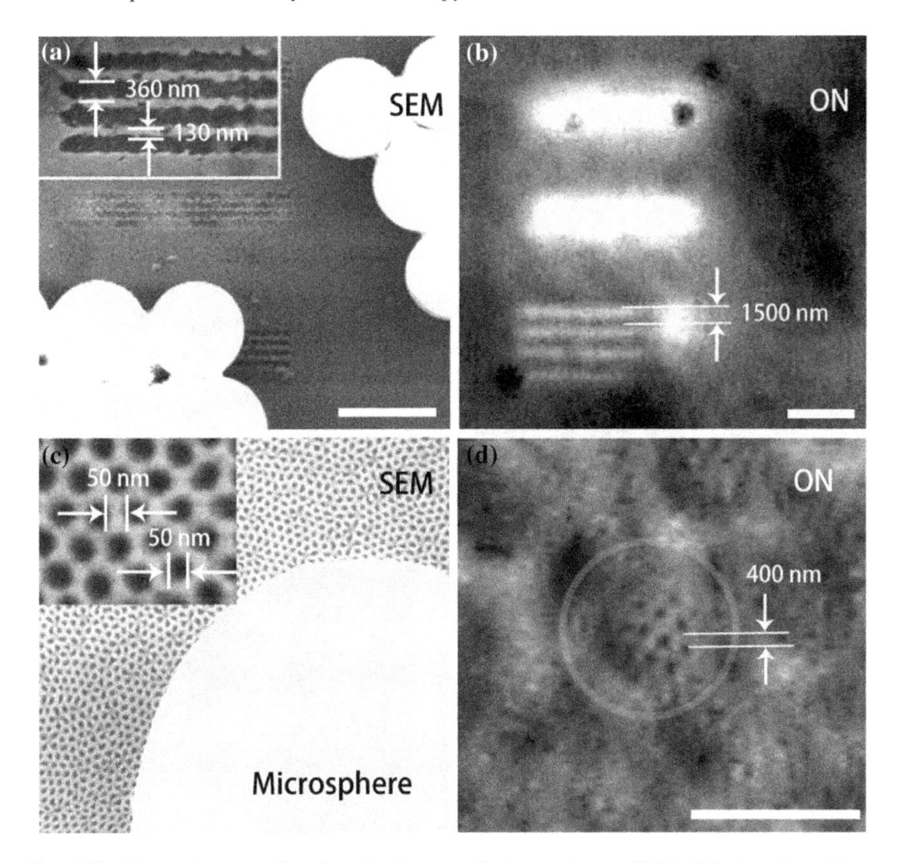

Fig. 6.27 Microsphere superlens imaging in transmission mode. **a**, **c** SEM of the imaging objects. 360-nm-wide lines spaced 130 nm apart (**a**) and a gold-coated fishnet AAO sample (**c**). The optical nanoscope (ON) image clearly resolves the pores that are 50 nm in diameter and spaced 50 nm apart (**c**). **b**, **d** Microsphere superlens imaging results. The size of the optical image between the pores within the image plane is 400 nm. It corresponds to a magnification factor of $M \approx 8$. Scale bar, 5 μm. Reproduced from [72] with permission. Copyright 2011, Macmillan Publishers Limited

solid immersion lens, such dielectric microspheres can be positioned closer to the investigated surface because they have smaller contact regions. Despite the great scientific interests [73–75], it requires the deposition of microspheres directly onto the sample surface, which is a random process and difficult to be well controlled. The view field is also restricted to the location of the microsphere, so the imaging application is greatly limited.

Recently, a remote-mode microsphere microscopy platform was proposed. As indicated in Fig. 6.29, a transparent microsphere as a tiny optical lens was positioned above the sample surface to form an enlarged virtual image. A conventional microscope system is used to capture the virtual image provided by the microsphere. A special adapter is designed to mount the microsphere setup on the objective lens. The

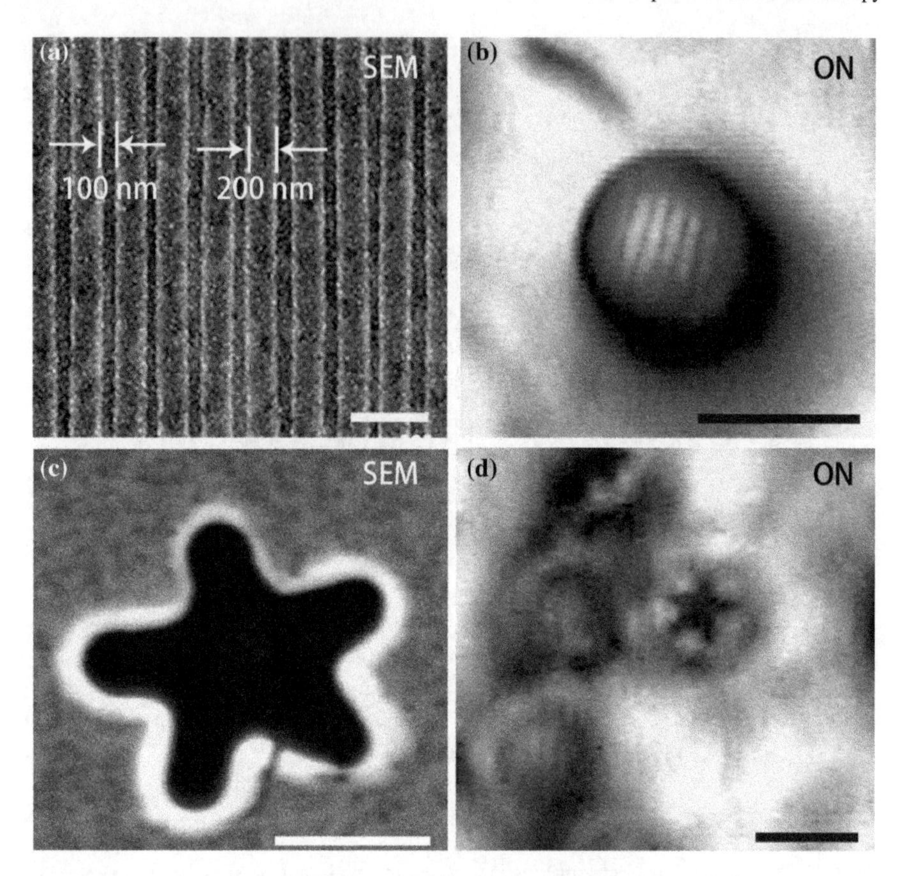

Fig. 6.28 Microsphere nanoscope reflection mode imaging. **a**, **b** Microsphere superlens reflection mode imaging of a commercial Blu-ray DVD disk. The 100-μm-thick transparent protection layer of the disk was peeled off before using the microsphere ($a = 2.37$ μm). The sub-diffraction-limited 100-nm lines **a** are resolved by the microsphere superlens **b**. **c** Reflection mode imaging of a star structure made on GeSbTe thin film for DVD disk. The complex shape of the star including 90-nm corner was clearly imaged. **d** Scale bar: SEM (500 nm), ON (5 μm). Reproduced from [72] with permission. Copyright 2011, Macmillan Publishers Limited

distance between the objective lens and the microsphere can be controlled precisely by a stepper motor on the mechanical frame [16].

First, a microsphere with 400 μm diameter was applied to a 20× conventional objective lens, which is located at 65 μm away from the object, ~100 times larger than the central wavelength of the incidence. Without sacrificing the observation capability and imaging quality, the view field is greatly extended to 125 μm × 125 μm via this image integration (theoretically there is no limitation on the scanning area while a larger area takes longer time to scan) [16]. Such a broad view field is sufficient to study the in vivo reactions between cells and multicellular mechanisms. From the microscopy results shown in Fig. 6.30, one can see that the 20× conventional

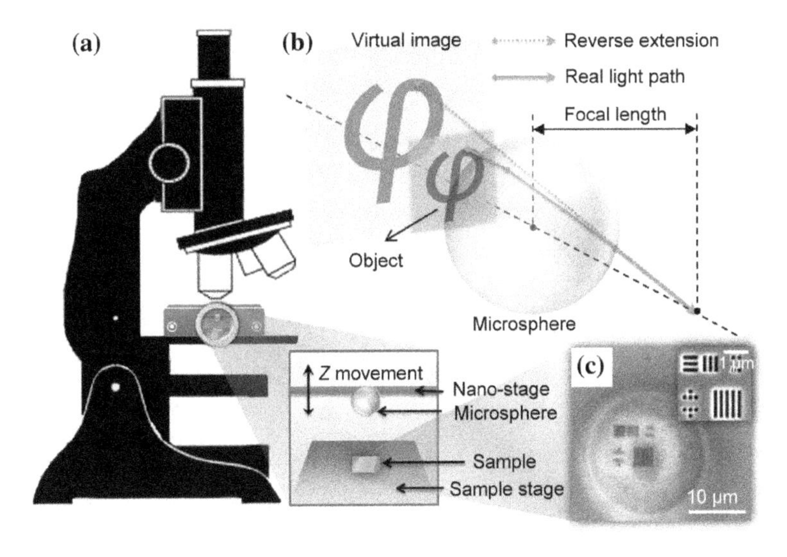

Fig. 6.29 **a** Schematic diagram of the remote-mode optical microsphere setup. **b** Mechanism to illustrate the enlarged virtual image by the microsphere. **c** Optical image captured by this system (sample: semiconductor testing sample; scale bar: 10 μm; imaged by a 20 μm silica microsphere compiled to an oil-immersion optical microscope with a 100× objective lens, NA = 1.4). Inset: SEM image (scale bar: 1 μm). Reproduced from [16] with permission. Copyright 2018, Institute of Optics and Electronics, Chinese Academy of Sciences

objective lens with the microsphere achieves the observation power similar to a 50× objective lens [16].

To demonstrate the resolving power of the system, a microsphere of 20 μm diameter was applied to a 100× conventional objective lens under the oil-immersion mode. Nanodots with 23 nm feature size and magnetic head with a nanogap of 77 nm were clearly observed. Compared with the conventional optical microscope with 200 nm resolution (such as Nikon Eclipse Ni-E), such configuration observes nanoscale features ~1/9 smaller, as indicated in Fig. 6.31 [16].

Although superlensing microsphere and microfibers can be manufactured via engineering processes including chemical synthesis and photolithography, these processes, however, are complex and require sophisticated engineering processes. Recently, a naturally occurring superlens in spider silks, which are transparent in nature and have micron-scale cylinder structure, has been utilized as biological superlens provided by nature [76]. Figure 6.32 shows the spider, its silk, and the corresponding imaging experimental setup. The silk used in experiments were minor ampullate glands filaments reeled directly from *Nephila edulis* spiders (Fig. 6.32a), which has a diameter of 6.8 μm and refractive index of 1.55. To facilitate manipulation and precision positioning of the silk at desired location, the silk was top-encapsulated using a transparent cellulose-based tape and directly placed on top of the imaging sample surface (Fig. 6.32b) [76].

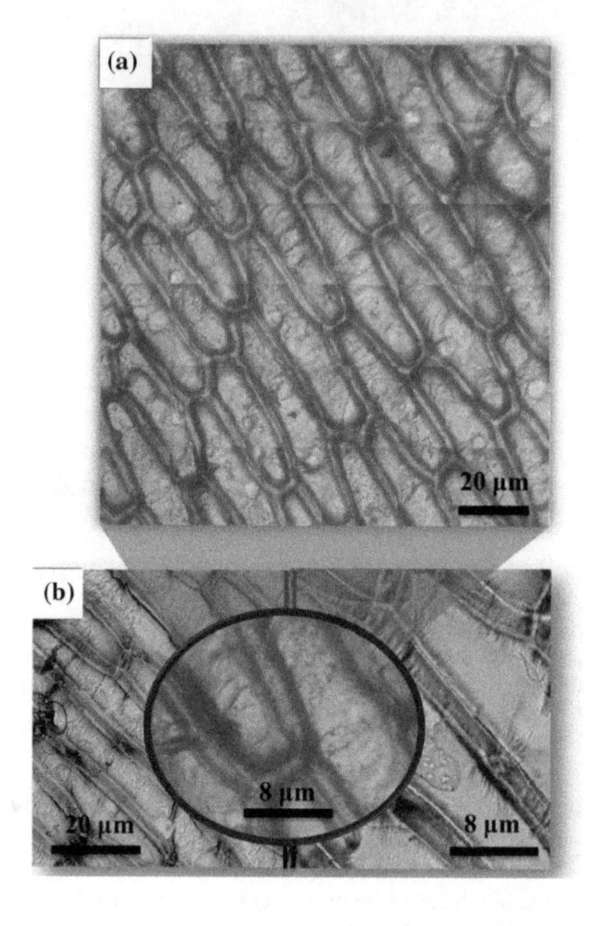

Fig. 6.30 **a** Integrated image of onion cells (scale bar: 20 μm). **b** Comparison of the optical images by three optical lenses: the 20× objective lens (left); 20× objective lens with the microsphere (middle); and 50× objective lens (right). Reproduced from [16] with permission. Copyright 2018, Institute of Optics and Electronics, Chinese Academy of Sciences

Two samples were used for the imaging experiment: a semiconductor chip with features about 400–500 nm (above diffraction limit) and a commercial Blu-ray disk with 100/200 nm features (below diffraction limit). The 1D imaging results are demonstrated in Fig. 6.33, in which a semiconductor chip sample with line features 400–500 nm (Fig. 6.33a) was imaged with silk lens placed on top (Fig. 6.33b) [76]. One can clearly see the silk generates a clear image of underlying line objects, but with a twisted angle stemming from anisotropic magnification effect of cylindrical lens. Then, a sub-diffraction-limited Blu-ray disk (Fig. 6.33c) containing 200 and 100 nm features was used as the imaging objects. Without silk, the microscope cannot resolve these feature due to the optical diffraction limit (600/2NA = 333.3 nm, where NA = 0.9). However, clearly magnified (M ~ 2.1) super-resolution images could be obtained after carefully adjusting the illumination angle of the lighting (Fig. 6.33d) [76].

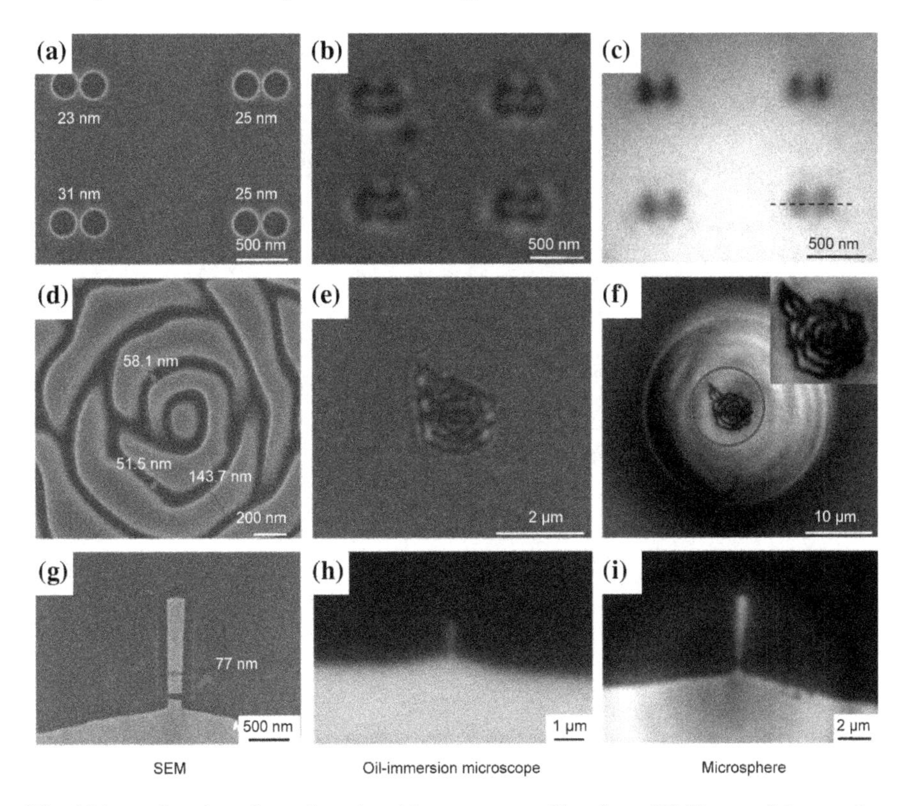

Fig. 6.31 **a–c** Imaging of nanodot pairs with nanogap on a Si wafer. **a** SEM image of the samples, showing sizes of nanogaps between each pair of nanodots. **b** Imaging of the samples by an oil-immersion microscope. **c** Neighboring separated nanodots are resolved clearly by a microsphere with 20 μm diameter. **d–f** Imaging of samples with complex features. **d** Zoomed-in SEM image with size notations. **e** Imaging result by the oil-immersion optical microscope. **f** Image under the 27 μm microsphere in scanning mode. **g–i** Imaging of a magnetic head in a hard disk drive from the production line. Reproduced from [16] with permission. Copyright 2018, Institute of Optics and Electronics, Chinese Academy of Sciences

6.8 Super-oscillation and Supercritical Microscopy

6.8.1 Super-oscillation Microscopy

Although some novel concepts including superlens and hyperlens are proposed and deliver revolutionary ideas for super-resolution microscopy by recovery of the non-propagating evanescent fields, far-field optical super-resolution can actually be achieved without making use of the evanescent waves. In recent years, "super-oscillation", a concept stems from super-directive antennas in the microwave community [77], has been proposed to achieve sub-diffraction-limited spot size and overcome the short working distance and fluorescence labeling constraints. Systematic

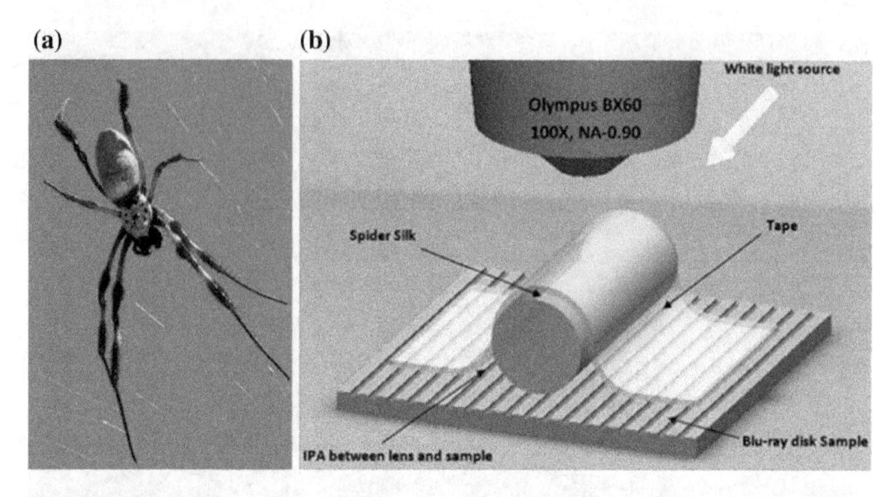

Fig. 6.32 **a** *Nephila edulis* spider in its web. **b** Schematic drawing of reflection mode silk biosuperlens imaging. The spider silk was placed directly on top of the sample surface by using a soft tape. The gaps between silk and sample were filled with IPA which improves imaging contrast. The silk lens collects the underlying near-field object information and projects a magnified virtual image into a conventional objective lens (100×, NA: 0.9). Adapted with permission from [76]. Copyright 2016, American Chemical Society

study of optical super-oscillations was recently revived in the context of quantum mechanics after Aharonov [78] showed that weak quantum mechanical measurements can have values outside the spectrum of the corresponding operator. In other words, a "local" measurement of a value, such as the wave number of an optical wave, can be outside the range seen when a global measurement is taken.

Seemingly counterintuitive, this anomalous effect may be understood by investigating the difference between the complex amplitude and intensity [79]. As depicted in Fig. 6.34, if one constructs an intensity function as $|\sin(2\pi x)+ 0.99|^2$, a small peak can be obtained with very small width. Though this example is very simple, it reveals some important aspects of super-oscillation interference. First, although the local intensity may oscillate more rapidly than the highest Fourier components, the complex amplitude often does not have this property. Second, the super-oscillation is weak and accompanied by strong side lobes.

According to the super-oscillation theory, one can construct waveforms which contain spatial variations more rapid than the operational wavelength and hence perform focusing or imaging with a resolution beyond the diffraction limit. In principle, there is no physical limitation on resolution and only a trade-off in transfer of intensity to sidebands and reduced energy concentration in the hot spot. Metasurface offers a flexible method to tune the phase and amplitude of transmission or reflection coefficients by locally tuning the subwavelength structures [80]. Moreover, benefiting from the nearly dispersionless feature of phase modulations with metasurfaces, a super-oscillatory metasurface filter was proposed for broadband super-resolution

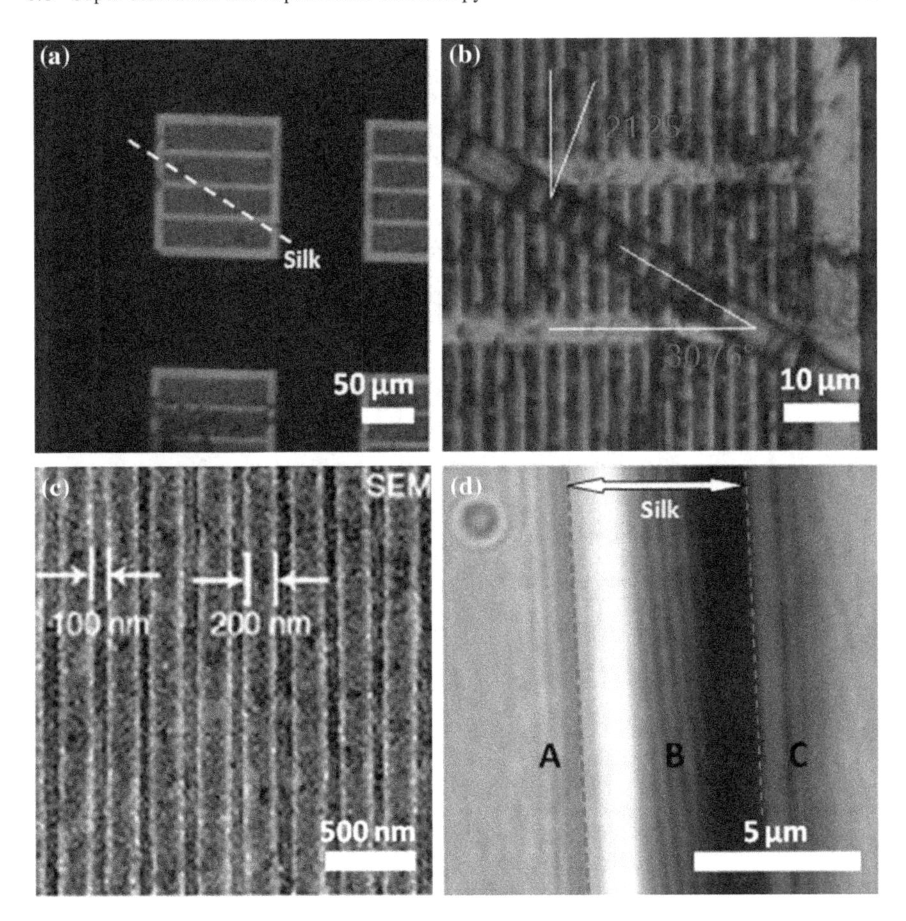

Fig. 6.33 A typical imaging example of silk superlens nanoscope in reflection mode imaging a surface of an integrated circuit and commercial Blu-ray disk. Microdimensional integrated surface pattern **a** is magnified by the spider silk **b**. The sub-diffraction-limited 100-nm channels SEM image **c** are resolved by the spider silk superlens **d** and correspond to a magnification fact of 2.1× Reproduced from [76] with permission. Copyright 2016, American Chemical Society

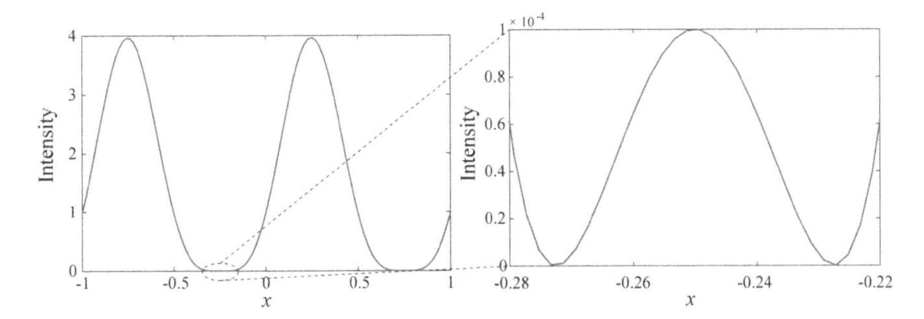

Fig. 6.34 A 1D super-oscillatory function (solid blue line) and its magnified view. Reproduced from [79] with permission. Copyright 2018, Optical Society of America

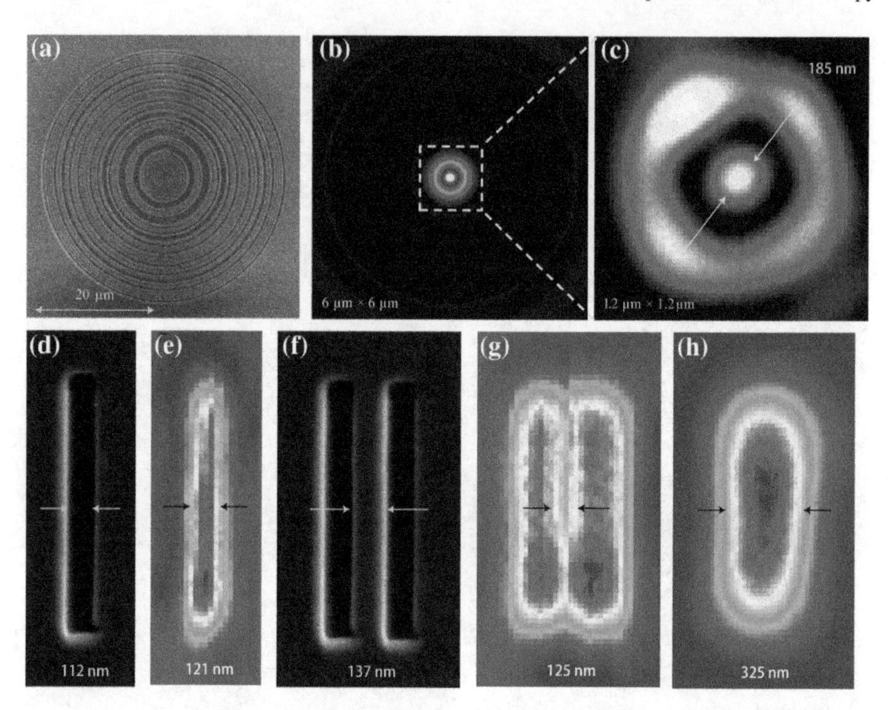

Fig. 6.35 **a** SEM image of the SOL. **b** Calculated energy distribution of the super-oscillatory lens at 10.3 μm from the lens. **c** The actual focal point, $\lambda = 640$ nm. **d–g** SEM image of a 112-nm slit (**d**) and its SOL image (**e**) a double slit (**f**) and its SOL image (**g**). **h** The image of the same double slit is not resolved using a conventional lens of NA = 1.4. Reproduced from [82] with permission. Copyright 2012, Macmillan Publishers Limited

imaging ranging from 400 to 700 nm [81]. This method is expected to potentially promote the development of super-resolving microscopes.

A super-oscillatory lens (SOL) was constructed by a binary mask with a sequence of concentric rings of different widths and diameters optimized by an iterative algorithm [82]. The ring pattern of the SOL (Fig. 6.35a) with outer diameter 40 μm was manufactured by focused ion-beam milling of a 100-nm-thick aluminum film supported on a round glass substrate and mounted as a microscope illuminating lens. Experiment results in Fig. 6.35c indicated the optimized SOL could generate a focal hotspot of 185 nm in diameter, located at a distance of 10.3 μm from the lens, when illuminated with a laser at $\lambda = 640$ nm. The intensity of light in the SOL hotspot was ~25 times the intensity of the incident light [82].

To demonstrate the optical resolution of the test objects, single and double nanoslits were manufactured in a 100 nm titanium film. A single slit (Fig. 6.35d) of 112 nm wide is seen by the SOL as a 121-nm slit (Gaussian fit, Fig. 6.35e). A 137 nm gap between two slits (Fig. 6.35f) is well resolved and seen as a 125 nm gap (double Gaussian fit, Fig. 6.35g). These slits are not resolved by a conventional liquid immersion lens with NA = 1.4 (Fig. 6.35h). These measurements demonstrate a resolution of better than $\Delta \approx 140$ nm, or $\lambda/4.6$ [82].

To demonstrate the ability of the super-oscillatory lens to image complex objects, a cluster of eight nanoholes is fabricated with widely varying hole separations (Fig. 6.36a) [82]. Details of the structure are blurry and not resolved in a conventional microscope image (Fig. 6.36b), but on the image taken with the super-oscillatory lens, all the major features of the cluster are sharp and resolved (Fig. 6.36c): Holes separated by 105 nm ($\lambda/6$) are seen as clearly distinct and separated. Moreover, two holes spaced by 41 nm ($\lambda/15$) are nearly resolved [82].

6.8.2 Supercritical Microscopy

Although the above technology achieves pure optical super-resolution imaging in far field, which could be used at any wavelength, and does not depend on the luminescence of object, it is still struggling against several intractable issues [83]: (i) image distortion and small field of view caused by the inevitable side lobes of a super-oscillatory spot; (ii) complicated nanolithographic technique due to the subwavelength features sizes; (iii) short working distance which will lead to a low tolerance of optical misalignment and environmental disturbance; (iv) short depth of focus (DOF), then consequently, a tiny bit of deviation in the object position resulting in huge degradation of super-resolution property. To surmount these inherent barriers and bring the academic concept into practical applications, Qin et al. [83] proposed a noninvasive microscopy based on a concept of supercritical lens (SCL).

The "supercritical" means that the focusing ability of such lens intervenes between the Rayleigh criterion (RC) of traditional Fresnel zone plate (FZP) [84, 85] that focuses light into an Airy spot with the spot size of $0.61\lambda/\text{NA}$ and the super-oscillation criterion with the spot size of $0.38\lambda/\text{NA}$. Therefore, SCLs are featured by the sub-diffraction-limited focal spot, weak side lobe, ultra-long focal length, and needle-like focal region [83]. A focal spot with its full FWHM of 165 nm (0.407λ) in lateral size is experimentally obtained at a distance of 55 μm ($\approx 135\lambda$) away from the SCL, one-order improvement compared with the reported SOL results. Based on the SCL, a customized microscope system working in scanning scheme was built for the demonstration of super-resolution imaging. The above benefits lead to a high-quality image for a large-area object at a high scanning speed [83].

As indicated in Fig. 6.37, such SCL can clearly distinguish a pattern with the minimal feature size of 65 nm in air with a 55 μm working distance. The imaging process is purely physical and captured in real time, which does not need any preprocessing to the samples and mathematical post-processing to the imaging results. Such SCL microscopy was able to map the horizontal details of a 3D object at one-time scanning, which is impossible for other planar lenses (Fig. 6.38). Another distinct property of such SCL microscopy system is the high-speed image reconstruction process, about 16 times faster than the reported SOL imaging [82].

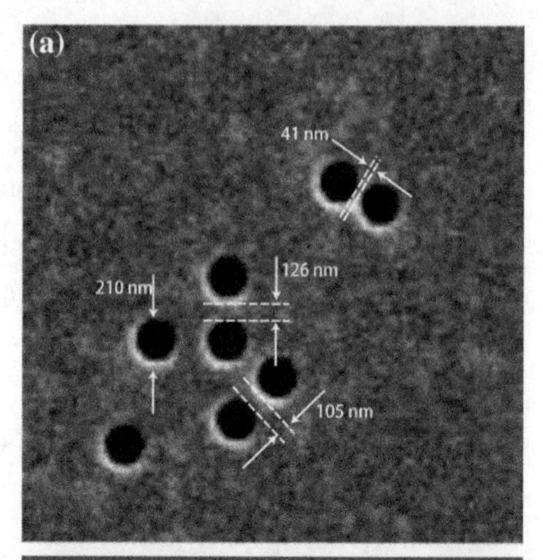

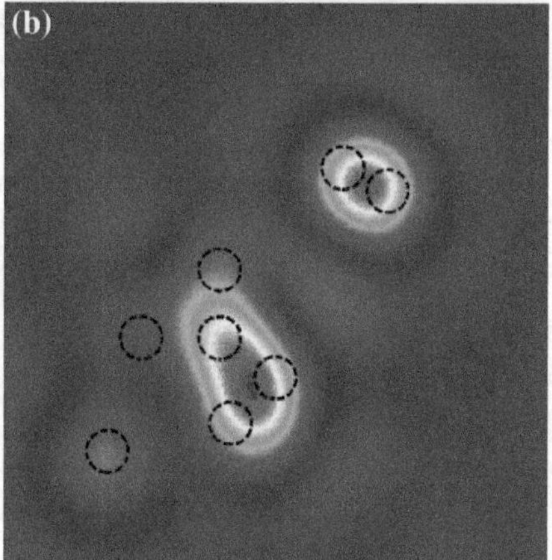

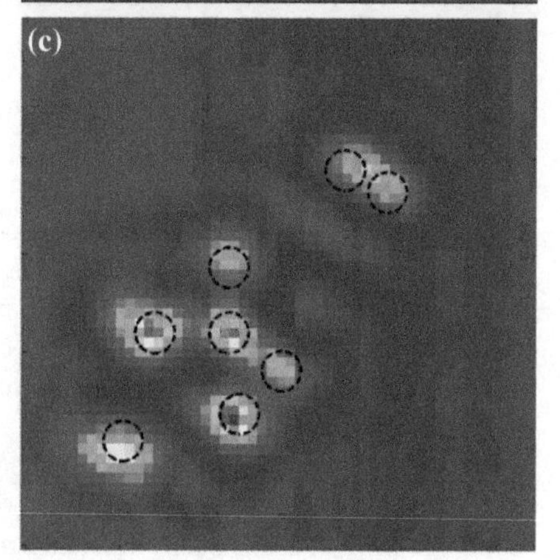

Fig. 6.36 **a** 2.75 μm × 2.755 μm SEM image of a cluster of nanoholes in a metal film. **b** The image of the cluster is not resolved with a conventional lens of NA = 1.4. **c** The SOL image resolves all the main features of the cluster. Reproduced from [82] with permission. Copyright 2012, Macmillan Publishers Limited

6.8.3 Bessel-Beam Microscopy

Bessel beam is one kind of non-diffractive beams that commonly generated by conical shaped lens, i.e., axicons, which could convert Gaussian beams into nondiffracting Bessel beams. Cascading lens and axicon have been utilized for the Bessel-beam microscopy (BBM) [86, 87]. Although BBM has been proved to achieve super-resolution imaging, the existing solutions based on cascading lens and axicon suffer from many drawbacks. For example, cascading configuration impedes the integration and increases the difficulties of practical implementation.

Recently, a planar "Bessel-lens" which superimposes the phase of focusing lens and axicon simultaneously in single metasurface was proposed for super-resolution imaging [88]. The total phase profile of the geometric phase should include two ingredients. The first one is a focusing wavefront (Fig. 6.39a, b), which can be written as:

$$\phi_{\text{LENS}}(r) = \overline{P_S P_L} * k = \left(\sqrt{r^2 + f^2} - f\right) * k \tag{6.8.1}$$

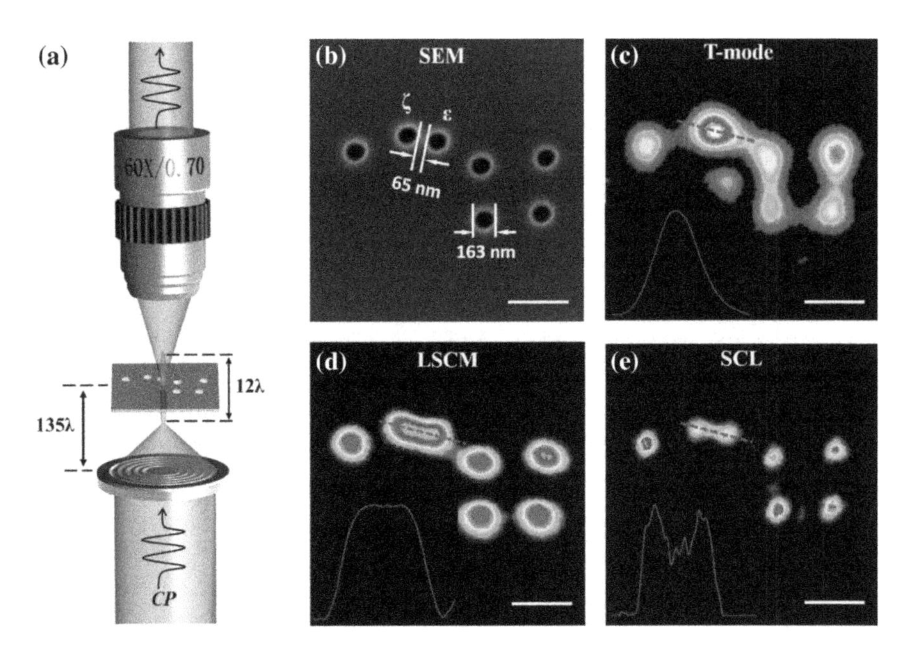

Fig. 6.37 **a** Schematic of SCL microscopy. **b** SEM of nanoscale big dipper used for the resolution demonstration. **c** Imaging result by a normal transmission-mode (T-mode) microscope. **d** Imaging result by the laser scanning confocal microscopy (LSCM) with the same wavelength and objective lens. **e** Imaging result by SCL microscopy. Scale bars: 500 nm. Reproduced from [83] with permission. Copyright 2016, WILEY-VCH Verlag GmbH & Co. KGaA, Weinheim

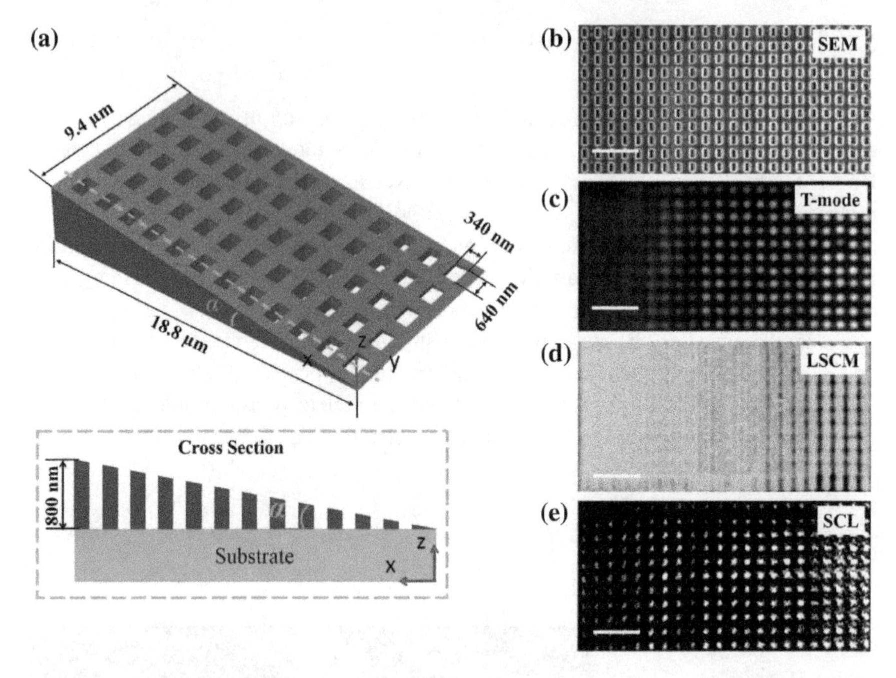

Fig. 6.38 **a** Sketch of a 3D fishnet wedge composed of an etched array of rectangular holes. **b** Top-view (x-y plane) SEM image of the fishnet wedge with a size of $18.8 \times 9.4\ \mu m^2$. **c–e** The imaging results of this wedge by transmission-mode microscopy (T-mode), LSCM, and SCL microscopy. Scale bar: 3 μm. Reproduced from [83] with permission. Copyright 2016, WILEY-VCH Verlag GmbH & Co. KGaA, Weinheim

where f is the focal length, and $k = 2\pi/\lambda$ represents the wave number. The other one is a Bessel wavefront (Fig. 6.39a, c), which mimics the function of an axicon lens [88]:

$$\phi_{\mathrm{AXICON}}(r) = \overline{P_S P_A} * k = r * k * \sin\beta \tag{6.8.2}$$

where $\beta = \tan^{-1}(R/f)$, and R is the radius of the metasurface. Considering the case of $f \gg R$, Eqs. (6.8.1) and (6.8.2) can be, respectively, simplified into [88]:

$$\phi_{\mathrm{LENS}}(r) = \frac{r^2}{2f} * k \tag{6.8.3}$$

and

$$\phi_{\mathrm{AXICON}}(r) = \frac{R}{f} * r * k \tag{6.8.4}$$

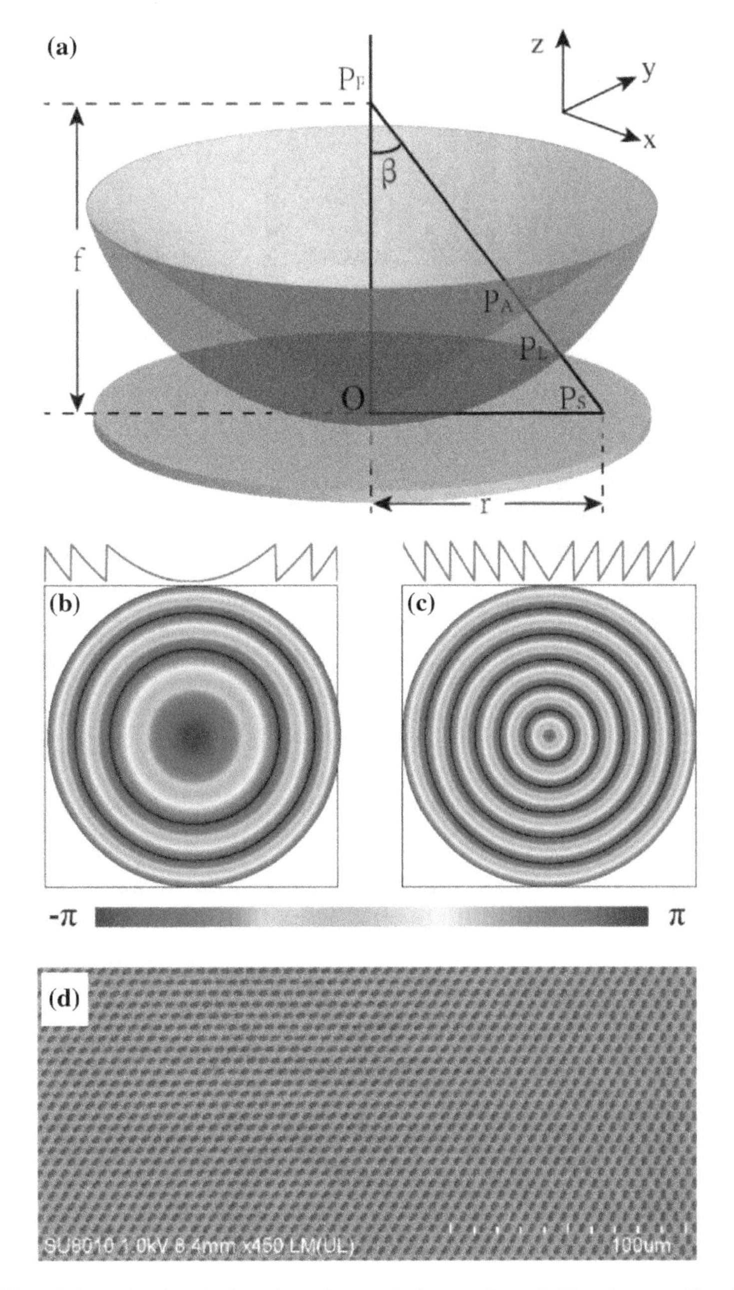

Fig. 6.39 **a** Schematic of designing planar lens and planar axicon. **b** The phase profile of planar lens after modulo operation by 2π. **c** The phase profile of planar axicon after modulo operation by 2π. **d** The SEM image of the metasurface. Reproduced from [88] with permission. Copyright 2017, Optical Society of America

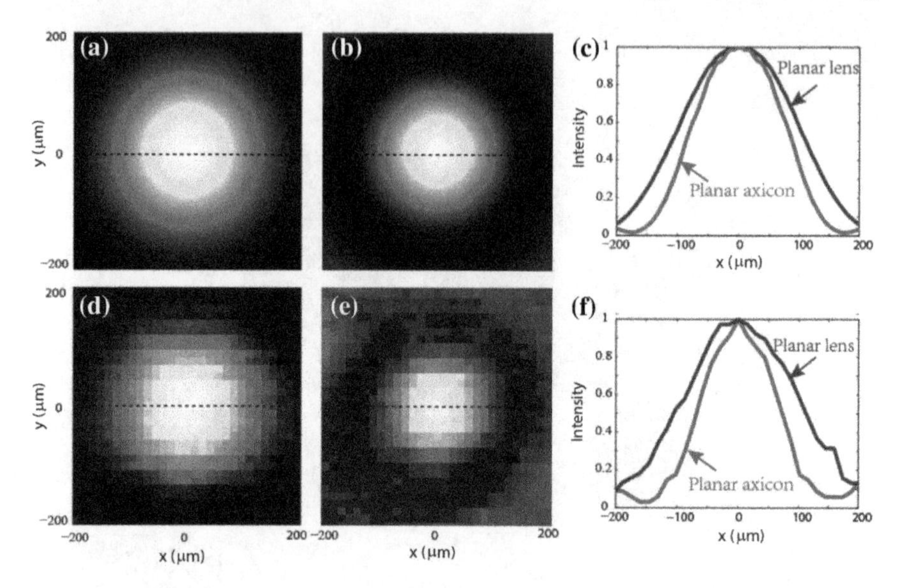

Fig. 6.40 Focusing results of planar lens and planar axicon. **a** The simulated focus result of planar lens by vector diffraction method. **b** The simulated focus result of planar axicon by vector diffraction method. **c** The transversal line of y =0 μm in **a** and **b**. **d** The experimental focus result of planar lens. **e** The experimental focus result of planar axicon. **f** The transversal line of y =0 μm in **d** and **e**. Adapted with permission from [88]. Copyright 2017, Optical Society of America

Therefore, the phase distribution of planar Bessel-lens can be derived as [88]:

$$\phi_{\text{Bessel-lens}}(r) = \phi_{\text{LENS}}(r) + \phi_{\text{AXICON}}(r) = \left(\frac{r^2}{2f_1} + \frac{R}{f_2}r\right) * k \tag{6.8.5}$$

The phase profile in Eq. (6.8.5) was realized by geometric metasurface consisting of rotating metallic apertures, as indicated in Fig. 6.39d.

In order to make a comparison between the planar lens and planar axicon, the focusing property of them is characterized by both vector diffraction simulation and experimental measurement [88]. From the results shown in Fig. 6.40, one can see that the focal spot of planar axicon is smaller than that of planar lens and approaches to the diffraction limit. This unique property will help to obtain a super-resolution imaging when the planar axicon is utilized for microscopy [88].

As a proof of concept, two-dimensional 5 × 5 aperture arrays with period of 110 μm were utilized as imaging object. The simulated and measured imaging results by planar lens and planar Bessel-lens are in Fig. 6.41 [88]. From the simulated results shown in Fig. 6.41a, b, it can be seen that planar lens could not distinguish apertures with each other, but planar Bessel-lens can obtain trenchant image. The experimental results presented in Fig. 6.41c, d agree well with simulations [88].

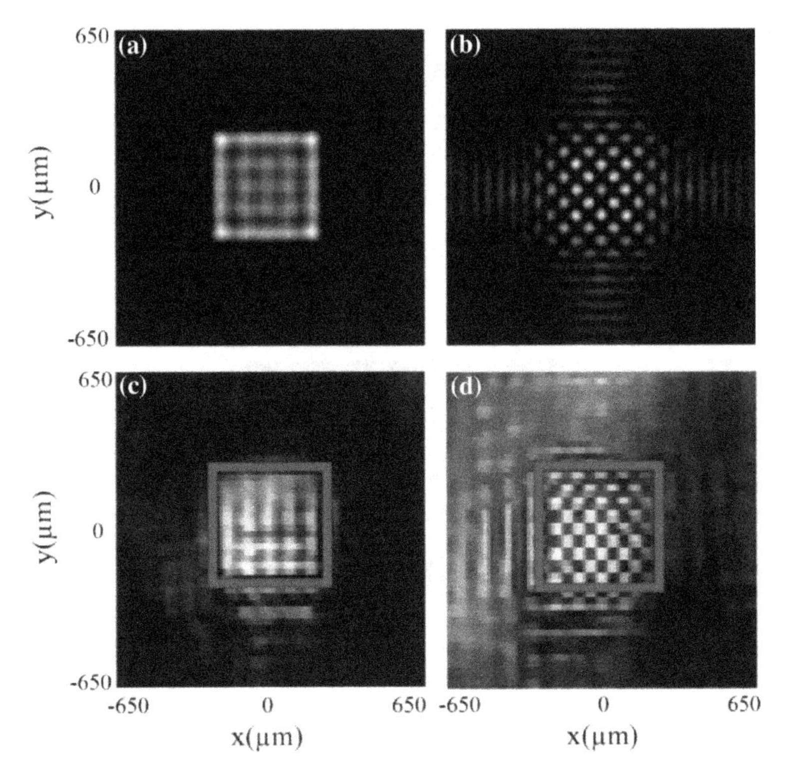

Fig. 6.41 Imaging results of planar lens and planar Bessel-lens. **a** Simulated imaging result of planar lens. **b** Simulated imaging result of planar Bessel-lens. **c** Experimental imaging result of planar lens. **d** Experimental imaging result of planar Bessel-lens. Reproduced from [88] with permission. Copyright 2017, Optical Society of America

Furthermore, randomly distributed apertures with different separation distances (shown in Fig. 6.42a) were imaged. Figure 6.42b shows the simulated imaging results of planar lens. A typical area is zoomed in and presented as an inset which includes a pair of adjacent apertures that cannot be resolved by planar lens [88]. In contrary, the simulation results in Fig. 6.42c show that the planar Bessel-lens can distinguish them. Theoretically, the optical resolution of traditional lens is about $R_{\text{lens}} = 0.61\lambda/\text{NA}$ due to the diffraction limit, i.e., $R_{\text{lens}} = 129.3\ \mu\text{m}$ when NA= 0.05. Thus, the planar lens cannot distinguish two point source spacing at 92 μm, while planar Bessel-lens can distinguish them, as shown in Fig. 6.42d–f [88].

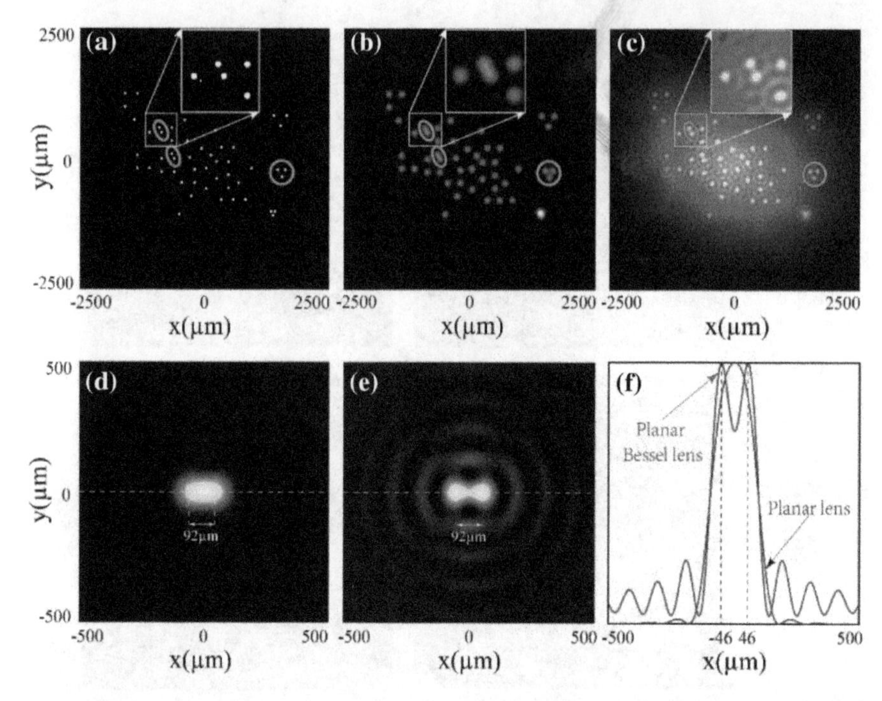

Fig. 6.42 Simulated imaging results of planar lens and planar Bessel-lens illuminated by incoherent light. The typical parts are circled by blue ellipse. **a** The object is composed of random distributed apertures whose radius is 25 μm. **b** The simulated imaging result of planar lens. **c** The simulated imaging result of planar Bessel-lens. **d, e** The imaging result of two point sources spaced at 92 μm for planar lens with NA = 0.05 and planar Bessel-lens with NA = 0.05. **f** The transversal lines of y =0 μm in **d** and **e**. Reproduced from [88] with permission. Copyright 2017, Optical Society of America

References

1. E. Abbe, Beiträge zur Theorie des Mikroskops und der mikroskopischen Wahrnehmung. Arch. Für Mikrosk. Anat. **9**, 413–418 (1873)
2. E.H. Synge, XXXVIII. A suggested method for extending microscopic resolution into the ultra-microscopic region. Lond. Edinb. Dublin Philos. Mag. J. Sci. **6**, 356–362 (1928)
3. E. Betzig, A. Lewis, A. Harootunian, M. Isaacson, E. Kratschmer, Near field scanning optical microscopy (NSOM): development and biophysical applications. Biophys. J. **49**, 269–279 (1986)
4. E. Betzig, J.K. Trautman, Near-field optics: microscopy, spectroscopy, and surface modification beyond the diffraction limit. Science **257**, 189–195 (1992)
5. E. Betzig, R.J. Chichester, Single molecules observed by near-field scanning optical microscopy. Science **262**, 1422–1425 (1993)
6. F. Lu, W. Zhang, L. Huang, S. Liang, D. Mao, F. Gao, T. Mei, J. Zhao, Mode evolution and nanofocusing of grating-coupled surface plasmon polaritons on metallic tip. Opto-Electron. Adv. **1**, 180010 (2018)
7. S.W. Hell, J. Wichmann, Breaking the diffraction resolution limit by stimulated emission: stimulated-emission-depletion fluorescence microscopy. Opt. Lett. **19**, 780–782 (1994)

8. M. Saxena, G. Eluru, S.S. Gorthi, Structured illumination microscopy. Adv. Opt. Photonics **7**, 241–275 (2015)
9. T. Horio, H. Hotani, Visualization of the dynamic instability of individual microtubules by dark-field microscopy. Nature **321**, 605 (1986)
10. S. Chowdhury, A.-H. Dhalla, J. Izatt, Structured oblique illumination microscopy for enhanced resolution imaging of non-fluorescent, coherently scattering samples. Biomed. Opt. Express **3**, 1841–1854 (2012)
11. K. Carlsson, P.E. Danielsson, R. Lenz, A. Liljeborg, L. Majlöf, N. Åslund, Three-dimensional microscopy using a confocal laser scanning microscope. Opt. Lett. **10**, 53–55 (1985)
12. M.J. Rust, M. Bates, X. Zhuang, Sub-diffraction-limit imaging by stochastic optical reconstruction microscopy (STORM). Nat. Methods **3**, 793–796 (2006)
13. E. Betzig, G.H. Patterson, R. Sougrat, O.W. Lindwasser, S. Olenych, J.S. Bonifacino, M.W. Davidson, J. Lippincott-Schwartz, H.F. Hess, Imaging intracellular fluorescent proteins at nanometer resolution. Science **313**, 1642 (2006)
14. S.T. Hess, T.P.K. Girirajan, M.D. Mason, Ultra-high resolution imaging by fluorescence photoactivation localization microscopy. Biophys. J. **91**, 4258–4272 (2006)
15. Y.M. Sigal, R. Zhou, X. Zhuang, Visualizing and discovering cellular structures with super-resolution microscopy. Science **361**, 880 (2018)
16. L.-W. Chen, Y. Zhou, M.-X. Wu, M. Hong, Remote-mode microsphere nano-imaging: new boundaries for optical microscopes. Opto-Electron. Adv. **1**, 170001 (2018)
17. V.G. Veselago, The electrodynamics of substances with simultaneously negative values of ε and μ. Sov. Phys. USPEKHI **10**, 509–514 (1968)
18. X. Zhang, Z. Liu, Superlenses to overcome the diffraction limit. Nat. Mater. **7**, 435 (2008)
19. J.B. Pendry, Negative refraction makes a perfect lens. Phys. Rev. Lett. **85**, 3966–3969 (2000)
20. J. Pendry, A. Holden, W. Stewart, I. Youngs, Extremely low frequency plasmons in metallic mesostructures. Phys. Rev. Lett. **76**, 4773–4776 (1996)
21. J.B. Pendry, A.J. Holden, D.J. Robbins, W.J. Stewart, Magnetism from conductors and enhanced nonlinear phenomena. IEEE Trans. Microw. Theory Tech. **47**, 2075–2084 (1999)
22. R. Shelby, D. Smith, S. Schultz, Experimental verification of a negative index of refraction. Science **292**, 77–79 (2001)
23. H.J. Lezec, J.A. Dionne, H.A. Atwater, Negative refraction at visible frequencies. Science **316**, 430 (2007)
24. T. Xu, A. Agrawal, M. Abashin, K.J. Chau, H.J. Lezec, All-angle negative refraction and active flat lensing of ultraviolet light. Nature **497**, 470–474 (2013)
25. H. Lee, Y. Xiong, N. Fang, W. Srituravanich, S. Durant, M. Ambati, C. Sun, X. Zhang, Realization of optical superlens imaging below the diffraction limit. New J. Phys. **7** (2005)
26. S.A. Ramakrishna, J.B. Pendry, D. Schurig, D.R. Smith, S. Schultz, The asymmetric lossy near-perfect lens. J. Mod. Opt. **49**, 1747–1762 (2002)
27. X. Luo, T. Ishihara, Surface plasmon resonant interference nanolithography technique. Appl. Phys. Lett. **84**, 4780–4782 (2004)
28. N. Fang, H. Lee, C. Sun, X. Zhang, Sub-diffraction-limited optical imaging with a silver superlens. Science **308**, 534–537 (2005)
29. T. Taubner, D. Korobkin, Y. Urzhumov, G. Shvets, R. Hillenbrand, Near-field microscopy through a SiC superlens. Science **313**, 1595 (2006)
30. M. Fehrenbacher, S. Winnerl, H. Schneider, J. Döring, S.C. Kehr, L.M. Eng, Y. Huo, O.G. Schmidt, K. Yao, Y. Liu, M. Helm, Plasmonic superlensing in doped GaAs. Nano Lett. **15**, 1057–1061 (2015)
31. Z. Liu, S. Durant, H. Lee, Y. Pikus, N. Fang, Y. Xiong, C. Sun, X. Zhang, Far-field optical superlens. Nano Lett. **7**, 403–408 (2007)
32. Y. Xiong, Z. Liu, C. Sun, X. Zhang, Two-dimensional Imaging by far-field superlens at visible wavelengths. Nano Lett. **7**, 3360–3365 (2007)
33. T. Xu, H.J. Lezec, Visible-frequency asymmetric transmission devices incorporating a hyperbolic metamaterial. Nat. Commun. **5**, 4141 (2014)

34. H. Liu, Y. Luo, W. Kong, K. Liu, W. Du, C. Zhao, P. Gao, Z. Zhao, C. Wang, M. Pu, X. Luo, Large area deep subwavelength interference lithography with a 35 nm half-period based on bulk plasmon polaritons. Opt. Mater. Express **8**, 199–209 (2018)
35. B. Wood, J.B. Pendry, D.P. Tsai, Directed subwavelength imaging using a layered metal-dielectric system. Phys. Rev. B **74**, 115116 (2006)
36. C. Wang, P. Gao, X. Tao, Z. Zhao, M. Pu, P. Chen, X. Luo, Far field observation and theoretical analyses of light directional imaging in metamaterial with stacked metal-dielectric films. Appl. Phys. Lett. **103**, 31911 (2013)
37. A. Salandrino, N. Engheta, Far-field subdiffraction optical microscopy using metamaterial crystals: theory and simulations. Phys. Rev. B **74** (2006)
38. I.I. Smolyaninov, Y.-J. Hung, C.C. Davis, Magnifying superlens in the visible frequency range. Science **315**, 1699–1701 (2007)
39. Z. Jacob, L.V. Alekseyev, E. Narimanov, Optical hyperlens: far-field imaging beyond the diffraction limit. Opt. Express **14**, 8247–8256 (2006)
40. L. Liu, K. Liu, Z. Zhao, C. Wang, P. Gao, X. Luo, Sub-diffraction demagnification imaging lithography by hyperlens with plasmonic reflector layer. RSC Adv. **6**, 95973–95978 (2016)
41. J. Rho, Z. Ye, Y. Xiong, X. Yin, Z. Liu, H. Choi, G. Bartal, X. Zhang, Spherical hyperlens for two-dimensional sub-diffractional imaging at visible frequencies. Nat. Commun. **1**, 143 (2010)
42. D. Lee, Y.D. Kim, M. Kim, S. So, H.-J. Choi, J. Mun, D.M. Nguyen, T. Badloe, J.G. Ok, K. Kim, H. Lee, J. Rho, Realization of wafer-scale hyperlens device for sub-diffractional biomolecular imaging. ACS Photonics **5**, 2549–2554 (2018)
43. Y. Xiong, Z. Liu, X. Zhang, A simple design of flat hyperlens for lithography and imaging with half-pitch resolution down to 20 nm. Appl. Phys. Lett. **94**, 203108 (2009)
44. W. Wang, H. Xing, L. Fang, Y. Liu, J. Ma, L. Lin, C. Wang, X. Luo, Far-field imaging device: planar hyperlens with magnification using multi-layer metamaterial. Opt. Express **16**, 21142–21148 (2008)
45. B.H. Cheng, Y.Z. Ho, Y.C. Lan, D.P. Tsai, Optical hybrid-superlens hyperlens for superresolution imaging. IEEE J. Sel. Top. Quantum Electron. **19**, 4601305 (2013)
46. B.H. Cheng, Y.-C. Lan, D.P. Tsai, Breaking optical diffraction limitation using optical hybrid-super-hyperlens with radially polarized light. Opt. Express **21**, 14898–14906 (2013)
47. X. Tao, C. Wang, Z. Zhao, Y. Wang, N. Yao, X. Luo, A method for uniform demagnification imaging beyond the diffraction limit: cascaded planar hyperlens. Appl. Phys. B **114**, 545–550 (2014)
48. F. Zernike, Luneburg lens for optical waveguide use. Opt. Commun. **12**, 379–381 (1974)
49. N. Yao, C. Wang, X. Tao, Y. Wang, Z. Zhao, X. Luo, Sub-diffraction phase-contrast imaging of transparent nano-objects by plasmonic lens structure. Nanotechnology **24**, 135203 (2013)
50. L. Wang, C. Vasilev, D.P. Canniffe, L.R. Wilson, C.N. Hunter, A.J. Cadby, Highly confined surface imaging by solid immersion total internal reflection fluorescence microscopy. Opt. Express **20**, 3311–3324 (2012)
51. D.S. Johnson, J.K. Jaiswal, S. Simon, Total internal reflection fluorescence (TIRF) microscopy illuminator for improved imaging of cell surface events. Curr. Protoc. Cytom. **61**, 12.29.1–12.29.19 (2012)
52. D. Axelrod, Total internal reflection fluorescence microscopy in cell biology. Traffic **2**, 764–774 (2001)
53. B. Rothenhäusler, W. Knoll, Surface plasmon microscopy. Nature **332**, 615 (1988)
54. G. Stabler, M.G. Somekh, C.W. See, High-resolution wide-field surface plasmon microscopy. J. Microsc. **214**, 328–333 (2004)
55. K. Watanabe, K. Matsuura, F. Kawata, K. Nagata, J. Ning, H. Kano, Scanning and non-scanning surface plasmon microscopy to observe cell adhesion sites. Biomed. Opt. Express **3**, 354–359 (2012)
56. J.S. Shumaker-Parry, C.T. Campbell, Quantitative methods for spatially resolved adsorption/desorption measurements in real time by surface plasmon resonance microscopy. Anal. Chem. **76**, 907–917 (2004)

57. B. Huang, F. Yu, R.N. Zare, Surface plasmon resonance imaging using a high numerical aperture microscope objective. Anal. Chem. **79**, 2979–2983 (2007)
58. B.K. Singh, A.C. Hillier, Surface plasmon resonance imaging of biomolecular interactions on a grating-based sensor array. Anal. Chem. **78**, 2009–2018 (2006)
59. W. Kong, W. Du, K. Liu, C. Wang, L. Liu, Z. Zhao, X. Luo, Launching deep subwavelength bulk plasmon polaritons through hyperbolic metamaterials for surface imaging with a tuneable ultra-short illumination depth. Nanoscale **8**, 17030–17038 (2016)
60. W. Kong, W. Du, K. Liu, H. Liu, Z. Zhao, M. Pu, C. Wang, X. Luo, Surface imaging microscopy with tunable penetration depth as short as 20 nm by employing hyperbolic metamaterials. J. Mater. Chem. C **6**, 1797–1805 (2018)
61. Z. Chen, A. Taflove, V. Backman, Photonic nanojet enhancement of backscattering of light by nanoparticles: a potential novel visible-light ultramicroscopy technique. Opt. Express **12**, 1214–1220 (2004)
62. S. Lecler, Y. Takakura, P. Meyrueis, Properties of a three-dimensional photonic jet. Opt. Lett. **30**, 2641–2643 (2005)
63. A. Heifetz, K. Huang, A.V. Sahakian, X. Li, A. Taflove, V. Backman, Experimental confirmation of backscattering enhancement induced by a photonic jet. Appl. Phys. Lett. **89**, 221118 (2006)
64. H. Guo, Y. Han, X. Weng, Y. Zhao, G. Sui, Y. Wang, S. Zhuang, Near-field focusing of the dielectric microsphere with wavelength scale radius. Opt. Express **21**, 2434–2443 (2013)
65. E. Mcleod, C.B. Arnold, Subwavelength direct-write nanopatterning using optically trapped microspheres. Nat. Nano **3**, 413–417 (2008)
66. G. Gu, R. Zhou, Z. Chen, H. Xu, G. Cai, Z. Cai, M. Hong, Super-long photonic nanojet generated from liquid-filled hollow microcylinder. Opt. Lett. **40**, 625–628 (2015)
67. Y. Shen, L.V. Wang, J.-T. Shen, Ultralong photonic nanojet formed by a two-layer dielectric microsphere. Opt. Lett. **39**, 4120–4123 (2014)
68. S.-C. Kong, A. Taflove, V. Backman, Quasi one-dimensional light beam generated by a graded-index microsphere. Opt. Express **17**, 3722–3731 (2009)
69. Z. Hengyu, C. Zaichun, C.T. Chong, H. Minghui, Photonic jet with ultralong working distance by hemispheric shell. Opt. Express **23**, 6626–6633 (2015)
70. M.X. Wu, B.J. Huang, R. Chen, Y. Yang, J.F. Wu, R. Ji, X.D. Chen, M.H. Hong, Modulation of photonic nanojets generated by microspheres decorated with concentric rings. Opt. Express **23**, 20096–20103 (2015)
71. M. Wu, R. Chen, J. Soh, Y. Shen, L. Jiao, J. Wu, X. Chen, R. Ji, M. Hong, Super-focusing of center-covered engineered microsphere. Sci. Rep. **6**, 31637 (2016)
72. Z. Wang, W. Guo, L. Li, B. Luk'yanchuk, A. Khan, Z. Liu, Z. Chen, M. Hong, Optical virtual imaging at 50 nm lateral resolution with a white-light nanoscope. Nat. Commun. **2**, 218 (2011)
73. H. Yang, R. Trouillon, G. Huszka, M.A.M. Gijs, Super-resolution imaging of a dielectric microsphere is governed by the waist of its photonic nanojet. Nano Lett. **16**, 4862–4870 (2016)
74. A. Darafsheh, C. Guardiola, A. Palovcak, J.C. Finlay, A. Cárabe, Optical super-resolution imaging by high-index microspheres embedded in elastomers. Opt. Lett. **40**, 5 (2015)
75. A. Darafsheh, N.I. Limberopoulos, J.S. Derov, D.E. Walker, V.N. Astratov, Advantages of microsphere-assisted super-resolution imaging technique over solid immersion lens and confocal microscopies. Appl. Phys. Lett. **104**, 61117 (2014)
76. J.N. Monks, B. Yan, N. Hawkins, F. Vollrath, Z. Wang, Spider silk: mother nature's bio-superlens. Nano Lett. **16**, 5842–5845 (2016)
77. G.T. di Francia, Super-gain antennas and optical resolving power. G Suppl. Nuovo Cimento **9**, 426–438 (1952)
78. Y. Aharonov, D. Bohm, Significance of electromagnetic potentials in the quantum theory. Phys. Rev. **115**, 485–491 (1959)
79. X. Luo, D. Tsai, M. Gu, M. Hong, Subwavelength interference of light on structured surfaces. Adv. Opt. Photonics **10**, 757–842 (2018)
80. D. Tang, C. Wang, Z. Zhao, Y. Wang, M. Pu, X. Li, P. Gao, X. Luo, Ultrabroadband superoscillatory lens composed by plasmonic metasurfaces for subdiffraction light focusing. Laser Photonics Rev. **9**, 713–719 (2015)

81. Z. Li, T. Zhang, Y. Wang, W. Kong, J. Zhang, Y. Huang, C. Wang, X. Li, M. Pu, X. Luo, Achromatic broadband super-resolution imaging by super-oscillatory metasurface. Laser Photonics Rev. **12**, 1800064 (2018)
82. E.T.F. Rogers, J. Lindberg, T. Roy, S. Savo, J.E. Chad, M.R. Dennis, N.I. Zheludev, A super-oscillatory lens optical microscope for subwavelength imaging. Nat. Mater. **11**, 432–435 (2012)
83. F. Qin, H. Kun, J. Wu, J. Teng, C. Qiu, M. Hong, A supercritical lens optical label-free microscopy: sub-diffraction resolution and ultra-long working distance. Adv. Mater. **29**, 1602721 (2017)
84. G. Cao, X. Gan, H. Lin, B. Jia, An accurate design of graphene oxide ultrathin flat lens based on Rayleigh-Sommerfeld theory. Opto-Electron. Adv. **1**, 180012 (2018)
85. S. Wang, X. Ouyang, Z. Feng, Y. Cao, M. Gu, X. Li, Diffractive photonic applications mediated by laser reduced graphene oxides. Opto-Electron. Adv. **1**, 170002 (2018)
86. C. Snoeyink, Imaging performance of Bessel beam microscopy. Opt. Lett. **38**, 2550–2553 (2013)
87. S.W. Hell, Far-field optical nanoscopy. Science **316**, 1153 (2007)
88. H. Gao, M. Pu, X. Li, X. Ma, Z. Zhao, Y. Guo, X. Luo, Super-resolution imaging with a Bessel lens realized by a geometric metasurface. Opt. Express **25**, 13933–13943 (2017)

Chapter 7
Sub-Diffraction-Limited Nanolithography

Abstract Sub-diffraction-limited nanolithography is one of the main applications in EO 2.0. In this chapter, we first give a brief introduction about the diffraction-limited lithography and the significance of breaking diffraction limit. Then we would like to summarize the research achievements of plasmonic lithography in the manners of interference, imaging, and direct writing. Some representative techniques are described in detail. The key aspects in evaluating the performance of plasmonic lithography are also discussed, such as resolution, fidelity, and the aspect ratio of nanopatterns. Some new physics and materials accompanying plasmonic devices design as well as lithography are presented. Subsequently, we discuss the engineering aspects of plasmonic lithography, like depth amplification and pattern transfer, resolution enhancement, and precision systems. In addition, practical applications of plasmonic lithography are introduced. The remaining problems and outlooks of plasmonic lithography are given in the end.

Keywords Plasmonic lithography · Diffraction limit · Evanescent waves

7.1 Introduction

Photolithography is a vital stage in the manufacture of micro-/nanoelectronics. In 1965, Gordon Moore predicted that the number of transistors per integrated circuit (IC) chip would continue to double in every 18–24 months. So the progress in semiconductor manufacturing is all about shrinkage in features, which allows faster and more advanced ICs that consume lesser power and can be produced at lower cost. Photolithography uses light to transfer a pattern of features from a mask to a light-sensitive chemical photoresist on a semiconductor wafer.

Suffering from the so-called diffraction limit, the minimum resolution of conventional photolithography is limited. In 1873, Ernst Abbe discovered a fundamental "diffraction limit" in optics: Whenever an object is imaged by an optical system, the features smaller than half the wavelength of the light are permanently lost in the image. The loss of information arises because high spatial frequency components that carry object's fine features (e.g., evanescent waves) exponentially decay, resulting

X. Luo, *Engineering Optics 2.0*, https://doi.org/10.1007/978-981-13-5755-8_7

in an imperfect image. Therefore, making a perfect optical lens that produces flawless images has been a fundamental question for centuries. In 1879, Lord Rayleigh defined the separation between two airy patterns in order to distinguish them as separate entities. In photolithography, the attainable structure size is a straightforward consequence of the diffraction-limited resolution of the projection optics. The lateral optical resolution CD is given by the quotient of the illumination wavelength, λ, and the numerical aperture, NA, of the projection optics according to the famous Rayleigh formula [1]:

$$\mathrm{CD} = k_1 \frac{\lambda}{\mathrm{NA}} \tag{7.1.1}$$

where k_1 is the process factor determined by the exact details of the optical system. To improve the resolution, and thus enable Moore's Law, all three parameters in Eq. (7.1.1) have been improved over time. For process factor k_1, the main developments are using image enhancement, feedback loops, and multiple exposure or spacer technologies. Progresses in parameters of λ and NA are summarized in Table 7.1. One can observe that over the years λ has decreased from UV (436 nm, g-line of mercury) to DUV (193 nm, ArF laser). Also the NA was increased from 0.28 to 1.35 (using a water-based immersion lens).

In this table, the pixel count, N, is defined as the number of pixels of size (CD × CD) that fit within the step/scan field, S_{SF}, or:

$$N \approx \frac{S_{\mathrm{SF}}}{\mathrm{CD}^2} \tag{7.1.2}$$

Note the dramatic increase in N over time, step/scan area, and CD; as the need for higher resolution and image area grew, dramatic increases in lens size and complex-

Table 7.1 Evolution of wavelength, NA, and k_1 over the years (ASML)

	David Mann (GCA) 4800	ASML/40	ASML/300	ASML/1100	ASML/19x0i
Year of prototype	1975	1987	1995	2000	2007
Wavelength (nm)	436	365	248	193	193
NA	0.28	0.4	0.57	0.75	1.35
k_1	0.90	0.77	0.57	0.39	0.27
CD (nm)	1400	700	250	100	38
Step/scan field	10 × 10	14 × 14	22 × 27	26 × 33	26 × 33
No. of pixels/field	50×10^6	400×10^6	10×10^9	86×10^9	600×10^9
Weight (kg)	2	20	250	400	1080

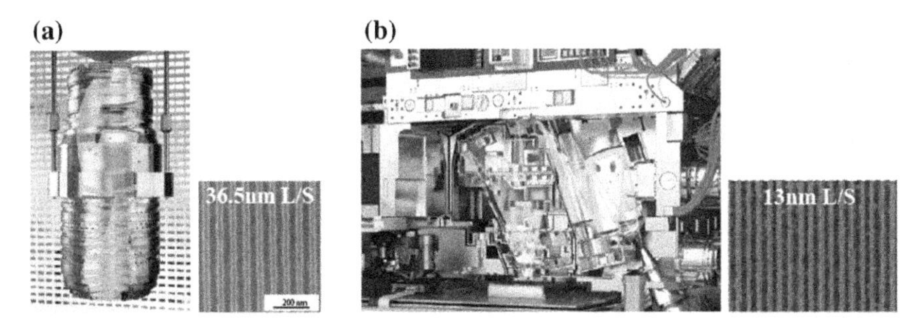

Fig. 7.1 **a** Lens of Starlith 1900 from Carl Zeiss for DUVL. The height of the lens is more than 1 m. The optical design and ray path are schematics and given only as an illustration. The right depicts a resist structure of 36.5-nm half-pitch, obtained with the lens. **b** Lens of NXE: 3300B EUVL with NA = 0.33. The right depicts a resist structure of 13-nm half-pitch, obtained with the lens (ASML). **a** Reprinted from [1] with permission. Copyright 2007, Nature Publishing Group

ity were required. The latest 193-nm immersion lithography with single exposure has approached the resolution limit. Figure 7.1a shows the lens DUV lithography (DUVL) with the NA = 1.35, comprising both the optical design and ray path. The inset shows that the limit resolution has reached 36.5 nm with polarized illumination [1].

Although EUV lithography is a promising next-generation lithography (NGL) with resolution beyond the current DUVL, it is a significant departure from previous lithography technologies. First, all the optical elements responsible for the imaging capabilities of the scanner must use reflective lenses (mirrors) rather than refractive lenses. Second, owing to matter's propensity to absorb EUV light, the entire optical path from the light source to the wafer must be in vacuum [2]. EUV lithography development has progressed over 30 years and has been applied in industrialization for full-scale high-volume manufacturing (HVM) recently. Owing to the limitation of NA, the resolution of EUVL with single exposure is 13 nm (near the illumination wavelength), as indicated in Fig. 7.1b.

Lithography patterning methods have critical dimension "cliffs", where a given patterning method can no longer be used to produce features with a half-pitch below that value. As shown in the international technology roadmap for semiconductors (ITRS) 2.0, roughly 40 nm is the cliff for ArF immersion single exposure line and space patterning, so 20 nm is the cliff for pattern doubling and 10 nm is the cliff for pattern quadrupling using ArF immersion patterning. The cliffs also exist in EUV lithography.

Therefore, it is significant to find an effective way to break the diffraction limit under the conventional optical lithography architecture. In recent years, many valuable experiments were carried out to beat the diffraction limit of optical lithography. These approaches include two-photon photopolymerization [3], sub-diffraction-limited lithography based on stimulated emission depletion (STED) [4] and absorbance modulation. Unfortunately, these methods do not break the limit at the optical level. Some scientists also proposed to improve the resolution by the use of two or more quantum entangled photons, but the implementation is too complex and the efficiency is rather low.

Recently, a couple of extraordinary optical phenomena associated with surface plasmon polaritons (SPPs) have been found, including extraordinary optical transmission (EOT) [5], beaming effects and extraordinary Young's interferences [6–7]. These counter-intuitive phenomena reveal the intriguing features of SPPs, including short wavelength property and local field enhancement property, which inspire researchers to devote to the revolutionary nanofabrication technology, i.e., plasmonic lithography [8]. As its name implies, plasmonic lithography utilizes various SPPs, including surface plasmon waves, bulk plasmon polaritons (BPPs), and localized surface plasmons (LSP), as avenues to surpass the traditional lithography resolution limited by diffraction.

Subwavelength plasmonic lithography experiments were initially carried out in 2003 and about 50-nm line width patterns were firstly obtained with a g-line mercury lamp at 436 nm via extraordinary Young's interference experiment [9]. The interference pattern was imaged from the top of the mask to the image plane, demonstrating that the metallic mask can be simultaneously used to generate interference pattern and form super-resolution images. It was predicted that sub-25 nm feature size can be allowed by SPPs. Furthermore, symmetric mode of thin metallic film that possesses shorter wavelength than the antisymmetric one was utilized to obtain higher lithography resolution. To decrease the stringent requirement of sub-diffraction-limited plasmon masks, hyperbolic metamaterials that can squeeze the bulk SPPs were introduced to plasmonic interference lithography. Moreover, enhanced nonlinear effects by the local field of SPPs were taken as an effective method to improve the lithography resolution.

Inspired by the concept of perfect lens and its sub-diffraction-limited imaging ability, various plasmonic lenses encompassing superlenses, plasmonic reflective lenses, and plasmonic cavity lenses were designed and employed to push the resolution of imaging lithography down to 22 nm at a wavelength of 365 nm, about 1/17 light wavelength [10]. Some functional devices like metalenses and nanopolarizers have been fabricated by plasmonic imaging lens lithography [11]. Furthermore, lithography quality involves exposure depth, contrast ratio, uniformity, and aspect ratio that have been greatly improved with the help of evanescent waves amplification and resolution-enhanced technologies by engineering the catenary optical fields in the imaging process [8, 12].

Finally, some plasmonic nanofocusing structures to localize SPPs energy to a subwavelength point were proposed and successfully implemented for point-to-point scanning nanolithography [13]. Regardless of plasmonic lithography operating in interference, imaging or direct writing manner, only low-cost, long-wavelength light sources, fully commercialized and developed materials of resists are employed. Therefore, it provides a potentially promising nanofabrication lithography method characterized by low-cost, large-area advantages. Moreover, SPP imaging and interference lithography are believed to deliver much higher throughput than some conventional nanofabrication tools like electron beam lithography (EBL) and focused ion beam milling (FIB).

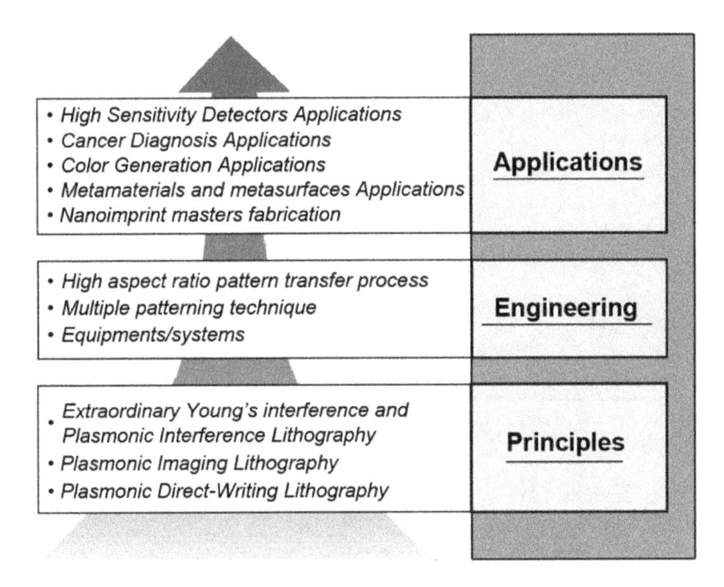

Fig. 7.2 System-level focus on plasmonic sub-diffraction-limited lithography

With the development of plasmonic sub-diffraction-limited lithography in recent years, it gradually formed an independent discipline including the principle, engineering, and applications, as shown in Fig. 7.2. Next, we will describe these three aspects in detail.

7.2 Plasmonic Interference Lithography

7.2.1 Extraordinary Young's Interference and Catenary Optical Fields

Originally observed in 1801 by Thomas Young, the double slits interference experiment was thought as one of the most beautiful physical experiments in history. It has been used to validate the wave physics of both photons and electrons. Even nowadays, there are still enormous researches on the quantum physics of multiple photon interference experiments with such a simple configuration. In this section, we show that many new physics would take place when the scale of light–matter interaction is scaled down to smaller than the wavelength [7, 8]. Particularly, this has been proposed to realize nanolithography as early as in 2003 [14, 15].

Figure 7.3 shows a schematic of the development of plasmonic lithography based on extraordinary Young's interference. In 2004, the patent [14] applied by Chen et al., was experimentally validated [9]. Subsequently, the lithographical structures were constantly optimized to realize much smaller critical dimensions and more complex

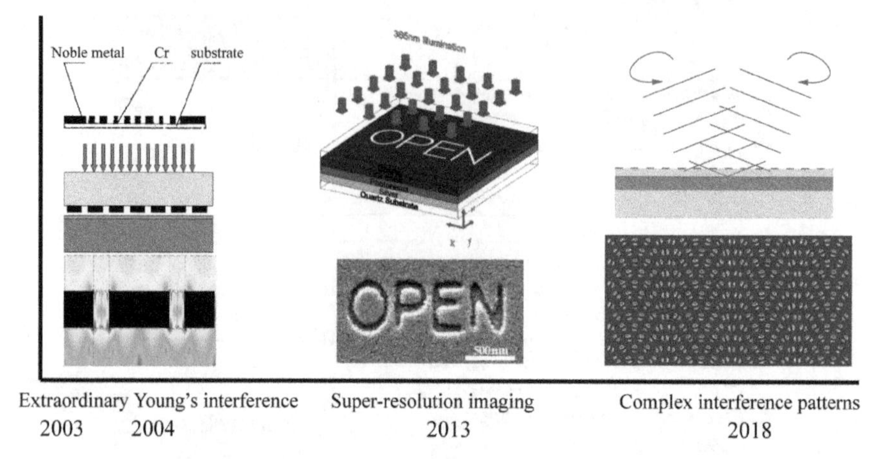

Fig. 7.3 From extraordinary Young's interference to nanolithography based on catenary plasmons [9, 7, 14, 16]

interference patterns (According to Abbe's imaging theory, the imaging process consisted of many interference effects). For example, as experimentally demonstrated in 2013, by using a reflective slab, the catenary plasmons could be constructed to realize a better image [16]. More recently, by employing the catenary optical fields, we demonstrate that spatially variant patterns could be defined by combining the extraordinary Young's interference and photonic spin-orbit interaction [7].

We note that classical Young's double slits experiment has made some assumptions. First of all, the material of screen is often assumed to be a perfect absorber or perfect electric conductor (PEC); second, the mutual coupling of the slits is ignored; third, the influence of slit width and polarization of incident light on the interference fringes is often neglected.

For a typical standing wave created by two counter-propagating waves, the period of the intensity peaks is generally known to be half of the wavelength. As a consequence, there will be no observable interference patterns when two slits are placed with a distance smaller than half a wavelength, and this condition is also required for effective medium theory to be applicable. To investigate Young's interference on deep-subwavelength scale, i.e., the geometric parameters are much smaller than the wavelength, the light field distribution for a pair of double slits separated by 100 nm and perforated in a 20-nm-thick silver film was investigated [7]. A very interesting phenomenon can be seen: At a wavelength of 365 nm, the peak-to-peak distance is close to 100 nm, and no interference pattern is observed; while at a longer wavelength of 385 nm, the peak-to-peak distance is decreased to about 25 nm, which is almost $\lambda/15$. The electric field distribution implies that surface plasmon resonance has been excited. If the localized fields are taken as an interference pattern, a more counter-intuitive conclusion may be drawn: The increase in wavelength has led to a much smaller interference pattern. This extraordinary Young's interference effect means that one need not reduce the wavelength to achieve higher resolution; thus

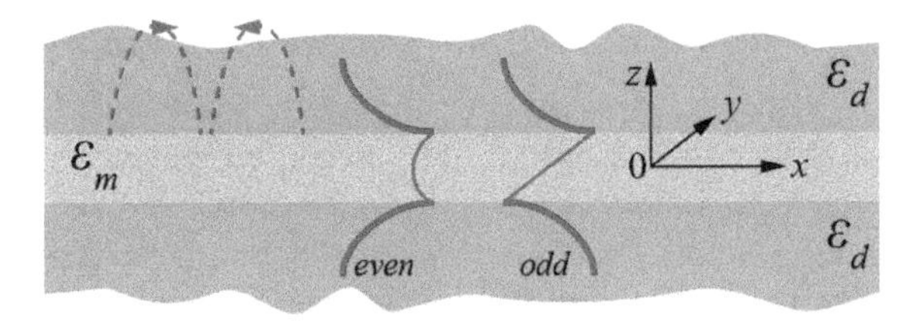

Fig. 7.4 Coupled plasmonic modes in metal–dielectric stacks. The dashed lines show the electric force lines with a shape of "catenary of equal phase gradient." The solid curves show the amplitude of E_x for even and odd modes. For even mode, E_x follows a catenary function. For odd mode, H_y and E_z have a catenary shape. Reprinted from [7] with permission. Copyright 2018, American Chemical Society

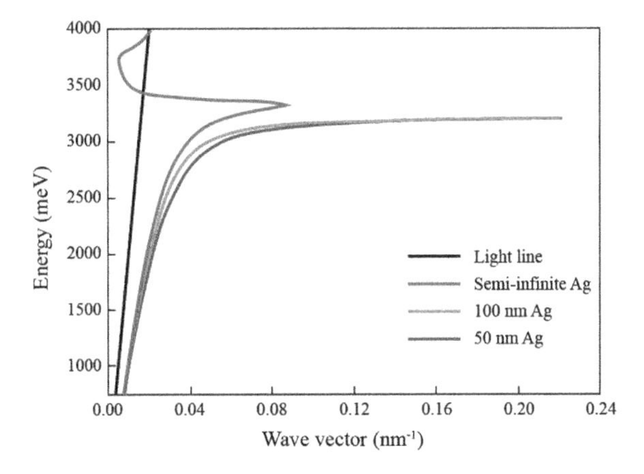

Fig. 7.5 Dispersion of the coupled SPPs (even mode) for different thickness silver film. Reprinted from [17] with permission. Copyright 2004, Optical Society of America

the classical diffraction limit could be overcome. Note that the above phenomena can only be observed for transverse magnetic (TM) but not transverse electric (TE) polarization, which is also distinct from the macroscopic experiment.

The extraordinary Young's interference effect can be well explained using the catenary optical fields, i.e., the coupled plasmons in thin metallic films. As shown in Fig. 7.4 [7], when the core metallic film is thin enough, the SPPs at the two metal–dielectric interfaces couple with each other. Using the dispersion relations of even modes derived from boundary conditions, it is straightforward to show that the effective wavelength of SPP, defined by the vacuum wavelength divided by the mode index, is dependent on the film thickness (Fig. 7.5) [17]. Consequently, one can increase the resolution of nanolithography by reducing the thickness and increasing the coupling strength of the catenary plasmons.

7.2.2 Principle of Plasmonic Interference Lithography

Similar to laser interference lithography, plasmonic interference lithography is a promising nanofabrication method due to its low cost, parallel and large-area features. By designing suitable structures, two or more surface plasmon waves are excited to interfere with each other and produce periodic nanopatterns. More generally, according to the Abbe's imaging theory, the interference of multiple spatial components would restore for images of the original complex objects.

What distinguishes SPPs from "regular" photons is that they have a much smaller wavelength than that in free space at the same frequency, owing to its dispersion curve lying below the light line. Theoretical investigation indicates that by decreasing the metallic film thickness, the effective wavelength of SPPs can be further shrunk to below one-tenth of the incidence wavelength. This unique feature could be readily explored for subwavelength interference lithography (Fig. 7.6).

For grating coupler, when light ($k = \omega/c$) hits a thick grating with periodicity Λ, at an angle θ_0, its component in the surface would have wave vectors $(\omega/c)(\varepsilon_d)^{1/2}\sin\theta_0 \pm nk_\Lambda$ where n is an integer and $k_\Lambda = 2\pi/\Lambda$. The dispersion relation satisfies:

$$\frac{\omega}{c}\sqrt{\varepsilon_\mathrm{d}}\sin\theta_0 \pm nk_\Lambda = \frac{\omega}{c}\sqrt{\frac{\varepsilon_\mathrm{m}\varepsilon_\mathrm{d}}{\varepsilon_\mathrm{m}+\varepsilon_\mathrm{d}}} \equiv k_\mathrm{SP},\ n = 1, 2, 3, \ldots \tag{7.2.1}$$

where ω is the angular frequency of light, c is the light velocity in vacuum, ε_m and ε_d are the permittivity of metal and dielectric, respectively. The propagation length along the surface can be obtained by theoretical prediction $1/\mathrm{Im}(k_\mathrm{SP})$. The decay length of SP waves along the z-direction (determines the exposure depth in the photoresist), can be theoretically estimated by [18]

$$\frac{1}{|k_z|} = \frac{\lambda}{2\pi}\left|\sqrt{\frac{\varepsilon'_\mathrm{m}+\varepsilon_\mathrm{d}}{\varepsilon_\mathrm{d}^2}}\right| \tag{7.2.2}$$

The ratio E_z/E_x affects the interference contrasts as discussed later and is determined by the permittivity of metal and dielectric:

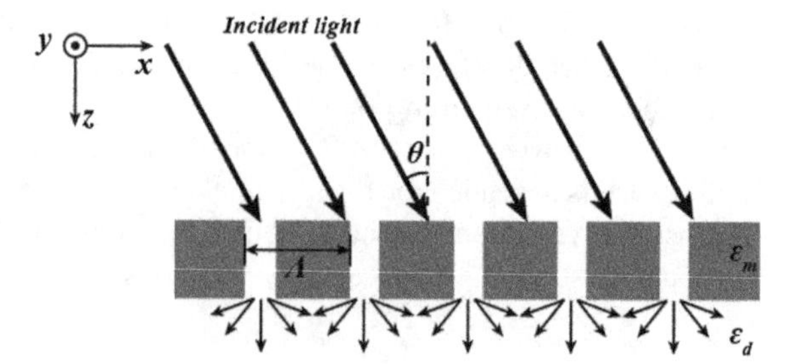

Fig. 7.6 Schematic of SPPs interference lithography with grating coupler

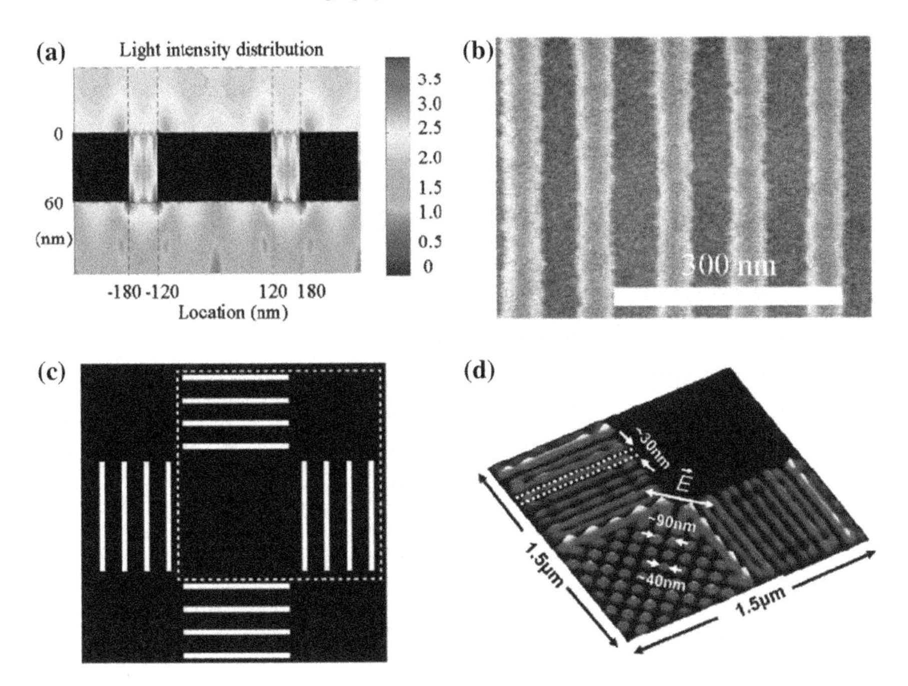

Fig. 7.7 **a** Simulated intensity distribution in the near-field of surface plasmon lithography. **b** SEM of resist recording results. **c** Schematic of 2D plasmon interference lithography, and **d** corresponding electric intensity distribution. **a** and **b** Reprinted from [9] with permission. Copyright 2004, American Institute of Physics. **d** Reprinted from [18] with permission. Copyright 2005, American Chemical Society

$$\frac{E_z}{E_x} = -\frac{k_x}{k_z} = -i\sqrt{\frac{|\varepsilon_{\mathrm{m}}'|}{\varepsilon_{\mathrm{d}}}} \tag{7.2.3}$$

The periodicity d of the interference fringes is half of the surface plasmon wavelength

$$d = \lambda_{\mathrm{SP}}/2 = \lambda_0 k_0/2k_{\mathrm{SP}} \tag{7.2.4}$$

Surface plasmon resonant interference nanolithography technique was firstly reported in 2004. Experimental and numerical results showed that high-resolution features of 50 nm could be readily created by mercury g-line (436 nm), where the metallic grating simultaneously performs as mask and superlens [9] (Fig. 7.7a and b). To get SPP interference patterns beneath the grating structures with acceptable uniformity, the grating structure for SPPs' excitation should be carefully designed to be compatible with the SPP mode (Fig. 7.7c and d).

It should be noted that, due to the spatial symmetry of the excitation grating, normal incident light generates SPP fields along two opposite directions on the metal,

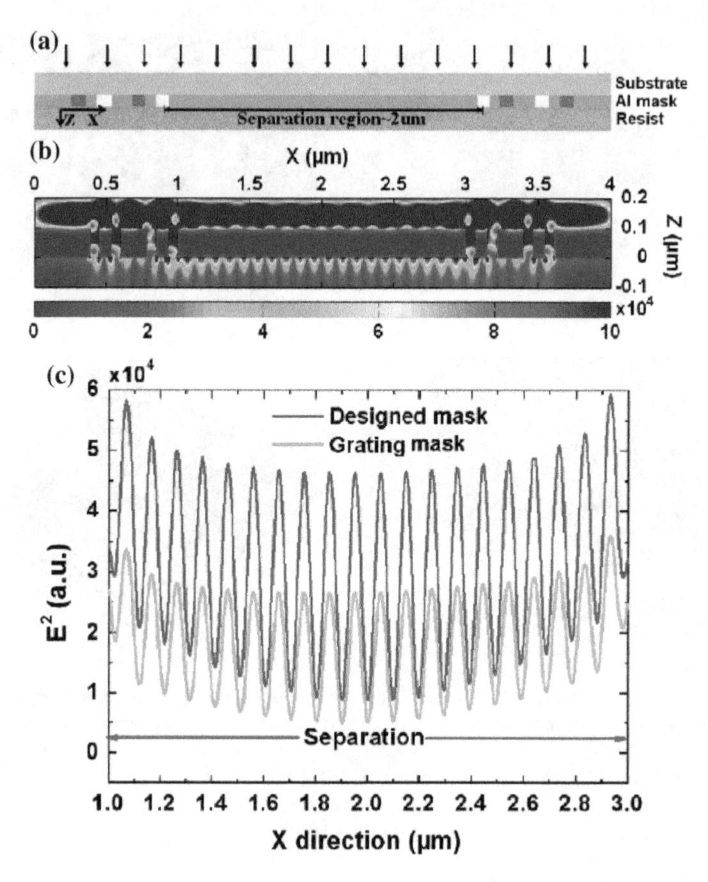

Fig. 7.8 **a** Schematic configuration of interference lithography based on unidirectional excitation of surface plasmons. **b** Simulated electric intensity distribution. **c** Cross section of the total electrical field along the x-direction, 30 nm away from the bottom of the aluminum mask. Reprinted from [20] with permission. Copyright 2009, IOP Publishing Ltd.

which means that half of the energy are wasted in the previous schema. In order to overcome this problem, unidirectional excitation of surface plasmons was utilized in SPP interference lithography (Fig. 7.8a and b), which is based on the extraordinary Young's interference observed in double metallic slits [19]. The directional excitation effect on the unilluminated surface requires that SPs interfere constructively along one direction while destructively along the opposite direction, i.e., the relative phases of SPPs at two exit apertures take the forms [20]:

$$\phi_1 + d\frac{2\pi}{\lambda_{SP}} = \phi_2 + 2N\pi \tag{7.2.5}$$

$$\phi_2 + d\frac{2\pi}{\lambda_{SP}} = \phi_1 + (2N+1)\pi \tag{7.2.6}$$

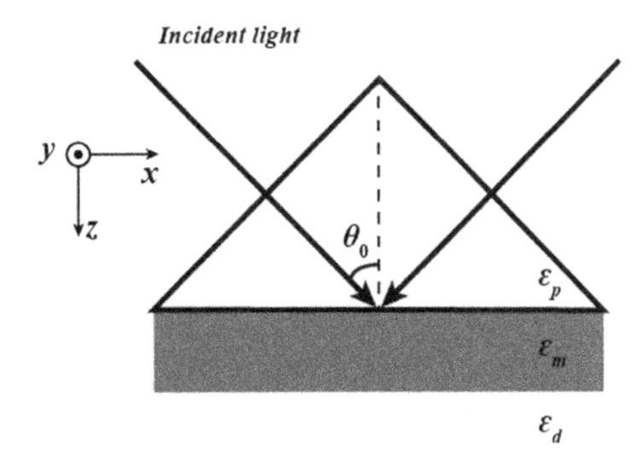

Fig. 7.9 Schematic of SPPs interference lithography with ATR coupler

where ϕ_1 and ϕ_2 are the relative phases of the generated SPPs at the exit apertures, d is the separation distance between them, and N is an arbitrary integer. According to the equations above, it can be deduced that

$$d = (4N + 1)\lambda_{\text{SP}}/4 \tag{7.2.7}$$

$$\phi_1 - \phi_2 = \frac{\pi}{2} \tag{7.2.8}$$

Based on the above principle, modified mask composed of nanoslits filled with different dielectric was proposed for interference lithography. The numerical results shown in Fig. 7.8c demonstrated that the proposed scheme could generate interference pattern with higher intensity and contrast ratio.

Subsequently, several modified configurations were proposed including multiple grating interferences to generate more complex 2D patterns, utilizing a sharp edge coupling mechanism [21], and adding reflective cladding metal slab below the resist layer to generate high contrast interference fringes [22].

When SPPs are excited by a grating or a subwavelength structure, the SPPs interference pattern area is greatly restricted or nonuniform due to the inevitable absorption and scattering of SPPs when they propagate along rough metal film. In contrast, large-area, uniform SPPs interference field is easily achieved with high-refractive-index prisms (Fig. 7.9).

For attenuated total reflection (ATR) coupler, if light is reflected at a metal surface covered with a dielectric medium ($\varepsilon_{\text{p}} > 1$), e.g., with a prism, its momentum becomes $(\hbar\omega/c)(\varepsilon_{\text{p}})^{1/2}$ instead of $\hbar\omega/c$ and its projection on the surface is

$$k_x = \sqrt{\varepsilon_{\text{p}}}\frac{\omega}{c}\sin\theta_0 \tag{7.2.9}$$

SPPs' prism excitation in the Kretschmann and Otto schemes could be introduced for large-area SPP interference lithography. Simulation result indicated that a high contrast large-area SPP interference features as small as 60 nm could be produced with the illumination wavelength of 441 nm by using an attenuated total reflection-coupling mode. Then, large-area patterns of grating lines and pillars with feature size ~90 nm were realized experimentally by employing a custom-made prism layer configuration at 364-nm illumination (Fig. 7.10a and b) [23]. Subsequently, high-density sub-50 nm periodic structures were realized by using Al film. It is found that the obtained periodic feature shows good exposure depth and high contrast (Fig. 7.10c and d) [24], but the resolution of interference pattern is limited due to the finite refractive index of the prism material.

7.2.3 Odd SPPs Mode Interference Lithography

In practical applications, two types of modes in Fig. 7.11 are commonly encountered with a metal (dielectric) film sandwiched between two semi-infinite dielectric (metallic) mediums. These waveguide and cavity-like structure are usually used to tailor SPPs modes for modifying both transverse wave vector and electromagnetic

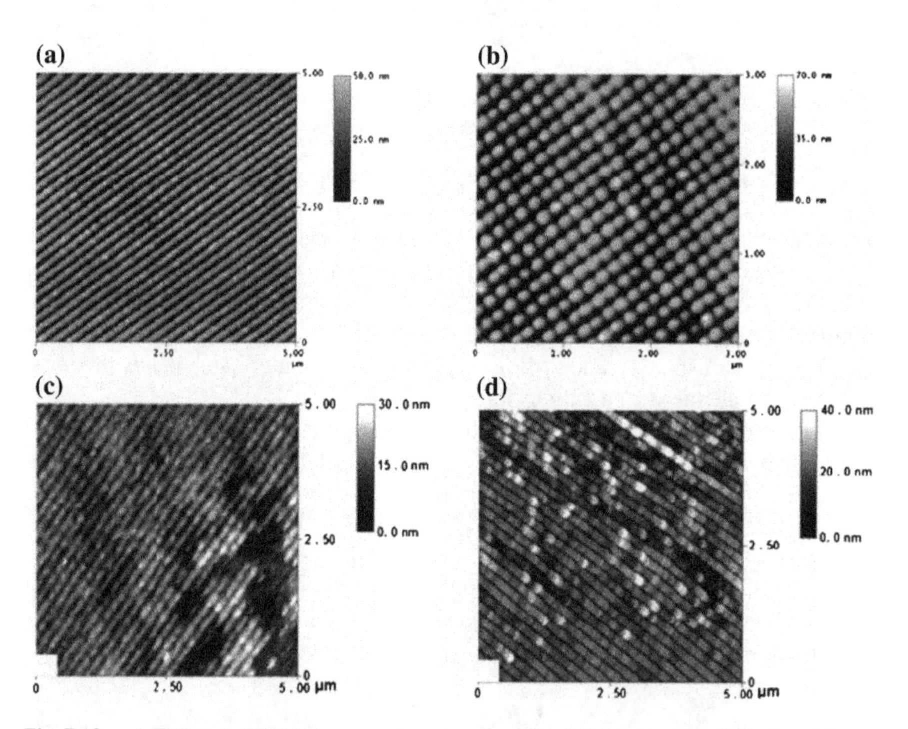

Fig. 7.10 **a** AFM image of the exposure pattern transferred onto the photoresist film with minimum line width of 89 nm and a 172-nm period. **b** 2D AFM images of the exposure pattern transferred onto the photoresist film with minimum spot size of 93 nm and 173 nm period. AFM images of exposure pattern transferred on the photoresist film using **c** Ag film and **d** Al film with 100-nm periodicity and 50-nm line width. **a** and **b** Reprinted from [23] with permission. Copyright 2010, Optical Society of America. **c** and **d** Reprinted from [24] with permission. Copyright 2010, Springer-Verlag

field distribution. In these two modes, the SPPs dispersion relation equation is written in a similar form as [25]:

$$\tanh k_1 \frac{d}{2} = -\frac{k_2 \varepsilon_1}{k_1 \varepsilon_2}, \tag{7.2.10}$$

$$\tanh k_1 \frac{d}{2} = -\frac{k_1 \varepsilon_2}{k_2 \varepsilon_1}, \tag{7.2.11}$$

where ε_1 and ε_2 are permittivities of media I and II, respectively; $k_1^2 = k_x^2 - \varepsilon_1 k_0^2$; $k_2^2 = k_x^2 - \varepsilon_2 k_0^2$ are the transverse wave vectors of SPPs wave.

SPPs in metal-insulator-metal (MIM) or insulator-metal-insulator (IMI) structure are split into two modes. This phenomenon could be explained by the coupling of two SPP modes at the two interfaces of MIM or IMI structures. Usually, they are termed as odd and even modes according to the electric field or magnetic field distribution along the direction normal to the film. In some investigations, they are labeled with long-range and short-range SPPs to characterize the traveling length feature. The difference mainly lies in the light confinement ratio inside the metal medium with variant geometrical parameters. It is worth noting that the magnitudes of transverse and normal electric field components E_x and E_z are different, especially for the two SPPs modes inside the MIM structure. In the case of odd electric field distribution, the E_z component dominates the SPPs field inside insulator region. For the even mode, however, the E_x component magnitude is much larger than that of E_z. This feature helps to get sharper imaging and interference fringes, compared with that of simple SPP modes.

Since the phase difference of $\pi/2$ between electric components E_x and E_z of TM polarized SPPs would deteriorate the interference fringe contrast, metal–photoresist–metal structure was utilized to tailor the ratio between E_x and E_z components. In principle, the up and below metal claddings form a MIM waveguide structure. Therefore, by using proper mask to excite diffraction that matches the antisymmetric mode of MIM waveguide, one can obtain interference patterning with higher spatial resolution. Besides, exposure depth can be greatly improved, which can be

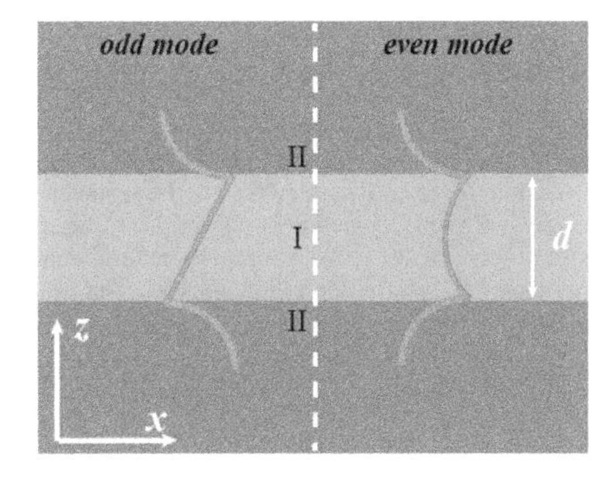

Fig. 7.11 Schematic structure of MIM or IMI, with d as the thickness of core layer. The core layer and claddings are represented by Roman numbers "I" and "II," respectively. The odd and even modes are represented by the orange-colored curve

attributed to the following reasons. On the one hand, the ratio r of E_z/E_x, calculated by $r = -k_x/k_z = -i(|\varepsilon_m|/\varepsilon_{PR})^{1/2}$, is increased by mismatched permittivities between the metal and photoresist [26]. On the other hand, the multi-reflection within the MIM cavity can further improve the contrast ratio.

MIM structure helps to realize modification of SPP mode with much larger transverse wave vector k_x and tailor the ratio between E_x and E_z components. This feature could be utilized for much smaller SPPs' interference period by launching the odd SPP mode and improve pattern contrast by inhibiting the transverse electric field E_x component (Fig. 7.12a and d) [26]. The SPPs' interference lithography using the odd mode in MIM structures is even extended to 193-nm DUV light and demonstrated numerically to push sub-diffraction-limited resolution beyond 15 nm [27]. And then it has been experimentally demonstrated with SiO_2-Al-photoresist-Al structure and 363.8-nm wavelength, yielding 45-nm half-pitch, 50 nm depth and area size up to 20 mm × 20 mm (Fig. 7.12b and e) [28]. Further experiments realized 61.25-nm half-pitch lines with large-area pattern uniformity and depth exceeding 100 nm, demonstrating that this approach can overcome two major drawbacks (poor uniformity and limited depth) in previous plasmonic lithography (Fig. 7.12c and f) [29]. Due to the strong resonance in the waveguide structure, the interference patterns exhibit large area, high aspect, and high resolution advantages.

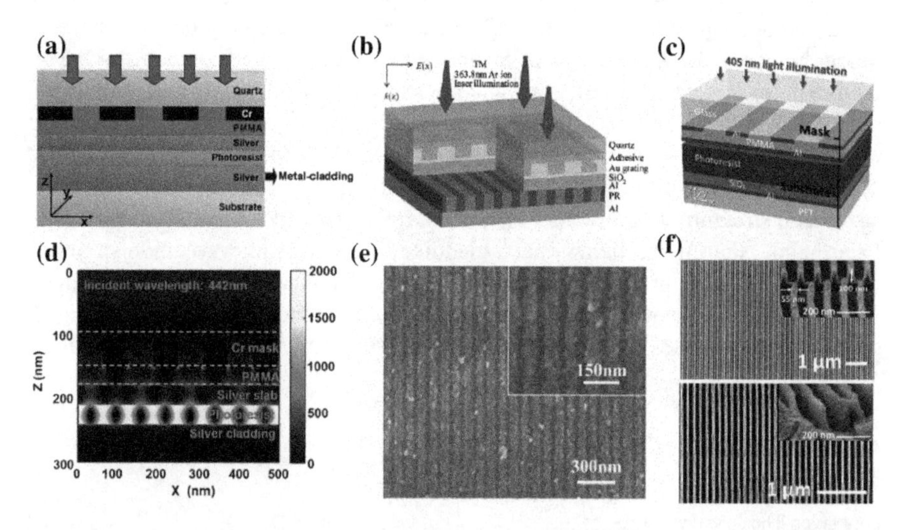

Fig. 7.12 Schematic structures of the odd SPP mode interference lithography at the wavelength of **a** 442 nm; **b** 363.8 nm; and **c** 405 nm. **d** Calculated total electric field distribution of **a**, the grating is made by Cr with 50-nm thickness and 124-nm period. The thickness of silver slab and photoresist are both equal to 30 nm. **e** Lithography patterns with 45-nm half-pitch of **b**. **f** SEM images of 100-nm (top) and 200-nm (bottom) thick PR produced by **c**. **a** and **d** Reprinted from [26] with permission. Copyright 2009, Springer-Verlag. **b** and **e** Reprinted from [28] with permission. Copyright 2016, The Authors. **c** and **f** Reprinted from [29] with permission. Copyright 2016, American Chemical Society

7.2.4 Spatial Frequency Filtering and BPPs Interference Lithography

Although the lithography resolution and quality can be greatly improved based on the antisymmetric mode of MIM waveguide structure, one critical issue is still not be resolved, i.e., sub-diffraction-limited plasmon masks are always required to form sub-diffraction-limited patterns to realize high excitation efficiency and large-area pattern uniformity. As a consequence, the assumed low cost advantages of SPPs interference lithography are diminished to some extent due to the expensive mask fabrication process. In order to solve this problem, it is needed to filter the high-order diffraction from BPPs, where the SPPs modes propagate across space instead of being confined to and propagating along the metal–dielectric interface. In order to excite BPPs, alternatively stacked subwavelength metal and dielectric films with a hyperbolic dispersion relation are utilized.

The BPPs modes are formed by the mutual coupling of plasmonic polaritons fields between adjacent metal–dielectric film interfaces, resulting in super plasmonic modes which show "propagation" behavior in the bulk space of multiple films system and exponentially decay outside. To address this point, the effective medium theory (EMT) could give good guidance by assuming that the effective permittivities of multiple metal and dielectric films could be approximated with anisotropic tensor elements like [30]

$$\varepsilon_x = \varepsilon_y = f \cdot \varepsilon_{\mathrm{m}} + (1 - f) \cdot \varepsilon_{\mathrm{d}} \tag{7.2.12}$$

$$\varepsilon_z = \varepsilon_{\mathrm{d}} \cdot \varepsilon_{\mathrm{m}} / [(1 - f) \cdot \varepsilon_{\mathrm{m}} + f \varepsilon_{\mathrm{d}}] \tag{7.2.13}$$

where ε_x, ε_y, and ε_z are the effective dielectric permittivity along the x-, y-, and z-directions, respectively, and f is the filling factor of metal films. Thus, the dispersion relation of light inside this effective medium could be written as $\left(k_x^2 + k_y^2\right)/\varepsilon_z + k_z^2/\varepsilon_x = k_0^2$ for TM polarization, where k_x, k_y, and k_z means the wave vector components along the x-, y-, and z-directions, and k_0 represents the magnitude of wave vector in vacuum.

Clearly, the dispersion contour of multi-layers is hyperbolic if $\varepsilon_x \varepsilon_y < 0$, which can be classified into type I ($\varepsilon_x = \varepsilon_y > 0$, $\varepsilon_z < 0$) and type II ($\varepsilon_x = \varepsilon_y < 0$, $\varepsilon_z > 0$) hyperbolic metamaterials [31]. The two dispersion types can be switched by tuning the permittivities of metal and dielectric (ε_{m} and ε_{d}), as well as the thicknesses f. Typically, the former is utilized for imaging lithography, which will be discussed in the following section, while the latter is explored in interference lithography due to its spatial filtering ability.

In order to illustrate the spatial filtering ability of type II hyperbolic materials, a metamaterial composed of alternatively stacked 5 pairs of SiO_2/Al films with sub-wavelength thickness 20 nm/10 nm is considered here. The corresponding three dimensional (3D) equi-frequency contour (EFC) calculated from EMT is shown in Fig. 7.13a [32]. As the EFC is open and oriented along the transverse vector k_x and k_y, any spatial frequency component with transversal wavevectors higher than $(\varepsilon_z)^{1/2}k_0$ would go through the metamaterial in principle. However, more strict analysis based on rigorous coupled wave analysis (RCWA) show that the maximum spatial wavevector can be supported is limited by the film thickness of metal–dielectric composite layer, forming a spatial bandpass filter, which is illustrated by the optical transfer function (OTF) in Fig. 7.13b. The bandwidth and center frequency of the bandpass filter can be adjusted by changing the dielectric constant and thicknesses.

In order to experimentally characterize the spatial filtering property of alternatively stacked metal–dielectric layers (ASMDLs), two gratings with unequal periods are introduced at the two sides of films structure to excite SPPs and generate moiré fringes observable in the far-field, as shown in Fig. 7.14 [33]. With the incidence of TM polarized light, different orders of diffracted evanescent waves are generated through the gratings in the laser illumination side by the equation

$$k_x = k_0 \sin\theta + \frac{2\pi}{P_{\text{in}}} n \; (n = 0, \pm 1, \pm 2, \ldots) \tag{7.2.14}$$

where θ is the incidence angle, n is the diffraction order, and P_{in} is the period of the gratings on the illumination side. These evanescent wave orders could be coupled to the SPPs modes in ASMDLs, which only allow the transmission of excited SPPs with specific transverse wave vectors and screen all the others. Then, the transmitted

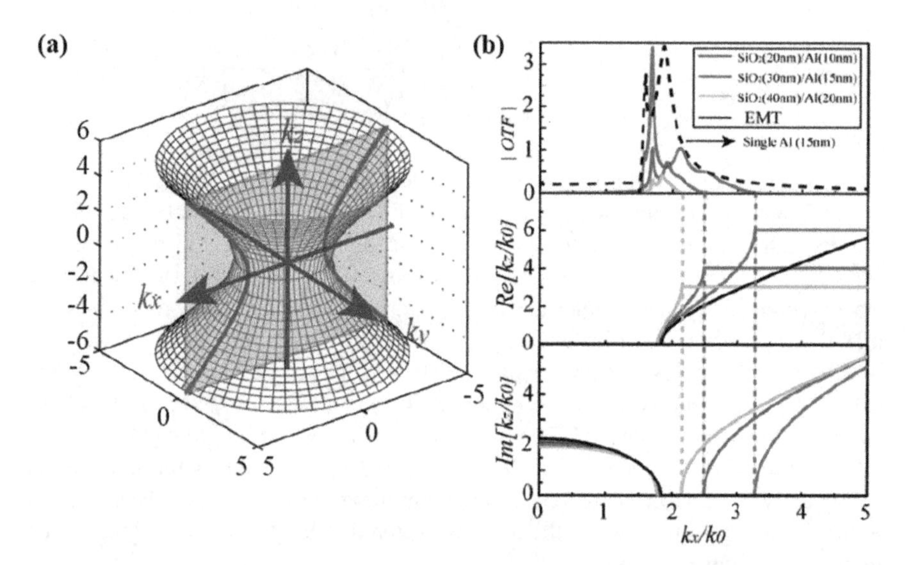

Fig. 7.13 **a** 3D plots of EFC surface in EMT approximation for metamaterial structure. **b** In top panel is OTF for SiO_2/Al films with variant thickness, calculated by RCWA and EMT. The middle and bottom panels are the real and imaginary parts of k_z for variant k_x calculated using Bloch theorem. The imaginary part of Al permittivity is ignored here for the visualization EFC curves. Reprinted from [32] with permission. Copyright 2015, WILEY-VCH Verlag GmbH & Co. KGaA, Weinheim

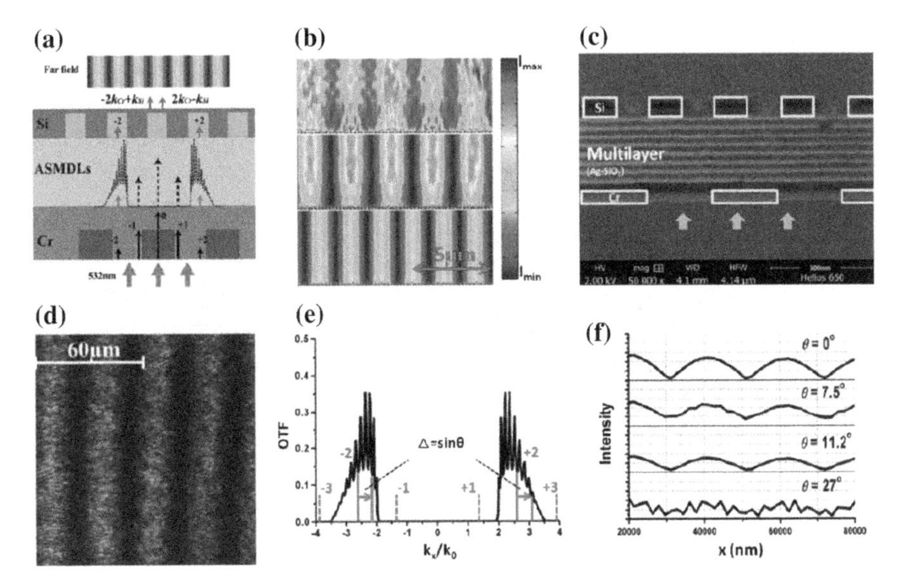

Fig. 7.14 **a** Schematic of characterization of spatial filtering property through far-field moire fringes. **b** Simulated moire fringes with different grating periods. **c** SEM of 8 pairs of Ag–SiO_2 multilayers film. **d** CCD observed image of moire fringes. **e** Transmission of diffraction orders of layered films with various incidence angles. **f** Corresponding moire fringes with different incident angles. Reprinted from [33] with permission. Copyright 2014, American Institute of Physics

SPPs through ASMDLs hit the gratings on the other side and generate light moiré fringes with period defined by

$$P_{\text{moire}} = \frac{P_{\text{in}} \times P_{\text{out}}}{|m \times P_{\text{out}} - n \times P_{\text{in}}|} \tag{7.2.15}$$

where P_{moire} is the period of the interference fringes and P_{out} is the period of gratings on the output side. In order to observe the moiré fringes in the far-field, special attention should be paid in two ways. One is the ASMDLs' geometries should be carefully designed to ensure there are only two diffraction orders of the input grating are supported by the ASMDLs. Otherwise, multiple orders of the grating at the exit side would deliver blurred moiré fringes. The other is the beat frequencies of moiré fingers are supported by the free space. Following the above principle, ASMDLs composed of 8 pairs of 20-nm-Ag/30-nm-SiO_2 films were fabricated with the periods of input grating and output grating, respectively, being 400 nm and 210 nm [33]. Clear fringes are observed by CCD camera with the measured period agreeing well with the theoretical predictions. By changing the incidence angle, the excitation spatial frequency will shift and thus the transmittance of beat frequency will be changed correspondingly. As a consequence, one can retrieve the filtering property of ASMDLs from the fluctuation of the contrast of fingers with the incidence angle.

Based on the unique spatial filtering property of metal–dielectric multilayers, deep-wavelength interference lithography was developed, where metal–dielectric multilayers that can filter the ±3 diffraction orders from the mask were utilized. In contrast to the aforementioned SPPs interference methods, the grating period of the plasmon mask can be several times larger than that of interference patterns, promising that the grating structures could be readily obtained by some simple and low-cost methods like large-area laser interference lithography.

Inspired by these pioneering numerical investigations, the experimental investigation was implemented recently, which is schematically illustrated in Fig. 7.15a [32]. The ±2nd diffraction orders of the mask were squeezed out layer by layer, and the uniformity was improved correspondingly through the hyperbolic metamaterial with spatial filtering property (Fig. 7.15b). By utilizing two and four SPPs interference lithography, 1D and 2D periodic patterns with half-pitch of 45 nm (~$\lambda/8$) were experimentally demonstrated (Fig. 7.15c). Numerical investigation further shows that higher lithography resolution up to 22.5 nm (~$\lambda/16$) is feasible by this method. But due to the great absorption of Au, the coupling between Au grating and HMM is weak. Further improvement of BPPs excitation was realized by TiO_2 grating, which could greatly improve the coupling between the diffraction light and the HMM compared to the Au grating. Consequently, the lithography energy efficiency was increased by more than one order of magnitude (Fig. 7.16a) [34]. This method was applied to achieve 20 mm × 20 mm and uniform periodic subwavelength lines with a half-pitch of 35 nm (~$\lambda/10$) (Fig. 7.16b).

7.2.5 *Two-Surface Plasmon Polaritons Interference Lithography*

So far, the nonlinear processes such as two-photon absorption and two-photon polymerization have been employed for 3D microfabrication and microscopes to obtain higher resolution. Recently, this idea was introduced into two-SPPs' absorption interference (TSPPA) lithography to improve the exposure pattern quality (Fig. 7.17a and b) [35]. Just like two-photon absorption, two-SPPs' quanta are simultaneously absorbed for the exposure of resist. Thus, the electric field intensity $|E|^2$ response is replaced by $|E|^4$ and a higher contrast or narrower SPPs' hot spot response would be expected. At the same time, the SPPs mode helps to confine light within a small volume and increase the absorption of two SPPs. Experimentally, TSPPA lithography uses two sets of side grating with 480 nm period and 800 nm femtosecond laser to achieve 120-nm line width interference fringes (Fig. 7.17c and d). Subsequently, resist patterns with the period of ~138 nm and line width of 70 nm have been obtained by using Al instead of Au to excite the SPPs, which explores the ability of the TSPPA-based lithography at the short wavelength (Fig. 7.17e and f) [36].

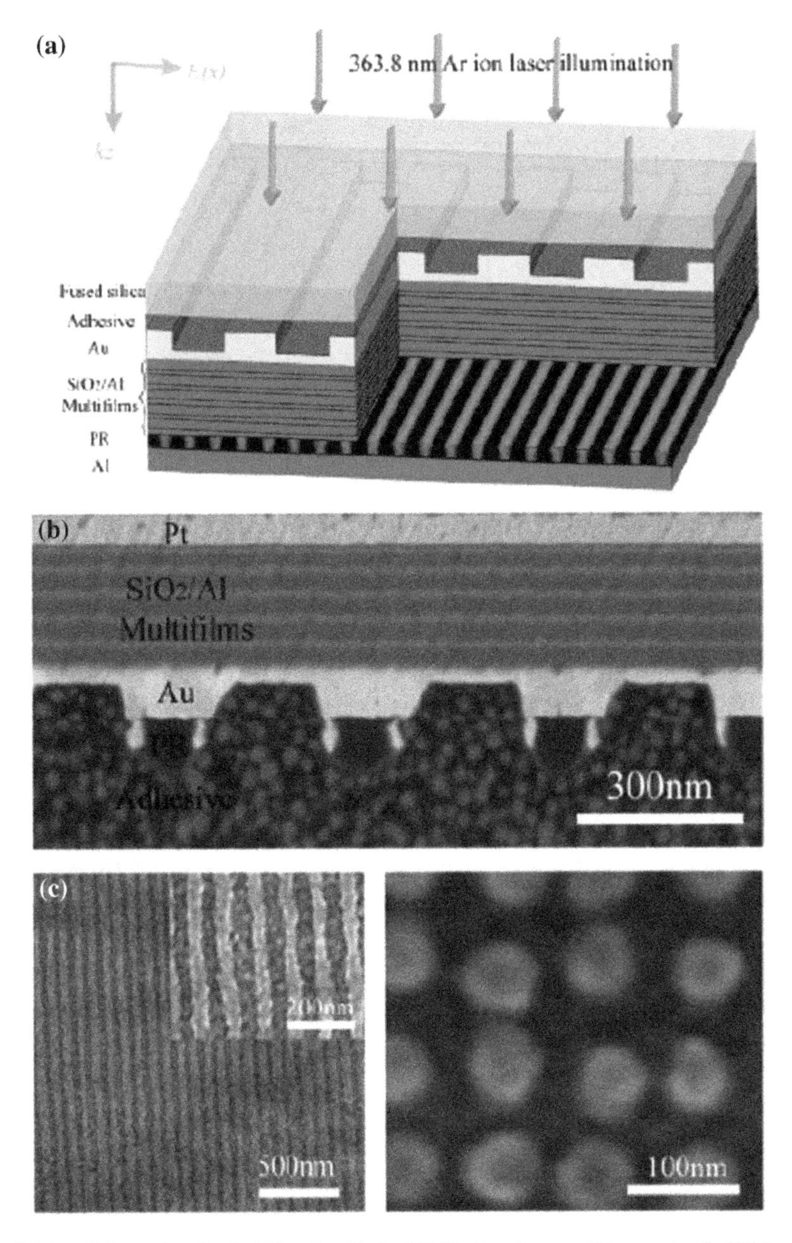

Fig. 7.15 **a** Schematic of sub-diffraction-limited BPPs interference lithography. **b** SEM cross-sectional picture of BPPs interference structure. **c** 1D and 2D periodic interference patterns recorded in PR layer. Reprinted from [32] with permission. Copyright 2015, WILEY-VCH Verlag GmbH & Co. KGaA, Weinheim

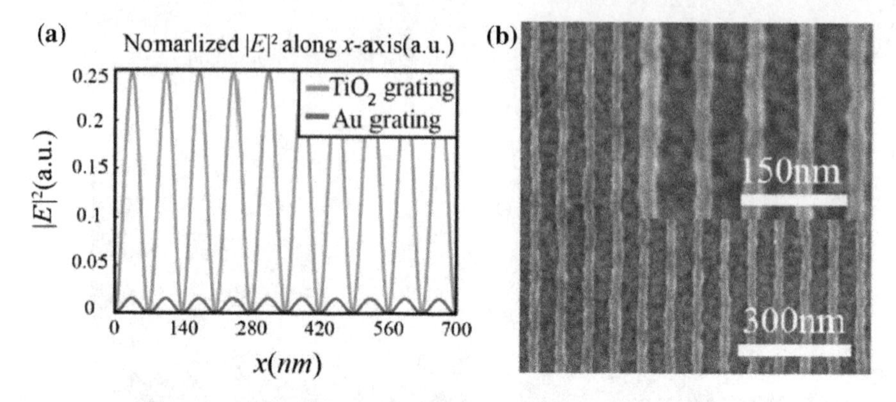

Fig. 7.16 **a** Normalized $|E|^2$ electric field intensity along the horizontal lines at the middle of PR layer for TiO_2 grating and Au grating. **b** SEM images of BPP interference fringes on PR layer based on TiO_2 grating. Reprinted from [34] with permission. Copyright 2018, Optical Society of America under the terms of the OSA Open Access Publishing Agreement

7.3 Plasmonic Imaging Lithography

For conventional projection optical lithography, the evanescent wave that carries subwavelength information about the object decays exponentially in a medium and cannot reach the image plane, resulting in the so-called diffraction limit. By exciting the SPPs in the metal slab, the evanescent wave can be coupled and amplified, and thus be restored in the imaging plane. In order to achieve a wider wave vector range and let more evanescent waves participate in super-resolution imaging, SPPs should be excited and manipulated effectively. Therefore, it is essential to design plasmonic imaging lithography structures properly, which determine the resolution, fidelity, depth, working distance, and efficiency.

7.3.1 Superlens Lithography

Superlens, a simplified case of a perfect lens, was first proposed by Pendry in 2000 [37]. The operation principle is as follows: When the permittivity of the metal slab and the surrounding environment meets the matching conditions (equal and of opposite sign), the TM wave excites SPPs at metal/dielectric interface. The evanescent field is then significantly enhanced and coupled with the other side of the metal layer. Then, the original field in front of the mask is restored and recorded by photoresist [17].

One of the unique properties of SPPs is the enhancement of evanescent wave, which can be expressed in a closed form. Considering a thin slab with negative refractive medium of thickness d, dielectric constant ε_2 and magnetic permeability μ_2 is sandwiched by medium 1 (ε_1 and μ_1) at object plane and medium 3 (ε_3 and

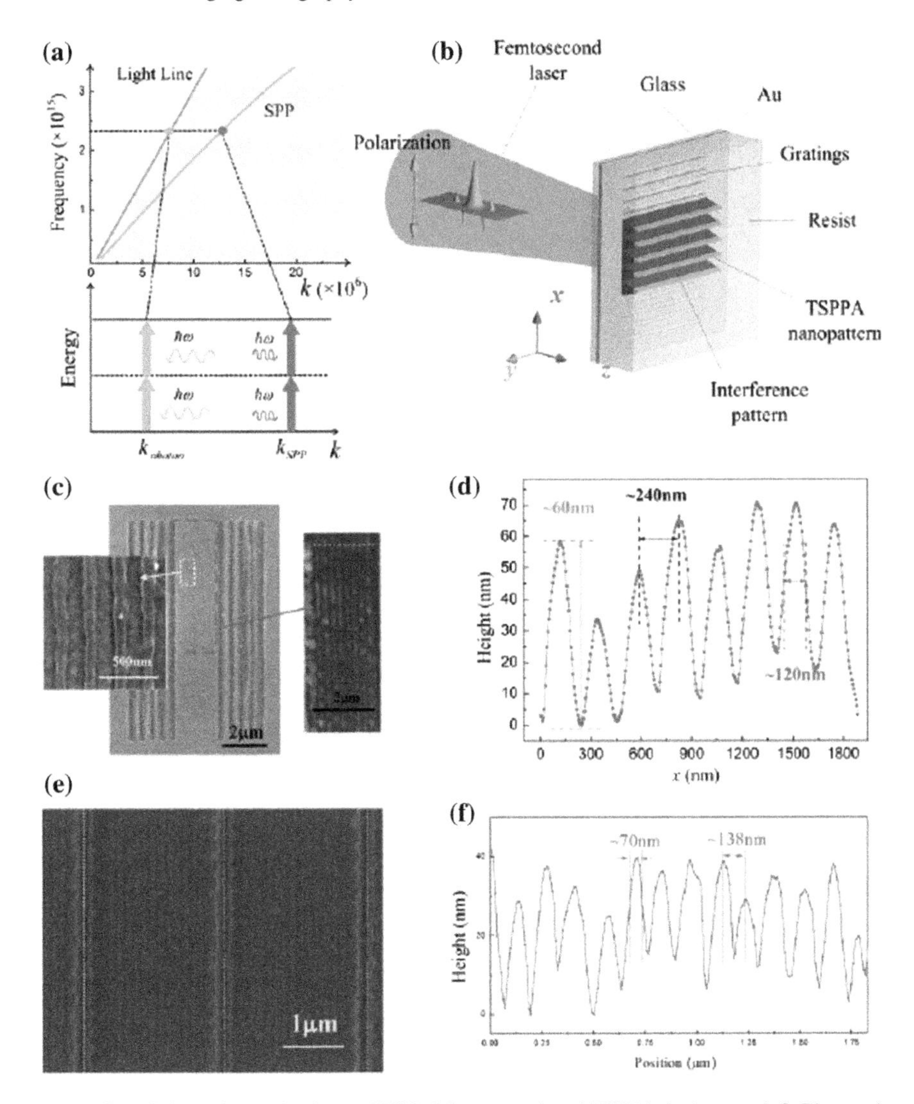

Fig. 7.17 **a** Schematic mechanisms of TPA (blue arrows) and TSPPA (red arrows). **b** Plasmonic interference structure with photoresist is illuminated by a femtosecond laser. **c** SEM and AFM images of the resist pattern, and **d** corresponding surface profile line with Au. SEM picture (**e**) and the surface profile (**f**) of the resist pattern with Al. **a**–**d** Reprinted from [35] with permission. Copyright 2013, American Institute of Physics. **e**–**f** Reprinted from [36] with permission. Copyright 2014, AIP Publishing LLC

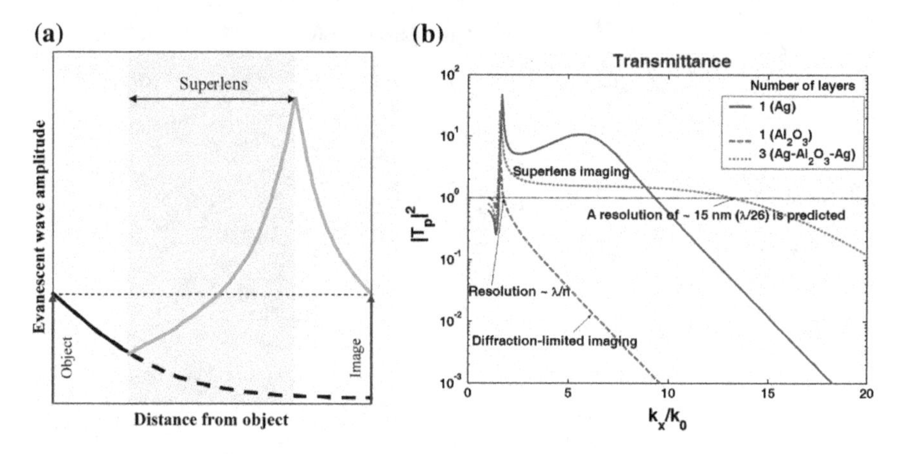

Fig. 7.18 **a** Evanescent field enhancement by superlens. The black curve shows the decay of evanescent wave in positive refractive index materials such as dielectrics. The blue curve shows the enhancement of evanescent waves by a superlens, which decays again on the other side restoring the original amplitude at the image plane. The black dotted curve represents a decaying field without a superlens. **b** Performance evaluation of diffraction-limited lens (dashed line), single-layer superlens (solid line), and multilayer superlens (dotted line). **a** Reprinted from [38] with permission. Copyright 2005, IOP Publishing Ltd. and Deutsche Physikalische Gesellschaft. **b** Reprinted from [39] with permission. Copyright 2006, Materials Research Society

μ_3) at imaging plane. Object plane and imaging plane are all at the distance d from the center of the slab.

For the electrostatic approximation and let the incident wave be TM polarized, the transmission coefficient through the slab no longer depends on μ and is only relevant to the dielectric function:

$$T_p(k_x) = \frac{4\varepsilon_2\varepsilon_3 \exp(-k_x d)}{(\varepsilon_1 + \varepsilon_2)(\varepsilon_2 + \varepsilon_3) - (\varepsilon_2 - \varepsilon_1)(\varepsilon_2 - \varepsilon_3)\exp(-2k_x d)} \tag{7.3.1}$$

The enhancement of the evanescent wave transmission depends on the condition of the dielectric constant. When $\varepsilon_1 \approx -\varepsilon_2 \approx \varepsilon_3$, $T_p(k_x) = \exp(k_x d)$ shows exponential growth of the evanescent field (Fig. 7.18a) [38]. This surface-to-surface coupling of this metal slab enables the transmission of enhanced evanescent field through the slab for high-resolution imaging, which is called superlensing. Ideally, with optimized design of multilayered superlens, the resolution of the superlens can reach 15 nm at 387.5 nm wavelength illumination (Fig. 7.18b) [39].

Figure 7.19a gives a rudimental demonstration of super-resolution nanolithography through a planar silver layer. However, different from conventional near-field lithography, the mask with dielectric spacers and silver lens is then brought into vacuum contact with photoresist for exposure. However, the resolution merely achieved 72.5-nm half-pitch with poor quality (Fig. 7.19b and c), probably resulting from the use of thick silver layers and additional SiO_2 spacer layer that embarrass the

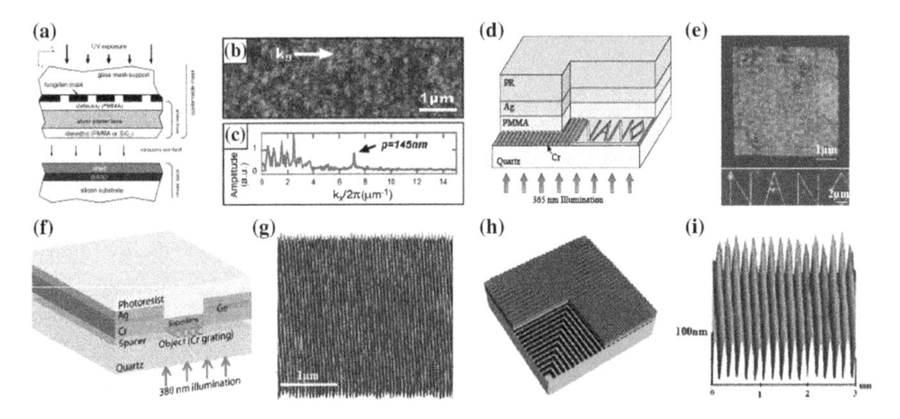

Fig. 7.19 **a** Schematic diagram of super-resolution nanolithography through a planar silver layer. **b** AFM image of a 145-nm period grating exposed by the lens of **a**. **c** Fourier transform in the direction of the grating vector k_g in **b**. Schematic of the silver superlens (**d**) and AFM image of grating and "NANO" patterns recorded on the photoresist (**e**). Schematic of smooth superlens (**f**) and AFM image of grating pattern recorded on the photoresist (**g**). Schematic of high aspect superlens (**h**) with high aspect, AFM image of grating pattern recorded on the photoresist (**i**). **a**–**c** Reprinted from [40] with permission. Copyright 2005, Optical Society of America. **d** and **e** Reprinted from [41] with permission. Copyright 2005, America Association for the Advancement of Science. **f** and **g** Reprinted from [42] with permission. Copyright 2010, American Institute of Physics. **h** and **i** Reprinted from [43] with permission. Copyright 2012, American Chemical Society

enhancement of evanescent waves [40]. In order to maximize the enhancement of evanescent waves, optical superlens lithography with sub-diffraction-limited resolution was proposed (Fig. 7.19d). Experimentally, 60-nm half-pitch gratings and "NANO" characters with ~89-nm line width were recorded in photoresist by utilizing a 35-nm Ag film under 365-nm mercury lamp illumination (Fig. 7.19e) [41].

The performance of superlens-based lithography is determined by many factors such as metal slab thickness and the parameter mismatches. To further improve the superlens lithography resolution, researchers optimized the structural parameters and preparation processes. Considering the information loss due to the rough surface morphology of silver film that hinders the resolution improvement, a "smooth superlens" with 30-nm half-pitch resolution at 380-nm illumination was proposed (Fig. 7.19f and g) [42]. Nanoimprint technique was utilized to reduce the thickness of the spacer layer down to 6-nm, and 1-nm-thick Ge was used as the wetting layer to fabricate a 15-nm-thick "smooth" silver superlens. By further reducing the PMMA spacer layer down to 20 nm, 50-nm half-pitch lines with high aspect profile of ~45 nm and fidelity ratio of 0.6 through 35-nm-thick Ag superlens were experimentally achieved (Fig. 7.19h and i) (Table 7.2) [43].

On the other hand, researchers paid special attention to the damping losses and surface roughness of metal slabs and derived some interesting and even seemingly opposite conclusions. Some viewpoints deem that small surface roughness can enhance higher spatial wave vector components, suppress the surface plasmon resonance

Table 7.2 Key features of the superlens, smooth superlens, and high aspect superlens

	"Superlens"	"Smooth superlens"	"High aspect superlens"
Illumination wavelength (nm)	365	380	365
Thickness of spacer (nm)	35	6	20
Thickness of Ag (nm)	35	15	35
Resolution (nm)	60	30	Sub-50
Depth (nm)	5–10	<6	45

peaks, and finally improve the resolution [44]. And some combined sinusoidal roughness with the additional loss and achieved flatter transfer function and 86% reduction of beam width compared to the lossless flat superlens [45]. Surface roughness mainly affects the imaging quality (e.g., uniformity and deviation), so interferential noise can be suppressed by adding suitable loss. However, an experiment demonstrated that the loss reduction in Ag superlens is a critical factor to achieve high-performance superlens when surface roughness is less than 2 nm in comparison with Ag/Ni and Ag/Ge lens [46].

7.3.2 Plasmonic Reflective Lens Lithography

Although traditional superlens can enhance the evanescent waves, they will decay again once depart from superlens. As a consequence, traditional plasmonic imaging lithography methods usually suffer from poor quality of resist patterns and are featured by shallow profiles, low contrast, and great aberrations compared with masks. Previous study presented numerical and experimental demonstrations of SPPs assisted nanoscale near-field photolithography by introducing a Ti shield slab below the photoresist [47, 48]. It is shown that evanescent waves amplification in reflection manner by appropriate metal shield helps to enhance the imaging performance in the near-field. However, only 500-nm periodicity pattern can be recorded in the photoresist with Ti shield slab. Further theoretically research proposed that a plasmonic lens would enhance evanescent waves in a reflective manner and help to relieve this effect. A clear analysis can be found in Ref. [49], where the imaging transfer process is approximated as a virtual light source in photoresist at a distance away from the metal reflector, as shown in the inset of Fig. 7.20. Interestingly, the combination of the counter-propagating waves would lead to a catenary-shaped intensity distribution.

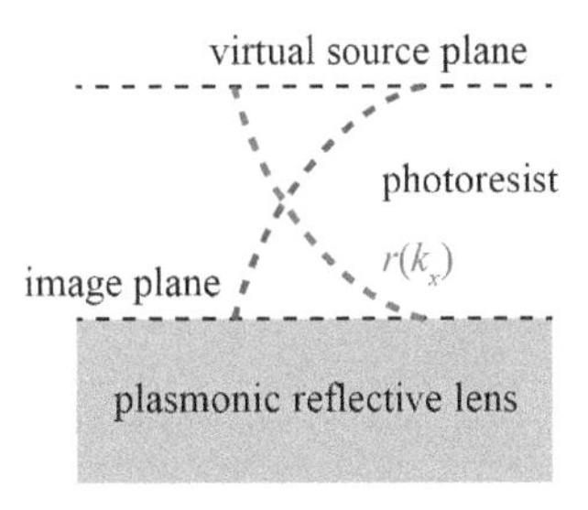

Fig. 7.20 Schematic illustration of plasmonic reflective lens lithography

For the imaging plane close to the photoresist–metal interface, the OTF for the plasmonic slab structure is defined as

$$T(k_x) = (1 + r(k_x))\exp(ik_{z,\mathrm{d}}d) = T^{'}(k_x)\exp(ik_{z,\mathrm{d}}d) \tag{7.3.2}$$

where k_x is the transversal wave vector, d is the thickness of the photoresist layer, and $r(k_x)$ is the reflection coefficient at the photoresist–metal interface. $k_{z,\mathrm{d}} = (\varepsilon_\mathrm{d}k_0^2 - k_x^2)^{1/2}$ is the perpendicular wave vector in the resist layer, where $k_0 = 2\pi/\lambda$. For TM polarized light illumination, the reflection at the PR–metal interface is:

$$r(k_x) = (k_{z,\mathrm{d}}/\varepsilon_\mathrm{d} - k_{z,\mathrm{m}}/\varepsilon_\mathrm{m})/(k_{z,\mathrm{d}}/\varepsilon_\mathrm{d} + k_{z,\mathrm{m}}/\varepsilon_\mathrm{m}) \tag{7.3.3}$$

where $k_{z,\mathrm{m}} = (\varepsilon_\mathrm{m}k_0^2 - k_x^2)^{1/2}$, ε_m and ε_d are the permittivities for the metal and resist layers. To get a convenient understanding of evanescent wave amplification, Eq. (7.3.2) is further approximated for large $k_x(k_x/k_0 \gg 1)$ as

$$T^{'} \approx 2\varepsilon_\mathrm{m}/(\varepsilon_\mathrm{m} + \varepsilon_\mathrm{d}) = 2(\varepsilon_\mathrm{m}^{'} + i\varepsilon_\mathrm{m}^{''})/\left[(\varepsilon_\mathrm{m}^{'} + \varepsilon_\mathrm{d}) + i\varepsilon_\mathrm{m}^{''}\right] \tag{7.3.4}$$

where $\varepsilon_\mathrm{m} = \varepsilon_\mathrm{m}^{'} + i\varepsilon_\mathrm{m}^{''}$. Consequently, Eq. (7.3.4) has a large value $-2i\left(\varepsilon_\mathrm{m}^{'} + i\varepsilon_\mathrm{m}^{'}\right)/\varepsilon_\mathrm{m}^{''}$ when the condition $\varepsilon_\mathrm{d} = -\varepsilon_\mathrm{m}^{'}$ is satisfied and $\varepsilon_\mathrm{m}^{''}$ is negligible. Therefore, the evanescent components of large k_x are amplified in the dielectric layer and image quality could be improved. For TE-polarized light case, however

$$r(k_x) = \left(k_{z,\mathrm{d}} - k_{z,\mathrm{m}}\right)/\left(k_{z,\mathrm{d}} + k_{z,\mathrm{m}}\right) \tag{7.3.5}$$

And $T' = (1 + r(k_x))$ approximates to 1 when $k_x/k_0 \gg 1$. The reflected evanescent components just decay exponentially without any amplification.

Considering the permittivity match condition, silver film is a good plasmonic reflective lens. By using such a plasmonic reflective lens, deep sub-diffraction-limited imaging lithography is numerically and experimentally demonstrated (Fig. 7.21a) [16]. The OTF for plasmonic reflective lens is the superposition of the transmission function of mask and the reflection function of lens. The performance of plasmonic reflective lens mainly depends on the excitation of SP resonance under dielectric constant matching. When this condition is satisfied, resolution, depth of exposure, and fidelity can be improved significantly. Consequently, nanocharacters with about 50-nm line width and 40 nm depth (Fig. 7.21b–e) can be obtained by using a plasmonic reflective lens at 365-nm illumination. Compared with superlens, plasmonic reflective lens can suppress the broadening of nanocharacters imaging which have a broad band of Fourier spectrum (Table 7.3).

From the description above, one can see that plasmonic reflective lens supplies new plasmonic imaging lithography with high fidelity and aspect ratio for isolated patterns. But for dense lines, the highest half-pitch resolution of experimental reports is 32 nm, probably resulting from the proximity effect of density patterns.

7.3.3 Plasmonic Cavity Lens Lithography

Regardless of either superlens imaging lithography or plasmonic reflective lens imaging lithography, the amplification of evanescent only occurs at one side of the photoresist. Therefore, two unwelcome features are encountered, especially in the superlens scheme. One is the ultra-shallow exposure depth, less than 10 nm, and the other is the low contrast of images due to the great decay of evanescent waves and the E_z

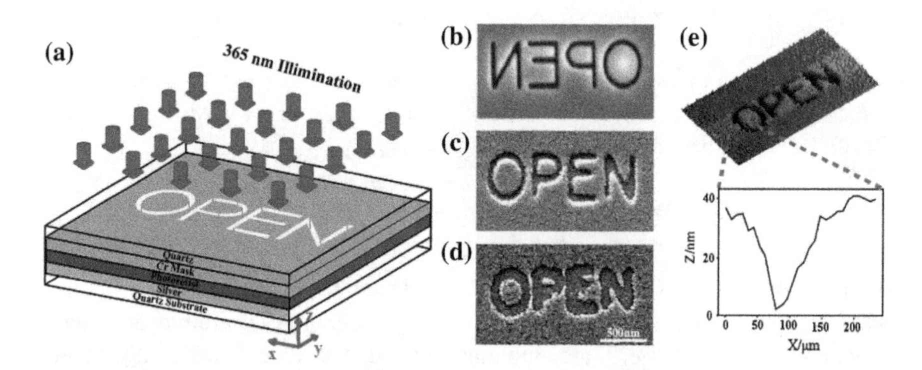

Fig. 7.21 **a** Schematic of the plasmonic reflective lens lithography. **b** SEM image of mask. **c** SEM image of the recorded resist patterns by plasmonic reflective lens. **d** SEM image of the recorded resist patterns by conventional near-field lithography. **e** AFM profile of "OPEN" and cross section at the depicted position. Reprinted from [16] with permission. Copyright 2013, Optical Society of America

Table 7.3 Key features of nanocharacters fabricated by superlens and plasmonic reflective lens

	Nanocharacters	Line width of object (nm)	Line width of imaging (nm)	Broadening (nm)
Superlens [41]	"NANO"	40	89	49
Plasmonic reflective lens [16]	"OPEN"	36	50	14

component's negative contributions. The combination of a superlens and plasmonic reflective lens forms a plasmonic cavity, which could significantly amplify evanescent waves in the entire photoresist region due to the multiple reflections inside the cavity [10].

Figure 7.22a schematically illustrates the amplitude profile of H_y transmitted through the Ag superlens. In analogy with superlens-assisted evanescent wave amplification, the objects' spatial frequency spectra could be amplified once again due to plasmonic reflection lens. In this case, the amplitude distribution of H_y in plasmonic cavity lens includes the transmitted (black curve) and multiple reflected fields (red curve) as shown in Fig. 7.22a. For an incident plane wave with a transversal wavevector k_x, the multi-reflection approximation *OTF* of plasmonic cavity lens for different electromagnetic field components inside the photoresist is given by [50]

$$\tilde{H}_y(k_x) = t(k_x)(1 + r_z(k_x)e^{ik_{z,\mathrm{Pr}}(d_{\mathrm{Pr}} - d_{\mathrm{Pr},1})}) \tag{7.3.6}$$

$$\tilde{E}_x(k_x) = t(k_x)(1 - r_x(k_x)e^{ik_{z,\mathrm{Pr}}(d_{\mathrm{Pr}} - d_{\mathrm{Pr},1})})\frac{k_{z,\mathrm{Pr}}\varepsilon_{\mathrm{SiO_2}}}{\varepsilon_{\mathrm{Pr}}k_{z,\mathrm{SiO_2}}} \tag{7.3.7}$$

$$\tilde{E}_z(k_x) = t(k_x)(1 + r_z(k_x)e^{ik_{z,\mathrm{Pr}}(d_{\mathrm{Pr}} - d_{\mathrm{Pr},1})})\frac{\varepsilon_{\mathrm{SiO_2}}}{\varepsilon_{\mathrm{Pr}}} \tag{7.3.8}$$

In Eqs. (7.3.6), (7.3.7), and (7.3.8), $t(k_x)$ is the *OTF* of superlens for H_y, while $t(k_x)(k_{z,\mathrm{Pr}}\varepsilon_{\mathrm{SiO_2}})/(\varepsilon_{\mathrm{Pr}}k_{z,\mathrm{SiO_2}})$ and $t(k_x)\varepsilon_{\mathrm{SiO_2}}/\varepsilon_{\mathrm{Pr}}$ are the *OTF* for E_x and E_z. The second terms in these equations should be eliminated during the *OTF* calculation of superlens in spite of reflection amplifying. However, the multi-reflection coefficients of

$$r_{x(z)}(k_x) = \frac{r_{\mathrm{Ag}}(k_x)e^{ik_{z,\mathrm{Pr}}d_{\mathrm{Pr}}}(e^{-ik_{z,\mathrm{Pr}}d_{\mathrm{Pr},1}} \mp r_{\mathrm{SL}}(k_x)e^{ik_{z,\mathrm{Pr}}d_{\mathrm{Pr},1}})}{1 - r_{\mathrm{Ag}}(k_x)r_{\mathrm{SL}}(k_x)e^{2ik_{z,\mathrm{Pr}}d_{\mathrm{Pr}}}} \tag{7.3.9}$$

contribute to *OTF* modification of plasmonic cavity lens due to Ag reflection lens, where $r_{\mathrm{Ag}}(k_x)$ defined as:

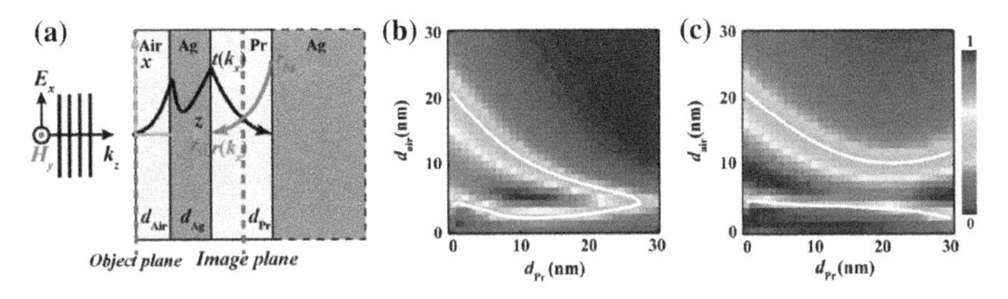

Fig. 7.22 **a** Schematic diagram of OTF calculation for plasmonic lens system. Simulation of hp 32-nm intensity contrast versus photoresist depth and air spacer with **b** superlens and **c** plasmonic cavity lens. The white solid curves are the intensity contrast contour of 0.4. **a** Reprinted from [50] with permission. Copyright 2016, Optical Society of America. **b** and **c** Reprinted from [10] with permission. Copyright 2015, AIP Publishing LLC

$$r_{\mathrm{Ag}}(k_x) = \frac{\varepsilon_{\mathrm{Ag}} k_{z,\mathrm{Pr}} - \varepsilon_{\mathrm{Pr}} k_{z,\mathrm{Ag}}}{\varepsilon_{\mathrm{Ag}} k_{z,\mathrm{Pr}} + \varepsilon_{\mathrm{Pr}} k_{z,\mathrm{Ag}}} \tag{7.3.10}$$

and $r_{\mathrm{SL}}(k_x)$ are the reflection coefficients of the Pr-Ag reflection lens interface and the Pr-Ag supelens-Air scheme. $\varepsilon_{\mathrm{Ag}}$ and $\varepsilon_{\mathrm{Pr}}$ are the permittivities of the Ag and photoresist layers. $k_{z,\mathrm{Ag}} = (\varepsilon_{\mathrm{Ag}} k_0^2 - k_x^2)^{1/2}$ and $k_{z,\mathrm{Pr}} = (\varepsilon_{\mathrm{Pr}} k_0^2 - k_x^2)^{1/2}$ are the longitudinal wave vectors in the Ag and Pr layers. d_{Pr} is the photoresist thickness, and $d_{\mathrm{Pr},1}$ is the selected depth position of observation line in photoresist. In the analytical expression of $r_{x(z)}(k_x)$, the subtraction and plus signs in the bracket of numerator term are corresponding to the *OTF* for E_x and E_z, respectively. Figure 7.22b and c show the dependence of simulated intensity contrast of the superlens and the plasmonic cavity lens on air spacer thickness and photoresist depth position. It has been shown that plasmonic cavity lens helps to achieve $V > 0.4$ in the whole photoresist depth direction.

Plasmonic cavity lens can effectively localize the surface plasmons for projecting deep-subwavelength patterns, through which fringes of 31 nm (smaller than 1/14 incident wavelength) can be achieved with greatly improved intensity contrast. Besides, by optimizing the thickness of resist and silver, the resolution can be further improved.

By solving the conundrum for the fabrication of the plasmonic cavity lens, the enhancement of aspect profile of ultra-deep subwavelength lithography has been demonstrated experimentally. The plasmonic cavity lens was in the form of a 20 nm thick top Ag layer-30 nm thick photoresist layer-50 nm thick bottom Ag layer. The profile depth of 32-nm half-pitch patterns was improved up to 23 nm (Fig. 7.23a and b), nearly 4 times that of the smooth superlens. In addition, the resolution of plasmonic cavity lens up to 22-nm half-pitch (~1/17 illumination wavelength) was experimentally demonstrated (Fig. 7.23c and d). The prediction (Surface-plasmon polaritons allow features to 25 nm) in 2004 was well verified by this result [51]. Plasmonic cavity lens provides a cost-effective, high aspect profile, and potentially promising way for nanolithography of 22-nm half-pitch and beyond.

7.3.4 Demagnification Through Hyperlens

The principle of hyperlens imaging and lithography can be explained by using dispersion relation of electromagnetic waves in cylindrical coordinates:

$$\frac{k_r^2}{\varepsilon_\theta} - \frac{k_\theta^2}{|\varepsilon_r|} = \left(\frac{\omega}{c}\right)^2 = k_0^2 \tag{7.3.11}$$

where k_r and k_θ are wave vectors, ε_r and ε_θ are the permittivities of the medium in radial and tangential directions, respectively, and k_0 is the wavevector in free space. A hyperlens is designed to be an effective anisotropic medium with negative ε_r and positive ε_θ so that waves with arbitrarily large tangential wavevector k_θ can

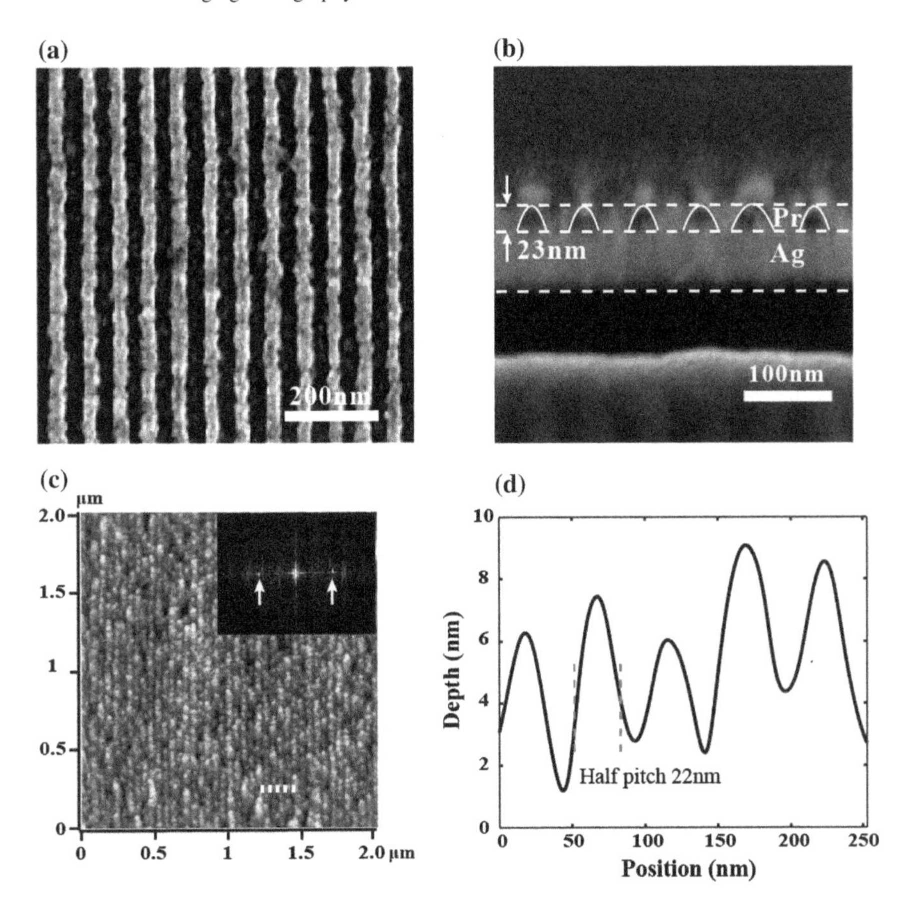

Fig. 7.23 **a** SEM image of the recorded 32-nm half-pitch resist pattern by plasmonic cavity lens. **b** Cross-sectional view of pattern in **a**. **c** AFM image of 22-nm half-pitch resist pattern, inset shows its two-dimensional Fourier analysis diagram. **d** Profile plot at position as depicted by the dashed lines in **c**. Reprinted from [10] with permission. Copyright 2015, AIP Publishing LLC

propagate in the hyperlens. For lithography purposes, the diffraction-limited pattern is projected onto the outer surface of a hyperlens from a diffraction-limited mask. As the waves propagate inward along the radial direction, tangential wavevectors are gradually increased while the waves can still propagate in the hyperlens [52]. As a result, a diffraction-limited pattern of the mask can be reduced to a sub-diffraction-limited pattern at the inner surface of the hyperlens. Numerical simulation showed that half-pitch resolution down to 20 nm was possible from a mask with 280-nm period at a working wavelength of 375 nm [53].

The hyperlens demagnifying imaging lithography experiment was firstly demonstrated in 2016. The hyperlens was composed of multiple Ag/SiO_2 films (Fig. 7.24a) [54], and a plasmonic Ag reflector was deposited behind the multi-films to enhance the exposure intensity and depth. Experiment results demonstrated a sub-diffraction-limited resolution of about 55-nm line width, where 1.8 demagnification factor was obtained at 365-nm light wavelength (Fig. 7.24c). A similar demagnifying imaging lithography experiment was carried at 405 nm, with a resolution of 170-nm line width (Fig. 7.24d, e) [55].

It should be mentioned that the curved geometry of a demagnifying hyperlens above causes difficulty in the practical implementation and applications. A simple way for planar hyperlens profile is to cut and polish hyperlens on one or both sides (Fig. 7.25a and b) [53]. However, this method seems challenging for fabrication and usually causes the deformation of images. As an alternative, hybrid lens combining hyperlens with planar superlens was proposed. However, the demagnification ratio is not uniform on the entire input plane. Subsequently, a conformal transformation theory was utilized to make planar input and output surfaces with specially designed and complicated mathematic functions for uniform demagnification, as indicated in Fig. 7.25c and d [56].

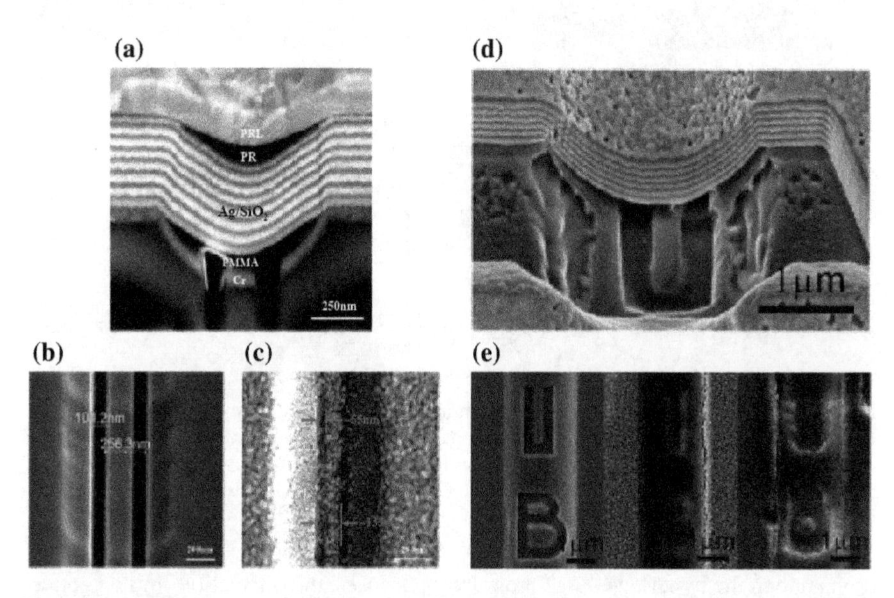

Fig. 7.24 **a** SEM picture of the cross section of demagnifying hyperlens. **b** Two slits mask pattern with 100-nm line width and 250 nm center-to-center distance. **c** Lithography result of resist pattern with hyperlens and PRL with 55-nm line width at 135-nm pitch. **d** Another experimental demonstration of demagnifying hyperlens. **e** (left) Cr mask pattern, (middle) subwavelength pattern on the photoresist recorded using hyperlens and (right) control experimental result. **a**–**c** Reprinted from [54] with permission. Copyright 2016, The Royal Society of Chemistry. **d** and **e** Reprinted from [55] with permission. Copyright 2016, American Chemical Society

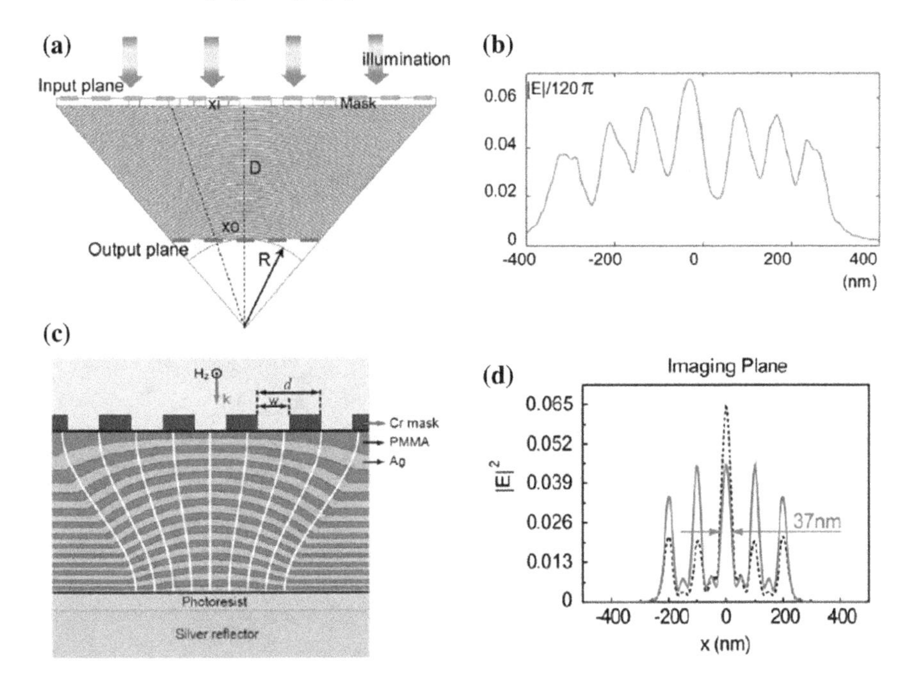

Fig. 7.25 Planer demagnified hyperlens. **a** A schematic of a flat interface hyperlens. **b** |E| field distribution along the plane that is 10 nm below the output plane. **c** Demagnification imaging lithography model with modified hyperlens. **d** Electric field distribution on the plane 10 nm below the output plane. The solid and dashed lines refer to the proposed structure and control structure, respectively. **a** and **b** Reprinted from [53] with permission. Copyright 2009, American Institute of Physics. **c** and **d** Reprinted from [56] with permission. Copyright 2013, Springer-Verlag Berlin Heidelberg

7.3.5 Wavefront Engineering

Figure 7.26 shows the schematic diagram of wavefront engineering in plasmonic imaging lithography [50]. The air spacer ridge is used to avoid mask degradation. This asymmetric lens consisting of a silver slab bounded by air and photoresist is applied for SPPs lithography and high-density data storage. Super-resolution imaging model is established to analytically describe plasmonic imaging process. To simplify the discussions, this model has two assumptions: (a) The coupling interaction between plasmonic lens and mask is neglected. (b) The slit-like objects are taken as uniform line sources with TM polarization. The electric fields in half ($z > 0$) 3D imaging space are then generally given according to vectorial angular spectrum theory by

$$\begin{bmatrix} E_x(x,y,z) \\ E_y(x,y,z) \\ E_z(x,y,z) \end{bmatrix} = \int_{-\infty}^{\infty}\int_{-\infty}^{\infty} \boldsymbol{G}(k_x,k_y)\tilde{\boldsymbol{E}}_{\text{obj}}(k_x,k_y)\boldsymbol{OTF}(k_x,k_y)$$

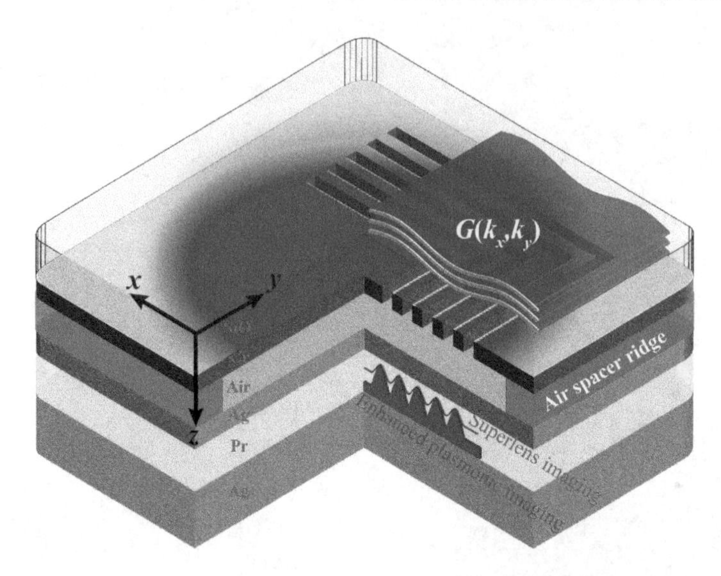

Fig. 7.26 Schematic diagram of plasmonic imaging enhancement with wavefront engineering technique. Reprinted from [50] with permission. Copyright 2016, Optical Society of America

$$\exp(ik_x x + ik_y y)\mathrm{d}k_x \mathrm{d}k_y \tag{7.3.12}$$

where $\boldsymbol{G}(k_x,k_y)$ is the illumination function of the plasmonic lens for different directions. $\tilde{\boldsymbol{E}}_{\mathrm{obj}}(k_x,k_y)$ is the E-field component of objects, which is given by

$$\tilde{\boldsymbol{E}}_{\mathrm{obj}}(k_x, k_y) = \left[\tilde{\boldsymbol{E}}_{x,\mathrm{obj}}, \tilde{\boldsymbol{E}}_{y,\mathrm{obj}}, \frac{k_x \tilde{E}_{x,\mathrm{obj}} + k_y \tilde{E}_{y,\mathrm{obj}}}{-k_z}\right] \tag{7.3.13}$$

$\boldsymbol{OTF}(k_x,k_y)$ is the optical transfer function of the plasmonic lens for different spatial frequency. Thus, the imaging result is expressed by the product of illumination function $\boldsymbol{G}(k_x,k_y)$, object spectrum $\tilde{\boldsymbol{E}}_{\mathrm{obj}}(k_x,k_y)$ and optical transfer function $\boldsymbol{OTF}(k_x,k_y)$. The object spectrum and ***OTF*** of plasmonic imaging lithography are the main factors of super-resolution imaging performance, but the role of wavefront engineering should be also considered. To simplify the analysis, we only consider the incidence transverse wavevector $k_{x,\mathrm{inc}}$, then $\boldsymbol{G}(k_x,k_y) = \boldsymbol{G}(k_x) = \exp(ik_{x,\mathrm{inc}}\, x_{\mathrm{inc}})$. Therefore, if we use the illumination with a high transverse wavevector, it can help to enhance plasmonic lens optical transfer ability of patterns' information even with a large air working distance. It is attributed to the fact that Fourier components of high spatial frequency spectrum are shifted to that with smaller wavevectors. The working distance elongation could be approximately evaluated by assuming that the magnitude of the transferred Fourier components k_{g} in plasmonic cavity lens under NI is compensated by the enhanced magnitude of the shifted Fourier components $k_{x,\mathrm{inc}}$ or $k_{x,\mathrm{inc}} - k_{\mathrm{g}}$. This compensation yields the distance in the form [57]

$$L = \frac{1}{i\sqrt{k_0^2 - k_s^2}} \ln \frac{|\mathrm{OTF}(k_g, 0)|}{|\mathrm{OTF}(k_s, 0)|} + \frac{\sqrt{k_0^2 - k_g^2}}{\sqrt{k_0^2 - k_s^2}} L_{\mathrm{NI}} \quad (7.3.14)$$

where L_{NI} and L represent the air distance under NI and high spacial frequency spectrum off-axis illumination (OAI), respectively; OTF(k_g,0) and OTF(k_s,0) are the optical transfer function values of plasmonic cavity lens with a zero gap distance and $|\mathrm{OTF}(k_s,0)| = \min\{|\mathrm{OTF}(k_i,0)|, |\mathrm{OTF}(k_i - k_g,0)|\}$. So far, there are two ways to excite a high transverse wavevector, one is OAI, and the other is surface plasmon illumination (SPI).

OAI is widely used to improve the resolution and depth of focus (DOF) in conventional projecting optics lithography. Also OAI can be applied to enhance the sub-diffraction-limited imaging performance of plasmonic lens. Both theoretical investigation and numerical simulation showed that the off-axis illumination with designed high numerical aperture and tailored electric field components could greatly improve the performance of plasmonic imaging lithography, which was subsequently experimentally demonstrated [57, 58]. The patterns with 60-nm half-pitch were well resolved by high spatial frequency spectrum OAI ($k_{x,\mathrm{inc}} = 1.5k_0$) at 80-nm working distance, approximately fourfold longer than that in the conventional superlens scheme (Fig. 7.27). Higher resolution with 45-nm half-pitch was also demonstrated with working distance of 40 nm.

Besides OAI, surface plasmon illumination (SPI) based on metal–dielectric multilayers can generate uniform field with a large wave vector [59]. SPI delivers the shifted spatial spectra components of mask, which helps to enhance the optical transferability. By using SPI, it significantly improves the working distance of ultra-deep subwavelength plasmonic imaging lithography (Fig. 7.28a). Numerical simulations showed that the maximum working distance could reach a length of 60 nm for patterns with 32-nm half-pitch, which is about six times that for the conventional superlens under normal illumination (NI) (Fig. 7.28b). It also can resolve patterns with 22-nm half-pitch under SPI ($3.6k_0$) with a working distance of 20 nm (Fig. 7.28c).

In general, imaging contrast depends on the ratio of peak and valley intensity of two overlapping spots. As for normal light illumination, the constructive interference between the neighboring slits with same phase difference may degrade the imaging contrast. These phenomena could be clearly seen from the blue curves in Fig. 7.29. With phase-shifting mask design, the π phase shift is alternatively induced in the object fields of $\tilde{\boldsymbol{E}}_{\mathrm{obj}}(k_x,k_y)$ in Eqs. (7.3.12) and (7.3.13). The red-dashed curves in Fig. 7.29 show that the imaging results of different patterns are fully resolved and imaging contrast under phase-shifting mask is significantly improved. For phase-shifting mask with π phase difference, the destructive interference between neighboring slits cancels the valley light intensity of two neighboring slits and contributes to the imaging contrast improvement [50].

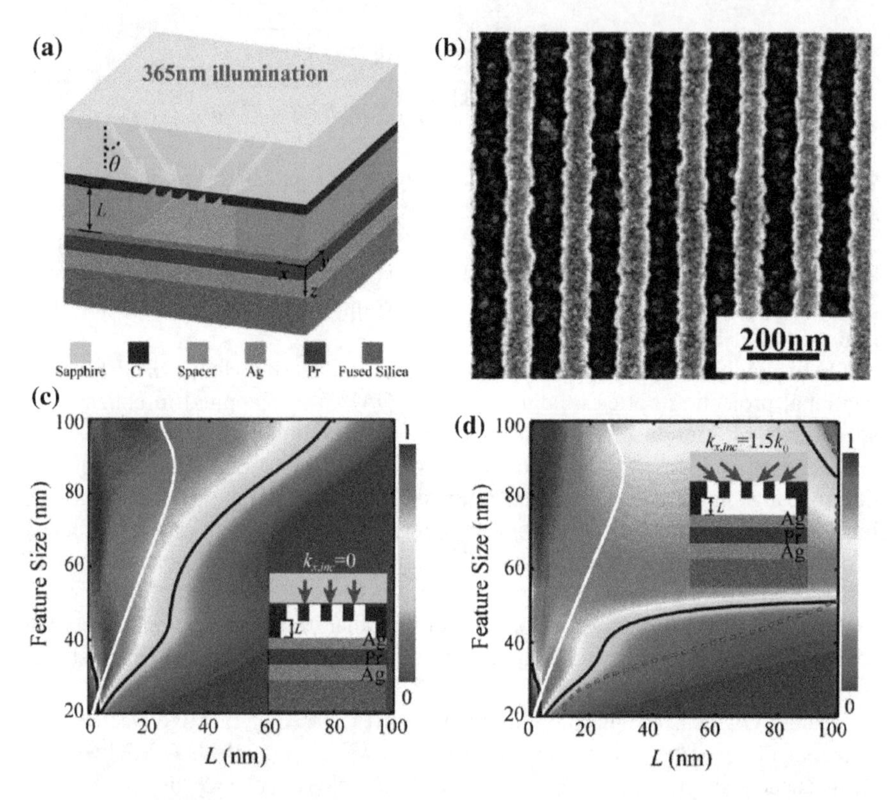

Fig. 7.27 **a** Schematic of plasmonic cavity lens lithography with off-axis illumination. **b** SEM image of the recorded resist patterns by plasmonic cavity lens lithography with off-axis illumination at 80-nm working distance. Imaging contrast as a function of the air distance (L) and half-pitch size of dense-line object for **c** the plasmonic cavity lens under normal incidence, and **d** the plasmonic cavity lens under high spatial frequency off-axis illumination ($k_{x,\text{inc}} = 1.5k_0$) from two sides. Reprinted from [57] with permission. Copyright 2015, The Authors

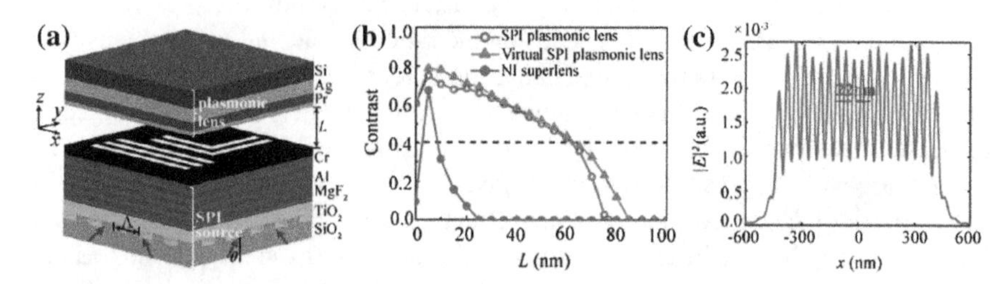

Fig. 7.28 **a** Schematic of plasmonic cavity lens lithography with surface plasmon illumination. **b** Imaging contrast for variant working distances with fixed 32-nm half-pitch. **c** Distribution of electric field intensity in the center of photoresist layer for image of 22-nm half-pitch grating with the working distance of 20 nm. Reprinted from [59] with permission. Copyright 2014, Springer Science+Business Media New York

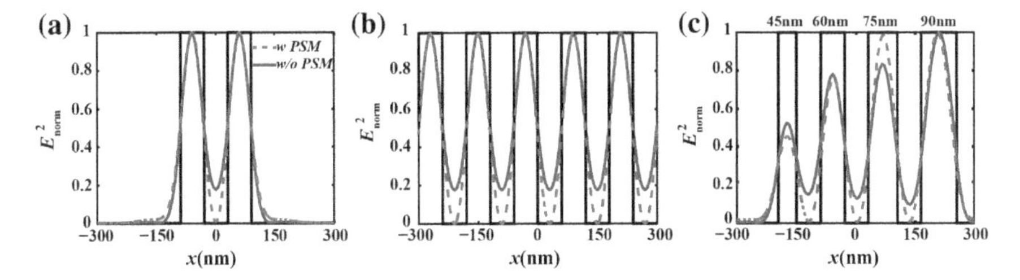

Fig. 7.29 Calculated imaging results of hp 60-nm line pair (**a**), dense lines (**b**), and multi-lines with plasmonic cavity imaging under phase-shifting mask design (**c**). Reprinted from [50] with permission. Copyright 2016, Optical Society of America

7.4 Plasmonic Direct Writing Lithography

LSPs are employed in nanooptical lithography associated with sharp metallic tips, nanometal particles and specifically designed nanoapertures-like bowtie structures, in which the large LSPs fields deliver nanoscale lithography resolution. Localized plasmonic lithography is usually performed in a point-to-point writing manner, and it is easy to do in parallel by integrating multiple plasmonic lenses over one writing head to increase throughput.

7.4.1 Nanofocusing with SPPs and BPPs

A simple plasmonic focusing structure is a transparent ring on a metallic film, in which subwavelength focus appears at the ring center due to the convergence of SPP excited over the metallic ring. Furthermore, a two concentric half-ring structure with a radius difference of about half of the SPPs wavelength is proposed to generate a solid focus spot [60]. SPPs focusing can be realized not only by changing the structural geometries but also by filling different medium into a semicircular slot. Based on the phase difference through the nanoslits, a tightly confined SPPs' spot can be also achieved under linear polarization illumination. In order to enhance the SPPs' excitation efficiency, a spiral slot incorporating a spiral triangle array lens is designed, which could couple the azimuthal polarization component into SPPs [61, 62].

Recently, an alternative nanofocusing approach using hybrid plasmonic Fano resonances in particular structures was proposed to achieve the subwavelength spot and elongate the focal length of focusing lens (Fig. 7.30) [63].

Particularly, hyperbolic metamaterial with multiple metal–dielectric films provides a volume space in which photons with large transverse wavevector k_x propagate inside in the form of BPPs. Thus, a plasmonic Fresnel plate and a hypergrating structure could be designed for generating nanofocusing effects inside or at the edge of BPP material by appropriately engineering BPP phases [64].

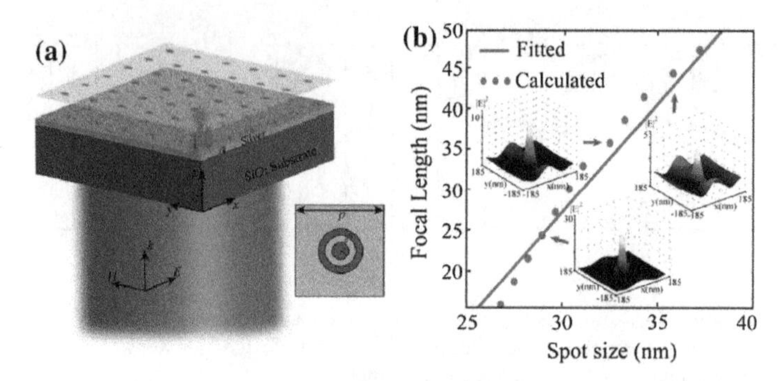

Fig. 7.30 **a** Nanofocusing by plasmonic Fano resonance lens. **b** Nanofocus length with variant focus size. Reprinted from [63] with permission. Copyright 2016, The Royal Society of Chemistry

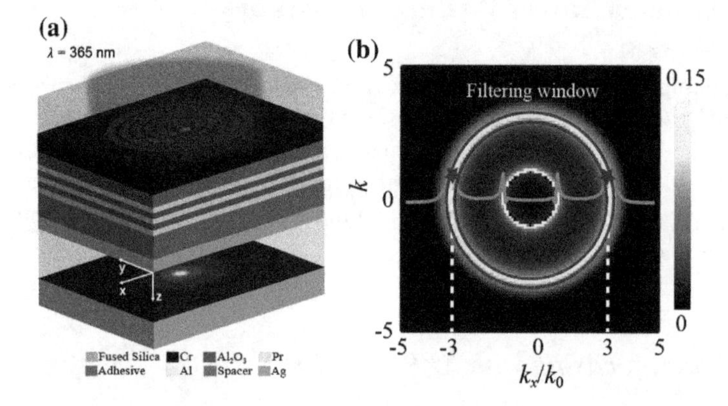

Fig. 7.31 **a** Schematic of the Al/Al_2O_3 hyperbolic metamaterial and Ag/Photoresist/Ag plasmonic imaging lithography. **b** Two-dimensional OTF versus the tangential wavevector k_x and k_y of HMM. The olive curve in **b** is the OTF distribution along the position of $k_y = 0$. The stars in **b** correspond to the center position of specific BPPs coupling window in multilayer metamaterial. Reprinted from [13] with permission. Copyright 2017, The Royal Society of Chemistry

Subsequently, a miniature evanescent Bessel beam generator has been proposed by utilizing the combination of metasurfaces based on concentric grating and hyperbolic metamaterials composed of alternative metal/dielectric multilayer (Fig. 7.31a) [13]. Bessel beam is described by the superposition of a set of plane waves with the wavevectors lying on the surface of cone in the Fourier space. The diffraction-free characteristic of Bessel beam emanating from the intensity distribution in the plane normal to z-axis is proportional to $J_0^2(k, r)$, where J_0 is the zero-order Bessel function of the first kind, k_r is the radial wavevector and $k_r = \left(k_0^2 - k_z^2\right)^{1/2}$. Accordingly, the full width at half maximum (FWHM) of Bessel beam is only decided by the spatial wavevector k_r.

Compared with SP modes, the BPP modes generated by HMM could provide higher spatial frequency. The HMM based on the Al/Al_2O_3 multilayer structure acts as a homogeneous medium with a highly anisotropic, hyperbolic spatial frequency dispersion. The dispersion relation of the multilayer structure represents the relationship between tangential wavevector k_x and longitudinal wavevector k_z, which exhibits a hyperbolic profile with no cut-off spatial frequency and thus evanescent waves with large wavevector could be launched by HMM. However, due to the quick growth of the imaginary part of k_z, the ultra-high spatial frequencies could not be supported, leading to the fact that light modes with a specified k_x range of spatial frequencies would go through the metamaterial. A filtering window with low absorption for the specific BPP waves around the wavevector $3k_0$ is created through the multilayer metamaterial, as shown in Fig. 7.31b, making other diffraction waves outside the window range being damped. The high transmission efficiency is mainly caused by the interaction of modes on the surface and the metamaterial medium.

Higher radial wavevector promises a narrower FWHM of Bessel beam. The simulation result shows that increasing the permittivity or decreasing the thickness could push the center wavevector of filtering window toward higher spatial frequency. The polarization of incident light plays another crucial role in the generation of deep subwavelength evanescent Bessel beam, which could be clarified by

$$\begin{aligned} E_x &= -\frac{1}{2} i |t(k_r)| k_z \big[\exp(i2\varphi) J_2(k_r r) - J_0(k_r r)\big] \exp(ik_z z) \\ E_y &= -\frac{1}{2} |t(k_r)| k_z \big[\exp(i2\varphi) J_2(k_r r) + J_0(k_r r)\big] \exp(ik_z z) \\ E_z &= -|t(k_r)| k_r \exp(i\varphi) J_1(k_r r) \exp(ik_z z) \end{aligned} \tag{7.4.1}$$

where r, φ, and z are cylindrical coordinates, k_r and k_z are the radial and longitudinal wavevectors with $k_r^2 + k_z^2 = k_0^2$, J_m is the mth-order Bessel function of the first kind, and $t(k_r)$ is the transmission coefficient of diffraction wave.

The vectorial electric field components have different contributions to the evanescent Bessel beam. The radial component intensity $|E_r|^2$, defined as $|E_r|^2 = |E_x|^2 + |E_y|^2$, presents a spot satisfying the zero-order Bessel function, while the longitudinal component intensity $|E_z|^2$ gives negative contribution with a donut distribution.

The recording patterns of Pr layer with the optimum exposure dose at different working distances in lithography processing are shown in Fig. 7.32a. Furthermore, as an example, the effect of exposure dose on the focusing spot sizes in the lithography experiment with the fixed distance of 40 nm. The experimental results with the working distance of 0 nm, shown in Fig. 7.32b, exhibits the spot size of 67 nm on the Pr layer. When the distance is elongated to 40 nm, the size of spot recorded on Pr layer reaches 65 nm, as depicted in Fig. 7.32c. Figure 7.32d indicates that even the distance arises to 80 nm, the spot could maintain the size of about 70 nm.

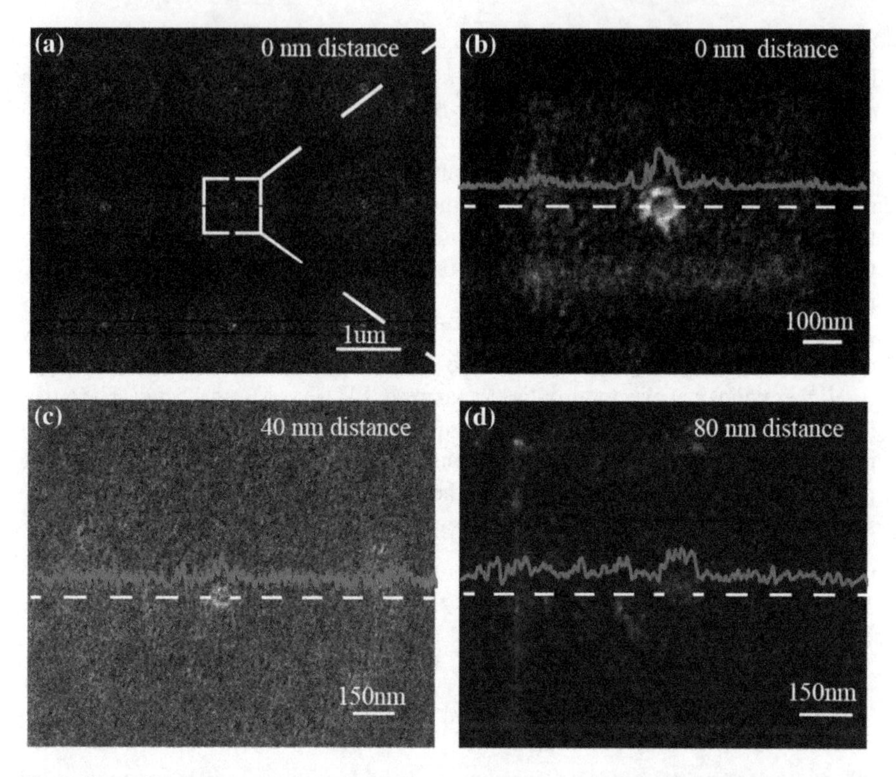

Fig. 7.32 **a** Lithography result by Bessel-BPPs focusing structure combined with MIM waveguide plasmonic imaging lens at a distance of 0 nm. **b** Zoom-out SEM picture of **a**. **c** and **d** Lithography results obtained at a distance of 40 and 80 nm. Reprinted from [13] with permission. Copyright 2017, The Royal Society of Chemistry

7.4.2 LSPs Lithography with Nanoaperture

Unlike SPPs and BPPs where light field is mainly confined to two-dimensional space, LSPs exist in a point-like region with near zero dimensions. The typical LSPs mode is in a nanometallic sphere, which would excite free electrons oscillation inside the sphere when illuminated. The strong resonance of light yields a great field enhancement and could be directly used for nanolithography.

Obtaining a light spot with subwavelength size could be done by transmitting it through a small hole. For example, a metallic hole on the end of a fiber is usually employed in near-field scanning microscopes and used for scanning nanolithography as well. One of the technical challenges of this method is low transmission efficiency, even in the case of employing metal coating. Another often used light localizing structure is a sharp metallic tip with ultra-small curvature, which behaves like a lightning rod and facilitates concentrating light energy around the tip apex. In this case, the light-enhancement factor could be large enough in comparison with the

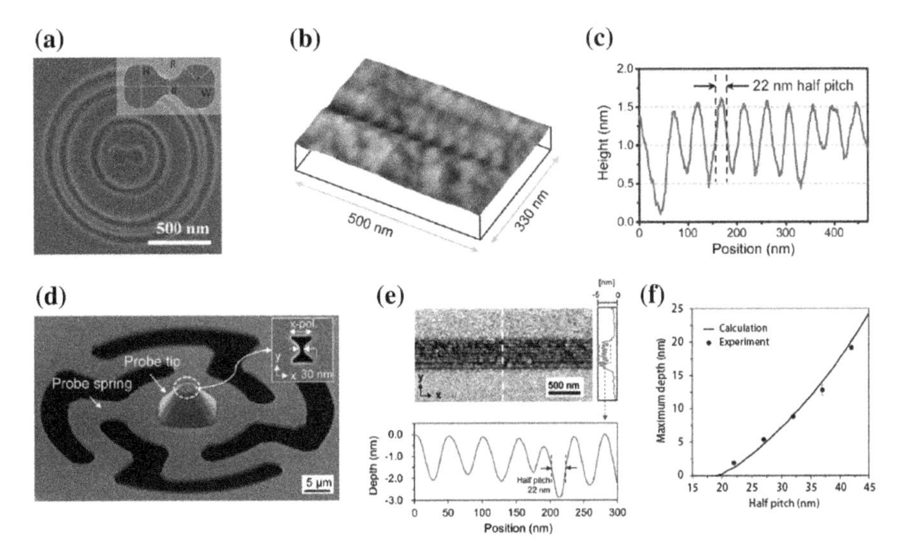

Fig. 7.33 **a** SEM picture of a multistage plasmonic lens. **b** 3D topography of the lithography patterns. **c** Cross-sectional profile of **b**. **d** SEM picture of circular contact probe with bowtie aperture for plasmonic lithography. **e** AFM image of resist pattern with half-pitch 22 nm. **f** Maximum depth of line array pattern as a function of half-pitch. Filled circles are experimental data and a solid line is obtained with the theoretical model. The error bar is a typical standard deviation for the experimental data points. **a–c** Reprinted from [67] with permission. Copyright 2011, The Authors. **d–f** Reprinted from [68] with permission. Copyright 2012, WILEY-VCH Verlag GmbH & Co. KGaA, Weinheim

light illumination. LSP mode in some specially designed structures, like a bowtie, helps to greatly enhance output light intensity and generates a nano hot spot. In addition, by introducing a number of slits or grooves surrounding the metallic hole or bowtie, excited plasmonic light energy would converge to the central part and is superimposed constructively to greatly enhance the central focus intensity [65].

Some point-to-point scanning nanolithography methods are performed by the use of LSPs, including bowtie, metallic small hole, and bull's eye structures, etc. The physical origin of nano-optical spots for those methods is light resonance manipulation and confinement in nanospaces with specially designed LSPs or SPPs resonance nanostructures. Compared with the nanoholes on a fiber end, one important improvement is the high efficiency of LSPs excitation. This is due to the partly opening geometrical feature of bowties and/or light concentrating into nanoapertures with controlled surface plasmons on the boundary. Recent experimental reports show nanolithography with critical dimensions of about 30–80 nm [66] and even 22 nm [67, 68] by combining SPPs, LSPs, and the threshold effect of material under femosecond laser exposure (Fig. 7.33).

Further, some improvements have been proposed for bowtie lithography with considerable confinement of focus and enhanced exposure depth. By combining a plasmonic cavity lens with bowtie, it is demonstrated that both resolution and focus depth could be significantly enhanced [69].

7.4.3 LSPs Lithography with Tips

Usually, two issues associated with LSPs aperture structure deliver some inconveniences in their applications [70]. Firstly, LSPs are mainly confined in the transversal direction, which make the emitted light from the apertures decays exponentially and becomes quickly emanative in the z-direction. This inevitably results in considerably decreased focal depth and shallow feature profiles in the photoresist. The second concern is related to the strong dependence of light spot size on the nanoaperture geometrical shape. For instance, the partly opening aperture of a bowtie usually yields a non-circular light spot and hot points around the aperture edges. Apertureless metallic tips can be employed for optical fabrication in the nanodimension as

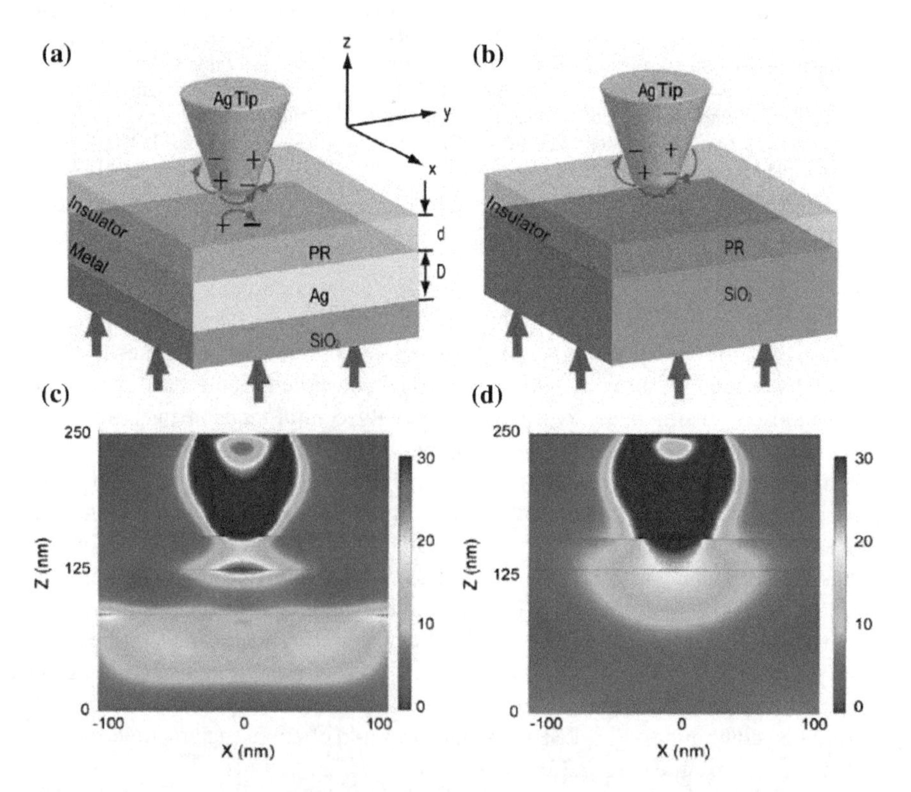

Fig. 7.34 Schematic of nanolithography with **a** tip-insulator-metal and **b** tip-insulator structure. Simulated electric field intensity distribution of **c** tip-insulator-metal and **d** tip-insulator structures. Reprinted from [71] with permission. Copyright 2013, Springer Science+Business Media New York

well by the help of hot spot effect, thermal and chemical reactions around the metal tips. The hot spot usually delivers considerable enhancement of light concentrated at the tip apex. The other useful advantage of apertureless tips is ease of fabrication. Unfortunately, spot enlargement and intensity decaying exist in this method as well.

Some modified structures for apertureless tip lithography have been proposed to solve the above problems, as shown in Fig. 7.34a and b [71]. The tip, photoresist and metallic layer form a tip-insulator-metal (TIM) structure, in which a highly confined mode of plasmonic light both in the transversal and longitude directions occurs, with normal light illumination beneath the substrate. In addition to the resolution improvement, the spot intensity inside the photoresist along the longitudinal direction becomes nearly uniformly distributed and results in an elongated depth of focus. Numerical simulations show that FWHM of the spot size could reach sub-10 nm by optimizing geometrical parameters of the TIM structure. Moreover, circularly polarized light illumination in the normal direction beneath the metal layer delivers a regular circular shape (Fig. 7.34c and d).

7.5 Engineering Aspects of Plasmonic Nanolithography

Since the amplified evanescent waves decay again after leaving the lens, the resolution of plasmonic nanolithography is limited by the working distance and the depth of exposure. The decay length depends on the Fourier components of the objects; the higher Fourier components decay faster, which results in deteriorated lithography quality. Thus far, some resolution enhancement techniques have been proposed to improve the aspect ratio and resolution of plasmonic nanolithography.

7.5.1 High Aspect Ratio Pattern Transfer Process

In practical applications, fabrication of nanodevice with plasmonic imaging lithography faces two major several challenges: the shallow depth of the lithographic patterns and the sloped sidewalls of the resist pattern. Therefore, it is important to develop a high aspect ratio pattern transfer process for depth amplification.

Plasmonic lithography can be utilized in combination with conventional hard-mask technology for the amplification of the depth of the lithographic pattern. Fig. 7.35a shows the schematic of the pattern transfer process flow [10]. In the first step of pattern transfer, ion beam etching (IBE) was utilized to transfer photoresist pattern into the SiO_2 layer. Then the residual Ag was removed by chrome remover. Finally, the SiO_2 pattern was transferred into bottom photoresist. Consequently, the patterns with 32-nm half-pitch were transferred into 80-nm-thick resist with angle of side >80° (Fig. 7.35b, c).

Figure 7.35d shows a similar process for plasmonic direct writing lithography, which uses 5-nm-thick sputtered SiO_2 layer as a hard mask. After plasmonic direct-

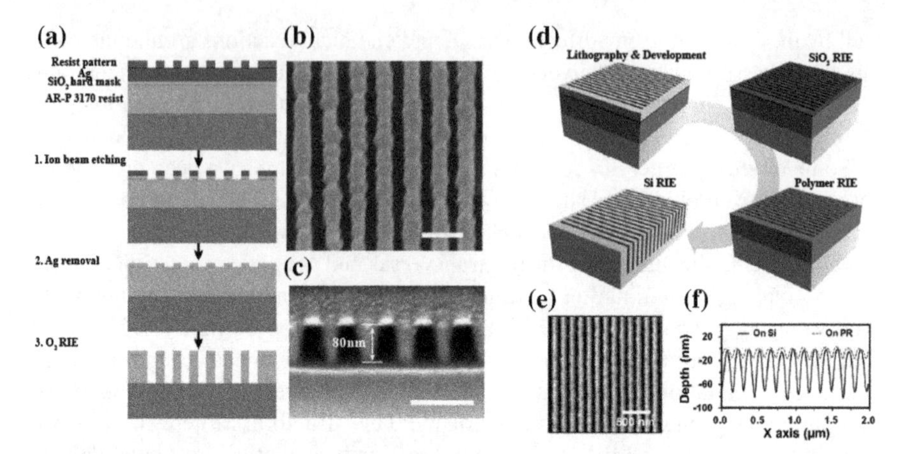

Fig. 7.35 **a** Schematic of the process of high aspect ratio pattern transfer for plasmonic imaging lithography. **b** Top view of the transferred 32-nm half-pitch pattern. **c** Cross-sectional view in **b**. Scale bar in **b** and **c**, 100 nm. **d** Schematic of the multilayer pattern transfer for plasmonic direct writing lithography. **e** AFM image of transferred silicon pattern with 70-nm half-pitch. **f** Pattern profile of 70-nm half-pitch before and after pattern transfer. **a–c** Reprinted from [10] with permission. Copyright 2015, AIP Publishing LLC. **d–f** Reprinted from [72] with permission. Copyright 2015, IOP Publishing Ltd.

writing lithography and development, the depth of an exposed pattern is shallow, less than 20 nm with 70 nm half-pitch [72]. Subsequently, the pattern is transferred onto silicon substrate with a depth of 75 nm, as shown in Fig. 7.35e and f.

7.5.2 Multiple-Patterning Technique

There are two major challenges in plasmonic imaging lithography for patterning with half-pitch beyond 22 nm at 365 nm: First, the excitation efficiency is too low for an evanescent wave with large wave number ($\geq 8.3k_0$); second, the intensity exponentially decays in the medium. The development of self-aligned multiple-patterning gives an opportunity to improve the resolution and quality of plasmonic imaging lithography.

A process flow of the plasmonic cavity lens lithography based multiplication process is given in Fig. 7.36a [73]. One of the key steps in this procedure is to pattern the core to form the spacer. Thus the line width of the core pattern needs to be equal to the targeted critical dimension. A trimming process is also required after transferring the high aspect ratio pattern. In the subsequent conformal deposition of a SiO_2 layer, the deposition temperature should be lower than the glass transition temperature of the resist-core. Finally, the top of spacer layer and resist-core are etched by plasma etching. Nanograting with 16-nm half-pitch (Fig. 7.36b) could be obtained by this technique. Similarly, nanoring with 9 nm feature size was achieved.

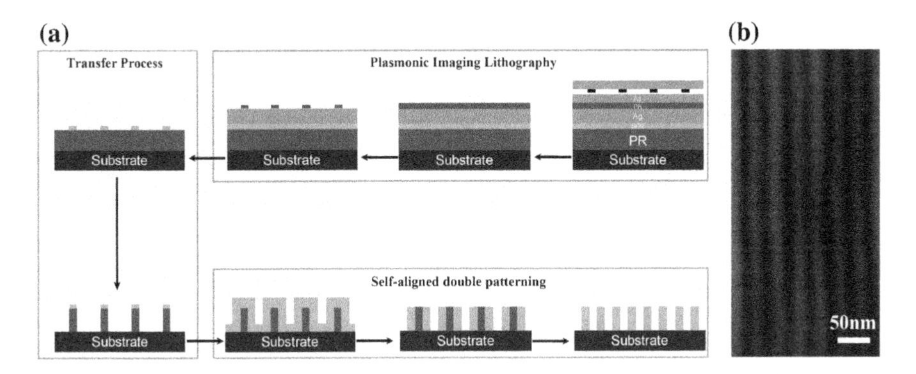

Fig. 7.36 **a** Process flow of the combination of plasmonic imaging lithography and self-aligned double patterning. **b** 16-nm half-pitch nanograting obtained by multiple-patterning technique. Reprinted from [73] with permission. Copyright 2017, Elsevier B.V.

7.5.3 Equipment and Systems

7.5.3.1 Plasmonic Imaging Lithography Systems

As mentioned above, plasmonic imaging lithography could be an alternative nanolithography technique. It is thus necessary to develop lithography equipment or instrument for practical applications.

In 2008, researchers from Center for Scalable and Integrated Nano Manufacturing (SINAM) have designed an ultra-precise positioning system with an interchangeable module, named Multi-Scale Alignment Positioning System (MAPS), to meet the requirements of SINAM's investigations (Fig. 7.37a) [74]. The module was first designed to perform nanoimprint lithography (NIL), and modules for plasmonic imaging lithography (PIL), Field-Assisted Parallel Nanoassembly (FAPNA), and AFM would be added later according to their plan. Figure 7.37b shows that the system has been shipped and now resides at the University of California Los Angeles [75]. Unfortunately, from the current public reports, they only integrated the NIL module and only got the imprint results. Moreover, the imprinted feature size was not at the nanoscale (Fig. 7.37c).

A prototype system for plasmonic imaging lithography has been reported by Liu et al. The system consists of illumination, mask holder, leveling unit, motion stages, substrate stage, power supplier, controller, vacuum, and air pressure (Fig. 7.38a–d) [76]. Furthermore, the authors introduced a step exposure method to fabricate a 5 $\times$ 5 array of grating patterns with a step length of 300 μm and the uniform patterns cover the whole area of about 2 $\times$ 2 mm^2. In the whole process of step exposure, line width roughness (LWR) and line edge roughness (LER) of the grating patterns reasonably remain constant, and the results show that little distortion exists from the first point to the last point (Fig. 7.38e, f).

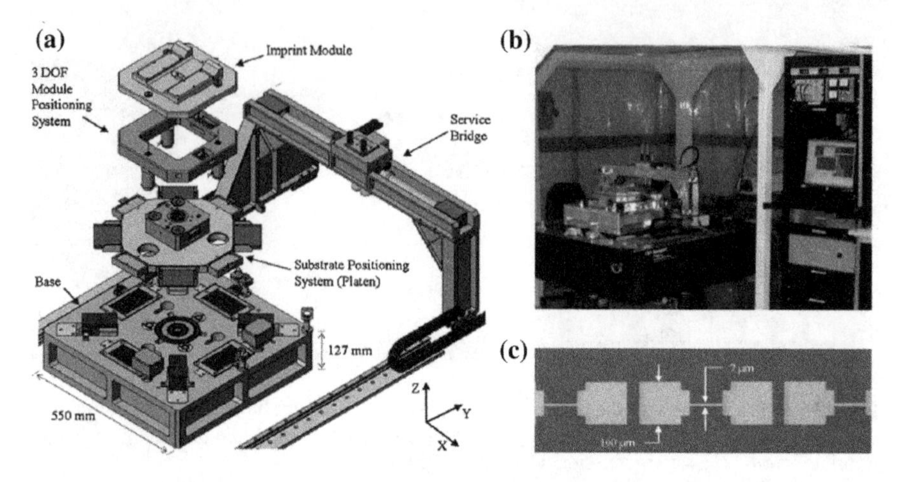

Fig. 7.37 **a** Schematic of MAPS with interchangeable module. **b** Image of MAPS at UCLA. **c** Dog bone imprint result. Reprinted from [75] with permission. Copyright 2012, Elsevier Inc.

Subsequently, a high-precision system that integrates the nanometer resolution gap measurement and control unit, ultra-high precision stage, and nanoscale alignment unit has been set up (Fig. 7.38g), which allows to fabricate nanostructures on wafers from 4 to 6 in. [73]. The gap measurement accuracy is better than ±4 nm. In order to increase the travel range and improve the accuracy of the stage, air bearing has been used to provide smooth and high speed motion. The travel range and maximum speed can reach 180 mm × 180 mm and 500 mm/s, respectively. The repeatability of each direction is better than 6 nm by using double-pass heterodyne laser interferometers for the displacement measurement. A Moiré fringe-based alignment technique is also demonstrated to achieve an accuracy of sub-100 nm for overlay.

7.5.3.2 Plasmonic Direct-Writing Lithography Systems

Plasmonic direct-writing lithography is a flexible fabrication tool, which can form nanostructures with arbitrary shape. Thus, designing and manufacturing different kinds of systems are necessary to push the development of new nanomanufacturing technologies.

Figure 7.39a shows the schematic of a high-speed flying plasmonic lithography system [66]. A pulse laser, controlled by a high-speed optical modulator according to the signals from a pattern generator, is utilized for illumination. The writing position is referred to the angular position of the disk from the spindle encoder and the position of a nanostage along the radial direction. To achieve high-speed scanning while maintain the nanoscale gap, a novel air-bearing slider is utilized to fly the plasmonic lens arrays at a height of 20 nm above the substrate at speeds of between 4 and 12 m/s (Fig. 7.39b). The rotation of the substrate creates an air flow along the bottom surface of the plasmonic flying head, known as the air-bearing surface

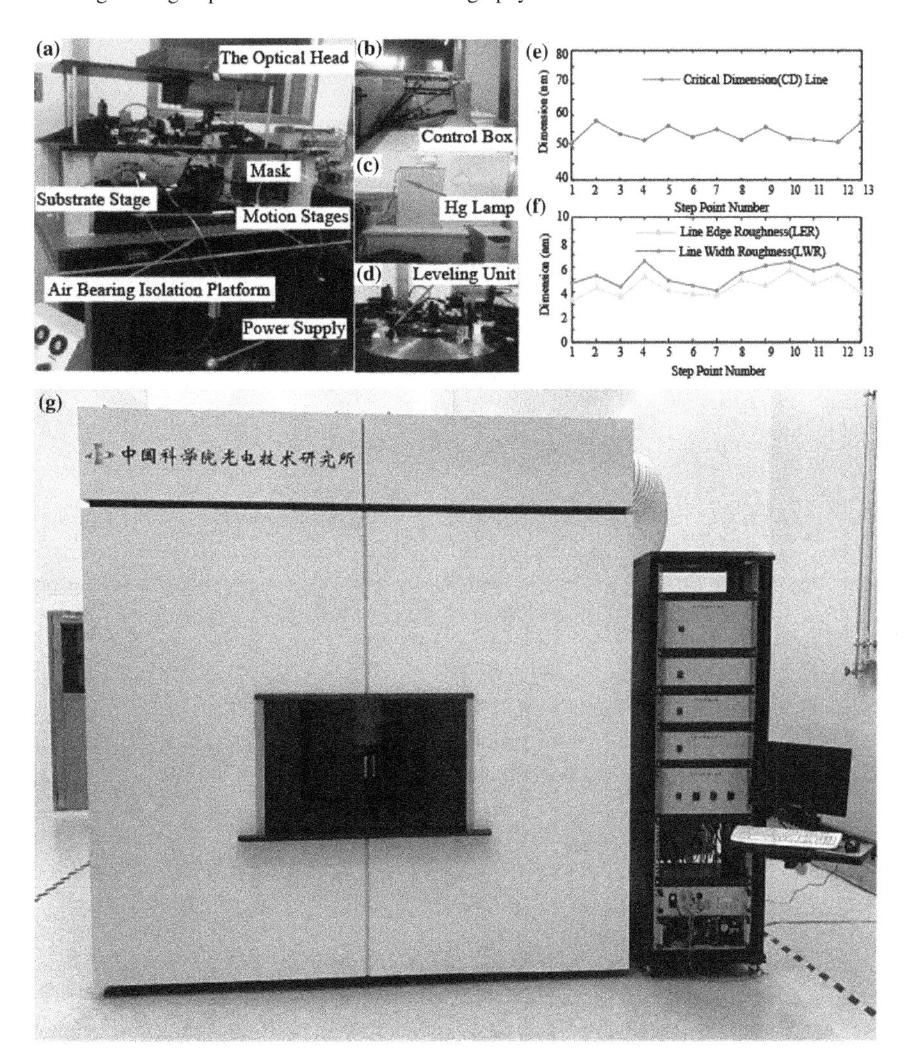

Fig. 7.38 **a** Prototype system of plasmonic imaging lithography. **b** The control box. **c** The light source. **d** The leveling component. **e** The variation of CD of grating patterns at different step point numbers. **f** The variation of LWR (3δ) and LER (3δ) at different step point numbers. **g** A new high-precision system for plasmonic imaging lithography. Figures are reproduced from **a**–**f** and **g**. **a**–**f** Reprinted from [76] with permission. Copyright 2016, American Vacuum Society. **g** Reprinted from [73] with permission. Copyright 2017, Elsevier B.V.

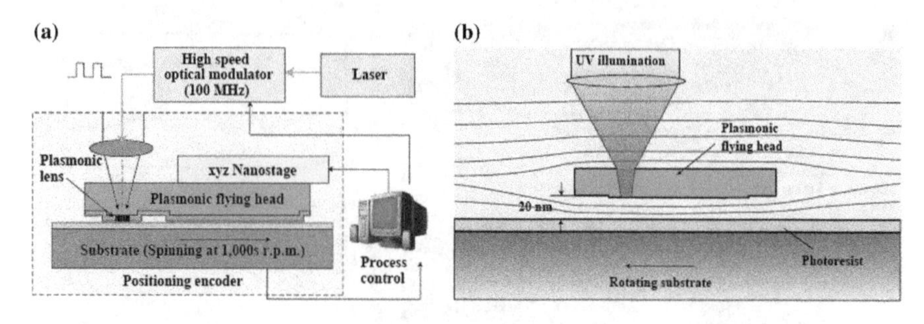

Fig. 7.39 **a** Process control system of flying plasmonic maskless nanolithography. **b** Cross-sectional schematic of the plasmonic head flying 20 nm above the rotating substrate which is covered with photoresist. Reprinted from [66] with permission. Copyright 2008, Macmillan Publishers Limited

(ABS), which generates an aerodynamic lift force to balance with the force supplied by the suspension arm for precisely regulating a nanoscale gap between the plasmonic lens arrays and the rotating substrate covered with photoresist. With the high bearing stiffness and small actuation mass, this self-adaptive method can provide an effective bandwidth up to 120 kHz. The use of an ABS eliminates the need for a feedback control loop and therefore overcomes the major technical barrier for high-speed scanning.

Figure 7.40a shows the contact-probe-based laser direct writing system [77]. The expose source is a continuous-wave laser with 405 nm wavelength, which is focused onto the contact probes by an objective lens (NA = 0.8). In order to form the minimal spot, a ridge aperture is used to localize the focused laser light, and controlled to maintain within a few nanometers. A well designed circular contact probe is applied to improve the positioning accuracy and throughput. Compared with a cantilever-type probe, circular contact probe can decrease the positioning errors in magnitude (Fig. 7.40b and c) [68]. For practical use of contact-probe-based laser direct writing system, alignment of recorded patterns for the overlay process can be a critical issue for lithography processes and outcome quality. To solve this problem, a new scheme based on near-field scanning optical microscope is employed, as shown in Fig. 7.40d [78]. A ridge aperture has high transmission originating from Fabry-Perot resonance at λ_L for plasmonic lithography and high sensitivity due to plasmonic resonance at λ_A for detecting a mark pattern on a substrate. A large shift of the plasmonic resonance for variation of the refractive index of the material underneath the ridge aperture is plotted in the lower area of the figure, leading to a large signal for detecting the mark pattern. The mark pattern is measured at wavelength λ_A, which is not sensitive to exposure of the photoresist. Figure 7.40e shows the image obtained in the experiment and the process defining the center. The overlay alignment is shown to be less than 2 nm.

Another scheme of the contact-probe-based laser direct writing system with ISPI gap detection is shown in Fig. 7.41a [79]. The system consists of two subsystems: the lithography system and the interferometric-spatial-phase-imaging (ISPI) system.

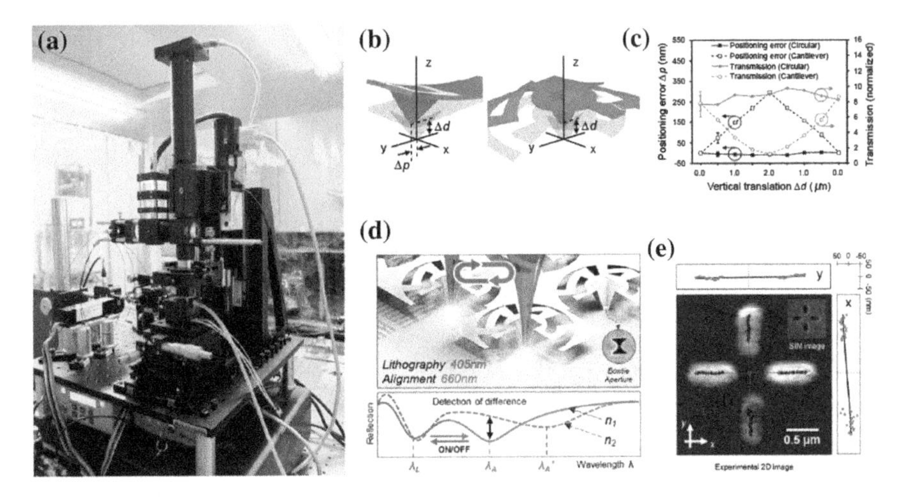

Fig. 7.40 **a** Pictorial view of the contact-probe-based laser direct writing system. **b** Schematic view of the positioning error induced by the vertical translation of cantilever- and circular-type probes. **c** The experimental results of the positioning error Δp and transmission with varying Δd. The error bar shows a typical standard deviation of the data points. **d** Concept of a plasmonic multifunctional probe. **e** Experimental 2D image and the corresponding reference point of the mark obtained by the algorithm, accompanied by the line profiles and curve fittings of each bar. **a** Reprinted from [77] with permission. Copyright 2012, SPIE. **b** and **c** Reprinted from [68] with permission. Copyright 2012, WILEY-VCH Verlag GmbH & Co. KGaA, Weinheim. **d** and **e** Reprinted from [78] with permission. Copyright 2013, Optical Society of America

Figure 7.41b shows the photograph of the system setup. As shown in Fig. 7.41c, the ISPI grating pattern consists of a pair of two-dimensional checkerboards, whose *y*-periodicity is uniform and the *x*-periodicity is chirped oppositely. The *y*-periodicity is designed for oblique angle to avoid interfering with the lithography processing beam. The *x*-periodicity creates interference fringes that are sensitive to the mask-substrate gap. Since the *x*-periodicity is chirped, the transmitted-diffracted angle of the incident beam will vary along the *x*-direction as shown in Fig. 7.41d. All the re-diffracted beams will then interfere and display a set of interference fringes at the imaging plane of the ISPI microscope. The sensitive of this gapping technique can reach 0.15 nm. With the use of ISPI and a dynamic feedback control system, it can precisely align the mask and the substrate and keep variation of the gap distance below 6 nm to realize parallel nanolithography.

7.5.3.3 Flexible Manufacturing

Plasmonic imaging lithography is applicable not only for rigid substrates but also for flexible substrates, e.g., membranes. Figure 7.42a shows the procedure of plasmonic imaging lithography on polyimide (PI) membranes [73]. The membrane was first assembled in a frame, and then the surface of membrane was planarized with

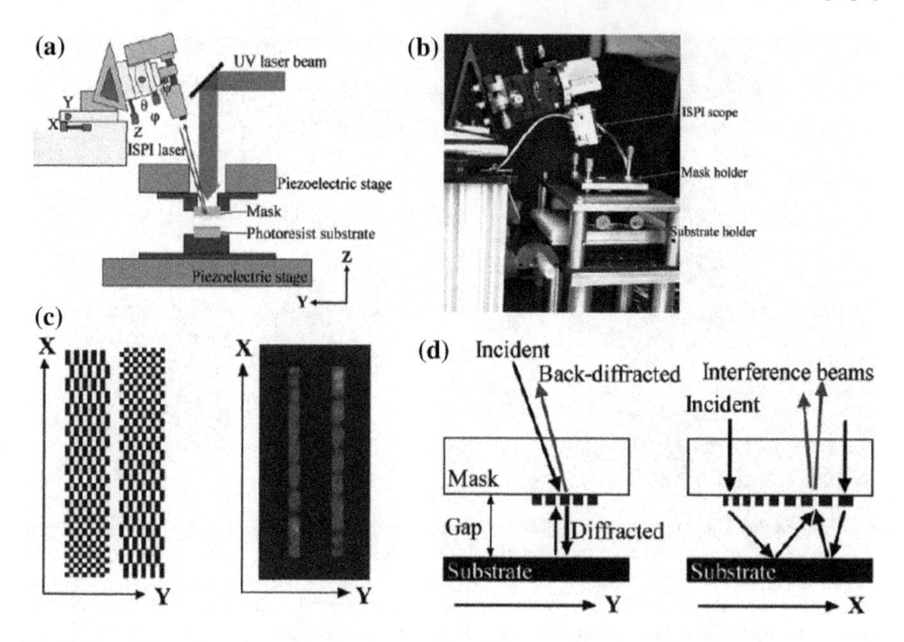

Fig. 7.41 **a** Illustration of the experimental setup. **b** Photograph of the experimental setup. **c** Sketch of ISPI checkerboard gratings and typical fringe image captured by the ISPI camera. **d** Illustration of the working mechanism of ISPI. Reprinted from [79] with permission. Copyright 2013, American Vacuum Society

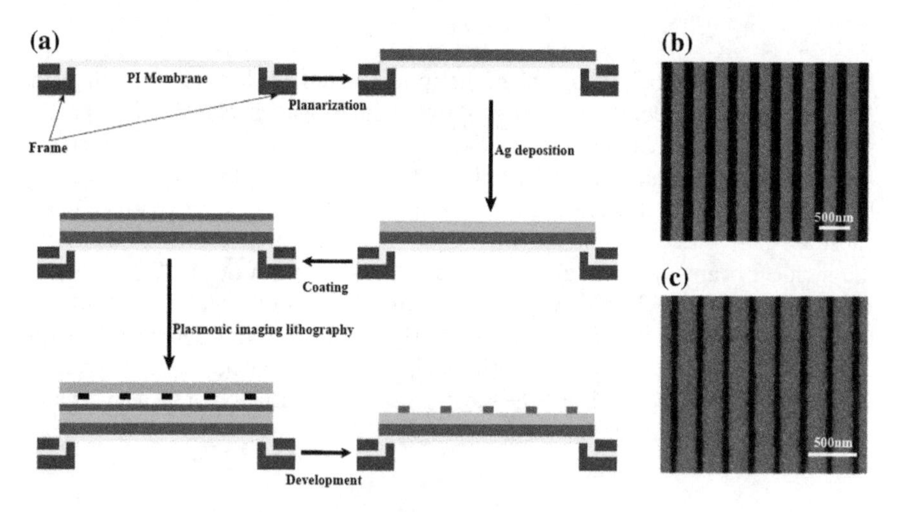

Fig. 7.42 **a** Procedure of plasmonic imaging lithography on PI membranes. **b** and **c** SEM images of the recorded resist patterns on a PI membrane by plasmonic reflective lens. The periods of the grating in **b** and **c** are 350 and 280 nm, respectively. Reprinted from [73] with permission. Copyright 2017, Elsevier B.V.

photoresist. In order to get a smooth reflective lens, a silver layer with a thickness of 50 nm was deposited on it, which serves as a plasmonic mirror. After coating, exposing and developing, the patterns were recorded in the resist. Figure 7.42b and c show the SEM images of the recorded resist patterns on a membrane at different local regions. The periods of the grating in Fig. 7.42b and c are 350 and 280 nm, respectively. The critical dimension of the grating shown in Fig. 7.42c is about 80 nm. Clearly, the reflective plasmonic lithography offers a new way to fabricate lightweight and conformable nanodevices, which can be applied in wearable devices, artificial intelligence, flexible displays, light sources, and large space optical telescopes, etc.

7.6 Applications of Plasmonic Nanolithography

Not only the feasibility of plasmonic lithography as an alternative nanofabrication tool has been widely proven with high-resolution patterns on the resist, but also the practical applications have been demonstrated in recent years.

7.6.1 High Sensitivity Detectors

Superconducting nanowire single-photon detectors (SNSPDs) are made up of a long and submicron-wide meander-type stripe of ultrathin superconducting film. In comparison with semiconducting single-photon detectors such as avalanche photodiodes (APDs) and photomultiplier tubes (PMTs), SNSPDs have high count rates, low dark counts and low timing jitter. These properties make the SNSPDs suitable for many applications. It was reported that SNSPDs have been used as single-photon detectors in quantum key distribution and imagers of defects in very large-scale integrated circuits [80, 81]. The standard nanofabrication process for thin superconducting films is based on EBL, because it requires highly uniform and smooth nanowire. But EBL is low efficiency through a point-by-point writing manner. Such a SNSPD can be fabricated using plasmonic lithography (Fig. 7.43) [11, 25]. The pattern quality is comparable to that fabricated using standard EBL.

7.6.2 Cancer Diagnosis

The early detection and monitoring of infectious diseases and cancer through affordable and accessible healthcare will significantly reduce the disease burden. Progress in nanotechnology has enjoyed exponential growth in the past couple of decades. We have seen the design and synthesis of metal nanoparticles (NPs) tailored specifically for biomedical diagnosis [82]. In particular, noble metals have attracted lots of attention. Because of their unique optical and electronic properties, Au and Ag NPs have

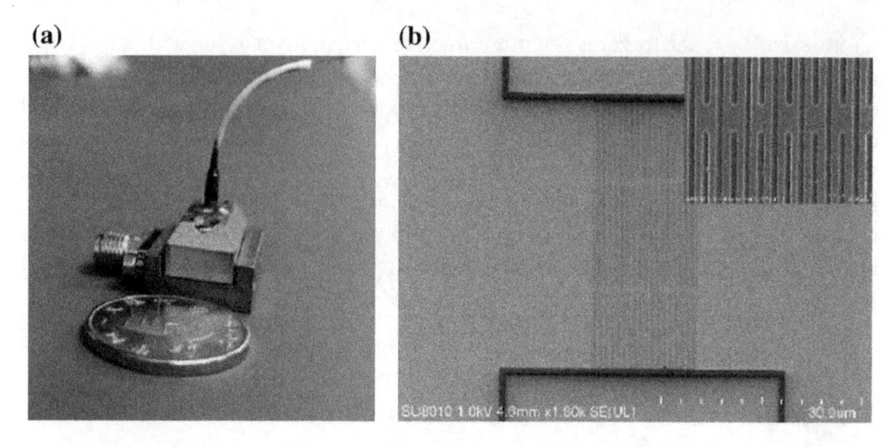

Fig. 7.43 **a** Photography of the fabricated device for superconducting nanowire single-photon detectors. **b** SEM images of the fabricated SNSPD nanowire structure. **a** Reprinted from [25] with permission. Copyright 2018, WILEY-VCH Verlag GmbH & Co. KGaA, Weinheim

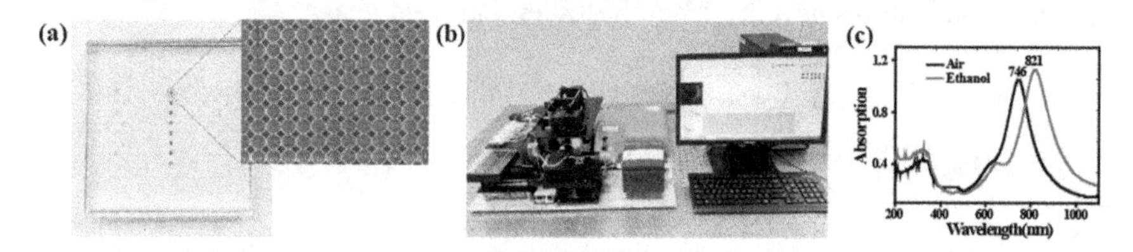

Fig. 7.44 **a** Photograph of the 8 individual channels with periodic rhombic NPs (insert SEM picture). **b** Photograph of home-build multi-channel biosensor system. **c** LSPR spectra of periodic rhombic NPs in air and ethanol

been exploited in the fabrication of LSPR chips for detection of biomolecules. They impart increased sensitivity and also allow the development of analytical platforms for label-free detection. These metal NPs show specific changes in their absorbance responses in the visible region of the spectrum upon binding with various molecules such as nucleic acids or proteins. In addition, the electronic properties, in particular, of Au and Ag NPs have been employed as labels for detection of proteins and other target molecules [82].

Considering the capability of plasmonic lithography and application, a periodic rhombic metal NPs with two more hot spots than the traditional Ag triangular NPs is introduced for the purpose of achieving high sensitivity and selectivity (Fig. 7.44a). The measured peak wavelength shift is about 207 nm/RIU (Fig. 7.44c), which is increased by 1.38 times over the whole detection area in comparison with that of the traditional triangular NPs fabricated by nanosphere lithography (NSL) in the same experimental conditions. The corresponding refractive index sensitivity of the rhombic Ag nanosensor is greatly improved.

7.6.3 *Color Generation Applications*

Colors and decorations are important for the perception and identification of both natural and artificial objects. Colors arise from the interaction of light with structure materials. Plasmonics is known to offer a platform for transformative applications in optical frequency. The engineering of plasmonic colors is a promising, rapidly emerging research field that could have a large technological impact. Plasmonic lithography can also be used to plasmonic structural color. Figure 7.45a displays a fabricated sample composed of metallic grating for structural color display, which is the symbol of the Institute of Optics and Electronics (IOE) [25]. Figure 7.45b shows the colorful images of this structural color.

7.6.4 *Metasurfaces Fabrication*

Metasurfaces have been demonstrated to be able to achieve full control of the amplitudes, phases, and polarization states of electromagnetic waves. Since such metasurfaces are much thinner and consequently easier to fabricate than metamaterials, it is believed that the first generation of practical meta-devices will utilize this scheme [83]. So far, most metasurfaces are fabricated in a direct writing manner with EBL or FIB, which imposes a serious cost barrier with respect to practical applications. Near-field optical lithography seemingly provides a high-resolution and low-cost way. However, the depth of the obtained resist patterns was ultra-shallow due to the exponential decay of evanescent waves. It is shown that by utilizing plasmonic reflective lens lithography, metasurfaces can be produced with high fidelity and contrast (Fig. 7.46a) [84]. The Ag metalens with 50-nm thickness over a thin silica layer was successfully fabricated as illustrated in Fig. 7.46b. The size of the nanoslot is

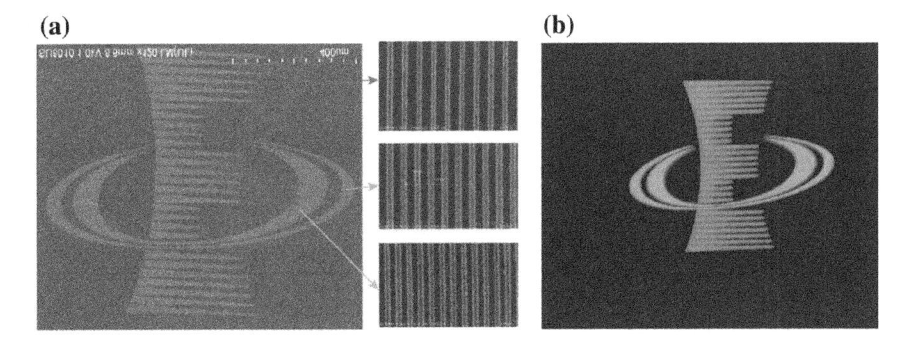

Fig. 7.45 **a** SEM and optical images of a large area sample for structural color. The periods of three gratings are 420, 350, and 280 nm. **b** Optical images of the colorful symbol of the Institute of Optics and Electronics (IOE). Reprinted from [25] with permission. Copyright 2018, WILEY-VCH Verlag GmbH & Co. KGaA, Weinheim

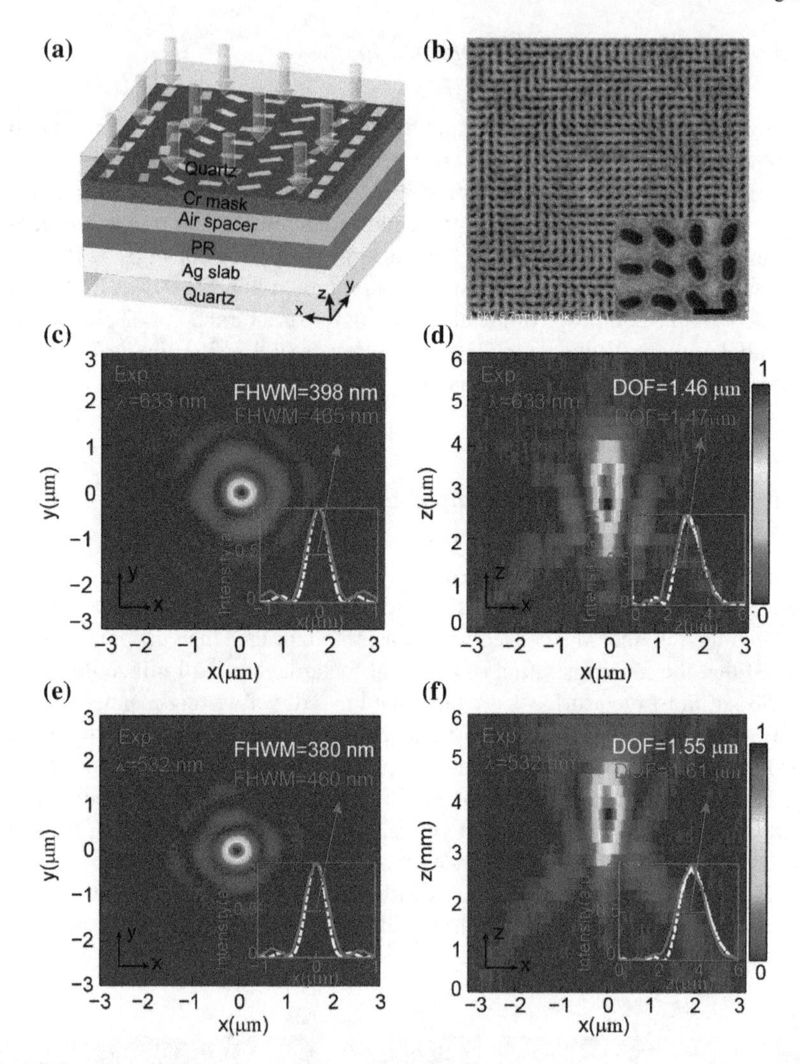

Fig. 7.46 **a** Schematic of plasmonic reflective lens lithography structure for nanoslots metasurface fabrication. **b** SEM image of fabricated metalens. **c** and **d** Experimental intensity distributions in the *xy*- and *xz*-plane for the circularly polarized light with 633-nm wavelength, respectively. **e** and **f** Experimental intensity distributions in the *xy*- and *xz*-plane for the circularly polarized light with 532-nm wavelength, respectively. Reprinted from [84] with permission. Copyright 2015, The Royal Society of Chemistry

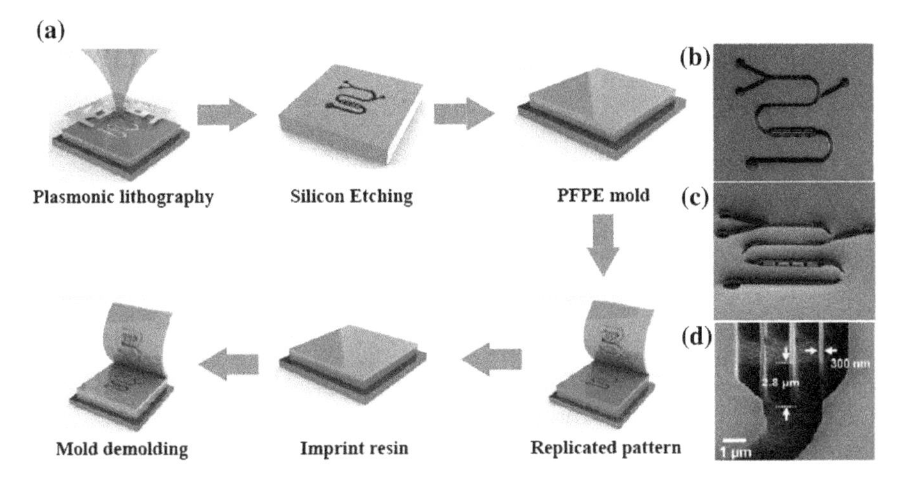

Fig. 7.47 **a** Fabrication process of nanoimprint master, mold, and replicated pattern. **b–d** Fabrication results of replicated pattern from the mold. Reprinted from [72] with permission. Copyright 2015, IOP Publishing Ltd.

about 75 nm × 150 nm. Figure 7.46c–f show the measured focusing efficiency of metalens fabricated by plasmonic reflective lens lithography (~11% for 633 nm and 5% for 532 nm), which is a bit smaller than the designed values (~15% for 633 nm and 9.7% for 532 nm), due to the fabrication errors such as width increase and shape degradation, etc. It is believed that this method would find potentially promising applications in the fabrication of a variety of metasurfaces employed in flat lenses, meta-holograms, and vortex phase plates, etc.

7.6.5 Nanoimprint Masters Fabrication

Nanoimprint lithography is a powerful technique for the manufacture of nanoscale devices. As feature size in devices decreases, high-resolution patterning is required to fabricate imprint masters. Although, EBL provides high-resolution patterning down to tens of nanometers, the lithography tool is expensive and requires a high vacuum system that adds complexity to use and maintenance. Compared with EBL, plasmonic direct-writing lithography is an alternative technology for practical manufacturing with simple system and low cost. Combined with the effects of localized surface plasmons, the resolution of plasmonic direct-writing lithography reaches below 22 nm half-pitch. Furthermore, it can pattern arbitrary shape nanostructures on a resist. Figure 7.47a gives a schematic of nanoimprint master, mold, and replicated pattern fabrication [72]. The arbitrary fluidic channel pattern is first recorded on a resist by plasmonic direct-writing lithography, and then the resist pattern is transferred onto the silicon substrate through deep reactive ion etching (DRIE). After that, use perfluoropolyether (PFPE) mold resin to fill the silicon pattern, a mold is formed by demolding. Subsequently, the replicated pattern can be fabricated from the mold by using PUA replication resin. Figure 7.47b–d show the replicated results of the fluidic channel with an aspect ratio of 7.2. So, plasmonic direct-writing lithography can be applied to the fabrication of imprint masters and photo masks.

7.7 Outlook

In summary, by utilizing the sub-diffraction-limited feature of SPPs, a variety of plasmonic structures including the illumination module, mask, and lens have been designed to improve the resolution and fidelity of plasmonic lithography. The latest experiment of plasmonic lithography demonstrated that patterns with 22-nm half-pitch can be achieved. By introducing the self-aligned multiple-patterning technique to plasmonic imaging lithography, nanograting with 16-nm half-pitch and nanoring with 9-nm feature size have been achieved. Based on these achievements, plasmonic lithography has demonstrated its ability to be a promising tool for nanofabrication. There are also some issues that need to be further addressed:

1. Although the resolution of plasmonic lithography has been extended to be a half-pitch of 22 nm, the exposure depth and quality need to be further improved.
2. Resists of plasmonic lithography must show considerable improvement in resolution, LER, and sensitivity.
3. The interest and use of plasmonic lithography in nanofabrication are expanding, but more applications still need to be fulfilled.
4. Uniformity for large-area patterns fabrication should be improved.

Nevertheless, we believe that plasmonic lithography offers an effective way to break the diffraction limit of the conventional optical lithography and gives potentially promising access to high resolution, efficient, large area, and low-cost nanolithography in EO 2.0.

References

1. M. Totzeck, W. Ulrich, A. Göhnermeier, W. Kaiser, Pushing deep ultraviolet lithography to its limits. Nat. Photonics **1**, 629 (2007)
2. C. Wagner, N. Harned, Lithography gets extreme. Nat. Photonics **4**, 24 (2010)
3. S. Kawata, H.-B. Sun, T. Tanaka, K. Takada, Finer features for functional microdevices. Nature **412**, 697 (2001)
4. Z. Gan, Y. Cao, R.A. Evans, M. Gu, Three-dimensional deep sub-diffraction optical beam lithography with 9 nm feature size. Nat. Commun. **4**, 2061 (2013)
5. T.W. Ebbesen, H.J. Lezec, H.F. Ghaemi, T. Thio, P.A. Wolff, Extraordinary optical transmission through sub-wavelength hole arrays. Nature **391**, 667 (1998)
6. H.J. Lezec, A. Degiron, E. Devaux, R.A. Linke, L. Martin-Moreno, F.J. Garcia-Vidal, T.W. Ebbesen, Beaming light from a subwavelength aperture. Science **297**, 820–822 (2002)

7. M. Pu, Y. Guo, X. Li, X. Ma, X. Luo, Revisitation of extraordinary Young's interference: from catenary optical fields to spin-orbit interaction in metasurfaces. ACS Photonics **5**, 3198–3204 (2018)
8. X. Luo, D. Tsai, M. Gu, M. Hong, Subwavelength interference of light on structured surfaces. Adv. Opt. Photonics **10**, 757–842 (2018)
9. X. Luo, T. Ishihara, Surface plasmon resonant interference nanolithography technique. Appl. Phys. Lett. **84**, 4780–4782 (2004)
10. P. Gao, N. Yao, C. Wang, Z. Zhao, Y. Luo, Y. Wang, G. Gao, K. Liu, C. Zhao, X. Luo, Enhancing aspect profile of half-pitch 32 nm and 22 nm lithography with plasmonic cavity lens. Appl. Phys. Lett. **106**, 093110 (2015)
11. X. Luo, Plasmonic metalens for nanofabrication. Natl. Sci. Rev. **5**, 137–138 (2018)
12. X. Luo, *Catenary Optics* (Springer, Singapore, 2019)
13. L. Liu, P. Gao, K. Liu, W. Kong, Z. Zhao, M. Pu, C. Wang, X. Luo, Nanofocusing of circularly polarized Bessel-type plasmon polaritons with hyperbolic metamaterials. Mater. Horiz. **4**, 290–296 (2017)
14. X. Chen, X. Luo, H. Tian, J. Shi, Contact or proximity nanolithography system using normal or long wavelength light, Chinese Patent Office Patent ZL03123574.3, 29 May 2003
15. X. Luo, D. Tsai, M. Gu, M. Hong, Extraordinary optical fields in nanostructures: from sub-diffraction-limited optics to sensing and energy conversion. Chem. Soc. Rev. (2019)
16. C. Wang, P. Gao, Z. Zhao, N. Yao, Y. Wang, L. Liu, K. Liu, X. Luo, Deep sub-wavelength imaging lithography by a reflective plasmonic slab. Opt. Express **21**, 20683–20691 (2013)
17. X. Luo, T. Ishihara, Subwavelength photolithography based on surface-plasmon polariton resonance. Opt. Express **12**, 3055–3065 (2004)
18. Z.-W. Liu, Q.-H. Wei, X. Zhang, Surface plasmon interference nanolithography. Nano Lett. **5**, 957–961 (2005)
19. H. Shi, X. Luo, C. Du, Young's interference of double metallic nanoslit with different widths. Opt. Express **15**, 11321–11327 (2007)
20. T. Xu, L. Fang, B. Zeng, Y. Liu, C. Wang, Q. Feng, X. Luo, Subwavelength nanolithography based on unidirectional excitation of surface plasmons. J. Opt. A Pure Appl. Opt. **11**, 085003 (2009)
21. Z. Liu, Y. Wang, J. Yao, H. Lee, W. Srituravanich, X. Zhang, Broad band two-dimensional manipulation of surface plasmons. Nano Lett. **9**, 462–466 (2009)
22. W. Ge, C. Wang, Y. Xue, B. Cao, B. Zhang, K. Xu, Tunable ultra-deep subwavelength photolithography using a surface plasmon resonant cavity. Opt. Express **19**, 6714–6723 (2011)
23. K.V. Sreekanth, V.M. Murukeshan, Large-area maskless surface plasmon interference for one- and two-dimensional periodic nanoscale feature patterning. J. Opt. Soc. Am. A **27**, 95–99 (2010)
24. K.V. Sreekanth, V.M. Murukeshan, Effect of metals on UV-excited plasmonic lithography for sub-50 nm periodic feature fabrication. Appl. Phys. A **101**, 117–120 (2010)
25. X. Luo, Subwavelength optical engineering with metasurface waves. Adv. Opt. Mater. **6**, 1701201 (2018)
26. T. Xu, L. Fang, J. Ma, B. Zeng, Y. Liu, J. Cui, C. Wang, Q. Feng, X. Luo, Localizing surface plasmons with a metal-cladding superlens for projecting deep-subwavelength patterns. Appl. Phys. B **97**, 175–179 (2009)
27. J. Dong, J. Liu, G. Kang, J. Xie, Y. Wang, Pushing the resolution of photolithography down to 15 nm by surface plasmon interference. Sci. Rep. **4**, 5618 (2014)
28. L. Liu, Y. Luo, Z. Zhao, W. Zhang, G. Gao, B. Zeng, C. Wang, X. Luo, Large area and deep sub-wavelength interference lithography employing odd surface plasmon modes. Sci. Rep. **6**, 30450 (2016)
29. X. Chen, F. Yang, C. Zhang, J. Zhou, L.J. Guo, Large-area high aspect ratio plasmonic interference lithography utilizing a single high-k mode. ACS Nano **10**, 4039–4045 (2016)
30. B. Wood, J.B. Pendry, D.P. Tsai, Directed subwavelength imaging using a layered metal-dielectric system. Phys. Rev. B **74**, 115116 (2006)

31. C. Wang, Y. Zhao, D. Gan, C. Du, X. Luo, Subwavelength imaging with anisotropic structure comprising alternately layered metal and dielectric films. Opt. Express **16**, 4217–4227 (2008)
32. G. Liang, C. Wang, Z. Zhao, Y. Wang, N. Yao, P. Gao, Y. Luo, G. Gao, Q. Zhao, X. Luo, Squeezing bulk plasmon polaritons through hyperbolic metamaterials for large area deep subwavelength interference lithography. Adv. Opt. Mater. **3**, 1248–1256 (2015)
33. Z. Guo, Z.Y. Zhao, L.S. Yan, P. Gao, C.T. Wang, N. Yao, K.P. Liu, B. Jiang, X.G. Luo, Moiré fringes characterization of surface plasmon transmission and filtering in multi metal-dielectric films. Appl. Phys. Lett. **105**, 141107 (2014)
34. H. Liu, Y. Luo, W. Kong, K. Liu, W. Du, C. Zhao, P. Gao, Z. Zhao, C. Wang, M. Pu, X. Luo, Large area deep subwavelength interference lithography with a 35 nm half-period based on bulk plasmon polaritons. Opt. Mater. Express **8**, 199–209 (2018)
35. Y. Li, F. Liu, L. Xiao, K. Cui, X. Feng, W. Zhang, Y. Huang, Two-surface-plasmon-polariton-absorption based nanolithography. Appl. Phys. Lett. **102**, 063113 (2013)
36. Y. Li, F. Liu, Y. Ye, W. Meng, K. Cui, X. Feng, W. Zhang, Y. Huang, Two-surface-plasmon-polariton-absorption based lithography using 400 nm femtosecond laser. Appl. Phys. Lett. **104**, 081115 (2014)
37. J.B. Pendry, Negative refraction makes a perfect lens. Phys. Rev. Lett. **85**, 3966–3969 (2000)
38. H. Lee, Y. Xiong, N. Fang, W. Srituravanich, S. Durant, M. Ambat, C. Sun, X. Zhang, Realization of optical superlens imaging below the diffraction limit. New J. Phys. **7**, 255 (2005)
39. P. Chaturvedi, N.X. Fang, Molecular scale imaging with a multilayer superlens. MRS Proc. **919**, 0919-J04-07 (2006)
40. D.O.S. Melville, R.J. Blaikie, Super-resolution imaging through a planar silver layer. Opt. Express **13**, 2127–2134 (2005)
41. N. Fang, H. Lee, C. Sun, X. Zhang, Diffraction-limited optical imaging with a silver superlens. Science **308**, 534 (2005)
42. P. Chaturvedi, W. Wu, V. Logeeswaran, Z. Yu, M.S. Islam, S.Y. Wang, R.S. Williams, N.X. Fang, A smooth optical superlens. Appl. Phys. Lett. **96**, 043102 (2010)
43. H. Liu, B. Wang, L. Ke, J. Deng, C.C. Chum, S.L. Teo, L. Shen, S.A. Maier, J. Teng, High aspect subdiffraction-limit photolithography via a silver superlens. Nano Lett. **12**, 1549–1554 (2012)
44. S. Huang, H. Wang, K.-H. Ding, L. Tsang, Subwavelength imaging enhancement through a three-dimensional plasmon superlens with rough surface. Opt. Lett. **37**, 1295–1297 (2012)
45. H. Wang, J.Q. Bagley, L. Tsang, S. Huang, K.-H. Ding, A. Ishimaru, Image enhancement for flat and rough film plasmon superlenses by adding loss. J. Opt. Soc. Am. B **28**, 2499–2509 (2011)
46. H. Liu, B. Wang, L. Ke, J. Deng, C.C. Choy, M.S. Zhang, L. Shen, S.A. Maier, J.H. Teng, High contrast superlens lithography engineered by loss reduction. Adv. Funct. Mater. **22**, 3777–3783 (2012)
47. D.B. Shao, S.C. Chen, Numerical simulation of surface-plasmon-assisted nanolithography. Opt. Express **13**, 6964–6973 (2005)
48. D.B. Shao, S.C. Chen, Surface-plasmon-assisted nanoscale photolithography by polarized light. Appl. Phys. Lett. **86**, 253107 (2005)
49. M.D. Arnold, R.J. Blaikie, Subwavelength optical imaging of evanescent fields using reflections from plasmonic slabs. Opt. Express **15**, 11542–11552 (2007)
50. Z. Zhao, Y. Luo, N. Yao, W. Zhang, C. Wang, P. Gao, C. Zhao, M. Pu, X. Luo, Modeling and experimental study of plasmonic lens imaging with resolution enhanced methods. Opt. Express **24**, 27115–27126 (2016)
51. http://www.laserfocusworld.com/articles/print/volume-40/issue-8/world-news/surface-plasmon-polaritons-allow-features-to-25-nm.html
52. W. Wang, H. Xing, L. Fang, Y. Liu, J. Ma, L. Lin, C. Wang, X. Luo, Far-field imaging device: planar hyperlens with magnification using multi-layer metamaterial. Opt. Express **16**, 21142 (2008)
53. Y. Xiong, Z. Liu, X. Zhang, A simple design of flat hyperlens for lithography and imaging with half-pitch resolution down to 20 nm. Appl. Phys. Lett. **94**, 203108 (2009)

54. L. Liu, K. Liu, Z. Zhao, C. Wang, P. Gao, X. Luo, Sub-diffraction demagnification imaging lithography by hyperlens with plasmonic reflector layer. RSC Adv. **6**, 95973–95978 (2016)
55. J. Sun, T. Xu, N.M. Litchinitser, Experimental demonstration of demagnifying hyperlens. Nano Lett. **16**, 7905–7909 (2016)
56. X. Tao, C. Wang, Z. Zhao, Y. Wang, N. Yao, X. Luo, A method for uniform demagnification imaging beyond the diffraction limit: cascaded planar hyperlens. Appl. Phys. B **114**, 545–550 (2014)
57. Z. Zhao, Y. Luo, W. Zhang, C. Wang, P. Gao, Y. Wang, M. Pu, N. Yao, C. Zhao, X. Luo, Going far beyond the near-field diffraction limit via plasmonic cavity lens with high spatial frequency spectrum off-axis illumination. Sci. Rep. **5**, 15320 (2015)
58. W. Zhang, N. Yao, C. Wang, Z. Zhao, Y. Wang, P. Gao, X. Luo, Off Axis illumination planar hyperlens for non-contacted deep subwavelength demagnifying lithography. Plasmonics **9**, 1333–1339 (2014)
59. W. Zhang, H. Wang, C. Wang, N. Yao, Z. Zhao, Y. Wang, P. Gao, Y. Luo, W. Du, B. Jiang, X. Luo, Elongating the air working distance of near-field plasmonic lens by surface plasmon illumination. Plasmonics **10**, 51–56 (2015)
60. Z. Fang, Q. Peng, W. Song, F. Hao, J. Wang, P. Nordlander, X. Zhu, Plasmonic focusing in symmetry broken nanocorrals. Nano Lett. **11**, 893–897 (2011)
61. W. Chen, R.L. Nelson, Q. Zhan, Efficient miniature circular polarization analyzer design using hybrid spiral plasmonic lens. Opt. Lett. **37**, 1442–1444 (2012)
62. S. Yang, W. Chen, R.L. Nelson, Q. Zhan, Miniature circular polarization analyzer with spiral plasmonic lens. Opt. Lett. **34**, 3047–3049 (2009)
63. M. Song, C. Wang, Z. Zhao, M. Pu, L. Liu, W. Zhang, H. Yu, X. Luo, Nanofocusing beyond the near-field diffraction limit via plasmonic Fano resonance. Nanoscale **8**, 1635–1641 (2016)
64. C. Ma, Z. Liu, A super resolution metalens with phase compensation mechanism. Appl. Phys. Lett. **96**, 183103 (2010)
65. Y. Wang, W. Srituravanich, C. Sun, X. Zhang, Plasmonic nearfield scanning probe with high transmission. Nano Lett. **8**, 3041–3045 (2008)
66. W. Srituravanich, L. Pan, Y. Wang, C. Sun, D.B. Bogy, X. Zhang, Flying plasmonic lens in the near field for high-speed nanolithography. Nat. Nanotechnol. **3**, 733 (2008)
67. L. Pan, Y. Park, Y. Xiong, E. Ulin-Avila, Y. Wang, L. Zeng, S. Xiong, J. Rho, C. Sun, D.B. Bogy, X. Zhang, Maskless plasmonic lithography at 22 nm resolution. Sci. Rep. **1**, 175 (2011)
68. S. Kim, H. Jung, Y. Kim, J. Jang, J.W. Hahn, Resolution limit in plasmonic lithography for practical applications beyond 2x-nm half pitch. Adv. Mater. **24**, OP337–OP344 (2012)
69. Y. Wang, N. Yao, W. Zhang, J. He, C. Wang, Y. Wang, Z. Zhao, X. Luo, Forming sub-32-nm high-aspect plasmonic spot via bowtie aperture combined with metal-insulator-metal scheme. Plasmonics **10**, 1607–1613 (2015)
70. C. Wang, W. Zhang, Z. Zhao, Y. Wang, P. Gao, Y. Luo, X. Luo, Plasmonic structures, materials and lenses for optical lithography beyond the diffraction limit: a review. Micromachines **7**, 118 (2016)
71. J. Zhou, C. Wang, Z. Zhao, Y. Wang, J. He, X. Tao, X. Luo, Design and theoretical analyses of tip–insulator–metal structure with bottom–up light illumination: formations of elongated symmetrical plasmonic hot spot at sub-10 nm resolution. Plasmonics **8**, 1073–1078 (2013)
72. H. Jung, S. Kim, D. Han, J. Jang, S. Oh, J.-H. Choi, E.-S. Lee, J.W. Hahn, Plasmonic lithography for fabricating nanoimprint masters with multi-scale patterns. J. Micromech. Microeng. **25**, 055004 (2015)
73. P. Gao, X. Li, Z. Zhao, X. Ma, M. Pu, C. Wang, X. Luo, Pushing the plasmonic imaging nanolithography to nano-manufacturing. Opt. Commun. **404**, 62–72 (2017)
74. O. Ozturk, Multi-scale alignment and positioning system II, Doctor, The University of North Carolina, 2008
75. R. Fesperman, O. Ozturk, R. Hocken, S. Ruben, T.-C. Tsao, J. Phipps, T. Lemmons, J. Brien, G. Caskey, Multi-scale alignment and positioning system—MAPS. Precis. Eng. **36**, 517–537 (2012)

76. M. Liu, C. Zhao, Y. Luo, Z. Zhao, Y. Wang, P. Gao, C. Wang, X. Luo, Subdiffraction plasmonic lens lithography prototype in stepper mode. J. Vac. Sci. Technol. B **35**, 011603 (2016)
77. H. Jung, Y. Kim, S. Kim, J. Jang, J.W. Hahn, High-resolution laser direct writing with a plasmonic contact probe, in *Proceedings of SPIE 8323, Alternative Lithographic Technologies IV*, vol. 8323 (2012), pp. 83232A-8323–7
78. S. Oh, T. Lee, J.W. Hahn, Multifunctional bowtie-shaped ridge aperture for overlay alignment in plasmonic direct writing lithography using a contact probe. Opt. Lett. **38**, 2250–2252 (2013)
79. X. Wen, L.M. Traverso, P. Srisungsitthisunti, X. Xu, E.E. Moon, High precision dynamic alignment and gap control for optical near-field nanolithography. J. Vac. Sci. Technol. B **31**, 041601 (2013)
80. H. Takesue, S.W. Nam, Q. Zhang, R.H. Hadfield, T. Honjo, K. Tamaki, Y. Yamamoto, Quantum key distribution over a 40-dB channel loss using superconducting single-photon detectors. Nat. Photonics **1**, 343 (2007)
81. X.-F. Shen, X.-Y. Yang, L.-X. You, Performance of superconducting nanowire single-photon detection system. Chin. Phys. Lett. **27**, 087404 (2010)
82. A. Tuantranont, *Applications of Nanomaterials in Sensors and Diagnostics*. Springer series on chemical sensors and biosensors (Springer, Berlin Heidelberg, 2014)
83. N. Meinzer, W.L. Barnes, I.R. Hooper, Plasmonic meta-atoms and metasurfaces. Nat. Photonics **8**, 889 (2014)
84. J. Luo, B. Zeng, C. Wang, P. Gao, K. Liu, M. Pu, J. Jin, Z. Zhao, X. Li, H. Yu, X. Luo, Fabrication of anisotropically arrayed nano-slots metasurfaces using reflective plasmonic lithography. Nanoscale **7**, 18805–18812 (2015)

Chapter 8
Sub-Diffraction-Limited Telescopies

Abstract Telescopy is one of the most important applications in engineering optics. In order to improve the observation ability, the optical aperture of telescopes is becoming much larger, which has nearly reached the state-of-the-art technique limit. Sub-diffraction-limited telescopies, which have improved observation ability without enlarging the size of the telescopes, are pursued for a long time. In this chapter, two kinds of sub-diffraction-limited telescopies are introduced for EO 2.0. First, a brief introduction of the telescopy in EO 1.0 is given. Then in Sect. 8.2, we introduce the telescopy based on the super-oscillation in detail. Super-oscillation telescopy with dielectric pupil filter (DPF), metasurfaces, as well as achromatic super-oscillation telescopy are discussed. In Sect. 8.3, a review on another super-resolution telescopy based on orbital angular momentum is presented.

Keywords Optical telescope · Sub-diffraction-limited · Super-oscillation

8.1 Introduction

Telescopes are optical imaging systems used to gather light from a distant object and then form a real image viewed by using an eyepiece or captured by optoelectronic sensors, such as a CCD or CMOS camera. Since it was invented about four hundred years ago, the optical telescope has always been the most important tool for applications requiring optical information of targets long-distance away, like astronomy observation, optical surveillance, and remote sensing. The telescopes are typically characterized by large apertures, long focal lengths, and comparatively narrow fields of view.

Suffering from the diffraction of light, the resolution of a telescope is theoretically limited by the diameter D of the objective aperture and light wavelength λ in the form of Rayleigh criterion $1.22\lambda/D$. In order to improve the resolution of telescopes, the main strategy at present is to increase the aperture of the telescope both for ground-based and space-based telescopes. Figure 8.1a, b illustrate two ground-based telescopes, i.e., the Extremely Large Telescope (ELT) and the William Herschel Telescope (WHT). The WHT is a 4.2-m optical/near-infrared reflecting telescope

X. Luo, *Engineering Optics 2.0*, https://doi.org/10.1007/978-981-13-5755-8_8

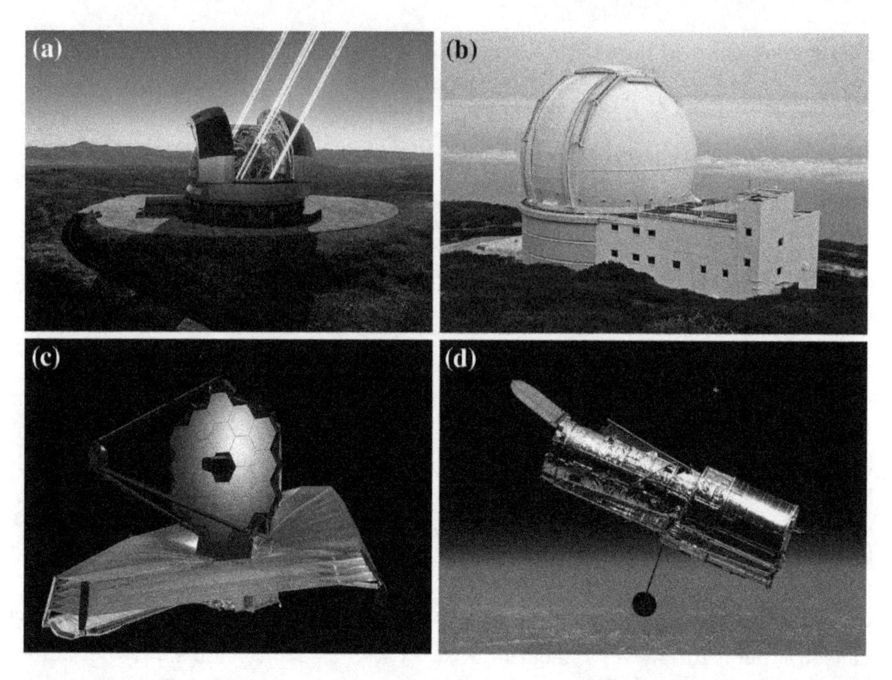

Fig. 8.1 **a** Artist's rendering of the Extremely Large Telescope, which is the world's largest optical/near-infrared extremely large telescope now under construction with a 39.3-m segmented primary mirror. **b** The William Herschel Telescope, which is a single optical telescope with a 4.2-m primary mirror. **c** The James Webb Space Telescope, which is a space telescope with a 6.5-m diameter mirror composed of 18 hexagonal mirror segments. **d** The Hubble Space Telescope, which entered service in 1990 with a 2.4-m diameter mirror. **a** Reproduced from Wikipedia [1]. **b** Reproduced from Wikipedia [2]. **c** Reproduced from Wikipedia [3]. **d** Reproduced from Wikipedia [4]

located at the Observatoriodel Roque de los Muchachos on the island of La Palma in the Canary Islands, Spain. Numerous discoveries have been made with the WHT including, for the first time, confirmation of the existence of a black hole in the galaxy. To further enhance the observation ability, extremely large telescopes, such as Thirty Meter Telescope (TMT) and ELT, which have diameters of the primary mirror larger than 30 m are under construction. However, the larger size will dramatically increase the cost. For example, in 2017, the construction cost for ELT was estimated to be €1.15 billion, in contrast to £15M cost (in 1984, equivalent to £44M in 2016) of the WHT [1]. Compared with the ground-based large telescope, building space telescope with lager aperture is much more costly and technically complex. Figure 8.1c, d show the James Webb Space Telescope (JWST) with a 6.5-m diameter mirror and the Hubble Space Telescope (HST) with a 2.4-m diameter mirror. Due to the huge cost and the extremely technical difficulty, the planned launch date for JWST has been pushed back many times from 2010 to 2021 [2].

How to improve the resolution at a fixed telescope aperture, i.e., enhance the observation ability without significantly increasing the technical complexity, is the goal always pursued all over the world. This resolution obstacle, so-called diffraction limit, has been fully realized for centuries and interpreted with the uncertainty principle from the viewpoint of quantum mechanics [5, 6]. Consequently, the main access to enhance the resolution of telescope relies on the increase of the aperture size of an objective lens, besides some complementary but important efforts, including adaptive optics for air distortion compensation [7] and recording devices with high sensitivity. On the other hand, to reduce the complexities and challenges to build an ultra-large telescope, some innovative concepts are proposed, such as Fourier transform telescope [8] and aperture synthesis [9]. Unfortunately, those methods still suffer from the diffraction limit and are usually restricted to specific applications due to the concerns for the posterior data processing and active illumination.

Recently, some novel concepts are proposed and deliver revolutionary ideas for super-resolution optics [10]. Perfect lens composed of a negative refractive index (NRI) metamaterial slab was proved in theory to be capable of image infinite small features of targets [11]. This idea was further extended to the superlens and hyperlens, which has been applied in nanolithography and helped to improve the resolution of an optical microscope. Special optical beams, such as non-diffracting Bessel beams, have also been used to realize super-resolution imaging [12, 13]. Based on fluorescence radiation manipulations, some super-resolution microscopes, including stimulated emission depletion microscopy (STED) [14], stochastic optical reconstruction microscopy (STORM) [15], and photoactivated localization microscopy (PALM) [16], delivered impressive achievements in observing nanoscale fluorescence targets. In contrast to the great interests and achievements in super-resolution optics, especially in the applications of microscopes and nanolithography, few efforts are made to break the diffraction barrier of telescopes. This could be mainly attributed to the fact that remote objects, like celestial targets, would not be accessible for artificial radiation manipulation and the great size of telescopes hamper the possibility of super-resolution lenses with metamaterial slabs. Some new super-resolution telescopies have been proposed and experimentally demonstrated recently, taking advantages of novel mechanisms including super-oscillation technique and orbital angular momentum modulation, which we will discuss in details in the following sections.

8.2 Super-Oscillation Telescopy

In 1952, Toraldo proposed a concept of super-gain antenna in microwave and optics to improve the resolution of imaging in the far-field [17], which is regarded as the well-known pupil filtering technology [18]. Berry and Popescu attributed this high-resolution technique to the super-oscillation phenomenon [19]. They indicated that the super-oscillation behavior occurs in a region where the band-limited functions are able to oscillate faster than their highest Fourier components, which is somewhat similar to the weak measurements in quantum mechanics, and this phenomenon usually arises from complex destructive interference of light with different spatial frequencies and variant phases. It was experimentally demonstrated that a super-oscillatory lens (SOL) generating a sub-diffraction-limited focusing spot could be applied for a super-resolution microscope in confocal scanning manner with coherent light illumination. The obtained resolution is better than $\lambda/6$ [20]. Recently, SOL has been extensively studied for its applications in the sub-diffraction-limited focusing and imaging [21–23] as well as the heat-assisted magnetic recording in confocal scanning system [24]. The key point of super-oscillation imaging relies on the generation and optimization of complex light fields, including their phase and amplitude distributions, which is usually realized by binary amplitude/phase elements [20, 25–27], spatial light modulators [28, 29], or optical eigenmodes methods [30].

Figure 8.2 shows the schematic of the super-oscillation telescope (SRT) system. An incoherent light illuminates on the target, which is positioned at the front focal plane of an optical collimator. Therefore, the target can be seen as an effective object from infinite distant. An objective lens with finite aperture sizes at the entrance-pupil plane is used to image the target. A small region of diffraction-limited optical images, restricted by a field diaphragm at the imaging plane of the objective lens, is relayed through a local diffraction shrinkable optics consisting of a $4f$ system with a specially SOL at the exit-pupil plane. The generated diffraction pattern is recorded with a CCD camera. The SOL can be a phase or amplitude modulation plate. Different from the super-oscillation sub-diffraction-limited microscope system, where the SOLs are utilized to scan the object so as to shrink the illuminating spot, the SOLs in telescope enable the resolution enhancement through the modulation of the optical transfer function (OTF) at the back end of the optical imaging system. It should also be noted that the phase modulation is more frequently adopted for SOL-based sub-diffraction-limited microscope systems. In the following sections, we will introduce three kinds of phase modulation-based SOL, including DPF, metalens, and achromatic metasurface.

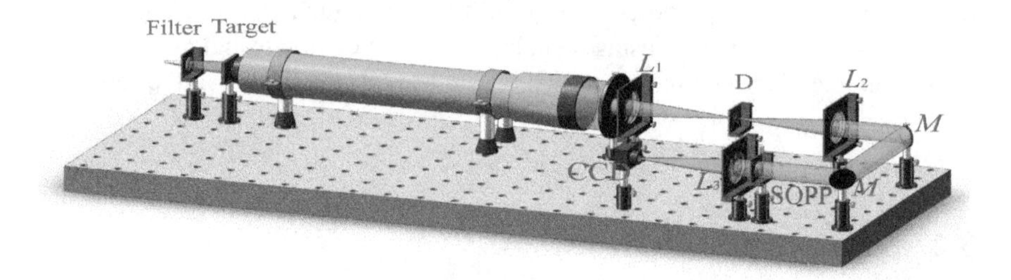

Fig. 8.2 Schematic of the optical setup of a SRT system, including a halogenated lamp, a filter, an optical collimator, an entrance pupil, an objective lens (L_1), and a $4f$ system consisting of a field diaphragm, two mirrors (M), two focusing lenses (L_2 and L_3), a designed super-oscillation lens, and a CCD camera. Reproduced from [31] with permission. Copyright 2015, The Authors

8.2.1 Super-Oscillation Telescopy with Dielectric Pupil Filter

The SRT system benefits from the shrinkable lateral full width of local point spread function (LPSF) beyond the diffraction limit due to the introduction of phase modulation. This feature, from the viewpoint of defined local optical transfer function (LOTF), enables higher local Fourier frequency components surpassing the cutoff frequency determined by the finite aperture size of the objective lens. The point spread function (PSF) of the SRT system is the diffraction pattern of the phase plate when the light point source is positioned at the central front focal plane of the optical collimator. The PSF can be approximately expressed as [32]

$$I(\rho) \propto \left(\frac{1}{\lambda f}\right)^2 \left| \int_0^R \exp[i\varphi_{\text{binary}}(r)] J_0\left(\frac{2\pi r\rho}{\lambda f}\right) r\,\mathrm{d}r \right|^2, \tag{8.2.1}$$

where f is the focal length of the last lens L_3 in the $4f$ system, and R is the aperture radius of the exit pupil. The modulation phase function $\varphi_{\text{binary}}(r)$ is assumed to be a circular symmetrical binary phase function (0 or π) with finite phase-jump positions. It can be calculated and optimized based on a desired resolution factor G and field of view (FOV) by linear programming method [32]. Here, resolution factor G is defined as the ratio between the full width of central spot size where the intensity of the central spot falls to minimum, and that of an ideal Airy spot; the FOV is defined as a region where intensities of side lobes normalized by the central peak intensity are controlled below a constant.

Figure 8.3a, b show the diffraction-limited pattern and its spatial Fourier spectrum with limited frequency, determined by the aperture size of the objective lens for the case without phase modulation at the wavelength of 532 nm. The PSF of the objective lens at the imaging plane and in the SRT system turn to be the Airy spot.

When a DPF is inserted as a phase plate, the diffractive pattern and its LOTF are totally modulated. A super-oscillation is observed as shown in Fig. 8.4. As clearly seen in Fig. 8.4a, the diffraction shrinkable phenomenon for SOL spot could be observed for the central spot with greatly reduced lateral size. The LPSF is designed to obtain a local diffraction shrinkable effect, which has 0.6 times the full width

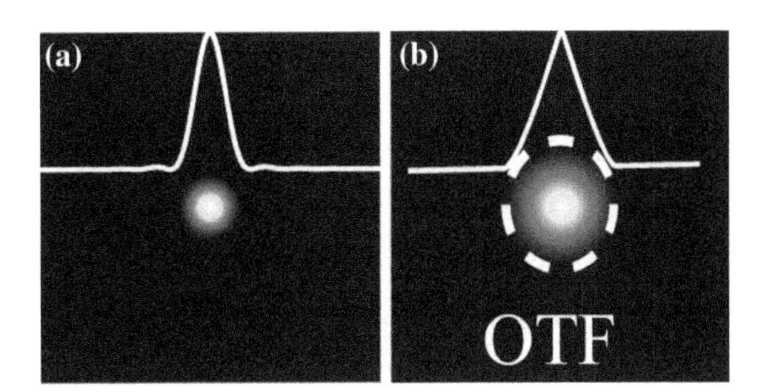

Fig. 8.3 Simulated diffraction-limited pattern (**a**) and corresponding OTF (**b**) at diaphragm plane. Dashed inset rings are the cutoff frequency of diffraction-limited OTF. Reproduced from [31] with permission. Copyright 2015, The Authors

of Airy pattern for sub-diffraction-limited imaging. Inside the FOV, the maximum intensity M_1 of side lobes normalized by the central peak intensity is assumed to be small enough to bring no significant influence for super-resolution imaging in this local FOV (here, M_1 is set as 0.1). Detailed parameters are presented in Table 8.1. The normalized π-phase-jump positions of the phase plate are optimized at $r_1 = 0.297$, $r_2 = 0.594$, and $r_3 = 0.85$.

The sub-diffraction-limited pattern defined by Eq. (8.2.1) would be attributed to the destructive interference of high and low Fourier components. It should be noted that the access to Fourier components beyond the cutoff frequency would not be possible from the definition of Fourier components in the total focus pattern region, as shown in Fig. 8.4b of the conventional OTF. Different conclusions could be obtained as we focus on the small region around the central diffraction shrinkage spot and ignore those great side lobes outside the FOV. To show this point, the concept of LOTF is proposed. In the given analyses, the key point is that the Fourier components of LOTF are not contributed from the total diffraction patterns, but only accounts for the sub-diffraction-limited pattern within the local FOV. As a result, local Fourier components beyond the cutoff frequency of conventional OTF could be seen in Fig. 8.4c. It is reasonable to rely on the tool of LOTF for the SRT imaging analysis, providing that the imaging field is constrained to the local FOV and the side lobes are neglected in the imaging process. Consequently, a remote target, as

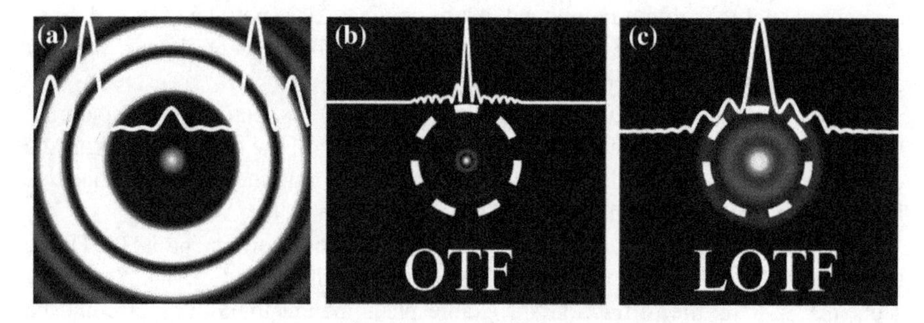

Fig. 8.4 Simulated sub-diffraction-limited pattern (**a**) and corresponding OTF (**b**) and LOTF **c** at CCD plane. Dashed inset rings are the cutoff frequency of diffraction-limited OTF. Reproduced from [31] with permission. Copyright 2015, The Authors

Table 8.1 Simulated (*S*) and experimental (*E*) parameters for the airy spots and the sub-diffraction-limited spots

Parameter (μm)		Airy	Spot 1
G	*S*	81.13	48.68
	E	82.8	51.75
FOV	*S*	∞	154.15
	E	∞	151.8
M_1	*S*	0.0175	0.1
	E	0.0256	0.16

the ensemble of a great deal of light spots, could be observed with more details and enhanced resolution, depending on the extent of the diffraction shrinkage.

To demonstrate the resolving ability of the SRT, two transparent holes with 20 μm diameter and 55 μm center-to-center distance, corresponding to 0.68 of equivalent Rayleigh criterion distance 81.13 μm, are used for the evaluation of super-resolution imaging (Fig. 8.5a). As expected, the two holes are completely unresolved at the imaging plane of the SRT's objective lens (Fig. 8.5b). The information seems to be lost forever at the entrance of the objective lens, from the viewpoint of conventional Fourier optics. However, at the CCD plane of the relayed optics with phase modulation, the two holes are clearly distinguished with a resolution beyond the Rayleigh criterion (Fig. 8.5c). The measured result in Fig. 8.5d shows good agreement with the simulation performed by convolving the objective function with LPSF. Figure 8.5e plots the experimental and simulated imaging contrasts for variant center-to-center distances of these two holes. Both the imaging resolution and contrast could be remarkably improved in the SRT system with SOL.

It should be noted that, the imaging results indicate that the proposed SRT method is different from the inverse filter method [33], in which incoherent imaging is restricted and image information beyond the diffraction limit could not be restored. The super-resolution realized here originates from the super-oscillation in a region where a band-limited function oscillates arbitrarily quickly, faster than its highest Fourier component. Importantly, it gives counterintuitive evidence to the common knowledge that the primary diffraction-limited images of an objective lens would not be observed in more details by the following relayed optics, as stated in the *Principles of Optics* [34].

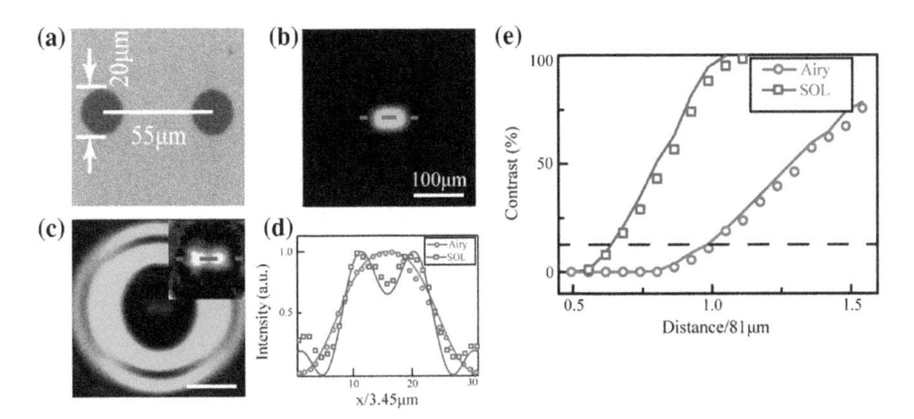

Fig. 8.5 **a** Microscope image of a two-hole target. **b** Experimental diffraction-limited imaging pattern without phase modulation. **c** Super-resolution imaging pattern with SOL. False-color inset shows the magnified central field distribution. **d** Experimental (symbolic lines) and simulated field (solid lines) distributions along horizontal lines in **b** and **c**. **e** Experimental (symbolic lines) and simulated (solid lines) imaging contrast for variant center-to-center distances of two-hole targets. Dashed line indicates the resolvable contrast of Rayleigh criterion. Reproduced from [31] with permission. Copyright 2015, The Authors

The SRT imaging could be obtained for complex targets in a single local FOV, instead of just focusing, which was realized with the help of a relayed optics to a telescope system. As an illustrative example, the character "*E*" target shown in Fig. 8.6a is employed, which could not be resolved for diffraction-limited imaging due to the structural size is smaller than the diffraction limit. However, the details of the character could be identified without ambiguity in the SRT imaging system as shown in Fig. 8.6b. Figure 8.6c presents the calculated Fourier spectra of local image of the character "*E*" in the SRT system, which clearly shows much higher local spatial Fourier components beyond the cutoff frequency. The retrieved high-frequency components enable the sub-diffraction-limited imaging of the SOL SRT system. The measured and simulated results of the details of the image show good agreement (Fig. 8.6d). This SOL imaging could also be used for large targets by superimposing multiple local fields of views when they are beyond a single FOV of the SOL. For large targets, a viewing field diaphragm, positioned at the diffraction-limited imaging plane of the objective lens, could be employed to avoid distortions of huge side lobes outside the FOV from nearby targets and enhance the overall FOV. SRT imaging could be performed by scanning the diaphragm and superimposing together multiple local fields of views at different positions. Figure 8.6e–h show a group of closely positioned holes could be stitched and clearly imaged with six local fields.

Local light diffraction shrinkable phenomenon can be applied for real-time, on-line, and incoherent SRT imaging. In theory, the diffraction shrinkable spot could be arbitrarily small and yields infinitely extended bandwidth of LOTF. Figure 8.7 gives the imaging properties of a SOL with 0.32 times resolvable distance of Rayleigh

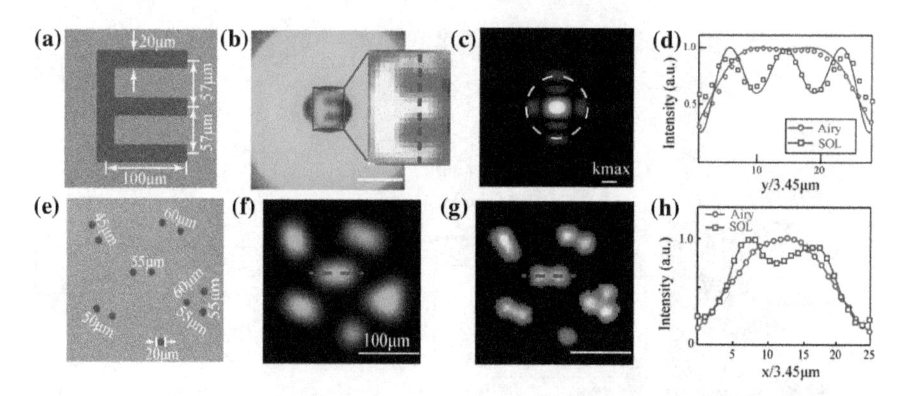

Fig. 8.6 **a** Microscope image of an "*E*" target. **b** Experimental super-resolution imaging pattern with SOL. Magnified false-color pattern shows the central field distribution. **c** Local Fourier spectrum of "*E*" in **b**. **d** Experimental (symbolic lines) and simulated (solid lines) field distributions along vertical lines as indicated in **b**. **e** Microscope images of a group of points. **f** Experimental diffraction-limited imaging patterns. **g** Super-resolution imaging pattern with SOL by superimposing together variant local fields of views. **h** Field distributions along horizontal lines in **f** and **g**. False-color insets show magnified central field distributions. Reproduced from [31] with permission. Copyright 2015, The Authors

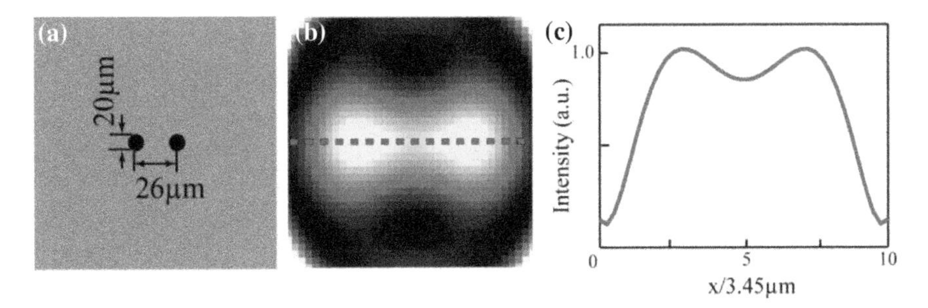

Fig. 8.7 Super-resolution super-oscillatory imaging for a two-hole target with 26 μm center-to-center distance. Reproduced from [31] with permission. Copyright 2015, The Authors

criterion. The designed G, L, and M_l are 24.34 μm, 40.57 μm, and 0.1, respectively. It can be seen that two holes with a center-to-center distance of 26 μm could be resolved as shown in Fig. 8.7b, c. However, the reduced central spot of the SOL system delivers decrease of central focusing energy. The calculated Strehl ratio, defined as the quotient of the central intensities of a sub-diffraction-limited spot and Airy spot, is 0.71%. Unlike the case in conventional imaging optics, where the reduced Strehl ratio indicates the degradation of the imaging quality, the ultra-small Strehl ratio leads to a great improvement of the imaging resolution. On the other hand, the ultra-low central intensity of sub-diffraction-limited spots would impose some obstacles for light signal detection of CCD, especially for the low-contrast targets imaging. There were also some investigations reporting the super-oscillation focal spot with relatively large Strehl ratio and small side lobes at the sacrifice of sub-diffraction-limited spot size [22], which are called supercritical imaging. But the side lobes, more than 20% of the central intensity, would seriously degrade the super-resolution imaging contrast and resolution. So one should consider the trade-off among those factors for practical applications. Regardless of the low focusing efficiency, noise, and effects of the huge side lobes in detecting, there seems no theoretical limit for a super-oscillatory spot and much smaller feature is possible.

8.2.2 Super-Oscillation Telescopy with Metasurfaces

The sub-diffraction-limited imaging of the super-oscillation telescope with DFP is nearly monochromatic with a narrow spectrum width due to the step phase changes fast with the wavelength. To analyze the influence of SRT's resolution on the phase change, a DPF is designed to work on a resolving ability of 0.6 times of Rayleigh criterion (assuming λ is 532 nm, R is 4.5 mm, and f is 500 mm), with the normalized radius of each phase-jump position $r_1 = 0.297$, $r_2 = 0.594$, and $r_3 = 0.85$. The etched depth Δd of the DPF can be calculated to be 578 nm according to:

$$\frac{2\pi}{\lambda}\Delta d(n-1)=\pi, \tag{8.2.2}$$

where n means the refraction index of the glass, which is 1.46 here. From Eq. (8.2.2), we can see that the step phase generated by the DPF will be strongly dependent on the wavelength. For a wide light spectrum range, the inherent chromatic aberration of imaging optics and dispersion feature of phase retardation would destroy the light diffraction shrinkage behavior without further design.

Figure 8.8 shows the resolution factor G for light wavelength ranging from 400 to 750 nm [35]. Obviously, the DPF method shows strong dependence on the light wavelength, where the super-oscillation effect occurs closely at the working wavelength 532 nm and G increases greatly as wavelength goes slightly away from it. This phenomenon indicates that super-oscillation is sensitive to the change of phase modulation, even for 10% variance generated by a binary phase structure with about 50 nm wavelength deviation [36]. Therefore, the DPF can only be applied to the imaging within a narrow bandwidth.

Figure 8.8 demonstrates clearly the great fragility of super-oscillation phenomenon to the slight change of light fields, as it arises from the delicate light interference behaviors, especially for the spot size far beyond Abbe diffraction limit. Figure 8.8b, c give the simulation images of "*E*" object at two wavelengths of 532 and 600 nm, which obviously demonstrate the narrow working band of DPF-based SRT system. Recently, an ultra-broadband super-oscillatory lens (UBSOL) has been proposed based on geometric metasurfaces [37]. The ultra-broadband feature mainly arises from the nearly dispersionless phase profile of transmitted light through spin–orbit interaction in the metasurfaces. It is demonstrated in experiment that sub-diffraction-limited light focusing behavior holds well with nearly unchanged

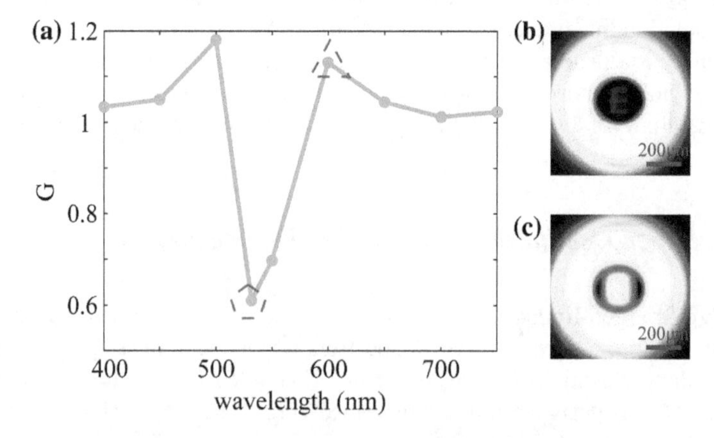

Fig. 8.8 **a** Resolution factor G distribution as the wavelengths range from 400 to 750 nm for DPF. **b**, **c** Simulation results for imaging of "*E*," corresponding wavelength points are indicated in square and triangle, respectively. The "*E*" is selected to be 100 μm × 124 μm. Reproduced from [35] with permission. Copyright 2018, WILEY-VCH Verlag GmbH & Co. KGaA, Weinheim

focal patterns for wavelengths spanning across visible and near-infrared light. This method is believed to find promising applications in super-resolution microscope or telescope, high-density optical data storage, etc.

Figure 8.9 illustrates the proposed UBSOL, which is realized by specially designed metasurface structure inscribed on a metal film [37]. The metasurface consists of a great number of transparent rectangular nano-apertures with variant orientations. A plane wave with left or right circular polarization is normally incident on the UBSOL. Based on the geometric phase modulation principle, the transmitted light for cross-polarization with respect to the incidence possesses a specific geometric phase profile, which shows nearly dispersionless feature [36, 38, 39] and allows the nearly same sub-diffraction-limited light focusing patterns at variant focal planes for broadband light wavelengths [40, 41].

The unique dispersionless phase manipulation of a metasurface structure for light with circular polarization plays a key role in designing an UBSOL. To show this point, light response and dispersion analysis are presented in Fig. 8.10. Full-wave simulations are performed by CST MICROWAVE STUDIO to calculate the response of nano-apertures for circular polarization incidence. As a representative structure, the metasurface is assumed to be periodically arrayed rectangular nanometallic apertures with fixed orientation, 200 nm × 200 nm period, 140 nm length, 60 nm width, and 120 nm thickness, as shown in Fig. 8.10a.

As light with circular polarization impinges at the nanorectangular metallic aperture, localized plasmonic modes within the aperture (in the form of catenary plasmon) would be excited and help to squeeze light through the nano-aperture with subwavelength dimension. As a reasonable approximation, each nano-aperture could be considered as a plasmonic dipole antenna, through which the transmitted light field is partially converted to its opposite helicity with an abrupt phase change Φ, governed by $\Phi = \pm 2\varphi$, where φ is the orientation angle ranging from 0° to 180°, ± sign represents right circular polarization RCP (LCP) incidence converted to LCP (RCP) transmission. This phase modulation could be demonstrated in Fig. 8.10b for LCP incidence at the wavelength of 632.8 nm, where the phase shift changes

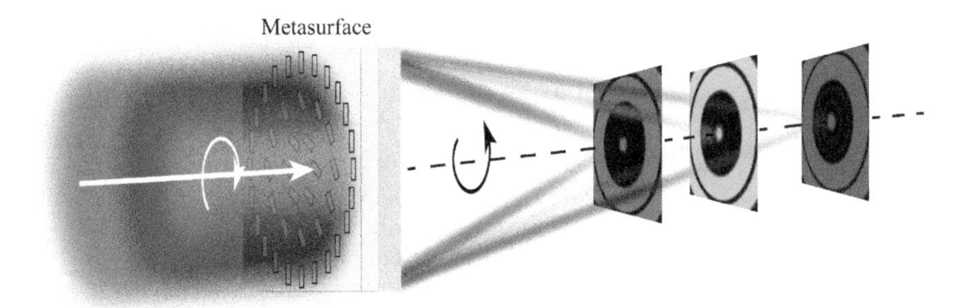

Fig. 8.9 Schematic of ultra-broadband sub-diffraction-limited focusing with super-oscillatory plasmonic metasurface. Reproduced from [37] with permission. Copyright 2015, WILEY-VCH Verlag GmbH & Co. KGaA, Weinheim

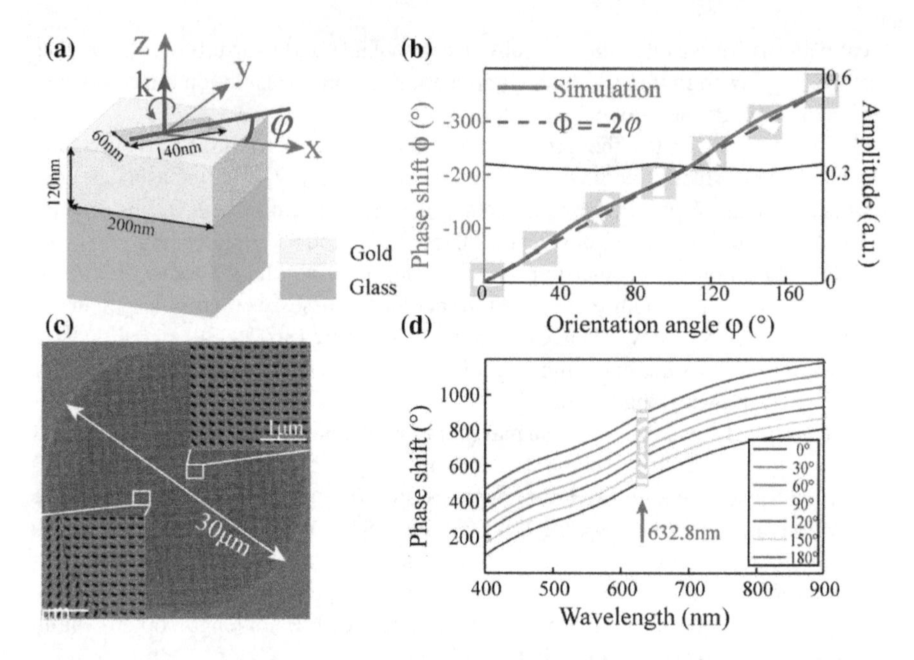

Fig. 8.10 **a** Schematic of the unit cell nano-aperture. **b** Phase shift and amplitude as a function of orientation angle at the wavelength of 632.8 nm. **c** Scanning electron microscopy image of the sample. The inset shows the section with varying orientations and the red-dashed line indicates phase-jump positions. **d** Phase shift of light through nano-aperture with variant orientation for incident wavelength from 400 to 900 nm. Reproduced from [37] with permission. Copyright 2015, WILEY-VCH Verlag GmbH & Co. KGaA, Weinheim

from 0° to −360° with a linear relation with aperture orientation angles. The slight deviation is mainly attributed to the angle dependence of light interaction between neighboring apertures. As the phase shift is independent of light wavelength, the phase-change values are approximately fixed under a broad spectrum from 400 to 900 nm, which ensures a nearly dispersionless phase profile generated by a metasurface. The bandwidth can be further increased by using the continuous catenary structure, which eliminates the resonance of nanoresonators and guarantees the conversion efficiency in an ultra-broadband frequency range [36].

The design of UBSOL begins at determining the phase profile required for sub-diffraction-limited focusing at a single wavelength. The phase of the UBSOL is combined by a hyperboloidal phase profile $\Phi_{\text{lens}}(r)$ for light focusing and a step phase modulation $\Phi_{\text{binary}}(r)$ for super-oscillatory behavior beyond Abbe diffraction limit. The focusing phase could be written as [42]:

$$\Phi_{\text{lens}}(r) = \frac{2\pi}{\lambda_0}(-\sqrt{f_0^2 + r^2} + f_0) + 2m\pi, \tag{8.2.3}$$

where λ_0 is the light wavelength, f_0 the focal length, and m an integer.

For the extra step phase part $\Phi_{\text{binary}}(r)$, it is assumed to be a circular symmetrical binary phase function (0 or π) with finite phase-jump positions, similar with that described in Sect. 8.2.1. The positions of the phase jump are optimized by linear programming method [32] to obtain a sub-diffraction-limited focal spot. The intensity distribution around the focal plane can be given by the Fresnel diffraction integral equation [34]

$$I(\lambda, \rho, z) \propto \left(\frac{1}{\lambda z}\right)^2 \left| \int_0^R \exp\left[i\Phi_{\text{binary}}(r) + i\Phi_{\text{lens}}(r)\right] \exp\left(i\frac{\pi r^2}{\lambda z}\right) J_0\left(\frac{2\pi r\rho}{\lambda z}\right) r\mathrm{d}r \right|^2. \tag{8.2.4}$$

To some extent, this could be considered as a generalization of traditional Fresnel's diffraction equation. In the paraxial region, the hyperboloidal phase profile of Eq. (8.2.3) could be approximately expressed as

$$\Phi_{\text{lens}}(r) \approx -\frac{\pi r^2}{\lambda_0 f_0} + 2m\pi. \tag{8.2.5}$$

Thus, Eq. (8.2.4) can be rewritten as

$$I(\lambda, \rho, z) \propto \left(\frac{1}{\lambda z}\right)^2 \left| \int_0^R \exp\left[i\Phi_{\text{binary}}(r)\right] \exp\left[i\pi r^2\left(\frac{1}{\lambda z} - \frac{1}{\lambda_0 f_0}\right)\right] J_0\left(\frac{2\pi r\rho}{\lambda z}\right) r\mathrm{d}r \right|^2. \tag{8.2.6}$$

Figure 8.11 plots three design examples with different $\Phi_{\text{binary}}(r)$. Sample *A* is designed without the binary phase modulation, i.e., $\Phi_{\text{binary}}(r) = 0$. So the design turns to be a conventional focusing lens with an Airy spot at the focal plane. The other two designs of $\Phi_{\text{binary}}(r)$ account for UBSOLs with different full width at half maximum (FWHM) beyond Abbe diffraction limit. Sample *B* is designed to generate a spot 0.807 times size of Abbe diffraction limit hot spot with the intensity of side lobes less than 20% of the peak intensity. Sample *C* is designed to generate a smaller focusing spot with spot size being 0.678 of Abbe diffraction limit. It is achieved with a limited FOV and much higher side lobes around it. The simulated focal patterns are plotted in Fig. 8.11b. The FOV in Sample *C*, defined as the area of low intensity less than 10% of the central intensity, could be used for real-time super-resolution imaging.

It should be noted that, although the metasurface delivers a nearly dispersionless phase distribution of transmitted light with opposite circular polarization for variant light wavelengths, a focus shift would occur owing to the axial chromatic aberration of phase profile defined in Eq. (8.2.3). Under paraxial approximation, the chromatic focal shift follows the rule that the product λz is kept to be nearly a constant. For the wavelengths away from the designed central wavelength, the modulated phase

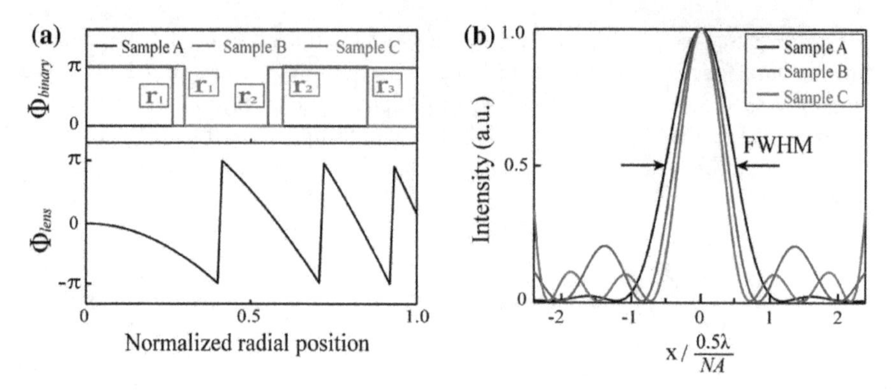

Fig. 8.11 **a** Designed phase profiles for UBSOLs with Sample *A*, *B*, and *C*. Bottom: hyperboloidal phase profile localized in $[-\pi, \pi]$. Top: optimized extra phase-jump functions. Sample *A*: without π-phase jump; Sample *B*: π-phase jump at positions $r_1 = 0.262$ and $r_2 = 0.556$; Sample *C*: at positions $r_1 = 0.297$, $r_2 = 0.594$, and $r_3 = 0.85$. **b** Theoretical intensity distributions at the focal plane for Sample *A*, *B*, and *C*. Reproduced from [37] with permission Copyright 2015, WILEY-VCH Verlag GmbH & Co. KGaA, Weinheim

function of transmitted light defined by Eq. (8.2.3) is not a perfect hyperboloidal phase distribution for the shifted focus position. However, this phase aberrations are usually small under paraxial approximation and would allow nearly unchanged super-oscillatory focal patterns around the shifted focal plane for a wide range of light wavelengths, which could be well understood by substituting $\lambda z \approx \lambda_0 f_0$ into Eq. (8.2.6), and we have

$$I(\rho) \propto \left(\frac{1}{\lambda_0 f_0}\right)^2 \left| \int_0^R \exp\left[i\Phi_{\text{binary}}(r)\right] J_0\left(\frac{2\pi r\rho}{\lambda_0 f_0}\right) r\,\mathrm{d}r \right|^2. \tag{8.2.7}$$

Based on the theoretical design and the phase modulation function shown in Fig. 8.10, three metasurface structures, with outer diameter $D = 30$ μm and a focal length $f_0 = 60$ μm, were designed at the wavelength $\lambda_0 = 632.8$ nm for LCP incidence. The structures were fabricated by focused ion beam (FIB) milling on a 120-nm-thick gold film deposited on a glass substrate. Figure 8.9c shows the scanning electron microscopy image for Sample *C*. The right-top inset with magnification shows the central section with varying orientations, and the red-dashed line in the left-bottom inset indicates phase-jump positions nearly corresponding to the nano-apertures' orientation angle shift of 90°.

In Fig. 8.12, the measured light distributions at the focal plane of $z = 60$ μm are presented and show good agreements with theoretical calculations. The measured focusing FWHM for Sample *A* is 1.364 μm, very close to the calculated Abbe diffraction limit of 0.5 λ/*NA* = 1.304 μm, with calculated numerical aperture *NA* = 0.2425. The focal FWHMs for Sample *B* and Sample *C* are measured to be 1.1 and 0.88 μm, about 0.843 and 0.674 times spot size of Abbe diffraction limit, respectively.

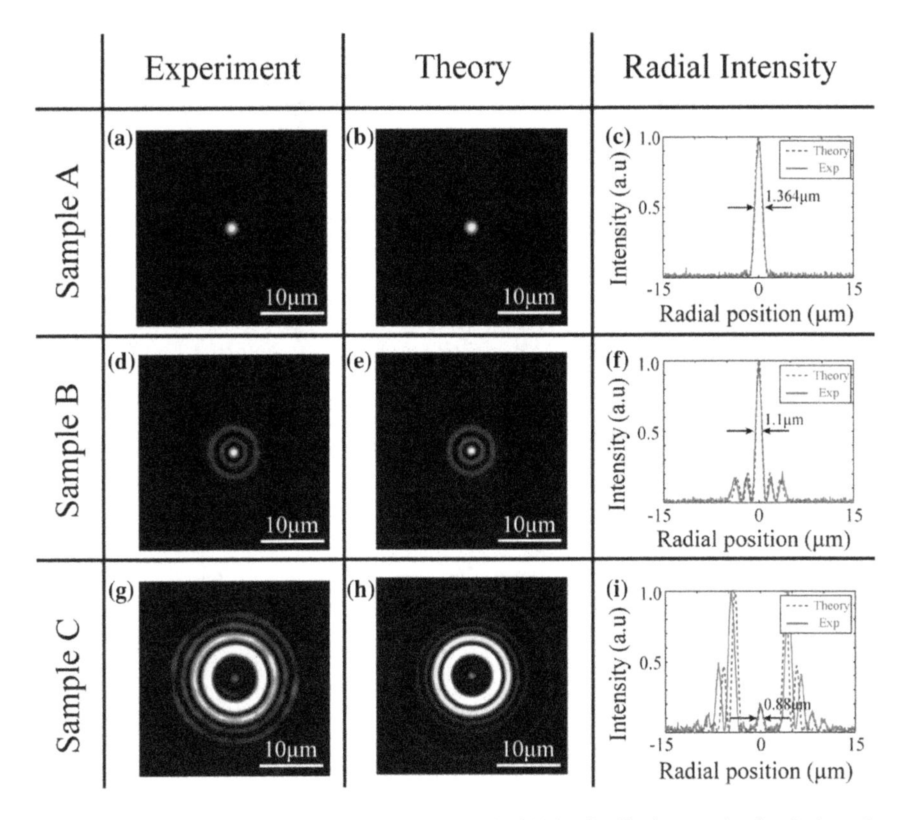

Fig. 8.12 **a**, **b**, **d**, **e**, **g**, **h** Experimental and theoretical light distributions at the focal plane for Sample *A*, *B*, and *C* with fixed wavelength of 632.8 nm, respectively. **c**, **f**, **i** Normalized light distributions along radial direction for the three samples, respectively. Reproduced from [37] with permission. Copyright 2015, WILEY-VCH Verlag GmbH & Co. KGaA, Weinheim

The slight discrepancy of side lobes between the experimental and theoretical results for the Sample *C* may be attributed to the enhanced fragility to fabrication errors for deep sub-diffraction-limited focusing.

Figure 8.13 shows the measured light distributions for the three samples with light wavelengths of 405, 532, 632.8, and 785 nm, respectively. Clearly, the focal patterns show almost similar optical field distributions, except for the great chromatic focus shift from 93 to 48 μm and the change of focusing intensity. The light distributions at variant focal planes show nearly same profiles shown in the bottom panel of Fig. 8.13. This invariance of diffraction patterns, as discussed above, benefits from the chromatic focal shift defined by the constant λz under paraxial approximation, which enables the focal patterns defined by Eq. (8.2.7) to show invariance with respect to light wavelengths. In other words, the change of focal patterns for variant light wavelengths is compensated by the axial chromatic focus shift. The invariant feature of light distributions for variant wavelengths still holds for a wide z-axis positions region, where the approximation of Eq. (8.2.6) keeps its validity. In addition, Sample

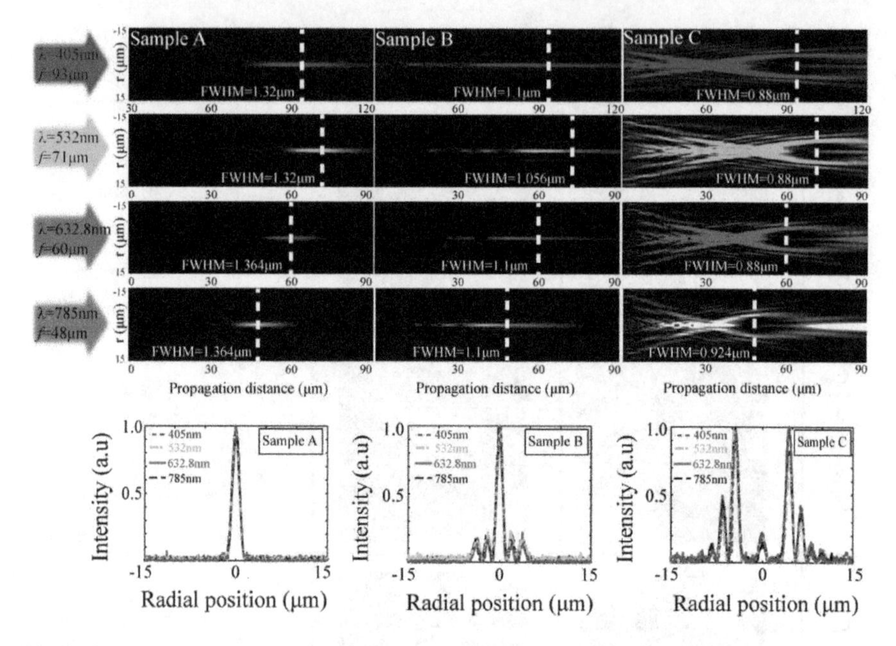

Fig. 8.13 Top: experimental light distributions along the *z*-direction for Sample *A*, *B,* and *C* at wavelengths of 405, 532, 632.8, and 785 nm, respectively. Bottom: normalized intensity distributions at the focal plane for the three samples with variant wavelengths. Note that the first row for 405 nm light wavelength plots with distance from 30 to 120 μm and the other rows from 0 to 90 μm. Reproduced from [37] with permission. Copyright 2015, WILEY-VCH Verlag GmbH & Co. KGaA, Weinheim

B shows a needle-like behavior [43, 44] with greatly elongated depth of focus in contrast to the Airy spot in Sample *A*. The needle-like focusing behavior is related to the diffraction pattern invariance feature of non-diffracting Bessel beams, and in the design of Sample *B,* the modulation of phase profile happens to generate the similar effect, which could be seen as well in some references using diffractive optical elements under radially or circularly polarized light [44]. For Sample *C*, the focal pattern, exhibiting greatly reduced spot size in the central dark region, is featured with a much shallower depth of focus due to its fragility of super-oscillatory spot.

Low efficiency is a common concern for both transmission-type metasurfaces and super-oscillatory focus. It is worth to note that most of light energy of SOL is contributed to the great side lobes around the central sub-diffraction-limited hot spot, especially for those SOL spots with very small lateral size, and the cost of energy is inevitable in super-oscillatory phenomenon, but there is a trade-off between sub-diffraction-limited spot size and focusing efficiency.

It is hard to give a definite wavelength range for the proposed UBSOL. As for light with much shorter wavelength and enlarged focal length, the finite size of nano-apertures would bring greater phase aberration due to light interactions between neighboring apertures and the aberration may contribute negatively to the super-oscillatory focusing. Theoretically, the sub-diffraction-limited focusing performances of Sample *B* and *C* may be observed for much larger wavelength in the near-infrared region. But this would deliver a shorter focal length and a greater focal pattern variance, as the approximation of Eq. (8.2.6) would not hold perfectly. On the other hand, the longitudinal electrical component would be increased obviously and deliver wider lateral focus size in the case of high numerical aperture [45, 46]. So the sub-diffraction-limited focusing feature would be greatly diminished for focus size evaluation with total electric field intensity [47].

8.2.3 Achromatic Super-Oscillation Telescopy

Section 8.2.2 has demonstrated a broadband super-oscillatory lens taking advantage of the nearly dispersionless, i.e., wavelength-independent, feature of phase modulations with metasurface. However, it should be noted that a common metasurface SOL can work in broadband, but not achromatically, which is a serious roadblock for broadband or white light super-resolution imaging. Recently, Li et al. combined the metasurface-based achromatic stepper phase engineering with a traditional achromatic lens and realized a super-oscillatory sub-diffraction-limited imaging for visible light ranging from 400 to 700 nm [35]. A resolving ability about 0.64 times of the Rayleigh criterion was obtained. This novel method is expected to potentially promote the development of the super-resolving telescopes or microscopes based on the super-oscillation optics.

As illustrated in Fig. 8.14, the metasurface filter is positioned at the exit-pupil plane of the imaging optical system, where objects illuminated with white incoherent light could be imaged at the image plane away from the metasurface filter. For the simplicity of analysis, the optical system could be regarded as an achromatic lens combined with the metasurface filter with phase function given in Eq. (8.2.1). The aim of designing metasurface filter with specific phase-change positions is to yield a PSF with minimum width of main lobe and small side lobe levels within the local field of FOV. The key point of the metasurface filter (MF) method is the employment of metasurface composed of nanoscale gratings with vertical orientations in concentric-ring regions. Instead of using phase retardations in conventional binary micro-structures etched on transparent substrates or spatial light modulators, the phase modulation of MF arises from the Pancharatnam–Berry phase or so-called photonic spin–orbit interaction in structures. This occurs as light in certain polarization state transmits through the nanograting structure and will be partly converted to the crossed polarization with phase variation dependent on the orientations of nanograting with respect to the incident light polarization. Thus, the transmitted light of cross-polarization possesses a specific geometric phase profile, which exhibits a nearly dispersionless feature in broadband spectrum.

The metasurface structure contains arrays of customized rectangular metallic gratings with $\varphi = 45°$ and $\varphi = 135°$ fabricated on a glass substrate, where φ is the angle between axis x and the orientation of the gratings. The polarization of the incident

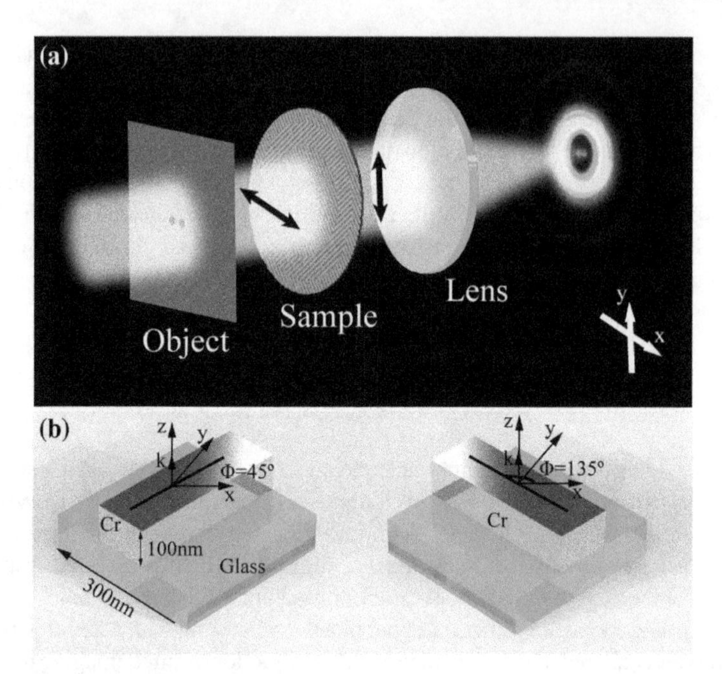

Fig. 8.14 Schematic of achromatic broadband super-resolution imaging by super-oscillatory metasurface. Reproduced from [35] with permission. Copyright 2018, WILEY-VCH Verlag GmbH & Co. KGaA, Weinheim

light is along the *x*- or *y*-directions. The rectangular nanometallic gratings with two orthogonal orientations introduce a π phase shift in broadband based on the geometric phase principle for linear polarized light. The chromium (Cr) is used as the grating material. The periodic of the Cr grating is 300 nm, and thickness is 100 nm as shown in Fig. 8.14b. The width of the Cr grating is optimized to be 100 nm for achieving a higher energy efficiency in a broad spectrum from 400 to 1000 nm. The metasurface is designed to contain four regions with the normalized radius of each phase-jump position $r_1 = 0.297$, $r_2 = 0.594$, and $r_3 = 0.85$, and these parameters have been demonstrated to obtain a 0.6 times the full width of Airy pattern in Sects. 8.2.1 and 8.2.2. For such a unique metasurface, the orthogonal gratings as shown in Fig. 8.15 are arranged alternately in neighboring annulus.

Figure 8.16 shows the simulated resolution factor *G* for light wavelength ranging from 400 to 750 nm. Obviously, the MF method shows independence of the resolution factor on the light wavelength. Figure 8.16b, c give the simulated images of "*E*" object through this telescope system at two wavelengths of 532 and 600 nm. The images are both clearly resolved which obviously demonstrates the achromatic working band of MF-based SRT system. Due to its dispersionless phase manipulation behavior, the MF successfully overcome the wavelength dependence feature of pupil filter as shown in Fig. 8.16. It is worth noting that the resolution enhancement factor *G* is scaled with the Airy spot size with correspondent light wavelength. Thus, the

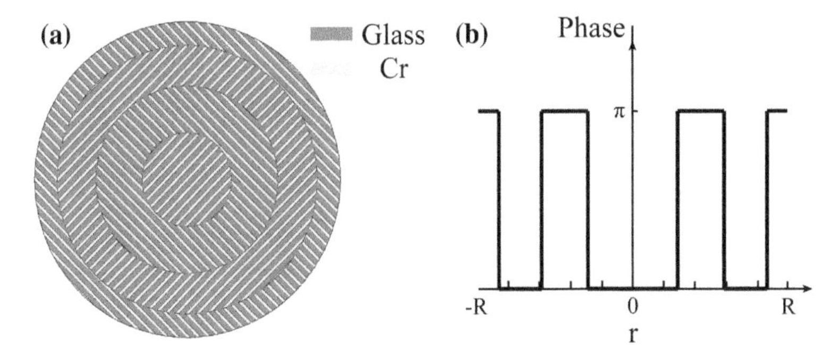

Fig. 8.15 **a** Phase manipulation of unit cell with different orientation. **b** Radial phase manipulation of metasurface. Reproduced from [35] with permission. Copyright 2018, WILEY-VCH Verlag GmbH & Co. KGaA, Weinheim

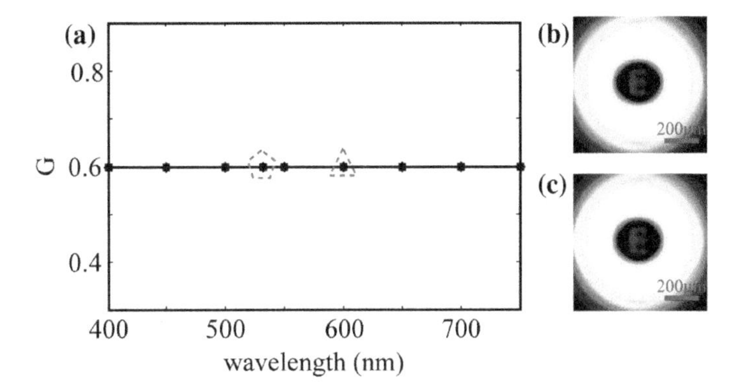

Fig. 8.16 **a** Resolution factor G distribution as the wavelengths range from 400 to 750 nm for MF. **b**, **c** Simulation results for imaging of "*E*," corresponding wavelength points in square and triangle, respectively. The "*E*" is selected to be 100 μm × 124 μm. Reproduced from [35] with permission. Copyright 2018, WILEY-VCH Verlag GmbH & Co. KGaA, Weinheim

resolving ability would scale linearly with wavelength and the resolution of a MF imaging system is a sum result for a wide light wavelength range, which could be approximately equal to the central wavelength.

In order to experimentally test the achromatic performance of the metasurface-based SRT system, a white light illumination is adopted. As shown in Fig. 8.17, the object, illuminated by a white light source of Xe-lamp, is placed at the front focal plane of optical collimator (compound achromatic lens, AL1), acting as the infinite distant target. Firstly, it is imaged at the focal plane of lens (AL2) with finite aperture and the image here is diffraction-limited. Following it is the relayed super-oscillatory imaging section, composed by a 4*f* system with two lenses AL3 and AL4. At the exit pupil of the system, near the lens AL4 is positioned with a pupil filter group with the metasurface filter sandwiched by two linear polarizers in crossed manner. The super-oscillatory image is recorded in the CCD at the second imaging plane.

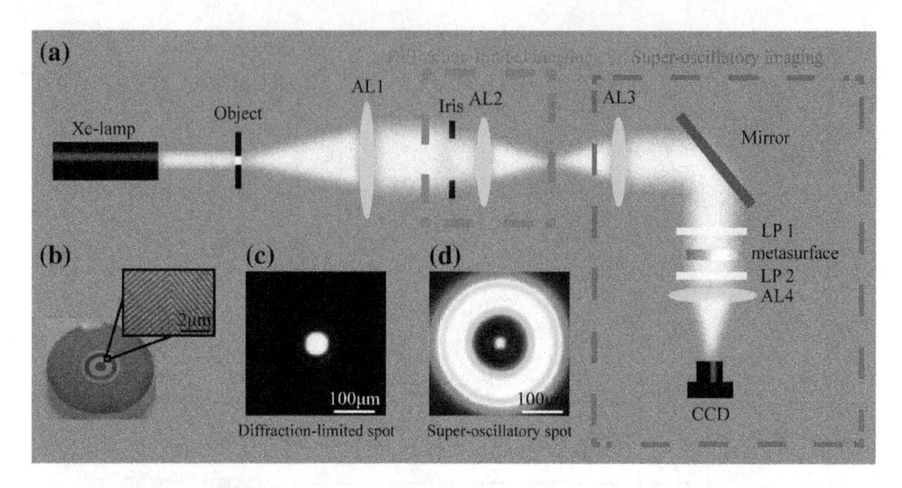

Fig. 8.17 **a** Schematic of experimental setup, including a Xe-lamp, an optical collimator (AL1, $f_1 = 1000$ mm), an iris, three same achromatic lenses (AL2, AL3, AL4, $f = 500$ mm), a mirror, two linear polarizers (LP1, LP2), a designed metasurface, and a CCD (ICL-B2520C). **b** The proposed metasurface. The inset indicates the distribution of the Cr gratings between two annuluses. **c** Diffraction-limited image of a 20 μm hole. **d** Super-oscillatory image of a 20 μm hole. AL: achromatic lens. Here, the positions and distances of elements are not in accurate scale with the experiment optic setup for a good visualization. Reproduced from [35] with permission. Copyright 2018, WILEY-VCH Verlag GmbH & Co. KGaA, Weinheim

The PSF of the MF-based SRT system was measured in the experiment by using a transparent circular hole (20 μm diameter) on an opaque screen. For the case without the filter group, the measured PSF at CCD plane is an Airy spot with full width of 82.8 μm, as shown in Fig. 8.17c. As the filter group is inserted in the optics, the PSF in Fig. 8.17d exhibits an obvious super-oscillation PSF pattern with a much smaller bright central spot surrounded by a wide lobe ring pattern about few spot size away and with much higher brightness. The super-oscillation central spot size is about 51.75 μm, being 0.625 times of the Airy spot and slight larger than the design result $G = 0.6$ because of some inevitable errors like fabrication and aberration in optics system. The central spot with reduced width provides the super-resolution imaging power of objects located inside the great lobe ring region of the super-oscillation PSF.

Figure 8.18 presents the SRT imaging for two transparent holes with 20 μm diameter and 60 μm center-to-center distance beyond diffraction-limited resolving power to demonstrate the resolving ability of the MF (Fig. 8.18a). As expected, the two point objects are totally unresolved for the case without MF (Fig. 8.16b) and are clearly distinguished with MF (Fig. 8.18c). The cross-sectional profiles of two images are plotted in Fig. 8.18d, showing the super-oscillatory image of clearly resolved two holes. The super-oscillation imaging also holds its super-resolution ability for extended objects as demonstrated in Fig. 8.18e–h. In this case, a target object "*E*" with 55 μm center-to-center and 100 μm width was used (Fig. 8.18e). Fig. 8.18f

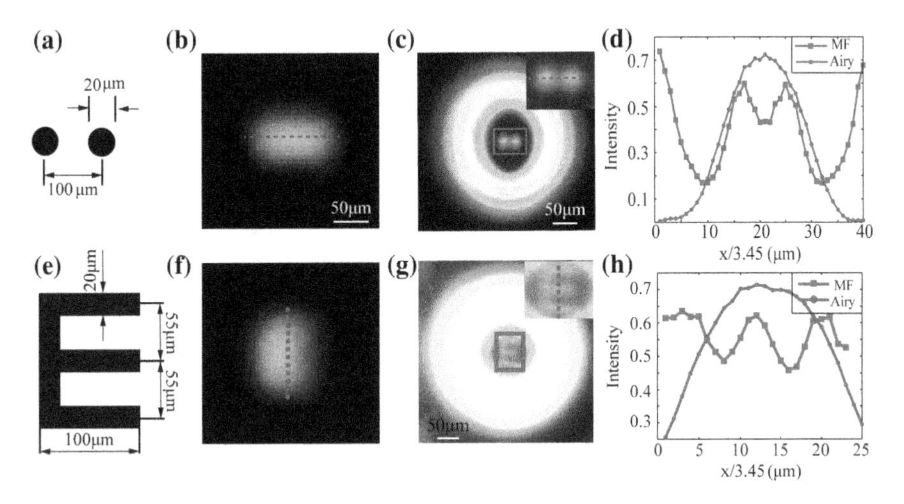

Fig. 8.18 **a** Diffraction-limited image. **b** Super-oscillatory image. **c** Intensity distribution of the central place marked by dashed lines in **a** and the insert of **b**. **d** Experiment imaging contrast for variant center-to-center distances of two-hole target. The dashed line is the contrast of Rayleigh criterion. Here, the contrast is defined as $(Ip-Iv)/(Ip+Iv)$, where Ip is the peak intensity and Iv the central valley intensity. When the two point objects are unresolved, the Ip is equal to Iv. **e** Target object "*E*." **f** Diffraction-limited imaging of "*E*." **g** Super-resolution imaging of "*E*." **h** Intensity distribution of the place labeled by red dashed in the **f** and the insert of **g**. Reproduced from [35] with permission. Copyright 2018, WILEY-VCH Verlag GmbH & Co. KGaA, Weinheim

shows that the "*E*" could not be resolved by the diffraction-limited imaging system, as predicted by the Rayleigh criterion. By employing MF, it could be imaged and resolved perfectly in Fig. 8.18g. The comparison of intensity distribution between the diffraction-limited pattern and the super-resolution one along the red line further verifies the super-resolving ability of MF in Fig. 8.18h. It is obvious in Fig. 8.18h that the image of "*E*" is surrounded by the high-intensity sidebands, which is the typical feature of super-oscillation phenomenon. Besides, compared to the SRT in Sect. 8.2.2, this method could realize the sub-diffraction-limited imaging in the whole visible light, instead of several given wavelengths.

In principle, there always exists inevitable side lobes around the sub-diffraction-limited hot spot in super-oscillatory phenomenon, which are detrimental to imaging and render low efficiency and limited FOV. Fortunately, it is the sub-diffraction-limited imaging feature in the far-field that we are mainly concerned about. According to the constraints on PSF, the special super-oscillatory pupil filter could be designed to make the FOV of PSF large enough so that the target objects are within the FOV. However, the efficiency would become lower with the increase of FOV. To solve this problem, some other methods have been used in super-oscillatory imaging, such as confocal scanning imaging [20, 22]. On the other hand, the low efficiency may be tolerable for some specific applications, such as observing target object with high intensity. Besides, high-sensitivity detectors and long exposing time are also advantageous.

8.3 Optical Telescope Based on Orbital Angular Momentum

Another way to overcome the diffraction limit in the far-field is to employ optical vortices (OV) in the imaging process [48, 49]. It has been demonstrated that OV is closely related to the orbital angular momentum (OAM) of light. In 1992, Allen and co-workers recognized that helically phased beam with $\Phi = l\varphi$ has quantized OAM of $l\hbar$, where l is known as topological charge and $\hbar$ is the Dirac constant. The details on the properties and generation of OAM will be introduced in Chap. 10.

The idea to utilize OV to break the Rayleigh criterion limit was proposed in 2001, and the schematic image is shown in Fig. 8.19a [48]. The original purpose of this design is to attenuate the intense glare of a bright coherent beam of light (as shown in the center of the left part in Fig. 8.19a) to enhance the detection of an incoherent background signal or a weak nearly collinear source (as shown in the bottom of the left part in Fig. 8.19a). The OV beam was produced by a transmission phase mask near a lens of focal length f and diameter D. After analyzing the amplitude profiles in the focal plane based on vector diffraction theory, the radius between the amplitude peaks with different l is plotted in Fig. 8.19b. The case $l = 0$ (the Airy disk) with $r = R_{\text{diff}} = 1.22\ \lambda f/D$ is also graphed in Fig. 8.19b to demonstrate the relative shifts of the amplitude peaks: $R_{l=1} = 0.64R_{\text{diff}}$, $R_{l=2} = 1.03R_{\text{diff}}$, $R_{l=3} = 1.37R_{\text{diff}}$, and $R_{l=4} = 1.71R_{\text{diff}}$. Interestingly, when $l = 1$, the radius between amplitude peaks are much smaller than that predicted by Rayleigh criterion; thus, sub-Rayleigh separation can be achieved.

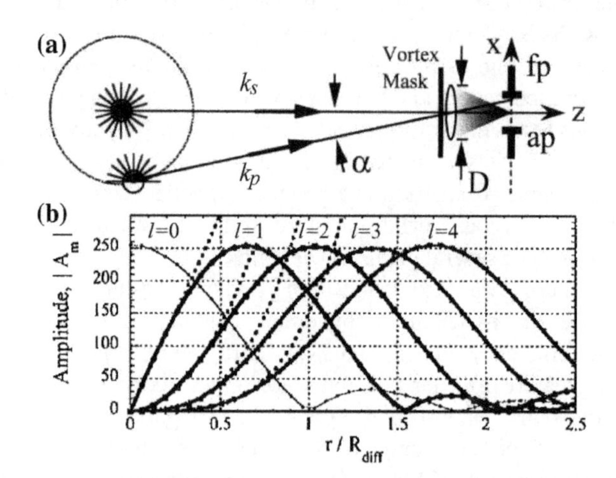

Fig. 8.19 **a** Schematic of the OV spatial filter system. k_s and k_p indicate light emitted from a bright source and a reflective plane. The right part of the system can be treated as a telescope, and the transmission phase mask near the lens is used to generate OV beam. **b** Line plots of the field amplitude between the peaks for values of topological charge l = 1~4. Adapted from [48] with permission. Copyright 2001, Optical Society of America

In the aforementioned work, it only discussed the separability of two identical overlapping OVs with the same integer OAM values. In 2006, a detailed research concerning the OV-assisted sub-Rayleigh separability was proposed with equally luminous monochromatic and white light sources [49]. The optical setup is shown in Fig. 8.20 where one beam was always coinciding with the optical center of the hologram (the hologram is encoded with OAM phase), thus generating an OV with integer topological charge $l = 1$. The other beam spanned the hologram in different positions starting from the optical center.

Figure 8.21a shows the simulated (upper row) and experimental (central row) results at different separations between two OV beams. The corresponding Airy patterns are given in the bottom row of Fig. 8.21a for comparison. It can be inferred from these figures that different Laguerre–Gaussian (LG) patterns can be obtained if the second beam is off centered because it acquires a non-integer OAM value producing an asymmetric pattern. Figure 8.21b plots the intensity of the central section of the combined profiles of the LG patterns along the direction connecting the two sources. Obviously, as the separation between the two beams increases, the initially symmetric profile is distorted and the intensity of the two maxima assumes different values.

To further validate the sub-Rayleigh separability of the proposed method, the ratio between the intensities of the peaks of the superposed LG modes vs the off-axis shift of the spot in units of the Rayleigh radius is plotted in Fig. 8.22. As the Rayleigh criterion is assumed that two identical sources are just resolved when the

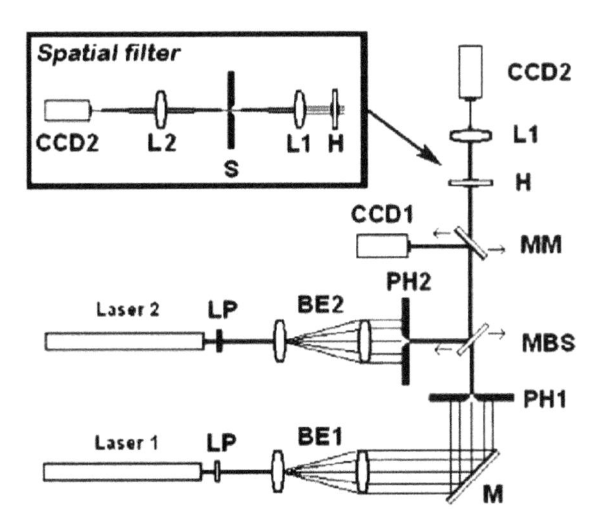

Fig. 8.20 Optical setup: LP's are neutral polarizing filters, BE1 and BE2 beam expanders, *M* a fixed mirror, MBS a moving beam splitter, MM a moving plane mirror, L1 a biconvex lens, H the fork hologram, CCD1 and CCD2 two CCD cameras. The inset represents the spatial filtering for the white light: L2 is a camera lens and S a narrow slit (aperture and 1 mm). Reproduced from [49] with permission. Copyright 2006, American Physical Society

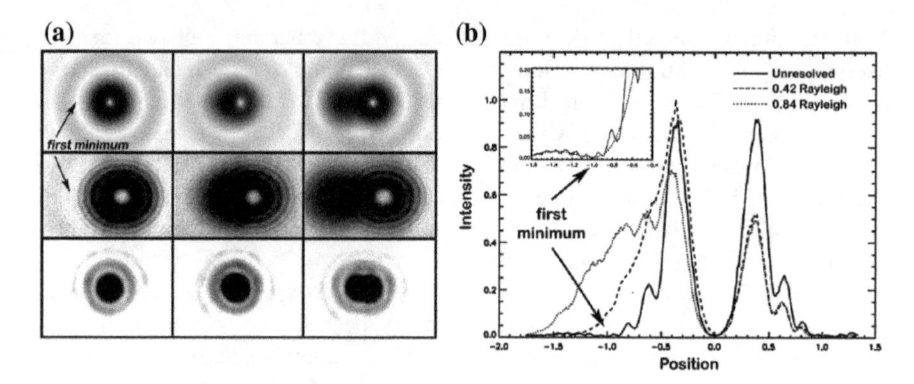

Fig. 8.21 **a** Images of the separation of two nearby monochromatic sources having same intensity. Left, middle and right columns correspond to the sources separated by 0, 0.42 and 0.84 times the Rayleigh criterion radius. **b** Experimental intensity profiles of the superposed LG modes at different separations. Reproduced from [49] with permission. Copyright 2006, American Physical Society

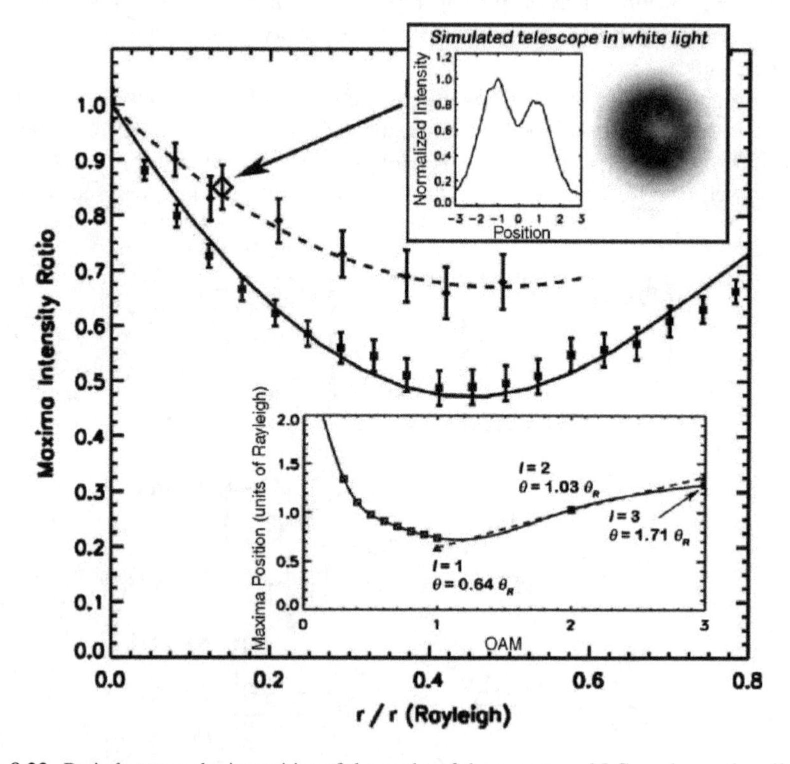

Fig. 8.22 Ratio between the intensities of the peaks of the superposed LG modes vs the off-axis shift of the spot in units of the Rayleigh radius. Reproduced from [49] with permission. Copyright 2006, American Physical Society

intensities of the asymmetric peaks differ by at least 5%, the theoretical separability is 50 times better than the Rayleigh limit in the monochromatic case. Analogously, the results obtained in white light illumination suggest a separability about 10 times better than the Rayleigh limit. This far-field sub-Rayleigh method is promising in several techniques in engineering optics and astronomy, especially in space missions and in telescopes with adaptive optics.

References

1. Extremely Large Telescope. https://en.wikipedia.org/wiki/Extremely_Large_Telescope
2. William Herschel Telescope. https://en.wikipedia.org/wiki/William_Herschel_Telescope
3. James Webb space telescope. https://en.wikipedia.org/wiki/James_Webb_Space_Telescope
4. Hubble space telescope. https://en.wikipedia.org/wiki/Hubble_Space_Telescope
5. X. Luo, Principles of electromagnetic waves in metasurfaces. Sci. China. Phys. Mech. Astron. **58**, 594201 (2015)
6. E.H.K. Stelzer, S. Grill, The uncertainty principle applied to estimate focal spot dimensions. Opt. Commun. **173**, 51–56 (2000)
7. J.Y. Wang, J.K. Markey, Modal compensation of atmospheric turbulence phase distortion. J. Opt. Soc. Am. **68**, 78–87 (1978)
8. M. Tegmark, M. Zaldarriaga, Fast fourier transform telescope. Phys. Rev. D **79**, 083530 (2009)
9. A.B. Meinel, Aperture synthesis using independent telescopes. Appl. Opt. **9**, 2501–2504 (1970)
10. L.W. Chen, Y. Zhou, M.X. Wu, M.H. Hong, Remote-mode microsphere nano-imaging: new boundaries for optical microscopes. Opto-Electron. Adv. **1**, 170001 (2018)
11. J.B. Pendry, Negative refraction makes a perfect lens. Phys. Rev. Lett. **85**, 3966–3969 (2000)
12. H. Gao, M. Pu, X. Li, X. Ma, Z. Zhao, Y. Guo, X. Luo, Super-resolution imaging with a Bessel lens realized by a geometric metasurface. Opt. Express **25**, 13933–13943 (2017)
13. C. Snoeyink, Imaging performance of bessel beam microscopy. Opt. Lett. **38**, 2550–2553 (2013)
14. S.W. Hell, J. Wichmann, Breaking the diffraction resolution limit by stimulated emission: stimulated-emission-depletion fluorescence microscopy. Opt. Lett. **19**, 780–782 (1994)
15. B. Huang, W. Wang, M. Bates, X. Zhuang, Three-dimensional super-resolution imaging by stochastic optical reconstruction microscopy. Science **319**, 810–813 (2008)
16. E. Betzig, G.H. Patterson, R. Sougrat, O.W. Lindwasser, S. Olenych, J.S. Bonifacino, M.W. Davidson, J. Lippincott-Schwartz, H.F. Hess, Imaging intracellular fluorescent proteins at nanometer resolution. Science **313**, 1642–1645 (2006)
17. G.T. Di Francia, Super-gain antennas and optical resolving power. Il Nuovo. Cimento. **9**, 426–438 (1952)
18. M. Martínez-Corral, P. Andrés, C.J. Zapata-Rodríguez, M. Kowalczyk, Three-dimensional superresolution by annular binary filters. Opt. Commun. **165**, 267–278 (1999)
19. M.V. Berry, S. Popescu, Evolution of quantum superoscillations and optical superresolution without evanescent waves. J. Phys. A: Math. Gen. **39**, 6965 (2006)
20. E.T.F. Rogers, J. Lindberg, T. Roy, S. Savo, J.E. Chad, M.R. Dennis, N.I. Zheludev, A super-oscillatory lens optical microscope for subwavelength imaging. Nat. Mater. **11**, 432–435 (2012)
21. C. Hao, Z. Nie, H. Ye, H. Li, Y. Luo, R. Feng, X. Yu, F. Wen, Y. Zhang, C. Yu, J. Teng, B. Luk'yanchuk, C-W. Qiu, Three-dimensional supercritical resolved light-induced magnetic holography. Sci. Adv. **3**, e1701398 (2017)
22. F. Qin, K. Huang, J. Wu, J. Teng, C.W. Qiu, M. Hong, A supercritical lens optical label-free microscopy: sub-diffraction resolution and ultra-long working distance. Adv. Mater. **29**, 1602721 (2016)

23. Y. Eliezer, L. Hareli, L. Lobachinsky, S. Froim, A. Bahabad, Breaking the temporal resolution limit by superoscillating optical beats. Phys. Rev. Lett. **119**, 043903 (2017)
24. G. Yuan, E.T.F. Rogers, T. Roy, Z. Shen, N.I. Zheludev, Flat super-oscillatory lens for heat-assisted magnetic recording with sub-50 nm resolution. Opt. Express **22**, 6428–6437 (2014)
25. F.M. Huang, T.S. Kao, V.A. Fedotov, Y. Chen, N.I. Zheludev, Nanohole array as a lens. Nano. Lett. **8**, 2469–2472 (2008)
26. F.M. Huang, N.I. Zheludev, Super-resolution without evanescent waves. Nano. Lett. **9**, 1249–1254 (2009)
27. K. Huang, H. Ye, J. Teng, S.P. Yeo, B. Luk'yanchuk, C.W. Qiu, Optimization-free superoscillatory lens using phase and amplitude masks. Laser. Photonics. Rev. **8**, 152–157 (2014)
28. A.M.H. Wong, G.V. Eleftheriades, An optical super-microscope for far-field, real-time imaging beyond the diffraction limit. Sci. Rep. **3**, 1715 (2013)
29. S. Kosmeier, M. Mazilu, J. Baumgartl, K. Dholakia, Enhanced two-point resolution using optical eigenmode optimized pupil functions. J. Opt. **13**, 105707 (2011)
30. J. Baumgartl, S. Kosmeier, M. Mazilu, E.T.F. Rogers, N.I. Zheludev, K. Dholakia, Far field subwavelength focusing using optical eigenmodes. Appl. Phys. Lett. **98**, 181109 (2011)
31. C. Wang, D. Tang, Y. Wang, Z. Zhao, J. Wang, M. Pu, Y. Zhang, W. Yan, P. Gao, X. Luo, Super-resolution optical telescopes with local light diffraction shrinkage. Sci. Rep. **5**, 18485 (2015)
32. H. Liu, Y. Yan, Q. Tan, G. Jin, Theories for the design of diffractive superresolution elements and limits of optical superresolution. J. Opt. Soc. Am. A **19**, 2185–2193 (2002)
33. W.K. Pratt, F. Davarian, Fast computational techniques for pseudoinverse and wiener image restoration. IEEE Trans. Comput. C **26**, 571–580 (1977)
34. M. Born, E. Wolf, *Principle of Optics*, 7th edn. (Pergamon, Oxford, UK, 2007)
35. Z. Li, T. Zhang, Y. Wang, W. Kong, J. Zhang, Y. Huang, C. Wang, X. Li, M. Pu, X. Luo, Achromatic broadband super-resolution imaging by super-oscillatory metasurface. Laser. Photonics. Rev. **12**, 1800064 (2018)
36. M. Pu, X. Li, X. Ma, Y. Wang, Z. Zhao, C. Wang, C. Hu, P. Gao, C. Huang, H. Ren, X. Li, F. Qin, J. Yang, M. Gu, M. Hong, X. Luo, Catenary optics for achromatic generation of perfect optical angular momentum. Sci. Adv. **1**, e1500396 (2015)
37. D. Tang, C. Wang, Z. Zhao, Y. Wang, M. Pu, X. Li, P. Gao, X. Luo, Ultrabroadband super-oscillatory lens composed by plasmonic metasurfaces for subdiffraction light focusing. Laser Photonics Rev. **9**, 713–719 (2015)
38. X. Li, L. Chen, Y. Li, X. Zhang, M. Pu, Z. Zhao, X. Ma, Y. Wang, M. Hong, X. Luo, Multicolor 3D meta-holography by broadband plasmonic modulation. Sci. Adv. **2**, e1601102 (2016)
39. X. Li, M. Pu, Z. Zhao, X. Ma, J. Jin, Y. Wang, P. Gao, X. Luo, Catenary nanostructures as compact bessel beam generators. Sci. Rep. **6**, 20524 (2016)
40. L. Huang, X. Chen, H. Mühlenbernd, H. Zhang, S. Chen, B. Bai, Q. Tan, G. Jin, K.W. Cheah, C.W. Qiu, J. Li, T. Zentgraf, S. Zhang, Three-dimensional optical holography using a plasmonic metasurface. Nat. Commun. **4**, 2808 (2013)
41. M. Khorasaninejad, W.T. Chen, R.C. Devlin, J. Oh, A.Y. Zhu, F. Capasso, Metalenses at visible wavelengths: Diffraction-limited focusing and subwavelength resolution imaging. Science **352**, 1190–1194 (2016)
42. J. Lin, S. Wu, X. Li, C. Huang, X. Luo, Design and numerical analyses of ultrathin plasmonic lens for subwavelength focusing by phase discontinuities of nanoantenna arrays. Appl. Phys. Express **6**, 022004 (2013)
43. G. Yuan, E.T.F. Rogers, T. Roy, G. Adamo, Z. Shen, N.I. Zheludev, Planar super-oscillatory lens for sub-diffraction optical needles at violet wavelengths. Sci. Rep. **4**, 6333 (2014)
44. H. Wang, L. Shi, B. Lukyanchuk, C. Sheppard, C.T. Chong, Creation of a needle of longitudinally polarized light in vacuum using binary optics. Nat. Photon. **2**, 501–505 (2008)
45. Q. Zhan, Properties of circularly polarized vortex beams. Opt. Lett. **31**, 867–869 (2006)
46. G.M. Lerman, U. Levy, Effect of radial polarization and apodization on spot size under tight focusing conditions. Opt. Express **16**, 4567–4581 (2008)

47. F. Qin, K. Huang, J. Wu, J. Jiao, X. Luo, C. Qiu, M. Hong, Shaping a subwavelength needle with ultra-long focal length by focusing azimuthally polarized light. Sci. Rep. **5**, 9977 (2015)
48. G.A. Swartzlander, Peering into darkness with a vortex spatial filter. Opt. Lett. **26**, 497–499 (2001)
49. F. Tamburini, G. Anzolin, G. Umbriaco, A. Bianchini, C. Barbieri, Overcoming the Rayleigh criterion limit with optical vortices. Phys. Rev. Lett. **97**, 163903 (2006)

Chapter 9
Metalenses and Meta-mirrors

Abstract Lenses are the fundamental optical components and play the key roles in most of the optical systems, including cameras, microscopes, telescopes, projective lithographic machines, and spectrometers. Traditional lenses are made from materials such as glass or plastic and are polished or molded to desired shapes. However, the traditional refractive/reflective or diffractive lenses have their intrinsic limits in integration, weight, chromatic aberration, among others. The newly emerging metalenses may be promising alternatives to overcome these limits for practical applications. In this chapter, we will start with a brief review of the traditional lens in Sect. 9.1. Then, the design methods of the planar metalens and meta-mirror in EO 2.0 are introduced in Sect. 9.2. In Sects. 9.3 and 9.4, planar lenses with large numerical aperture (NA) and wide field of view, which are extremely difficult to realize in traditional optics with compact volume, are discussed in detail. Important technologies and the latest developments in metalenses, including achromatic or super-chromatic imaging, and tunable imaging, are elaborated and highlighted in Sects. 9.5 and 9.6. At last, we also give a brief introduction of nonlinear metalens in Sect. 9.7.

Keywords Flat optics · Flat lens · Snell's law · Active lens

9.1 Lenses in Traditional Optics

The word lens comes from lēns, the Latin name of the lentil, because a double-convex lens is lentil-shaped. The functions that lenses perform include the convergence or divergence of light beams and the formation of virtual or real images. Lenses are also used to implement the Fourier transform optically which is extremely important in frequency spectra processing in information optics. Traditional lenses in EO 1.0 are widely constructed based on the laws of refraction and reflection in geometric optics and the classical diffraction theory in physical optics. Refractive lenses have been maturely used in optical imaging systems as shown in Fig. 9.1; however, they are often bulky, heavy, and costly. The mass of the lens usually increases approximately by the cube of its diameter. Thus, for example, growth of refractive telescope's aper-

X. Luo, *Engineering Optics 2.0*, https://doi.org/10.1007/978-981-13-5755-8_9

Fig. 9.1 **a** Singlet refractive lenses. **b** A cross section of the objective lens system of Olympus E-30 DSLR camera. **c** Starlith 1900 (from Carl Zeiss) high-NA projection system based on lenses and curved mirrors. **a** Reproduced from Wikipedia [1]. **b** Reproduced from Wikipedia [2]. **c** Reproduced from [3] with permission. Copyright 2007, Nature Publishing Group

ture is extremely hard when the diameter becomes larger than about 1 m. Although large telescopes based on reflective scheme have much smaller weight compared with those refractive lenses at the same aperture, the precise fabrication and measurement are still very challenging. On the other hand, the mass and dimension of the lens systems are also strongly dependent on their performances, which require very precise environment control. Taking the lithographic objective lens as an example, the high numerical aperture (NA) Starlith 1900 objective lens (from Carl Zeiss) for deep-ultraviolet (DUV) photolithography has a height more than 1 m and a weight of around 1000 kg.

The surfaces of both the refractive and reflective lenses are shaped to be curved according to the Fermat's principle. The curved profile and large weight of refractive/reflective lenses hinder the development of the next-generation optical systems, especially in large-aperture space telescopes and wearable optical devices and systems. Planar components based on diffractive optics (Fig. 9.2) are supposed to overcome these limits and have been developed for decades. However, the constitutive elements in diffractive lenses are spaced on a wavelength scale, thus suffering from the high-order diffractions. These spurious orders not only degrade the efficiency of

Fig. 9.2 **a** Lighthouse with Fresnel lens. **b** Photon sieve. **c** Solar imaging system with photon sieves. **a** Reproduced from Wikipedia [4]. **b** Reproduced from [5] with permission. Copyright 2007, Optical Society of America. **c** Reproduced from NASA [6]

diffractive lenses but also give rise to undesired effects such as virtual focal spots, halos, and ghost images. Another great limit of the traditional diffractive lenses, including Fresnel lenses and photon sieves (Fig. 9.2a, b), is the strong chromatic aberrations that are hard to compensate. Metalenses, thin and flat lenses composed of gradient subwavelength structures, circumvent the formation of spurious diffraction orders, which generally prevails in conventional diffractive components. High-performance optical systems with high-level integration can be realized by taking advantages of the super design flexibility and ability of the subwavelength structures.

Furthermore, in EO 1.0, fundamental limits of the performances in refractive/diffractive optical lenses and imaging system widely exist, especially for singlet lenses. For examples, the NA and the field of view (FOV) of the traditional refractive/diffractive lenses are limited, even for imaging systems with lens group; the singlet diffractive lens suffers from axial chromatic aberration, which is determined by the Abbe number of the lens material and focal ratio or f-number (FN) of the lens, which is a key barrier for the widely application of diffractive optics [7, 8]. Apart from the basic design principles and strategies, in this chapter, we will give an introduction of the recent development of the singlet metalenses for high NA, large

FOV, and also achromatic aberration. Aberration correction approach adopted from traditional design in EO 1.0 will also be analyzed for its application in metalenses.

9.2 Planar Metalens and Meta-mirror

Planar and thin lenses are highly desirable for both the compact and large-aperture optical imaging systems. Figure 9.3 shows a schematic plot of a planar metalens. According to the Fermat's principle, the required phase distribution of the transmission light at the output surface is:

$$\phi(x, y) = 2m\pi + \frac{2\pi f}{\lambda} - \frac{2\pi\sqrt{f^2 + x^2 + y^2}}{\lambda}, \quad (9.2.1)$$

where m is an arbitrary integer number, f the focal length, and λ the wavelength.

The design of metalenses is to construct this phase distribution at the output surface using subwavelength structures in a planar way. In general, there are three categories of localized phase modulation with subwavelength structures according to the physical mechanism as introduced in Sect. 2.3. The first kind is the propagation phase engineering, such as surface plasmon polariton (SPP)-based phase modulation using metallic waveguide array made of subwavelength slits and their variations, and also dielectric propagation phase modulation. The second localized phase modulation scheme is based on the circuit resonance in complex metallic surfaces. The third approach for phase modulation relies on the geometric phase, a frequency-

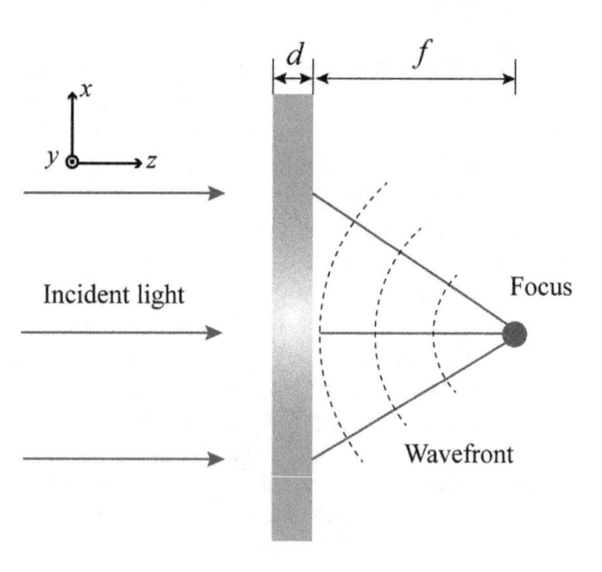

Fig. 9.3 Schematic of a thin planar lens

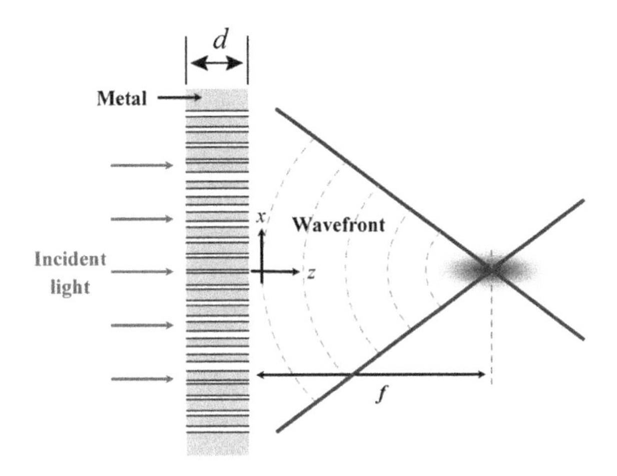

Fig. 9.4 Schematic of plasmonic metalens based on nanoslits array. Reproduced from [9] with permission. Copyright 2007, American Institute of Physics

independent phase originating from the photonic spin–orbit interaction during the process of polarization conversion.

Figure 9.4 illustrates the schematic of a one-dimensional metalens based on SPP propagation phase modulation, where a metallic film is perforated with a great number of nanoslits with specifically designed widths; thus, the transmitted light from the slits is modulated and converges in free space [9].

Recalling the dispersion equation of metal–insulator–metal (MIM) waveguide, the wave number of the fundamental mode for transverse magnetic (TM) illumination k_{sp} can be calculated from [9, 10],

$$\tanh(\sqrt{k_{\mathrm{sp}}^2 - k_0^2\varepsilon_{\mathrm{d}}}w/2) = \frac{-\varepsilon_{\mathrm{d}}\sqrt{k_{\mathrm{sp}}^2 - k_0^2\varepsilon_{\mathrm{m}}}}{\varepsilon_{\mathrm{m}}\sqrt{k_{\mathrm{sp}}^2 - k_0^2\varepsilon_{\mathrm{d}}}}, \tag{9.2.2}$$

where k_0 is the wavevector of light in free space, ε_m and ε_d are the respective permittivity of the metal and dielectric material inside the slits, and w is the slit width. The real and imaginary parts of k_{sp} determine the phase velocity and the propagation loss of SPPs inside the metallic slit, respectively. The phase retardation of light transmitted through the slit is expressed as,

$$\phi = \mathrm{Re}(k_{\mathrm{sp}})d + \theta, \tag{9.2.3}$$

where θ is the factor stemming from the multiple reflections of light between the entrance and exit interfaces and can be calculated with the following equation:

$$\theta = \arg\left[1 - \left(\frac{1 - k_{\mathrm{sp}}/k_0}{1 + k_{\mathrm{sp}}/k_0}\right)^2 \exp(i2k_{\mathrm{sp}}d)\right] \tag{9.2.4}$$

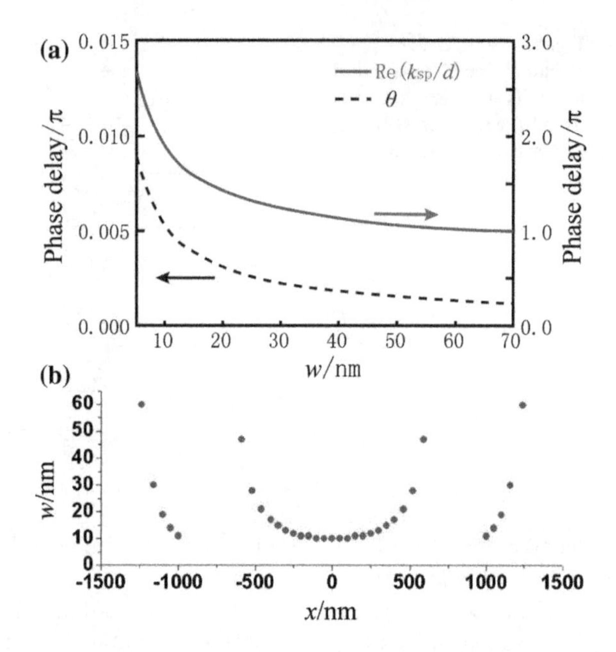

Fig. 9.5 **a** Dependence of phase delay on the slit width. Red and black tags represent the contributions for phase delay from the real part of propagation constant and multiple reflections. **b** Distribution of silt widths and positions on the designed metallic slab lens. Reproduced from [14] with permission. Copyright 2007, American Institute of Physics

The dependence of phase retardation on slit width is shown in Fig. 9.5a. Both the propagation phase and the multiple reflection phase increase steadily with decreased slit widths and grow rapidly for slit width below 20 nm. And it can be clearly seen that the phase retardation is dominantly determined by the real part of propagation constant. Therefore, $\Delta\phi$ can be approximated as $\mathrm{Re}(k_{\mathrm{sp}}d)$, which can be tuned by varying the slit width. In the calculation, silver is chosen as the metal material, and its permittivity $\varepsilon_{\mathrm{m}} = -29.26 + i1.348$ is used at the incident wavelength of 810 nm; the dielectric is air with $\varepsilon_{\mathrm{d}} = 1$. Notably, the evanescent coupling of SPPs in the metallic slits results in a catenary-shaped intensity distribution (hyperbolic cosine function), which not only explains the width-dependent propagation constant but also enables high-contrast nanolithography with plasmonic cavity [11–13].

Combining the width-dependent phase delay with the phase distribution of a lens, i.e., Eq. (9.2.1), a series of Ag slits are designed with predefined phase retardations, as shown in Fig. 9.5b. The thickness of the Ag slab is 300 nm. The space between every two adjacent slits is assumed to be larger than the metal's skin depth to prevent the coupling effect. Finite-difference time-domain (FDTD) simulations are performed to illustrate the validity of this metalens. Figure 9.6 shows the simulation results for point-to-point imaging with metalens. The point source is localized at $x = 0$ and $z = 0.2$ μm. The metallic slab lens ranges from $z = 1.2$ μm to $z = 1.5$ μm. The radiated light from the source is TM-polarized with a wavelength of 810 nm. Obvious image spot can be seen at the position around $z = 2.43$ μm with the full width at half maximum (FWHM) of 396 nm, which is approximately half of the incident wavelength. The curve at the right side represents the cross section of image plane at

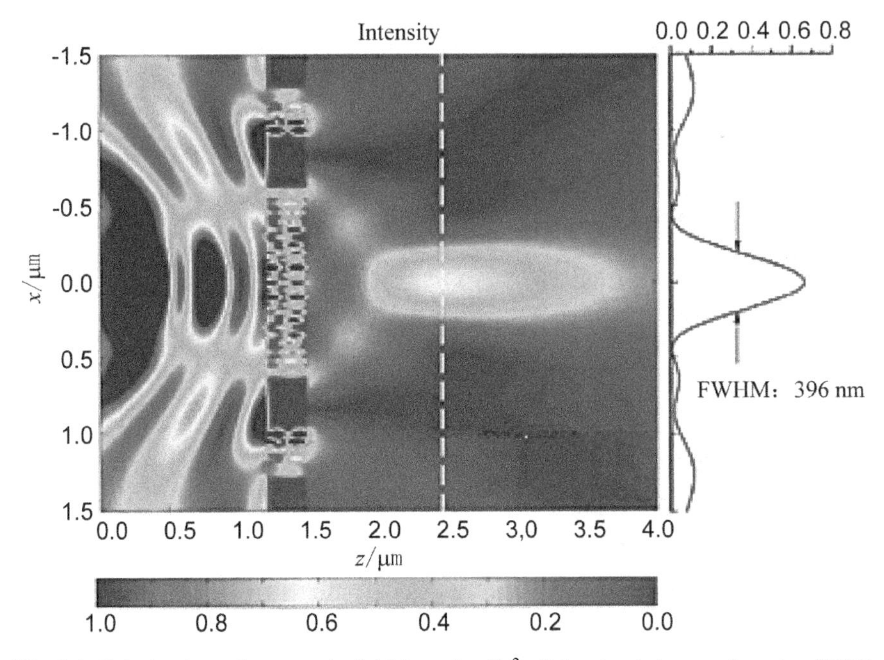

Fig. 9.6 Calculated steady magnetic field intensity Hy^2 of the simulation results using FDTD method. Reproduced from [14] with permission. Copyright 2007, American Institute of Physics

$z = 2.435$ μm. The nearest adjacent side lobe level is about 0.07, nearly approaching the theoretical value of sinc function.

The first experimental demonstration of far-field one-dimensional lensing using a plasmonic slit array was performed by Fan's group in 2009 [15]. They implemented a planar nanoslit lens using a combination of thin-film deposition and focused ion-beam milling. Geometry of the lens consists of a 400-nm optically thick gold film with air slits of different widths (80–150 nm) milled therein on a fused silica substrate as shown in Fig. 9.7. A *p*-polarized wave with a wavelength of 637 nm was focused at the focal length of 5.3 μm with a FWHM of 0.88 μm through the plasmonic variable lens.

To realize polarization-independent 2D focusing, the nanoslits were replaced by circular or square plasmonic holes with variable radius or dielectric nanorods to generate polarization-independent phase modulation [16, 17]. Simultaneous control of light polarization and phase distributions of light was also achieved using more complex nanoslits [18]. It should be noted that the focusing mechanism of these holes is completely different from the so-called Fresnel lenses or photon sieves [19, 20], which do not change the local phase front of the incoming waves.

Dielectric-based propagation phase engineering is an alternative way to realize high-efficiency metalens with comparatively larger thickness [21–25]. Metalens consisting of nanoposts with different materials, such as silicon, silicon nitride, gallium nitride, and titanium dioxide can be designed through modulating the propaga-

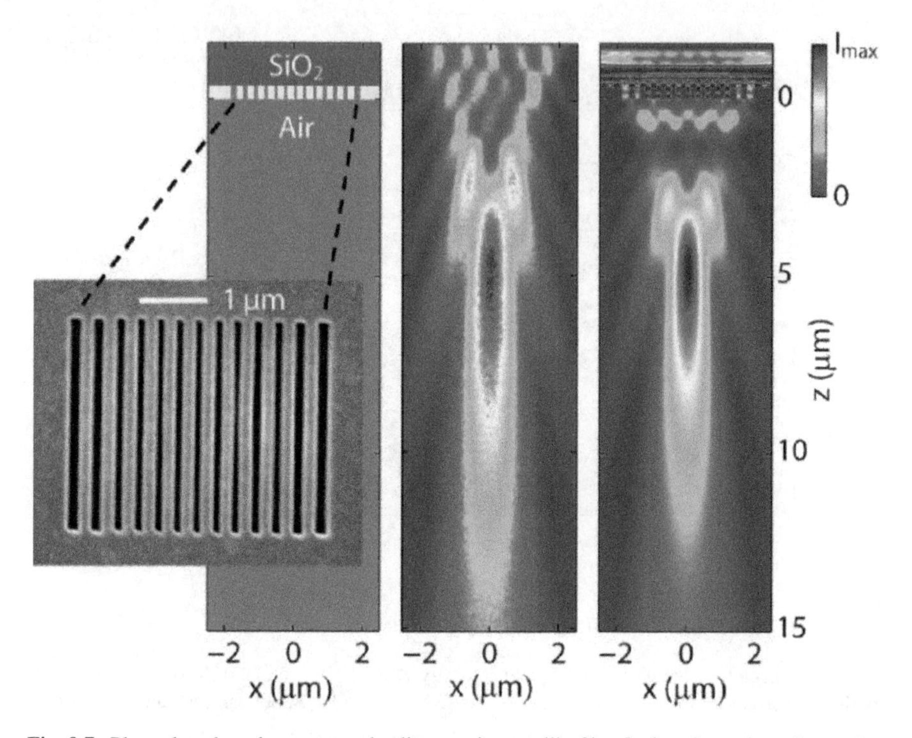

Fig. 9.7 Planar lens based on nanoscale slit array in metallic film. Left: schematic and scanning electron micrograph of the 1D metalens. Middle: focusing pattern measured by confocal scanning optical microscopy (CSOM). Right: finite-difference frequency domain (FDFD) simulated focusing pattern of the field intensity through the center of the slits. Reproduced from [15] with permission. Copyright 2009, American Chemical Society

tion constant by varying their diameters. Silicon-based metalens has been successfully applied in endoscope, termed nano-optic endoscope, for high-resolution optical coherence tomography (OCT) in vivo [21]. Figure 9.8a, b illustrates the schematic and the real photograph of the nano-optic endoscope with metalens. Light is delivered and subsequently collected through a length of single-mode fiber housed within a drive cable and a ferrule with an 8° angle-polished end facet to reduce reflections. A prism attached to the ferrule redirects the light toward the metalens, for a side-viewing catheter suitable for imaging luminal organs. The transmissive metalens is comprised of amorphous silicon (a-Si) nanopillars on a 170-μm-thick cover glass that focuses and collects light from a target object. The schematic and scanning electron microscope (SEM) images of the Si nanopost metalens are shown in Fig. 9.8c, d. The a-Si nanopillars featuring a height of 750 nm and varying diameters are used to locally impart the required phase.

Due to its ability to modify the propagation phase at subwavelength level, the nano-optic endoscope can achieve near diffraction-limited imaging through negating non-chromatic aberrations. The imaging on fruit flesh (grape) and ex vivo swine airways

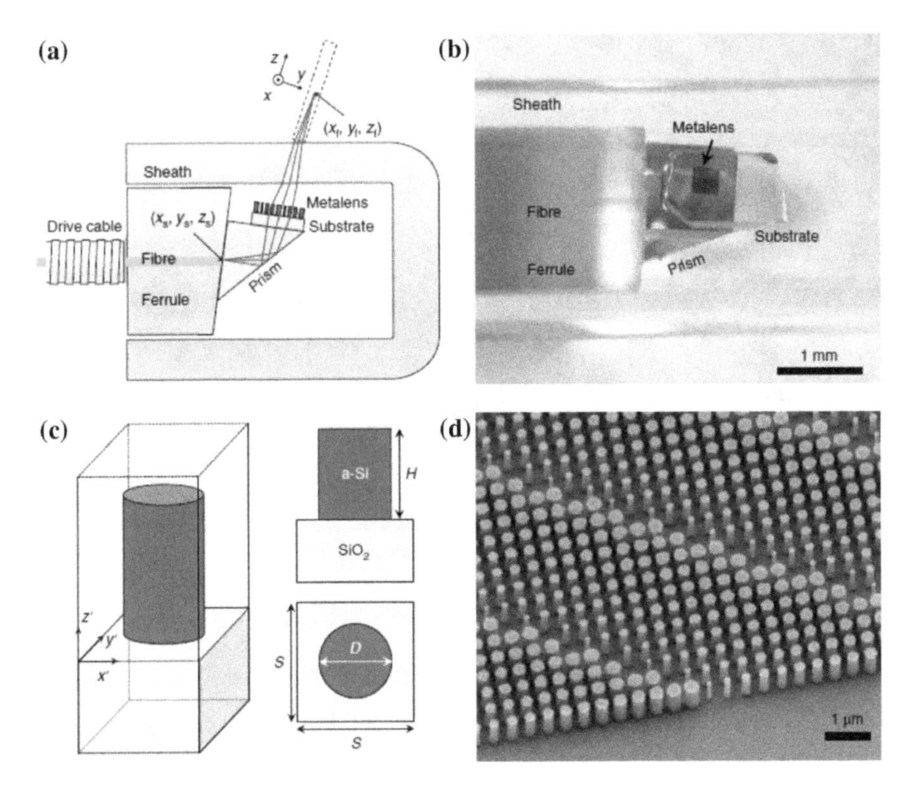

Fig. 9.8 **a** Schematic of the nano-optic endoscope. The metalens was designed to image a point source at (x_s, y_s, z_s) to a diffraction-limited spot at (x_f, y_f, z_f) with working distance WD = 0.5 mm. **b** Photographic image of the distal end of the nano-optic endoscope. **c** Schematic of an individual metalens building block consisting of an amorphous silicon (a-Si) nanopillar on a glass substrate. The nanopillars have height $H = 750$ nm and are arranged in a square lattice with unit cell size $S = 400$ nm. Phase imparted by a nanopillar is controlled by its diameter (D). **d** Scanning electron micrograph image of a portion of a fabricated metalens. Reproduced from [21] with permission. Copyright 2018, The Authors

with nano-optic endoscope and a conventional OCT catheter with a ball lens is shown in Fig. 9.9. Compared with image captured by the conventional OCT catheter, the image captured using the nano-optic endoscope is notably of superior quality for fruit flesh imaging (Fig. 9.9a, b). The cellular walls are more clearly visualized, and small-sized cells can be identified in the magnified images obtained using the nano-optic endoscope. Figure 9.9c, d shows the comparison of the images obtained in the swine airway using a conventional catheter and the nano-optic endoscope. The clear delineation of the layers of the airway wall and the visualization of fine glands in the bronchial mucosa further highlight the superior image quality of the nano-optic endoscope. The endoscopic imaging in resected human lung specimens and in sheep airways in vivo has also been demonstrated. The combination of the superior

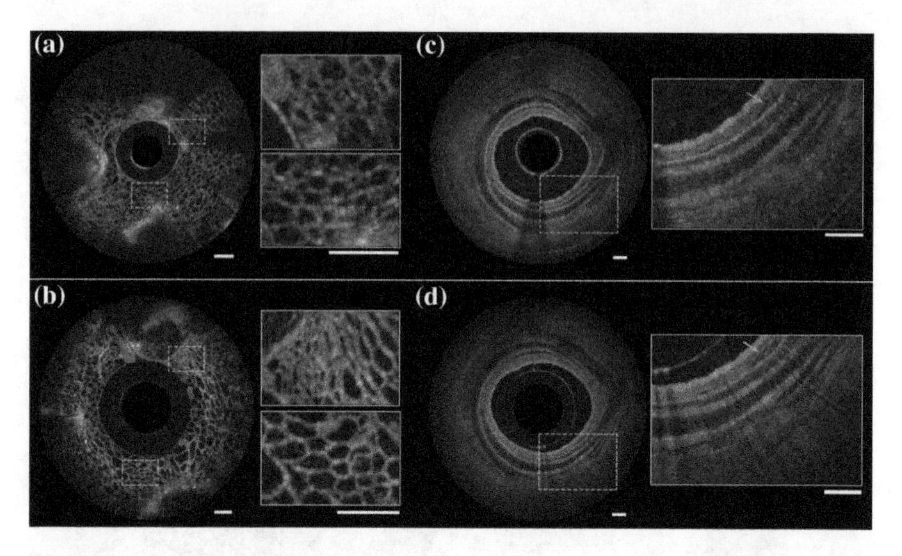

Fig. 9.9 Comparison of OCT images acquired using the nano-optic endoscope and a conventional OCT catheter. **a**, **b** OCT images of fruit flesh (grape) obtained using a ball lens catheter (**a**) and the nano-optic endoscope (**b**). **c**, **d** Ex vivo images of swine airway using a ball lens catheter (**c**) and the nano-optic endoscope (**d**). The arrows indicate fine glands in bronchial mucosa. All scale bars, 500 μm. Reproduced from [21] with permission. Copyright 2018, The Authors

resolution and higher imaging depth of focus (DOF) of the nano-optic endoscope is likely to increase the clinical utility of endoscopic optical imaging.

For the purpose of realistic application, large-aperture thin-film metasurface lenses were fabricated on flexible polyimide (PI) substrate [26]. The measured focal point of the large-aperture metalens in a telescope system reaches the diffraction limit at wavelengths ranging from 450 to 650 nm as depicted in Fig. 9.10. For white light, the measured focal spot is 52.5 μm, which is just a bit larger than the theoretical result (51.6 μm). Due to the use of light PI substrate, the areal density of this lens is reduced to be only 0.066 kg/m^2.

Similar design strategy of planar metalens has also been used with circuit-type phase engineering and geometric phase engineering [27–32]. In some cases, the circuit-type phase engineering-based metalens is also referred as Huygens' metasurface. The circuit-type phase modulation can be equivalent to the engineering of the localized effective electric and magnetic current. So a metalens can be designed by building an electrometric sheet with corresponding electric and magnetic current distribution. Geometric phase-based dielectric metalens has attracted great attention due to its broadband characteristics, especially in optical band [29–32]. Figure 9.11 illustrates a typical geometric phase-based chiral metalens with spatial rotated titanium dioxide dielectric nanofins [30]. Figure 9.11a shows the SEM image of the dielectric metalens, and its chiral imaging for the beetle, Chrysina gloriosa, is shown in Fig. 9.11b. The design principle of this metalens is taking advantage of the geometric phase modulation through rotating the orientation directions of the nanofins.

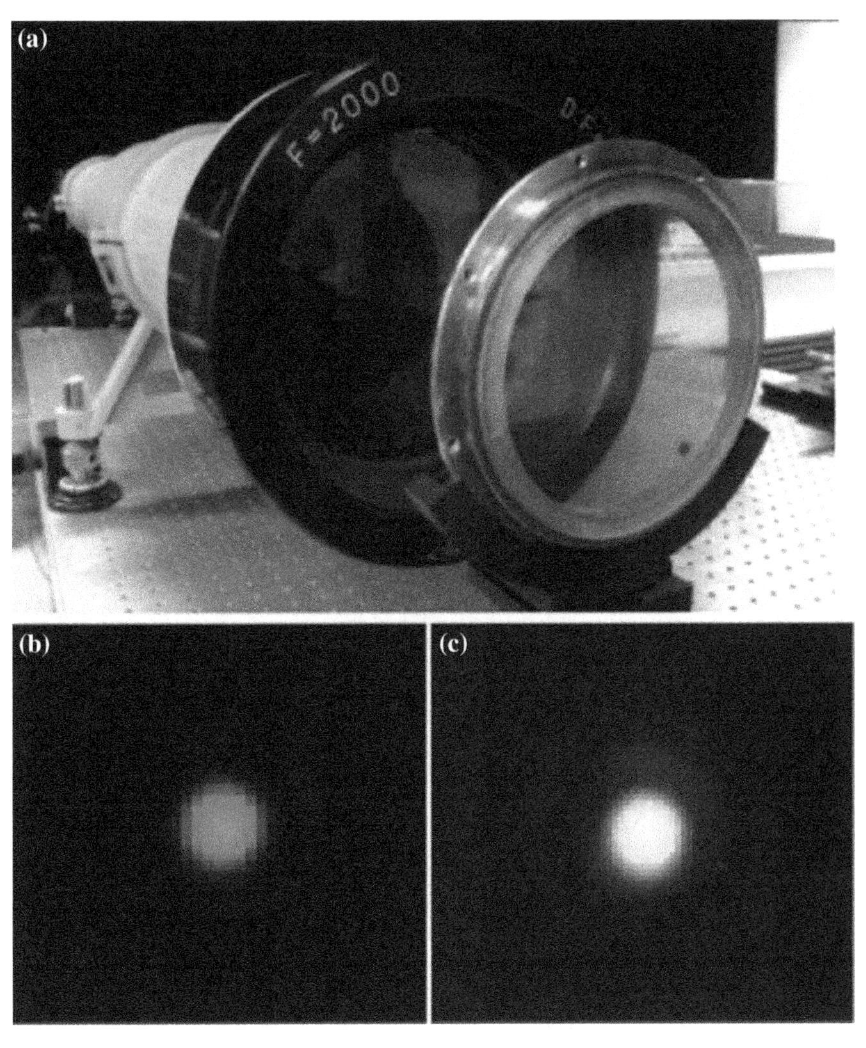

Fig. 9.10 **a** Photograph of the thin-film metasurface lens. Focal spots for red light (**b**) and white (**c**) light. Reproduced from [26] with permission. Copyright 2015, Science China Press and Springer-Verlag Berlin Heidelberg

And a single nanofin can be treated as a half-wave plate to transform a circularly polarized beam to its cross-polarization effectively.

Another kind of important planar metalens is meta-mirror, which works in reflective mode similar to the traditional concave mirror such as the parabolic mirror. Parabolic mirror plays a vital role in the design of microwave antennas and various integrated optical systems including IR spectrometers and the lithographic objective lens. Different from the concave mirror, meta-mirror focuses the light by shaping the phase with planar subwavelength structures instead of shaping its curved pro-

Fig. 9.11 **a** SEM micrograph of a portion of the chiral metalens showing the high-aspect-ratio TiO_2. Scale bar: 600 nm. **b** Imaging with the chiral metalens for the beetle, Chrysina gloriosa. The left image is formed by focusing left circularly polarized light reflected from the beetle and the right image is from right circularly polarized light reflected from the beetle. Illumination was provided by green LEDs paired with 10 nm bandpass filter centered at 532 nm. Reproduced from [30] with permission. Copyright 2016, American Chemical Society

file. Architecture combing H-shaped, also called I-shaped subwavelength structures, with a metallic ground is commonly used in design of the meta-mirrors [33–35]. Figure 9.12a illustrates the unit cell of I-shaped metasurface composed of two metallic wires being connected with a pair of metallic patches. When the electric field is along the y-direction, the structure can be treated as a combination of an effective inductor L and capacitor C. The corresponding effective sheet impedance is $Z = j\omega L + 1/(j\omega C)$. The capacitor C can be simply tuned by changing the gap width g; thus, the phase of the reflection light is manipulated for y-polarization illumination. The phase shift for different gap widths is illustrated in Fig. 9.12b. It is nearly linear in the whole frequency range considered here (8–12 GHz). This design strategy was adopted to construct a 1D meta-mirror in the microwave frequency [35].

To simplify the fabrication of the meta-mirror in optical frequency, the I-shaped structures can be modified to rectangular patches as shown in Fig. 9.13. This metal–insulator–metal configuration, whose top metal layer consists of a periodic arrangement of differently sized metal nanobricks, was designed to function as broadband focusing flat meta-mirror [36].

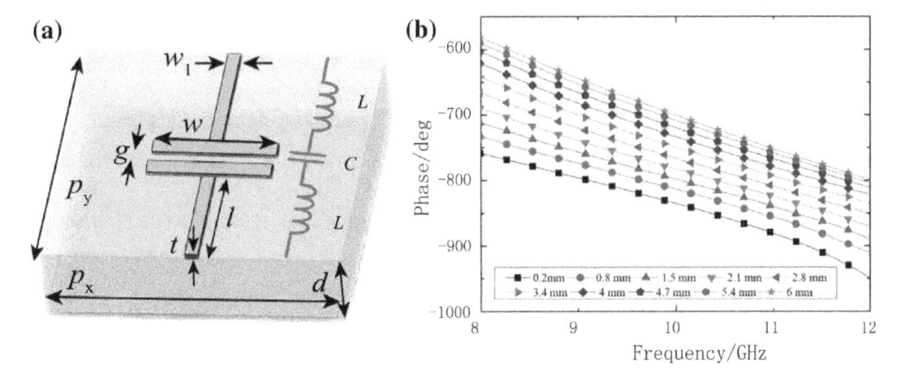

Fig. 9.12 **a** Schematic of the I-shaped unit cell of meta-mirror. **b** Reflection phases for different values of gap width g for thickness of $d = 6$ mm. The other geometric parameters are chosen as $l = 4$ mm and $w = 0.2$ mm. Reproduced from [35] with permission. Copyright 2013, The Authors

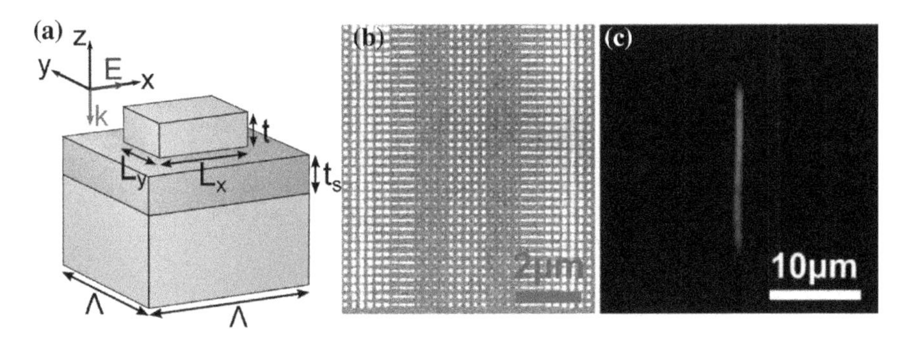

Fig. 9.13 **a** Schematic of the considered gold–glass–gold unit cell with fixed parameters $\Lambda = 240$ nm, $t_s = 50$ nm, and $t = 50$ nm. The normally incident wave ($\lambda = 800$ nm) is x-polarized. **b** The SEM image of the metalens with circuit-type phase engineering. **c** The measured focusing result. Reproduced from [36] with permission. Copyright 2013, American Chemical Society

9.3 High-NA Metalens

The NA is one of the most basic parameters of a lens, which determines the ability to focus light and the resolving capability. The NA of a lens is defined as,

$$\text{NA} = n \sin\theta, \tag{9.3.1}$$

where n is the refractive index in which the image lies and θ is the slope angle of the marginal ray exiting the lens. High NA is desired for applications requiring small light–matter interaction volumes or large angular collections, including high-resolution microscope, lithography, and high-density data recording. Traditionally, a refractive lens with large NA requires precision bulk optic. The high-NA lens demands the implementation of optical elements with rapidly varying phase profiles, which corresponds rapid thickness change for refractive lens and multi-steps grayscale lithography for traditional diffractive lens. Both of these two strategies are very difficult to be accomplished. Currently, the maximum free-space NA of optical objectives available in the market is typically limited to 0.9–0.95 with complex optical system. In contrast, metasurfaces allow the lens designer to circumvent those issues that producing high-NA lenses in an ultra-flat fashion [32, 37–42]. There are two kinds of design strategies to obtain high-NA metalenses so far. The first one is making use of the phase engineering at nanoscale, and the second one is based on the diffraction energy redistribution of the subwavelength high-contrast gratings.

For the first configuration with phase engineering, the phase modulation should vary fast enough, especially at the edge of the lens. Silicon nanoposts with varying diameters were developed to construct a high-NA micro-lens based on the propagation phase engineering [38]. Polarization-insensitive, micron-thick, high-contrast transmitarray micro-lenses with focal spots as small as 0.57λ were obtained. Figure 9.14a shows the simulated transmission coefficient and phase for periodic hexagonally arranged Si nanoposts with a lattice constant of 800 nm, and varying post diameters from 200 to 550 nm. The height of the posts is 940 nm. The phases spanning the entire 0–2π range can be covered with large transmission amplitudes higher than 92%. Light is concentrated inside the posts that behave as weakly coupled low-quality factor resonators. This behavior is fundamentally different from the low-contrast gratings operating in the effective medium regime whose diffractive characteristics are mainly determined by the duty cycle and the filling factor. Figure 9.14b shows the results for a micro-lens that focuses at $d = 25$ μm away from the lens. The FWHM of the focal spot is 1.06 μm or 0.68λ ($\lambda = 1550$ nm).

Figure 9.15a shows the measured FWHM spot size, transmission, and focusing efficiency for the lenses with different focusing distances. The micro-lens shows >82% focusing efficiency for the lenses designed for focusing distance $d = 500$ μm. The normalized measured intensity profile at the focal plane for a micro-lens with the focusing distance of $d = 50$ μm is shown in Fig. 9.15b. A spot with a FWHM of 0.57λ was obtained, which agrees well with the designed result.

Recently, a silicon nanofins-based metalens with even higher NA (NA = 0.98) was experimentally demonstrated by making use of the geometric phase engineering [40]. Micro-lenses of this type, composed of nanoposts or nanofins, are fabricated in one lithographic step that could be performed with high-throughput photo- or nano-imprint lithography, thus enabling widespread adoption. The high index of the dielectric scatters results in negligible mutual coupling, so the light scattered at each grating is dominated by the scatter properties rather than by the collective behavior of multiple coupling. In principle, there is no limit for designing the high-NA lenses with localized phase engineering, except for the focusing or imaging efficiency.

The local phase modulation approach introduced above is challenging to implement in the visible spectrum, for which the constituent elements are not deep subwavelength, thus facing the discretization issues. As an alternative, recently subwavelength high-contrast gratings were adopted to construct lenses with high NA larger than 0.99, which corresponds to an ability to efficiently bend light at angles as large as 82° [41]. The design principle is using the basis of asymmetric scattering from nano-

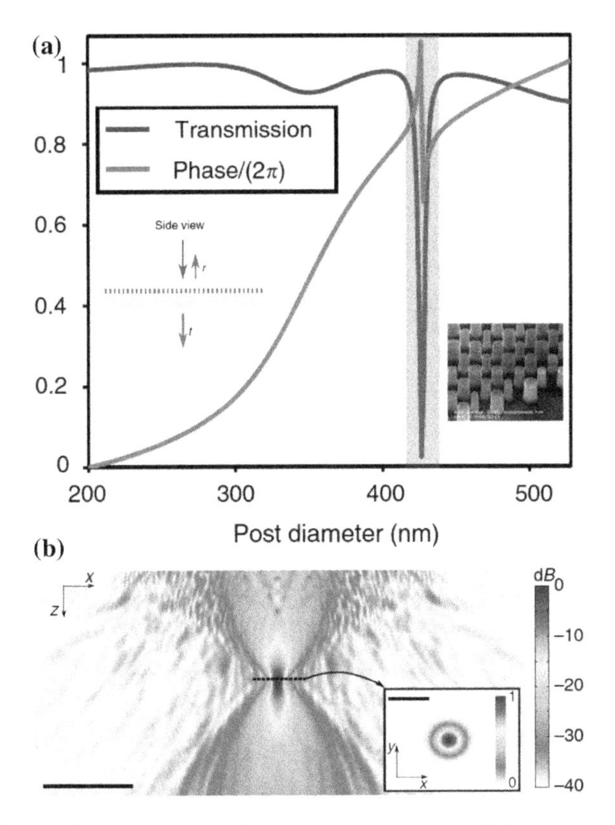

Fig. 9.14 **a** Simulated transmission and phase of the transmission coefficient for a family of periodic hexagonal high-contrast transmitarrays with lattice constant of 800 nm and varying post diameters. The shaded part of the graph is excluded when using this graph to map transmission phase to post diameter. In all these simulations, the posts made of amorphous silicon ($n = 3.43$) are 940 nm tall and the wavelength is $\lambda = 1550$ nm. Inset: SEM image of the silicon posts forming the micro-lens. Scale bars, 1 μm. **b** Logarithmic scale electric energy density in the *xz* cross section. The inset shows the real part of the *z* component of the Poynting vector at the plane of focus. Scale bars, 20 μm in the main figure and 2 μm in the inset. Reproduced from [38] with permission. Copyright 2015, Macmillan Publishers Limited

antenna patterns. The thickness of the metalens can be subwavelength (~λ/3) at an operating wavelength of 715 nm. The operating principle is schematically depicted in Fig. 9.16a. The light beam is diffracted by a simple square lattice. The number of diffraction orders and their corresponding angles can be tuned by adjusting the periods of the unit cell. Assuming that the diffraction energy is designed to the T_{+1} order, the scattering from each nano-antenna must be inhibited in the directions of the other diffraction orders that are opened, i.e., T_{-1}, R_0, R_{+1}, and R_{-1}, as schematically shown in Fig. 9.16a. Figure 9.16b is the optimized unit cell for large bending angle of 82° for normally incident light. It consists of an asymmetric dimer nano-antenna made of Si nanodisks with the height $H = 250$ nm placed on top of a silicon dioxide (SiO_2)

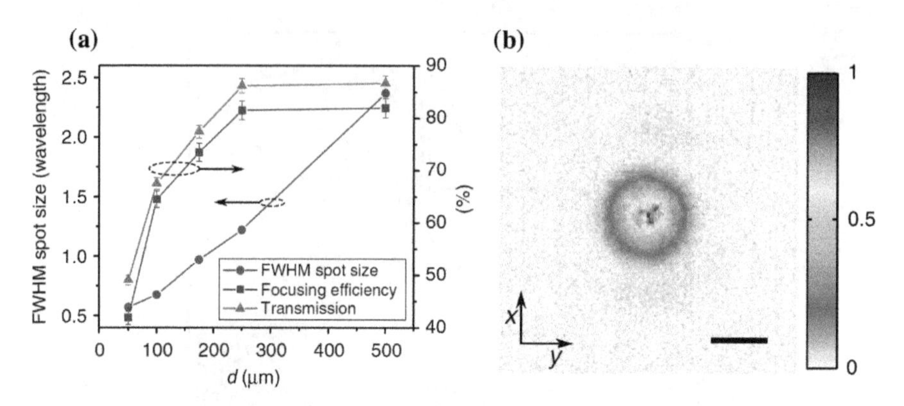

Fig. 9.15 **a** FWHM spot size, transmission, and focusing efficiency of the micro-lenses as a function of their focusing distance. **b** Measured 2D intensity profile at the focal plane for a micro-lens with $d = 50$ μm. Scale bar, 1 μm. Reproduced from [38] with permission. Copyright 2015, Macmillan Publishers Limited

substrate and surrounded by air. The diameters of these nanodisks are $D_1 = 150$ nm and $D_2 = 190$ nm and the gap $g = 50$ nm. The diffractive and non-diffractive periods of the square lattice are, respectively, $P_d = 721$ nm and $P_{nd} = 260$ nm. The former is selected in such a way to provide the first diffractive order going to air at 82° to the normal, and thus, it is fixed for a given wavelength and angle of diffraction. The latter is chosen to be small enough to avoid diffraction in air in this direction and keep a high array density. Figure 9.16c–e illustrates the simulated spectral dependencies of diffraction efficiencies of this array, defined as the amount of power channeled into a particular order divided by the incident one. The diffraction efficiency into the T_{+1} order for *p*-polarized and *s*-polarized light are around 35 and 32%, respectively, at 715 nm illumination. Meanwhile, around 75% (*p*-polarized) and 70% (*s*-polarized) of the transmitted power are bended to the T_{+1} order.

Using the design method introduced above, unit cells for different diffraction angles can be obtained. Then, a near-unity NA lens can be constructed. Figure 9.17a, b is the SEM images of the fabricated high-NA lens. The lens has a total diameter of $d = 600$ μm and a focal distance of $f \sim 42$ μm. The nano-antenna inclusions and unit cells on different areas of the lens are designed to generate different bending angles corresponding to their radial positions within the lens, thus approximating the ideal parabolic phase profile to a piecewise linear one, sampled in one-degree steps. A slightly elongated focal spot at a distance of approximately 42–47 μm from the lens was experimentally demonstrated, which implies a maximum experimental bending angle of around 82°, corresponding to a maximum NA of ~0.99. The measurements also reveal a highly symmetric intensity distribution in the focal plane (*xy*-plane), shown in Fig. 9.17c, d, and a very low level of background. The cut across the *x*-axis passing through the focus is closed to that of a diffraction-limited system. The measured FWHM along the cut is 385 nm, corresponding to an NA of around 0.935, which is at the limit of resolution of the measurement system with Nikon 100x,

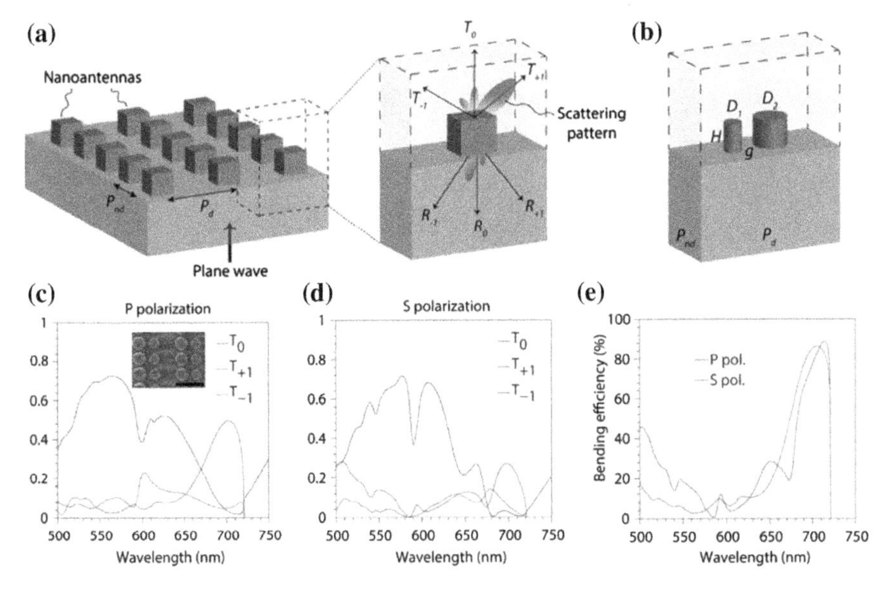

Fig. 9.16 **a** Schematic description of the nano-antenna arrays with controlled energy distribution among the supported diffraction orders. The number and angle of these orders is determined by the diffractive period of the array, P_d. The energy redistribution is determined by the scattering pattern of each nano-antenna in the array. **b** Schematic description of the array of asymmetric dimers producing energy concentration into the T_{+1} diffraction order leading to light bending at 82° for plane waves normally incident from the substrate side at a wavelength of 715 nm. Simulated diffraction efficiencies under normally incident light illumination with **c** p- or **d** s-polarization. **e** Simulated bending efficiency of the array, representing the amount of power into the desired order relative to the total transmitted one. Reproduced from [41] with permission. Copyright 2018, American Chemical Society

NA = 0.95 lens as the collection objective. The designed value of 0.99 has been further proved by using the metalens to collect light from a single sub-diffractive scatterer located at its focus.

Table 9.1 illustrates the representative summary of previously reported transmissive metalenses with high NA. Dry lens with NA = 0.99 and oil-immersion lens with NA = 1.48 have been obtained so far. These values outperform the conventional singlet lenses or even lens group, although there is plenty of room to be improved in other important optical performances, such as the efficiency, chromatic aberration, and spherical aberration for their practical applications.

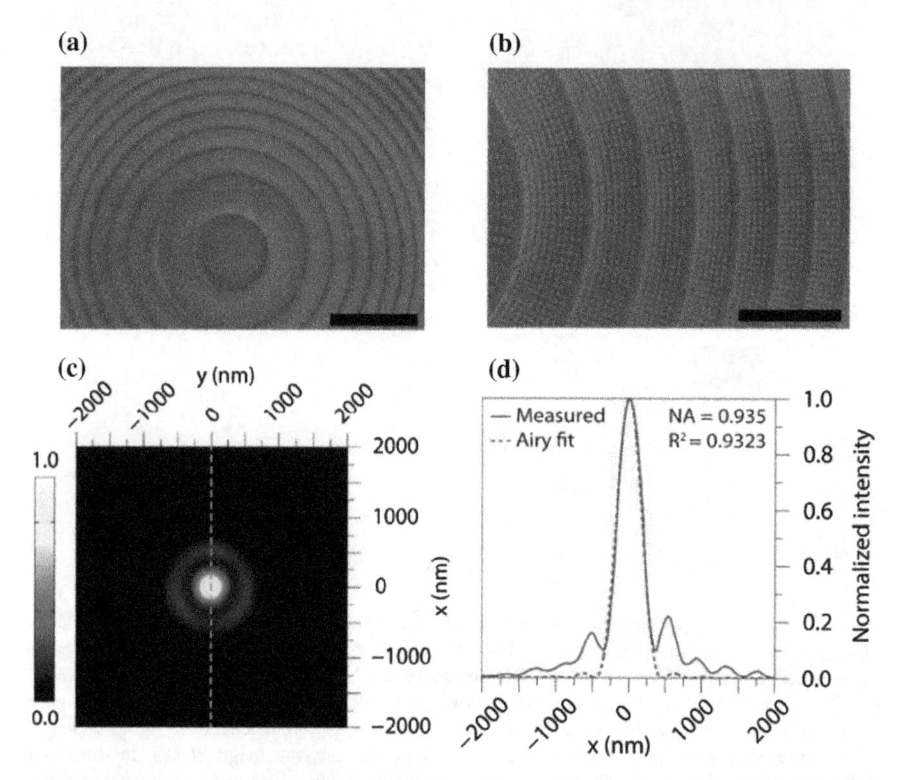

Fig. 9.17 **a** SEM image of the fabricated sample showing a low magnification image of the central part. **b** SEM image of the low angle bending part of the lens. **c** Normalized measured intensity map in the focal plane ($z = 45$ μm). **d** Intensity profile along the x-axis, passing through the focus, together with the fitting to the Airy profile. Reproduced from [41] with permission. Copyright 2018, American Chemical Society

Table 9.1 Previously reported transmissive metalenses with high NA [40]

References [32, 38–41]	Materials	Wavelength (nm)	FWHM spot size (λ)	NA	Efficiency (%)
Chen et al.	TiO_2	532	0.57	1.1 ($n =$ 1.52)	50
Khorsaninejad et al.	TiO_2	405, 532, 660	0.67, 0.63, 0.73	0.8 (in air)	86, 73, 66
Paniagua-Dominguez et al.	a-Si	715	0.54	0.99 (in air)	35
Arbabi et al.	a-Si	1550	0.57	0.97 (in air)	42
Liang et al.	c-Si	532	0.52	0.98 (in air)	67
Liang et al.	c-Si	532	0.4	1.48 ($n =$ 1.512)	48

9.4 Wide Field-of-View Metalens

Apart from the chromatic aberration, most planar lenses surfer from the off-axis aberration at large incident angle, since the ideal phase profile is dependent on the incidence angle [43, 44]. In order to increase the FOV, many methods have been investigated. Luneburg lens is a typical gradient index (GRIN) component free from such aberration because of its rotational symmetry, which can be also be used as scanning antenna with large FOV and high gain. The GRIN requirement can be addressed by using metamaterial homogenization techniques. For example, by utilizing subwavelength holes etched into silicon on insulator (SOI), flattened all-dielectric Luneburg lens at telecommunication wavelengths was realized, which exhibits beam forming (focusing) from a planar focal surface over a wide FOV of 67° [45, 46]. Subsequently, a 3D approximate transformation optics lens was proposed in the microwave frequency band [47], which was fabricated by multilayered dielectric plates with inhomogeneous holes. By locating and shifting a planar array of feeding sources on the flattened focal plane, the radiation beam scanned in a range of 50°. A semispherical lens was then fabricated in GHz frequencies, to resolve the sources in subwavelength scale with high resolution that could be captured directly by a conventional microwave imaging device [48]. As another example, a 3D version of Luneburg lens at optical frequencies was fabricated by using the ultra-fine femtosecond laser direct-writing technique [49]. The effective refractive index was spatially and gradually modified by tailoring the volume-filling fraction of simple cubic metamaterial structures.

Although Luneburg lens that possesses spherical symmetry in refractive index distribution has a large FOV, the rotational symmetry of these near-optimal lenses makes them incompatible with flat imaging optics as well as the current planar fabrication technologies. Theoretical analysis shows that this aberration, known as coma, can be greatly reduced compared with conventional bulk spherical lenses by building the flat lens on a curved substrate or stacking several flat lenses together [43]. It is also commonly known that cascade lenses can be optimized to function as a fisheye lens with significantly reduced monochromatic aberration. This design strategy may also be adopted in integrated metalens doublets [50]. Figure 9.18 schematically shows the focusing property by a spherical-aberration-free singlet metalens and doublet lens. The singlet metalens exhibits diffraction-limited focusing for normal incidence and significant aberrations for incident angles as small as a few degrees. In contrast to the singlet metalens, the doublet lens acts as a fisheye objective, so the FOV can be greatly extended. The phase profile of each surface of the doublet can be designed and optimized using ray tracing approach using commercial optical design software (i.e., Zemax OpticStudio, Zemax LLC). The phase profiles were defined as even-order polynomials of the radial coordinate r as,

$$\phi(r) = \sum_{n=1}^{5} a_n \left(\frac{r}{R}\right)^{2n}, \tag{9.4.1}$$

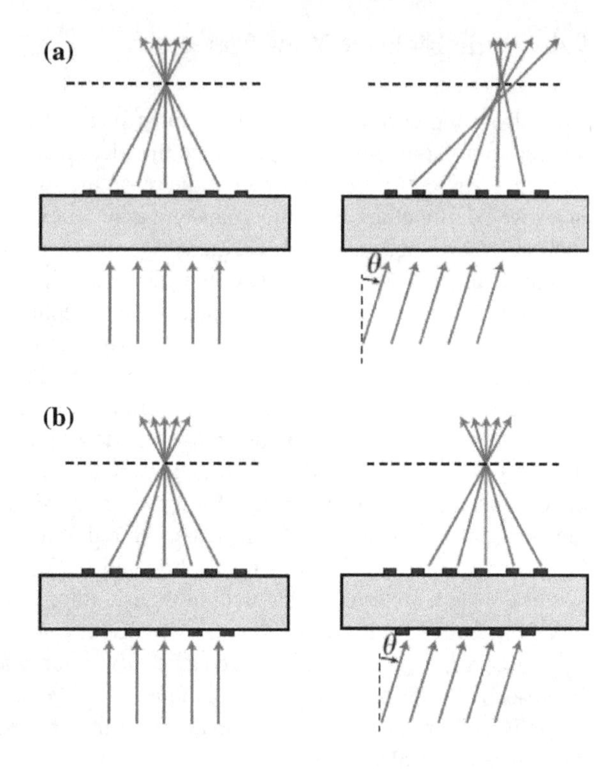

Fig. 9.18 **a** Schematic illustration of focusing of on-axis and off-axis light by a spherical-aberration-free metasurface singlet lens. **b** Similar illustration in **a** but for a metasurface doublet lens corrected for monochromatic aberrations. Both lenses have aperture diameter of 800 mm and focal length of 717 mm (f-number of 0.9), the wavelength is 850 nm. Reproduced from [50] with permission. Copyright 2016, The Authors

where R is the radius of the metasurface, and a_n is the coefficients, which are optimized for minimizing the focal spot size at different incident angles. In the optimum design, the first metasurface operates as a corrector plate and the second one performs the significant portion of focusing; thus, they can be referred as correcting and focusing metasurfaces, respectively (Fig. 9.18).

A flat wide FOV metalens with a monolithic metasurface doublet was demonstrated at an operating wavelength of 850 nm [50]. The doublet lens consists of a hexagonal array of 600-nm-tall amorphous Si nanoposts with spatially varying diameters on the top and bottom surfaces of a 1-mm-thick transparent fused silica substrate, which has a small f-number of 0.9, a FOV larger than 60° × 60°, and ~70% focusing efficiency. The doublet metalens is then integrated to a CMOS image sensor to produce a miniature camera of total dimensions 1.6 mm × 1.6 mm × 1.7 mm. Figure 9.19 schematically shows the miniature camera composing of the metasurface doublet lens and a low-cost color CMOS image sensor (OmniVision OV5640, pixel size: 1.4 μm) with a cover glass thickness of 445 ± 20 μm. An air gap of 220 μm was set between the metasurface doublet lens and the image sensor to facilitate the assembly of the camera. Image captured by the camera and its insets depicting the zoomed-in views of the images at 0°, 15°, and 30° view angles are shown in Fig. 9.19c, which verified its ability of large-FOV imaging. The modulation transfer function (MTF) represents the relative contrast of the image versus the spatial details of the

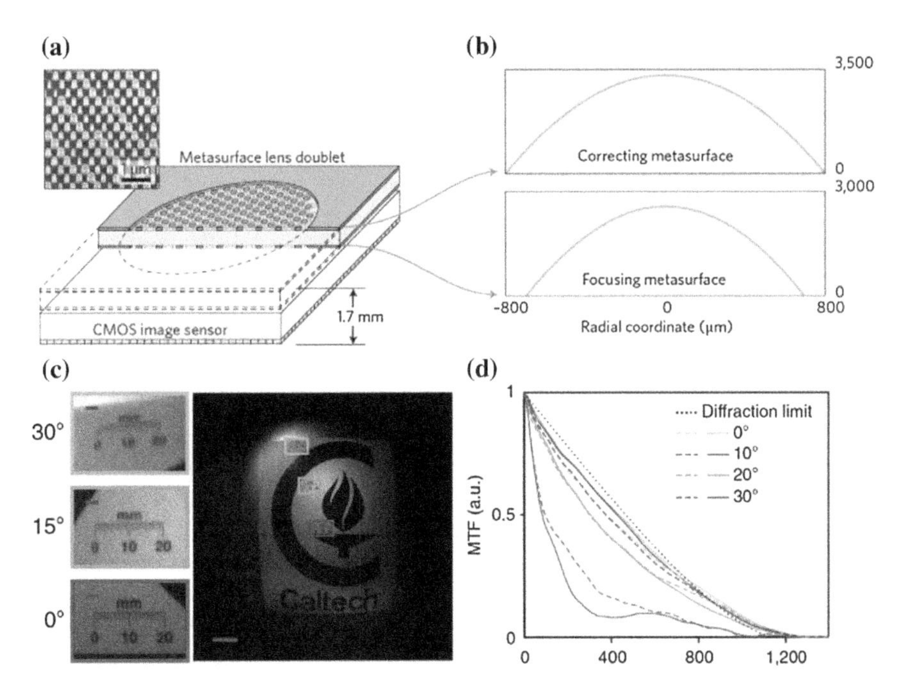

Fig. 9.19 **a** Schematic illustration of the miniaturized camera, consisting of a metasurface lens doublet integrated on the surface of a CMOS image sensor, with total thickness of 1.7 mm. The planar metasurface lens doublet consists of a correcting metasurface and a focusing metasurface, fabricated on both sides of a transparent substrate. The inset shows a magnified view of the metasurface lens, consisting of a 2D array of nanoposts with spatially varying dimensions. **b** Phase profile of the correcting and focusing metasurfaces. **c** Image recorded using the miniaturized camera. **a** and **b** Reproduced from [51] with permission. Copyright 2016, Macmillan Publishers Limited, part of Springer Nature. **c** and **d** Reproduced from [50] with permission. Copyright 2016, The Authors

object, which can be obtained by computing the Fourier transform of the focal spot intensity of each point in the FOV. The MTFs for the doublet metalens were computed using the measured focal spots and are shown in Fig. 9.19d. Both the images and the MTFs shown in Fig. 9.19c, d demonstrate the effectiveness of correction achieved by cascading two metasurfaces and the diffraction-limited performance of the doublet metalens over a wide FOV. It is worth noting that stacks of this kind of planar lenses can be monolithically fabricated without additional alignment steps, and they can be further integrated with image sensors during the alignment process.

Similar doublet metalenses have been also demonstrated in the visible region using the geometric phase engineering based on TiO_2 nanofins [52]. Compared with multi-micro-lens objective fabricated by two-photon laser direct writing [53], the doublet metalens is more compact and easier to scale up. As has been demonstrated, the above doublet metalens is inherently a wide-angle Fourier transform lens. The measured relative location of the doublet lens focal spot is a function of incident angle along with the $f\sin(\theta)$ curve. This kind of lens can transform the rotational symmetry to the translational symmetry associated with the off-axis incident light, allowing a large FOV as well as optical Fourier transformation. However, these metasurface doublets' FOVs are still no larger than 60°. How to further enlarge the FOV remains a challenge. The Defense Advanced Research Projects Agency (DARPA) indicated the question "What advances are needed to open the FOV up to 2π steradians without increasing the size and weight of the system" as one of the key questions in extreme challenges in optics and imaging [54].

Recently, this question has been addressed to some extent by virtue of the symmetric conversion method [55]. A strategy that is able to achieve full control of such symmetry in 2D single flat lens with rapid phase gradient, the so-called super-symmetric lens, has been proposed [55], which could enable perfect conversion from rotational symmetry to translational symmetry, resulting in an almost perfect wide-angle lensing performance with an extraordinary FOV of larger than 160° × 160°. Figure 9.20a schematically shows the compound eyes possessing spherical symmetry in refractive index distribution, and therefore, light rays coming from different orientation angles could be perfectly directed to predefined spherical focal surfaces. Nevertheless, the rotational symmetry of these near-optimal lenses makes them to be not compatible with flat optics as well as current planar fabrication technologies.

The key to construct a wide FOV metalens is the realization of perfect conversion from the rotational symmetry to translational symmetry in light fields. For this purpose, the following relation should be met:

$$k_0 \sin\theta_x x + k_0 \sin\theta_y y + \Phi_m(x, y) = \Phi_m(x + \Delta_x, y + \Delta_y), \tag{9.4.2}$$

where k_0 is the wave number in free space, $\Phi_m(x, y)$ is the phase shift profile carried by the flat lens, and Δ_x and Δ_y correspond to the translational shift of $\Phi_m(x, y)$ at incidence angles of θ_x and θ_y.

With some mathematical manipulations, Eq. (9.4.2) can be rewritten as:

$$\Phi_m(x + \Delta_x, y + \Delta_y) - \Phi_m(x, y) = k_x x + k_y y, \tag{9.4.3}$$

and

$$\begin{aligned} \frac{\partial\Phi_m}{\partial x} &= \frac{k_x}{\Delta_x}x, \\ \frac{\partial\Phi_m}{\partial y} &= \frac{k_y}{\Delta_y}y, \end{aligned} \tag{9.4.4}$$

where $k_x = k_0\sin\theta_x$ and $k_y = k_0\sin\theta_y$. Consequently, the phase profile of the flat lens should satisfy:

$$\Phi_m = \frac{k_x x^2}{2\Delta_x} + \frac{k_y y^2}{2\Delta_y}. \tag{9.4.5}$$

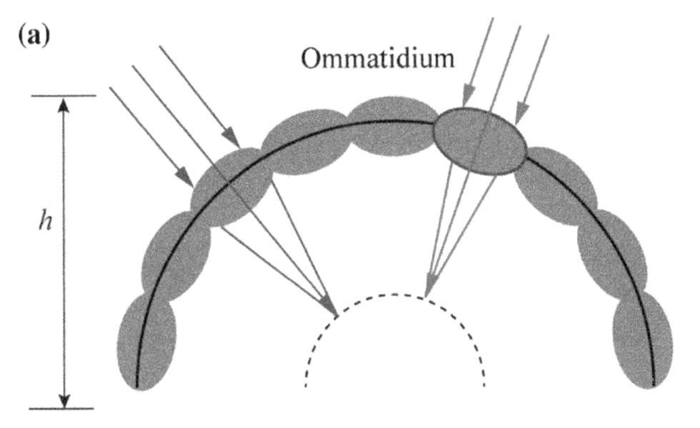

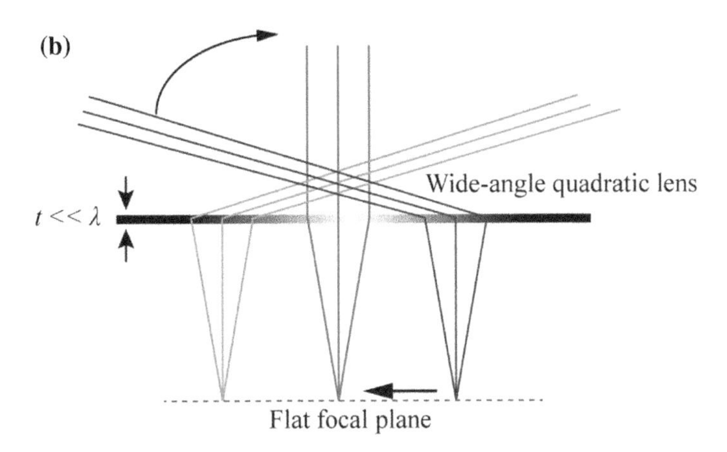

Fig. 9.20 **a** Compound eyes composed of many ommatidia. **b** Wide-angle flat lens. The thickness t is far smaller than the wavelength. The red, blue, and yellow lines represent the light rays with different incidence angles. Reproduced from [55] with permission. Copyright 2017, Optical Society of America

If the lens is circular symmetric, there are

$$\Delta_x = \Delta_y = \Delta, \quad k_x = k_y = k_0 \sin\theta. \tag{9.4.6}$$

Then, we have

$$\Phi_m = \frac{k_0 \sin\theta}{2\Delta}(x^2 + y^2) = k_0 \sin\theta \frac{r^2}{2\Delta}. \tag{9.4.7}$$

This is just the phase profile for a normal thin lens in the paraxial regime:

$$\Phi(r) = k_0 \frac{r^2}{2f} = \frac{\pi r^2}{\lambda f}, \tag{9.4.8}$$

where f is the focal length and the horizontal shift can be written as $\Delta = f\sin\theta$. For inclined illumination beams that lie in the xz-plane with an arbitrary angle of θ to the normal axis of the lens, the phase carried by the outgoing light should be:

$$\Phi(r) = k_0 \frac{r^2}{2f} + k_0 x \sin\theta = \frac{k_0}{2f}\left((x + f\sin\theta)^2 + y^2\right) - \frac{f k_0 \sin^2\theta}{2}. \tag{9.4.9}$$

Since the last term in the right hand of Eq. (9.4.9) is independent of r and can be neglected, there is only a transversal shift of $f\sin\theta$ in the x-direction with respect to the Eq. (9.4.8), as indicated in Fig. 9.20b. In this regard, the rotational effect of the oblique incidence light is perfectly converted to the translational symmetry of the output one. The phase of the super-symmetric lens follows a quadratic form. Therefore, the lens is also termed "quadratic lens." The quadratic lens could be realized using either discrete nano-antennas [56, 57] or semicontinuous catenary structures [58–60].

Figure 9.21a shows the quadratic lens constructed by semicontinuous catenary structures. The rotational shift of oblique incidence is converted to the translational movement of the focusing beam. As shown in Fig. 9.21b, the obliquely incident laser beam is horizontally shifted with a value of $f\sin\theta$, demonstrating an unprecedented improvement of FOV over traditional flat lens, which can be used in Fourier transform and wide-angle imaging. Moreover, such lens could operate in an achromatic way in the entire visible range and beat the classic diffraction limit to some extent. To increase the energy efficiency of single-layered metasurface, Liu et al. proposed a 1D Fourier metalens made of an array of dielectric waveguide resonators [61], which shows focusing efficiency as high as $\approx$50% for incidence angles of 0–60° and a broad bandwidth ranging from 1100 to 1700 nm (Fig. 9.22).

The translational symmetry ensures that the flat lens is easily integrated in lens antennas, whose radiation direction and side lobe can be readily tuned by adjusting the horizontal position and the distance between the source and the flat lens. Compared with previous beam steering techniques such as Luneburg lens and rotational prisms, this approach provides great advantages such as low profile, easy implementation, and dramatically reduced side lobe. As demonstrated recently [62], a high-efficiency (>80% even when the incidence angle is titled by 60°) and ultra-thin (0.127λ) metalens was realized using bilayer geometric metasurfaces. Wide-angle beam steering ability beyond ±60° was experimentally demonstrated in 16–19 GHz. Compared with previous beam steering techniques such as Luneburg lens and rotational prisms, this approach provides great advantages such as low profile, easy implementation, and dramatically reduced side lobe (Fig. 9.23).

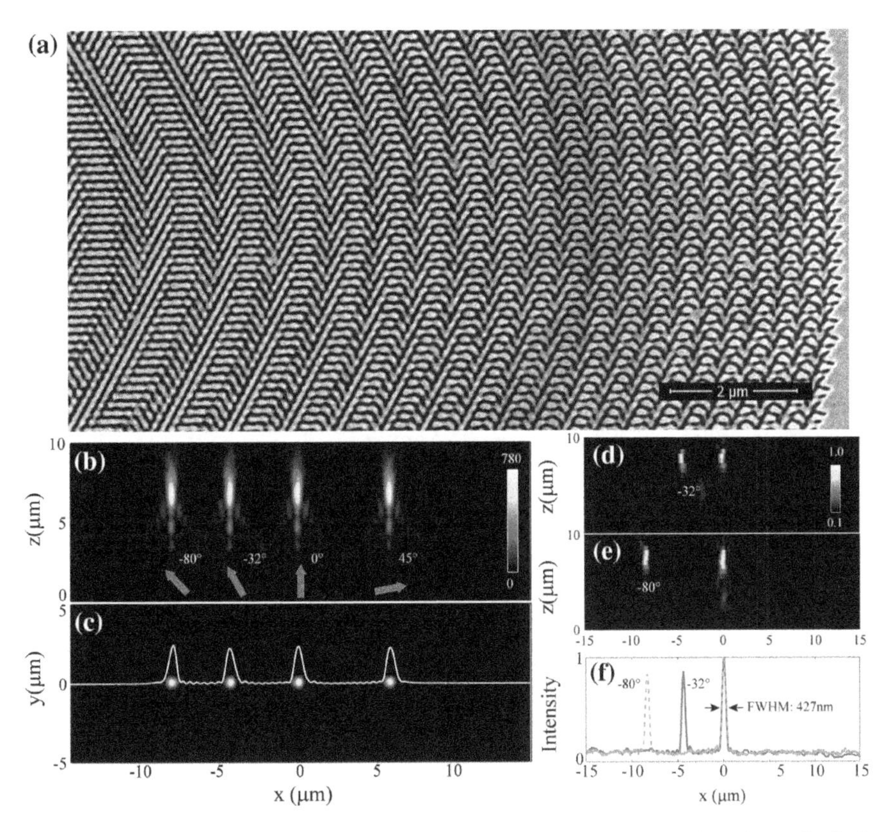

Fig. 9.21 Performance of the wide-angle flat lens. **a** SEM image of the fabricated sample. Scale bar: 2 μm. **b**, **c** Simulated light intensity distributed on the xz-plane ($y = 0$) and xy-plane ($z = 7.5$ μm) at a wavelength of 632.8 nm. **d**, **e** Experimental results of the intensity distribution on the xz-plane ($y = 0$) for $\theta = 0°, -32°$ (**d**) and $\theta = 0°, -80°$ (**e**). **f** Cross-sectional view ($z = 7.5$ μm) of the intensity distribution along the x-direction shown in (**d** and **e**). The FWHM is about 427 nm, a little larger than the numerical value (~380 nm). Reproduced from [55] with permission. Copyright 2017, Optical Society of America

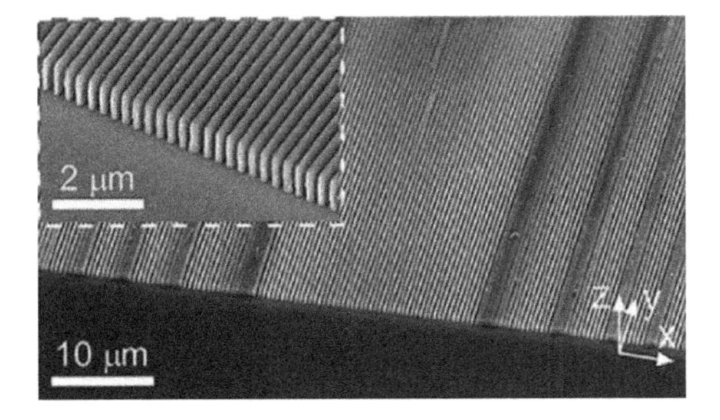

Fig. 9.22 Dielectric Fourier lens based on gradient silicon grating. Reproduced from [61] with permission. Copyright 2018, WILEY-VCH Verlag GmbH & Co. KGaA, Weinheim

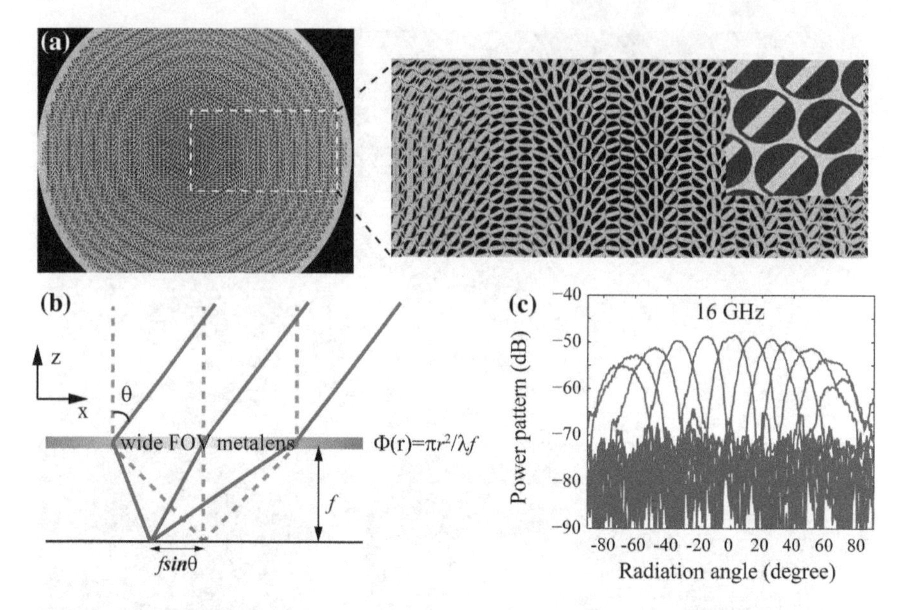

Fig. 9.23 **a** Perspective and zoom view of the wide-angle flat lens. **b** Illustration of the symmetry conversion from rotational symmetry of incidence to translational symmetry of outgoing wave based on a wide FOV metalens. **c** Measured far-field power patterns at different frequencies when a circularly polarized antenna is transversely shifted at a distance of $z = -87.5$ mm at frequency of 16 GHz. Reproduced from [62] with permission. Copyright 2018, WILEY-VCH Verlag GmbH & Co. KGaA, Weinheim

9.5 Achromatic and Super-dispersive Elements

Chromatism is a quantity describing the performance change of optical device when the operating light wavelength changes. It exists not only in traditional flat optical components such as Fresnel zone plates and photon sieves [20], but also in metasurface-based optical devices, which makes the deflection angle and focal length increase and decrease with increasing wavelength, respectively. The achromatic focusing requires the focal length to keep as a constant, which implies that the phase shift should vary with the wavelength. According to the Fermat's principle, the ideal wavelength-dependent phase profile for an achromatic lens could be written as [63]:

$$\Delta\Phi(r,\lambda) = -\frac{2\pi}{\lambda}\left(\sqrt{r^2+f^2} - f\right), \tag{9.5.1}$$

where r and f are the radius and focal length, respectively. Since metalens would behave the same with an arbitrary additional constant, Eq. (9.5.1) can be revised as:

$$\Delta\Phi(r,\lambda) = -\frac{2\pi}{\lambda}\left(\sqrt{r^2+f^2} - f\right) + C(\lambda). \tag{9.5.2}$$

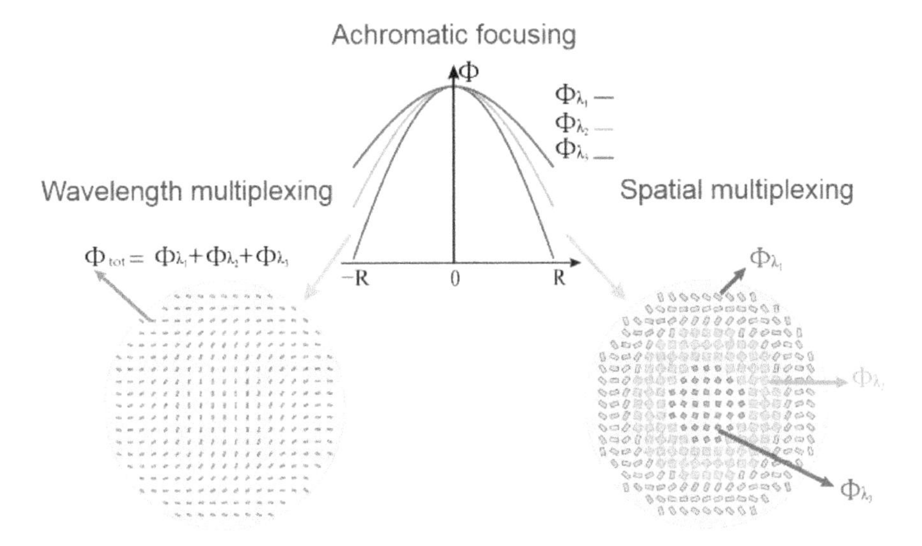

Fig. 9.24 Schematic of multi-wavelength achromatic lens based on wavelength and spatial multiplexing

Since the ideal phase is inversely proportional to the wavelength, it is difficult to achieve this goal using common subwavelength structures. However, with an additional term $C(\lambda)$ added in this equation, more degrees of freedom could be exploited to realize near-achromatic focusing performance. Since it is easy to realize metalens for a single wavelength, we focus our attention on the achromatic metalens, which utilizes intricate dispersion engineering methods to reduce the chromatic dispersion.

9.5.1 Multi-wavelength Achromatic Metalens

One straightforward method to optimize $C(\lambda)$ and eliminate the achromatism of focusing lens is the phase and spatial multiplexing by taking advantages of the design flexibility of metasurfaces, as shown in Fig. 9.24.

Phase multiplexing achromatic lens takes advantages of the complex amplitude superposition. Supposing that there are several point sources of light with different wavelength, the distribution of phase on the metasurface can be written as [64]:

$$\Phi(x, y) = \arg\left(\sum_{1}^{N} E_n\right) = \arg\left(\sum_{1}^{N} A_n \exp\left(i\frac{2\pi}{\lambda_n}\sqrt{x^2 + y^2 + f^2}\right)\right). \quad (9.5.3)$$

We assume that the amplitudes of these sources are equal. In the design and fabrication of the metalens, as an example, three discrete wavelengths, i.e., 532 nm, 632.8 nm, and 785 nm, are used.

The radius of the circular sample is 10 μm, and the focal length is set as f = 9 μm. Figure 9.25a shows the SEM image of the fabricated sample, where the nano-apertures are arranged based on Eq. (9.5.3). We simulated and measured the performance of the metasurface at the three discrete wavelengths. The left panels in Fig. 9.25b–d display the numerical intensity patterns in the xz-plane for $\lambda = 532$ nm, 632.8 nm, and 785 nm, which qualitatively agree with the experiment results given in the right panels. We compared the FWHM of the numerical and experimental results at $z = 9$ μm for all the three wavelengths. As shown in Fig. 9.26, this metalens has the ability of focusing close to the theoretically expected values (or the diffraction limit). By utilizing the metasurface-assisted diffraction theory, it is possible to further obtain achromatic super-resolution imaging at multi-wavelengths. In principle, the performance of the multi-wavelength metasurface could be further improved by combining the amplitude and phase modulation techniques. If we divide a metasurface into many regions, and each region is composed of a particular kind of nanostructures that permit only one wavelength to transmit, such metasurface can be used to control different light separately.

Spatial multiplexing is another direct approach to realize a multi-wavelength achromatic metalens as shown in Fig. 9.24. By combining multiple structures with different response wavelengths into a single metasurface, it is possible to realize achromatic performance at multiple wavelengths [22]. In this approach, the dispersion of different unit cells is designed to only respond to separate wavelengths. Based on the spectral filtering and space division, the focal spots for different wavelengths are designed to spatially overlap at the designated focal length; thus, a multi-wavelength achromatic metalens can be realized [65]. Figure 9.27a shows a specific design based on spatial multiplexing, where the red-, green-, and blue-colored rectangles represent the three nanocuboids responding to the blue (473 nm), green (532 nm), and red (632.8 nm) lights, respectively. In order to realize the wavelength-independent wavefront shaping, three dedicatedly designed nanocuboids with different geometries, denoted by Sb ($l = 80$ nm, $w = 42$ nm), Sg ($l = 91$ nm, $w = 67$ nm), Sr ($l = 131$ nm, $w = 102$ nm), are utilized to filter out the three wavelengths (473 nm, 532 nm, and 632.8 nm), respectively (Fig. 9.27b). The pixels are arranged with a period of $P = 200$ nm and a height of $h = 400$ nm. The focal length of the flat achromatic lens is $f = 10$ μm for all the three wavelengths, and the corresponding NA is calculated as 0.6294.

Figure 9.28a–c shows the normalized intensity distribution of achromatic metalenses in xz-plane at the wavelengths $\lambda = 473$, 532, and 632.8 nm. Obviously, this achromatic metalens is able to focus these waves to the same plane at designed focal length ($f = 10$ μm). The exact focal lengths in the three cases are 10.19 μm (blue), 10.19 μm (green), and 9.89 μm (red). The focusing efficiency of achromatic metalens is defined as the ratio of the optical power of the focused beam to the optical power of the incident beam. The calculated results are 21.13% (blue), 54.66% (green), and 31.49% (red) for three unit cells.

Spatial multiplexing achromatic lens can also be realized by stacking multilayered metasurfaces to form a multilayered composite structure with each layer responding

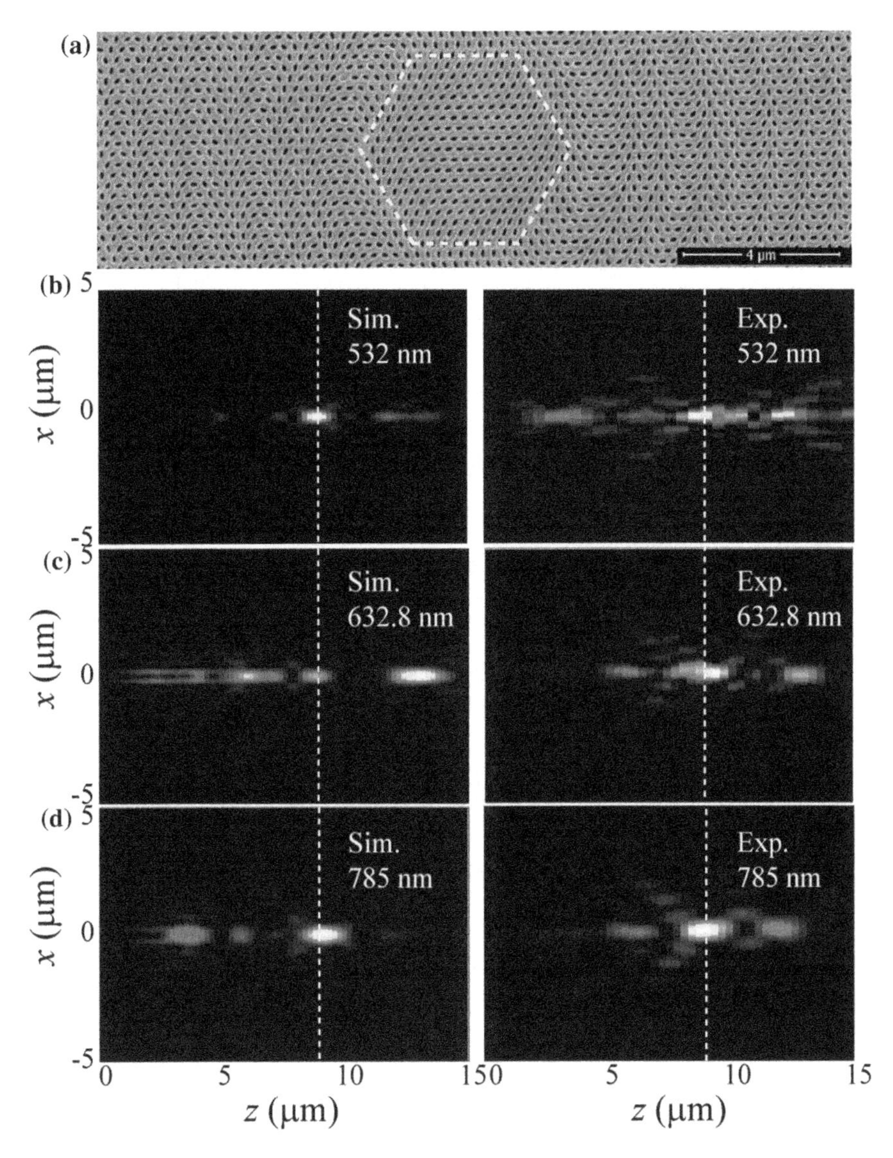

Fig. 9.25 **a** SEM image of the fabricated sample. Scale bar: 4 μm. **b–d** Numerical (left panels) and experimental (right panels) intensity maps for $\lambda = 532$ nm, 632.8 nm, and 785 nm in the *xz*-plane. The focal length is 9 μm as indicated by the white lines. Reproduced from [64] with permission. Copyright 2015, The Authors

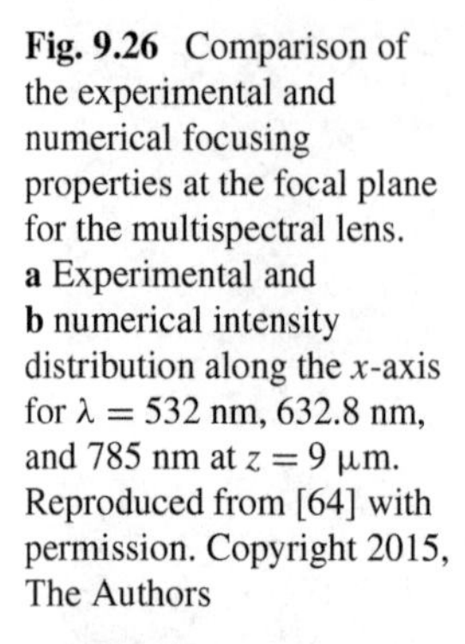

Fig. 9.26 Comparison of the experimental and numerical focusing properties at the focal plane for the multispectral lens. **a** Experimental and **b** numerical intensity distribution along the x-axis for $\lambda = 532$ nm, 632.8 nm, and 785 nm at $z = 9$ μm. Reproduced from [64] with permission. Copyright 2015, The Authors

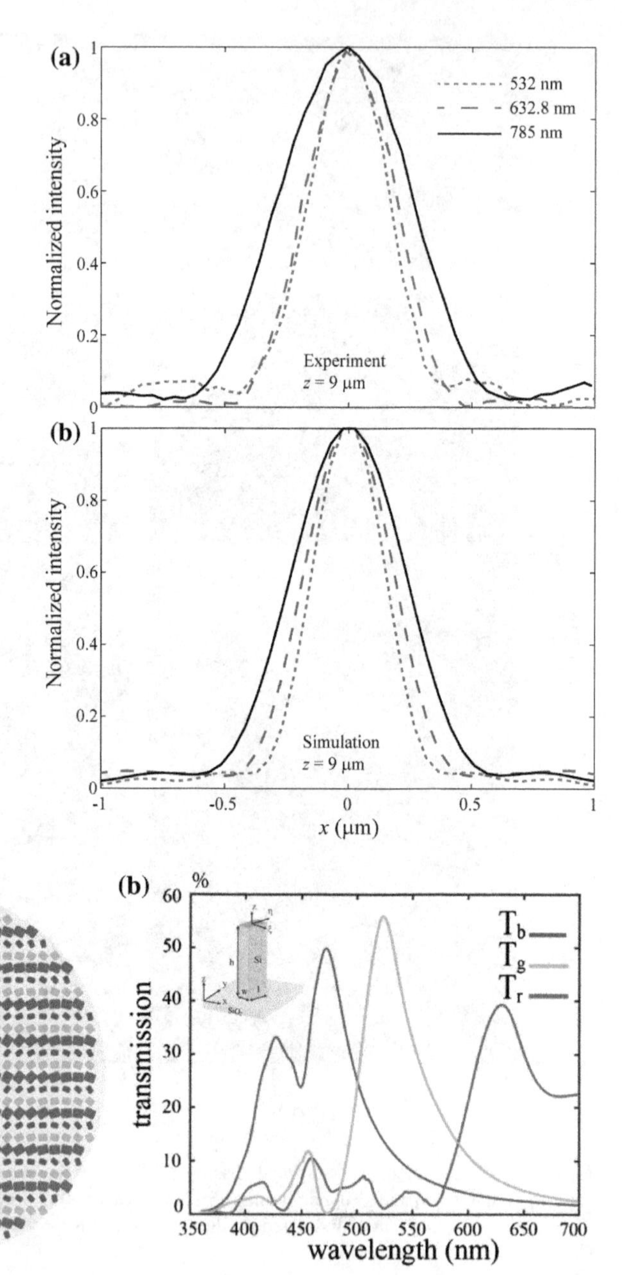

(a)

Fig. 9.27 **a** Top view of a spatial multiplexing achromatic metalens with a radius of 3 μm. **b** The simulation transmission (conversion efficiency) results of three individual designed nanocuboids illuminated by a normally incident LCP (left circularly polarized) light beam. Inset: 3D view of the basic unit cell. Reproduced from [65] with permission. Copyright 2017, Optical Society of America

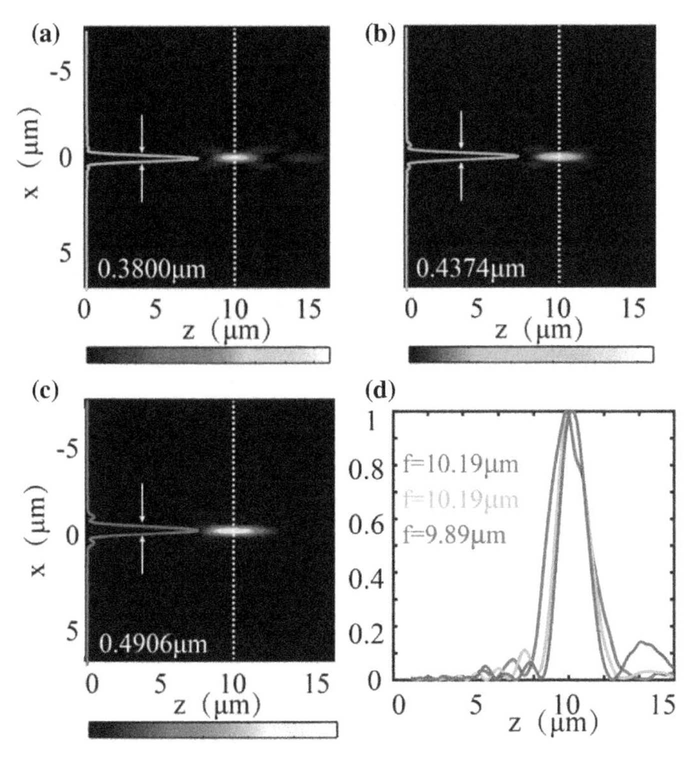

Fig. 9.28 **a**–**c** Simulated normalized intensities distribution of achromatic flat metalenses in *xz*-plane and (inset) the intensity profiles across the focal plane. **d** Comparison of the normalized intensity curves of three light along the *z*-axis. Reproduced from [65] with permission. Copyright 2017, Optical Society of America

to a single wavelength. The spectral response of this kind of metalens is spatially multiplexed in longitudinal direction, not in horizontal direction.

The concept of an aberration-corrected multilayer composite structure is illustrated in Fig. 9.29 [66]. The lens consists of three closely stacked metasurfaces, each composed of nano-antennas made of different metal: gold, silver, and aluminum, and is designed to optimally interact with light at wavelengths of 650 nm, 550 nm, and 450 nm, respectively. Each of the layers acts as a narrowband binary Fresnel zone plate (FZP) lens that focuses its targeted light to the common focal point. Within each layer, the nanoparticles are closely spaced to avoid diffraction-grating effects. For the presented lens, an interlayer distance of 200 nm was chosen to minimize the near-field crosstalk between the individual nano-antennas in the different layers (Fig. 9.30).

Owing to the importance of achromatism for optical imaging system, a lot of other explorations have been carried out with subwavelength structures. For example, dual-wavelength focusing was achieved by using a plasmonic lens composed of an annular slit and a concentric groove, and the groove was placed at the position where the nodes of the two interference patterns match [67]. The second way is by

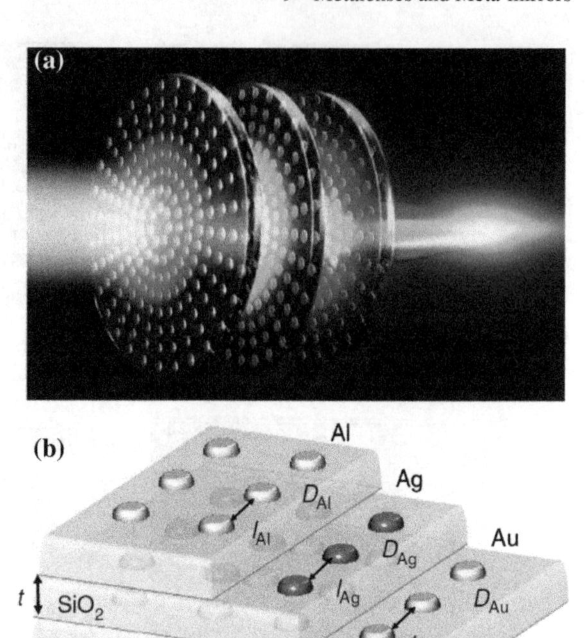

Fig. 9.29 **a** Artist's view of the three-layer lens. When illuminated with white light, each layer focuses its designated part of the spectrum to a distance of 1 mm along the optical axis. **b** Schematic illustration of the layered structure. Each layer consists of nanodisks with the following diameters D and separations I: $D_{Au} = 125$ nm, $I_{Au} = 185$ nm; $D_{Ag} = 85$ nm, $I_{Ag} = 195$ nm; $D_{Al} = 120$ nm, $I_{Al} = 150$ nm. Reproduced from [66] with permission. Copyright 2017, The Authors

combining two polarization-divided structures in one metasurface; each structure independently performs a similar function at a special wavelength. In this way, a metasurface with strong polarization and wavelength selectivity has been realized using tightly packed cross- and rod-shaped optical nano-antennas [68]. This selectivity allows to address any superposition of the two colors at the focus of the lenses by controlling the polarization of light. By combining multiple structures with different response wavelengths into a single metasurface, it is possible to realize achromatic performance at multiple wavelengths. For instance, a multi-wavelength achromatic metasurface comprised of multiple metallic nanogroove gratings was demonstrated [69]. To achieve achromatic diffraction, the ratio between the resonance wavelength and the period of each elementary grating was fixed. Incident light at those multiple resonance wavelengths can be efficiently diffracted into the same direction with near-complete suppression of the specular reflection. Based on a similar approach, a wide-angle off-axis achromatic flat lens was also realized for focusing light of different wavelengths into the same position.

Another method to optimize $C(\lambda)$ and eliminate the chromatism of focusing lens is the utilization of high-order resonant modes. By using an aperiodic arrangement of coupled dielectric resonators, Capasso's group proposed an achromatic metasurface lens (Fig. 9.31a–d) working at three different wavelengths of 1300 nm, 1550 nm, and 1800 nm [70, 71].

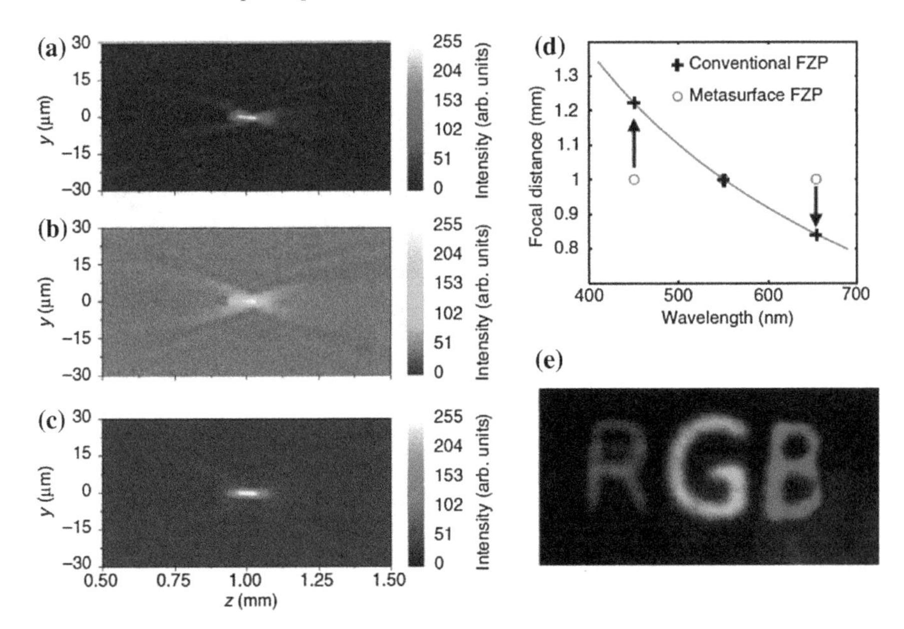

Fig. 9.30 **a** Measured light focusing with composite metalens. Images of the focal region for a conventional FZP illuminated by laser light at 450 nm (**a**), 550 nm (**b**), and 650 nm (**c**) for the composite metalens. **d** Theoretical calculation of the focal distance for a conventional FZP (red line) and the measured focal points at the RGB wavelengths of the conventional FZP (crosses) and metalens (circles). **e** Demonstration of color imaging using the fabricated composite metalens element. Reproduced from [66] with permission. Copyright 2017, The Authors

Taking advantage of the dispersion engineering of dielectric nanopillars, an achromatic reflective metalens operating in visible light was obtained [72]. A particle swarm optimization algorithm (PSOA) was utilized to optimize the geometric parameters and distribution of the nanopillars, so as to obtain an optimized $C(\lambda)$. Figure 9.32 shows the achromatic metalens operating over a continuous bandwidth in the visible from $\lambda = 490$–550 nm, accomplished via dispersion engineering of TiO_2 nanopillars tiled on a dielectric spacer layer above a metallic mirror. The PSOA provides a powerful tool to optimize the optical characteristics of designed unit elements; an optimal arrangement is determined to make the actual phase profile be close to the ideal phase profile. The working bandwidth of this meta-mirror is around 60 nm, which is not broad enough for practical purposes due to the limited design freedom of the structure.

Recently, by utilizing PSOA, a transmissive achromatic infrared metalens from 8 to 14 μm with a large diameter (18 cm) was also designed. The fabricated larger-scale singlet achromatic metalens was integrated into an IR imaging system, which moves an important step to its realistic application. In this design, high-aspect-ratio silicon (Si) nanoposts are chosen as the basic building blocks to realize anomalous dispersion engineering, as shown in Fig. 9.33a. The employment of a circular cross section guarantees polarization-insensitive operation of the achromatic metalens. Generally,

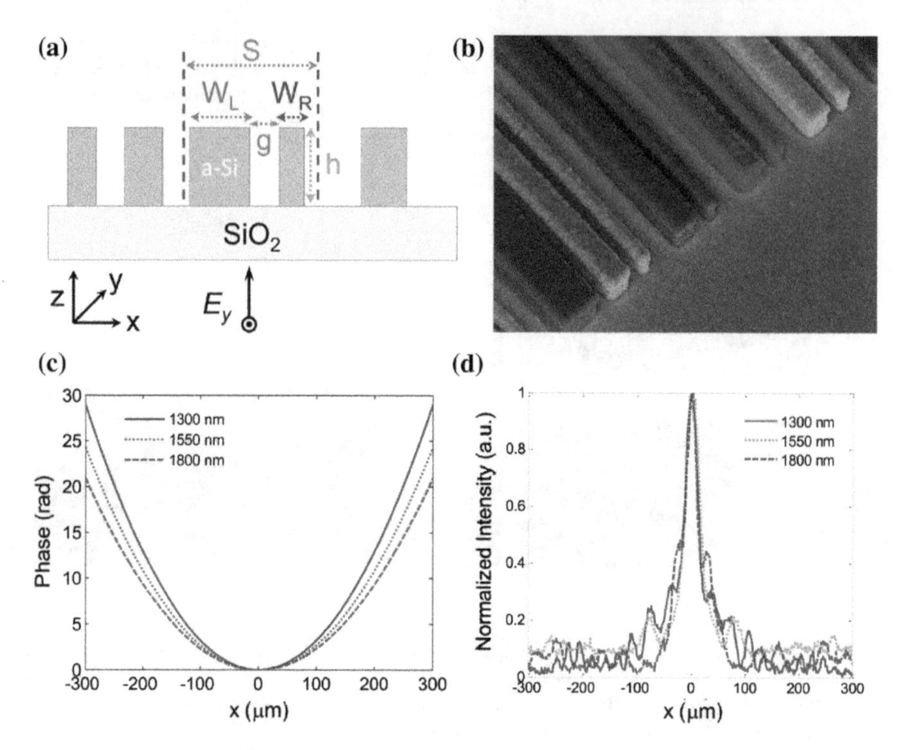

Fig. 9.31 **a** Schematic of the metasurface consisting of amorphous silicon (a-Si) rectangular dielectric resonators on a fused silica substrate. **b** False colored side-view SEM image of the metasurface lens. Each unit cell is identified by a different color. Scale bar is 400 nm. **c** Phase profiles of lens with numerical aperture NA = 0.04, diameter D = 600 μm, and focal length f = 7.5 mm for wavelengths of 1300 nm, 1550 nm, and 1800 nm as a function of the distance from the center of the lens. **d** Measured intensity profiles across the focal plane of the lens for three wavelengths of 1300 nm, 1550 nm, and 1800 nm. Reproduced from [70] with permission. Copyright 2015, AAAS

the required phase coverage is achieved by optimizing the diameter of nanopost with a fixed period. In large-area etching process, however, uneven structural gaps will lead to uneven etching depth, which can be solved by a varied period ($p = d$ + gap, gap = 1.3 μm). Figure 9.33b shows the phase shift as a function of the diameter of the nanopost at several different wavelengths. Note that the phase is folded between 0 and 2π.

The metalens can be obtained with only one step of lithography (laser direct writing) followed by a single etching process (inductive-coupled plasmonic) on a double-sided polished Si wafer (wafer thickness is 500 μm). Figure 9.34a–c shows the fabricated metalens, its SEM images, and the metalens singlet-based IR imaging system. The array of ceramic heating lamps (Fig. 9.34e) was employed as the target object for the object distance of 20 m, as shown in Fig. 9.34d. The lamps are well resolved by using the achromatic metalens, as shown in Fig. 9.34f. The uneven brightness is due to the inconsistent heating power of the lamps. This work provides

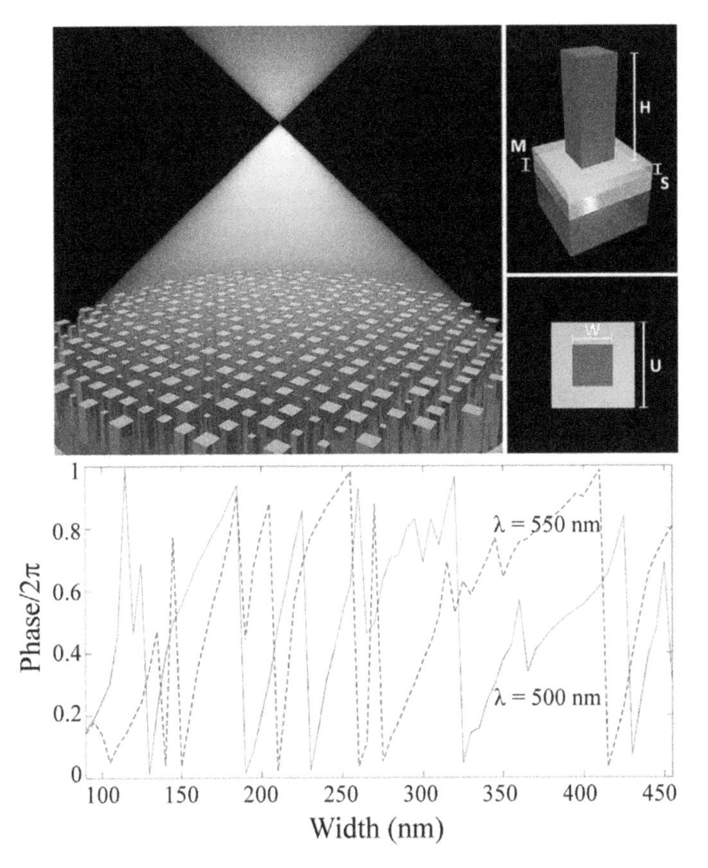

Fig. 9.32 Achromatic metalens in reflection mode. The building block consists of a titanium dioxide (TiO_2) nanopillar with height $H = 600$ nm on a substrate. The bottom shows reflection phase shift as a function of the nanopillar width at two different wavelengths of 500 and 550 nm. Adapted from [72] with permission. Copyright 2017, American Chemical Society

an efficient way for development of compact optical devices for long-wavelength infrared technology.

9.5.2 Broadband Achromatic Metalens

The essence of the multi-wavelength lens, however, is achromatic at some discrete wavelengths rather than in a continuously broadband. Recently, some methods have been proposed to improve the efficiency of the metalenses with a continuous working band, which is an important step to the practical application of metalens. As shown in Fig. 9.34, a novel method to design broadband achromatic plasmonic component was proposed based on metallic nanoslit lens. By making use of the compensation

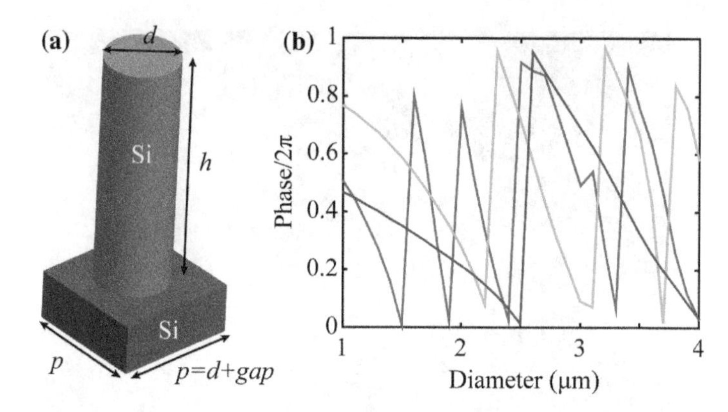

Fig. 9.33 **a** Schematic of a unit element, showing the geometric parameters, $h = 15$ μm, gap $= 1.3$ μm. **b** Simulated transmission phase shift as a function of the nanopost diameter at three different wavelengths of 8 μm (red), 11 μm (green), and 14 μm (blue)

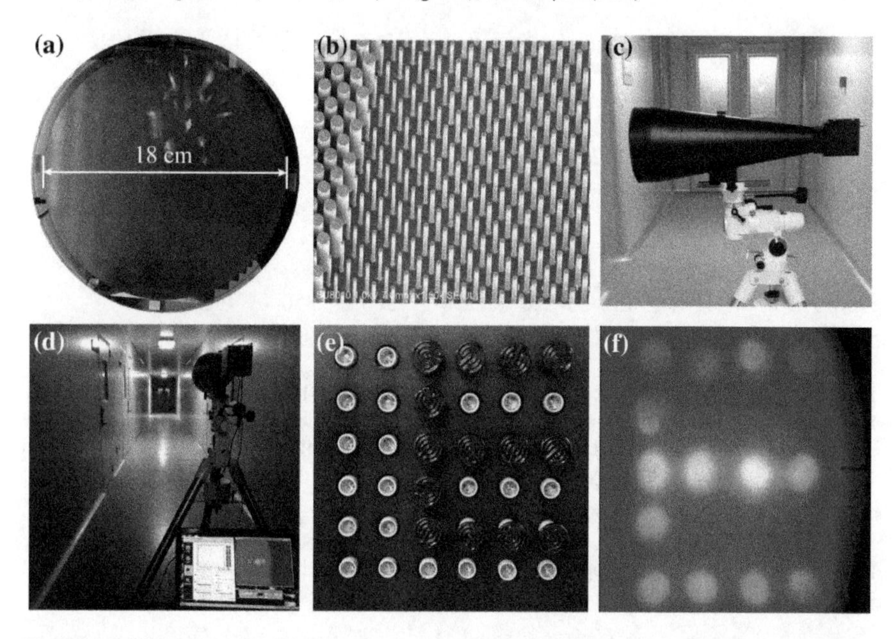

Fig. 9.34 **a** The image of fabricated achromatic metalens captured by a mobile phone. The focal length is 50 cm. **b** SEM image of the metalens. **c** The graph of imaging system object consisting of a metalens, an infrared CCD and lens cone. **d** Imaging scene with an object distance of 20 m. **e** The graph of the array of ceramic heating lamps with the diameter of ~5.5 cm and the center-to-center distance of 8 cm. **f** The images formed by the metalens

between the structural dispersion of MIM waveguide and material dispersion of metal, it was theoretically demonstrated that the plasmonic nanoslit structures can be used to construct broadband achromatic lenses [63]. When a subwavelength MIM slit is illuminated by TM-polarized beam, the SPP mode can exist, which produces phase retardation when it propagates through the subwavelength slit. The complex propagation constant β of the fundamental SPP mode in MIM waveguide is given by the eigenvalue equation [14]:

$$\tanh\left(\frac{\sqrt{\beta^2 - k_0^2\varepsilon_\mathrm{d}}w}{2}\right) = -\frac{\varepsilon_\mathrm{d}\sqrt{\beta^2 - k_0^2\varepsilon_\mathrm{m}}}{\varepsilon_\mathrm{m}\sqrt{\beta^2 - k_0^2\varepsilon_\mathrm{d}}}, \tag{9.5.4}$$

where w is the width of slit, k_0 is the wavevector of light in free space, and ε_d and ε_m are the permittivities of the dielectric medium filled in the slit and the metal, respectively.

The material dispersive behavior of metal can be approximated by the Drude model:

$$\varepsilon_\mathrm{m}(\omega) = \varepsilon_\infty - \frac{\omega_\mathrm{p}^2}{\omega^2 + i\omega\gamma}, \tag{9.5.5}$$

where ω is the angular frequency of the incident electromagnetic radiation, ε_∞ is the permittivity at infinite angular frequency, ω_p is the bulk plasma frequency which represents the natural frequency of the oscillations of free conduction electrons, and γ is the collision frequency. At the frequency $\omega \ll \omega_p$ and $\omega \gg \gamma$, $\varepsilon_\mathrm{m}(\omega)$ can be approximated to:

$$\varepsilon_\mathrm{m}(w) \approx -\frac{\omega_\mathrm{p}^2}{\omega^2} \ll -1. \tag{9.5.6}$$

Applying Eq. (9.5.6) to simplify Eq. (9.5.4), the $\beta\lambda$ can be written as a function of permittivity of dielectric ε_d and slit width w:

$$\beta\lambda = 2\pi\sqrt{\varepsilon_\mathrm{d}\left(\frac{2c}{\omega_\mathrm{p}w} + 1\right)} = \mathrm{const}. \tag{9.5.7}$$

When the SPP wave passes through the subwavelength metallic slit, the output phase retardation $\Delta\varphi$ of light transmitted through each slit can be expressed by [14]:

$$\Delta\varphi = 2m\pi + \mathrm{Re}(\beta h) + \theta, \tag{9.5.8}$$

where $\theta = \arg\left[1 - \left(\frac{1-\beta/k_0}{1+\beta/k_0}\right)^2 \exp(i2\beta h)\right]$ originates from multiple reflections between the entrance and exit surfaces [14], and h presents the length of the MIM waveguide. Both physical analysis and numerical simulations show that βh plays a dominating role in phase shift as we discussed in Sect. 9.1. Therefore, if we choose $m = 0$, $\Delta\varphi$ can be approximated as $\mathrm{Re}(\beta h)$. The imaginary part of propagation constant of the SPP in the MIM slit is usually ignorable ($\mathrm{Im}(\beta) \ll 1$) at the frequency ω much higher than collision frequency γ. The waveguide length h is a constant in components based on MIM slits, so:

$$\beta h\lambda \approx \Delta\varphi \cdot \lambda = \text{const.} \tag{9.5.9}$$

Thus, the MIM waveguide is theoretically proved achromatic in the frequency range $\omega \ll \omega_{\mathrm{p}}$ and $\omega \gg \gamma$.

Furthermore, the dispersive behavior of MIM waveguide is demonstrated in theoretical calculation and numerical simulation. The basic unit of the achromatic plasmonic component based on MIM waveguide is shown by the schematic cross section in the inset of Fig. 9.35a. The metallic slit width w is varied from 20 to 100 nm, and the length of waveguide h is fixed at 3 μm. The relative permittivity of the material filled in the slit is assumed to be $\varepsilon_{\mathrm{d}} = 1$ for air. Silver is chosen as the metal in this model due to its lower loss. In the near-infrared frequency, the Drude dispersion of the gold counteracts the dispersion of the propagation SPP mode, which makes the achromatic condition, i.e., Eq. (9.5.9), satisfied. The designed metalens can focus the near-infrared light over a continuous bandwidth broader of 1000–2000 nm (Fig. 9.35c).

The phase distribution of the achromatic plasmonic lens is designed for $f =$ 5 μm. The number of slits in our design is 51, and the period of the structure is chosen as 200 nm. As shown in Fig. 9.36a, the focal length is close to 5 μm at different wavelengths. Insets of Fig. 9.36a show the electric field intensity distribution of plasmonic lens illuminated by the light at wavelength of 1000 nm, 1500 nm, and 2000 nm, respectively. Although the sizes of focal spot are different, which is determined by diffraction limit, the focal lengths are the same so this flat lens is achromatic.

To analyze the achromatic performance, the simulated phase distributions at 200 nm above the output plane of the flat lens are shown as solid lines in Fig. 9.36b. The dashed lines in Fig. 9.36b stand for the theoretical achromatic distributions for target focal length at different wavelengths obtained from Eq. (9.5.9). The simulated phase distributions of output light with different wavelengths show good agreement with the theoretical prediction. Figure 9.36c shows the normalized $\Delta\varphi \cdot \lambda$ as a function of the x-position. Simulated $\Delta\varphi \cdot \lambda$ is almost the same, which leads to the same focal length. The slight focal shift can be explained by the little deviation of $\Delta\varphi \cdot \lambda$. The limitation of this method is the fabrication difficulty of the nanoslits with high aspect ratios. Similar tactic was later demonstrated in an all-dielectric flat focusing lens, which was composed of subwavelength silicon-air slit waveguide array with varied widths [73]. Such lenses can realize achromatic focusing in a wide spectral range from 8 to 12 μm by engineering the width of the silicon slits.

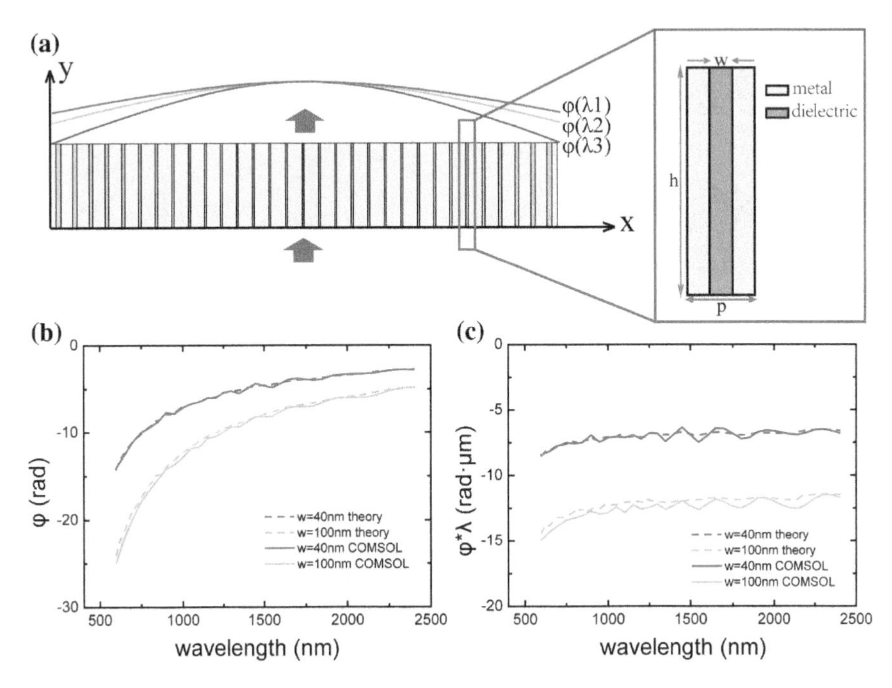

Fig. 9.35 Metallic nanoslits array as an achromatic lens. **a** Side view. **b** Relative phase shifts obtained by theoretical calculations and numerical simulations. **c** The product of φ and λ for each slit width. Adapted with permission from [63]. Copyright 2016, The Authors

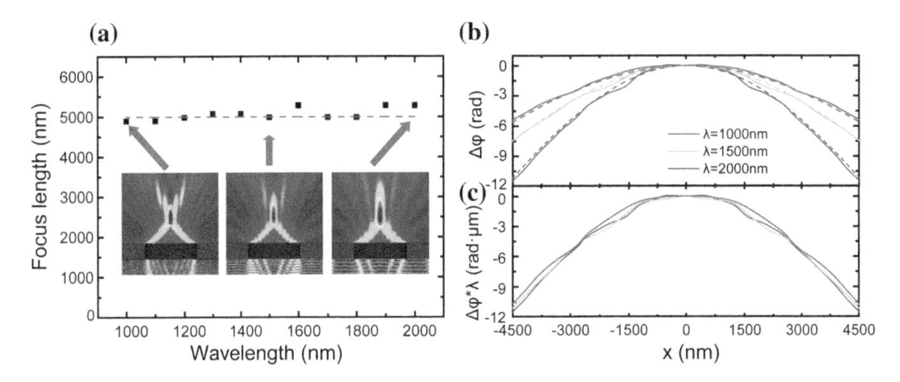

Fig. 9.36 **a** Achromatic plasmonic lens. **a** Focal lengths of the designed achromatic lens at different wavelengths. Electric field intensity distributions at the wavelength $\lambda = 1000$, 1500, and 2000 nm are shown in the inset. **b** Numerically (solid line) simulated and ideal (dashed line) phase distribution at 200 nm above the output surface at the wavelength $\lambda = 1000$ nm (blue), 1500 nm (green) and 2000 nm (red). **c** Simulated spatial distribution of $\Delta\varphi \cdot \lambda$ at the wavelength $\lambda = 1000$ nm (blue), 1500 nm (green), and 2000 nm (red). Adapted with permission from [63]. Copyright 2016, The Authors

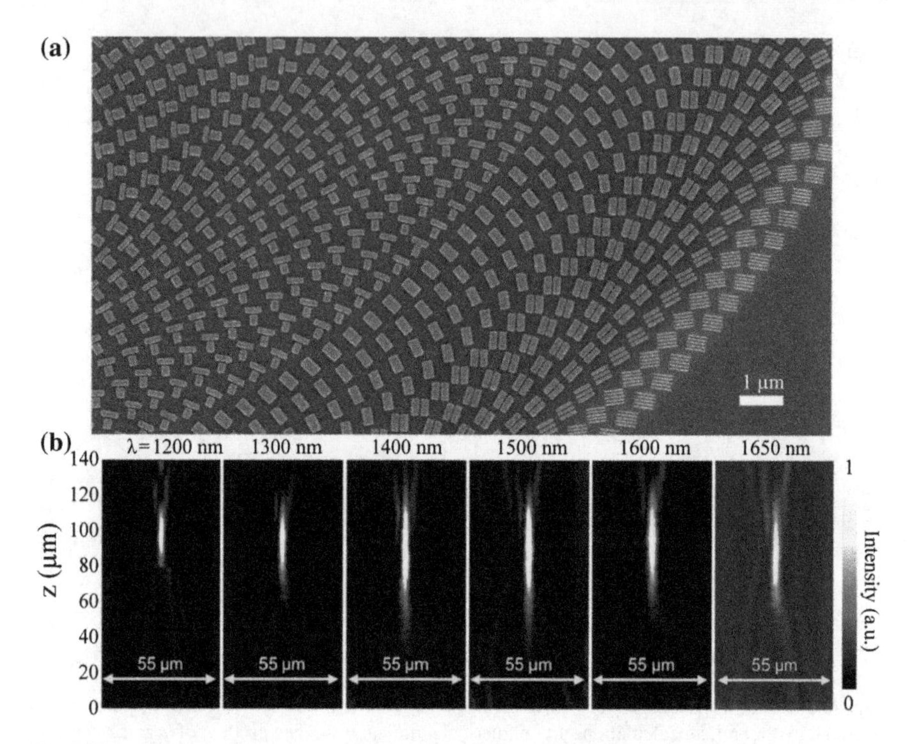

Fig. 9.37 **a** Zoom-in SEM image of fabricated metalens. **b** Experimental intensity profiles along axial planes at various incident wavelengths. Reproduced from [74] with permission. Copyright 2017, The Authors

Geometric phase provides another promising way to realize achromatic performance. For instance, an integrated resonant unit element combined of geometric phase and circuit-type phase shift was employed to design broadband achromatic flat optical components [74]. Achromatic converging metalens and beam deflector were demonstrated within a broad infrared band from wavelength of 1200–1680 nm, and broad visible band from 420 to 650 nm [75]. As a result of the limited phase shift coverage, the achromatic lens shown in Fig. 9.37 has a small NA of 0.268. For all the measured wavelengths, the focused light has a strong intensity at the focal length of 100 μm.

Achromatic transmissive metalens working in visible regime has also been designed based on dielectric resonators [76, 77]. As illustrated in Fig. 9.38, the achromatic metalens is constructed by varying both the sizes and orientation angles of two complementary structures (holes and pillars) [77]. With a diameter of 50 μm and NA of 0.106, the metalens is able to form achromatic images as shown in Fig. 9.38c–e. In order to increase the aperture size, while simultaneously maintaining the achromatic performance, larger phase shift and stronger dispersion correcting ability should be provided.

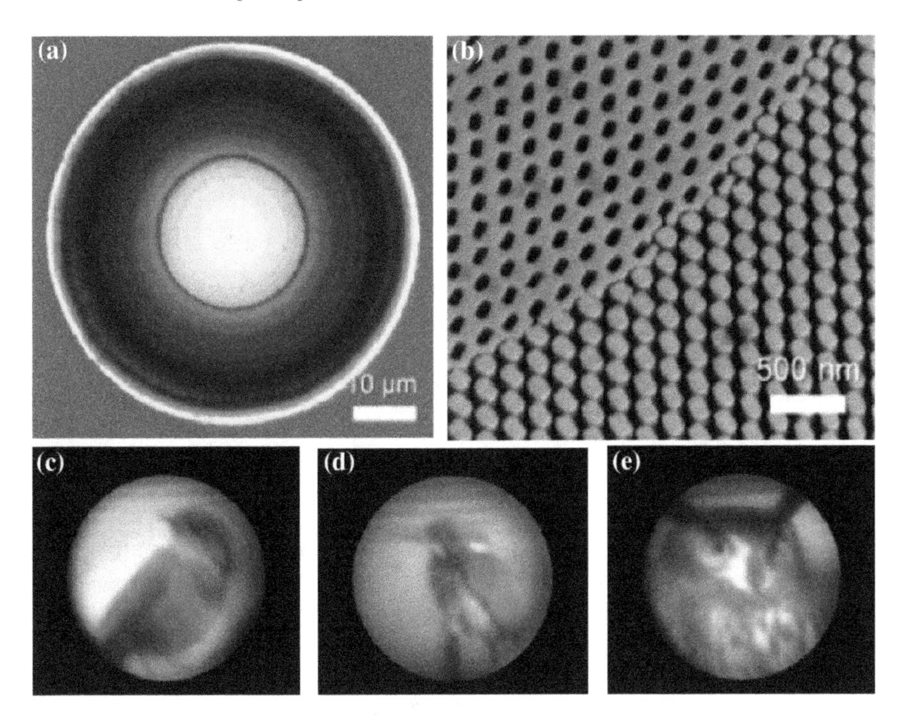

Fig. 9.38 **a** Optical image of broadband achromatic metalens. **b** Zoom-in SEM images at the boundary of nanopillars and Babinet GaN-based structures. **c**–**e** Colorfully captured images from achromatic metalens. Reproduced from [77] with permission. Copyright 2018, Macmillan Publishers Limited

9.5.3 Super-dispersive Metalens

Metasurfaces with zero and positive dispersion would be useful for making achromatic singlet and doublet lenses, and the larger-than-regular dispersion of hyperdispersive metasurface gratings would enable high-resolution spectrometers.

Figure 9.39a, c illustrates the super-dispersion positive and negative metalenses based on the silicon nanocuboids with spatially varying orientations. Their respective super-dispersion phase distribution is shown in Fig. 9.39b, d. The simulation results of super-dispersion positive and negative metalenses are given in Fig. 9.40. Figure 9.40a–c, e–g shows the normalized intensity distribution of positive and negative in xz-plane, where the intensities are normalized with respect to their maximum light intensity. Based on the vectorial diffraction method, the theoretical results (dotted line) are calculated and compared with the simulation results (solid line) in Fig. 9.40d, h, which show that super-dispersion positive and negative metalenses are able to separate three wavelengths completely. The slight deviations originate from the interaction between the unit cells where the wavefronts that metalenses shaped are not the spherical as desired. As shown in Fig. 9.40d, h, the focal lengths of posi-

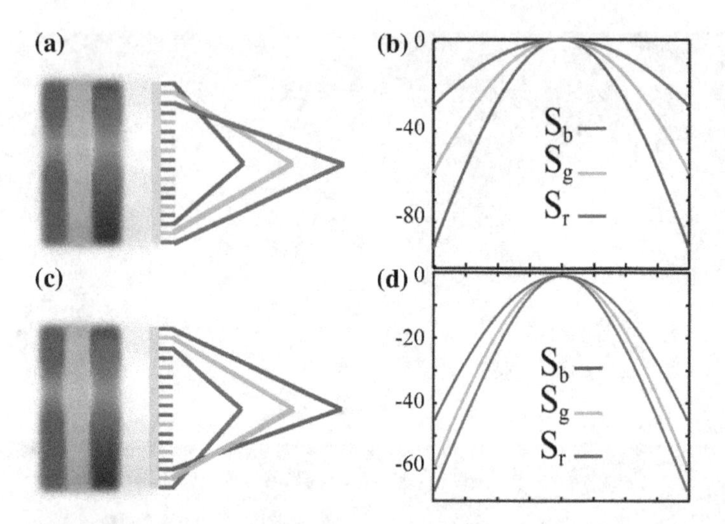

Fig. 9.39 **a** Schematic and **b** phase distribution of super-dispersion positive metalens designed to separate different wavelength light with normal dispersion. **f** Schematic and **g** phase distribution of super-dispersion negative metalens with anomalous dispersion. Reproduced from [65] with permission. Copyright 2017, The Authors

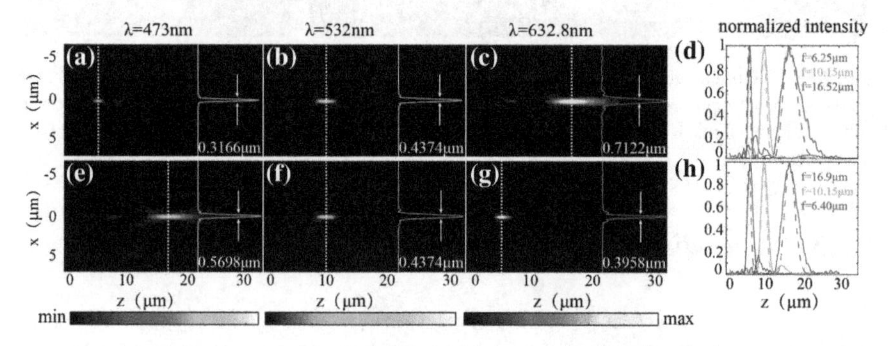

Fig. 9.40 Simulation results of super-dispersion metalenses. **a–c** Simulated normalized intensity distribution of super-dispersion positive metalenses in xz-plane and (inset) the intensity profiles across the focal plane upon illumination with $\lambda = 473$, 532, and 632.8 nm, respectively. **d** Comparison of theoretical results (dotted line) and simulated results (solid line) of the normalized intensity curves of three wavelengths along the z-axis. **e–g** Simulated normalized intensities distribution of super-dispersion negative metalenses in xz-plane and (inset) the intensity profiles across the focal plane upon illumination with $\lambda = 473$, 532, and 632.8 nm, respectively. **h** Comparison of theoretical results (dotted line) and simulated results (solid line) of the normalized intensity curves of three wavelengths along the z-axis. Reproduced from [65] with permission. Copyright 2017, The Authors

tive and negative lenses are close to their designed value, and FWHMs of them are very close to their diffraction limits, respectively.

9.6 Tunable Metalenses

Tunable or reconfigurable meta-devices, which rely on the ability to dynamically control the resonant properties of every individual subwavelength structure, are essential for applications in tunable electromagnetic devices, including beam deflectors, dynamic holograms, and flat lenses [58, 78]. Focal adjustment and zooming are widely used in cameras and advanced optical systems. It is typically performed along the optical axis by mechanical or electrical means. Traditional zoom lenses are regularly constructed by multi-element lenses which are bulky and difficult to integrate. Deformable solid- and liquid-filled lenses with mechanical, electromechanical, electrowetting, and thermal tuning mechanisms have been demonstrated. Although these devices are more compact than multi-element varifocal lenses, they are still bulky due to their refractive element-based configurations. Spatial light modulator (SLM) is one type of traditional dynamically tunable device with relatively compact volume. However, its pixel size is one order of magnitude larger than visible wavelength, which limits its tuning range and NA. The commonly used liquid crystal-based SLMs are polarization dependent and have limited response speed. In this section, we introduce three kinds of compact tunable metalens. The working mechanisms and their respective advantages/disadvantages are discussed.

9.6.1 Tunable Metalens with Mechanical Stretching and Moving

Metasurfaces with structural tenability and stretchable substrates have been developed for mechanical reconfigurable ability [79]. For instance, by changing the lattice constant of a complex Au nanorod array fabricated on a stretchable polydimethylsiloxane (PDMS) substrate (Fig. 9.41a) [80], a mechanically reconfigurable metasurface that can continuously tune the wavefront has been demonstrated in the visible frequency range. The anomalous refraction angle of visible light at 632.8 nm can be adjusted from 11.4° to 14.9° by stretching the substrate by ~30%. Furthermore, an ultra-thin flat 1.7× zoom lens whose focal length can be continuously changed from 150 to 250 μm has been demonstrated (Fig. 9.41b). Therefore, by applying an oscillating force to the stretchable substrate, it is possible to build metalens enabling axial scanning. In order to overcome the limited efficiency and polarization-dependent operation, Kamali et al. demonstrated highly tunable dielectric metasurface devices based on subwavelength thick silicon nanoposts encapsulated in a thin transparent elastic polymer (Fig. 9.41c) [81]. Measured optical intensities in the axial plane (Fig. 9.41d, left) and the focal plane (Fig. 9.41d, right) at 6 different strain values (0–50%) show a large focal distance tunability from 600 to 1400 μm through radial strain, while maintaining a diffraction-limited focus and a focusing efficiency above 50%.

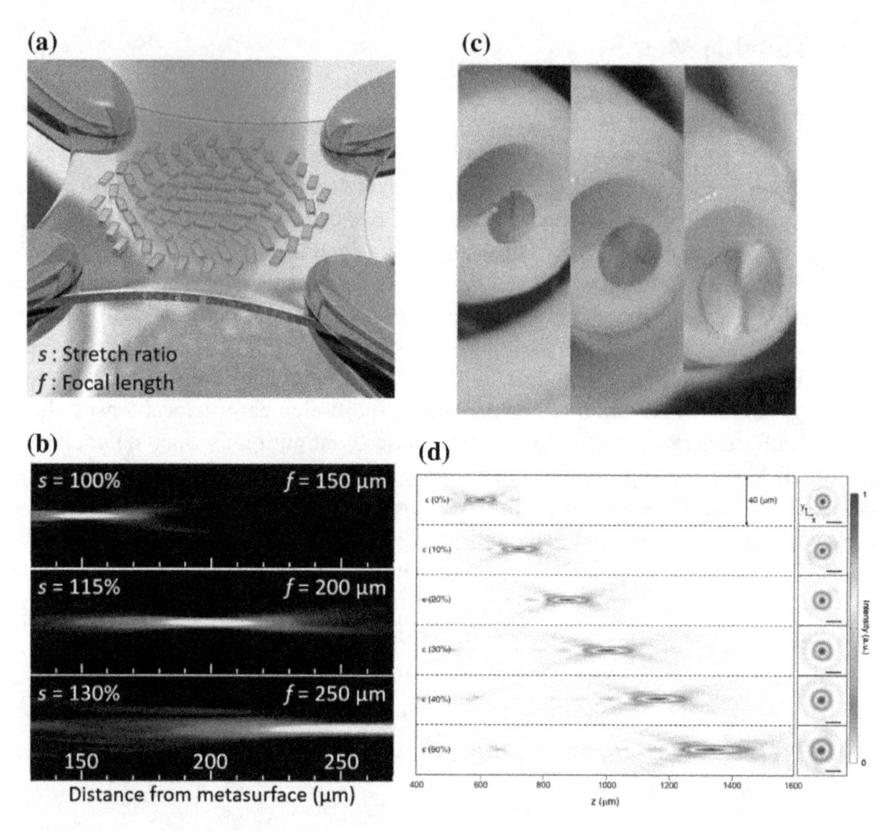

Fig. 9.41 **a** Schematic illustrations of a tunable lens based on stretched PDMS. **b** Measured longitudinal beam profiles generated on the transmission side of the flat zoom lens with different stretch ratio 100% (top), 115% (middle), and 130% (bottom). **c** An array of elastic metasurface micro-lenses clamped between two Teflon rings, under different strain. **d** Measured optical intensity profiles of a radially strained metasurface micro-lens in the axial plane (left) and the focal plane (right). Scale bars: 5 μm. **a**, **b** Reproduced from [80] with permission. Copyright 2016, American Chemical Society. **c**, **d** Reproduced from [81] with permission. Copyright 2016, WILEY-VCH Verlag GmbH & Co. KGaA, Weinheim

Another kind of mechanically tunable metalens was designed using a concept of Alvarez lens which was invented several decades ago [82]. The Alvarez lens includes two cascade inverse phase plates. The central concept of the Alvarez lens is the dependence of the focal length on the lateral displacement of these two phase plates with cubic profile. The first Alvarez phase plate obeys [83]:

$$\varphi_{\mathrm{Alv}}(x, y) = A\left(\frac{1}{3}x^3 + xy^2\right), \tag{9.6.1}$$

and the inverse phase plate obeys:

$$\varphi_{\mathrm{Inv}}(x, y) = -A\left(\frac{1}{3}x^3 + xy^2\right). \tag{9.6.2}$$

So the summed phase satisfies $\varphi_{\mathrm{Alv}}(x, y)+\varphi_{\mathrm{Inv}}(x, y) = 0$ for aligned phase plates. For a displacement d along the x-axis, the addition of the two surfaces produces a quadratic phase profile plus a constant phase offset:

$$\varphi_{\mathrm{sum}}(d) = \varphi_{\mathrm{Alv}}(x + d, y) + \varphi_{\mathrm{Inv}}(x - d, y) = 2Ad(x^2 + y^2) + \frac{2}{3}d^3. \tag{9.6.3}$$

By neglecting the constant phase offset, and setting $r^2 = (x^2 + y^2)$, we recognize the expression for a lens under the paraxial approximation:

$$\varphi_{\mathrm{Len}}(d) = 2Adr^2 = \frac{r^2}{2f}, \tag{9.6.4}$$

with focal length as a function of displacement:

$$f(d) = \frac{1}{4Ad}. \tag{9.6.5}$$

From Eq. (9.6.5), the Alvarez lenses show adjustable focus when a lateral displacement of the two-wave plate takes place. Many methods to realize cubic surfaces have been implemented so far. Unfortunately, it remains a great challenge with traditional approaches, due to the fabrication difficulties in freeform optics with the most state-of-the-art tools. Metasurface, as a new category of lithographically defined diffractive devices, may surmount this limit and enable thin and lightweight optical elements with precisely engineered phase profiles. Figure 9.42a plots the schematic of the setup used to measure the performance of the metasurface-based Alvarez lens. The Alvarez phase plate is mounted on the LED side, while the inverse phase plate is mounted on the microscope side. The Alvarez phase plate is allowed to move in the x-direction. The microscope is free to move along the z-axis, allowing us to image into and out of the focal plane for each displacement. The Alvarez lens was designed with $A = 1.17 \times 10^7\ \mathrm{m}^{-2}$, and each of the Alvarez metasurface phase plate was fabricated in 633 nm silicon nitride film deposited on top of a 500-μm fused quartz substrate with a length of 150 μm. Figure 9.42b shows the SEM image of half of the Alvarez lens. The focal lengths for displacements d of each metasurface from 2 to 50 μm was measured as shown in Fig. 9.42c. The focal distances change from a minimum of 0.5 mm to a maximum of 3 mm. This indicates that with a physical displacement of 100 μm, the focal length changes by 2.5 mm corresponding to a change in optical power (inverse focal length) of about 1600 diopters.

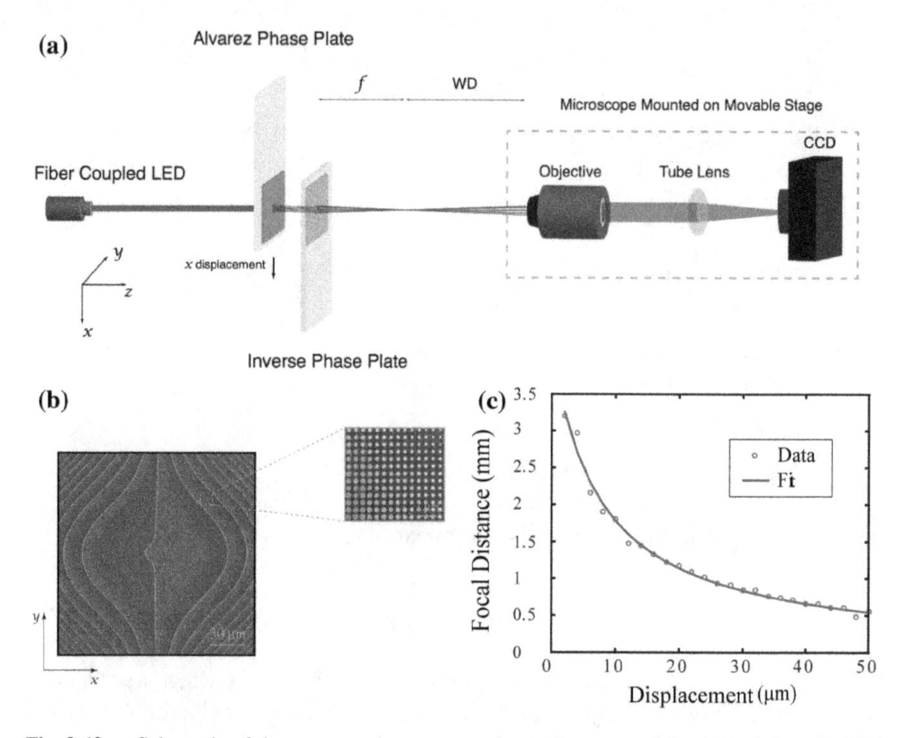

Fig. 9.42 **a** Schematic of the setup used to measure the performance of the Alvarez lens. **b** SEM images of half of the Alvarez lens. Inset: Zooms of specific locations of the metasurface showing the gradient in pillar sizes. **c** Measured focal distance of the Alvarez lens pair plotted against x displacement. The red line is a theoretical fit to the focal length data. Reproduced from [83] with permission. Copyright 2017, The Authors

9.6.2 Tunable Metalens with Nonlinear or Phase-Change Materials

As introduced above, highly tunable metasurface lenses based on stretchable substrates have already been demonstrated. However, they normally have low speeds and require a radial stretching mechanism that might increase the total device size. In 2007, Min et al. investigated a type of metallic tunable lens consisting of slits with variant widths, filled with Kerr nonlinear media [84]. Each slit transmits light with specific phase retardation controlled by the intensity of incident light, owing to the nonlinear response. This new lens can actively control the deflection angle and the focal length of output beam, as illustrated in Fig. 9.43.

Phase-change materials with tunable optical properties upon external stimuli are also crucial for the realization of versatile platforms with reconfigurable functionalities. For example, vanadium dioxide (VO_2)-based hybrid metamaterials have been shown to have tunable resonances resulting from the VO_2 phase transition at THz and IR frequencies. VO_2 can undergo a thermally driven insulator-to-metal transi-

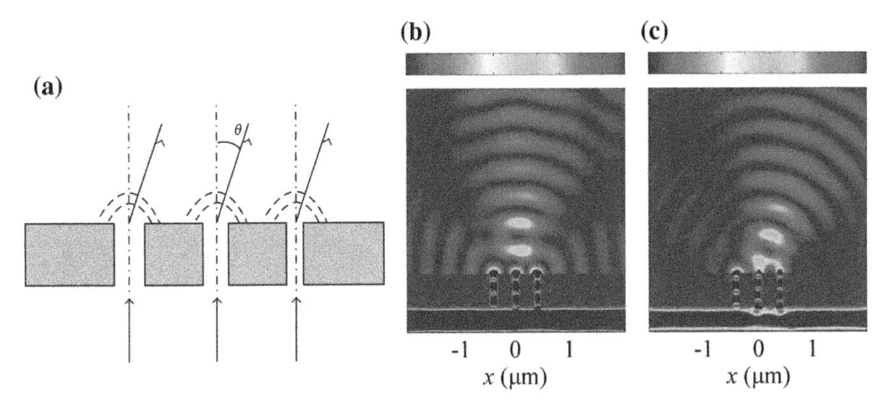

Fig. 9.43 Reconfigurable nanoslits lens. **a** Schematic of metallic nanoslits lens containing nonlinear media. The parameters of three-slit structure are as follows: The thickness of the Ag film is 560 nm, the distance between two silts is 400 nm (center to center), the slit width is 90, 70, and 60 nm in sequence from up to down. **b**, **c** The FDTD simulations of electric field intensity distribution at different incident amplitudes: **b** 1×10^8 V/m and **c** 2.5×10^8 V/m. Reproduced from [84] with permission. Copyright 2007, Optical Society of America

tion at ~67 °C, leading to large tuning range of reflectivity from ~80% to 0.25% at 11.6 μm [85]. It is also shown that the resonances of plasmonic elements can be tuned by utilizing the large change of the optical properties of the underlying VO_2 substrate in proximity to its phase transition [86]. As another phase-change material often adopted in optical recording, Ge2Sb2Te5 (GST) has recently been utilized to develop reconfigurable metamaterials and metasurfaces, owing to their good stability, high-speed, and reversible switching performance [87–89]. GST undergoes a phase change from amorphous state to crystalline state through pumping by ultrafast femtosecond laser beams or thermal heating, with remarkable dielectric property differences between the two states. In particular, the degree of crystallization can be gradually obtained by applying controllable amounts of external excitation, giving rise to greater flexibilities to access continuous refractive index control. The combination of these characteristics with specifically designed metasurfaces can provide new functionalities toward active electromagnetic perfect absorbers [87, 90], polarization manipulation [91], beam steering [92], and dynamic planar lenses [88, 93].

A hybrid planar lens composed of an array of slits filled with GST has been proposed to engineer the far-field focusing pattern, which exploits the large refractive index contrast between its amorphous and crystalline phases in the visible and IR regions. Figure 9.44 illustrates the GST-based tunable lens nanoslits scheme [88]. The GST material was filled into the slits through magnetron sputtering. By varying the crystallization level of GST from 0 to 90%, a transmitted electromagnetic phase modulation as large as 0.56π was realized at the working wavelength of 1.55 μm. Consequently, different phase fronts can be constructed spatially by assigning the designed GST crystallization levels to the corresponding slits, achieving various far-field focusing patterns.

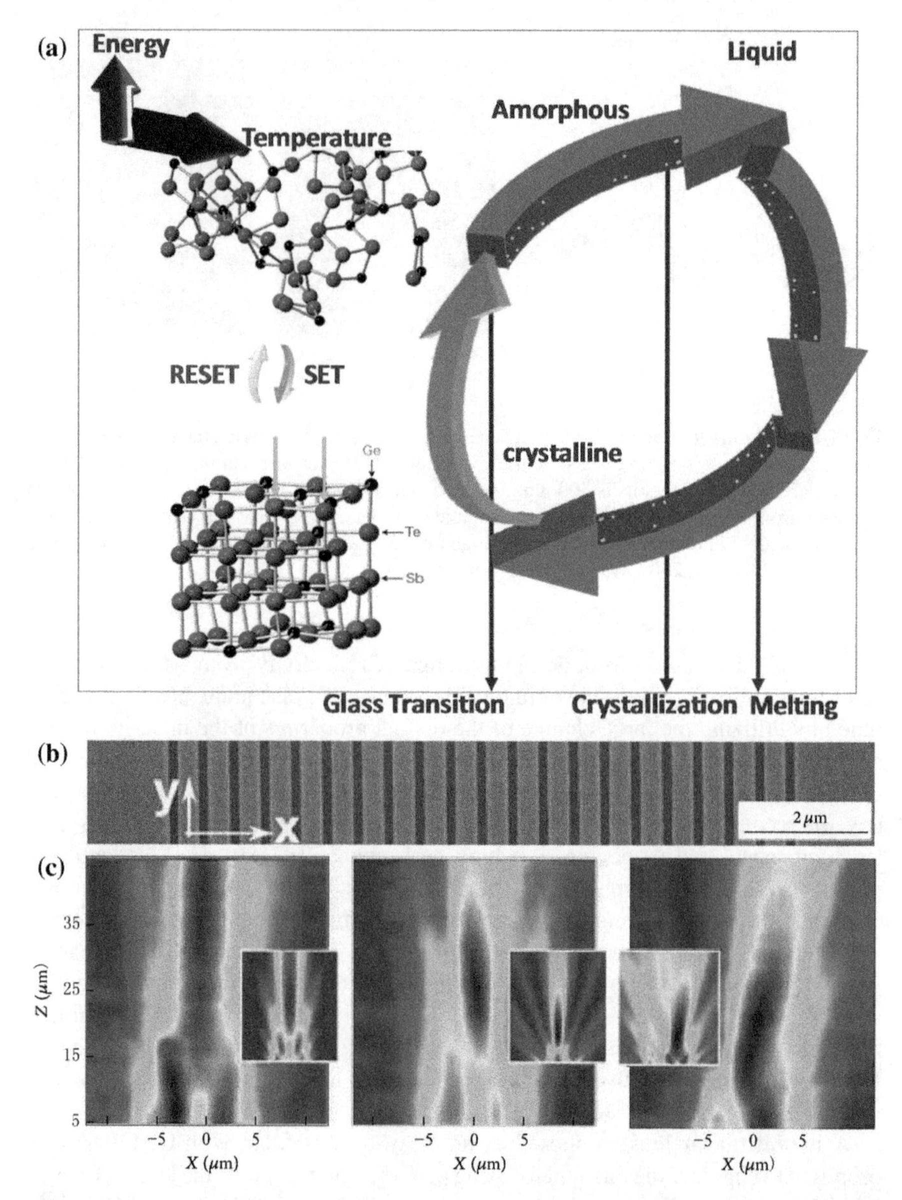

Fig. 9.44 **a** Schematic illustration of phase transition of GST between amorphous and crystalline phases. **b** SEM image of the fabricated planar lens before sputtering GST. **c** Focusing pattern measured in *xz*-plane by confocal scanning optical microscopy for different status of GST (lower panel). **a** Reproduced from [94] with permission. Copyright 2017, The Authors. **b**, **c** Reproduced from [88] with permission. Copyright 2015, The Authors

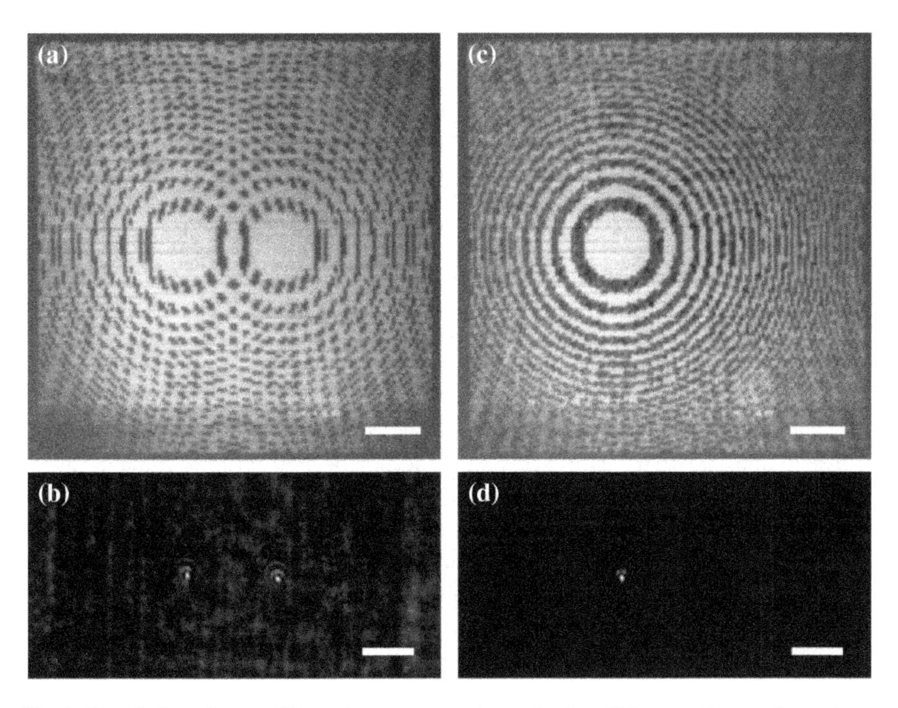

Fig. 9.45 **a**, **b** Superimposed Fresnel zone patterns imaged at $\lambda = 633$ nm as they are first written. **c**, **d** The second Fresnel zone pattern is erased. Reproduced from [93] with permission. Copyright 2015, Macmillan Publishers Limited

More recently, the introduction of the phase-change material with tailored trains of femtosecond pulse allows different functionalities to be written into the same optical materials without changing the structure of the optical system. As illustrated in Fig. 9.45, by a recycle of written, erased, and rewritten, a variety of devices were demonstrated, including a visible-range reconfigurable dichromatic and multi-focus Fresnel zone plate, a super-oscillatory lens with subwavelength focus, a grayscale hologram, and resonant metamaterials [93].

9.6.3 Tunable Metalens with MEMS

Using electrical means to tune metasurfaces in microwave regime is of great importance because they are suitable for current printed circuit board technologies [95]. Electrical tunability can be obtained through integrating PIN or varactor diodes in the metallic circuits of metasurfaces. Selectively connecting/disconnecting PIN diode can significantly modulate the electromagnetic response, in particular the resonant phase shift, of the metasurface [96, 97]. Different from PIN diodes, the capacitance of the varactor diode changes under different reverse bias voltages that may result in electromagnetic responses such as the shift of resonant frequency. In a preliminary demonstration, an active absorber using coupled resonators and a varactor diode was constructed, whose absorption frequency changes from 4.35 to 5.85 GHz while keeping an absorption efficiency more than 90% when the bias voltage changes from 0 to −19 V [98]. If both the PIN and varactor diodes are embedded in the absorbers, the resonant frequency and absorptance can be simultaneously dynamically controlled [99].

Besides the polarization conversion and electromagnetic absorption, active beam steering has also been demonstrated by incorporating varactor diodes into the texture [100]. There is also a subwavelength reconfigurable Huygens' metasurface realized by loading it with controllable active elements [101]. As one representative demonstration, a reconfigurable metalens at the microwave frequencies was experimentally validated, where multiple complex focal spots can be controlled simultaneously at distinct spatial positions and reprogrammable in a desired fashion. Limited by the sizes and operation speed, these diodes are restricted in radio, microwave, and possibly millimeter wave regimes. At higher frequencies, the Schottky diodes and photocarriers are widely adopted to introduce tunability [102, 103]. In addition, electrical tuning methods based on graphene have substantial technological potential in terms of response time, broadband operation, compactness, and compatibility with silicon technology as well as large-scale fabrication, because graphene has high electrical and thermal conductivity, broadband widely tunable electro-optical properties, and good chemical resistance [104–106].

In optical regimes, a tunable lens taking advantages of micro-electromechanical system (MEMS) technology has been proposed recently [107]. Figure 9.46 shows the tunable metalens, with MEMS to modulate the axial distance between multiple optical elements. The system consists of a stationary metasurface on a glass substrate and a moving metasurface on a SiNx membrane. The membrane can be electrostatically actuated to change the distance between the two metasurfaces. The lenses are designed such that a small change in the distance between them, $\Delta x \sim 1$ μm, leading to a large tuning of the focal length ($\Delta f \sim 36$ μm change in the front focal length from 781 to 817 μm when the lens separation is changed from 10 to 9 μm). The capacitor plates are shown in the inset of Fig. 9.46a. The contacts are configured to make two series capacitors. Each capacitor has one plate on the glass substrate and another one on the membrane, resulting in an attractive force between the membrane and the glass substrate. Figure 9.46b shows the SEM images of the fabricated capacitor plates.

The metasurfaces are constructed with high-contrast dielectric transmitarrays, which consist of arrays of high-index dielectric scatterers (nanoposts) with different shapes and sizes, as shown in Fig. 9.46c. With proper design, the nanoposts enable complete control of phase and polarization at the subwavelength scale.

Using MEMS control principle, the axial distance tunable doublet can be used for imaging with electrically controlled focusing. Figure 9.47a shows the schematic setup of the tunable system. A transmissive object is placed in front of the imaging system. A 1.8-mm-diameter pinhole is placed in front of the aspheric lens to reduce the aperture and increase contrast. The system images the object to a plane ~130 μm

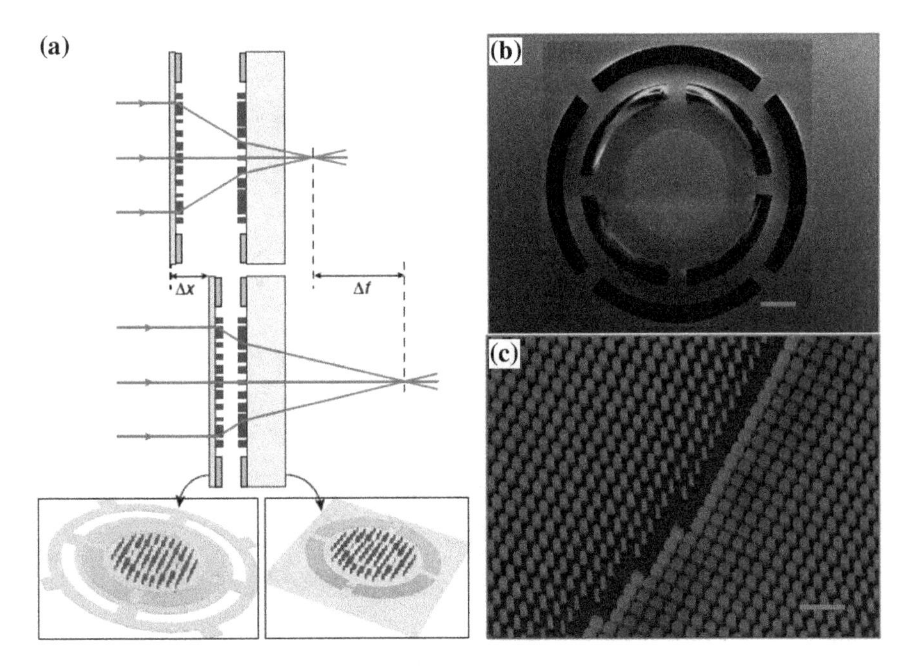

Fig. 9.46 **a** Schematic illustration of the MEMS-based tunable lens, comprised of a stationary lens on a substrate, and a moving lens on a membrane. With the correct design, a small change in the distance between the two lenses ($\Delta x \sim 1$ μm) results in a large change in the focal distance ($\Delta f \sim 35$ μm). (Insets: schematics of the moving and stationary lenses showing the electrostatic actuation contacts.) **b** SEM images of the lens on the membrane, and **c** nanoposts that form the lens. Scale bars are 100 μm. Reproduced from [107] with permission. Copyright 2018, The Authors

outside the stationary lens substrate. The imaging results at different voltages are plotted in Fig. 9.47b. When the object is $p \sim 15$ mm away and no voltage is applied, the image is out of focus. If the applied voltage is increased to 85 V in the same configuration, the image comes to focus. Changing the object distance to $p \sim 9.2$ mm, the voltage should also be changed to 60 V to keep the image in focus. At 0 V, the object should be moved to $p \sim 4$ mm to be in focus, and applying 85 V to the doublet will result in a completely out-of-focus image in this configuration. As observed here, by moving the membrane only about 4 μm, the effective focal length of the overall system changes from 44 to 122 mm, a ratio of about 1:2.8. It has been also demonstrated that this scheme can be integrated with a third metasurface to make compact microscopes (~1 mm thick) with a large corrected FOV (~500 μm or 40°) and fast axial scanning for 3D imaging. This may pave the way toward MEMS integrated metasurfaces as a platform for tunable and reconfigurable optics.

Recently, Capasso's group experimentally demonstrated electrically tunable large-area metalenses controlled by elastomers, with simultaneous larger focal length tuning (>100%) and real-time astigmatism and image shift corrections, which were only possible in electron optics [108]. The device's thickness is only 30 μm. These results demonstrate the possibility of future optical microscopes, which operate fully

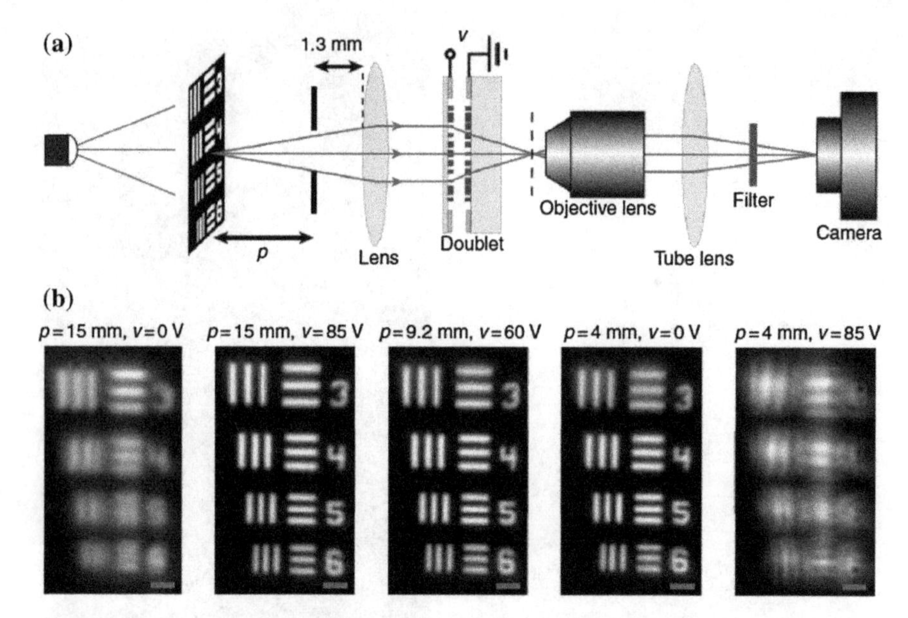

Fig. 9.47 **a** Schematic illustration of the imaging setup using a regular glass lens and the tunable doublet. The image formed by the doublet is magnified and re-imaged using a custom-built microscope with a $\times 55$ magnification onto an image sensor. **b** Imaging results, showing the tuning of the imaging distance of the doublet and glass lens combination with applied voltage. By applying 85 V across the device, the imaging distance p increases from 4 to 15 mm. The scale bars are 10 μm. Reproduced from [107] with permission. Copyright 2018, The Authors

electronically, as well as compact optical systems that employ the principles of adaptive optics to correct many orders of aberrations simultaneously.

9.7 Nonlinear Metalenses

In recent years, increasing attention has been focused on the nonlinear optical properties, particularly in the context of second and third harmonic generation and beam steering by phase gratings [109–111]. However, due to the somewhat limited phase and amplitude control over the nonlinearities of individual plasmonic element, fundamental issue of phase matching across metasurfaces has not yet been thoroughly addressed, which always depends on periodic structures (gratings) that impose specific angles of diffraction.

Nonlinear metalens acting as a nonlinear binary-phase Fresnel zone plate was firstly demonstrated by Segal et al. [112]. The nonlinear zone plate is constructed by annular zones, with adjacent zones composed of SRRs of opposite directions, leading to binary nonlinear phases of 0 and π. Recently, Almeida et al. [113] demonstrated full control over the nonlinear phase on the subwavelength scale in phase-gradient

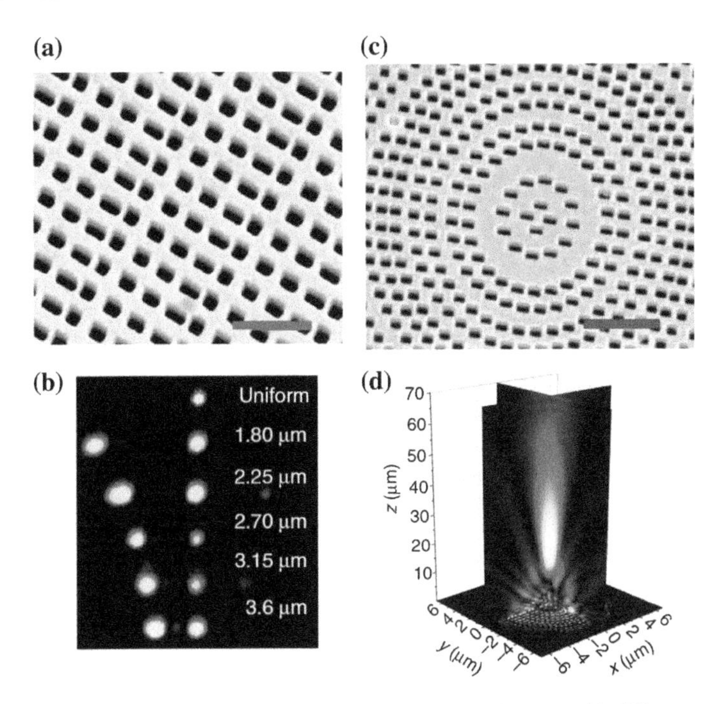

Fig. 9.48 **a** Blazed grating with unitary cell consisting of four elements with different aspect ratio AR = 1.1, 1.5, 1.9, 2.9. Scale bar, 1 μm. **b** CCD images for the zeroth and first diffraction orders for different periodicities. **c** SEM images of the fabricated ultra-thin nonlinear metalenses. Scale bars, 5 μm. **d** Experimental measurement of the field in focal region. Reproduced from [113] with permission. Copyright 2016, The Authors

metasurfaces. As in the linear regime, where Snell's law has to be modified, the phase control over the nonlinear nano-antennas leads to a modified manifestation of the laws governing nonlinear phenomena, such as nonlinear scattering, refraction, and frequency conversion. Four-wave mixing (FWM) from such metasurfaces revealed a new feature: The scattering from a phase-gradient unit cell enables anomalous phase-matching condition. This phase-matched FWM is efficient, and its phase-matched direction of propagation may be controlled by proper design of the phase gradients, enabling beam bending at any angle (Fig. 9.48a, b). Figure 9.48c shows a designed ultra-thin nonlinear metalenses with rectangular nanoholes. Tight focusing with focal lengths of several microns was demonstrated (Fig. 9.48d). This nonlinear metalens does not have any restrictions on the symmetry of the design which is characteristic to elements based on second harmonic generation (SHG) and can be integrated in light detectors based on frequency conversion to provide more sensitive detection.

Multi-quantum-well (MQW) semiconductor heterostructures, known to provide one of the largest nonlinear responses in condensed matter, respond only to the electric fields oriented normally to the semiconductor layers [114]. The highly confined and enhanced electric fields of plasmonic structures are promising to be used

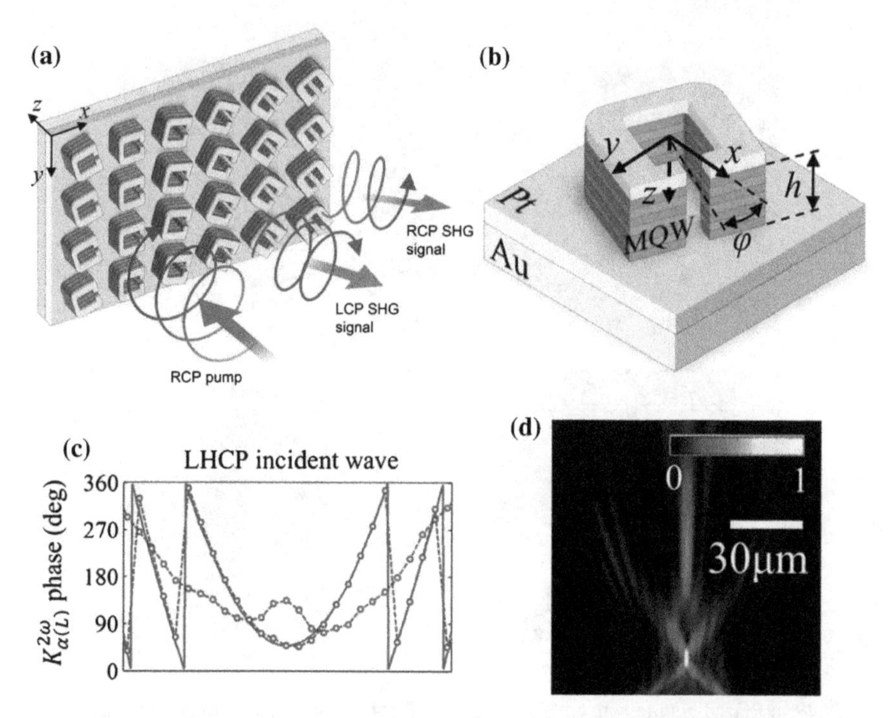

Fig. 9.49 **a** A sketch of PB nonlinear metasurface with a phase gradient in the x-direction. **b** Geometry of the PB metasurface element. **c** Analytically (solid lines) and numerically (dashed lines with markers) computed phases of RCP and LCP components of the effective nonlinear current induced on the metasurface by a LCP incident plane wave at ω. **d** The spatial distribution of the energy density above the metasurface illuminated by a 30-μm-wide LCP Gaussian beam. Reproduced from [115] with permission. Copyright 2015, American Physical Society

in enlarging the nonlinear responses of MQW heterostructures. The giant level of nonlinearities experimentally observed in plasmonic systems opens a new avenue in nonlinear optics due to their very large nonlinear response in deep subwavelength volumes, relaxing the necessity for phase matching, and providing a significant nonlinear response in a confined pixel. Plasmonic nonlinear metalens with PB phase engineering was demonstrated through highly nonlinear MQW substrates, establishing a platform to control the nonlinear wavefront based on giant localized nonlinear effects [115]. Figure 9.49a, b schematically show the nonlinear metasurface based on PB phase engineering, which works in reflective mode. A thin MQW substrate with layers is sandwiched between an array of suitably designed plasmonic resonators and a metallic ground plane. Each element is designed to ensure giant nonlinear response. A nonlinear metalens was designed to focus the LCP component of the generated beam at focal length $f = 20$ μm under LCP normal incidence. Figure 9.49c shows the analytical and numerical phase of the induced surface currents. The spatial distribution of the time-averaged energy density of the radiated field is given in Fig. 9.49d. Nearly perfect focusing of the radiated LCP wave is achieved at the desired point.

References

1. Lens. https://en.wikipedia.org/wiki/Lens_(optics)
2. Objective lens system of Olympus E-30 DSLR Camera. https://commons.wikimedia.org/wiki/File:E-30-Cutmodel.jpg
3. M. Totzeck, W. Ulrich, A. Göhnermeier, W. Kaiser, Pushing deep ultraviolet lithography to its limits. Nat. Photon. **1**, 629 (2007)
4. Fresnel lens, https://en.wikipedia.org/wiki/Fresnel_lens
5. G. Andersen, D. Tullson, Broadband antihole photon sieve telescope. Appl. Opt. **46**, 3706–3708 (2007)
6. Solar imaging system with photon sieves. https://www.nasa.gov/topics/technology/features/kitchen-optics.html
7. G. Cao, X. Gan, H. Lin, B. Jia, An accurate design of graphene oxide ultrathin flat lens based on Rayleigh-Sommerfeld theory. Opto-Electron. Adv. **1**, 180012 (2018)
8. S. Wang, X. Ouyang, Z. Feng, Y. Cao, M. Gu, X. Li, Diffractive photonic applications mediated by laser reduced graphene oxides. Opto-Electron. Adv. **1**, 170002 (2018)
9. H. Shi, C. Wang, C. Du, X. Luo, X. Dong, H. Gao, Beam manipulating by metallic nano-slits with variant widths. Opt. Express **13**, 6815–6820 (2005)
10. T. Xu, C. Wang, C. Du, X. Luo, Plasmonic beam deflector. Opt. Express **16**, 4753–4759 (2008)
11. L. Bourke, R.J. Blaikie, Genetic algorithm optimization of grating coupled near-field interference lithography systems at extreme numerical apertures. J. Opt. **19**, 095003 (2017)
12. P. Gao, N. Yao, C. Wang, Z. Zhao, Y. Luo, Y. Wang, G. Gao, K. Liu, C. Zhao, X. Luo, Enhancing aspect profile of half-pitch 32 nm and 22 nm lithography with plasmonic cavity lens. Appl. Phys. Lett. **106**, 093110 (2015)
13. M. Pu, Y. Guo, X. Li, X. Ma, X. Luo, Revisitation of extraordinary Young's interference: from catenary optical fields to spin-orbit interaction in metasurfaces. ACS Photonics **5**, 3198–3204 (2018)
14. T. Xu, C. Du, C. Wang, X. Luo, Subwavelength imaging by metallic slab lens with nanoslits. Appl. Phys. Lett. **91**, 201501 (2007)
15. L. Verslegers, P.B. Catrysse, Z. Yu, J.S. White, E.S. Barnard, M.L. Brongersma, S. Fan, Planar lenses based on nanoscale slit arrays in a metallic film. Nano Lett. **9**, 235–238 (2008)
16. S. Ishii, V.M. Shalaev, A.V. Kildishev, Holey-metal lenses: sieving single modes with proper phases. Nano Lett. **13**, 159–163 (2012)
17. Y. Chen, C. Zhou, X. Luo, C. Du, Structured lens formed by a 2D square hole array in a metallic film. Opt. Lett. **33**, 753–755 (2008)
18. J. Li, S. Chen, H. Yang, J. Li, P. Yu, H. Cheng, C. Gu, H.-T. Chen, J. Tian, Simultaneous control of light polarization and phase distributions using plasmonic metasurfaces. Adv. Funct. Mater. **25**, 704–710 (2015)
19. K. Huang, H. Liu, F.J. Garcia-Vidal, M. Hong, B. Luk'yanchuk, J. Teng, C.-W. Qiu, Ultrahigh-capacity non-periodic photon sieves operating in visible light. Nat. Commun. **6**, 7059 (2015)
20. L. Kipp, M. Skibowski, R.L. Johnson, R. Berndt, R. Adelung, S. Harm, R. Seemann, Sharper images by focusing soft X-rays with photon sieves. Nature **414**, 184–188 (2001)
21. H. Pahlevaninezhad, M. Khorasaninejad, Y.-W. Huang, Z. Shi, L.P. Hariri, D.C. Adams, V. Ding, A. Zhu, C.-W. Qiu, F. Capasso, M.J. Suter, Nano-optic endoscope for high-resolution optical coherence tomography in vivo. Nat. Photon. **12**, 540–547 (2018)
22. E. Arbabi, A. Arbabi, S.M. Kamali, Y. Horie, A. Faraon, Multiwavelength metasurfaces through spatial multiplexing. Sci. Rep. **6**, 32803 (2016)
23. Z.-B. Fan, Z.-K. Shao, M.-Y. Xie, X.-N. Pang, W.-S. Ruan, F.-L. Zhao, Y.-J. Chen, S.-Y. Yu, J.-W. Dong, Silicon nitride metalenses for close-to-one numerical aperture and wide-angle visible imaging. Phys. Rev. Appl. **10**, 014005 (2018)

24. A. She, S. Zhang, S. Shian, D.R. Clarke, F. Capasso, Large area metalenses: design, characterization, and mass manufacturing. Opt. Express **26**, 1573–1585 (2018)
25. M. Khorasaninejad, A.Y. Zhu, C. Roques-Carmes, W.T. Chen, J. Oh, I. Mishra, R.C. Devlin, F. Capasso, Polarization-insensitive metalenses at visible wavelengths. Nano Lett. **16**, 7229–7234 (2016)
26. X. Luo, Principles of electromagnetic waves in metasurfaces. Sci. China Phys. Mech. Astron. **58**, 594201 (2015)
27. X. Chen, L. Huang, H. Mühlenbernd, G. Li, B. Bai, Q. Tan, G. Jin, C.-W. Qiu, S. Zhang, T. Zentgraf, Dual-polarity plasmonic metalens for visible light. Nat. Commun. **3**, 1198 (2012)
28. X. Chen, M. Chen, M.Q. Mehmood, D. Wen, F. Yue, C.-W. Qiu, S. Zhang, Longitudinal multifoci metalens for circularly polarized light. Adv. Opt. Mater. **3**, 1201–1206 (2015)
29. F. Zhang, M. Pu, X. Li, P. Gao, X. Ma, J. Luo, H. Yu, X. Luo, All-dielectric metasurfaces for simultaneous giant circular asymmetric transmission and wavefront shaping based on asymmetric photonic spin-orbit interactions. Adv. Funct. Mater. **27**, 1704295 (2018)
30. M. Khorasaninejad, W.T. Chen, A.Y. Zhu, J. Oh, R.C. Devlin, D. Rousso, F. Capasso, Multispectral chiral imaging with a metalens. Nano Lett. **16**, 4595–4600 (2016)
31. X. Xie, X. Li, M. Pu, X. Ma, K. Liu, Y. Guo, X. Luo, Plasmonic metasurfaces for simultaneous thermal infrared invisibility and holographic illusion. Adv. Funct. Mater. **28**, 1706673 (2018)
32. M. Khorasaninejad, W.T. Chen, R.C. Devlin, J. Oh, A.Y. Zhu, F. Capasso, Metalenses at visible wavelengths: Diffraction-limited focusing and subwavelength resolution imaging. Science **352**, 1190–1194 (2016)
33. M. Pu, P. Chen, C. Wang, Y. Wang, Z. Zhao, C. Hu, C. Huang, X. Luo, Broadband anomalous reflection based on gradient low-Q meta-surface. AIP Adv. **3**, 052136 (2013)
34. X. Li, S. Xiao, B. Cai, Q. He, T.J. Cui, L. Zhou, Flat metasurfaces to focus electromagnetic waves in reflection geometry. Opt. Lett. **37**, 4940–4942 (2012)
35. M. Pu, P. Chen, Y. Wang, Z. Zhao, C. Huang, C. Wang, X. Ma, X. Luo, Anisotropic meta-mirror for achromatic electromagnetic polarization manipulation. Appl. Phys. Lett. **102**, 131906 (2013)
36. A. Pors, M.G. Nielsen, R.L. Eriksen, S.I. Bozhevolnyi, Broadband focusing flat mirrors based on plasmonic gradient metasurfaces. Nano Lett. **13**, 829–834 (2013)
37. A.B. Klemm, D. Stellinga, E.R. Martins, L. Lewis, G. Huyet, L. O'Faolain, T.F. Krauss, Experimental high numerical aperture focusing with high contrast gratings. Opt. Lett. **38**, 3410–3413 (2013)
38. A. Arbabi, Y. Horie, A.J. Ball, M. Bagheri, A. Faraon, Subwavelength-thick lenses with high numerical apertures and large efficiency based on high-contrast transmitarrays. Nat. Commun. **6**, 7069 (2015)
39. W.T. Chen, A.Y. Zhu, M. Khorasaninejad, Z.J. Shi, V. Sanjeev, F. Capasso, Immersion metalenses at visible wavelengths for nanoscale imaging. Nano Lett. **17**, 3188–3194 (2017)
40. H. Liang, Q. Lin, X. Xie, Q. Sun, Y. Wang, L. Zhou, L. Liu, X. Yu, J. Zhou, T.F. Krauss, J. Li, Ultrahigh numerical aperture metalens at visible wavelengths. Nano Lett. **18**, 4460–4466 (2018)
41. R. Paniagua-Domínguez, Y.F. Yu, E. Khaidarov, S. Choi, V. Leong, R.M. Bakker, X. Liang, Y.H. Fu, V. Valuckas, L.A. Krivitsky, A.I. Kuznetsov, A metalens with a near-unity numerical aperture. Nano Lett. **18**, 2124–2132 (2018)
42. F. Lu, F.G. Sedgwick, V. Karagodsky, C. Chase, C.J. Chang-Hasnain, Planar high-numerical-aperture low-loss focusing reflectors and lenses using subwavelength high contrast gratings. Opt. Express **18**, 12606–12614 (2010)
43. F. Aieta, P. Genevet, M. Kats, F. Capasso, Aberrations of flat lenses and aplanatic metasurfaces. Opt. Express **21**, 31530–31539 (2013)
44. A. Kalvach, Z. Szabó, Aberration-free flat lens design for a wide range of incident angles. J. Opt. Soc. Am. B **33**, A66–A71 (2016)

45. J. Hunt, T. Tyler, S. Dhar, Y.-J. Tsai, P. Bowen, S. Larouche, N.M. Jokerst, D.R. Smith, Planar, flattened Luneburg lens at infrared wavelengths. Opt. Express **20**, 1706–1713 (2012)
46. F. Zhang, M. Pu, J. Luo, H. Yu, X. Luo, Symmetry breaking of photonic spin-orbit interactions in metasurfaces. Opto-Electron. Eng. **44**, 319–325 (2017)
47. H. Ma, T. Cui, Three-dimensional broadband and broad-angle transformation-optics lens. Nat. Commun. **1**, 124 (2010)
48. W.X. Jiang, C.-W. Qiu, T.C. Han, Q. Cheng, H.F. Ma, S. Zhang, T.J. Cui, Broadband all-dielectric magnifying lens for far-field high-resolution imaging. Adv. Mater. **25**, 6963–6968 (2013)
49. Y.-Y. Zhao, Y.-L. Zhang, M.-L. Zheng, X.-Z. Dong, X.-M. Duan, Z.-S. Zhao, Three-dimensional Luneburg lens at optical frequencies. Laser Photonics Rev. **10**, 665–672 (2016)
50. A. Arbabi, E. Arbabi, S.M. Kamali, Y. Horie, S. Han, A. Faraon, Miniature optical planar camera based on a wide-angle metasurface doublet corrected for monochromatic aberrations. Nat. Commun. **7**, 13682 (2016)
51. C. Sun, Shrinking the camera size. Nat. Mater. **16**, 11 (2016)
52. B. Groever, W.T. Chen, F. Capasso, Meta-Lens doublet in the visible region. Nano Lett. **17**, 4902–4907 (2017)
53. T. Gissibl, S. Thiele, A. Herkommer, H. Giessen, Two-photon direct laser writing of ultra-compact multi-lens objectives. Nat. Photon. **10**, 554–560 (2016)
54. Extreme challenges in optics and imaging. https://www.fbo.gov/index?s=opportunity&mode=form&id=dc0f5e99441421af64f2048f696c5168&tab=core&_cview=0
55. M. Pu, X. Li, Y. Guo, X. Ma, X. Luo, Nanoapertures with ordered rotations: symmetry transformation and wide-angle flat lensing. Opt. Express **25**, 31471–31477 (2017)
56. X. Li, L. Chen, Y. Li, X. Zhang, M. Pu, Z. Zhao, X. Ma, Y. Wang, M. Hong, X. Luo, Multicolor 3D meta-holography by broadband plasmonic modulation. Sci. Adv. **2**, e1601102 (2016)
57. Y. Li, X. Li, L. Chen, M. Pu, J. Jin, M. Hong, X. Luo, Orbital angular momentum multiplexing and demultiplexing by a single metasurface. Adv. Opt. Mater. **5**, 1600502 (2017)
58. X. Li, M. Pu, Y. Wang, X. Ma, Y. Li, H. Gao, Z. Zhao, P. Gao, C. Wang, X. Luo, Dynamic control of the extraordinary optical scattering in semicontinuous 2D metamaterials. Adv. Opt. Mater. **4**, 659–663 (2016)
59. X. Li, M. Pu, Z. Zhao, X. Ma, J. Jin, Y. Wang, P. Gao, X. Luo, Catenary nanostructures as compact Bessel beam generators. Sci. Rep. **6**, 20524 (2016)
60. M. Pu, X. Li, X. Ma, Y. Wang, Z. Zhao, C. Wang, C. Hu, P. Gao, C. Huang, H. Ren, X. Li, F. Qin, J. Yang, M. Gu, M. Hong, X. Luo, Catenary optics for achromatic generation of perfect optical angular momentum. Sci. Adv. **1**, e1500396 (2015)
61. W. Liu, Z. Li, H. Cheng, C. Tang, J. Li, S. Zhang, S. Chen, J. Tian, Metasurface enabled wide-angle fourier lens. Adv. Mater. **30**, 1706368 (2018)
62. Y. Guo, X. Ma, M. Pu, X. Li, Z. Zhao, X. Luo, High-efficiency and wide-angle beam steering based on catenary optical fields in ultrathin metalens. Adv. Opt. Mater. **6**, 1800592 (2018)
63. Y. Li, X. Li, M. Pu, Z. Zhao, X. Ma, Y. Wang, X. Luo, Achromatic flat optical components via compensation between structure and material dispersions. Sci. Rep. **6**, 19885 (2016)
64. Z. Zhao, M. Pu, H. Gao, J. Jin, X. Li, X. Ma, Y. Wang, P. Gao, X. Luo, Multispectral optical metasurfaces enabled by achromatic phase transition. Sci. Rep. **5**, 15781 (2015)
65. K. Li, Y. Guo, M. Pu, X. Li, X. Ma, Z. Zhao, X. Luo, Dispersion controlling meta-lens at visible frequency. Opt. Express **25**, 21419–21427 (2017)
66. O. Avayu, E. Almeida, Y. Prior, T. Ellenbogen, Composite functional metasurfaces for multispectral achromatic optics. Nat. Commun. **8**, 14992 (2017)
67. P. Venugopalan, Q. Zhang, X. Li, L. Kuipers, M. Gu, Focusing dual-wavelength surface plasmons to the same focal plane by a far-field plasmonic lens. Opt. Lett. **39**, 5744–5747 (2014)
68. O. Eisenbach, O. Avayu, R. Ditcovski, T. Ellenbogen, Metasurfaces based dual wavelength diffractive lenses. Opt. Express **23**, 3928–3936 (2015)
69. Z.-L. Deng, S. Zhang, G.P. Wang, Wide-angled off-axis achromatic metasurfaces for visible light. Opt. Express **24**, 23118–23128 (2016)

70. M. Khorasaninejad, F. Aieta, P. Kanhaiya, M.A. Kats, P. Genevet, D. Rousso, F. Capasso, Achromatic metasurface lens at telecommunication wavelengths. Nano Lett. **15**, 5358–5362 (2015)
71. F. Aieta, M.A. Kats, P. Genevet, F. Capasso, Multiwavelength achromatic metasurfaces by dispersive phase compensation. Science **347**, 1342–1345 (2015)
72. M. Khorasaninejad, Z. Shi, A.Y. Zhu, W.T. Chen, V. Sanjeev, A. Zaidi, F. Capasso, Achromatic metalens over 60 nm bandwidth in the visible and metalens with reverse chromatic dispersion. Nano Lett. **17**, 1819–1824 (2017)
73. S. Wang, J. Lai, T. Wu, C. Chen, J. Sun, Wide-band achromatic flat focusing lens based on all-dielectric subwavelength metasurface. Opt. Express **25**, 7121–7130 (2017)
74. S. Wang, P.C. Wu, V.-C. Su, Y.-C. Lai, C.H. Chu, J.-W. Chen, S.-H. Lu, J. Chen, B. Xu, C.-H. Kuan, T. Li, S. Zhu, D.P. Tsai, Broadband achromatic optical metasurface devices. Nat. Commun. **8**, 187 (2017)
75. H.H. Hsiao, H. Chen Yu, J. Lin Ren, C. Wu Pin, S. Wang, H. Chen Bo, P. Tsai Din, Integrated resonant unit of metasurfaces for broadband efficiency and phase manipulation. Adv. Opt. Mater. **6**, 1800031 (2018)
76. W.T. Chen, A.Y. Zhu, V. Sanjeev, M. Khorasaninejad, Z. Shi, E. Lee, F. Capasso, A broadband achromatic metalens for focusing and imaging in the visible. Nat. Nanotechnol. **13**, 220 (2018)
77. S. Wang, P.C. Wu, V.-C. Su, Y.-C. Lai, M.-K. Chen, H.Y. Kuo, B.H. Chen, Y.H. Chen, T.-T. Huang, J.-H. Wang, R.-M. Lin, C.-H. Kuan, T. Li, Z. Wang, S. Zhu, D.P. Tsai, A broadband achromatic metalens in the visible. Nat. Nanotechnol. **13**, 227 (2018)
78. A. Nemati, Q. Wang, M. Hong, J. Teng, Tunable and reconfigurable metasurfaces and metadevices. Opto-Electron. Adv. **1**, 180009 (2018)
79. S. Song, X. Ma, M. Pu, X. Li, K. Liu, P. Gao, Z. Zhao, Y. Wang, C. Wang, X. Luo, Actively tunable structural color rendering with tensile substrate. Adv. Opt. Mater. **5**, 1600829 (2017)
80. H.-S. Ee, R. Agarwal, Tunable metasurface and flat optical zoom lens on a stretchable substrate. Nano Lett. **16**, 2818–2823 (2016)
81. S.M. Kamali, E. Arbabi, A. Arbabi, Y. Horie, A. Faraon, Highly tunable elastic dielectric metasurface lenses. Laser Photonics Rev. **10**, 1062 (2016)
82. L.W. Alvarez, Two-element variable-power spherical lens. US Patent, US3305294A (1967)
83. A. Zhan, S. Colburn, C.M. Dodson, A. Majumdar, Metasurface freeform nanophotonics. Sci. Rep. **7**, 1673 (2017)
84. C. Min, P. Wang, X. Jiao, Y. Deng, H. Ming, Beam manipulating by metallic nano-optic lens containing nonlinear media. Opt. Express **15**, 9541–9546 (2007)
85. M.A. Kats, D. Sharma, J. Lin, P. Genevet, R. Blanchard, Z. Yang, M.M. Qazilbash, D.N. Basov, S. Ramanathan, F. Capasso, Ultra-thin perfect absorber employing a tunable phase change material. Appl. Phys. Lett. **101**, 221101 (2012)
86. M.A. Kats, R. Blanchard, P. Genevet, Z. Yang, M.M. Qazilbash, D.N. Basov, S. Ramanathan, F. Capasso, Thermal tuning of mid-infrared plasmonic antenna arrays using a phase change material. Opt. Lett. **38**, 368–370 (2013)
87. Y. Chen, X. Li, X. Luo, S.A. Maier, M. Hong, Tunable near-infrared plasmonic perfect absorber based on phase-change materials. Photon. Res. **3**, 54–57 (2015)
88. Y. Chen, X. Li, Y. Sonnefraud, A.I. Fernandez-Dominguez, X. Luo, M. Hong, S.A. Maier, Engineering the phase front of light with phase-change material based planar lenses. Sci. Rep. **5**, 8860 (2015)
89. Y.G. Chen, T.S. Kao, B. Ng, X. Li, X.G. Luo, B. Luk'yanchuk, S.A. Maier, M.H. Hong, Hybrid phase-change plasmonic crystals for active tuning of lattice resonances. Opt. Express **21**, 13691–13698 (2013)
90. V.K. Mkhitaryan, D.S. Ghosh, M. Rudé, J. Canet-Ferrer, R.A. Maniyara, K.K. Gopalan, V. Pruneri, Tunable complete optical absorption in multilayer structures including Ge2Sb2Te5 without lithographic patterns. Adv. Opt. Mater. **5**, 1600452 (2016)
91. T. Li, L. Huang, J. Liu, Y. Wang, T. Zentgraf, Tunable wave plate based on active plasmonic metasurfaces. Opt. Express **25**, 4216–4226 (2017)

92. C.H. Chu, M.L. Tseng, J. Chen, P.C. Wu, Y.-H. Chen, H.-C. Wang, T.-Y. Chen, W.T. Hsieh, H.J. Wu, G. Sun, D.P. Tsai, Active dielectric metasurface based on phase-change medium. Laser Photonics Rev. **10**, 986–994 (2016)
93. Q. Wang, E.T.F. Rogers, B. Gholipour, C.-M. Wang, G. Yuan, J. Teng, N.I. Zheludev, Optically reconfigurable metasurfaces and photonic devices based on phase change materials. Nat. Photon. **10**, 60–65 (2016)
94. N. Raeis-Hosseini, J. Rho, Metasurfaces based on phase-change material as a reconfigurable platform for multifunctional devices. Materials **10**, 1046 (2017)
95. A.M. Shaltout, A.V. Kildishev, V.M. Shalaev, Evolution of photonic metasurfaces: from static to dynamic. J. Opt. Soc. Am. B **33**, 501–510 (2016)
96. H.-X. Xu, S. Sun, S. Tang, S. Ma, Q. He, G.-M. Wang, T. Cai, H.-P. Li, L. Zhou, Dynamical control on helicity of electromagnetic waves by tunable metasurfaces. Sci. Rep. **6**, 27503 (2016)
97. B.O. Zhu, K. Chen, N. Jia, L. Sun, J. Zhao, T. Jiang, Y. Feng, Dynamic control of electromagnetic wave propagation with the equivalent principle inspired tunable metasurface. Sci. Rep. **4**, 4971 (2014)
98. J. Zhao, Q. Cheng, J. Chen, M.Q. Qi, W.X. Jiang, T.J. Cui, A tunable metamaterial absorber using varactor diodes. New J. Phys. **15**, 043049 (2013)
99. X. Wu, C. Hu, Y. Wang, M. Pu, C. Huang, C. Wang, X. Luo, Active microwave absorber with the dual-ability of dividable modulation in absorbing intensity and frequency. AIP Adv. **3**, 022114 (2013)
100. D.F. Sievenpiper, J.H. Schaffner, H.J. Song, R.Y. Loo, G. Tangonan, Two-dimensional beam steering using an electrically tunable impedance surface. IEEE Trans. Antennas Propag. **51**, 2713–2722 (2003)
101. K. Chen, Y. Feng, F. Monticone, J. Zhao, B. Zhu, T. Jiang, L. Zhang, Y. Kim, X. Ding, S. Zhang, A. Alù, C.-W. Qiu, A reconfigurable active Huygens' metalens. Adv. Mater. **29**, 1606422 (2017)
102. H.T. Chen, J.F. O'Hara, A.K. Azad, A.J. Taylor, R.D. Averitt, D.B. Shrekenhamer, W.J. Padilla, Experimental demonstration of frequency-agile terahertz metamaterials. Nat. Photon. **2**, 295–298 (2008)
103. H.T. Chen, W.J. Padilla, J.M.O. Zide, A.C. Gossard, A.J. Taylor, R.D. Averitt, Active terahertz metamaterial devices. Nature **444**, 597–600 (2006)
104. O. Balci, E.O. Polat, N. Kakenov, C. Kocabas, Graphene-enabled electrically switchable radar-absorbing surfaces. Nat. Commun. **6**, 6628 (2015)
105. Z. Fang, Y. Wang, A.E. Schlather, Z. Liu, P.M. Ajayan, F.J. García de Abajo, P. Nordlander, X. Zhu, N.J. Halas, Active tunable absorption enhancement with graphene nanodisk arrays. Nano Lett. **14**, 299–304 (2014)
106. W. Li, B. Chen, C. Meng, W. Fang, Y. Xiao, X. Li, Z. Hu, Y. Xu, L. Tong, H. Wang, W. Liu, J. Bao, Y.R. Shen, Ultrafast all-optical graphene modulator. Nano Lett. **14**, 955–959 (2014)
107. E. Arbabi, A. Arbabi, S.M. Kamali, Y. Horie, M. Faraji-Dana, A. Faraon, MEMS-tunable dielectric metasurface lens. Nat. Commun. **9**, 812 (2018)
108. A. She, S. Zhang, S. Shian, D.R. Clarke, F. Capasso, Adaptive metalenses with simultaneous electrical control of focal length, astigmatism, and shift. Sci. Adv. **4**, eaap9957 (2018)
109. M. Rahmani, G. Leo, I. Brener, A. Zayats, S. Maier, C. De Angelis, H. Tan, V.F. Gili, F. Karouta, R. Oulton, Nonlinear frequency conversion in optical nanoantennas and metasurfaces: materials evolution and fabrication. Opto-Electron. Adv. **1**, 180021 (2018)
110. M. Kauranen, A.V. Zayats, Nonlinear plasmonics. Nat. Photon. **6**, 737–748 (2012)
111. S. Chen, G. Li, W. Cheah Kok, T. Zentgraf, S. Zhang, Controlling the phase of optical nonlinearity with plasmonic metasurfaces. Nanophotonics **7**, 1013–1024 (2018)
112. N. Segal, S. Keren-Zur, N. Hendler, T. Ellenbogen, Controlling light with metamaterial-based nonlinear photonic crystals. Nat. Photonics **9**, 180–184 (2015)

113. E. Almeida, G. Shalem, Y. Prior, Subwavelength nonlinear phase control and anomalous phase matching in plasmonic metasurfaces. Nat. Commun. **7**, 10367 (2016)
114. J. Lee, M. Tymchenko, C. Argyropoulos, P.-Y. Chen, F. Lu, F. Demmerle, G. Boehm, M.-C. Amann, A. Alu, M.A. Belkin, Giant nonlinear response from plasmonic metasurfaces coupled to intersubband transitions. Nature **511**, 65–69 (2014)
115. M. Tymchenko, J.S. Gomez-Diaz, J. Lee, N. Nookala, M.A. Belkin, A. Alù, Gradient nonlinear pancharatnam-berry metasurfaces. Phys. Rev. Lett. **115**, 207403 (2015)

Chapter 10
Generation and Manipulation of Special Light Beams

Abstract Special light beams satisfying the Helmholtz equation are of great importance in various applications ranging from high-resolution imaging, high data capacity optical communication to micromanipulation in EO 2.0. In this chapter, we summarize some typical special light beams emerging in recent decades and having attracted arising attentions. We especially review their generating and modulating methods with subwavelength structures, which are the core building blocks in EO 2.0. By suitably adjusting the shape, size, position, and orientation of the structures with high spatial resolution, one can control the basic properties of light (phase, amplitude, polarization) and thus engineer the beams' wavefront profile at will. This possibility greatly expands the frontiers of optical engineering with attendant reduction of thickness, size, and complexity.

Keywords Photonic spin Hall effect · Vortex beam · Bessel beam · Airy beam

10.1 Special Light Beams

The Gaussian beam is a fundamental beam and of great importance in EO 1.0, which satisfies the paraxial Helmholtz equation. In EO 2.0, various kinds of new beams, such as Laguerre–Gaussian (LG) beam, Bessel beam, and also Airy beam, can be generated using novel compact elements.

Helically phased light carries an OAM irrespective of the radial distribution of the beam. One typical beam carrying OAM is the LG mode set. These modes have amplitude distributions of [1]

$$\begin{aligned} \mathrm{LG}_{pl} = & \sqrt{\frac{2p!}{\pi(p+|l|)!}}\frac{1}{w(z)}\left[\frac{r\sqrt{2}}{w(z)}\right]^{|l|}\exp\left[\frac{-r^2}{w^2(z)}\right]L_p^{|l|}\left(\frac{2r^2}{w^2(z)}\right)\exp[il\varphi] \\ & \exp\left[\frac{ik_0r^2z}{2\left(z^2+z_R^2\right)}\right]\exp\left[-i(2p+|l|+1)\tan^{-1}\left(\frac{z}{z_R}\right)\right] \end{aligned}, \tag{10.1.1}$$

X. Luo, *Engineering Optics 2.0*, https://doi.org/10.1007/978-981-13-5755-8_10

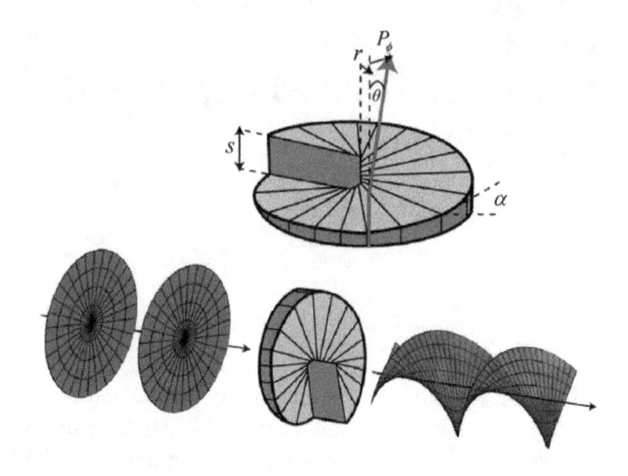

Fig. 10.1 A spiral phase plate can generate a helically phased beam from a Gaussian beam. Adapted from [1] with permission. Copyright 2011, Optical Society of America

where the $1/e$ radius of the Gaussian term is given by $w(z) = w(0)\left[\left(z^2 + z_R^2\right)/z_R^2\right]^{1/2}$ with $w(0)$ being the beam waist, z_R the Rayleigh range, and $(2p + |l| + 1)\tan^{-1}(z/z_R)$ the Gouy phase. $L_p^{|l|}(x)$ is an associated Laguerre polynomial, where l is the azimuthal index giving an OAM per photon, and p is the number of radial nodes in the intensity distribution. For $p = 0$, the complex LG mode can be simplified to

$$\mathrm{LG}_{0l} = A\exp[il\varphi], \quad (10.1.2)$$

and this is the expression of a simple vortex beam with only one radial node and topological charge of l. Figure 10.1 presents a common approach to the generation of a vortex beam by passing a plane wave through an optical element with a helical surface. The optical thickness of the helical phase plate increases with azimuthal position according to $l\lambda\theta/(2\pi(n-1))$, where n is the refractive index of the plate.

Bessel beam is another form of light beam. A Bessel beam gets its name from the description of such a beam using a Bessel function, which leads to a predicted cross-sectional profile of a set of concentric rings. In general, the complex amplitude of Bessel beam carrying optical vortex should be written as:

$$A(r, \varphi, z) = J_l(k_r r)\exp(il\varphi + ik_z z), \quad (10.1.3)$$

where J_l is the lth-order Bessel function, k_r and k_Z are the radial and longitudinal wavevectors with $k = \sqrt{k_r^2 + k_z^2} = 2\pi/\lambda$ (λ is the wavelength of the electromagnetic radiation making up the Bessel beam), r, φ, and z are the radial, azimuthal, and longitudinal components, respectively, and l is the topological charge of optical vortex. When $l = 0$, this reduced to the common Bessel beam, i.e., zero-order Bessel beam. Since the amplitude of the Bessel beam is independent of z, its horizontal shape would not change as that of Gaussian beam. Consequently, Bessel beam is a kind of non-diffractive beam. One way to understand Bessel beams is to think of them as plane waves traveling on a cone. A commonly used way to generate a Bessel beam is

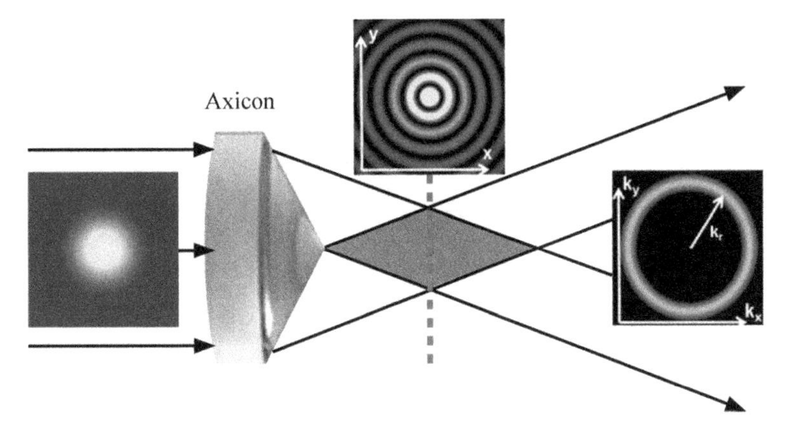

Fig. 10.2 Schematic of generation of Bessel beams with an axicon. Reproduced from [2] with permission. Copyright, 2013. Optical Society of America

passing a Gaussian beam through an axicon to obtain a radial wavevector k_r as shown in Fig. 10.2. Since the Bessel beam has the same radial wavevector, the interference pattern never changes; in momentum space, there is only one value for k_r. Apart from its "non-diffracting" property, one appealing property of the Bessel beam is "self-healing," which means that the beam shape would be recovered after scattered by an obstacle.

The Airy beam is another solution to the paraxial Helmholtz equation and exhibits the "self-accelerating" and "self-healing" properties. Considering the two-dimensional (2D) paraxial wave equation in free space, the dynamic equation which governs the propagation of the electric field associated with planar optical beams is given as follows

$$i\frac{\partial\phi}{\partial z}+\frac{1}{2}\frac{\partial^2\phi}{\partial s^2}=0, \tag{10.1.4}$$

where z is the coordinate along the propagation axis, $s=x/x_0$ represents a dimensionless transverse coordinate, and x_0 is an arbitrary transverse scale. Ignoring the time, the Airy beams' field distribution at the aperture plane is

$$\phi(s,z=0)=Ai(s), \tag{10.1.5}$$

where Ai is the Airy function. Like the Bessel beam, the ideal Airy beams carry an infinite energy and cannot be achieved in practice.

Considering the exponentially decaying window function, the initial optical field distribution at aperture can be written as

$$\phi(s,z=0)=Ai(s)\exp(\alpha s), \tag{10.1.6}$$

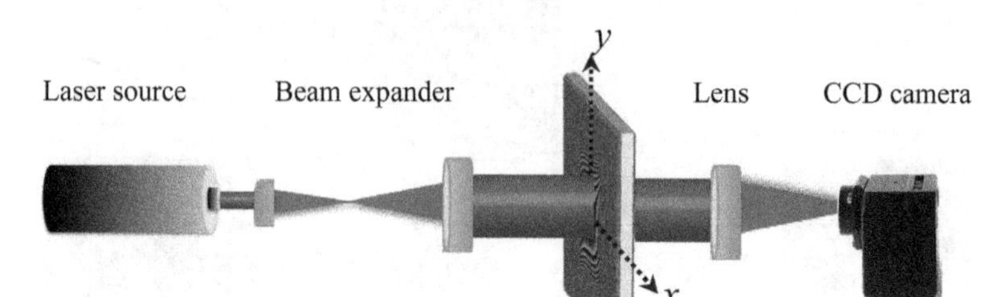

Fig. 10.3 Schematic of generation of Airy beams

Considering the one-dimensional (1D) case, the Fourier spectrum of the Airy beams at the aperture is

$$A_0(k_x, z = 0) \propto \exp(-\alpha k_x^2)\exp\left(\frac{i}{3}(k_x^3 - 3\alpha^2 k_x - i\alpha^3)\right), \tag{10.1.7}$$

where k_x is spatial frequency. It can be seen that the finite energy Airy beams have a cubic phase-modulated Gaussian power spectrum. Consequently, the phase mask to generate the Airy beams is generally cubic phase. So, Airy beams are commonly generated by imprinting a cubic phase onto a Gaussian beam in the Fourier plane and subsequently transforming it back to the real space by a lens as shown in the optical system in Fig. 10.3.

10.2 Interaction of Light Beam with Interface

Although the reflection and refraction of plane wave on the material interface can be easily calculated using Snell's law and Fresnel's equations, the situation is more complicated for a real optical beam which has a finite width. With respect to the plane of incidence, there may be in-plane and out-of-plane spatial shifts (i.e., lateral displacements) and similar angular shifts (i.e., deflections), as shown in Fig. 10.4 [3]. The spatial and angular shifts can also be regarded as coordinate and momentum shifts, respectively. The displacements of the beam centroids, $\langle X^a \rangle$ and $\langle Y^a \rangle$, $a = r, t$, represent spatial Goos–Hänchen and Imbert–Fedorov shifts, respectively. The deflection angles $\langle \Theta_X^a \rangle = \langle P_X^a \rangle / k^a$ and $\langle \Theta_Y^a \rangle = \langle P_Y^a \rangle / k^a$ are associated with the angular (or momentum space) Goos–Hänchen and Imbert–Fedorov shifts, respectively. In many cases, the transverse Imbert–Fedorov shifts are also called spin Hall effect (SHE) of photons.

As we know, each electron has a spin angular momentum (SAM) of $\pm 1/2\hbar$, where $\hbar$ is the reduced Plank constant. The spin of electron will accumulate at the opposite

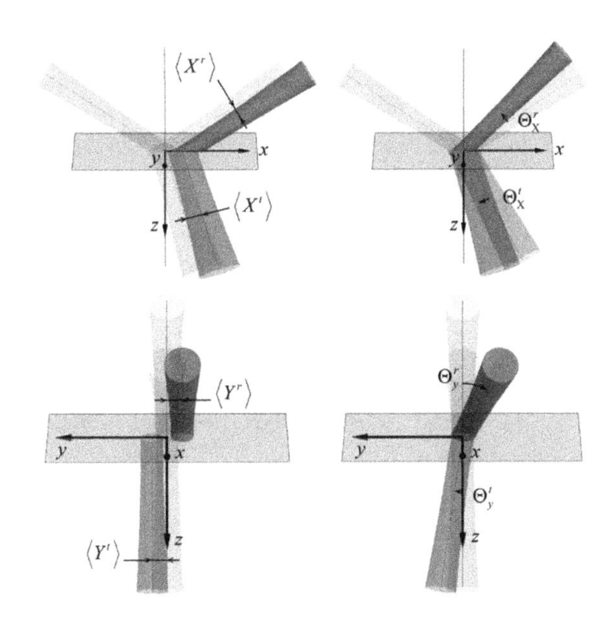

Fig. 10.4 Schematic picture of the beam shifts upon reflection and refraction at a plane interface ($z = 0$). The plane of incidence of the beam is (x; z). In-plane (upper panels) and out-of-plane (lower panels) spatial (left panels) and angular (right panels) shifts of the reflected (r) and transmitted (t) beams. Reproduced from [3] with permission. Copyright 2013, IOP Publishing Ltd.

sides of an electron flow, which is the famous SHE in an electric system [4, 5]. Similar to the electrons, Beth demonstrated that each photon of circularly polarized light (CPL) possesses SAM of $\pm\sigma\hbar$ per photon, where $\sigma = \pm 1$ [6]. In 2004, Onoda theoretically predicted that the SHE also existed in the optics as the counterpart of the electronic SHE [7]. It was predicted that, to conserve total angular momentum, an inhomogeneity of material's index of refraction can cause momentum transfer between the orbital and the spin angular momentum of light along its propagation trajectory, and will result in transverse splitting in polarizations. However, in the traditional cases, the photonic spin Hall effect (PSHE) is extremely week due to the limited optical spin–orbit interaction. Extremely precise metrology is required for the detection of such phenomenon. In 2008, Hosten and Kwiat successfully amplified and observed this effect occurring at an air–glass interface by employing the quantum weak measurement technology [8].

Based on the underlying physics of the SHE in spin–orbit interaction of light, there are two kinds of PSHE during geometric phase engineering [3, 9]. The first class of PSHE corresponds to shift of different spin states of photons in the direction perpendicular to the incident plane of light beam [3]. This kind of PSHE is also called Imbert–Fedorov shift, which is very weak because of the extremely small photon momentum and spin–orbit interaction. The mechanism of this PSHE originates from the induced spin-redirection phase during the variation of propagation trajectory in space. The spin-redirection phase is a special kind of geometric phase, which was firstly found by Rytov in 1938. The second class of PSHE originates from the Pancharatnam–Berry (PB) phase engineering during the interaction of the light beam with the anisotropic structures or materials. Different from the spin-redirection phase-based PSHE, this kind of PSHE is inspired during the variation of polarization (not propagation trajectory), and it corresponds to a angular shift of different spin states of photons, increases during beam propagation, and thus can be directly detected without using the weak measurement. Recently, a series of efforts have been devoted to enhance the PSHE by introducing a stronger or enlarged spin–orbit interaction. However, it still remains a big challenge to arbitrarily manipulate the PSHE. The recently emergent metasurfaces provide a possible approach to solve this problem [10]. Metasurfaces are extraordinary interfaces with exceptional abilities in controlling the light beams, and a rapid phase variation can be on-demand designed. Both types of PSHE have been demonstrated to be enhanced by planar metasurfaces.

10.2.1 PSHE with Spin-Redirection Phase

In 2013, Yin et al. obtained a giant photonic SHE by applying a metasurface to generate a rapid phase gradient [10]. The rapidly varying phase discontinuities along the metasurface breaking the axial symmetry of the system enable the direct observation of large transverse motion of CPL, even at normal incidence. The strong spin–orbit interaction deviates from the trajectory prescribed by the ordinary Fermat principle.

Figure 10.5 illustrates the metasurface consisting of a collection of V-shaped nanoantennas [11]. The period of the constituent V-shaped antenna is 180 nm, which is at a subwavelength scale. Eight antennas with different lengths, orientations, and spanning angles were chosen for a linear phase retardation with a phase gradient of 4.4 rad/mm. Transverse polarization splitting is induced by the metasurface with the strong phase gradient along the x-direction. The rapid phase retardation refracts light in a skewing direction and results in the PSHE. The mechanism of the PSHE is the momentum conservation at the metasurface. The induced effective circular birefringence is determined by the gradient of the in-plane phase change or the curvature of the ray trajectory. The strong spin–orbit interaction within the optically thin material leads to the accumulation of circular components of the beam in the transverse directions (y-directions) of the beam (Fig. 10.5a), even when the incident angle is normal to the surface. The faster the in-plane phase changes, the stronger the effect becomes. One can quantitatively evaluate the PSHE shift at metasurfaces by considering the total energy transport at the interfaces. Any transverse motion of light is captured by the transverse components of Poynting vectors. Integrating the Poynting vectors (including the evanescent fields) over the full half-space above the metasurface allows to evaluate the overall transverse displacement for a transmitted beam with finite size,

$$\Delta = \frac{1}{S_z} \int_0^\infty S_y(z) \mathrm{d}z, \tag{10.2.1}$$

where S_y denotes the Poynting vector component that is perpendicular to the plane of incidence, *i.e.*, along transverse y-direction in Fig. 10.5. S_z is the z component of

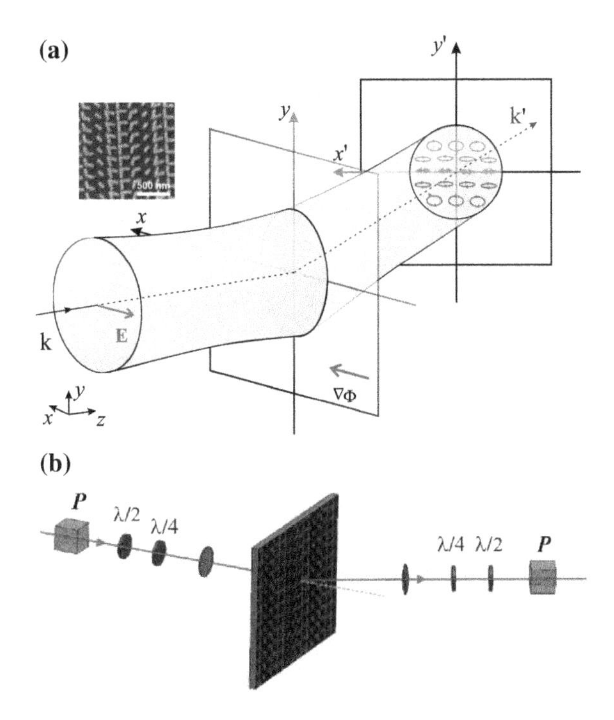

Fig. 10.5 **a** Schematic of the PSHE with a metasurface. Inset: scanning electron microscope (SEM) image of a metasurface with a rapid phase gradient in the horizontal (x) direction. Scale bar, 500 nm. **b** Experimental setup of the characterization system. Reproduced from [10] with permission. Copyright 2013, AAAS

the Poynting vector of the transmitted beam at the interface ($z = 0$). The procedure is general and applicable to both transmitted and reflected fields. A similar approach has been employed by Renard to analyze the Goos–Hänchen effect—a longitudinal beam displacement occurs at total internal reflection, by integrating the longitudinal components of Poynting vector [12].

The experimental test system is shown in Fig. 10.5b. Light from a broadband source is focused on the metasurface by a lens with a focal length of $f = 50$ mm. The polarization of the illumination is controlled through cascading an achromatic half-wave plate and a quarter-wave plate. The regularly and anomalously refracted far-field transmission through the metasurface was collected using a lens with $f =$ 50 mm and imaged on an InGaAs camera. The polarization state of the transmission is resolved by using an achromatic quarter-wave plate ($\lambda/4$), a half-wave plate ($\lambda/2$), and a polarizer (P) with a high extinction ratio greater than 10^5 for all wavelengths of interest.

Figure 10.6a and b show the PSHE effect for x- and y-polarized incidence, respectively. Figure 10.6c shows the numerically integrated PSHE shifts for a metasurface with a phase gradient of 4.4 rad/μm using the rigorous Fourier modal method for solving the Poynting vectors of the transmitted fields [13]. The incident polarization is circular, and the result agrees well with the experiments. Such a strong and broadband photonic SHE may provide a route for exploiting the spin and orbit angular momentum of light for information processing and communication.

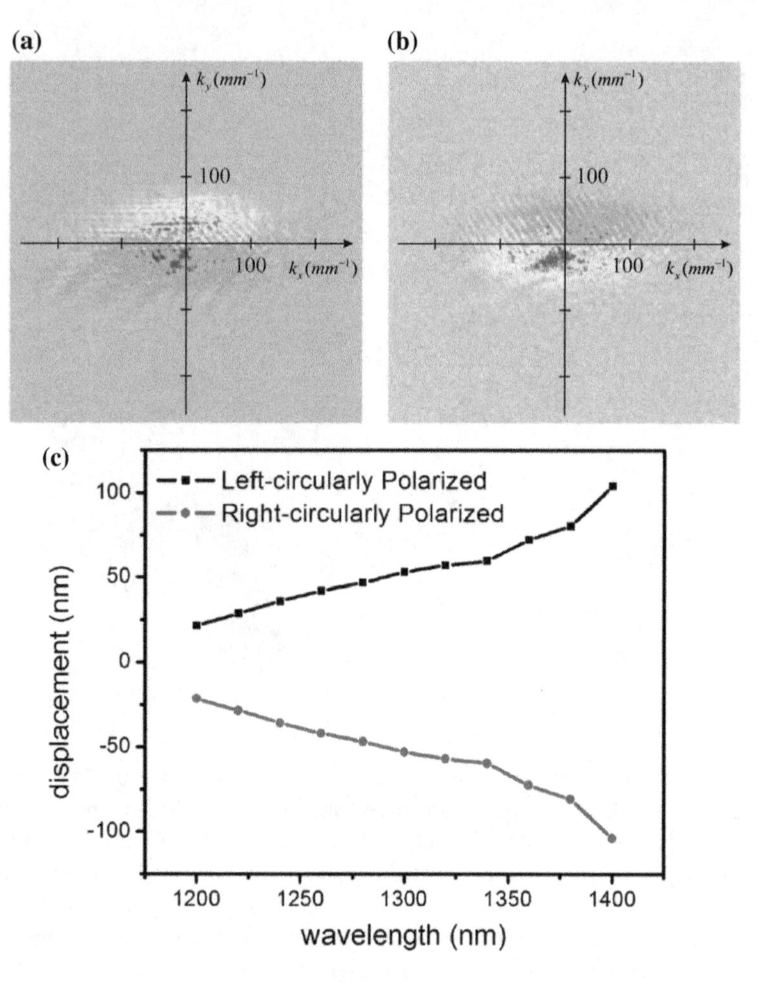

Fig. 10.6 **a** Observation of a giant SHE: the helicity of the anomalously refracted beam. The incidence is from the silicon side onto the metasurface and is polarized along the x-direction along the phase gradient. The incidence angle is at surface normal. Red and blue represent the right and left circular polarizations, respectively. **b** Helicity of the refracted beam with incidence polarized along the y-direction. **c** PSHE of CPL for a metasurface with a phase gradient of 4.4 rad/μm. Reproduced from [10] with permission. Copyright 2013, AAAS

10.2.2 PSHE with PB Phase

PB phase is another mechanism for generating PSHE. A spin-dependent splitting in k-space will be desired as the PB phase can be tailored by arranging the orientation of local optical axis in a space-variant manner. A shift of wavevector will arise due to the phase gradient,

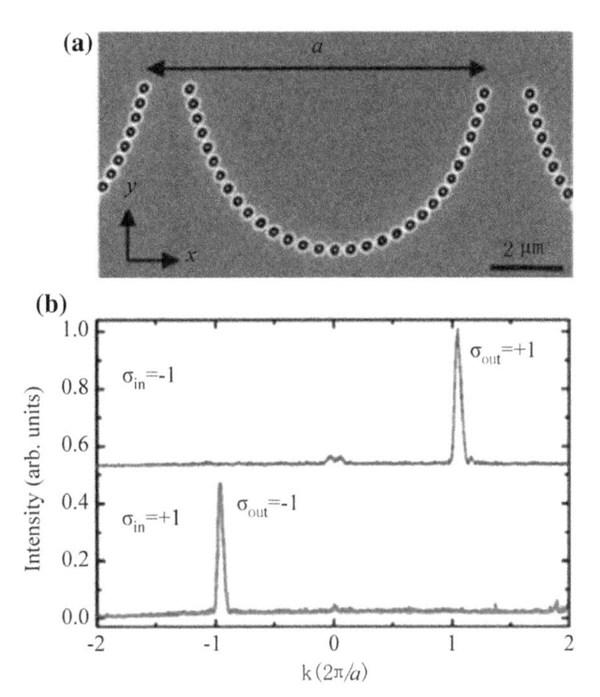

Fig. 10.7 **a** SEM image of a curved chain whose local orientation is varied linearly along the x-axis with a rotation period of $a = 9$ μm and a structure length of 135 μm. **b** The spin-dependent momentum deviation for the PSHE at a wavelength of 780 nm. The red and blue lines stand for incident right- and left-handed CPL, respectively ($\sigma_{\text{in}} = \pm 1$). σ_{out} denotes the spin state of the scattered light. Reproduced from [14] with permission. Copyright 2011, American Chemical Society

$$\Delta k(x, y) = \nabla \Phi(x, y) = \frac{\partial \Phi}{\partial x}\hat{e}_x + \frac{\partial \Phi}{\partial y}\hat{e}_y, \tag{10.2.2}$$

where $\hat{e}_x$ and $\hat{e}_y$ are the unit direction vectors in the x-axis and y-axis of space coordinate. The shift of wavevector in momentum space will cause an angular shift in real space after propagation.

In 2011, Shitrit et al. realized a PSHE based on the PB phase engineering in coupled localized plasmonic chains, which they called locally isotropic optical spin Hall effect [14]. This effect was regarded as the interaction between the optical spin and the path ξ of the plasmonic chain with an isotropic unit cell as shown in Fig. 10.7a. The local orientation of the curved chain induces local anisotropy variation in the scattered field. Figure 10.7b shows the measured PSHE observed from the chain. The intensity distribution was measured in the far-field, which corresponds to the momentum space. Polarization analysis shows that the scattering from the curved chain comprises two components: ballistic and spin-flip. The ballistic component maintains the polarization state of the incident beam, while the spin-flip component has an opposite spin state. Spin-dependent beam deflection is observed in the experiment via orthogonal circular polarizers and corresponds to a momentum shift of $\Delta k_x = -2\sigma\pi/a$, where $\sigma = \pm 1$ is the incident spin state. The scattering by the bent chain is most conveniently studied using a rotating reference frame, which is attached to the axis of the local anisotropy of the chain and follows the chain's route.

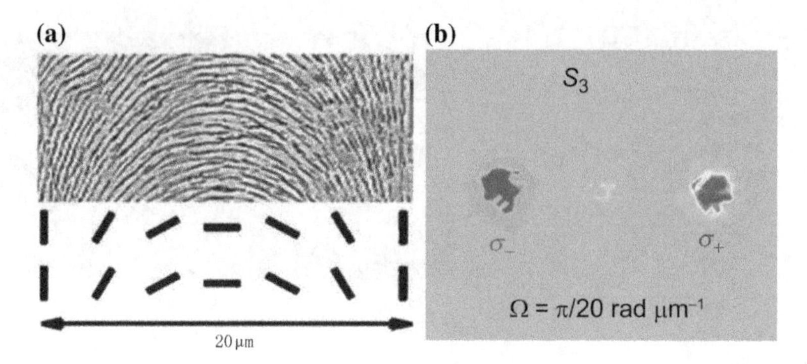

Fig. 10.8 **a** SEM image picture of the metasurface. **b** The intensity and S_3 parameter with a rotation rate of $\Omega = \pi/20$ rad μm^{-1}. The observation distance is 10 cm away from the metasurface. Reproduced from [15] with permission. Copyright 2011, Springer Nature

There is another type of PB phase-based PSHE, which was called locally anisotropic optical spin Hall effect [14, 15]. Different from the locally isotropic PSHE, which originates from the specified path and the anisotropy coupling between the isotropic structures, the locally anisotropic PSHE occurs due to the interaction between the optical spin and the local anisotropy of the unit cell, which is independent of the chain path. Figure 10.8a shows a metasurface constructed by anisotropic nanostructures which were fabricated by the femtosecond laser self-assembly in fused silica [15]. The rotated local optical axes in horizontal direction endowed the metasurface with a spin-dependent phase gradient. Thus, the angular splitting in real space can be detected directly after passing the glass (Fig. 10.8b).

More recently, a single achiral nano-aperture, named catenary aperture, was utilized as a meta-macromolecule enabling giant angular PSHE [16, 17]. The catenary is the curve that free-hanging chain assumes under its own weight, and thought to be a "true mathematical and mechanical form" in architecture by Robert Hooke in the 1670s, with nevertheless no significant phenomena observed in optics. However, it has been demonstrated that the catenary can serve as a unique building block of metasurfaces with continuous and linear phase shift covering $[0, 2\pi]$, a mission which is extremely difficult if not impossible with state-of-the-art technology.

The catenary aperture has a length much larger than the width, while the width is much smaller than the operating wavelength. It can be mathematically treated with two parameters, i.e., the inclination angle $\xi(x)$ with respect to the x-axis and the width Λ as shown in Fig. 10.9.

The catenary aperture can be regarded as a chain of space-variant-orientated slits with non-uniform phase shift under circularly polarized illumination. Here, ξ is defined as the angle between the slit and one specific coordinate, and Φ as the geometric phase. For the simplicity of discussion, left-handed circular polarization (LCP) illumination is assumed ($\sigma = 1$), i.e., $\Phi = 2\xi$, in the following deduction. In general, the design procedure of the catenary apertures consists of two steps. First, write explicitly the space-variant angle $\xi(x, y)$ according to the required phase

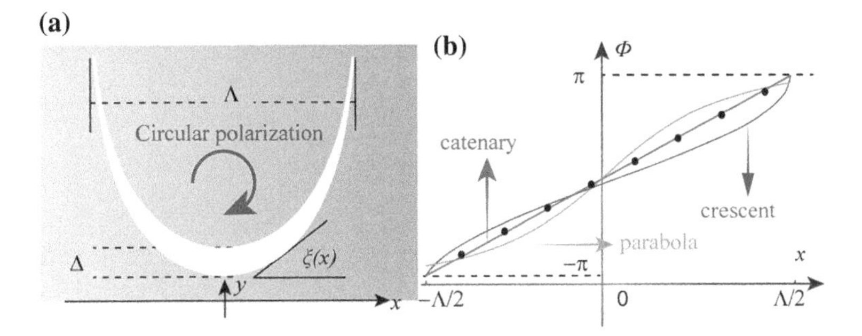

Fig. 10.9 **a** Sketch map of a catenary aperture illuminated at normal incidence by CPL. The inclination angle with respect to the x-axis is denoted as $\xi(x)$. **b** Phase distributions of the catenary (red), parabola (orange), crescent (blue), and discrete antennas (black dot) for LCP illumination ($\sigma = 1$). Reproduced from [17] with permission. Copyright 2015, The Authors

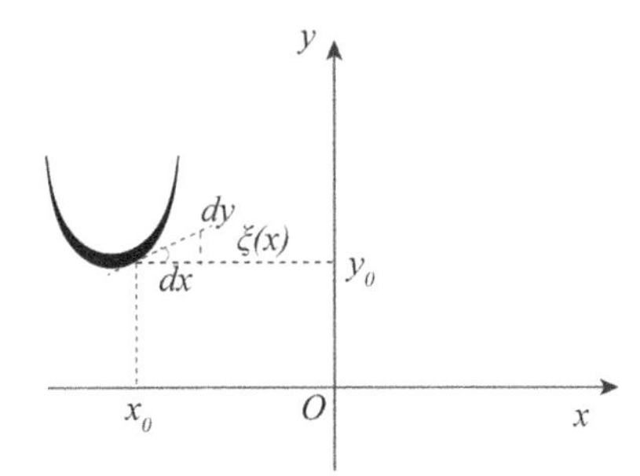

Fig. 10.10 Schematic of the integration procedures for Cartesian coordinate. Reproduced from [17] with permission. Copyright 2015, The Authors

distribution as given by the geometric relation $\xi(x, y) = \Phi(x, y)/2$; Then, integrate the tangent of the angle, i.e., $\tan\xi(x, y)$, over Cartesian or polar coordinates (Fig. 10.10). For a linear phase distribution along x-direction, $dy/dx = \tan(\pi x/\Lambda)$ can be directly integrated into the x-axis to obtain

$$y = \frac{\Lambda}{\pi}\ln(|\sec(\pi x/\Lambda)|), \tag{10.2.3}$$

where Λ is the horizontal length of a single catenary. In the catenary structure, the inclination angle between the curve tangent and x-axis, $\xi(x)$, is varying from $-\pi/2$ to $\pi/2$ between the left and right endpoints, yielding a linear geometric phase as a result of the geometric relation $\Phi(x) = 2\sigma\xi(x)$, where $\sigma = \pm 1$ denote the left- and right-handed circular polarizations (LCP and RCP). It should be noted that the phase in catenary is free of the discrete phase sampling as in the case of nano-antenna arrays. From the deduction, it is intrinsically linear while other deformed shapes such as circular crescent and parabola are characterized by large phase nonlinearity, as can be directly calculated using the tangent angles (Fig. 10.9b).

A single catenary aperture is then obtained by shifting Eq. (10.2.3) with Δ along y-direction and joined the four tips.

$$\begin{aligned} y_1 &= \frac{\Lambda}{\pi}\ln(|\sec(\pi x/\Lambda)|) \\ y_2 &= \frac{\Lambda}{\pi}\ln(|\sec(\pi x/\Lambda)|) + \Delta \end{aligned}. \tag{10.2.4}$$

Because the value of Eq. (10.2.4) is infinite for $x = \pm\Lambda/2 + n\Lambda$, where n is an integer, thus the curves are truncated at the two ends with δx in practical designs; i.e., the span of x is $(-0.5\Lambda + \delta x + n\Lambda, 0.5\Lambda - \delta x + n\Lambda)$. Owing to the geometric phase, the diffraction of the cross-polarization from the catenary aperture violates Kirchhoff's theory. Intuitively speaking, a single element of the catenary aperture can act as an artificial bias for the incident CPL and show dramatic angular PSHE. The theoretical deflection angle of the cross-polarized light at normal incidence is $\theta = -\sigma \sin^{-1}(\lambda/\Lambda)$, corresponding to an additional horizontal momentum of $\Delta k = -2\sigma\pi/\Lambda$.

Figure 10.11 shows the schematic of the characterization system. In the experiment, the incident light propagates perpendicularly to the sample plane. Under LCP and RCP illumination at $\lambda = 632.8$ nm, the cross-polarized (RCP and LCP) fields transmitted from the aperture were measured using a homemade microscope in transmission mode. An incident collimated beam was converted into RCP light through cascaded polarizer and quarter-wave plate, and then illuminated on the samples. The cross-polarization (LCP) component of the transmitted field through the samples was filtered by an additional quarter-wave plate and polarizer. The intensity distribution of the fields was imaged through a 100× objective and a tube lens, and then collected by a charge-coupled device (CCD, 1600 × 1200 pixels, WinCamD-UCD15, DataRay Inc.) camera. The inset of Fig. 10.11 gives the SEM of the single catenary aperture with $\Lambda = 2$ μm and $\Delta = 200$ nm, which was fabricated via focused ion-beam (FIB) milling on a 120-nm-thick gold film.

Figure 10.12a represents the cross-polarized near-field electric fields and far-field scattering of this aperture using a commercial electromagnetic software (CST MICROWAVE STUDIO). This far-field scattering corresponds well with the experimental results (Fig. 10.12b). In the simulations, the metal film was assumed to be a perfect electric conductor (PEC) because the excited surface plasmon polaritons (SPPs) do not contribute to the far-field pattern. It is also noted that the phase retardation originating from the SPPs propagation [18] can be neglected because it is much smaller than the geometric phase for such thin metallic film. As shown in Fig. 10.12b, the intensity distribution in the *xy*-planes reveals that the CPL has been deflected by the catenary aperture along the *x*-axis with an angle of ~ ±18° (the theoretical value is ±18.42°) for LCP and RCP, respectively.

Owing to the geometric nature of the spin–orbit conversion, the handedness scattering can be observed within a broadband frequency range. To demonstrate this, the far-field scattering power of the single catenary aperture illuminated by LCP light at different wavelengths was calculated. As shown in Fig. 10.13a, the cross-polarized scattering is larger than the co-polarization within the whole range. Because the amplitude of the electric field of the incident plane wave is set as 1 V/m, the power of the light incident on the aperture can be evaluated as 30×10^{-16} W. Clearly, the

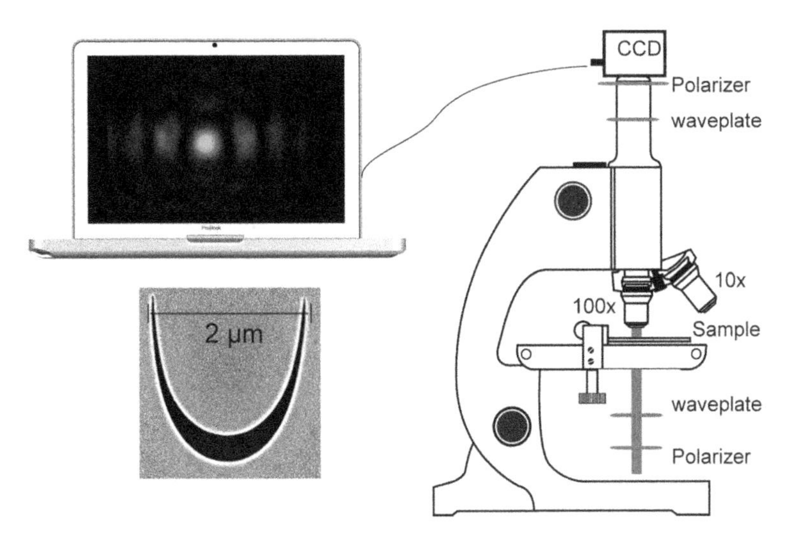

Fig. 10.11 Schematic of the experimental setup. Inset: SEM image of the fabricated catenary aperture. Reproduced from [16] with permission. Copyright 2017, The Authors

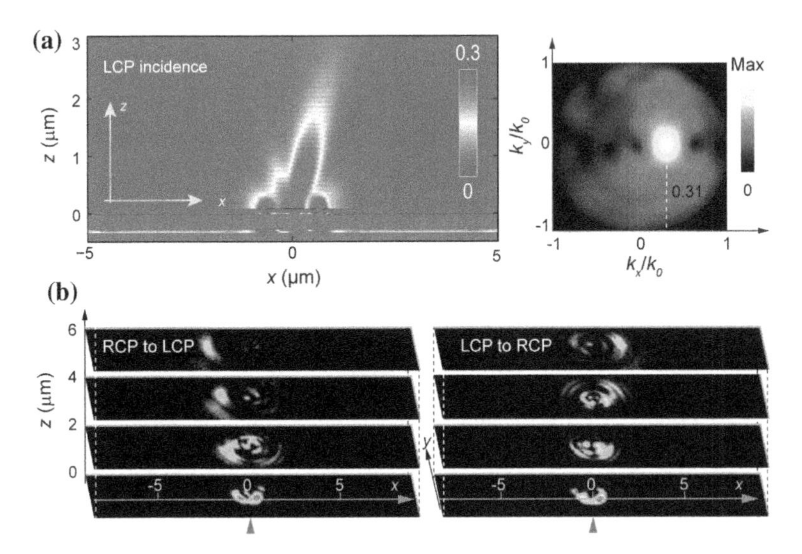

Fig. 10.12 **a** Simulation of a single catenary aperture under LCP incidence. The left panel represents the cross-polarized electric intensity in the xz-plane ($y = 1$ μm). The right panel shows the amplitude of the far-field angular spectrum in the k-space, where k_x and k_y are the horizontal wave numbers in the x- and y-directions. **b** Measured cross-polarized intensity patterns in the xy-planes for $z = 0, 2, 4, 6$ μm under RCP (left panel) and LCP (right panel) illumination ($\lambda = 632.8$ nm). The top surface of the sample is set at $z = 0$. Reproduced from [16] with permission. Copyright 2017, The Authors

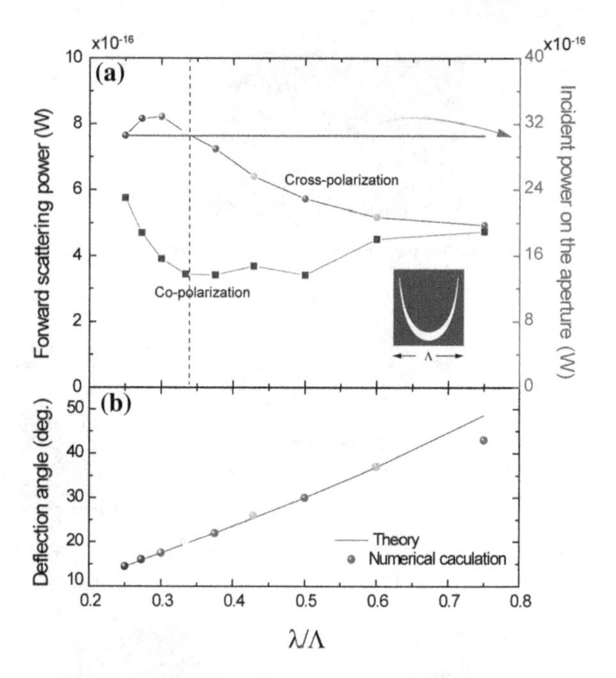

Fig. 10.13 Broadband light scattering by a single catenary aperture. **a** Calculated scattering powers in forward direction for co- and cross-polarized components (left axis) and the total input LCP power integrated into the aperture (right axis). The scattering powers are obtained by integrating the power pattern given by CST. **b** Theoretical ($\theta = \sin^{-1}(\lambda/\Lambda)$) and numerically calculated deflection angles at different wavelengths. Reproduced from [16] with permission. Copyright 2017, The Authors

conversion efficiency, defined as the ratio of the cross-polarization power to the input power, is typically below 25%, except for the range where $0.25 \leq \lambda/\Lambda \leq 0.35$. Moreover, the cross-polarized component becomes evanescent and does not contribute to the far-field when $\lambda/\Lambda > 1$. In the short wavelength region ($\lambda/\Lambda < 0.25$), the degradation of performance would be attributed to the increased diffraction because the wavelength is comparable to Δ. Figure 10.13b plots the deflection angles evaluated from CST MWS at different wavelengths, showing good agreement with the theoretical results obtained via the relation $\theta = -\sigma \sin^{-1}(\lambda/\Lambda)$. The deviation at large angles is mainly because the beam width increases with the deflection angle, which would in turn induce stronger scattering by the edge of the aperture.

10.3 Vortex Beam Generation

As a quantum mechanical description of photons, the optical angular momentum is crucial for various classical and quantum applications [19]. In 1992, Allen and co-workers [20] recognized that helically phased beam with $\Phi = l\varphi$ possesses quantized OAM of $l\hbar$, where l is known as topological charge. Featuring a helix phase front and zero intensity distribution at the beam center, beams carrying OAM are also called optical vortices and have drawn increasing attention from many realms, including optical communications [21, 22], super-resolution imaging [23], optical micromanipulation [24], and detection of rotating objects [25]. For optical communication, one of the critical issues in this area is that the exponential growth of data traffic will exhaust the available capacity in the near future. Avoiding the Shannon limit requires taking advantage of the remaining untapped spatial dimension. Recent progresses in modal basis that carry OAM facilitate space-division multiplexing due to the inherent orthogonality of the different OAM modes. However, traditional technologies to generate and manipulate the vortex beam suffer from bulky size and thus cannot be integrated into nanophotonic systems. By utilizing the localized phase-shifting mechanisms of metasurfaces, it is quite easy to obtain a compact element for vortex beam generation.

10.3.1 OAM Generator with Discrete Phase Engineering

Discrete phase engineering, also called digital phase engineering, is a common way to construct a variety of metasurfaces as introduced in previous chapters. In recent years, the OAM generators designed with different categories of discrete phase engineering at subwavelength scale have been realized [26–36]. Plasmonic local propagating phase modulation in metallic holes can be utilized to generate optical vortex for both linear polarization and circular polarization, as shown in Fig. 10.14 [36]. Similar to the 1D nanoslit, 2D metallic holes can modulate the phase retardation by varying the diameter based on the effective mode index manipulating as shown in Fig. 10.14a. In order to relax the fabrication requirement in visible band, one can fill high-refractive index materials into the metallic slits or holes to obtain stronger local phase engineering ability. In this way, a nanowaveguide radius change of approximately 56 nm results in a 2π phase change. The holes with spatially changed diameters are arranged along azimuth direction corresponding to the desired phase distribution of a vortex beam (Fig. 10.14b and c).

Geometric phase engineering [27, 31–33, 37] and circuit-type resonant phase engineering [38, 39] are other two typical ways to give rise to a helical phase front carrying orbital angular momentum. Figure 10.15a shows an OAM generator consisting of nanorod array with spatially varying orientations [27]. When illuminated by CPL, the subwavelength plasmonic rods will generate local abrupt phase changes for its cross-polarized component according to the geometric phase principle. An azimuthally linear phase gradient is abruptly introduced by arranging the nanorods with a constant orientation angle gradient. The evolution of the vortex beam for different incident wavelengths is shown in Fig. 10.15b, and a characteristic dark spot with zero intensity in the center is observed.

A vortex plate can also be built with the structure comprised of V-shaped antennas [26]. The varying length and spanning angles of the two arms will generate a circuit-type phase, and a rotation of 90° will generate a 180° geometric phase for its cross-polarized light, when illuminated with a linearly polarized beam. Four unit cells were designed to obtain a total π phase shift with $\pi/4$ phase increments. An extra π phase shift was obtained by rotating the initial four nanorods. By arranging the eight elements along the azimuthal direction, a helical phase profile could be generated.

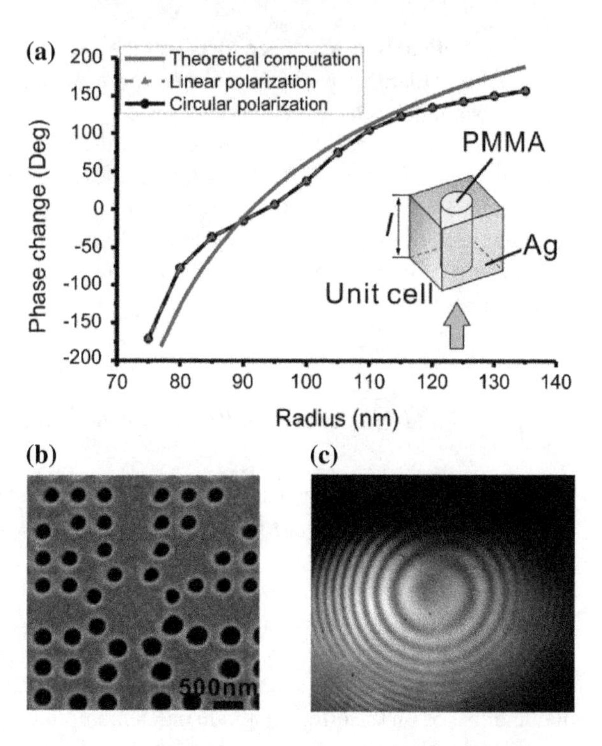

Fig. 10.14 **a** Propagating phase change as a function of the nanowaveguide radius at a wavelength of 532 nm. **b** SEM picture of the nanowaveguide array fabricated by FIB. **c** Measured spiral interference pattern. Reproduced from [36] with permission. Copyright 2014, American Chemical Society

By combining spiral wavefront along the azimuthal direction with a spherical wavefront along the radial direction, a focusing optical vortex can also be obtained [40]. The phase distributions of the OAM lens in the incident plane are

$$\Phi(r, \varphi) = -k\sqrt{r^2 + f^2} + l\varphi, \quad (10.3.1)$$

where $k = 2\pi/\lambda$ is the wavevector in vacuum. The OAM lens composing of ellipse nanorods is shown in Fig. 10.16. This metasurface lens can be seen as a composite of a positive lens and a polarization converter. Therefore, A point source radiating RCP light is transformed into light carrying OAM of $l = 2$ and the spherical wave is converted into plane wave (Fig. 10.16d). Based on the similar design principles with subwavelength geometric phase manipulation given in previous discussion, twisted focusing optical vortex [41] and deflecting vortex [42] have also been demonstrated with compact metasurfaces recently.

A nanoslit metasurface that can generate the multi-channel vortex light in equal energy at wavelength of 632.8 nm was proposed and experimentally demonstrated [32]. Different from the phase-only metasurface, both the phase and the amplitude manipulation are employed by varying the orientation and geometry of the antennas. Such approach can suppress the energy of the higher-order diffraction and achieve equal energy in different channels. The uniformity of the light array is 0.146 in experiment, approximating to the simulated one of 0.10. The phase and amplitude

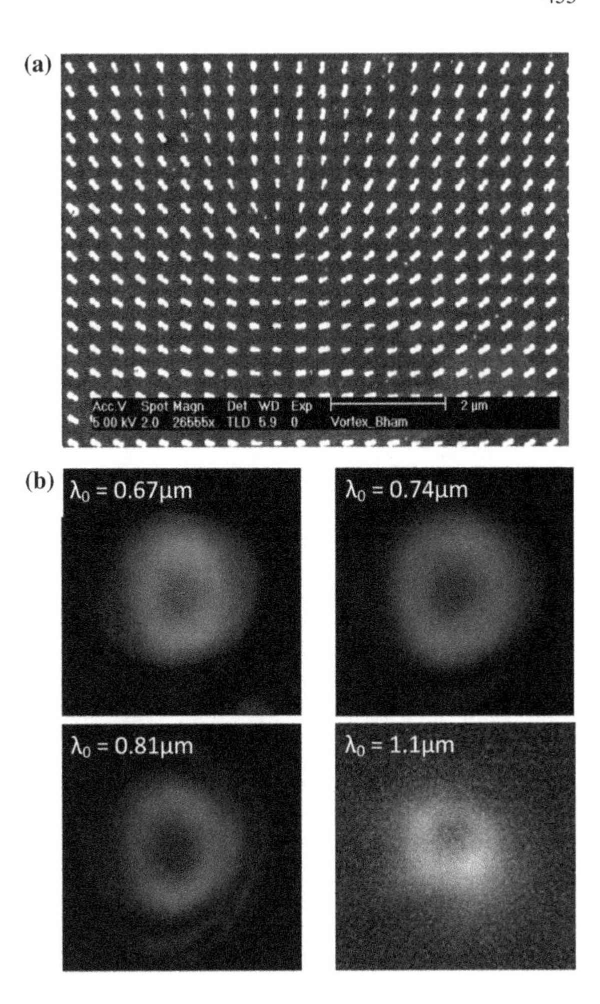

Fig. 10.15 **a** SEM image of the dipole array which was designed for generating an optical vortex beam. **b** Measured intensity distribution of the vortex beam patterns for different wavelengths from 670 to 1100 nm. Reproduced from [27] with permission. Copyright 2012, American Chemical Society

information of the multi-channel vortex light were obtained by employing the principle of holography and then encoded to the spatial orientation and geometry of the antenna. This approach shows the merits of ultra-thin, easy design and fabrication, multi-channel, multi-topological charge, and equal energy in different channels. In 2016, Ren et al. proposed an angular momentum multiplexing method using a nanoring aperture with a chip-scale footprint as small as $4.2 \times 4.2\ \mu m^2$, where nanoring slits exhibit a distinctive outcoupling efficiency on tightly confined plasmonic modes [43]. The mode-sorting sensitivity and scalability of this approach enable on-chip parallel multiplexing over a bandwidth of 150 nm in the visible wavelength range.

Compared with vortex beams featuring a homogeneous polarization state, vectorial vortex beams with an inhomogeneous polarization distribution (such as radial polarization) in the transverse plane have also attracted much attentions. Many approaches, including liquid crystal q-plates, spatial light modulators (SLMs), and

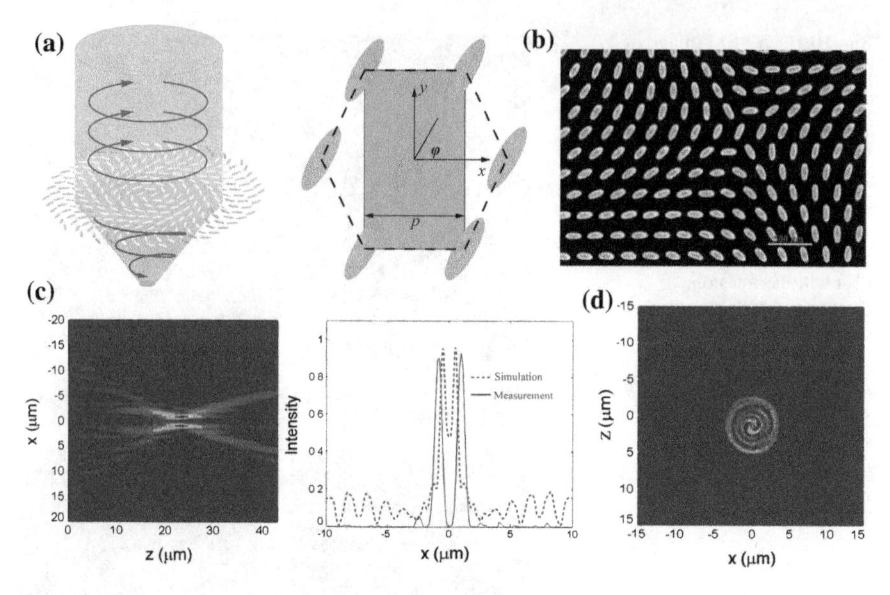

Fig. 10.16 **a** Schematic view of the metasurface for OAM converting and focusing. Inset: unit cells of the metasurface. **b** SEM image of the fabricated sample which converts the incident CP light into a focused light carrying OAM with topological charge $l = 2$. **c** Measured power intensity profile of the focused field in *xoz*-plane. Inset: the simulated (dashed lines) and measured (solid lines) intensity profile of the focused field, respectively, along *x*-axis. **d** Simulated results of beaming function for the chiral metasurface. Reproduced from [40] with permission. Copyright 2015, The Authors

optical elements using femtosecond laser direct writing technology, have been proposed to generate vector vortex beams [19]. However, these systems could not be straightforwardly downsized, preventing widespread applications in integrated optics. Recently, a new type of holographic interface is realized, which is able to manipulate the three fundamental properties of light (phase, amplitude, and polarization) over a broad wavelength range [44]. The design strategy relies on replacing the large openings of conventional holograms by arrays of subwavelength apertures, oriented to locally select a particular state of polarization. The resulting optical element can therefore be viewed as the superposition of two independent structures with very different length scales, that is, a hologram with each of its apertures filled with nanoscale openings to only transmit a desired state of polarization. Lately, an approach to generate vector vortex beams with a single metasurface was proposed and experimentally demonstrated by locally tailoring phase and transverse polarization distribution [45]. By maintaining the converted part and the residual part to be equal in amplitudes, the cylindrically polarized vortex beams carrying orbital angular momentum were experimentally demonstrated based on a single metasurface.

10.3.2 OAM Multiplexing and Demultiplexing with Metasurfaces

Recent progresses in modal basis that carries OAM facilitate space-division multiplexing due to the inherent orthogonality of the different OAM modes [22]. However, traditional technologies to generate and manipulate vortex beams suffer from bulky size and thus cannot be integrated into nanophotonic systems. Li et al. proposed a physical methodology for multiple OAMs multiplexing and demultiplexing by an off-axis designing principle to integrate all the space-division multiplexing (SDM), wavelength-division multiplexing (WDM), and polarization-division multiplexing (PDM) into one single ultra-thin metasurface [46].

Figure 10.17 illustrates the progresses of OAM multiplexing and demultiplexing for OAM with separable topological charges, by a single metasurface.

OAM is commonly described in LG modes. Therefore, it is possible to generate OAM beams with arbitrary topological charges at designed separable directions at a single component. The optical transmission function of this multi-OAM generation via off-axis integration (MOG-OAI) component can be expressed as $t(r,\theta) = \sum A_m(r)\exp\big(i(l\theta_m + k_{xm}x + k_{ym}y)\big)$, where k_{xm} and k_{ym} represent the transverse wavevector of the mth channel on x- and y-axes, respectively. $A_m(r)$ dominates the spatial intensity of the beam. r and θ refer to the radial position and the azimuthal coordinates, respectively. Due to the commutativity of k-spaces before and after the multiplexer, directions of independent OAM beams carrying different topological charges can be adjusted by modifying the k-space of incidence. With an off-axis beam incidence, $E_{\text{inc}} = e^{ik_{x0}x + ik_{y0}y}$, the diffraction field at the Fourier plane, which can be designed by integrating a lens, or monitoring the far-field, as Fraunhofer diffraction is given by:

$$E_F = F\{E_{\text{inc}} \cdot t\} = \sum_m F\big\{E_{\text{OAM}(l_m)}\big(k_{x0} + k_{xm}, k_{y0} + k_{ym}\big)\big\} \qquad (10.3.2)$$

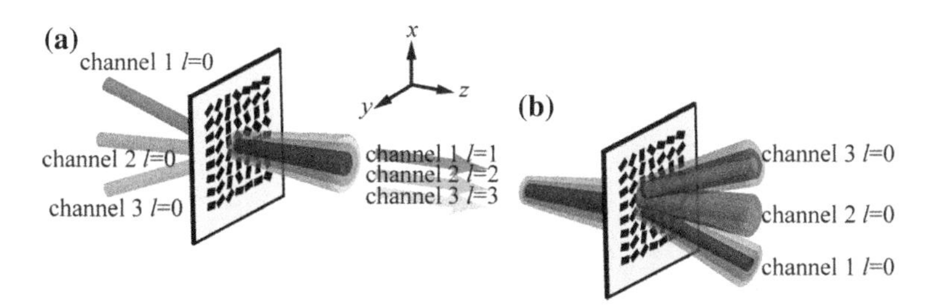

Fig. 10.17 Schematics of **a** off-axis incidence multi-OAM multiplexer and **b** off-axis multi-OAM demultiplexer. Reproduced from [46] with permission. Copyright 2017, Wiley-VCH Verlag GmbH & Co. KGaA, Weinheim

For OAM demultiplexing, when a plane wave illuminates the MOG-OAI component at the direction $(-k_{xm}, -k_{ym})$, a vortex beam with topological charge l_m can be generated at the direction normal to the component. Therefore, beams carrying independent information are multiplexed together in coaxial beams with separable OAMs by different incident angles as shown in Fig. 10.17b. Furthermore, the dispersion of wavevector k makes it possible to realize WDM of signals, although the geometric metasurface is intrinsic dispersionless. The dispersion of wavevector follows: $k_{xm} = k_0 \sin(\delta_{xm}) = 2\pi \sin(\delta_{xm})/\lambda = 2\pi \sin(\delta_{xm0})/\lambda_0$, where δ_{xm} is the deflection angle of the mth beam. Simultaneously, the dispersionless phase shift ensures unchanged performance in OAM modulation. Therefore, OAM beams with multi-wavelength can be separated without influencing the OAM demultiplexing function.

The functional metasurfaces are designed based on phase-only optical antennas. The phase distribution of phase-only component can be expressed as $\varphi_{\text{total}}(r, \theta) = \text{angle}(t)$. A multiplexer/demultiplexer generating the OAMs $l = -2, -1, 0, 1$, and 2 is designed for a LCP incidence ($\lambda = 632.8$ nm) dipole antenna metasurface. The radius of the component is set as 25 μm, the deflection angle is 9° and the size of pixel is 0.5 μm. As an OAM demultiplexer, the fundamental modes appear at the opposite directions for different input OAM beams as shown in Fig. 10.18a–c. Figure 10.18d–f indicates the simulated performance of the dipole antenna metasurface at another incident polarization (RCP). As shown in Fig. 10.18d, the beams propagate at opposite directions with desired opposite OAM. Comparing Fig. 10.18e with Fig. 10.18b, the RCP incident $l = 1$ beam is deflected into the position opposite to that of LCP $l = 1$. Therefore, the OAM beams carrying the same topological charges with different polarizations can be separated by the same component for OAM demultiplexing. Recently, linearly polarization controlled vortex beam generation has also been demonstrated [47]. Besides polarization demultiplexing, Fig. 10.18g–i shows demultiplexing of OAM beams at different wavelengths of 405 (blue), 532 (green), and 632.8 nm (red). The beams with different OAMs are demultiplexed into diverse directions with solid spots, while the beams with the same OAM can propagate at the same direction with diverse deflection angles for different wavelengths. The separation distances can be raised by enlarging the deflection angles. This response is similar to blazed grating, which can be applied to WDM in communication systems. The polarization and wavelength can be designed with prescribed range in k-space as simulated. Therefore, such metasurface based on dipole antenna can be a highly integrated demultiplexer for SDM, PDM, and WDM. Similarly, multiplexers can also support three multiplexing methods by a single component. This method can be applied to quantum OAM communication, which can operate two entangled photons symmetrically, and used for multi-channel quantum key distribution. Demultiplexing of SDM, WDM, and PDM can be done by a single metasurface in OAM communication systems. In addition, utilizing off-axis technique for optical integration has great potentials in complex optical systems with information on space, wavelength, and polarization.

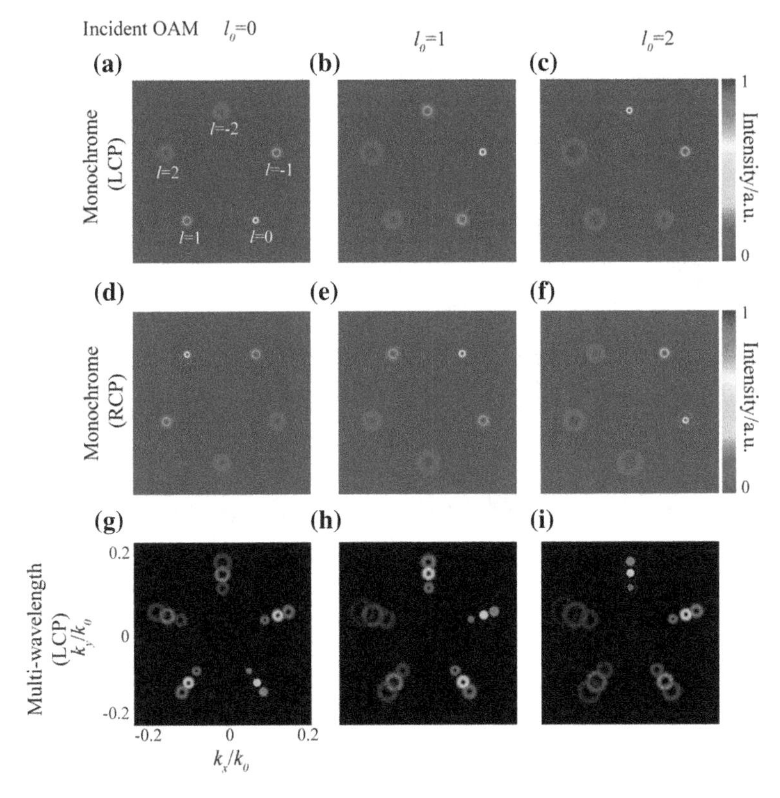

Fig. 10.18 Simulated performance of designed OAM demultiplexer. **a–c** The diffraction images of the metasurface illuminated by monochromatic ($\lambda = 532$ nm) LCP beams with OAM $l_0 = 0$, 1, and 2. **d–f** The diffraction images of the metasurface illuminated by monochromatic RCP beams with OAM $l_0 = 0$, 1, and 2. **g–i** The diffraction images of the metasurface illuminated by multi-wavelengths (red: $\lambda = 632.8$ nm; green: $\lambda = 532$ nm; blue: $\lambda = 405$ nm) LCP beams with OAM $l_0 = 0$, 1, and 2. Reproduced from [46] with permission. Copyright 2017, Wiley-VCH Verlag GmbH & Co. KGaA, Weinheim

10.3.3 OAM Generator with Continuous Phase Engineering

In principle, the gradient metasurfaces constructed by discrete meta-atoms still have shortages. For example, the discontinuous nature inevitably decreases the purity of generated OAM. Recently, it was shown that quasi-continuous design may overcome this problem [48], since the wavefront engineering accuracy is significantly improved by their nearly infinite small “pixel sizes” and thus the phase noise can be greatly suppressed. The early proposed computer-generated space-variant subwavelength dielectric gratings [49, 50] can be taken as a rudimentary quasi-continuous design in the long-wavelength infrared. In 2013, Brasselet et al. proposed that closed-path nanoslits and a discrete set of continuous deformation of a circular nanoslit [51] can also be utilized to generate and control optical vortices at the microscopic scale. By

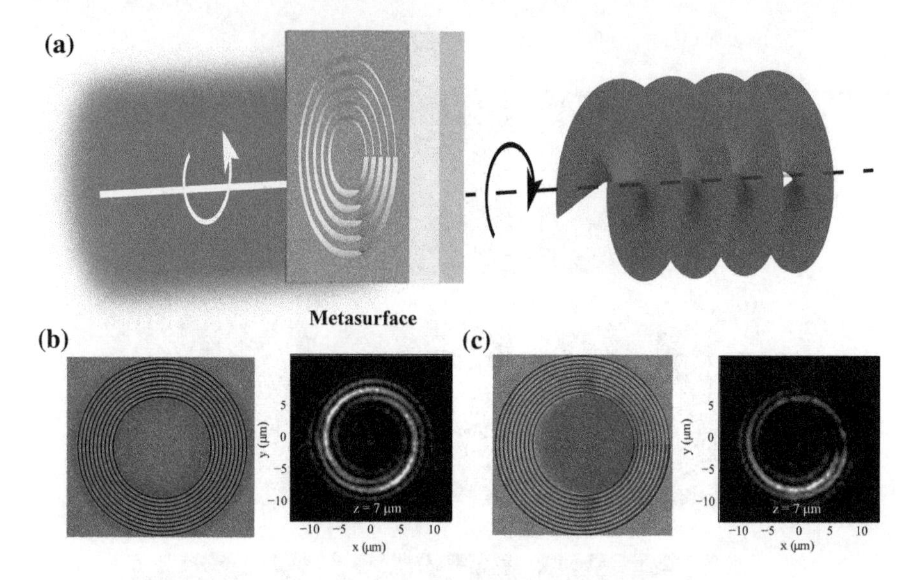

Fig. 10.19 **a** Schematic of the OAM generator with quasi-continuous structures. **b** SEM image and the interference fields of OAM generator for $l = 2$. **c** SEM image and the interference fields of OAM generator for $l = 1.5$. Reproduced from [48] with permission. Copyright 2016, American Chemical Society

virtue of the geometric phase engineering of a micro-ring or micro-helix together with the plasmonic propagating phase engineering with varying widths, an OAM beam with high purity can be generated by a continuously shaped plasmonic metasurface as illustrated in Fig. 10.19 [48]. One interesting effect of this configuration is the spin–orbit interaction becomes asymmetric [52], which has many novel applications in OAM communication and holography display [53–56].

Single catenary nano-aperture for giant spin Hall effect has been introduced in Sect. 10.2.2. Linear and continuous phase retardation can be intrinsically generated due to the unique trajectory of the equation. This continuous structure allows a higher diffraction efficiency. It is quite a straight way to obtain a vortex beam with continuously geometric phase engineering by arranging the catenary structures along the azimuthal direction (Fig. 10.20a). The catenary curve in polar coordinates (r and φ) can be derived to represent the corresponding surface topography.

For generation of OAM, the surface structure should be designed so that the phase distribution is $\Phi(r, \varphi) = l\varphi$. As illustrated in the geometric relation in Fig. 10.20b, the tangential angle of the curve at a given position (r, φ) with respect to the azimuthal direction is $\xi = l\varphi/2 - \varphi$. Consequently, the equation describing the surface profile can be written as

$$\frac{\mathrm{d}r}{r\mathrm{d}\varphi} = \tan\xi = \tan\left[\frac{(l-2)\varphi}{2}\right]. \tag{10.3.3}$$

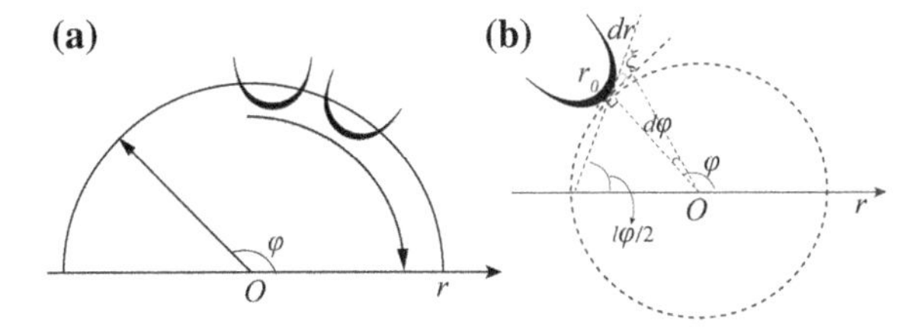

Fig. 10.20 **a** Sketch map of the arrangement of catenaries to obtain OAM beam. **b** Catenary equation derivation in polar coordinate frame. Reproduced from [17] with permission. Copyright 2015, The Authors

By some mathematical manipulation, the unique curve for generating OAM is obtained,

$$\begin{aligned} r &= (r_0 + n\Delta)\exp\left\{\frac{2}{l-2}\ln\left(\left|\sec[\frac{(l-2)\varphi}{2}]\right|\right)\right\}, n = 0, 1, 2, 3\ldots \\ &= (r_0 + n\Delta)\left(\left|\sec[\frac{(l-2)\varphi}{2}]\right|\right)^{\frac{2}{l-2}}, \end{aligned} \tag{10.3.4}$$

where $r_0 + n\Delta$ denotes the vertex of a single curve, n is the index of different curves, and Δ is the distance between the adjacent vertexes. The final aperture arrays are obtained by connecting the $2n$th and $(2n + 1)$th curves and truncating the curves properly. Obviously, Eq. (10.3.3) has a singularity at $l = 2$. In such a specific case, the direct integration of Eq. (10.3.2) leads to a set of concentric rings in accordance with the design mentioned above. Following Eq. (10.3.3), a sample with topological charge of $l = -6$, for LCP illumination (the topological charge is reversed from l to $-l$ for opposite circular polarized illuminations), was fabricated via FIB milling on a 120-nm-thick Au film (Fig. 10.21a). In this case, Δ and r_0 were chosen as 200 nm and 1.5 μm. Under circularly polarized illumination, the catenary apertures generate simultaneously two kinds of beams with approximately equal intensity, one uniformly phased and the other helically phased with opposite circular polarization. Based on the interference effect, the topological charge of the OAM can be directly identified, without the use of additional interference beam. Three laser sources at $\lambda =$ 532, 632.8, and 780 nm were adopted to demonstrate the broadband performances. Figure 10.21c shows the intensity patterns for different wavelengths and polarizations at a few microns away from the sample. The measured results (the first column in c) are in good agreement with the theoretical results obtained by vectorial diffraction theory (the second column in c). For the x- and y-polarized components, the intensity patterns are manifested by rotating petals encircling the beam centers, where the modulus and sign of l are determined by the number and twisting direction of petals. The conversion efficiency η, defined as the ratio between the power of the beam carrying OAM and the overall transmitted power, was measured. The mean values for $\lambda = 532$, 632.8, and 780 nm are 23.2, 39.8, and 54.4%, exhibiting at least 30-fold enhancement compared with that in previous work with circular apertures [51].

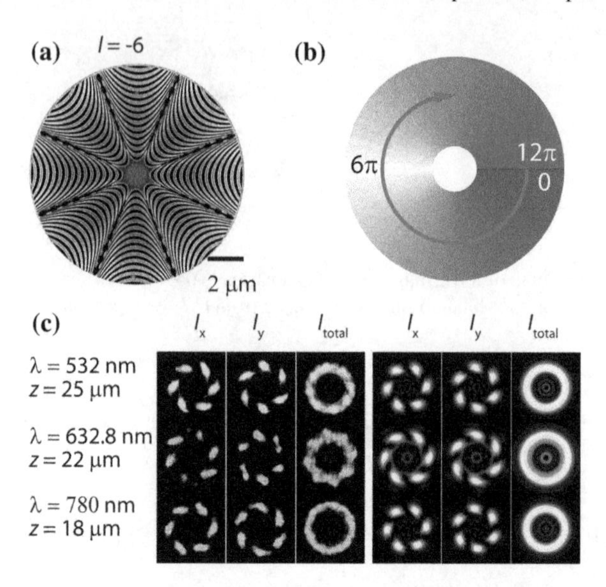

Fig. 10.21 **a** SEM images of the fabricated OAM generators based on the catenary arrays. The topological charge ($\sigma = 1$) is −6. **b** The corresponding spiral phase profiles. **c** The measured (the first column) and calculated (the second column) results of I_x, I_y, and I_{total} for $\lambda = 532$ nm (top row), $\lambda = 632.8$ nm (middle row), and $\lambda = 780$ nm (bottom row). The distances between the recording planes and the sample are indicated in each row. Reproduced from [17] with permission. Copyright 2015, The Authors

The most amazing property of the catenary-based OAM is its ability to realize achromatic performance which can be concluded with the following observations. On the one hand, the geometric phase is intrinsically independent on the operational frequency according to its definition. On the other hand, the conversion efficiency η is also achromatic. To demonstrate this, the conversion efficiencies for the unit cells of both the catenary apertures and discrete antennas are calculated using commercial software CST MICROWAVE STUDIO with Floquet ports (Fig. 10.22a and b). Note that the width w is changing all over the catenary; thus, the periods of unit cell ($p = 2w$) should be varied correspondingly. As illustrated in Fig. 10.22c, the efficiencies of continuous unit cells for $p = 100$–400 nm are depicted in a region shown in gray color (the lower envelope of this area corresponds to the lowest efficiency), while the results for discrete antennas are evaluated for only $p = 400$ nm. To explain the physical mechanism of the achromatic η, the electromagnetic modes in the unit cells of catenary and discrete antennas are shown in Fig. 10.22d and e. Resorting to the optical nanocircuit theory proposed by Engheta [57], there are two orthogonal modes in the catenaries, one inductive and one capacitive along the main axes. The inductive and capacitive modes are related to the achromatic rejection and transmission, which implies that η can be nearly a constant in ultra-broadband spectrum, although obvious fluctuations can be observed in the visible spectrum owing to the resonance at $\omega_0 = (\text{LC})^{-1/2}$ along the v-direction.

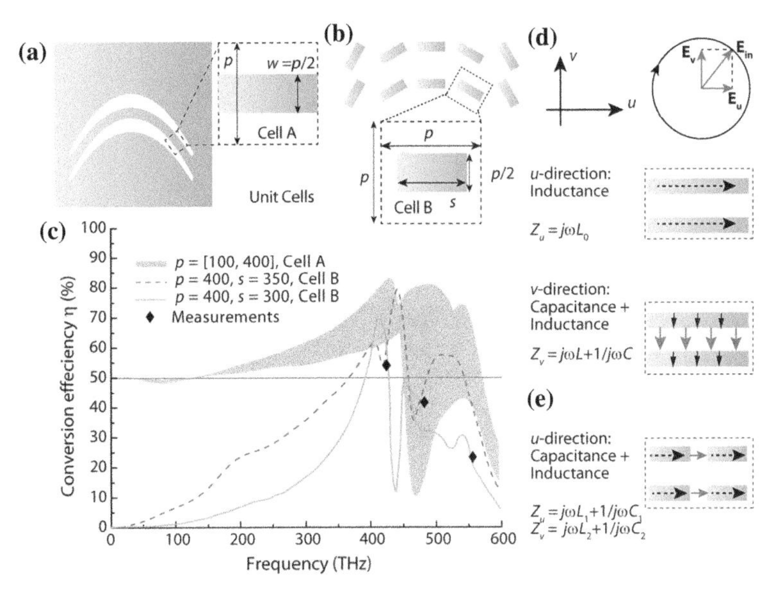

Fig. 10.22 Comparison of the conversion efficiencies and electromagnetic modes for the catenary apertures and discrete nano-antennas. Schematic of the unit cell in catenary apertures (**a**) and discrete antennas (**b**). The two cells differ from each other in the length of metallic bar with respect to the period. **c** Numerically calculated polarization conversion efficiency η for the catenary apertures ($l = p$) and nano-antennas ($l < p$). The vertical line depicts the resonant frequency ω_0 for the LC resonance along v-direction. **d** and **e** Schematic of the orthogonal modes in u-v-coordinates for the catenary and discrete nano-antennas. The surface impedances as a function of frequency are depicted by nanocircuit model. The resistances stemming from the ohmic loss are neglected here. Reproduced from [17] with permission. Copyright 2015, The Authors

Based on the phase profile in Eq. (10.3.1), a catenary aperture-based OAM lens can also be designed. Figure 10.23a illustrates a catenary OAM lens with $k_r = \arcsin(\lambda/2)$ k, $l = 2$, and $f = 40$ μm. The central and outer radii, defined as r_1 and r_2, are set as 10.6 and 20.8 μm, respectively. As shown in Fig. 10.23b and c, the intensity distributions along the propagation direction reveal the main two characteristics, i.e., the focusing property and intensity singularity. The measured center-to-center distance of the donut-shaped intensity patterns at $z = 40$ μm is 1.895 μm, which is a bit larger than the theoretically expected 1.56 μm, owing to the fabrication and measurement errors.

10.3.4 Vectorial Vortex Beam Generation

A radially polarized beam can be focused more sharply and give rise to a centered longitudinal field, thus ensuring higher resolution lithography and optical sensing. An azimuthally polarized beam with a helical phase front, which carries an orbital

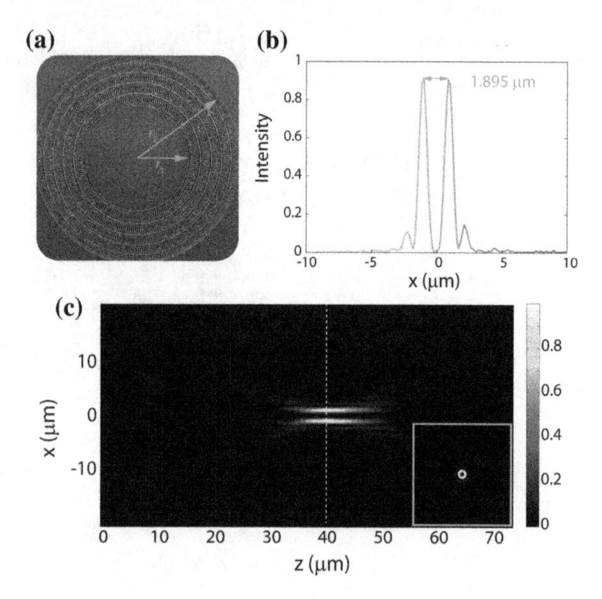

Fig. 10.23 **a** SEM images of the catenary arrays for focused OAM. **b** Measured cross-polarized intensities of the focused OAM in the *xz*-plane at a wavelength of $\lambda = 632.8$ nm. The intensities at the *xy*-plane ($x, y \in [-20, 20]$ mm, $z = 40$ μm) are plotted in the inset. **c** The intensity profile at the cross section of $z = 40$ μm and $y = 0$. Reproduced from [17] with permission. Copyright 2015, The Authors

angular momentum, can effectively achieve a significantly smaller spot size in comparison with that for a radially polarized beam with a planar wavefront in a higher-NA condition. Current technologies to generate vectorial beam suffer from low efficient energy use, poor resolution, low-damage threshold, and bulky size, preventing further practical applications.

Lin et al. [44] reported a holographic interface, which is able to manipulate the three fundamental properties of light (phase, amplitude, and polarization) over a broad wavelength range. The manipulation mechanism is illustrated in Fig. 10.24, where a detour phase of $k_0 \cdot \sin(\theta) \cdot \delta_{nm}$is generated at a special obtain angle θ with the phase shift determined by the offset distance with respect to the center of a unit cell. The widths of the apertures are fixed, while the lengths are changed to control the amount of light passing through the pixel. The polarization manipulation strategy relies on replacing the large openings of conventional holograms by arrays of subwavelength apertures, oriented to locally select a particular state of polarization.

The whole hologram that can generate radially polarized optical beams from circularly polarized incident light is shown in Fig. 10.25a, which can therefore be viewed as the superposition of two independent structures with very different length scales, that is, a hologram with each of its apertures filled with nanoscale openings to only transmit a desired state of polarization. The fabricated structure is tested with two different wavelengths (633 and 850 nm). A QWP converts the linearly polarized laser beam to a circularly polarized one. The far-field distribution is captured by a CCD camera after passing through a linear analyzer oriented at varying angles. These intensity patterns demonstrate that the $m = +1$ beam is radially polarized and the device is broadband.

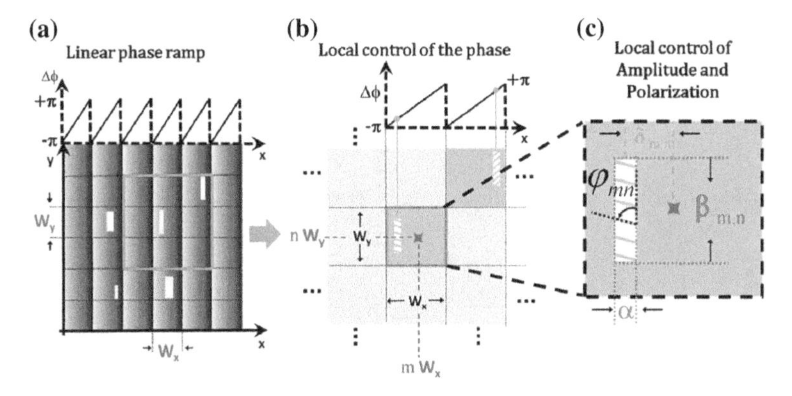

Fig. 10.24 **a** Illustration of the concept of the detour phase. **b** The overall complex amplitude of light emerging through each pixel is modulated by controlling both the position and the size of the slit in the pixel. Schematics (**a**) and (**b**), without the subwavelength apertures, describe a conventional detour-phase hologram. **c** By decorating each aperture with subwavelength features to form a wire grid polarizer oriented at some angle, linear polarization orthogonal to the wire grid is selected. Adapted with permission from [44]. Copyright 2013, American Chemical Society

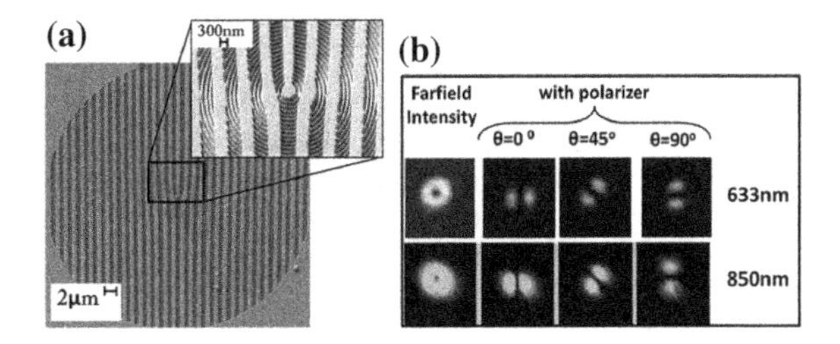

Fig. 10.25 **a** SEM micrograph of the device. **b** Measured far-field intensity distributions at two wavelengths, without and with analyzer, the latter oriented at different angles in front of the CCD camera. Reproduced from [44] with permission. Copyright 2013, American Chemical Society

Figure 10.26 shows another schematic of vector vortex beam generation. For an incident beam, the reflected light from the metasurface consists of two circular polarization states: One has the same handedness as the incident CPL but with an additional phase delay, and the other has the opposite handedness without the additional phase delay. The additional phase delay is known as Pancharatnam–Berry phase with a value of $\pm 2\phi$, where ϕ is the orientation angle of each nanorod. Mathematically, the relation between the CPL and the radially polarized light (RPL) can be expressed as:

$$\begin{pmatrix} \vec{E}_r \\ \vec{E}_\varphi \end{pmatrix} = \begin{pmatrix} \exp(-j\phi) & \exp(j\phi) \\ -j\exp(-j\phi) & j\exp(j\phi) \end{pmatrix} \begin{pmatrix} \vec{E}_{\mathrm{LCP}} \\ \vec{E}_{\mathrm{RCP}} \end{pmatrix} \tag{10.3.5}$$

where $\vec{E}_r$, $\vec{E}_\varphi$, $\vec{E}_{\mathrm{LCP}}$, and $\vec{E}_{\mathrm{RCP}}$ are the electric field of RPL, azimuthally polarized light, LCP, and RCP, respectively. Therefore, the superimposition of LCP and RCP

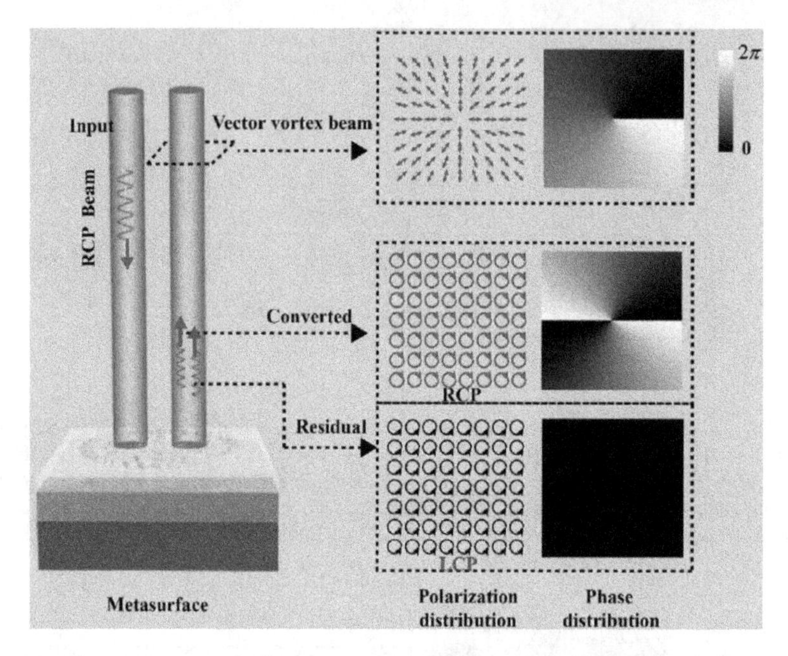

Fig. 10.26 Schematic of the vector vortex beam generation through a metasurface and the polarization and phase distributions of the generated beams. The resultant beam is a superposition of the converted part and the residual part. The converted part has the same circular polarization as that of the incident beam and an additional phase pickup, while the residual part has opposite helicity but no phase change. Reproduced from [45] with permission. Copyright 2016, American Chemical Society

with a phase difference of $\pm 2\phi$ after reflected by the metasurface can just be utilized to generate vectorial beam.

In order to generate spatial inhomogeneous vectorial field, a reflective metasurface composed of rotating metallic antennas has been fabricated with the orientation of each antenna defined by:

$$\alpha(r, \phi) = \phi + \alpha_0 + \frac{\pi}{4}, \tag{10.3.6}$$

where (r, ϕ) is the polar coordinate representation. α_0 is the initial angle related to the phase difference of two eigenstates. The generated structured beams from the fabricated metasurfaces are characterized and validated by passing through a linear polarizer with a different transmission angle. The appearance of "*s*"-shaped patterns is theoretically predicted and experimentally confirmed (Fig. 10.27). The observed patterns indicate that the resultant beams indeed have an inhomogeneous polarization distribution and a helical wavefront. Moreover, the twisted direction of the "*s*" shape varying with the helicity of the circular polarization is also experimentally confirmed from the obtained intensity patterns.

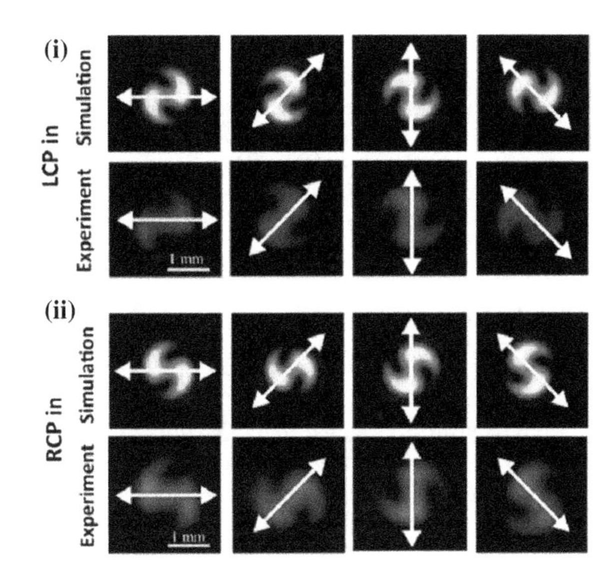

Fig. 10.27 Simulated and experimentally recorded intensity profile of the vector vortex beam after passing through a polarizer with different polarization angles including horizontal, diagonal, vertical, and anti-diagonal directions. The polarization angles are denoted by white double-headed arrows. **i** LCP light input and **ii** RCP light input. Reproduced from [45] with permission. Copyright 2016, American Chemical Society

10.4 Bessel Beam Generation

Besides the optical vortex, there is another interesting optical beam named Bessel beam, which is a kind of non-diffracting beam with a Bessel-type intensity distribution along the radial direction. Originally discovered in 1987 by Durnin [58], it was demonstrated to be an exact solution to the Helmholtz equation. The ideal Bessel beam requires the horizontal dimensions as well as the power carried by the beam to be infinite, which cannot be achieved in practical situations. Nevertheless, it was shown that even truncated with a Gaussian function, the Bessel beam can still maintain its unique properties such as diffraction-free and self-healing, which enable a variety of important applications in super-resolution imaging [59], optical micromanipulation, and nanofabrication [24, 60, 61].

A general description of electromagnetic fields at any cross section perpendicular to the propagation direction of a Bessel beam can be written as:

$$E(r, \varphi, z) = A_0 \exp(il\varphi + ik_z z) J_l(k_r r). \tag{10.4.1}$$

On the one hand, when $l = 0$, the Bessel beam does not carry orbital angular momentum. On the other hand, $l \neq 0$ corresponds to Bessel beams carrying orbital angular momentum with a topological charge of l, which are called high-order Bessel beams (HOBBs). Traditionally, zero-order Bessel beams (ZOBBs) can be generated by using either refractive optical elements, a conical prism (axicons) [62] or an objective paired with an annular aperture [63], as well as diffractive optical elements such as computer-generated holograms (CGHs) [64]. Similarly, HOBBs can be obtained by illuminating an axicon with a LG beams or propagating a ZOBB through a spiral phase plate. In recent years, more compact devices such as SLM were used to create

HOBBs. However, the pixel size of a typical SLM is at least 6.4 × 6.4 μm, which is more than one order of magnitude larger than the wavelength of visible light, which limited the performance of such devices. Metasurface, allowing one to sample the phase or amplitude distribution at the subwavelength scale, is another high-integrated scheme to obtain both ZOBBs and HOBBs, which provides a substantial improvement of the performances.

10.4.1 Bessel Beam Generator with Discrete Phase Engineering

Various metasurfaces with different kinds of discrete unit cells have been designed to obtain Bessel beams in both microwave/terahertz [65–68] and optical band [69–72]. In 2012, Aieta et al. designed and fabricated an optical axicon using V-shaped nano-antennas at telecom wavelengths [69]. The metallic antenna limits the efficiency of the devices due to the polarization conversion and the inevitable ohmic loss. In order to enlarge the efficiency, a dielectric Bessel beam generator comprising dense arrangement of Si nano-antennas was proposed by Lin et al. [70]. A non-diffracting Bessel beam at 550-nm wavelength was experimentally realized. The efficiency was around 50% and can be further increased. It should be noted that the use of semi-conductors can broaden the general applicability of gradient metasurfaces, as they offer facile integration with electronics and can be realized by mature semiconductor fabrication technologies.

Recently, Chen et al. demonstrated a meta-axicon with a high NA up to 0.9 capable of generating Bessel beams with full width at half maximum (FWHM) about as small as ~$\lambda/3$ ($\lambda = 405$ nm), which may enable advanced research and applications related to Bessel beams, such as laser fabrication, imaging, and optical manipulation [71]. The meta-axicon is constructed by TiO_2 nanofins (Fig. 10.28a and b), providing high conversion up to 90% in the visible range. What's more, it was demonstrated that the generated Bessel beams have transverse intensity profiles independent of wavelength across the visible spectrum (Fig. 10.29a–d). This achromatic property originates that the radial wavevector k_r is independent of the wavelength due to the intrinsic character of the geometric phase engineering. This performance exhibites advantages over conventional axicons, whose NA is almost constant within the visible region due to the weak dispersion of glass. Thus, the FWHM of the Bessel beam is proportional to wavelength and varies accordingly. Changing the wavelength from 400 to 700 nm results in a difference of 175% in the FWHM.

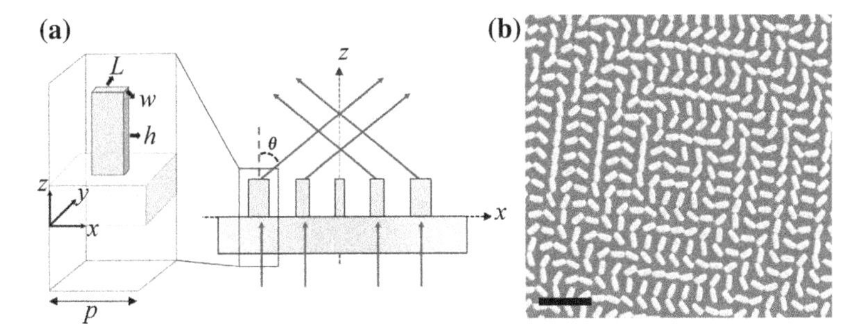

Fig. 10.28 **a** Schematic diagram of a meta-axicon. The meta-axicon deflects an incident collimated light to an angle θ toward its center to generate Bessel beams. The meta-axicon is composed of identical TiO_2 nanofins but different rotation angles. **b** SEM micrograph of a fabricated meta-axicon. Scale bar: 1 μm. Reproduced from [71] with permission. Copyright 2017, The Authors

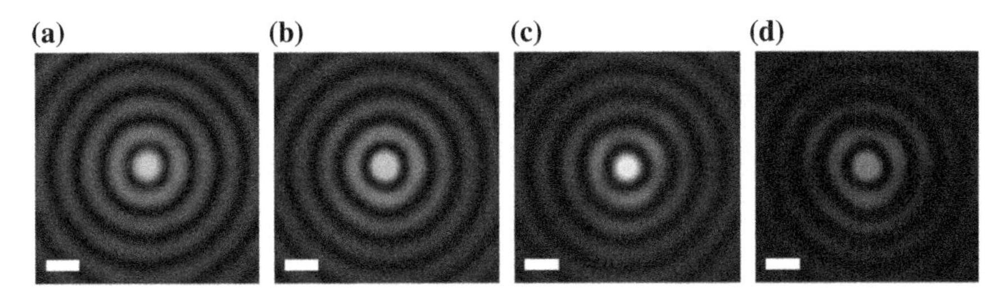

Fig. 10.29 **a–d** Measured intensity profiles at wavelengths $\lambda = 480, 530, 590$, and 660 nm, respectively. Intensity profiles are false colored for ease of visualization. Reproduced from [71] with permission. Copyright 2017, The Authors

10.4.2 Bessel Beam Generator with Continuous Phase Engineering

A general design procedure for continuous phase engineering-based Bessel beam generator can be obtained from the principle of holography. The complex amplitude of such a Bessel beam can be recorded and then used as a holographic Bessel beam generator [73]. Figure 10.30 illustrates an interesting design scheme to generate Bessel beams. This scheme combines micrometer rings or spirals, which act as a zone plate or spiral zone plate in traditional optics to generate vortex phases along the azimuthal direction, and a sub-micrometer grating with analytically designed catenary nanoslits to create conical wavefronts along the radial direction. As the quasi-continuous catenary nanostructures can generate continuous phase modulation between 0 and 2π, 2D arbitrary phase modulation was obtained by locating and rotating the catenary elements along a predefined trajectory, which offers a great flexibility for the generation of both ZOBBs and HOBBs. The quasi-continuous phase modulation makes its pixel to be extremely small, which helps to create Bessel beams at micrometer and even sub-micrometer size.

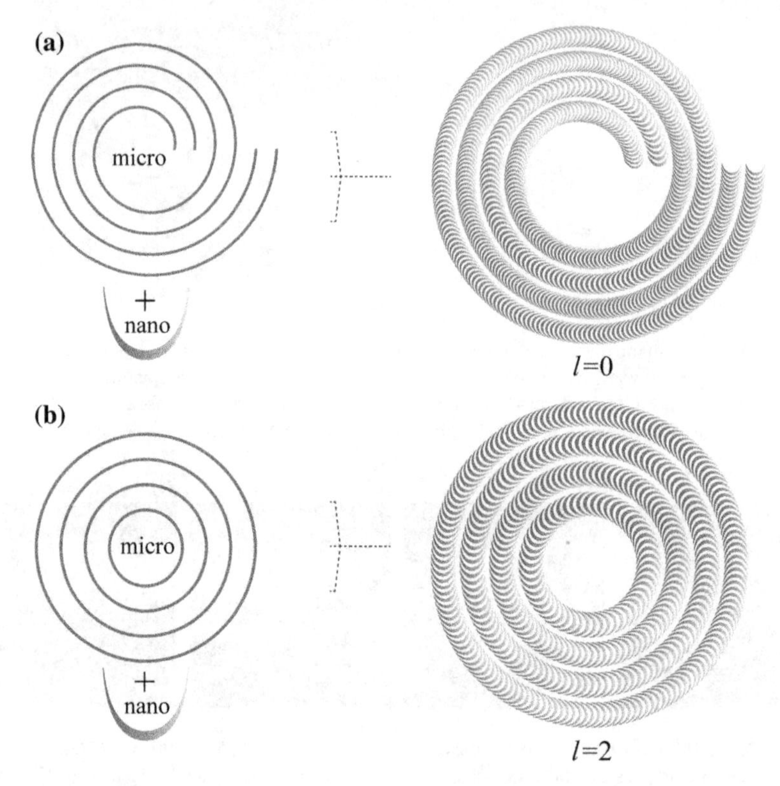

Fig. 10.30 Sketch of the Bessel beam generators combining micrometer/sub-micrometer gratings with $l = 0$ (**a**) and $l = 2$ (**b**)

The shape of the nanoslit can be described by the catenary curve, which is given in Eq. (10.2.3). As demonstrated in Sect. 10.2.2, the phase shifts through the catenary structures for circular polarization incidence are $\Phi = \sigma 2\pi x/\Lambda$ where $\sigma = \pm 1$ denotes the RCP light beam and LCP light beam, respectively. For the simplicity of discussion, we only consider the phase distribution of $\sigma = 1$ (for $\sigma = -1$, the phase profile will be reversed in sign). The single catenary can be arranged in the polar coordinates to achieve phase retardations along both the radial and the azimuthal directions. Obviously, the phase difference between the two ends of the catenary is $\pm 2\pi$; thus, the trajectory of the catenary ends forms the concentric or spiral gratings, which can be written as:

$$r = \frac{(l - 2)\varphi + (2m + 1)\pi}{k_r}, \tag{10.4.2}$$

where $m = 1, 2, 3, \ldots$

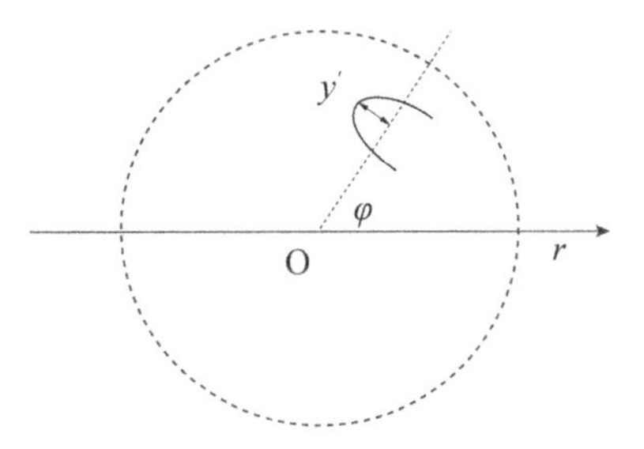

Fig. 10.31 Schematic of the integration procedures for polar coordinates

To obtain a HOBB, the corresponding catenary can be constructed by local approximation assuming that the radial and azimuthal component can be decoupled. Along each radial direction with constant φ, the governing equation becomes

$$y' = \frac{2}{k_r} \ln\left(\left|\sec\left(\frac{-k_r r + l_1 \varphi}{2}\right)\right|\right), \tag{10.4.3}$$

where y' is depicted in Fig. 10.31.

The sample was milled in an Au film on a quartz substrate. The thickness of the Au film is 120 nm. Figure 10.32a shows the SEM image of ZOBB generator. The diffraction-free propagation properties of the Bessel beams were performed by using a homebuilt microscope. Figure 10.32b and c shows the calculation and experiment results of the intensity distributions across the center of ZOBB. The propagation-invariant manner of the Bessel beams was validated in both theoretical simulations and experiment. The center of the zero-order Bessel beam ($l = 0$) is solid as shown in Fig. 10.32d. The theoretical calculation based on vectorial angular spectrum theory is found to be in good agreement with experimental data.

Following the analytic relationship in Eq. (10.4.3), HOBBs generators for topological charges $l = 2$, 3, and 4 were designed and fabricated. The corresponding SEM images are shown in Fig. 10.33a–c. The period p in the radial direction is 2 μm for all the samples, which determines the radial wavevector of the diffraction beam, by the relation $k_r = 2\pi/p$. HOBBs with hollow centers are generated after the illuminated light passing through these devices. As depicted in Fig. 10.33d–o, the spot size, defined as the radius of the innermost ring of the Bessel beam, increases with the topography charge l for the same radial wavevector and wavelength.

Meanwhile, the spot size of the HOBBs can also be manipulated by changing the period p of the microscale/sub-microscale gratings. Figure 10.34a–f plots the field distribution in the xz-plane for various p at the wavelength of 632.8 nm. The calculated and the tested spot sizes of the HOBBs for different p are also given in Fig. 10.34g. It is shown that the spot size approximately linearly depends on the period p. For reduced p, the radius of Bessel beams can even be subwavelength, which is not realizable by traditional SLM devices, due to the large pixel size.

Both the theoretically calculated and experimental interference patterns of HOBBs generators with $l = 2$, 3, and 4 are plotted in Fig. 10.35a–c and d-f, respectively.

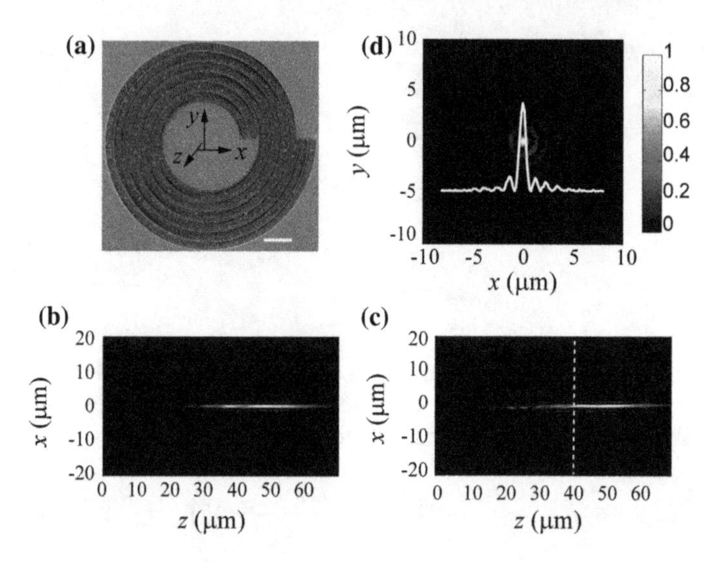

Fig. 10.32 **a** SEM images of ZOBB generator. Scale bar, 5 μm. Calculation (**b**) and experiment (**c**) results of the field distribution in the xz-plane ($y = 0$). **d** Experimental cross-sectional view of the intensity along the dashed line marked in (**c**). Insets show the normalized intensity at $y = 0$. Reproduced from [73] with permission. Copyright 2016, The Authors

The petallike intensity patterns originate from the interference between the beams carrying OAM with opposite signs. The number of the petals in the intensity patterns is twice of $|l|$. It should be noted that the proposed structure inherently works in a broadband since the geometric phase is intrinsically independent of the operational wavelength.

10.5 Airy Beam Generation

The non-spreading Airy wave packet with curved propagation profile was first predicted by Berry and Balazs in 1979. Within the framework of quantum mechanics, they demonstrated that the force-free Schrödinger equation could exhibit an Airy function form solution [74]. What sets the Airy wave apart from other known solution is its very ability to freely accelerate during propagation—even in the absence of any external potential. Due to the mathematical similarity between the Schrödinger equation and the paraxial Helmholtz equation, Siviloglou et al. introduced the Airy function solution in the domain of optics in 2007 [75, 76]. Account for its peculiar properties and applications, Airy beam has attracted spectacular research interest in forming optical bullet [76], optical micromanipulation [77], and producing curved plasma channels, among others [78]. According to the nature of the Airy wave packet, a 3/2-power phase modulation along the lateral dimension of beam is required. Such beams are commonly generated by imprinting a cubic phase onto a Gaussian beam

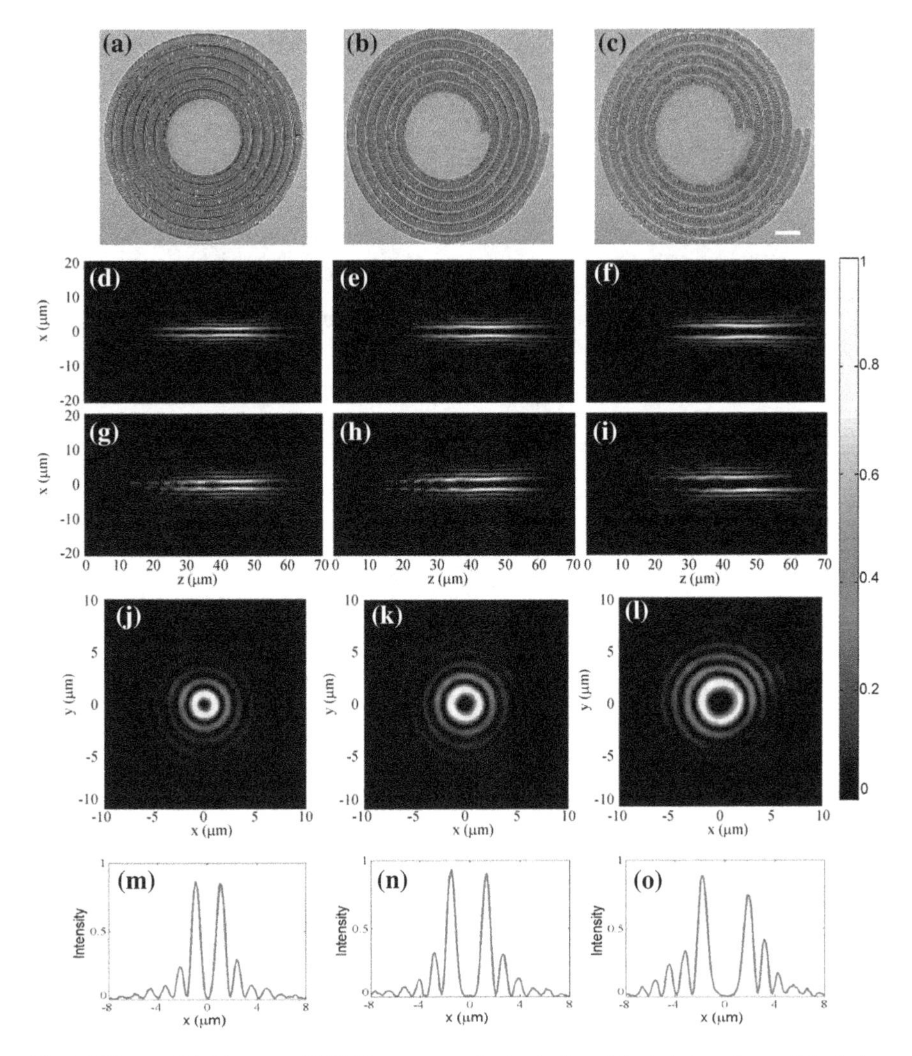

Fig. 10.33 **a–c** SEM images of HOBBs generator for $l = 2$, 3, and 4. Scale bar, 5 μm. Calculation (**d–f**) and experiment (**g–i**) results of the field distribution in the xz-plane ($y = 0$). **j–l** Experimental cross-sectional view of the intensity along the dashed line marked in (**g–i**). **m–o** shows the normalized intensity at $y = 0$. Reproduced from [73] with permission. Copyright 2016, The Authors

in the Fourier plane and subsequently transforming it back to the real space by a lens. Liquid crystal cell [79], three-wave mixing processes [80], phase plate [81], etc., have been adopted for generating the desired cubic phase distribution. However, traditional methods to generate Airy beam usually involve the Fourier transform of lens which needs at least one focal length and thus affects the compactness of optical system. Airy beam generation with a much compact way has attracted growing attention in recent years.

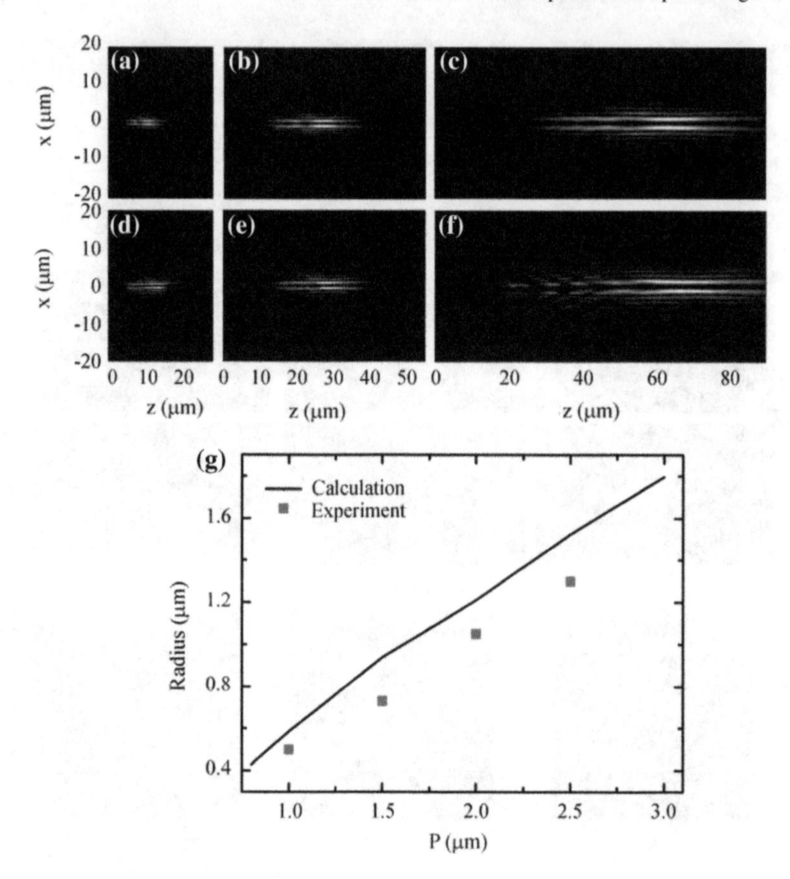

Fig. 10.34 **a–c** Calculation results of the field distribution in the xz-plane ($y = 0$) for $p = 1$, 1.5, 2.5 μm. **d–e** The corresponding experimental results. **g** Periodic p dependent spot sizes of the HOBBs. The topography charged l is 3, and all the data are obtained at the wavelength $\lambda = 632.8$ nm

10.5.1 Airy Plasmon Generator

Integrated Airy beam generators were initially proposed for surface Airy wave on a metal surface [82–85], so-called Airy plasmon, after its theoretical prediction by Salandrino and Christodoulides in 2010 [86]. Due to the phase mismatch between the free-space light and the plasmon, special designs are needed to excite the Airy plasmon. Figure 10.36 shows an approach to generate Airy plasmon by both amplitude and phase modulations of the illumination [83]. The generator was fabricated on an air–gold interface of a 150-nm-thick gold film deposited on a glass substrate. The envelope of the Airy packet is an oscillating function with slowly decaying amplitude, which is shown in Fig. 10.36b, and the phase distribution of the Airy function has values of 0 and π, as shown in Fig. 10.36c. This amplitude modulation is sampled by the length design of the rectangular slits in the transverse direction x, and the

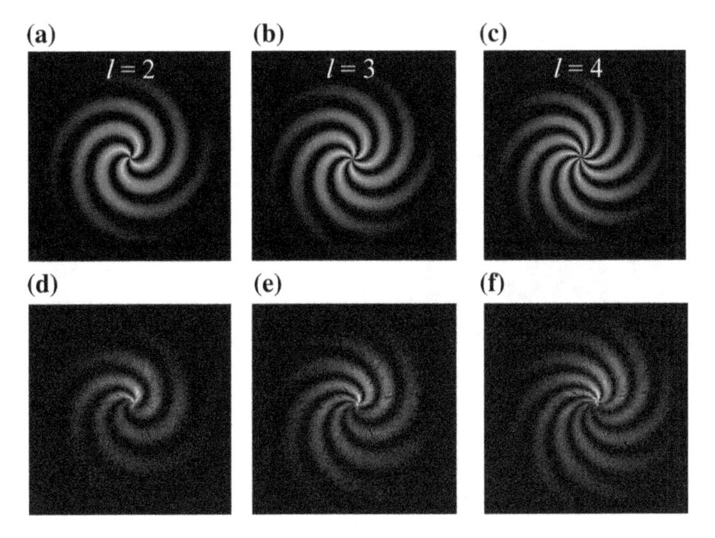

Fig. 10.35 Calculated (**a**–**c**) and experimental (**d**–**f**) interference patterns for HOBBs with $l = 2$, 3, and 4. Reproduced from [73] with permission. Copyright 2016, The Authors

phase modulation is realized by shifting of the next column of slits by half the SPP's wavelength (Fig. 10.36d). The slits are arranged periodically along the z-direction with a period of 764 nm (equal to the SPP wavelength) to effectively couple the illuminated light into the SPP wave at normal incidence. The generated Airy plasmon beam at the gold–air interface was characterized by a near-field scanning optical microscope (NSOM) as shown in the inset of Fig. 10.36a. The measured near-field distribution clearly shows that a surface plasmon is generated and propagates on the metal surface along a curved trajectory for more than 20 μm. The main lobe bends by more than two lobe widths before the wave loses its power due to the plasmon propagation losses.

In another approach to generate the Airy plasmon, a 3/2-power phase distribution was directly obtained by a non-periodically arranged nanocave array along the propagation direction of the SPPs [82]. Figure 10.37a shows the SEM image of generator with a coupling grating and a nanocave array, which is fabricated on a 60-nm-thick silver film deposited on a 0.2-mm-thickness SiO_2 substrate, by FIB milling. An in-plane propagating SPP wave is generated by the grating coupling of a He–Ne laser and then directly propagates to the non-periodically arranged nanocave array. The diffracted SPP waves from nanocaves will interfere and ultimately build a SPP Airy beam. Figure 10.37b and c shows both the experimentally detected and the theoretically calculated SPP beams. The experimental result obtained by a homebuilt leakage radiation microscope (LRM) system clearly manifests a set of Airy-like wave profiles which match well with the theoretical calculation based on the Huygens–Fresnel principle.

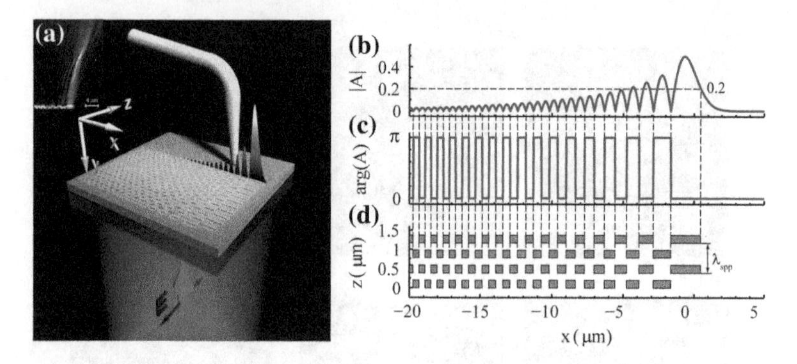

Fig. 10.36 **a** Schematic of the Airy plasmon generation with the engineered grating. The grating is composed of 11 periods of 200-nm-thick slits (in z-direction) and varying width (in x-direction). The grating is excited from the glass-substrate side with a broad Gaussian beam at 784 nm and polarization perpendicular to the slits. Inset: measured near-field intensity profile of the Airy plasmon (**b**), **c** absolute value and phase of the amplitude function of the Airy plasmon. The main lobe half width is $x_0 = 700$ nm. **d** Grating geometry for generation of Airy plasmons, $\lambda_{SPP} = 764$ nm, denotes the SPP wavelengths. Reproduced from [83] with permission. Copyright 2011, American Physical Society

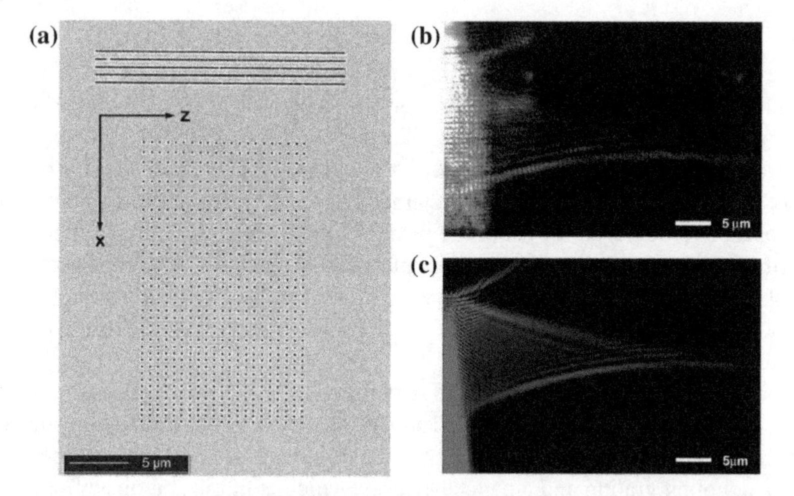

Fig. 10.37 **a** SEM image of Airy plasmon generator based on graded nanocave array fabricated by FIB milling, where the lattice parameter is graded in the x dimension (the period in the x dimension is a_x from 420 to 780 nm, grads $\Delta = 10$ nm), and the period in the z dimension is $p_z = 620$ nm. **b** Experimentally achieved SPP beam trajectories and **c** the calculated one. Reproduced from [82] with permission. Copyright 2011, American Physical Society

10.5.2 Free-Space Airy Beam Generator

Different kinds of integrated Airy beam generators for free-space light were also built recently [87–89]. One-dimensional Airy beam and vortical Airy beam were numerically demonstrated through phase-only modulation with spatially rotating elliptical amorphous TiO_2 nanofins [88]. Simultaneous phase and amplitude sampling were also realized for free-space Airy beam generation [89]. Figure 10.38a shows an Airy beam generator consisting of rotating C-shaped apertures, which is inscribed on a 150-nm-thick Au layer. For the cross-polarized component for a linearly polarized beam illumination, the amplitude of E-field can be continuously tuned by simply changing the rotated angle, and a stepped π phase shift can also be obtained for the rotating angles within $[-\pi/2, 0]$ and $[0, \pi/2]$. Figure 10.38b

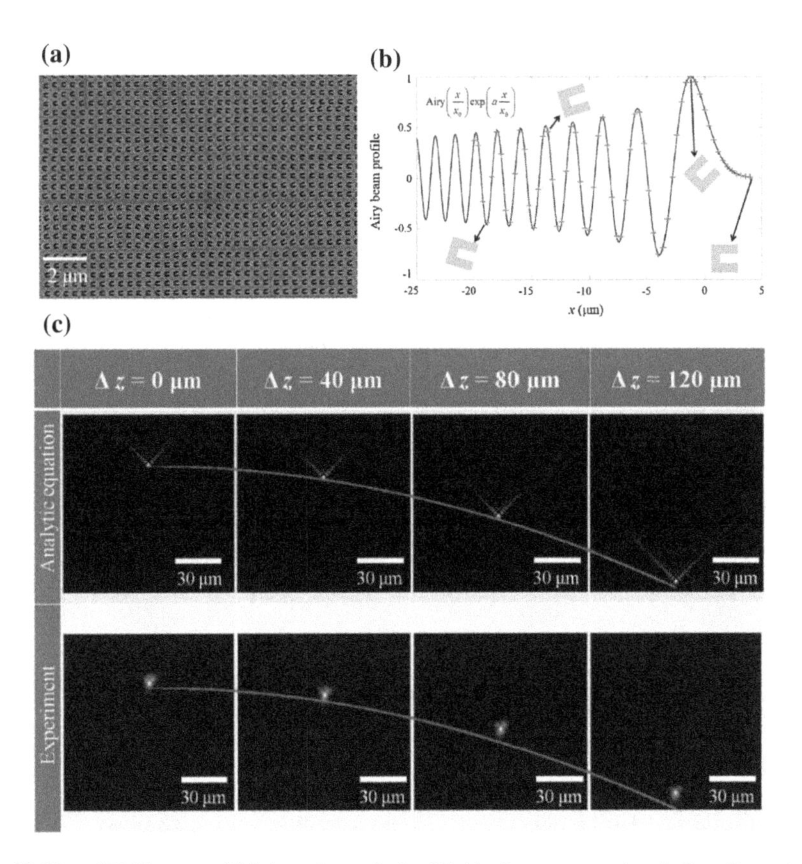

Fig. 10.38 **a** SEM images of fabricated sample for 2D Airy beam generation. **b** One-dimensional Airy beam profile. The locations of unit cells are marked as "+". At some selected unit cells, apertures with specific tilt angles were plotted. **c** Analytical and experimental results for $z = 0$, 40, 80, and 120 μm at wavelength $\lambda = 976$ nm. Reproduced from [89] with permission. Copyright 2017, Wiley-VCH Verlag GmbH & Co. KGaA, Weinheim

gives the 1D Airy beam profile. The envelope of the Airy beam is sampled by the C-shaped apertures with corresponding rotating angles. The theoretical and experimental results for the generated 2D Airy beams are presented in Fig. 10.38c. The field profiles at $z = 0$, 40, 80, and 120 μm, at incidence wavelengths $\lambda = 976$ nm, were plotted. The experimental results demonstrate the performance of the device with reasonable deviation from analytical results.

References

1. A.M. Yao, M.J. Padgett, Orbital angular momentum: origins, behavior and applications. Adv. Opt. Photon. **3**, 161–204 (2011)
2. A. Dudley, M. Lavery, M. Padgett, A. Forbes, Unraveling Bessel Beams. Opt. Photonics News **24**, 22–29 (2013)
3. K.Y. Bliokh, A. Aiello, Goos-Hanchen and Imbert-Fedorov beam shifts: an overview. J. Opt. **15**, 014001 (2013)
4. J.E. Hirsch, Spin Hall effect. Phys. Rev. Lett. **83**, 1834–1837 (1999)
5. J. Sinova, D. Culcer, Q. Niu, N.A. Sinitsyn, T. Jungwirth, A.H. MacDonald, Universal intrinsic spin Hall effect. Phys. Rev. Lett. **92**, 126603 (2004)
6. R.A. Beth, Mechanical detection and measurement of the angular momentum of light. Phys. Rev. **50**, 115–125 (1936)
7. M. Onoda, S. Murakami, N. Nagaosa, Hall effect of light. Phys. Rev. Lett. **93**, 083901 (2004)
8. O. Hosten, P. Kwiat, Observation of the spin Hall effect of light via weak measurements. Science **319**, 787–790 (2008)
9. X. Ling, X. Zhao, K. Huang, Y. Liu, C.-W. Qiu, H. Luo, S. Wen, Recent advances in the spin Hall effect of light. Rep. Prog. Phys. **80**, 066401 (2017)
10. X. Yin, Z. Ye, J. Rho, Y. Wang, X. Zhang, Photonic spin Hall effect at metasurfaces. Science **339**, 1405–1407 (2013)
11. N. Yu, P. Genevet, M.A. Kats, F. Aieta, J.-P. Tetienne, F. Capasso, Z. Gaburro, Light propagation with phase discontinuities: generalized laws of reflection and refraction. Science **334**, 333–337 (2011)
12. R.H. Renard, Total reflection: a new evaluation of the Goos-Hänchen shift. J. Opt. Soc. Am. **54**, 1190–1197 (1964)
13. L. Li, C.W. Haggans, Convergence of the coupled-wave method for metallic lamellar diffraction gratings. J. Opt. Soc. Am. A **10**, 1184–1189 (1993)
14. N. Shitrit, I. Bretner, Y. Gorodetski, V. Kleiner, E. Hasman, Optical spin Hall effects in plasmonic chains. Nano Lett. **11**, 2038–2042 (2011)
15. X. Ling, X. Zhou, X. Yi, W. Shu, Y. Liu, S. Chen, H. Luo, S. Wen, D. Fan, Giant photonic spin Hall effect in momentum space in a structured metamaterial with spatially varying birefringence. Light Sci. Appl. **4**, e290 (2015)
16. X. Luo, M. Pu, X. Li, X. Ma, Broadband spin Hall effect of light in single nanoapertures. Light Sci. Appl. **6**, e16276 (2017)
17. M. Pu, X. Li, X. Ma, Y. Wang, Z. Zhao, C. Wang, C. Hu, P. Gao, C. Huang, H. Ren, X. Li, F. Qin, J. Yang, M. Gu, M. Hong, X. Luo, Catenary optics for achromatic generation of perfect optical angular momentum. Sci. Adv. **1**, e1500396 (2015)
18. T. Xu, C. Wang, C. Du, X. Luo, Plasmonic beam deflector. Opt. Express **16**, 4753–4759 (2008)
19. S. Franke-Arnold, L. Allen, M. Padgett, Advances in optical angular momentum. Laser Photonics Rev. **2**, 299–313 (2008)
20. L. Allen, M.W. Beijersbergen, R.J.C. Spreeuw, J.P. Woerdman, Orbital angular-momentum of light and the transformation of Laguerre-Gaussian laser modes. Phys. Rev. A **45**, 8185–8189 (1992)

21. N. Bozinovic, Y. Yue, Y. Ren, M. Tur, P. Kristensen, H. Huang, A.E. Willner, S. Ramachandran, Terabit-scale orbital angular momentum mode division multiplexing in fibers. Science **340**, 1545–1548 (2013)
22. A.E. Willner, H. Huang, Y. Yan, Y. Ren, N. Ahmed, G. Xie, C. Bao, L. Li, Y. Cao, Z. Zhao, J. Wang, M.P.J. Lavery, M. Tur, S. Ramachandran, A.F. Molisch, N. Ashrafi, S. Ashrafi, Optical communications using orbital angular momentum beams. Adv. Opt. Photon. **7**, 66–106 (2015)
23. F. Tamburini, G. Anzolin, G. Umbriaco, A. Bianchini, C. Barbieri, Overcoming the Rayleigh criterion limit with optical vortices. Phys. Rev. Lett. **97**, 163903 (2006)
24. K. Dholakia, P. Reece, M. Gu, Optical micromanipulation. Chem. Soc. Rev. **37**, 42–55 (2008)
25. M.P.J. Lavery, F.C. Speirits, S.M. Barnett, M.J. Padgett, Detection of a spinning object using light's orbital angular momentum. Science **341**, 537–540 (2013)
26. P. Genevet, N. Yu, F. Aieta, J. Lin, M.A. Kats, R. Blanchard, M.O. Scully, Z. Gaburro, F. Capasso, Ultra-thin plasmonic optical vortex plate based on phase discontinuities. Appl. Phys. Lett. **100**, 013101–013103 (2012)
27. L. Huang, X. Chen, H. Mühlenbernd, G. Li, B. Bai, Q. Tan, G. Jin, T. Zentgraf, S. Zhang, Dispersionless phase discontinuities for controlling light propagation. Nano Lett. **12**, 5750–5755 (2012)
28. M.I. Shalaev, J. Sun, A. Tsukernik, A. Pandey, K. Nikolskiy, N.M. Litchinitser, High-efficiency all-dielectric metasurfaces for ultracompact beam manipulation in transmission mode. Nano Lett. **15**, 6261–6266 (2015)
29. Y. Yang, W. Wang, P. Moitra, I.I. Kravchenko, D.P. Briggs, J. Valentine, Dielectric meta-reflectarray for broadband linear polarization conversion and optical vortex generation. Nano Lett. **14**, 1394–1399 (2014)
30. X. Wang, Z. Nie, Y. Liang, J. Wang, T. Li, B. Jia, Recent advances on optical vortex generation. Nanophotonics (2018)
31. J. Jin, J. Luo, X. Zhang, H. Gao, X. Li, M. Pu, P. Gao, Z. Zhao, X. Luo, Generation and detection of orbital angular momentum via metasurface. Sci. Rep. **6**, 24286 (2016)
32. J. Jin, M. Pu, Y. Wang, X. Li, X. Ma, J. Luo, Z. Zhao, P. Gao, X. Luo, Multi-channel vortex beam generation by simultaneous amplitude and phase modulation with two-dimensional metamaterial. Adv. Mater. Technol. **2016**, 1600201 (2016)
33. K. Yang, M. Pu, X. Li, X. Ma, J. Luo, H. Gao, X. Luo, Wavelength-selective orbital angular momentum generation based on a plasmonic metasurface. Nanoscale **8**, 12267–12271 (2016)
34. A. Niv, G. Biener, V. Kleiner, E. Hasman, Propagation-invariant vectorial Bessel beams obtained by use of quantized Pancharatnam-Berry phase optical elements. Opt. Lett. **29**, 238–240 (2004)
35. A. Niv, G. Biener, V. Kleiner, E. Hasman, Rotating vectorial vortices produced by space-variant subwavelength gratings. Opt. Lett. **30**, 2933–2935 (2005)
36. J. Sun, X. Wang, T. Xu, Z.A. Kudyshev, A.N. Cartwright, N.M. Litchinitser, Spinning light on the nanoscale. Nano Lett. **14**, 2726–2729 (2014)
37. Z.E. Bomzon, V. Kleiner, E. Hasman, Pancharatnam-Berry phase in space-variant polarization-state manipulations with subwavelength gratings. Opt. Lett. **26**, 1424–1426 (2001)
38. X. Li, X. Ma, X. Luo, Principles and applications of metasurfaces with phase modulation. Opto-Electron. Eng. **44**, 255–275 (2017)
39. S. Chen, Z. Li, Y. Zhang, H. Cheng, J. Tian, Phase manipulation of electromagnetic waves with metasurfaces and its applications in nanophotonics. Adv. Opt. Mater. **6**, 1800104 (2018)
40. X. Ma, M. Pu, X. Li, C. Huang, Y. Wang, W. Pan, B. Zhao, J. Cui, C. Wang, Z. Zhao, X. Luo, A planar chiral meta-surface for optical vortex generation and focusing. Sci. Rep. **5**, 10365 (2015)
41. H. Liu, M.Q. Mehmood, K. Huang, L. Ke, H. Ye, P. Genevet, M. Zhang, A. Danner, S.P. Yeo, C.-W. Qiu, J. Teng, Twisted focusing of optical vortices with broadband flat spiral zone plates. Adv. Opt. Mater. **2**, 1193–1198 (2014)
42. J. Zeng, L. Li, X. Yang, J. Gao, Generating and separating twisted light by gradient–rotation split-ring antenna metasurfaces. Nano Lett. **16**, 3101–3108 (2016)

43. H. Ren, X. Li, Q. Zhang, M. Gu, On-chip noninterference angular momentum multiplexing of broadband light. Science **352**, 805–809 (2016)
44. J. Lin, P. Genevet, M.A. Kats, N. Antoniou, F. Capasso, Nanostructured holograms for broadband manipulation of vector beams. Nano Lett. **13**, 4269–4274 (2013)
45. F. Yue, D. Wen, J. Xin, B.D. Gerardot, J. Li, X. Chen, Vector vortex beam generation with a single plasmonic metasurface. ACS Photonics **3**, 1558–1563 (2016)
46. Y. Li, X. Li, L. Chen, M. Pu, J. Jin, M. Hong, X. Luo, Orbital angular momentum multiplexing and demultiplexing by a single metasurface. Adv. Opt. Mater. **5**, 1600502 (2017)
47. C. Yan, X. Li, M. Pu, X. Ma, F. Zhang, P. Gao, Y. Guo, K. Liu, Z. Zhang, X. Luo, Generation of polarization-sensitive modulated optical vortices with all-dielectric metasurfaces. ACS Photonics (2019)
48. Y. Guo, M. Pu, Z. Zhao, Y. Wang, J. Jin, P. Gao, X. Li, X. Ma, X. Luo, Merging geometric phase and plasmon retardation phase in continuously shaped metasurfaces for arbitrary orbital angular momentum generation. ACS Photonics **3**, 2022–2029 (2016)
49. E. Hasman, V. Kleiner, G. Biener, A. Niv, Polarization dependent focusing lens by use of quantized Pancharatnam-Berry phase diffractive optics. Appl. Phys. Lett. **82**, 328–330 (2003)
50. G. Biener, A. Niv, V. Kleiner, E. Hasman, Formation of helical beams by use of Pancharatnam-Berry phase optical elements. Opt. Lett. **27**, 1875–1877 (2002)
51. E. Brasselet, G. Gervinskas, G. Seniutinas, S. Juodkazis, Topological shaping of light by closed-path nanoslits. Phys. Rev. Lett. **111**, 193901 (2013)
52. Y. Guo, L. Yan, W. Pan, B. Luo, Generation and manipulation of orbit angular momentum by all-dielectric metasurfaces. Plasmonics 11, 337–344 (2016)
53. F. Zhang, M. Pu, X. Li, P. Gao, X. Ma, J. Luo, H. Yu, X. Luo, All-dielectric metasurfaces for simultaneous giant circular asymmetric transmission and wavefront shaping based on asymmetric photonic spin-orbit interactions. Adv. Funct. Mater. **27**, 1704295 (2018)
54. J.P.B. Mueller, N.A. Rubin, R.C. Devlin, B. Groever, F. Capasso, Metasurface polarization optics: independent phase control of arbitrary orthogonal states of polarization. Phys. Rev. Lett. **118**, 113901 (2017)
55. M. Khorasaninejad, F. Capasso, Metalenses: versatile multifunctional photonic components. Science **358**, eaam8100 (2017)
56. F. Zhang, M. Pu, J. Luo, H. Yu, X. Luo, Symmetry breaking of photonic spin-orbit interactions in metasurfaces. Opto-Electron. Eng. **44**, 319–325 (2017)
57. N. Engheta, Circuits with light at nanoscales: optical nanocircuits inspired by metamaterials. Science **317**, 1698–1702 (2007)
58. J. Durnin, Exact solutions for nondiffracting beams. I. The scalar theory. J. Opt. Soc. Am. A **4**, 651–654 (1987)
59. C. Snoeyink, S. Wereley, Single-image far-field subdiffraction limit imaging with axicon. Opt. Lett. **38**, 625–627 (2013)
60. J. Arlt, V. Garces-Chavez, W. Sibbett, K. Dholakia, Optical micromanipulation using a Bessel light beam. Opt. Commun. **197**, 239–245 (2001)
61. D. McGloin, V. Garcés-Chávez, K. Dholakia, Interfering Bessel beams for optical micromanipulation. Opt. Lett. **28**, 657–659 (2003)
62. R.M. Herman, T.A. Wiggins, Production and uses of diffractionless beams. J. Opt. Soc. Am. A **8**, 932–942 (1991)
63. T.A. Planchon, L. Gao, D.E. Milkie, M.W. Davidson, J.A. Galbraith, C.G. Galbraith, E. Betzig, Rapid three-dimensional isotropic imaging of living cells using Bessel beam plane illumination. Nat. Meth. **8**, 417–423 (2011)
64. D. McGloin, K. Dholakia, Bessel beams: diffraction in a new light. Contemp. Phys. **46**, 15–28 (2005)
65. Y. Meng, J. Yi, S.N. Burokur, L. Kang, H. Zhang, D.H. Werner, Phase-modulation based transmitarray convergence lens for vortex wave carrying orbital angular momentum. Opt. Express **26**, 22019–22029 (2018)
66. Z. Wang, S. Dong, W. Luo, M. Jia, Z. Liang, Q. He, S. Sun, L. Zhou, High-efficiency generation of Bessel beams with transmissive metasurfaces. Appl. Phys. Lett. **112**, 191901 (2018)

67. C. Pfeiffer, A. Grbic, Metamaterial Huygens' surfaces: tailoring wave fronts with reflectionless sheets. Phys. Rev. Lett. **110**, 197401 (2013)
68. Z. Ma, S.M. Hanham, P. Albella, B. Ng, H.T. Lu, Y. Gong, S.A. Maier, M. Hong, Terahertz all-dielectric magnetic mirror metasurfaces. ACS Photonics **3**, 1010–1018 (2016)
69. F. Aieta, P. Genevet, M.A. Kats, N. Yu, R. Blanchard, Z. Gaburro, F. Capasso, Aberration-free ultrathin flat lenses and axicons at telecom wavelengths based on plasmonic metasurfaces. Nano Lett. **12**, 4932–4936 (2012)
70. D. Lin, P. Fan, E. Hasman, M.L. Brongersma, Dielectric gradient metasurface optical elements. Science **345**, 298–302 (2014)
71. W.T. Chen, M. Khorasaninejad, A.Y. Zhu, J. Oh, R.C. Devlin, A. Zaidi, F. Capasso, Generation of wavelength-independent subwavelength Bessel beams using metasurfaces. Light Sci. Appl. **6**, e16259 (2017)
72. Y. Zhu, D. Wei, Z. Kuang, Q. Wang, Y. Wang, X. Huang, Y. Zhang, M. Xiao, Broadband Variable meta-axicons based on nano-aperture arrays in a metallic film. Sci. Rep. **8**, 11591 (2018)
73. X. Li, M. Pu, Z. Zhao, X. Ma, J. Jin, Y. Wang, P. Gao, X. Luo, Catenary nanostructures as compact Bessel beam generators. Sci. Rep. **6**, 20524 (2016)
74. M.V. Berry, N.L. Balazs, Nonspreading wave packets. Am. J. Phys. **47**, 264–267 (1979)
75. G.A. Siviloglou, J. Broky, A. Dogariu, D.N. Christodoulides, Observation of accelerating Airy beams. Phys. Rev. Lett. **99**, 213901 (2007)
76. G.A. Siviloglou, D.N. Christodoulides, Accelerating finite energy Airy beams. Opt. Lett. **32**, 979–981 (2007)
77. J. Baumgartl, M. Mazilu, K. Dholakia, Optically mediated particle clearing using Airy wavepackets. Nat. Photon. **2**, 675–678 (2008)
78. P. Polynkin, M. Kolesik, J.V. Moloney, G.A. Siviloglou, D.N. Christodoulides, Curved plasma channel generation using ultraintense Airy beams. Science **324**, 229–232 (2009)
79. D. Luo, H.T. Dai, X.W. Sun, H.V. Demir, Electrically switchable finite energy Airy beams generated by a liquid crystal cell with patterned electrode. Opt. Commun. **283**, 3846–3849 (2010)
80. T. Ellenbogen, N. Voloch-Bloch, A. Ganany-Padowicz, A. Arie, Nonlinear generation and manipulation of Airy beams. Nat. Photon. **3**, 395–398 (2009)
81. E. Abramochkin, E. Razueva, Product of three Airy beams. Opt. Lett. **36**, 3732–3734 (2011)
82. L. Li, T. Li, S.M. Wang, C. Zhang, S.N. Zhu, Plasmonic Airy beam generated by in-plane diffraction. Phys. Rev. Lett. **107**, 126804 (2011)
83. A. Minovich, A.E. Klein, N. Janunts, T. Pertsch, D.N. Neshev, Y.S. Kivshar, Generation and near-field imaging of airy surface plasmons. Phys. Rev. Lett. **107**, 116802 (2011)
84. P. Zhang, S. Wang, Y. Liu, X. Yin, C. Lu, Z. Chen, X. Zhang, Plasmonic Airy beams with dynamically controlled trajectories. Opt. Lett. **36**, 3191–3193 (2011)
85. A.E. Minovich, A.E. Klein, D.N. Neshev, T. Pertsch, Y.S. Kivshar, D.N. Christodoulides, Airy plasmons: non-diffracting optical surface waves. Laser Photonics Rev. **8**, 221–232 (2014)
86. A. Salandrino, D.N. Christodoulides, Airy plasmon: a nondiffracting surface wave. Opt. Lett. **35**, 2082–2084 (2010)
87. Z. Li, H. Cheng, Z. Liu, S. Chen, J. Tian, Plasmonic Airy beam generation by both phase and amplitude modulation with metasurfaces. Adv. Opt. Mater. **4**, 1230–1235 (2016)
88. Q. Fan, D. Wang, P. Huo, Z. Zhang, Y. Liang, T. Xu, Autofocusing Airy beams generated by all-dielectric metasurface for visible light. Opt. Express **25**, 9285–9294 (2017)
89. E.-Y. Song, G.-Y. Lee, H. Park, K. Lee, J. Kim, J. Hong, H. Kim, B. Lee, Compact generation of Airy beams with C-aperture metasurface. Adv. Opt. Mater. **5**, 1601028 (2017)

Chapter 11
Structural Colors and Meta-holographic Display

Abstract As a picture is worth a thousand words, display technology simplifies information sharing. In this chapter, the developments of structural colors based on various subwavelength structures are introduced, with some special attentions paid on the polarization-encoded and dynamic structural colors. Then, the meta-holography with polarization-independent/dependent, full-color, three-dimensional, high-efficiency, and broadband properties is discussed.

Keywords Structural colors · Meta-hologram · Meta-holography · Polarization encryption · Chiral holography

11.1 Introduction

Perception of color with our eyes is one of the major sources of information that we gain from our surroundings. The pioneering experiments on visible light and colors can be traced back to the seventeenth century, when Sir Isaac Newton observed different transmitted colored bands when a narrow beam of sunlight struck the face of a glass prism at an angle (Fig. 11.1a, b) [1]. This early experiment implies that a desirable color can be obtained by spectrally filtering the white light. Indeed, most of the modern color displays rely on spectrum filters to produce primary colors [e.g., red, green, and blue (RGB)]. Traditionally, such spectrum filters are based on colorant pigments (Fig. 11.1c, d) [2] that absorb light in certain wavelength range, which finds dominant usage in many display systems such as liquid crystal displays (LCDs). On the other hand, color can also be created by light interactions with certain nanostructures that strongly influence the light propagation through the structures. In fact, this is frequently observed in nature, for example, the wings of the morpho butterfly (a special southern American butterfly family) shine a magnificent blue color, which was found to be due to surface-relief volume-diffractive nanostructures on its wings (Fig. 11.1e, f) [3]. Structural colors are resistant to photobleaching and possess rainbow effect due to their complex viewing direction-dependent optical properties. Although structural colors have been mimicked by a couple of artificial structures [4–6] including thin-layer interference, diffraction grating, light scattering, photonic crystals, and so on, there are still some problems should be solved before

X. Luo, *Engineering Optics 2.0*, https://doi.org/10.1007/978-981-13-5755-8_11

Fig. 11.1 **a** Portrait of Sir Isaac Newton. **b** Illustration of a dispersive prism separating white light into the colors of the spectrum, as discovered by Newton. **c** Pigments for sale at a market stall in Goa, India. **d** Dyeing wool cloth, 1482: from a French translation of Bartholomeus Anglicus. **e** Morpho butterfly and **f** the scanning electron microscope (SEM) image of the nanostructures on its wings. Scale bar: 1 μm. **a–d** Reproduced from Wikipedia [1, 2] with permission. **e, f** Reproduced from [3] with permission. Copyright 1990, Optical Society of America

they can find widespread applications, e.g., the large pixel size, low color purity, and low tunability.

Holographic displays generate realistic three-dimensional (3D) images that can be viewed without the need for any visual aids. They operate by generating carefully tailored light fields that replicate how humans see an actual environment. Holography is a technique using the interference pattern between the scattered light of an object and a coherent reference beam to reconstruct the 3D features of the object. Owing to its potential in the cutting-edge optical technologies, holography has attracted continuous scientific interest. Traditional holography has got great achievements, and a lot of holographic productions can be found in our daily life.

11.2 Structural Colors

Coloring potential of metallic particles has been known for centuries and used for millennia, from the Roman Lycurgus cup in fourth century [7] (Fig. 11.2) to the coloration of ancient church windows. Subsequently investigations on the structural color generated from metallic structures enabled the discovery of plasmonics. For example, Michael Faraday's research in the 1850s performed pioneering, systematic studies on the optical properties of gold (Au) leaf that was beaten so thin that rendered translucent, transmitting the green and reflecting the yellow part of incident sunlight. He also synthesized and investigated Au colloidal particles and examined their distinct ruby red colors [8].

With the development of nanofabrication and characterization techniques, surface plasmons and related plasmonic nanostructures have attracted considerable interest in recent years [9]. One can exploit the geometries of the plasmonic nanostructures to manipulate light properties, including visible light wavelength selection and thereby the generation of structural color. Compared with the color produced by ordinary electronic absorption determined by the material's specific molecular spectrum, plasmon-based structural color is primarily determined by the geometry parameters of the plasmonic nanostructures. This characteristic brings in several advantages for the plasmon-based structural color [9]: (1) Different colors can be obtained by using the same material but only with different structural parameters, which could greatly simplify the manufacturing process for multiple colored units; (2) unlike traditional chemical pigment and dye that are affected by photobleaching, the plasmonic nanostructures are chemically stable, and therefore, the structural color is attractive for applications requiring high light intensities or continuous illuminations; (3) compared with other dielectric periodic structures, the highly confined surface plasmons in metallic nanostructures make the device dimension more compact.

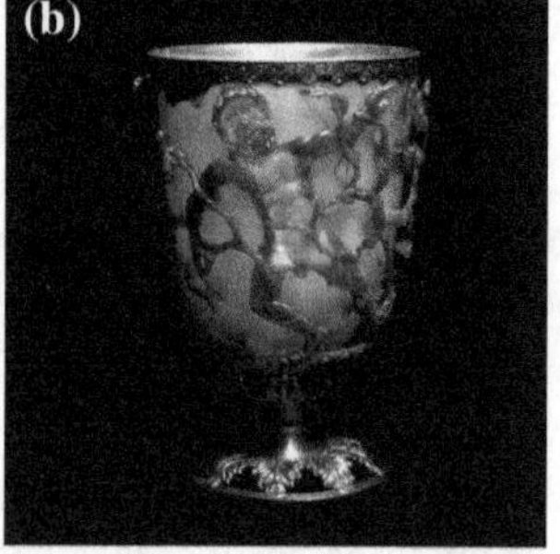

Fig. 11.2 Lycurgus cup showing different colors in reflected (**a**) and transmitted light (**b**). Reproduced from [7] with permission. Copyright 2004, Elsevier Ltd.

11.2.1 Transmissive Structural Colors

(1) Structural colors based on perforated metallic film

Extraordinary optical transmission of light through tiny apertures in optical metallic film [10] draws the attention of the optical community. Perforated metal film generates colors in transmission mode wherein the color generation (filtering) happens owing to the interference of surface plasmon polaritons (SPPs) between neighboring holes. For the subwavelength hole arrays with square lattice, the transmission peak wavelength in the normal incidence can be approximately calculated from the surface plasmon dispersion relation as

$$\lambda_{\max} \cong \frac{P}{\sqrt{i^2 + j^2}} \sqrt{\frac{\varepsilon_{\mathrm{m}} \varepsilon_{\mathrm{d}}}{\varepsilon_{\mathrm{m}} + \varepsilon_{\mathrm{d}}}}, \tag{11.2.1}$$

where P is the square lattice constant; the indices i and j are the scattering orders from the hole arrays; ε_{m} and ε_{d} are the permittivity for the metal and dielectric medium, respectively. From Eq. (11.2.1), one can see that the transmission peak wavelength $\lambda_{\max}$ is proportional to lattice constant of the hole arrays. Therefore, the subwavelength hole arrays in metal film can act as optical filters for which the transmitted color can be selected by simply adjusting the lattice.

Figure 11.3a, b show the lowest order transmission peak ($i, j = 1, 0$ in Eq. 11.2.1) of the perforated square lattice hole arrays on a free-standing 300-nm-thick silver film [11]. When the lattice constant of hole arrays increases from 300 to 550 nm, the transmission peak wavelength changes from 436 to 627 nm, covering primary RGB

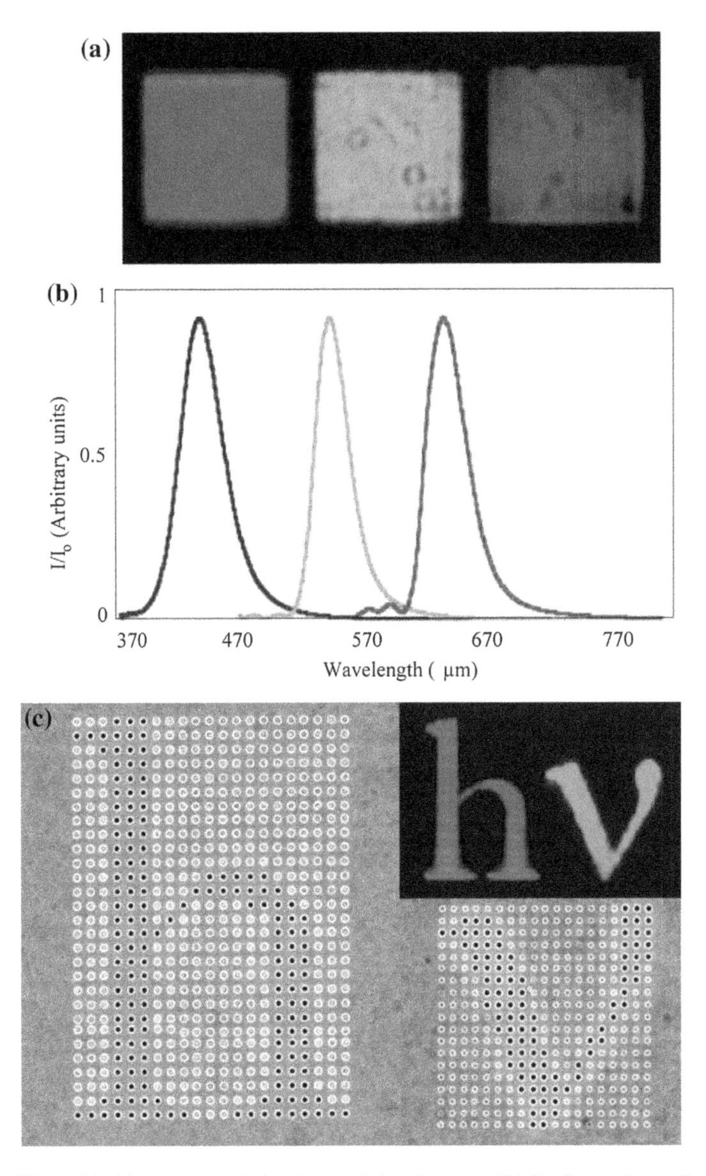

Fig. 11.3 Normal incidence transmission images (**a**) and spectra (**b**) for three subwavelength hole arrays. **c** Holes in an Ag dimple array generating the colored letters '*hv*' in transmission. Some of the dimples are milled through to the other sides so that light can be transmitted. The lattice constants were 550 and 450 nm to achieve the red and green colors. Reproduced from **a**, **b** [11] and **c** [12] with permissions. **a**, **b**. Copyright 2003, Nature Publishing Group. **c** Copyright 2007, Nature Publishing Group

colors in the visible spectrum. In addition, by arranging the perforated and unperforated subwavelength hole arrays in the metal film, arbitrary polarization-independent colored patterns can be realized [12], as indicated in Fig. 11.3c. Similarly, triangular lattice hole arrays have also been utilized to generate structural color [13]. Since triangular lattice subwavelength hole arrays have a larger wavelength interval between the first two transmission peaks than the square lattice hole arrays with the same lattice constant, lower cross-talk and high-purity color can be realized. Nevertheless, the transmission passbands of such filters are relatively broad and do not satisfy the requirement for the multi-band spectral imaging.

Further investigation shows that directional propagation is possible by surrounding a subwavelength aperture with periodic corrugations, the well-known beaming effect. The transmission peak wavelength λ_{SPP} of the aperture can be tuned by controlling the groove periodicity P as predicted by Eq. (11.2.2) for normal incidence illumination [14, 15].

$$\lambda_{\max} = P\sqrt{\frac{\varepsilon_m \varepsilon_d}{\varepsilon_m + \varepsilon_d}}. \tag{11.2.2}$$

(2) Structural colors based on metallic gratings

Diffraction grating is one of the primitive classes of man-made structural color devices. The grating as a periodic structure can split the white light into different colors like a dispersive medium. In order to realize localized structural color, metal–insulator–metal (MIM) grating structures with stronger field confinement were utilized [16], as shown in Fig. 11.4.

The basic principle of this device is to use MIM resonator arrays to realize the photon–plasmon–photon conversion efficiently at specific resonant wavelengths. For the transverse magnetic (TM)-polarized waves, arbitrary color in the visible spectrum can be filtered with this structure by selecting the proper stack array period, due to the linear dispersion of plasmon at specific resonant wavelengths (Fig. 11.4b) [16]. Compared with the aforementioned color filtering methods, this new design significantly improves the absolute transmission, bandwidth, and compactness. In addition, the filtered light is naturally polarized, making it attractive for direct integration in LCDs without a separate polarizer layer.

One can use different nanoresonator arrays to form arbitrary colored patterns on a micrometer scale. As a demonstration, a yellow character '*M*' in a navy blue background is shown in Fig. 11.5a [16]. The pattern size of the '*M*' logo measures only 20 μm × 12 μm and the pattern uses two periods: 310 nm for the yellow letter '*M*' and 220 nm for the navy blue background. The optical microscopy image of the pattern illuminated with the white light is shown in Fig. 11.5b [16]. A clear-cut yellow '*M*' sharply contrasts the navy blue background. It is important to note that the two distinct colors are well preserved even at the sharp corners and boundaries of the two different patterns, which implies that such color filter scheme can be extended to ultra-high-resolution color displays [16].

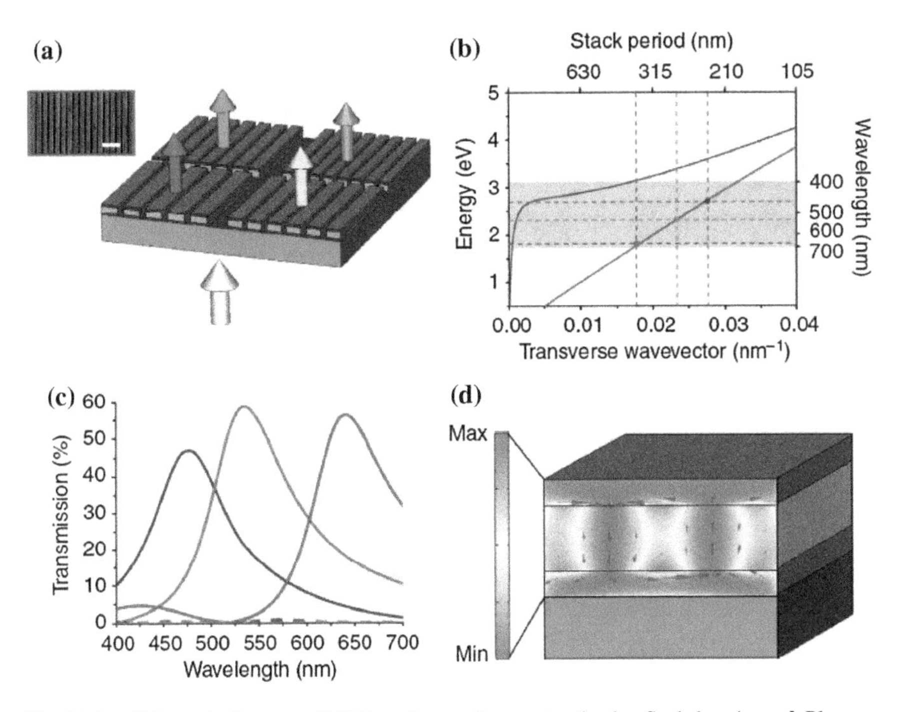

Fig. 11.4 **a** Schematic diagram of MIM stack array for structural color. Scale bar, 1 μm. **b** Plasmon dispersions in MIM stack array. Red, green, and blue dots correspond to the case of filtering primary RGB colors. Red and blue curves correspond to antisymmetric and symmetric modes, respectively. The shaded region indicates the visible range. **c** Simulated transmission spectra for the RGB filters. **d** Cross section of the time-averaged magnetic field intensity and electric displacement distribution (red arrow) inside the MIM stack. Reproduced from [16] with permission. Copyright 2010, Macmillan Publishers Limited

By gradually changing the periods of the plasmonic nanoresonator array, one can design and demonstrate a plasmonic spectroscope for spectral imaging. Figure 11.5c shows the fabricated device consisting of gradually changing periods from 200 to 400 nm that covers all colors in the visible range. When illuminated with white light, the structure becomes a rainbow stripe, with light emitting from the stack array, as shown in Fig. 11.5d [16]. Plasmonic spectroscopes can disperse the whole visible spectrum in just a few micrometer distances, which are orders of magnitude smaller than the dispersion of the conventional prism-based device. This feature indicates that the color pixels formed by these structures could provide extremely high spatial resolution for application in multi-band spectral imaging systems. Such thin-film stack structures can be directly integrated on top of focal plane arrays to implement high-resolution spectral imaging or to create chip-based ultra-compact spectrometers [16].

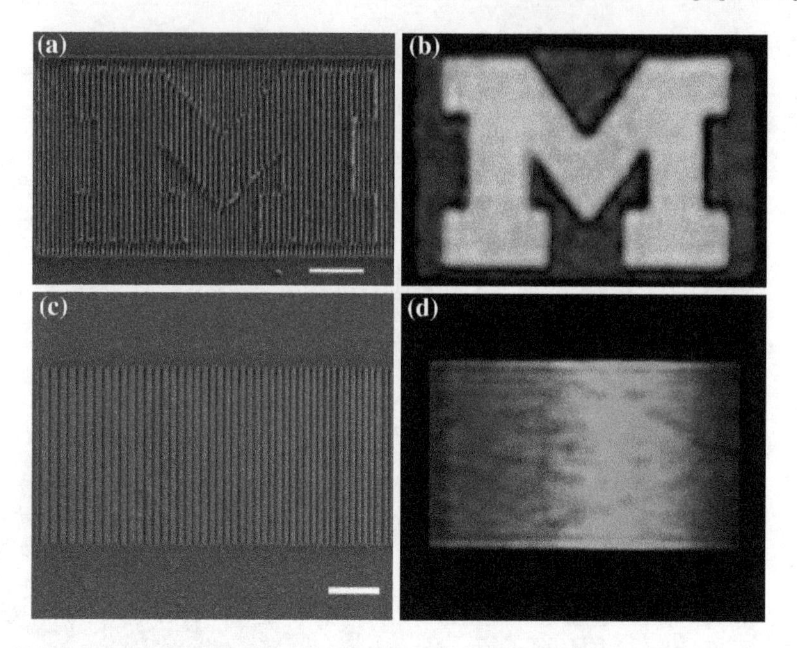

Fig. 11.5 **a** SEM image of the pattern '*M*' formed by two stack periods. Scale bar, 2 μm. **b** Optical microscopy image of the pattern illuminated with white light. **c** SEM image of the fabricated 1D plasmonic spectroscope with gradually changing periods from 400 to 200 nm (from left to right). Scale bar, 2 μm. **d** Optical microscopy image of the plasmonic spectroscope illuminated with white light. Reproduced from [16] with permission. Copyright 2010, Macmillan Publishers Limited

11.2.2 Reflective Structural Colors

(1) Structural colors based on Fabry–Perot cavity resonance

A typical configuration of reflective structural color filter is composed of metal-dielectric multilayer thin films based on Fabry–Perot (F-P) cavity resonances. F-P cavity-based reflective color filters consisting of a layer of thin nickel (Ni) film and a thick aluminum (Al) film, separated by a silicon dioxide (SiO_2) dielectric layer, are illustrated in Fig. 11.6a [17]. A representative cross-sectional SEM image of such a configuration is shown by Fig. 11.6b, where the thickness of top Ni metallic layer t and Al back-reflector h are, respectively, set to be 6 and 100 nm. SiO_2 is used as the spacer layer, and its optical thickness d determines the position of the absorption and reflectance peaks, as indicated in Fig. 11.6c at normal incidence [17]. The corresponding photographs of the fabricated samples ($d = 120$, 170, 220, and 270 nm) are displayed in Fig. 11.6d (top row), which show four different large-area color filters with high saturation and brightness including blue, green, yellow, and purple, demonstrating the broad color coverage of this F-P type configuration. While, for the reference samples without Ni layer (bottom row), all of them have high reflectivity in the entire visible range [17].

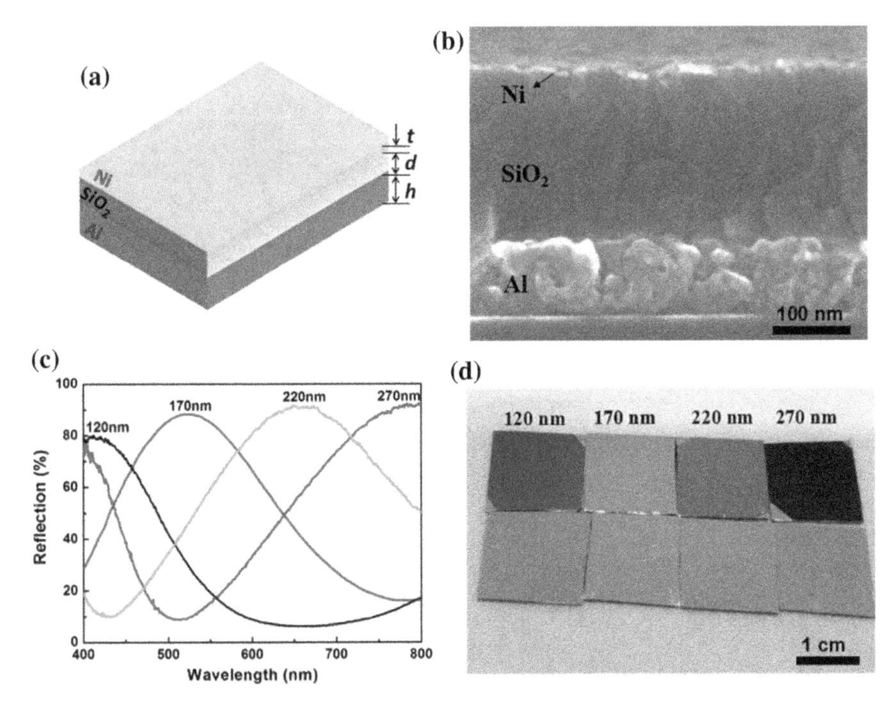

Fig. 11.6 **a** Schematic of a planar Ni/SiO_2/Al cavity. **b** The cross section SEM image of a representative Ni/SiO_2/Al tri-layer absorber on a Si substrate. **c** The measured reflection spectra for the F-P cavities with SiO_2 spacer thickness of d = 120, 170, 220, and 270 nm at normal incidence with unpolarized light. **d** The corresponding photograph of the fabricated devices with (top row) and without (bottom row) Ni layer. Reproduced from [17] with permission. Copyright 2016, WILEY-VCH Verlag GmbH & Co. KGaA, Weinheim

The presented method can be used in color printing; e.g., a large reddish rose and a large blue lotus with high brightness and saturation are shown in Fig. 11.7a, b [17]. Remarkably, the distinct color of the two images was solely attributed to the different thickness of the SiO_2 films. The SiO_2 thickness was 240 nm for the rose and 120 nm for the lotus. By resorting to high-resolution fabrication technology, e.g, electron-beam lithography (EBL), more complex color patterns can be realized, as illustrated in Fig. 11.7c, d [17].

(2) Structural colors based on magnetic dipole resonance

Figure 11.8a illustrates a schematic configuration of reflective structural color filters [18]. Each of the filters is based on a metasurface incorporating a c-Si nanopillar (NP) array that is integrated with an Al disk mirror (DM) at the top and an Al holey mirror (HM) at the bottom. The normal incidence may resonantly couple to the NPs in association with the Si metasurface, thus introducing a strong wavelength-selective suppression in the reflection spectra [18]. Such a reflection dip can be scanned across the entire visible spectral regime through the control of the NP diameter D, thus

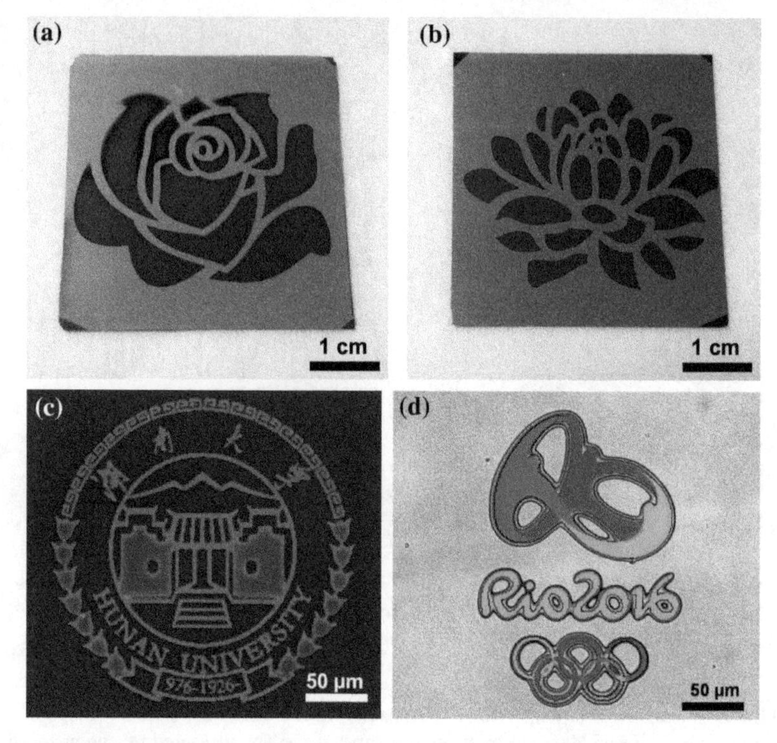

Fig. 11.7 Demonstration of colorful patterns fabricated by Ni/SiO_2/Al films on Si substrates with a different SiO_2 thickness: **a** reddish rose with 240 nm SiO_2 spacer, **b** blue lotus with 120 nm SiO_2 spacer, **c** a red microscale Logo of Hunan University, and **d** the full-color Logo of 2016 Rio de Janeiro Olympic Games. Reproduced from [17] with permission. Copyright 2016, WILEY-VCH Verlag GmbH & Co. KGaA, Weinheim

providing a vivid color output. Figure 11.8b shows the SEM images of the completed CMY color filters, having NP diameters of $D = 90$, 115, and 140 nm for a period of $P = 240$ nm. The bright-field microscope images of generated colors of yellow, magenta, and cyan, with dimensions of 30 μm × 30 μm are included in the inset of Fig. 11.8a [18].

11.2.3 Polarization-Encoded Structural Colors

The above designs of color filters are all based on the direct transmission or reflection of light wave through subwavelength periodic structures. EBL and focused ion beam (FIB) milling offer the platform to fabricate such nanostructures with small separations; however, these methods are expensive and inefficient [19]. Recently, interference lithography, a cost-effective approach to manufacture large-area mass-

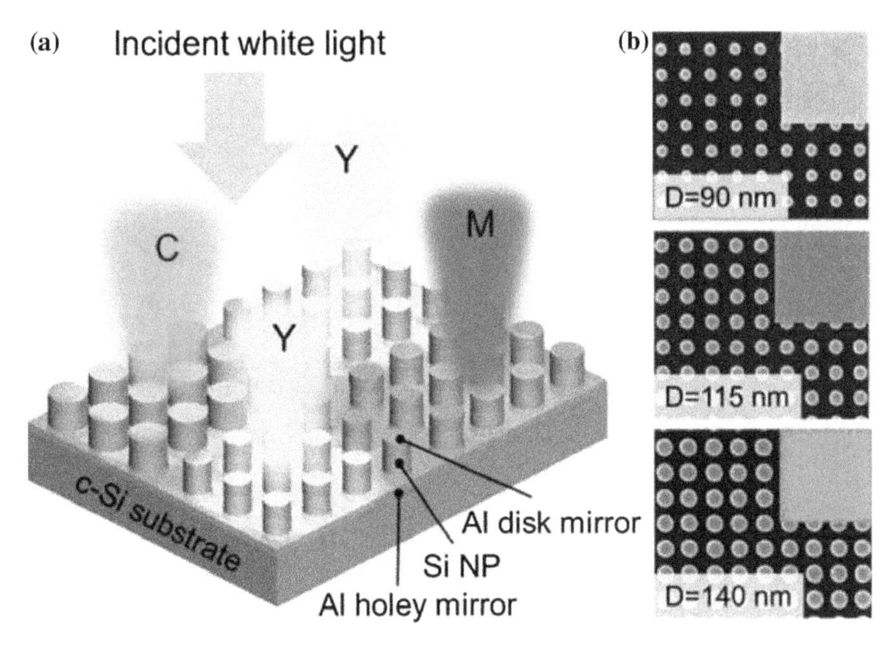

Fig. 11.8 **a** Configuration of subtractive color filters incorporating a metasurface that is comprised of an array of c-Si NPs. **b** SEM images of the completed filters of CMY colors for which the NP diameter is chosen to be $D = 140$, 115, and 90 nm for a constant period of $P = 240$ nm, respectively. The insets include the individual bright-field microscope images relating to the prepared devices with a footprint of 30 μm × 30 μm. Reproduced from [18] with permission. Copyright 2017, WILEY-VCH Verlag GmbH & Co. KGaA, Weinheim

production periodic nanostructures, is utilized to fabricate ultra-smooth silver grating for reflective structural color. Different from the principle of previous designs, such plasmonic shallow grating produces colors by photon spin restoration (Fig. 11.9a), which reflects a circularly polarized (CP) light to its co-polarized state at specific wavelengths (it is well known that the spin direction would be reversed when a CP light is reflected from a common mirror) [19]. Figure 11.9b depicts the experimentally measured reflective cross-polarization spectra. Three sharp peaks are observed at 627 nm (red), 525 nm (green), and 430 nm (blue), whereas the corresponding periods of plasmonic shallow gratings (PSGs) are 590, 470, and 300 nm, respectively. The reflective efficiency peaks reach up to 75%, with the smallest full width at half magnitude (FWHM) to be only ~16 nm. Furthermore, the magnitudes at off-resonant wavelengths are strongly suppressed, both of which help to improve the purity of the generated colors [19].

Figure 11.10a depicts the simulated and measured reflective peak wavelengths with respect to the grating period. Figure 11.10b reveals the experimentally measured optical images of the colors yielded by the PSGs with periods of 300, 390, 410, 470, 510, and 590 nm, respectively [19]. The corresponding chromatic coordinates are calculated and plotted in Commission Internationale de l'Eclairage (CIE)

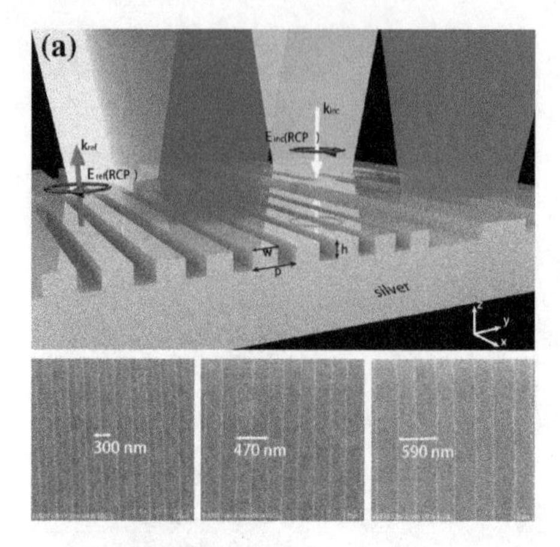

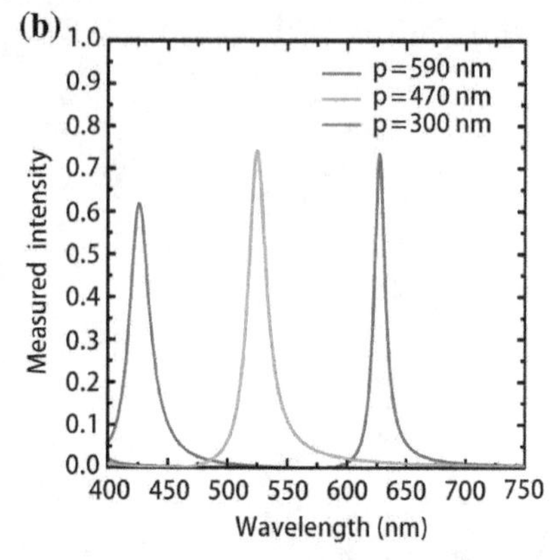

Fig. 11.9 **a** Schematic diagram of the light–structure interactions. The bottom row is SEM images of the prepared filters with periods of 300, 470, and 590 nm. **b** Experimentally measured reflection spectra of the cross-polarized scattered light for 45° polarized LP at normal incidence. Reproduced from [19] with permission. Copyright 2018, The Authors

1931 chromaticity diagram, as shown in Fig. 11.10c, d, which indicate the region of monochromatic colors. The chromatic coordinates for experimentally obtained spectra reasonably agree with their simulation counterparts [19].

Besides the realization of high-purity color filters, the feasibility of the high-resolution display of arbitrary chromatic patterns is investigated. As an example, three red characters "IOE" in a blue background are experimentally demonstrated, as shown in Fig. 11.11 [19]. The logo fabricated by the nested interference lithography spans 30 × 30 mm, with a group of silver grooves exhibiting two periods: 600 nm for the red characters "IOE" and 400 nm for the blue background. Figure 11.11a, b show the SEM images of the different periodic building blocks. The reflective cross-polarization photographs taken for incident light polarized at 45° are shown in Fig. 11.11c [19]. It can be seen that a bright red "IOE" sharply contrasts the blue background, and the two distinct colors are still observed even at the corners and edges of the logo. Furthermore, it is shown that when the linear polarizer in front of the CCD camera is rotated at −20° with respect to the grooves, a dim red acronym "IOE" and green background emerge as depicted in Fig. 11.11d [19]. Figure 11.11e indicates that the sample looks white when the reflective light and incident light possess the same polarization state. These fascinating phenomena offer platforms for security or encryption applications, which can be only unscrambled by a couple of specific incident polarization and/or reflective helicity [19].

Figure 11.12 presents a schematic illustration of polarization-encoded structural colors [20], where spatially rotating meta-atoms function as an ultra-compact half-wave plates. For incidence with horizontal polarization, they will rotate the polarization by 2θ (θ is the angle between the long axis and the horizontal direction, as shown in Fig. 11.12a) [20]. As a consequence, the metasurface can modulate an inci-

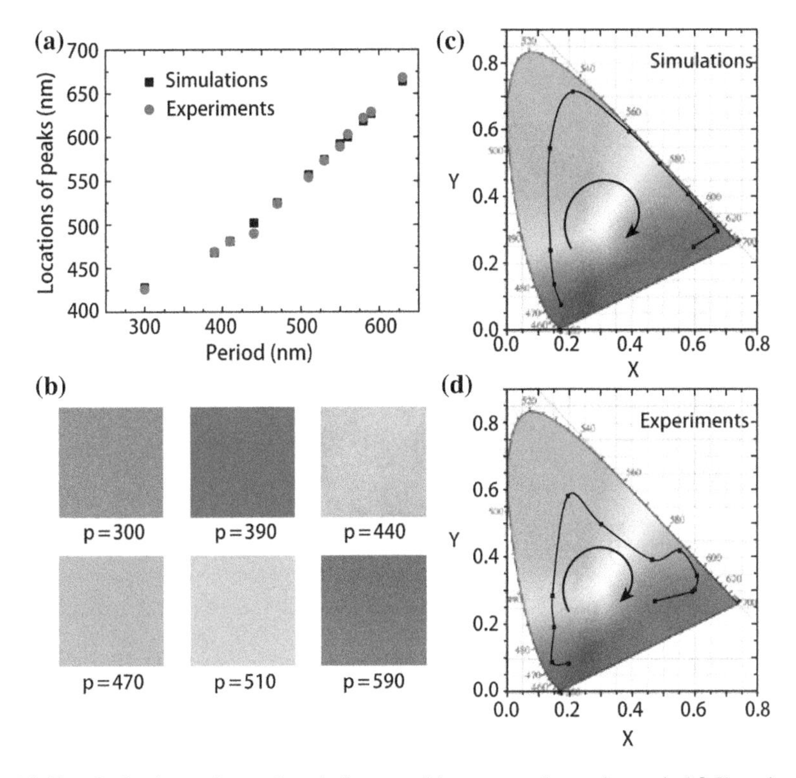

Fig. 11.10 **a** Reflective peak wavelength changes with respect to the grating period. **b** Experimental images of the colors captured by the CCD camera. A broad palette of color with high contrasts is realized. **c**, **d** Chromatic coordinates in the CIE 1931 diagram obtained from the simulated and measured spectra. Reproduced from [19] with permission. Copyright 2018, The Authors

dent light beam with uniform linear polarization into a vector beam with spatially varying polarization states. When the output light passes through an analyzer (linear polarizer) with a transmission axis along the vertical direction, the spatial polarization information will be transformed into spatial intensity information, forming the image pattern. When different meta-atoms that operate at different wavelengths are utilized (Fig. 11.12b), each atom will exhibit color and the brightness is determined by the intensity of transmitted light [20].

Following this principle, a high-resolution color image of rose is selected as a target image for polarization encoding (Fig. 11.12c) [20]. Figure 11.12d shows a selected area from the rose with 6×5 pixels, which contains both red and green pixels with spatial brightness variation. The corresponding polarization profile of a light beam composed of two different colors is shown in Fig. 11.12e [20]. The desired polarization profiles for different colors are realized by controlling the orientation angles and feature sizes of the nanoblocks (Fig. 11.12f). The corresponding SEM image of part of the fabricated sample is shown in Fig. 11.12g [20].

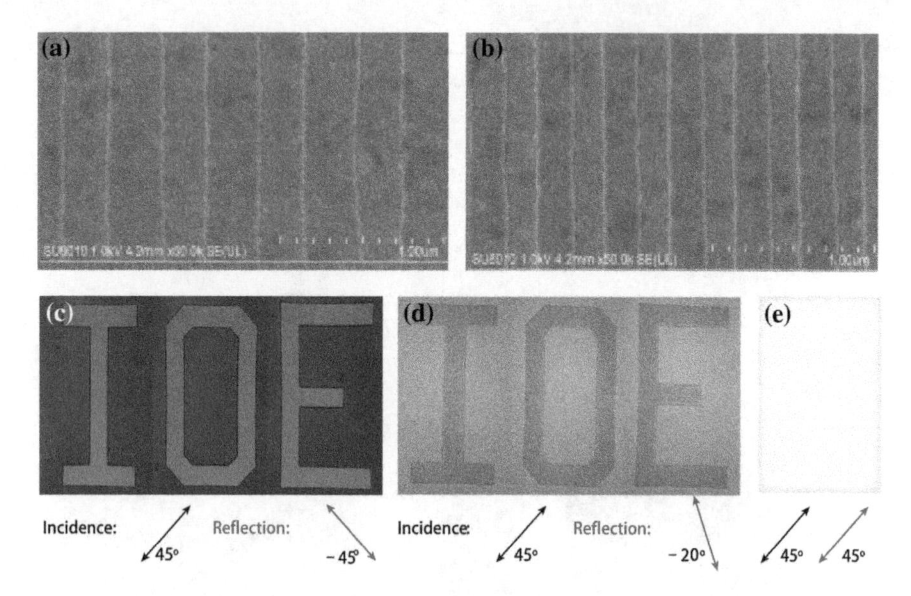

Fig. 11.11 **a** SEM images of the building blocks for the logo with period of **a** 600 nm and **b** 400 nm. **c–e** Microscopic images of the sample for 45° polarized incident light. The polarization direction of the reflection light is **c** −45°, **d** −20°, and **e** 45°. Reproduced from [19] with permission. Copyright 2018, The Authors

The experimental results are shown in Fig. 11.13, from which one can see by adding a polarization analyzer before the detector, the concealed images can be revealed, indicating the great potential of this technique for encryption application [20]. Furthermore, by merging different meta-atoms in a super-cell and adjusting their orientations, color mixing can be realized as illustrated. Consequently, a polarization-encoded full-color-like images without and with the polarization analyzer were realized [20].

11.2.4 Dynamic Structural Colors

Dynamic color tuning is a very important and fascinating direction in the field of structural colorations due to its possible applications in stealth, anti-counterfeiting, displaying techniques, etc. So far, dynamic structural colors have been realized based on different methods.

(1) Dynamic colors based on tensile substrate

Recently, a tensile substrate (e.g., polydimethylsiloxane, PDMS) was introduced into conventional plasmonic structures to demonstrate dynamic tunable structural colors [21]. The proposed idea is experimentally realized by fabricating samples via

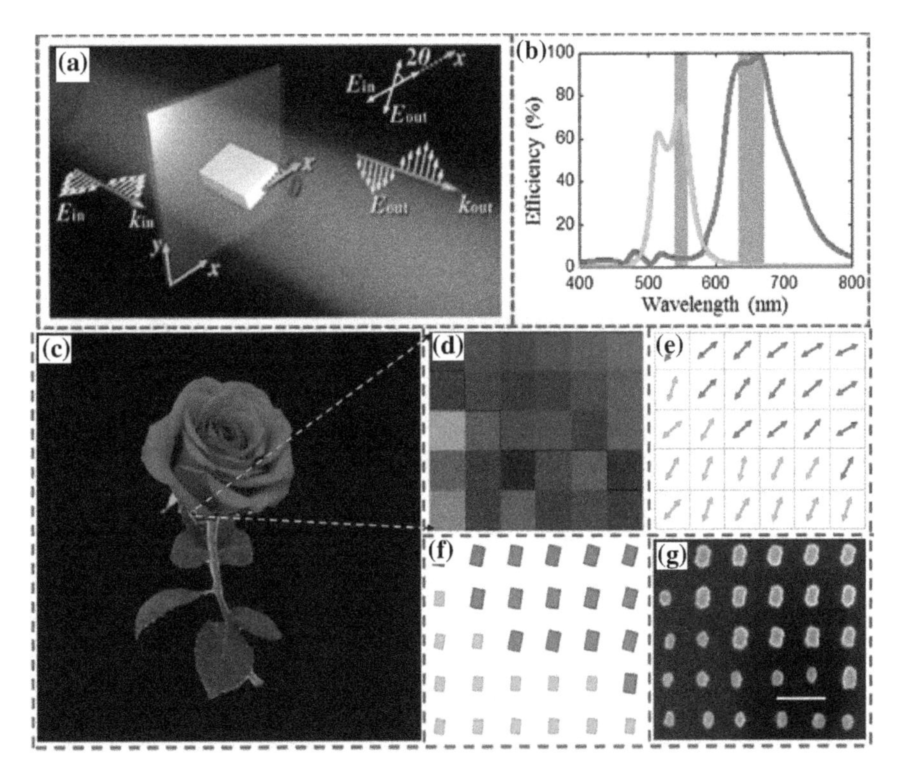

Fig. 11.12 **a** Schematic of the polarization rotating based on a single silicon nanoblock. **b** The calculated conversion efficiency for nanoblocks. **c** The target image of the red rose with green leaves. **d** An selected area which contains both red and green pixels with different gray levels. **e** The desired polarization distribution for encoding the detail of the selected area. **f** The corresponding nanoblock distributions of metasurfaces. **g** The SEM image of the fabricated sample. The scale bar is 500 nm. Reproduced from [20] with permission. Copyright 2018, WILEY-VCH Verlag GmbH & Co. KGaA, Weinheim

Fig. 11.13 Measured results without (**a**, **c**) and with (**b**, **d**) the required analyzer. Reproduced from [20] with permission. Copyright 2018, WILEY-VCH Verlag GmbH & Co. KGaA, Weinheim

interference photolithography, which was also subsequently demonstrated with EBL [22].

As shown in Fig. 11.14, the resonant wavelength shifts from 530 to 620 nm, indicating that the grating period changes under the external force. The SEM images show that when the PDMS is at the natural status ("0 mm"), the diameter of the nanoparticle and period are 151 and 322 nm, respectively. When the stretched length deviates from the natural status, the deformation of PDMS occurs. Interestingly, the resonant wavelengths increase nearly linearly with raising the stretched lengths [21]. As indicated in Fig. 11.15a, great CIE chromaticity gamut variations from green to fuchsia are presented. Although the CIE chromaticity could present the variations of the color, it cannot demonstrate the luminance of the spectra; i.e., it cannot represent the real colors as we see in the natural world. In order to have a better understanding of the actual colors, the corresponding colors of the spectra under D65 illumination (north daylight with sunshine) are depicted in Fig. 11.15b [21]. The captured colors via CCD camera are also exhibited as comparison. The hues of the captured colors via CCD camera under xenon lamp are almost the same with the colors calculated under D65 illumination. As shown in Fig. 11.15c, the photographs of the device with different stretched lengths are taken under the xenon lamp. The colors with different stretched lengths vary from green to fuchsia apparently [21].

(2) Dynamic colors based on electrochromic materials

Alternative method for dynamic and tunable structural color is introducing active materials into the plasmonic structures. Among them, electrochromic materials are particularly attractive for flexible display applications because of their vibrant col-

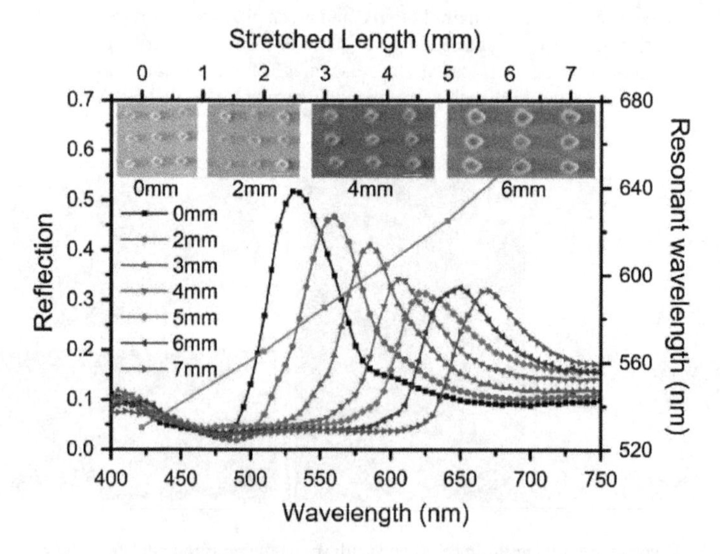

Fig. 11.14 Measured reflective spectra with different stretched lengths. The insets are the SEM images with different stretched lengths. Each scale bar is 200 nm. Reproduced from [21] with permission. Copyright 2017, WILEY-VCH Verlag GmbH & Co. KGaA, Weinheim

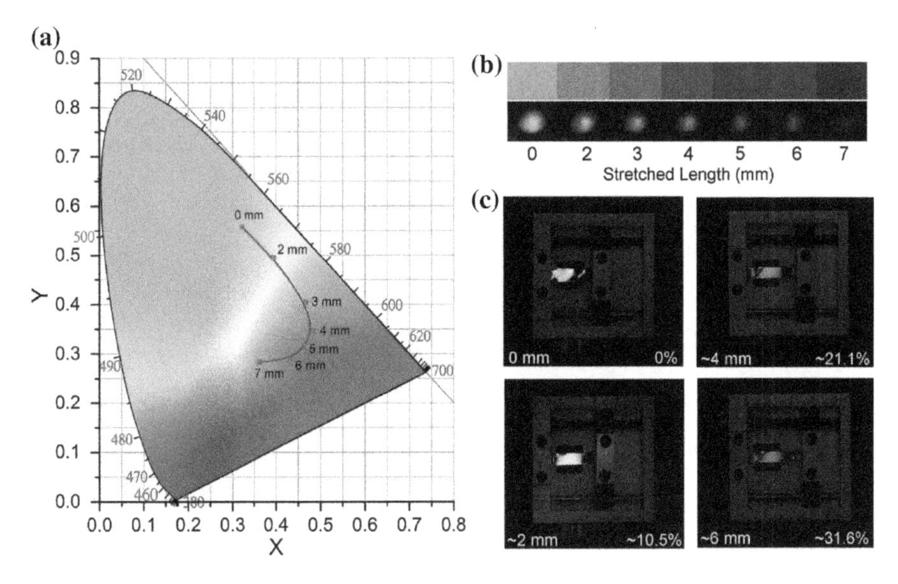

Fig. 11.15 **a** Calculated CIE chromaticity diagram of the reflective spectra. **b** The comparisons of measured colors captured via CCD camera under the illumination of xenon lamp and the calculated colors under D65 illuminated light. **c** The photographs of the structural colors with different stretched lengths under the xenon lamp. Reproduced from [21] with permission. Copyright 2017, WILEY-VCH Verlag GmbH & Co. KGaA, Weinheim

ors, low-cost, and relatively simple processing requirements. By incorporating electrochromic materials, such as polyaniline (PANI) and poly (2,2-dimethyl-3,4 propylenedioxythiophene) (PolyProDOT-Me_2), into plasmonic nanoslit arrays, Xu et al. [23] demonstrated a high-contrast, fast-monochromatic, and full-color switching. As shown in Fig. 11.16, the slits are covered sidewalls by the PANI and the whole structure is immersed into electrochemical cells (Fig. 11.16a). Applying various voltages to the cells causes the switching of the polymers between the reduced and oxidized states, causing changes of the refractive index of the environment surrounding the slits (Fig. 11.16b) [23].

Figure 11.17 shows the experimentally measured optical transmission spectra of the fabricated Al-nanoslit electrodes, along with corresponding optical micrographs, for both transmitting ON (applied voltage $V_{\mathrm{ON}} = 0.2$ V vs. Ag wire) and absorbing OFF (applied voltage $V_{\mathrm{OFF}} = 0.6$ V vs. Ag wire) states of the polymer [23]. Distinct with the uniformly dark colors exhibited in the OFF state, the Al-nanoslit electrodes in the ON state show various period-dependent colors covering the entire visible spectrum. The experimentally measured absolute transmission at filtered wavelengths in the ON state ranges from 13% to 18%, corresponding to a switching contrast ranging from 73% to 90% [23].

(3) Dynamic colors based on photonic doping

In order to generate dynamic structural color, photonic doping is proposed recently, in analogy to electrical/chemical doping, which is triggered by the interplay of structural

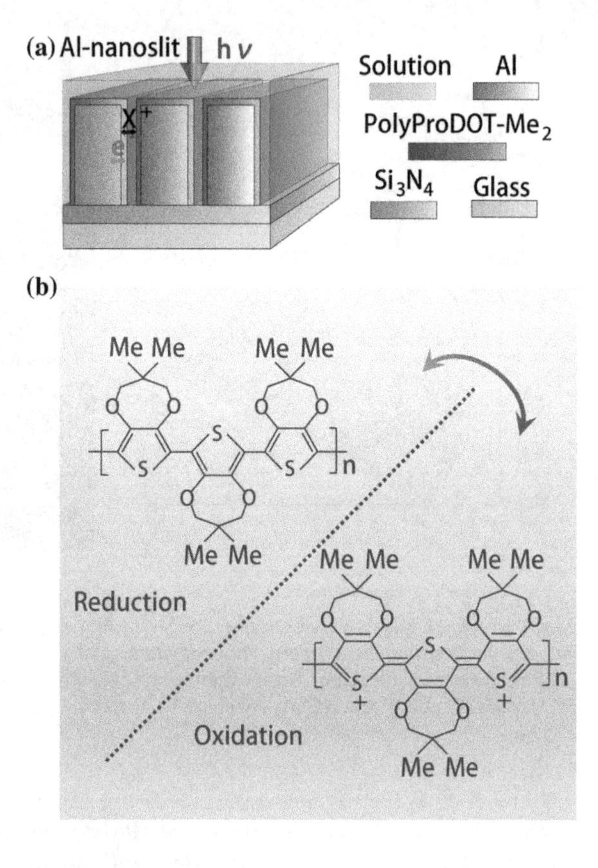

Fig. 11.16 **a** Schematic diagram of a plasmonic electrochromic electrode incorporating Al-nanoslit array. **b** Chemical structures of PolyProDOT-Me$_2$ in the oxidized and reduced form. Reproduced from [23] with permission. Copyright 2016, The Authors

colors and photon emission of methylammonium lead halide perovskite ($MAPbX_3$, where MA = $CH_3NH_3^+$ and X = Cl, Br, I, or their mixture) gratings [24]. The working principle is illustrated in Fig. 11.18, where both white light and laser light are incident on the sample simultaneously. On the one hand, the $MAPbX_3$ can generate high reflection. On the other hand, as a direct bandgap semiconductor, $MAPbBr_3$ can form emission colors with intensities determined by the excitation [24]. Consequently, a promising dynamical color tuning scheme could be realized by mixing the extrinsic structural color with intrinsic emission color, where the former color acts as the base and the latter color functions as a photonic impurity. Significantly, with the pumping density-dependent color, this scheme could realize in situ color tuning by controlling the excitation [24].

Based on the above results, a mixed color can be obtained when a $MAPbX_3$ grating is illuminated by white light and laser light simultaneously. What is more, by controlling the brightness of the green colors with the laser pumping density, the ratio of emission to reflection can be dynamically controlled [24]. Consequently, when the sample was illuminated by a white light source, a colorful image was successfully recorded by the CCD camera (see Fig. 11.19b). In contrast to the conventional gratings, the colors in the university logo can also be dynamically controlled via external

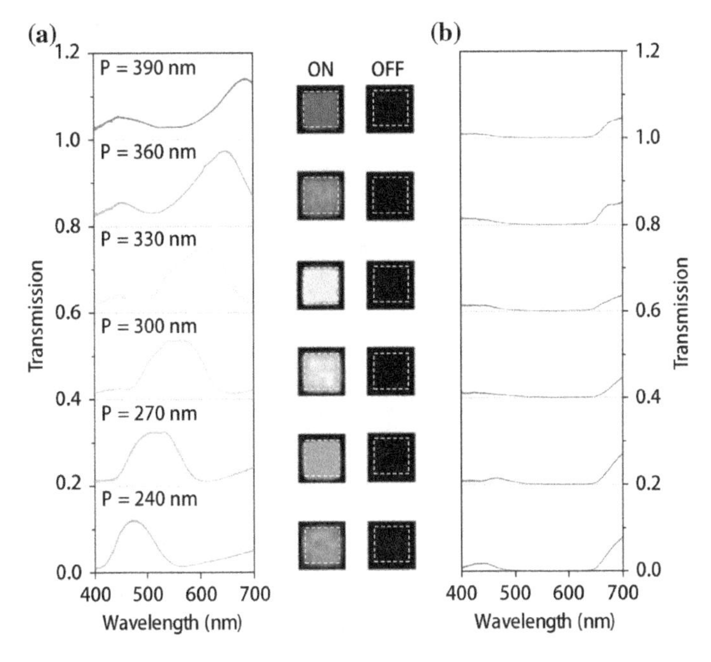

Fig. 11.17 Optical transmission spectra of PolyProDOT-Me_2-coated Al-nanoslit structures with respective values of slit period $P = 240$, 270, 300, 330, 360, and 390 nm, along with corresponding optical micrographs of device areas imaged in transmission. Transmission spectra and micrographs for **a** ON and **b** OFF states of the polymer are displayed, respectively. Reproduced from [23] with permission. Copyright 2016, The Authors

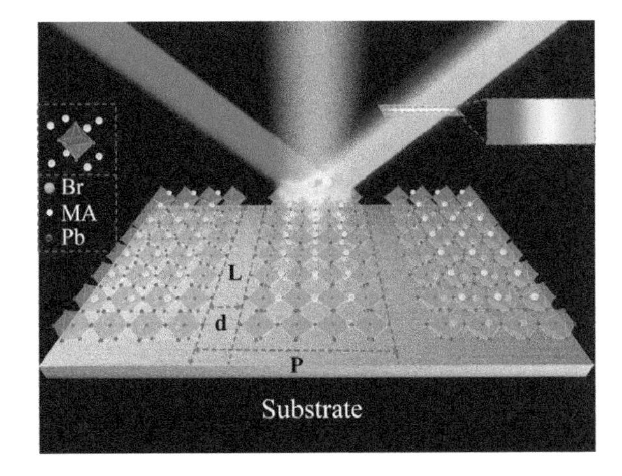

Fig. 11.18 Working principle. The schematic design of pixels and the in situ color generation by mixing extrinsic structural color and intrinsic emission color on $MAPbX_3$ perovskite gratings. Reproduced from [24] with permission. Copyright 2018, American Chemical Society

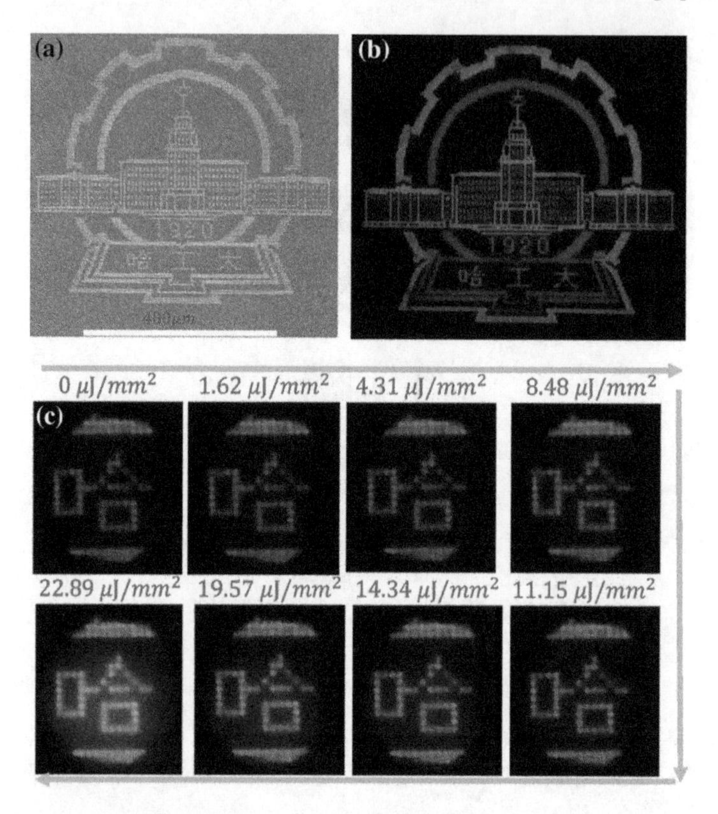

Fig. 11.19 **a** Top-view SEM image of the logo of Harbin Institute of Technology. **b** Microscope image of the university logo without photon doping. **c** Microscope images of part of the university logo at different pumping densities. Reproduced from [24] with permission. Copyright 2018, American Chemical Society

excitation. From Fig. 11.19c, one can see that the color of the Chinese character has been successfully tuned from red to green as the pumping density increased gradually [24].

11.3 Meta-holography Display

Another application of metasurface in the display area is realizing holography based on the localized phase or amplitude modulation. In traditional holography, the hologram is recorded by the interference of the scattered light from a real object and the reference light. Therefore, the traditional holography can only record and recur existing objects. The computer-generated holography (CGH), which simplifies the recording process by using numerical calculations, is a good method to solve this problem. Using the method of CGH to code the hologram by a certain material/equipment, such as a spatial light modulator (SLM), one can get the hologram of any objects, even though the object is inexistent in the world. But the minimum pixel size of the SLM limits the realization of a bigger viewing angle of the holographic image. Recently, the emergence of meta-hologram based on metasurfaces offers an effective method to overcome the problems above and may play important roles in future display technology.

11.3.1 Ultra-broadband Meta-holography

(1) Ultra-broadband polarization-independent meta-holography

A direct method to realize polarization-independent meta-holography is adopting meta-atoms with multi-fold or circular symmetry [25–27]. However, owing to the inherent dispersive property of propagation phase, the operation bandwidth of such meta-hologram is often limited. Recently, a broadband polarization-independent hologram is proposed based on the dispersionless geometric phase of anisotropic metallic apertures [28]. The meta-hologram is designed to synchronously generate the target picture and a conjugate image (i.e., reversed phase profile). As a consequence, when the handedness is changed, the preset target picture and the conjugate image will exchange their positions, but the whole output optical field will not be affected. Besides, since the linear polarization can be taken as a superimposition of circular polarizations with opposite handedness, the hologram will also not change.

In order to construct such hologram that can simultaneously generate target picture and its conjugate image, desired and reversed phase change should be generated simultaneously. That is to say, the total phase modulation should be $\exp(i\sigma\theta) + \exp(-i\sigma\theta)$, which leads to a amplitude modulation of $\cos\theta$, where θ is the orientation of the subwavelength metallic aperture [28]. Fortunately, such amplitude modulation can be realized by the polarization conversion of spatially rotating metallic apertures under the illumination of linear polarization illumination. The relation between the amplitude and elliptical nanohole orientation is in approximately accord with the line of $|\sin(2\theta)|$ [29]. Therefore, the amplitude of the cross-polarized light can be control continuously by changing the orientation of nanohole. So this elliptical nanohole can be used to code the above described hologram with the complex amplitude distribution [28].

The corresponding experimental results with the incident x-polarization, y-polarization, LCP and RCP light are shown in Fig. 11.20, which unambiguously indicate that the proposed design algorithm is feasible for realizing polarization-independent hologram [28]. It should be noted that, for non-centrosymmetric holographic imaging, only half of the energy is utilized, while the other contributes to the background noise. In order to overcome this problem, the method above is further utilized for centrosymmetric holographic imaging. Since the centrosymmetric picture and its conjugate imaging are totally superimposed, the wasted energy and high background noise can be avoided. In addition, owing to the dispersionless phase and

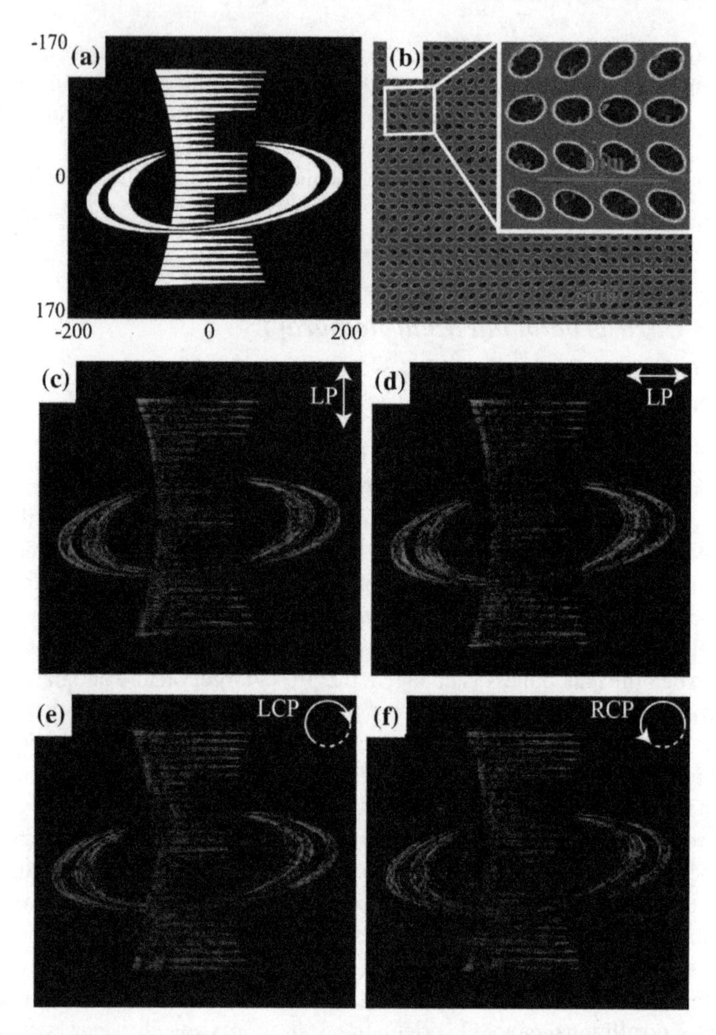

Fig. 11.20 **a** Target holographic image. The unit is μm. **b** SEM image of the fabricated elliptical nanoholes with the scale bar 5 μm. The up-right inset shows a higher resolution ratio SEM image, and the scale bar is 1 μm. Holographic images for **c** *y*-polarization, **d** *x*-polarization, **e** LCP and **f** RCP light. The incident wavelength is set as 632.8 nm. Reproduced from [28] with permission. Copyright 2018, The Royal Society of Chemistry

amplitude modulation, this method can realize broadband polarization-independent hologram, as indicated in Fig. 11.21 [28].

(2) Ultra-broadband chiral meta-holography

Chiral meta-holography means when a meta-hologram is illuminated by a different helicity of circularly polarized light (CPL), it will generate different holographic images, which is quite difficult to realize by conventional methods. Thanks to the

Fig. 11.21 Broadband holographic image at the wavelength of 632.8, 532, 473, and 405 nm. The incident x-polarization or y-polarization has been marked in the figure. Reproduced from [28] with permission. Copyright 2018, The Royal Society of Chemistry

recently emerged metasurface, a direct method to achieve this goal is exploiting the polarization-dependent geometric phase, which can be simply expressed as $\Phi = 2\sigma\theta$, where $\sigma = \pm 1$ denotes the left- and right-handed circular polarizations (LCP and RCP) and θ defines the orientation angle of the metallic rod [30]. By combining two groups of meta-atoms that, respectively, respond to the phase profiles of LCP and RCP into a single metasurface, chiral meta-holography can be easily realized. Nevertheless, this method has two disadvantages. One is the low efficiency because only part of the metasurface responds to the special helicity. The other is that it is quite difficult to realize a colorful holography, since the geometric phase is wavelength-independent.

Figure 11.22 shows a schematic of colorful chiral holography, where two different spin states for the wavelengths of 632.8 and 532 nm are used to code the red and green component of a color picture [31]. Generally, a conjugate image is generated when the helicity of incidence is reserved. In order to realize different holography, off-axis imaging is utilized by introducing extra gradient phase. As a consequence, the conjugate phase is avoided. Besides, benefiting from the off-axis imaging, the background noise stemming from unconverted light is also avoided.

The simulation and experiment results in Figs. 11.23 and 11.24 prove that the proposed method is feasible for realizing colorized holographic display with different spin state incident light [31]. Three different holographic images will be obtained by filtering the special polarization components of the output. Specifically, for a "Taiji" hologram, when it is illuminated by the linearly polarized incidence, the conjugated patterns will simultaneously emerge, which results in a homogenous pattern. Such polarization-dependent holography can be utilized for encrypted display [31].

Alternative method to realize chiral meta-holography is breaking the symmetry of spin–orbit interaction and realizing helicity-independent phase modulation [32]. To this purpose, one can merge this helicity-dependent geometric phase with helicity-

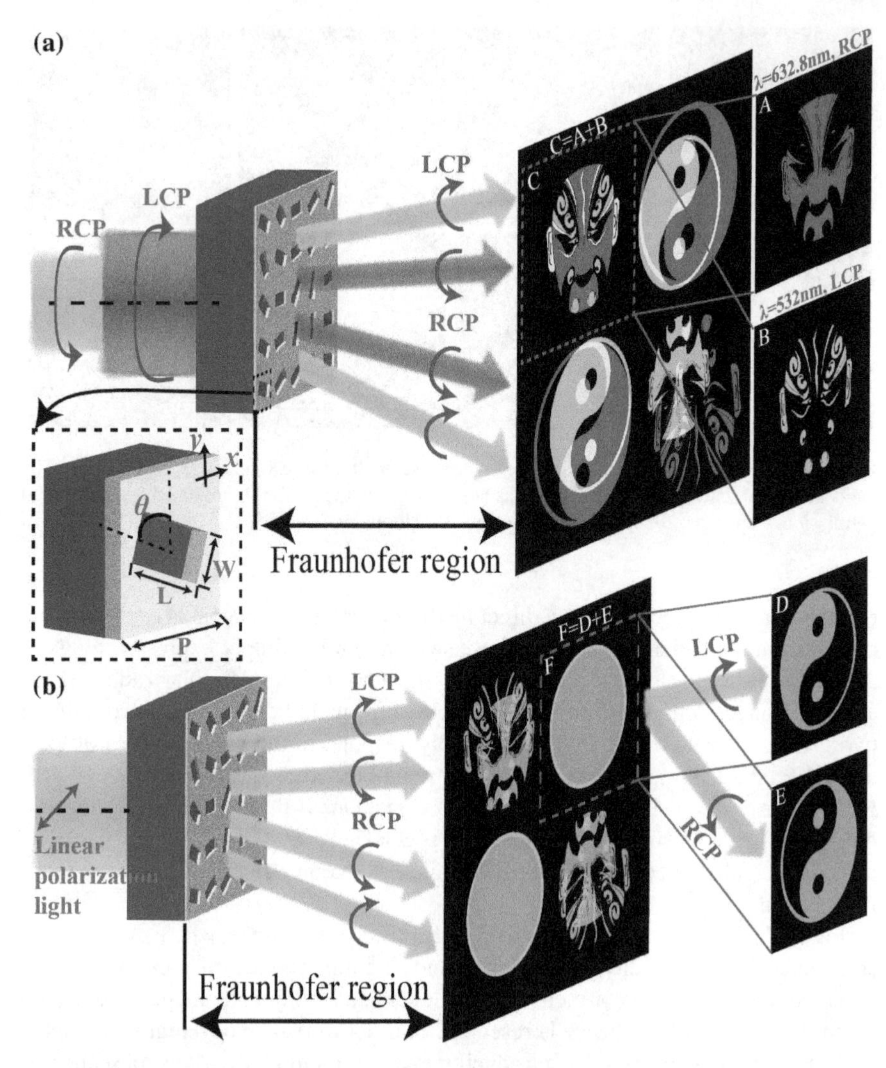

Fig. 11.22 Schematic diagrams to show the helicity-multiplexed metasurface holography. **a** When synchronously incident the LCP and RCP light at the wavelength of 632.8 and 532 nm, the mixed output light (window C) can form a colorized facial mask in the Fraunhofer region of the hologram. The results with single incident light of the wavelength 632.8 or 532 nm are shown in window A, B, respectively. **b** With linear polarization light normally impinges on the metasurface, three different pictures (window D–F) can be realized with detecting different polarizations of the output light, respectively. Reproduced from [31] with permission. Copyright 2017, WILEY-VCH Verlag GmbH & Co. KGaA, Weinheim

independent propagation phase into a single metasurface, where both the orientation and geometry of the unit cell change with the spatial location (Fig. 11.25) [33].

Assuming the orientation and propagation phase retardation of the unit cell are, respectively, θ and ϕ, in order to realize two asymmetric holography, the following relationship should be satisfied [33]:

$$\begin{aligned} 2\theta - \phi &= \psi_1 + 2n_1\pi \\ -2\theta - \phi &= \psi_{-1} + 2n_{-1}\pi \end{aligned}, \tag{11.3.1}$$

where ψ_1 and ψ_{-1} are independent phase distributions for $\sigma = \pm 1$, and n_1 and n_{-1} are integers. The above equation can be transformed to

$$\begin{aligned} \phi &= -\frac{\psi_1 + \psi_{-1}}{2} - (n_1 + n_{-1})\pi \\ \theta &= \frac{\psi_1 - \psi_{-1}}{4} + \frac{(n_1 - n_{-1})\pi}{2} \end{aligned}. \tag{11.3.2}$$

The obtained spin-independent phases are then quantized to different levels, ranging from 0 to 2π, realized by different geometries of nanopillars, and the spin-

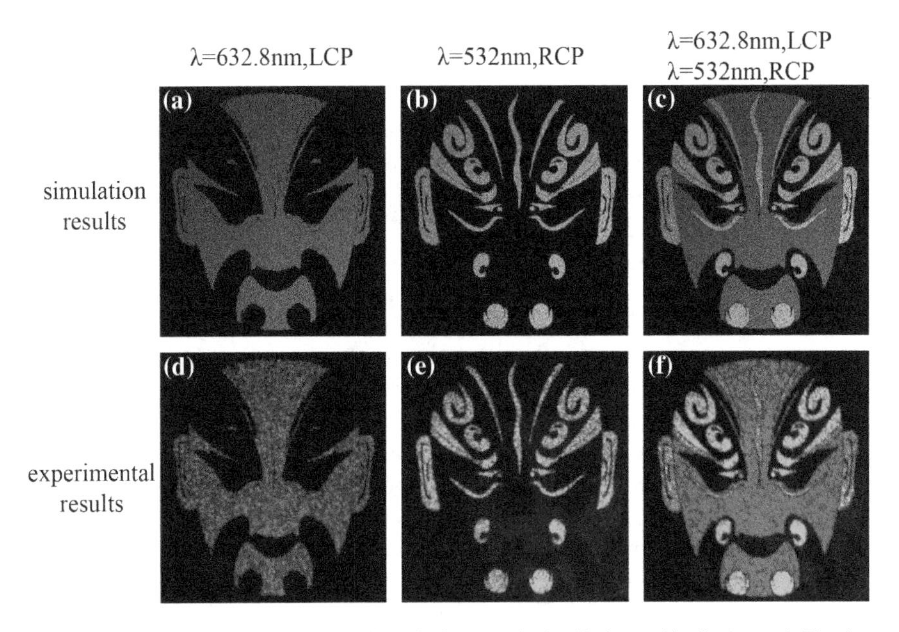

Fig. 11.23 Simulation and experimental results for the colorized holographic display. **a**–**b** The simulation results of the incident light at the wavelength of 632.8 nm with the LCP and the wavelength of 532 nm with the RCP, respectively. **c** The simulation multi-color holographic image with LCP light of the wavelength 632.8 nm and RCP light at the wavelength of 532 nm. **d**–**f** The corresponding experimental results. Reproduced from [31] with permission. Copyright 2017, WILEY-VCH Verlag GmbH & Co. KGaA, Weinheim

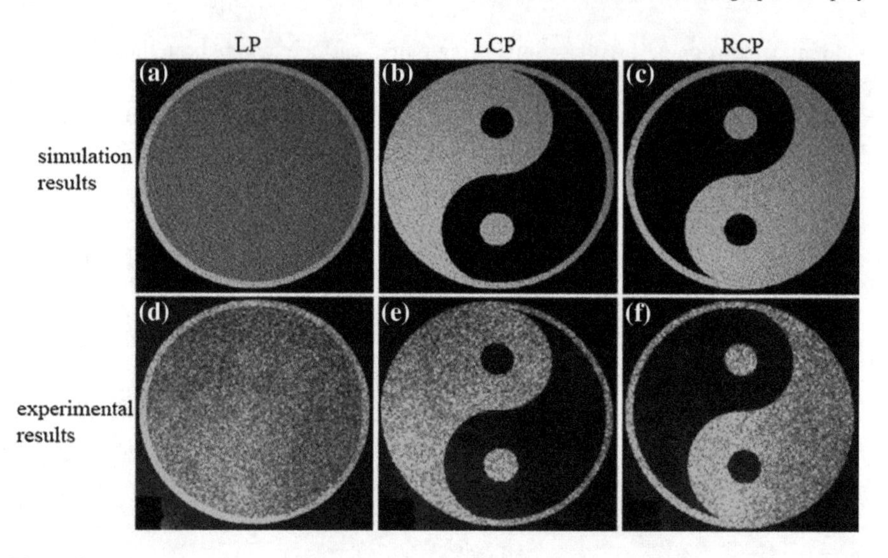

Fig. 11.24 Simulation and experimental results for the polarization encryption imaging. **a–c** The simulation results for the images with the transmission polarizations of LP, LCP and RCP, respectively (The incident wavelength is set as 532 nm). **d–f** The corresponding experimental results. Reproduced from [31] with permission. Copyright 2017, WILEY-VCH Verlag GmbH & Co. KGaA, Weinheim

dependent phases are generated by changing the orientations θ of nanopillars at each position. Following this principle, two independent groups [33, 34] simultaneously reported the asymmetric spin–orbit interactions in dielectric metasurfaces and demonstrated symmetry-breaking holography. In other words, the holographic images generated by circular polarizations with opposite spin are not mirror symmetric again, as indicated in Figs. 11.26 and 11.27 [33, 34].

In principle, this independent phase modulation method can be extended to any pair of orthogonal states of polarization, not only limited to linear and circular polarizations, and thus significantly expands the scope of metasurface polarization optics. Specifically, we can also simultaneously change the transmission phase and amplitude of the cross-circular polarization by utilizing a super-cell design [35], as shown in Fig. 11.28. For LCP incidence, the superimposition of propagation phase ϕ and geometric phase 2θ causes constructive interference at the transmission side. For RCP incidence, the superimposition of propagation phase ϕ and geometric phase -2θ generates a destructive interference at the transmission side [35].

Consequently, the chiral transmission gives rise to strong circular dichroism (CD) and asymmetric transmission (AT); i.e., the transmission is quite different when CPL is incident from different sides of the hologram, as indicated in Fig. 11.29 [35]. An extinction ratio of $\approx$10:1 and an AT parameter of $\approx$0.69 at the designed central wavelength of 9.6 μm were obtained, and the full width half maximum (FWHM) of AT parameter is about 2.9 μm ($\approx$30% of the peak wavelength). These items are much better than previous reports [35].

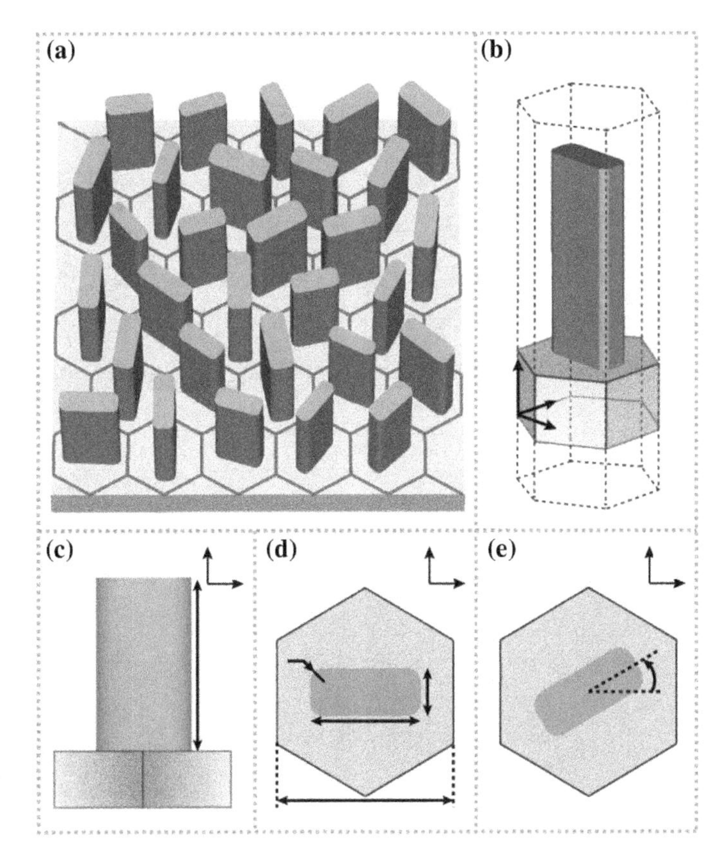

Fig. 11.25 Unit cell design. **a** Top view of the proposed metasurface consisting of chamfered nanofins on a substrate, with the same height and chamfering radius, but different sizes and orientations. **b** The nanofin is placed at the center of hexagonal unit cell. **c**–**e** Side and top views of hexagonal unit cell showing height H, width W, length L, chamfering radius R and rotation angle θ of the nanofin, with the lattice constant P. Reproduced from [33] with permission. Copyright 2017, Institute of Optics and Electronics, Chinese Academy of Sciences

In order to simultaneously realize asymmetric transmission and chiral holography, the orientation of each super-cell with asymmetric transmission was rotated according to the expected phase distribution (Fig. 11.30a) [35]. To characterize the performance of the holograms under different circular polarizations, the far-field diffraction pattern under the illumination of linear polarization was measured. Distinct from traditional geometric phase holograms that generate two conjugated images (Fig. 11.30b) under linearly polarized incidence, the chiral hologram only generates isolate holographic images (Fig. 11.30c) [35].

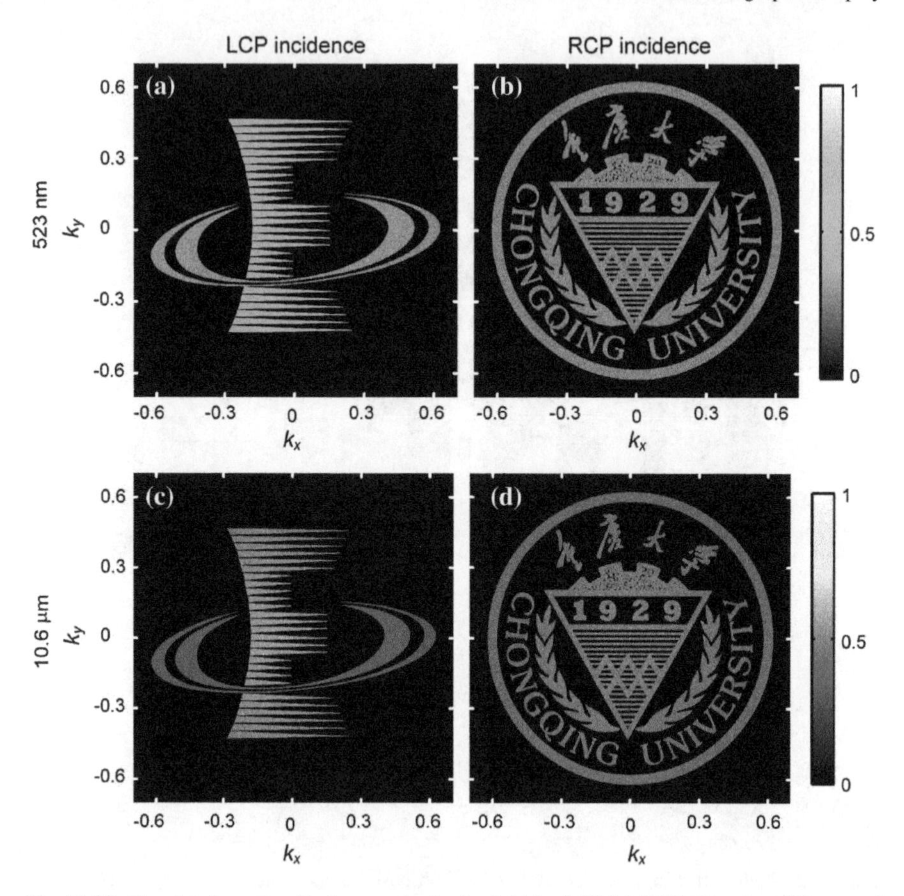

Fig. 11.26 Simulated results of holograms in the far-field for LCP (**a**), RCP (**b**) at $\lambda = 532$ nm and LCP (**c**), RCP (**d**) at $\lambda = 10.6\ \mu$m. The corresponding simulated sizes are 1 mm × 1 mm and 2 cm × 2 cm, respectively, for two working wavelengths (532 and 10.6 μm). Reproduced from [33] with permission. Copyright 2017, Institute of Optics and Electronics, Chinese Academy of Sciences

11.3.2 Vectorial Meta-holography

Recently, Deng et al. [36] demonstrated a new concept of vectorial holography based on diatomic metasurfaces consisting of meta-molecules formed by two orthogonal meta-atoms. Vectorial holography requires the simultaneous phase and polarization manipulation. The phase modulation relies on the detour phase of the 1st diffraction:

$$\Delta\varphi = 2\pi p / p_0. \tag{11.3.3}$$

Obviously, the phase shift is proportional to the displacement p and is independent of both the wavelength λ and the incident angle θ_0.

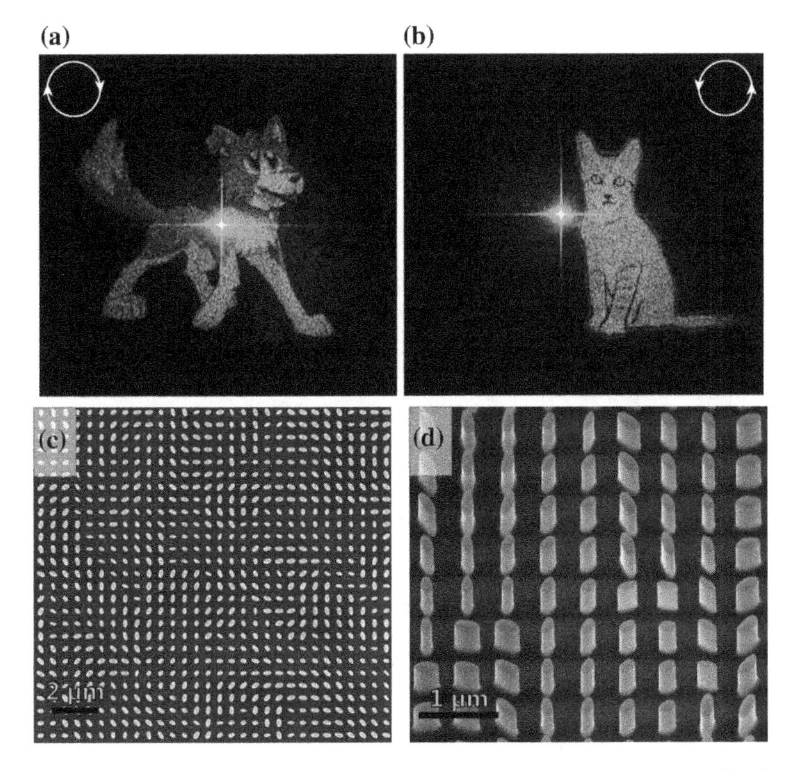

Fig. 11.27 **a, b** A single metasurface encodes two independent hologram phase profiles for each circular polarization at $\lambda = 532$ nm. When illuminated with RCP (LCP), the metasurface projects an image of a cartoon dog (cat) to the far field. **c** The metasurface encoding these holograms was 350×300 μm in size and contained 420,000 TiO_2 pillars of elliptical cross section. Shown is a SEM of the device. **d** Oblique view. Reproduced from [34] with permission. Copyright 2017, American Physical Society

In order to ensure full polarization control, meta-molecules composed of two orthogonally aligned meta-atoms with a displacement s are utilized, as shown in Fig. 11.31. The diffracted waves from each nanorod are cross-polarized, and they have the forms as follows [36]:

$$\begin{aligned} E_1 &= \vec{e}_1 C E_{\text{inc}} \sin \psi \\ E_2 &= \vec{e}_2 C E_{\text{inc}} e^{i2\pi s/p_0} \cos \psi \end{aligned}, \tag{11.3.4}$$

where $2\pi s/p_0$ is the displacement-targeted phase retardance. The superposition of the two orthogonal polarized field components can lead to arbitrary vectorial forms, which can be expressed as the Jones vector [36],

$$\begin{pmatrix} E_1 \\ E_2 \end{pmatrix} = C E_{\text{inc}} \begin{pmatrix} \sin \psi \\ \cos \psi \exp\left(i 2\pi \frac{s}{p_0} \right) \end{pmatrix}. \tag{11.3.5}$$

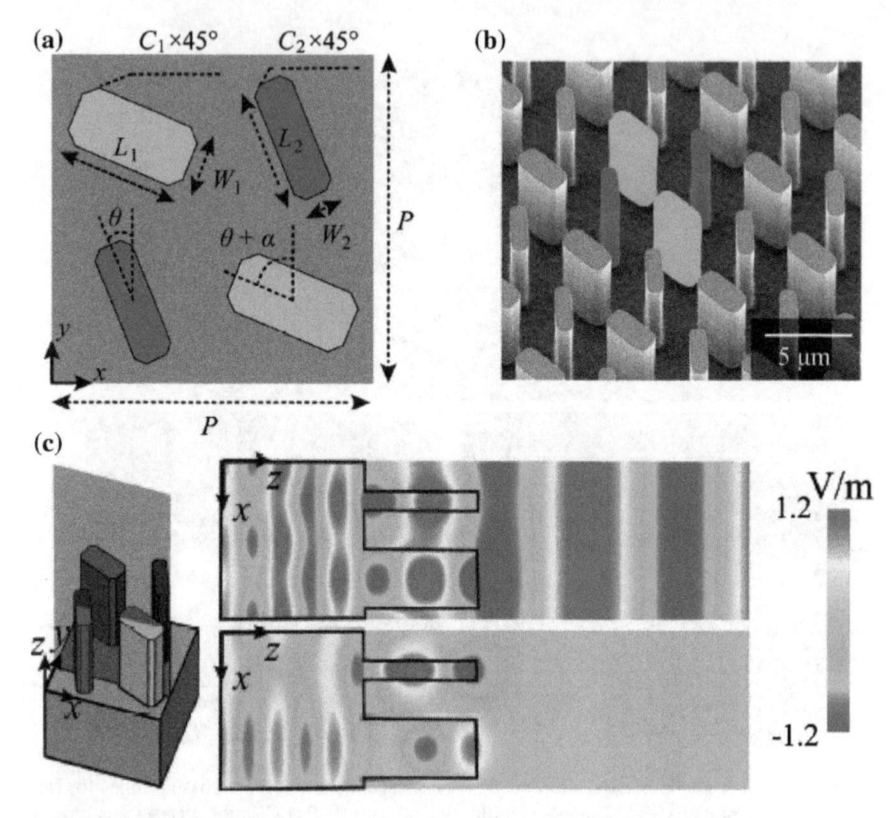

Fig. 11.28 **a** Top views of all-silicon super-cells showing the dimensions of $W_1 = 1.45$, $W_2 = 0.9$, $L_1 = 3.1$, $L_2 = 2.95$, $C_1 = 0.32$, $C_2 = 0.25$, $P = 8.1$, and $H = 6.0$ μm. **b** SEM image of the fabricated metasurface. **c** The distributions of electric field E_y in the periodic super-cell (left) under the illumination of LCP (top right) and RCP (bottom right) light from the substrate side at the resonant wavelength of 9.9 μm. Reproduced from [35] with permission. Copyright 2017, WILEY-VCH Verlag GmbH & Co. KGaA, Weinheim

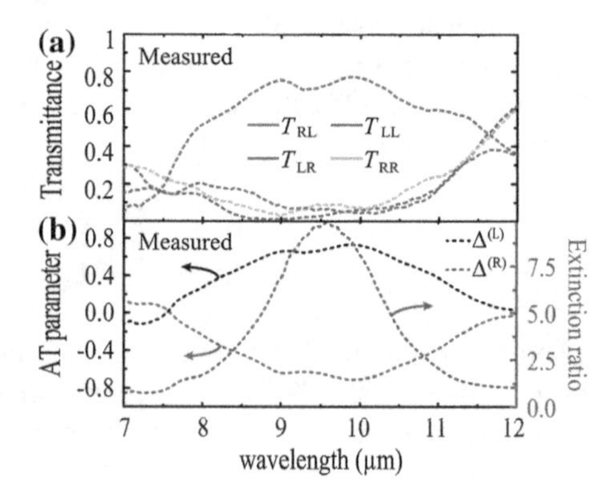

Fig. 11.29 **a** Measured transmittances of the sample. **b** Measured AT parameters and extinction ratios of the metasurface. Reproduced from [35] with permission. Copyright 2017, WILEY-VCH Verlag GmbH & Co. KGaA, Weinheim

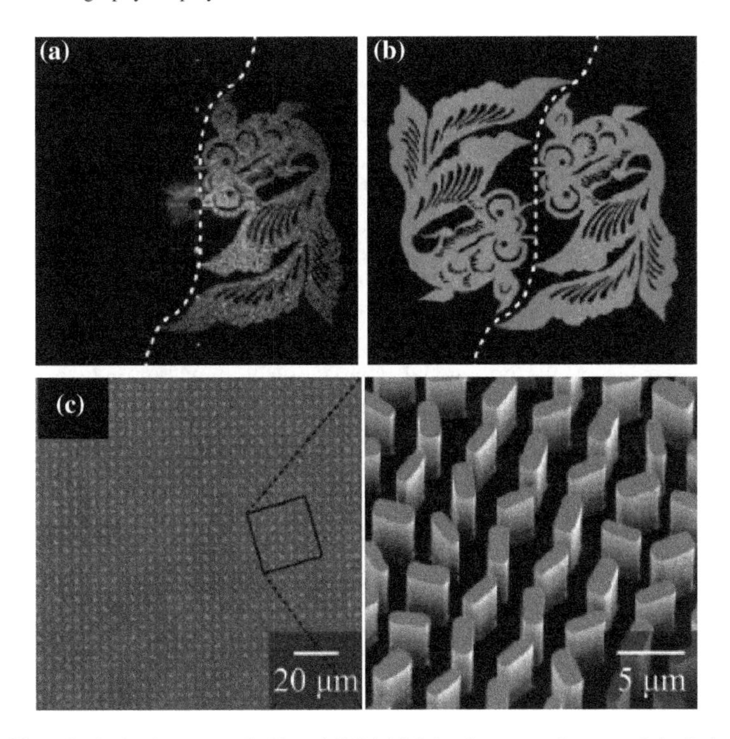

Fig. 11.30 **a** Optical microscope (left) and SEM (right) microscope images of the hologram. **b** Theoretical holographic images created with traditional PB-based phase hologram under the illumination of LP incidence. **c** The measured holographic images generated by the hologram with AT. Reproduced from [35] with permission. Copyright 2017, WILEY-VCH Verlag GmbH & Co. KGaA, Weinheim

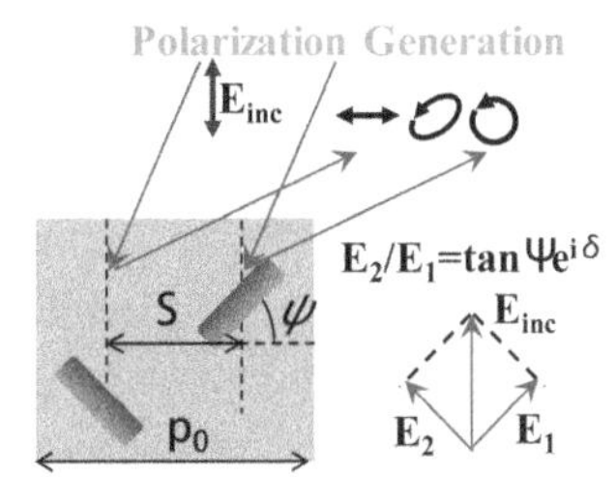

Fig. 11.31 Schematic illustration of the displacement-targeted diatomic meta-molecules for polarization control. Reproduced from [36] with permission. Copyright 2018, American Chemical Society

In other words, arbitrary polarization states of the diffracted light can be generated. Combining the two displacement-targeted phase control scheme above, the overall diffracted wave can be written as:

$$\begin{pmatrix} E_1 \\ E_2 \end{pmatrix} = CE_{\text{inc}} \begin{pmatrix} \sin\psi \\ \cos\psi \exp\left(i2\pi\frac{s}{p_0}\right) \end{pmatrix} \exp\left(i2\pi\frac{p}{p_0}\right). \tag{11.3.6}$$

As a consequence, both the phase and polarization of the 1st diffraction beam can be fully modulated by the three parameters (p, s, ψ), which are independent of wavelengths and incident angles, but only dependent on the relative displacements and orientation angles between consistent meta-atoms [36].

To realize vectorial hologram, one can arrange meta-molecules with multiple sets of parameters ($p_i(x, y)$, ψ_i, s_i), ($i = 1, 2, \ldots$) in sequence, and then, the ith hologram encoded by p_i (x, y) will have the predefined polarization states determined by ψ_i and s_i. The superposition of such multiple polarized holograms leads to the holographic image with spatially varying polarization states [36]. Owing to the linear wave plate nature of the diatomic meta-molecule, the vectorial meta-holograms can be switchable by the orthogonal polarization illumination, which allows for polarization-multiplexed hologram, as indicated in Fig. 11.32 [36].

11.3.3 Off-Axis Colorful Meta-holography

Alternative method to increase the information capacity is adopting off-axis meta-holography. The deep subwavelength building blocks of metasurface holograms mean the evanescent wave may appear in the operating frequency spectrum. But

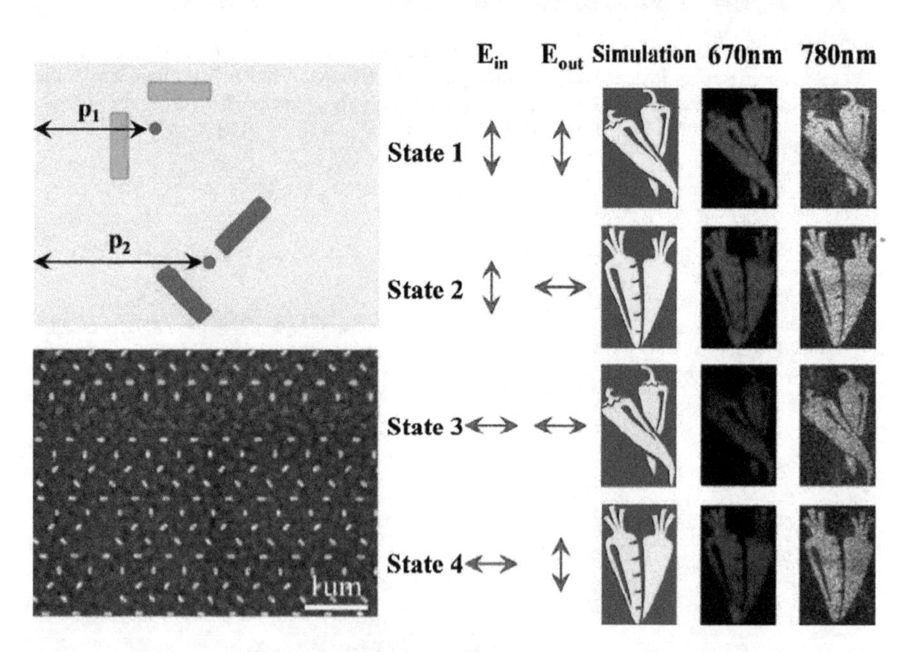

Fig. 11.32 Broadband vectorial meta-holograms for multistate-encrypted anti-counterfeiting by dual-way polarization switching. Two sets of polarizers at both input and output sides can independently switch the appearance of multistate-encrypted anti-counterfeiting images without mutual disturbance. Reproduced from [36] with permission. Copyright 2018, American Chemical Society

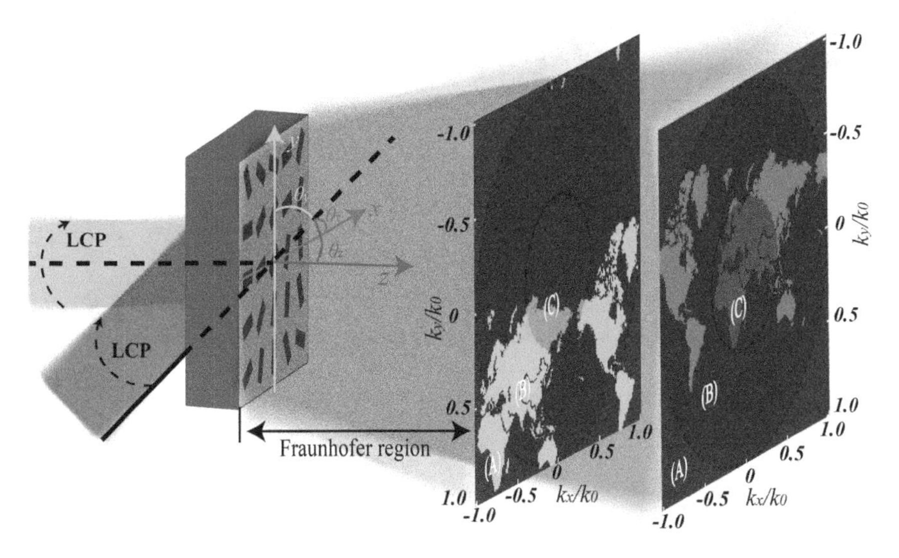

Fig. 11.33 Schematic of off-axis holography to increase the information capacity. The green color represents the light normal impinging on the sample, and the red color means the light incidence on the hologram with an oblique angle θ (θ_x, θ_y, θ_z), where θ_x, θ_y, and θ_z are the angles between the incident orientation and the positive directions of x-axis, y-axis, z-axis, respectively. The three regions signed in the frequency spectrum with a different color mean: (A) evanescent wave region; (B) distortion region; (C) distortion-free region. The incident wavelength is 405 nm. Reproduced from [37] with permission. Copyright 2017, The Royal Society of Chemistry

most metasurface holograms did not use the evanescent wave for far-field imaging. In other words, much information is discarded in current metasurface holographic technology [37].

In order to increase the information capacity, off-axis illumination is utilized to transfer the evanescent wave into the propagating wave, as illustrated in Fig. 11.33 [37]. When the light normally incidents on the sample, as shown by the green color light, the holographic image in the Fraunhofer region can be expressed as: $G(f_x, f_y) = F(g(x, y))$, where (x, y) represents the coordinate of the metasurface hologram, (f_x, f_y) is the coordinate in the corresponding frequency spectrum ($f_x = k_x/(\lambda k_0)$, $f_y = k_y/(\lambda k_0)$), the $g(x, y)$ describes the electromagnetic field just after the sample, and F means the fast Fourier transform algorithm. When the light incidents on the metasurface hologram with an angle θ (θ_x, θ_y, θ_z), the related far-field image can be written as [37]:

$$\begin{aligned} &G(f_x - \cos(\theta_x)/\lambda,\ f_x - \cos(\theta_x)/\lambda) \\ &\quad = F\{g(x, y)\exp[i2\pi(x\cos(\theta_x) + y\cos(\theta_y))/\lambda]\}' \end{aligned} \tag{11.3.7}$$

where θ_x and θ_y indicate the angle between the incident orientation and the positive directions of the x-axis and y-axis, respectively. As shown in Eq. (11.3.7), varying

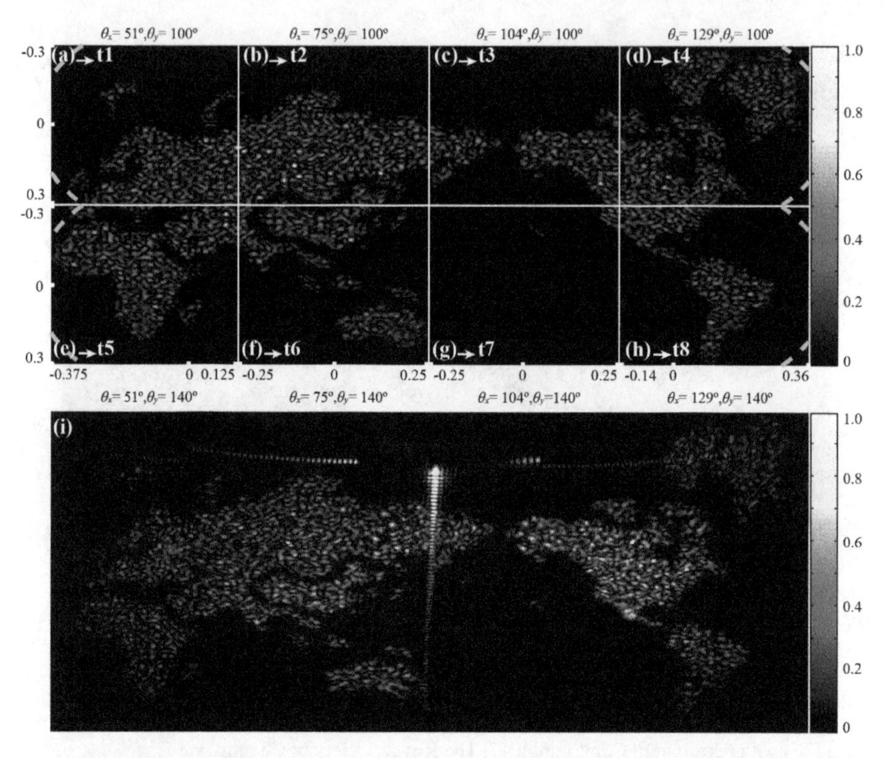

Fig. 11.34 **a–h** Simulation results with eight groups of incident angles, which are related to different time. The green dashed lines in **a**, **d**, **e**, and **h** denote the boundaries between the distortion region and distortion-free region. **i** The stitched holographic image of experimental results with different incident angles. The incident wavelength is 405 nm. The intensity has been normalized. Reproduced from [37] with permission. Copyright 2017, The Royal Society of Chemistry

incident angle could lead to the displacement of the frequency spectrum. As a consequence, when illuminating the sample with an oblique angle, the Africa part changes from the down-left region of the world map (green map) to the central (red map), but the American part seems to be shifted from right-down region to the left, as shown in Fig. 11.34. Mathematically, the whole frequency spectrum of the pixel hologram can be expressed as [37]:

$$G_w(f_x, f_y) = \sum_{m=-\infty}^{\infty} \sum_{n=-\infty}^{\infty} G(f_x - m/p,\ f_y - n/p), \tag{11.3.8}$$

where m and n are integers, and p is the period of the square pixels of the meta-hologram. This equation implies the whole frequency spectrum is periodic. Therefore, it is possible to recover all spectrum information by appropriate off-axis illumination [37].

In order to validate the scheme, a hologram of world map is realized by combining multiple holograms with a different off-axis illumination, as shown in Fig. 11.34 [37]. In theory, if the incidence wavelength is larger than twice of the period of the structure (>400 nm), the evanescent wave up to $2k_0$ can be recovered by off-axis illumination. Considering the distortion-free region is a circular area with radius of $0.417k_0$ in the frequency spectrum, several measurement windows with all the corresponding images in the distortion-free region are used in the experiment to transfer all the evanescent wave information into distortion-free region [37].

Although the meta-holography provides a sufficient viewing range, the new challenge is how to simultaneously multiplex multi-wavelengths into one metasurface and to eliminate the cross-talk among different wavelengths [37]. Although researchers have demonstrated multi-color spectral modulation by integrating three plasmonic pixels with different sizes into one metasurface, respond to RGB lights, further improvement of this method is challenging because of one fundamental limitation: Larger bandwidth leads to more cross-talks. Besides, because this method is based on the resonance of the nanorods, several nano-antennas need to be grouped to obtain an effective response, which in turn reduces the data density and the viewing angle of the holography [37].

Off-axis meta-holography has been utilized to overcome the above questions, and a schematic diagram of this methodology is shown in Fig. 11.35. Laser beams with different wavelengths are obliquely irradiated upon the metasurface formed by nanoslit antennas [38]. The outcoming light beams at designed angles are then superimposed to form the final multi-color image. Simultaneously, all the unwanted images shift to the evanescent wave region. By this method, the cross-talk among different wavelengths is eliminated. As shown in Fig. 11.36, two holographic images (a flower and China map) are generated at different off-axis angles with low cross-talk [38].

The reduced cross-talk also enables holography with more colors. With this new technique, a seven-color meta-hologram image was designed and demonstrated, as shown in Fig. 11.37 [38]. It follows the similar design principle of the RGB hologram except that lasers at seven different wavelengths were used to reconstruct the image. Thus, the color gamut is increased by 1.39 times. A larger color gamut describes a broader spectrum that the holography can be reproduced in the color space. Human eyes can capture a broader spectrum beyond the colors mixed by the RGB components. Seven-color mixing extends the range of the colors available for the holography and provides a much improved capability to display a more colorful and superior image [38].

11.3.4 Three-Dimensional Meta-holography

Using the above off-axis holography method, a multi-color 3D object was also demonstrated [38], as shown in Fig. 11.38. A modified point-source algorithm is reproduced to design the metasurface. The 3D object is represented by a collection

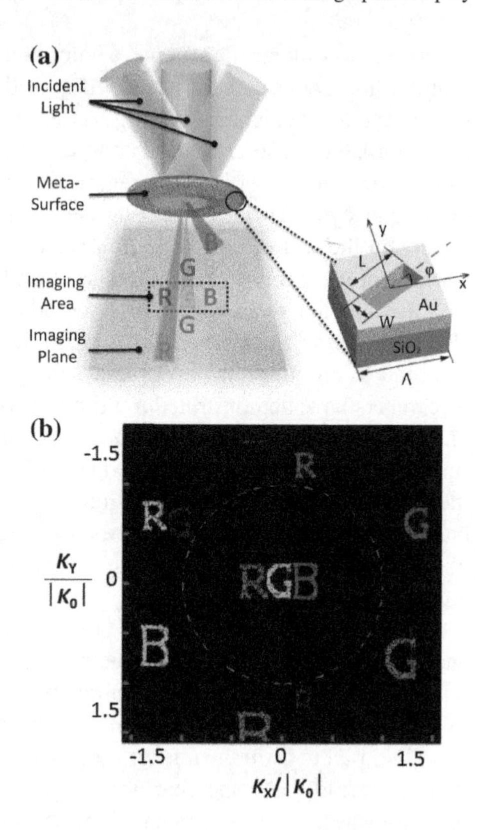

Fig. 11.35 **a** Schematic diagram of the off-axis illumination method for meta-holography and the building blocks of metasurface. **b** Simulation results for the hologram design that shifts all the unwanted images into the *k* space corresponding to evanescent waves. Reproduced from [38] with permission. Copyright 2016, The Authors

of the RGB point sources. The complex amplitude at the hologram plane is then calculated as the superposition of the light fields from the entire 3D object consisting of all point sources. Different from the 2D holographic characterization setup, the 3D object needs to scan through the vertical direction (*z* dimension) of the imaging space. At different positions along the *z*-axis, the holographic image is captured by the CCD camera. Points with varying colors can be identified in these images. These points are arranged in a space of 60 μm × 60 μm × 90 μm. The volume of this space is 324,000 μm^3 [38].

11.3.5 High-Efficiency and Broadband Meta-holography

In traditional phase-only CGH designs, the phase profile is controlled by etching different depths into a transparent substrate. Generally, two-level binary CGHs have been widely used due to their ease of fabrication. Such CGHs have a theoretical diffraction efficiency of only 40.5%, and the issue of twin-image generation cannot be avoided. Although multi-level-phase CGHs can alleviate the problem of low efficiency and twin-image generation, they pose a great challenge to the fabrication.

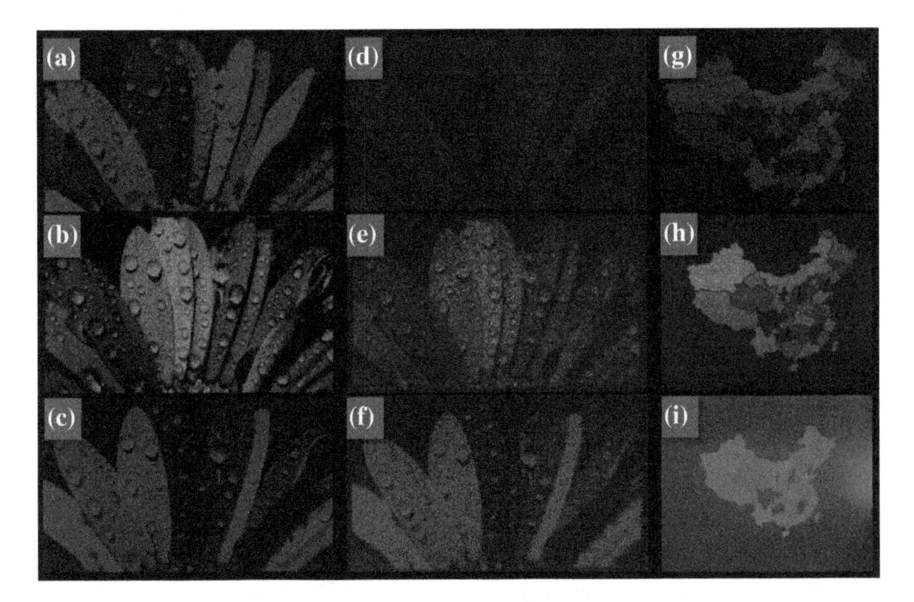

Fig. 11.36 **a–c** Simulation results for the flower holography for the RGB image patterns. **d–f** Experimental results corresponding to the red, green, and blue image patterns, respectively. **g–i** RGB holographic images of map of China, which are reconstructed by the same hologram metasurface of the flower image with different light incident angles. Reproduced from [38] with permission. Copyright 2016, The Authors

Compared with traditional two-level binary CGHs, geometric metasurfaces provide an alternative approach to achieving high-efficiency holograms without complicated fabrication procedures.

To increase the efficiency of geometric metasurfaces, a reflective metasurface hologram consists of three layers—a ground metal plane, a dielectric spacer layer, and a top layer of antennas—was utilized [39, 40]. Experiment results show that the polarization conversion efficiency of the geometric metasurfaces is high up to 80% at 825 nm and maintains larger than 50% in a broad range from 630 to 1050 nm (Fig. 11.39b). With the above structure as building block, a phase-only CGH was designed according to the classical GS algorithm. Because the phase delay is determined solely by the orientation of the nanorod antennas, 16 phase levels (Fig. 11.39a) were used to obtain the high performance from the CGH. Consequently, a broadband and high-efficiency metasurface hologram was realized, as shown in Fig. 11.39c–e.

More recently, all-metallic geometric metasurface hologram was proposed for simultaneous thermal invisibility and holographic illusion [41], as shown in Fig. 11.40. On the one hand, the all-metallic geometric metasurface hologram can abnormally reflect most infrared light into predesigned directions, which makes it invisible to the infrared thermal detector (Fig. 11.40a) [41]. On the other hand, the broadband (8–14 μm) and high efficiency (>80%) spin–orbit interactions of the metallic grating make it can realize broadband wavefront shaping by geometric phase

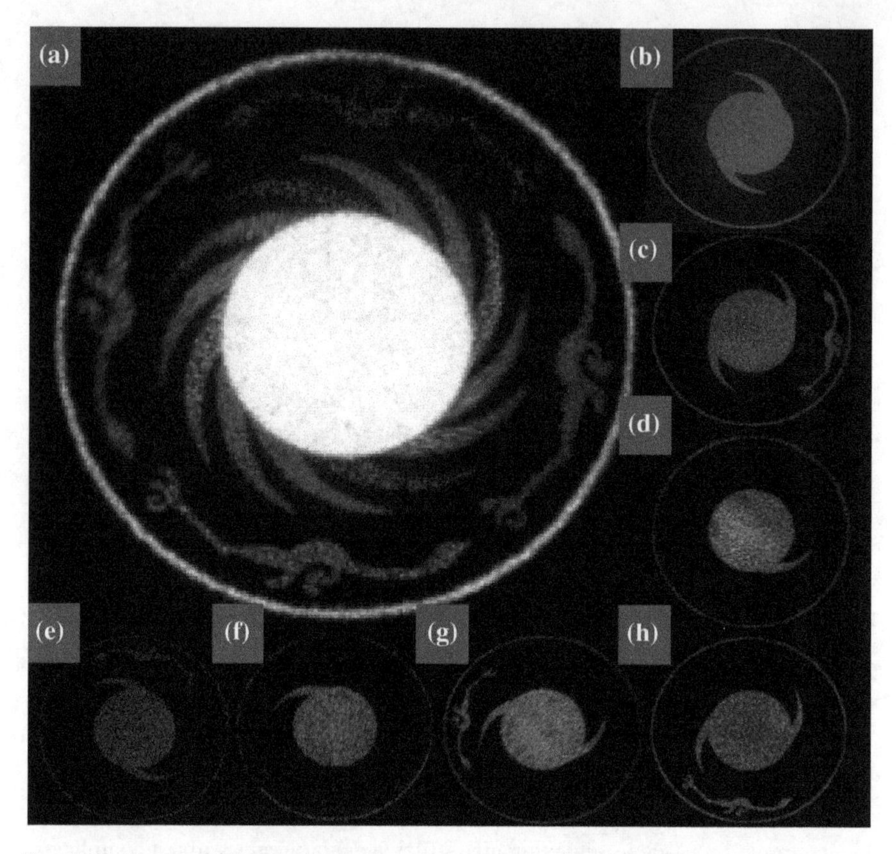

Fig. 11.37 Seven-color holographic image for the Sun Phoenix, a pattern discovered on an ancient artifact gold coil, made in the Chinese Shang dynasty 3000 years ago. **b–h** Patterns corresponding to purple, blue, cyan, red, orange, yellow, and green colors, respectively. Reproduced from [38] with permission. Copyright 2016, The Authors

metasurface (Fig. 11.40b). Based on this principle, broadband and high-efficiency holographic illusion have been demonstrated (Fig. 11.40c) [41].

11.3.6 Dynamic Holography

To date, a number of metasurface holograms have been accomplished in the microwave, terahertz, infrared, and visible spectral regimes. However, most of the reconstructed holographic images have been restricted to be static because of the fixed phase and amplitude profiles possessed by the metasurfaces once fabricated [42]. Figure 11.41 illustrates a switchable geometric phase metasurface hologram. The unit cell is composed of Au-nanorod/GST-225/Au-mirror, which can realize switchable

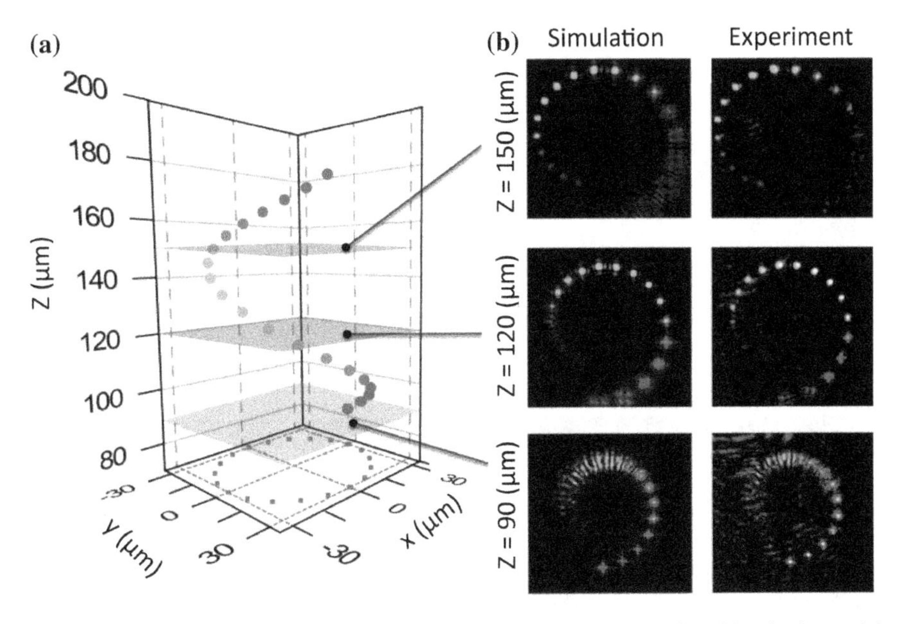

Fig. 11.38 Multicolor 3D meta-holography. **a** Schematic diagram of the 3D object in the spatial coordinate. **b** Experimental results for the cross sections at different z positions. Reproduced from [38] with permission. Copyright 2016, The Authors

spin–orbit interactions (Fig. 11.41a) [43]. The measured polarization conversion ratio (PCR) is shown in Fig. 11.41b, from which one can see that the PCR is higher than 80% in amorphous state and lower than 10% in crystalline state in a broadband spectral range from 8.5 to 10.5 μm, demonstrating the excellent switching performance. Based on the tunable spin–orbit interactions, switchable hologram was constructed just by rotating the metallic nanorod to realize the desired wavefront modulation calculated by point-source algorithm (Fig. 11.41c and d) [43]. The measured results in Fig. 11.41e and f show that when the GST is in amorphous state (i.e., "ON state" of spin–orbit interactions, and most light is converted to the cross-polarization), the holographic image can be detected, while it cannot be detected when the GST is changed into crystalline state (i.e., "OFF state" of spin–orbit interactions, and most light is directly reflected) [43]. Note that the catenary optical fields are very efficient to explain these results [44].

Similar to the dynamic structural color presented above, dynamic hologram can be realized by exploiting the phase transition of Mg from metal to dielectric upon hydrogen loading, forming magnesium hydride (MgH_2) [45]. This transition is reversible through dehydrogenation using oxygen. As a result, the plasmonic response of an Mg nanorod can be reversibly switched on and off, constituting a dynamic plasmonic pixel. Such plasmonic pixels can be used to dynamically control the phase of CPL via the PB phase by changing the orientation of the Mg nanorod [45]. Figure 11.42 shows a metasurface containing nanorods as dynamic (P_1, Mg/Pd) and static pixels

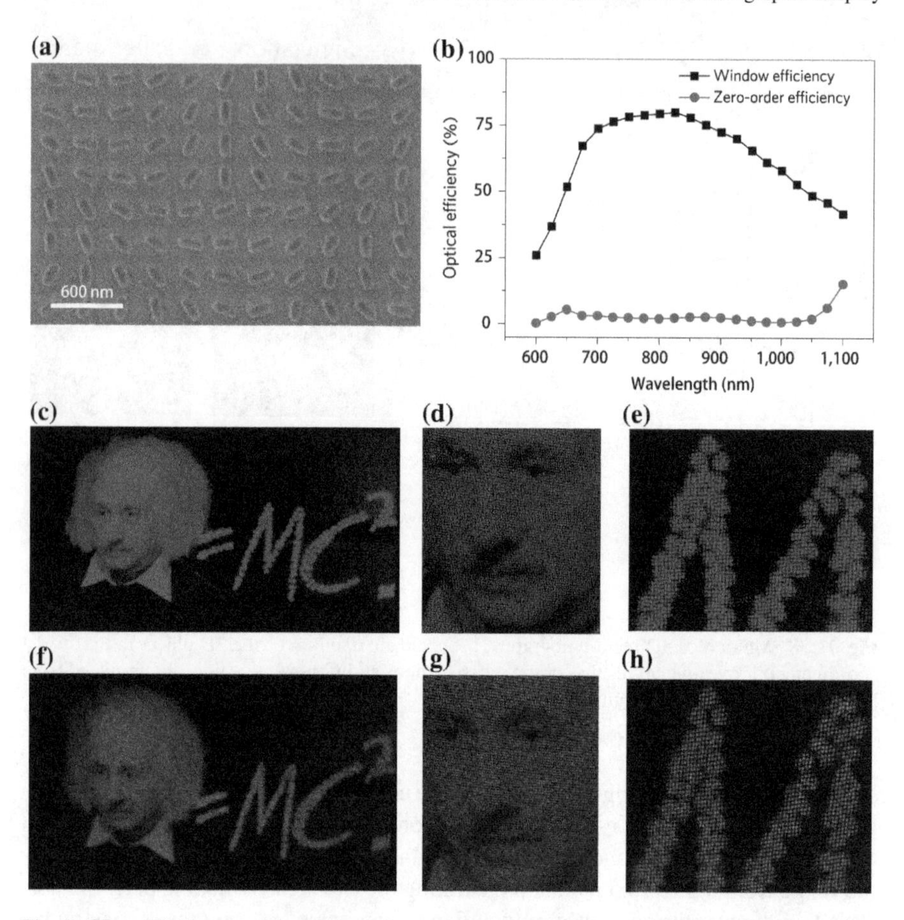

Fig. 11.39 **a** SEM image of the fabricated nanorod array (partial view). **b** Experimentally obtained optical efficiency for both the image and the zeroth-order beam. **c**–**e** Experimentally obtained images captured by a 'visible' camera in the far-field. The operating wavelength is 632.8 nm. **f**–**h** Experimentally obtained images captured by an infrared camera in the far-field, with an operating wavelength of 780 nm. Reproduced from [39] with permission. Copyright 2015, Macmillan Publishers Limited

(P_2, gold (Au)), respectively. The two sets of pixels are interpolated into each other to generate two off-axis holographic images with Chinese words, "harmony" (left) and "peace" (right) [45]. Figure 11.42b shows the representative snapshots of the holographic images during hydrogenation and dehydrogenation. At the initial state, two high-quality holographic patterns, harmony (left), and peace (right), are simultaneously observed by illumination of RCP light onto the metasurface sample [45]. Upon hydrogenation, harmony gradually diminishes because P_1 instantly decreases and reaches the "OFF" state at t_2, whereas peace remains unchanged. Upon dehydrogenation, harmony can be recovered so that both holographic patterns are at the ON state again.

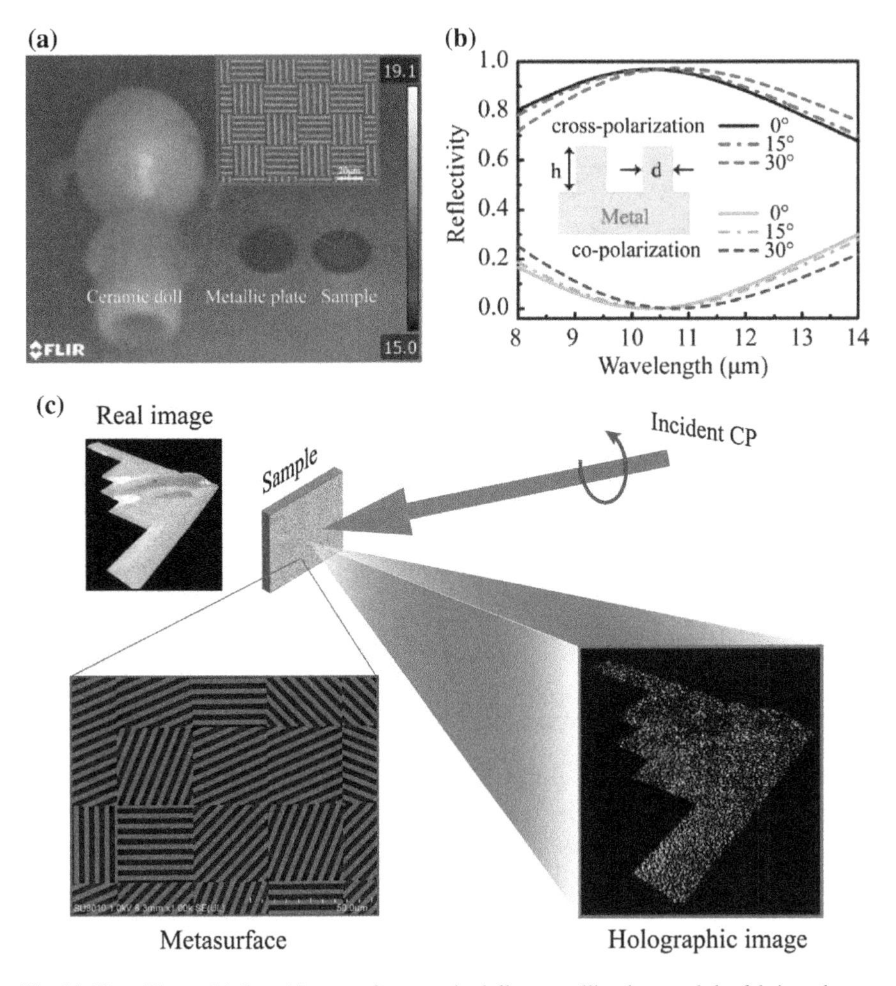

Fig. 11.40 **a** Thermal infrared image of a ceramic doll, a metallic plate, and the fabricated sample. Inset is the SEM image of the fabricated metasurface. Scale bar: 20 μm. **b** Simulated cross-polarization and co-polarization reflectivities of metallic subwavelength grating under circularly polarized illumination for different incident angles. **c** SEM image of the all-metallic metasurface hologram and the measured holographic image at a wavelength of 10.6 μm. Reproduced from [41] with permission. Copyright 2018, American Chemical Society

11.4 Simultaneous Structural Color and Hologram

Although conventional metasurfaces have demonstrated many promising functionalities in light control by tailoring either phase or spectral responses of subwavelength structures, simultaneous control of both responses is still a great challenge. Recently, a dual-mode metasurface was proposed, which enables simultaneous control of phase and spectral responses for two kinds of operation modes of transmission and reflec-

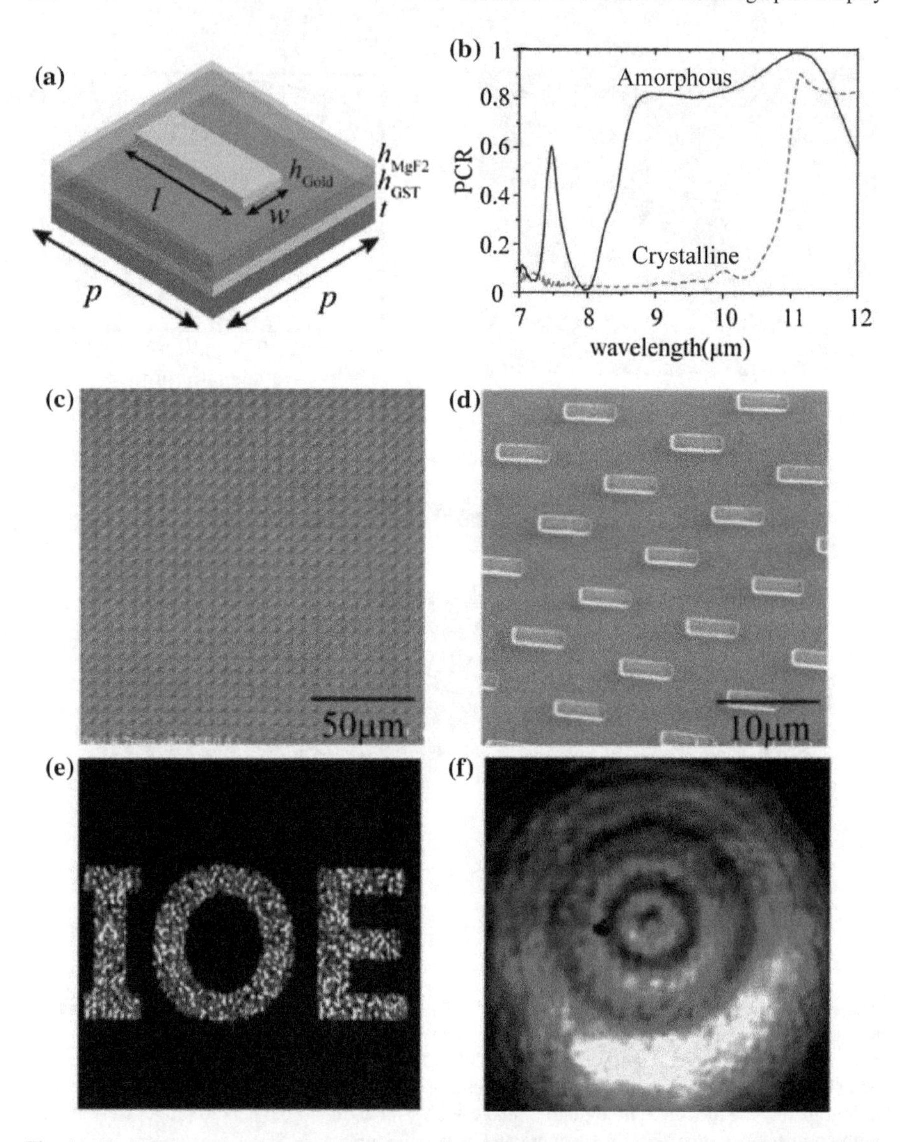

Fig. 11.41 **a** Schematic of the unit cell for switchable spin–orbit interactions. **b** Measured PCR with respect to the wavelength with amorphous and crystalline state, respectively. **c**, **d** Partial SEM image of the dynamic hologram with different scaling ratio. Measured holographic images with amorphous **e** and crystalline **f** states. Reproduced with permission from [43]. Copyright 2018, WILEY-VCH Verlag GmbH & Co. KGaA, Weinheim

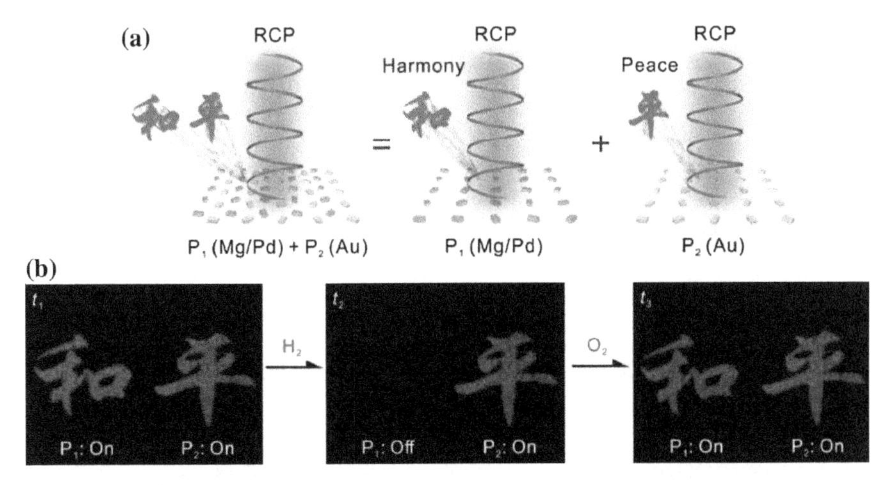

Fig. 11.42 **a** Two holographic patterns, harmony, and peace, reconstructed from two independent phase profiles containing Mg/Pd (P_1) and Au (P_2) nanorods as dynamic and static pixels, respectively. Ti is omitted for conciseness. **b** Representative snapshots of the holographic images during hydrogenation and dehydrogenation. Harmony (left) can be switched off/on using H_2/O_2, whereas peace (right) stays still. Reproduced with permission from [45]. Copyright 2018, The Authors

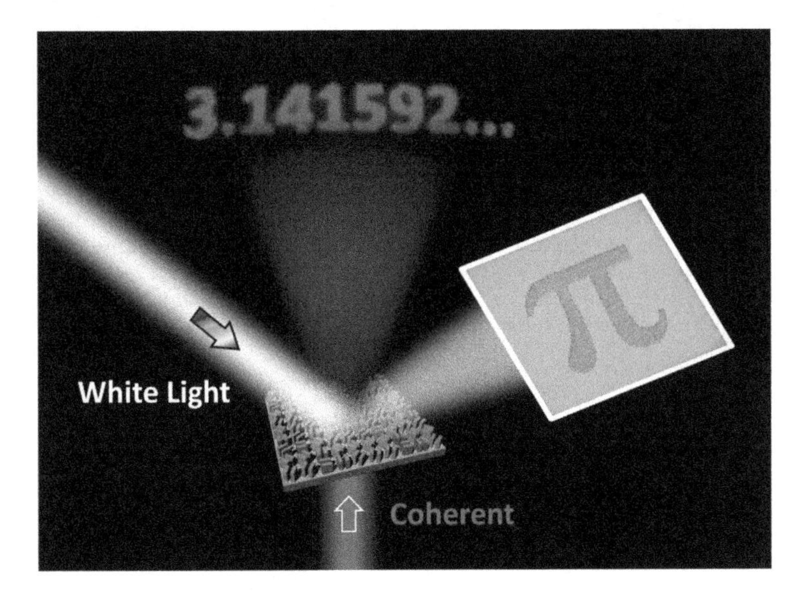

Fig. 11.43 Operation schematic of the crypto-display. The transmission mode uses single-wavelength coherent light to produce the holographic image of "3.141,592..." in the image plane, whereas the reflection mode uses white light to represent a reflected colored image of "π". The transmission mode requires the spatial coherence of the light source for its operation, but the reflection mode does not require any coherence of the light source; i.e., the reflection mode can be activated using incoherent white light such as sunlight. Reproduced from [46] with permission. Copyright 2018, American Chemical Society

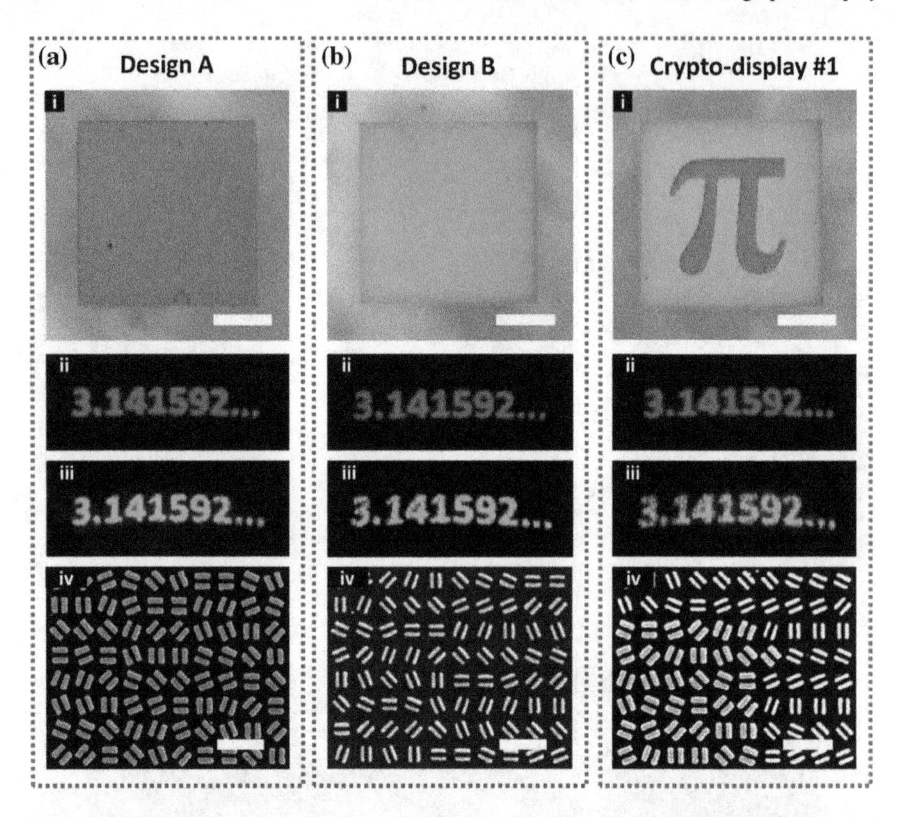

Fig. 11.44 Experimental demonstration of the crypto-display. Each figure set contains (i) an optical microscope image with the white point correction (scale bar = 100 μm), (ii) generated holographic images under 635 nm and (iii) 532 nm laser illumination, and (iv) a SEM image (scale bar = 1 μm). **a, b** Metasurfaces composed of either design A or B nano-antennas, respectively. **c** The demonstrated crypto-display consists of both design A and B nanoantennas. Reproduced from [46] with permission. Copyright 2018, American Chemical Society

tion, respectively [46]. In the transmission mode, the dual-mode metasurface acts as conventional metasurfaces by tailoring phase distribution of incident light. In the reflection mode, a reflected colored image is produced under white light illumination, as indicated in Fig. 11.43 [46].

In order to achieve this goal, two different units are simultaneously utilized in a metasurface. On the one hand, the two unit cells have the same cross-polarization transmittance (CPT) at 630 nm; thus, they can generate the same holography at the coherent illumination with a wavelength of 630 nm. On the other hand, the two unit cells exhibit different reflection spectra, and thus, they can generate different structural colors when they are illuminated by the incoherent light [46]. As expected, it will generate the same holography in the transmission side, but different structural colors in the reflection color, as indicated in Fig. 11.44a and b. When both of the two unit cells are utilized in a metasurface (one unit call generates a background

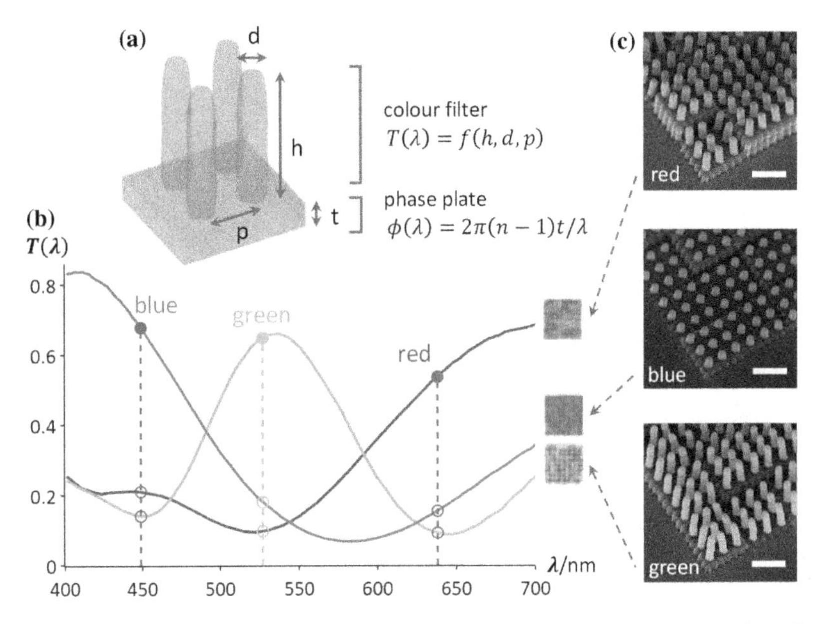

Fig. 11.45 **a** Schematic of a holographic color pixel that provides combined phase and amplitude control, comprising an array of pillars (color filter) integrated on top of a block (phase plate) in a dielectric with refractive index n. The color filter controls the amplitude of light based on its transmission spectrum $T(\lambda) = f(h, d, p)$, which depends on the pillar array dimensions (h, d, p) (height, diameter, and pitch). The phase plate controls the phase of transmitted light where the phase shift arises from path length differences that depend on the block thickness. **b** Transmission spectra and corresponding optical micrographs of pillar arrays with red, green, and blue colors. **c** False-color tilt-view SEMs of pillar arrays with dimensions optimized to give the best selectivity for RGB light. Reproduced from [47] with permission. Copyright 2019, The authors

color while the other unit cell generates a character "π"), not only a holography can be generated in the transmission but also a two-color pattern will be generated in the reflection side (Fig. 11.44c) [46]. Consequently, the crypto-display looks a normal reflective display under white light illumination, but generates a hologram that reveals the encrypted phase information under single-wavelength coherent light illumination. Since two operation modes do not affect each other, the crypto-display can have applications in security techniques [46].

An alternative method to realize simultaneous structural color and hologram is shown in Fig. 11.45 [47]. The top layer color filter comprises an array of pillar, and the bottom layer is composed of dielectric plates with thickness t and refractive index n. With coherent monochromatic illumination, the incident light is filtered ($T(\lambda) = f(h, d, p)$) such that only the relevant phase plates (written as $\phi(\lambda) = 2\pi(n-1)t/\lambda$) with color filters that closely match the illumination wavelength are selected for a given holographic projection, while the other phase plates do not form a projection as their color filters are mismatched [47]. Therefore, the multiplexed holographic color print will show different holographic projections when illuminated by red, green,

Fig. 11.46 **a** Optical characterization of the print: Transmission optical micrograph of a two-tone multiplexed hologram in which the pixels are arranged to form a QR code color image (left). Holographic projections in transmission, photographed on a white wall in a dark room: The image of a Chinese seal is shown under 638 nm red laser illumination (center), and the image of a penny black stamp is shown under 449 nm blue laser illumination (right). **b** SEMs of the print at various scales. Reproduced from [47] with permission. Copyright 2019, The authors

and blue lasers. Under incoherent white light illumination, the phase modulation of the holograms is effectively ignored and the color filters act as amplitude-modulating color pixels that together show the desired color image [47].

A two-tone holographic color print is shown in Fig. 11.46. Obviously, a QR code (blue and yellow pattern) is generated through the optical transmission of incoherent light (Fig. 11.46a), which is attributed to special arrangement of different geometric parameters of the pixel (Fig. 11.46b). For coherent light illumination, the sample will generate different holography, as indicated in Fig. 11.46a [47].

References

1. Isaac Newton. https://en.wikipedia.org/wiki/Isaac_Newton
2. Pigment. https://en.wikipedia.org/wiki/Pigment
3. J.J. Cowan, Aztec surface-relief volume diffractive structure. J. Opt. Soc. Am. A **7**, 1529–1544 (1990)
4. P. Jiang, D.W. Smith, J.M. Ballato, S.H. Foulger, Multicolor pattern generation in photonic bandgap composites. Adv. Mater. **17**, 179–184 (2005)
5. Z.-Z. Gu, H. Uetsuka, K. Takahashi, R. Nakajima, H. Onishi, A. Fujishima, O. Sato, Structural color and the lotus effect. Angew. Chem.-Int. Ed. **42**, 894–897 (2003)
6. S. Kinoshita, S. Yoshioka, J. Miyazaki, Physics of structural colors. Rep. Prog. Phys. **71**, 076401 (2008)
7. I. Freestone, N. Meeks, M. Sax, C. Higgitt, The lycurgus cup—a Roman nanotechnology. Gold Bull. **40**, 270–277 (2007)

8. M.L. Brongersma, Introductory lecture: nanoplasmonics. Faraday. Discuss. **178**, 9–36 (2015)
9. T. Xu, H. Shi, Y.-K. Wu, A.F. Kaplan, J.G. Ok, L.J. Guo, Structural colors: from plasmonic to carbon nanostructures. Small **7**, 3128–3136 (2011)
10. T.W. Ebbesen, H.J. Lezec, H.F. Ghaemi, T. Thio, P.A. Wolff, Extraordinary optical transmission through sub-wavelength hole arrays. Nature **391**, 667–669 (1998)
11. W.L. Barnes, A. Dereux, T.W. Ebbesen, Surface plasmon subwavelength optics. Nature **424**, 824–830 (2003)
12. C. Genet, T.W. Ebbesen, Light in tiny holes. Nature **445**, 39–46 (2007)
13. D. Inoue, A. Miura, T. Nomura, H. Fujikawa, K. Sato, N. Ikeda, D. Tsuya, Y. Sugimoto, Y. Koide, Polarization independent visible color filter comprising an aluminum film with surface-plasmon enhanced transmission through a subwavelength array of holes. Appl. Phys. Lett. **98**, 093113 (2011)
14. C. Wang, C. Du, X. Luo, Refining the model of light diffraction from a subwavelength slit surrounded by grooves on a metallic film. Phys. Rev. B **74**, 245403 (2006)
15. C. Wang, C. Du, Y. Lv, X. Luo, Surface electromagnetic wave excitation and diffraction by subwavelength slit with periodically patterned metallic grooves. Opt. Express **14**, 5671–5681 (2006)
16. T. Xu, Y.-K. Wu, X. Luo, L.J. Guo, Plasmonic nanoresonators for high-resolution colour filtering and spectral imaging. Nat. Commun. **1**, 59 (2010)
17. Z. Yang, Y. Zhou, Y. Chen, Y. Wang, P. Dai, Z. Zhang, H. Duan, Reflective color filters and monolithic color printing based on asymmetric fabry-perot cavities using nickel as a broadband absorber. Adv. Opt. Mater. **4**, 1196–1202 (2016)
18. W. Yue, S. Gao, S.-S. Lee, E.-S. Kim, D.-Y. Choi, Highly reflective subtractive color filters capitalizing on a silicon metasurface integrated with nanostructured aluminum mirrors. Laser Photonics Rev. **11**, 1600285 (2017)
19. M. Song, X. Li, M. Pu, Y. Guo, K. Liu, H. Yu, X. Ma, X. Luo, Color display and encryption with a plasmonic polarizing metamirror. Nanophotonics **7**, 323 (2018)
20. X. Zang, F. Dong, F. Yue, C. Zhang, L. Xu, Z. Song, M. Chen, P.-Y. Chen, G.S. Buller, Y. Zhu, S. Zhuang, W. Chu, S. Zhang, X. Chen, Polarization encoded color image embedded in a dielectric metasurface. Adv. Mater. **30**, 1707499 (2018)
21. S. Song, X. Ma, M. Pu, X. Li, K. Liu, P. Gao, Z. Zhao, Y. Wang, C. Wang, X. Luo, Actively tunable structural color rendering with tensile substrate. Adv. Opt. Mater. **5**, 1600829 (2017)
22. M.L. Tseng, J. Yang, M. Semmlinger, C. Zhang, P. Nordlander, N.J. Halas, Two-dimensional active tuning of an aluminum plasmonic array for full-spectrum response. Nano Lett. **17**, 6034–6039 (2017)
23. T. Xu, E.C. Walter, A. Agrawal, C. Bohn, J. Velmurugan, W. Zhu, H.J. Lezec, A.A. Talin, High-contrast and fast electrochromic switching enabled by plasmonics. Nat. Commun. **7**, 10479 (2016)
24. Y. Gao, C. Huang, C. Hao, S. Sun, L. Zhang, C. Zhang, Z. Duan, K. Wang, Z. Jin, N. Zhang, A.V. Kildishev, C.-W. Qiu, Q. Song, S. Xiao, Lead halide perovskite nanostructures for dynamic color display. ACS Nano **12**, 8847–8854 (2018)
25. Q.-T. Li, F. Dong, B. Wang, F. Gan, J. Chen, Z. Song, L. Xu, W. Chu, Y.-F. Xiao, Q. Gong, Y. Li, Polarization-independent and high-efficiency dielectric metasurfaces for visible light. Opt. Express **24**, 16309–16319 (2016)
26. K.E. Chong, L. Wang, I. Staude, A.R. James, J. Dominguez, S. Liu, G.S. Subramania, M. Decker, D.N. Neshev, I. Brener, Y.S. Kivshar, Efficient polarization-insensitive complex wavefront control using Huygens' metasurfaces based on dielectric resonant meta-atoms. ACS Photonics **3**, 514–519 (2016)
27. K. Huang, H. Liu, F.J. Garcia-Vidal, M. Hong, B. Luk/'yanchuk, J. Teng, C.-W. Qiu, Ultrahigh-capacity non-periodic photon sieves operating in visible light. Nat. Commun. **6**, 7059 (2015)
28. X. Zhang, X. Li, J. Jin, M. Pu, X. Ma, J. Luo, Y. Guo, C. Wang, X. Luo, Polarization-independent broadband meta-holograms via polarization-dependent nanoholes. Nanoscale **10**, 9304–9310 (2018)

29. L. Liu, X. Zhang, M. Kenney, X. Su, N. Xu, C. Ouyang, Y. Shi, J. Han, W. Zhang, S. Zhang, Broadband metasurfaces with simultaneous control of phase and amplitude. Adv. Mater. **26**, 5031–5036 (2014)
30. M.V. Berry, Quantal phase factors accompanying adiabatic changes. Proc. R. Soc. Lond. Math. Phys. Eng. Sci. **392**, 45–57 (1984)
31. X. Zhang, M. Pu, J. Jin, X. Li, P. Gao, X. Ma, C. Wang, X. Luo, Helicity multiplexed spin-orbit interaction in metasurface for colorized and encrypted. Ann. Phys. **529**, 1700248 (2017)
32. Y. Guo, M. Pu, Z. Zhao, Y. Wang, J. Jin, P. Gao, X. Li, X. Ma, X. Luo, Merging Geometric phase and plasmon retardation phase in continuously shaped metasurfaces for arbitrary orbital angular momentum generation. ACS Photonics **3**, 2022–2029 (2016)
33. F. Zhang, M. Pu, J. Luo, H. Yu, X. Luo, Symmetry breaking of photonic spin-orbit interactions in metasurfaces. Opto-Electron. Eng. **44**, 319–325 (2017)
34. J.P. Balthasar Mueller, N.A. Rubin, R.C. Devlin, B. Groever, F. Capasso, Metasurface polarization optics: independent phase control of arbitrary orthogonal states of polarization. Phys. Rev. Lett. **118**, 113901 (2017)
35. F. Zhang, M. Pu, X. Li, P. Gao, X. Ma, J. Luo, H. Yu, X. Luo, All-dielectric metasurfaces for simultaneous giant circular asymmetric transmission and wavefront shaping based on asymmetric photonic spin-orbit interactions. Adv. Funct. Mater. **27**, 1704295 (2017)
36. Z.-L. Deng, J. Deng, X. Zhuang, S. Wang, K. Li, Y. Wang, Y. Chi, X. Ye, J. Xu, G.P. Wang, R. Zhao, X. Wang, Y. Cao, X. Cheng, G. Li, X. Li, Diatomic metasurface for vectorial holography. Nano Lett. **18**, 2885–2892 (2018)
37. X. Zhang, J. Jin, M. Pu, X. Li, X. Ma, P. Gao, Z. Zhao, Y. Wang, C. Wang, X. Luo, Ultrahigh-capacity dynamic holographic displays via anisotropic nanoholes. Nanoscale **9**, 1409–1415 (2017)
38. X. Li, L. Chen, Y. Li, X. Zhang, M. Pu, Z. Zhao, X. Ma, Y. Wang, M. Hong, X. Luo, Multicolor 3D meta-holography by broadband plasmonic modulation. Sci. Adv. **2**, e1601102 (2016)
39. G. Zheng, H. Mühlenbernd, M. Kenney, G. Li, S. Zhang, Metasurface holograms reaching 80% efficiency. Nat. Nanotechnol. **10**, 308–312 (2015)
40. M. Pu, Z. Zhao, Y. Wang, X. Li, X. Ma, C. Hu, C. Wang, C. Huang, X. Luo, Spatially and spectrally engineered spin-orbit interaction for achromatic virtual shaping. Sci. Rep. **5**, 9822 (2015)
41. X. Xie, X. Li, M. Pu, X. Ma, K. Liu, Y. Guo, X. Luo, Plasmonic metasurfaces for simultaneous thermal infrared invisibility and holographic illusion. Adv. Funct. Mater. **28**, 1706673 (2018)
42. A. Nemati, Q. Wang, M. Hong, J. Teng, Tunable and reconfigurable metasurfaces and metadevices. Opto-Electron. Adv. **1**, 180009 (2018)
43. M. Zhang, M. Pu, F. Zhang, Y. Guo, Q. He, X. Ma, Y. Huang, X. Li, H. Yu, X. Luo, Plasmonic metasurfaces for switchable photonic spin-orbit interactions based on phase change materials. Adv. Sci. **5**, 1800835 (2018)
44. M. Pu, Y. Guo, X. Li, X. Ma, X. Luo, Revisitation of extraordinary Young's interference: from catenary optical fields to spin–orbit interaction in metasurfaces. ACS Photonics **5**, 3198–3204 (2018)
45. J. Li, S. Kamin, G. Zheng, F. Neubrech, S. Zhang, N. Liu, Addressable metasurfaces for dynamic holography and optical information encryption. Sci. Adv. **4**, eaar6768 (2018)
46. G. Yoon, D. Lee, K.T. Nam, J. Rho, "Crypto-display" in dual-mode metasurfaces by simultaneous control of phase and spectral responses. ACS Nano **12**, 6421–6428 (2018)
47. K.T. Lim, H. Liu, Y. Liu, J.K. Yang, Holographic colour prints for enhanced optical security by combined phase and amplitude control. Nat. Commun. **10**, 25 (2019)

Chapter 12
Polarization Manipulation, Detection, and Imaging

Abstract Among all the properties of electromagnetic wave, polarization plays an important role. Polarization sensitivity has been utilized to substantially enhance the functionality of optical technologies, such as in polarization spectroscopy, microscopy, and imaging systems. Manipulation and detection of polarized light wave have been a hot topic for quite a long time but limited by the following two reasons. On the one hand, the anisotropy and dichroism in naturally occurring material are quite weak. On the other hand, different from the intensities and frequencies that are easy to assess using suitable detectors and spectrometers, the polarization state is difficult to probe experimentally since the inherent vectorial information is completely lost in the conventional detection process. As a consequence, the optical systems involving polarization applications have generally been of complexity and bulky size. Miniaturization of these devices and systems is highly desirable for the development of polarization-sensitive systems. The recent development of metamaterials and metasurfaces provides new opportunities to achieve polarization manipulation, measurement, and imaging with ultra-thin artificial structures. In this chapter, we shall give a detailed discussion about the polarization manipulation, detection, and imaging in EO 2.0.

Keywords Polarization conversion · Form birefringence · Anisotropic · Chiral · Polarization imaging

12.1 Introduction

It is well known in physical textbooks that there are two basic kinds of polarization states, i.e., the linear polarization and circular polarization. They are often utilized as an expansion basis in anisotropic and chiral materials. When a wave incidents on an anisotropic slab with its fast and slow optical axes parallel to the surface, it can be decomposed into two linearly polarized waves aligned along the optical axes. After passing through this slab, the two components would attain different phase shifts and thus combine to be a new wave with its polarization state determined by the phase difference. Similarly, for chiral materials where circular polarizations are

X. Luo, *Engineering Optics 2.0*, https://doi.org/10.1007/978-981-13-5755-8_12

normal modes, any input wave could be decomposed into two circular waves, which would combine together after passing from this material. In the following, the Jones Matrix formalism is used to explain the above effect. Although it cannot describe unpolarized or partially polarized light and is limited to treating only completely polarized light, the Jones Matrix is "simpler" than the Mueller matrix [1].

In general, the Jones vector for a wave could be written as:

$$\mathbf{J} = \begin{bmatrix} A_x \\ A_y \end{bmatrix}, \tag{12.1.1}$$

where A_x and A_y are the complex amplitudes along the x- and y-direction, respectively. For linear polarization in x-direction, the Jones vector can be written as:

$$\mathbf{J} = \begin{bmatrix} 1 \\ 0 \end{bmatrix}. \tag{12.1.2}$$

When the linearly polarized wave has a rotation angle of θ with respect to x, there is

$$\mathbf{J} = \begin{bmatrix} \cos\theta \\ \sin\theta \end{bmatrix}. \tag{12.1.3}$$

For right-handed circular polarization (RCP) and left-handed circular polarization (LCP), there are

$$\mathbf{J} = \frac{1}{\sqrt{2}}\begin{bmatrix} 1 \\ j \end{bmatrix} \tag{12.1.4}$$

and

$$\mathbf{J} = \frac{1}{\sqrt{2}}\begin{bmatrix} 1 \\ -j \end{bmatrix}. \tag{12.1.5}$$

For any polarization system, its optical performance could be described by a 2×2 matrix $\mathbf{T}$, so that the output vector $\mathbf{J}_2$ could be connected with the input one $\mathbf{J}_1$ via

$$\mathbf{J}_2 = \mathbf{T}\mathbf{J}_1. \tag{12.1.6}$$

For an anisotropic waveplate which induces a phase shift of $\varphi/2$ along the x-direction and $-\varphi/2$ along the y-direction, the matrix is

$$\mathbf{T} = \begin{bmatrix} e^{j\varphi/2} & 0 \\ 0 & e^{-j\varphi/2} \end{bmatrix}, \tag{12.1.7}$$

which can be recast into

$$\mathbf{T}_1 = \begin{bmatrix} 1 & 0 \\ 0 & e^{-j\varphi} \end{bmatrix}. \tag{12.1.8}$$

The Jones matrices for a quarter-wave plate (QWP) $\varphi = \pi/2$ and half-wave plate (HWP) $\varphi = \pi$ are,

$$\mathbf{T}_1 = \begin{bmatrix} 1 & 0 \\ 0 & -j \end{bmatrix} \tag{12.1.9}$$

and

$$\mathbf{T}_1 = \begin{bmatrix} 1 & 0 \\ 0 & -1 \end{bmatrix}. \tag{12.1.10}$$

The Jones matrix for a rotator is

$$\mathbf{R} = \begin{bmatrix} \cos\theta & \sin\theta \\ -\sin\theta & \cos\theta \end{bmatrix}. \tag{12.1.11}$$

The Jones matrix $\mathbf{T}$ of an optical system is transformed into $\mathbf{T}'$ by

$$\mathbf{T}' = \mathbf{R}(\theta)\mathbf{T}\mathbf{R}(-\theta). \tag{12.1.12}$$

12.1.1 Birefringent Crystals

The propagation of light in crystals under general conditions is rather complex. However, the rule is relatively simple if the plane wave is traveling in a direction parallel to the principal axes of the crystal. In this case, the difference of phase shift for the two orthogonal components can be written as:

$$\varphi = \varphi_y - \varphi_x = (n_2 - n_1)k_0 d \tag{12.1.13}$$

where n_2 and n_1 are the refractive indexes along the slow and fast optical axes, $k_0 = 2\pi/\lambda_0$ is the wave number in vacuum, and d is the thickness of the sample. From the above equation, it can be seen that if the difference in refractive index is a constant, the phase shift is inversely proportional to the wavelength, which means that the operational bandwidth is intrinsically limited. Furthermore, since the refractive index is often small for many naturally occurring materials, as can be seen in Table 12.1, the thickness of the sample should be much larger than the wavelength. In many other

Table 12.1 Refractive indexes of uniaxial crystals, at $\lambda = 590$ nm [2]

Material	Crystal system	n_o	n_e	Δn
Barium borate, BaB_2O_4	Trigonal	1.6776	1.5534	−0.1242
Beryl, $Be_3Al_2(SiO_3)_6$	Hexagonal	1.602	1.557	−0.045
Calcite, $CaCO_3$	Trigonal	1.658	1.486	−0.172
Ice, H_2O	Hexagonal	1.309	1.313	+0.004
Lithium niobate, $LiNbO_3$	Trigonal	2.272	2.187	−0.085
Magnesium fluoride, MgF_2	Tetragonal	1.380	1.385	+0.006
Quartz, SiO_2	Trigonal	1.544	1.553	+0.009
Ruby, Al_2O_3	Trigonal	1.770	1.762	−0.008
Rutile, TiO_2	Tetragonal	2.616	2.903	+0.287
Sapphire, Al_2O_3	Trigonal	1.768	1.760	−0.008
Silicon carbide, SiC	Hexagonal	2.647	2.693	+0.046
Tourmaline (complex silicate)	Trigonal	1.669	1.638	−0.031
Zircon, high $ZrSiO_4$	Tetragonal	1.960	2.015	+0.055
Zircon, low $ZrSiO_4$	Tetragonal	1.920	1.967	+0.047

applications, these materials should be made to be amorphous or polycrystalline to reduce the polarizing effects.

12.1.2 *Optical Activity*

The phenomenon of optical activity, that is, the ability to rotate the linear polarization state of light, is a fundamental effect of electrodynamics which is traditionally associated with mirror asymmetry (chirality) of organic molecules. For optical active materials, the normal modes are circularly polarized waves and the propagation properties are determined by their corresponding refractive indexes. For linearly polarized incidence, the two circular components would experience different phase delay; thus, the output wave is still linearly polarized but with a rotation of the polarization direction. The rotatory power (rotation angle per unit length) is written as:

$$\rho = \frac{\pi}{\lambda_0}(n_- - n_+) \tag{12.1.14}$$

where the subscripts "+" and "−", respectively, stand for the RCP and LCP. Similar to the anisotropic case, the optical rotation is dependent on both the length and wavelength. Moreover, the two factors are often linked to each other. For example, if one chooses a larger refractive index to shorten the thickness of length, the dispersion of the refractive index would increase. When the difference in refractive index becomes large enough, the interfacial reflection must be considered. As demonstrated by the gold-helix metamaterials, one circular polarization can pass through, while the other is almost totally reflected [3, 4].

12.1.3 Interfacial Polarization Effect

Besides the polarization effects in bulky anisotropic or chiral materials, the reflection and refraction at a single interface could also induce strong polarization splitting or conversion. For a homogeneous and isotropic interface, the reflection and refraction are determined by the Fresnel's equations (see Chap. 2). At normal incidence, the two linear components have equal reflection and transmission. However, the difference is obvious for oblique incidence [5], as highlighted by the Brewster effect for lossless materials and pseudo-Brewster effect for lossy material [6]. If a subwavelength metallic grating is placed at the interface, only one polarization with electric field perpendicular to the metallic lines can pass; thus, it could operate as a linear polarizer at a wide angle of incidence. More interestingly, with proper subwavelength surface structures, the reflection and transmission rules could be extended to the so-called generalized Fresnel's equations [7]. By cascading anisotropic metasurfaces, both birefringent and chiral artificial materials could be constructed to realize broadband polarization control.

12.2 Artificial Anisotropic and Chiral Materials

Typical artificial anisotropic and chiral subwavelength structures are shown in Fig. 12.1, mainly encompassing artificial anisotropic and chiral subwavelength structures. In principle, there are two main methods to get the polarization rotation or conversion, so-called phase and amplitude routes [8]. The former exploits the phase difference in birefringent media with approximately equal transmitted amplitudes. The latter relies on different transmission coefficients, also termed as dichroism of material. In such medium, the unnecessary polarization is filtered out by the higher absorption and/or higher reflection and the remains dominate the output polarization.

12.2.1 Anisotropic Phase Engineering

According to the operation manner, polarization converters can be classified into two categories, i.e., transmissive and reflective types. For transmissive polarization converter, it requires simultaneous control of the transmission amplitude and phase along the two axes of anisotropic medium. Early investigations mainly focus on the microwave band and subsequently extend to higher frequency band including tera-

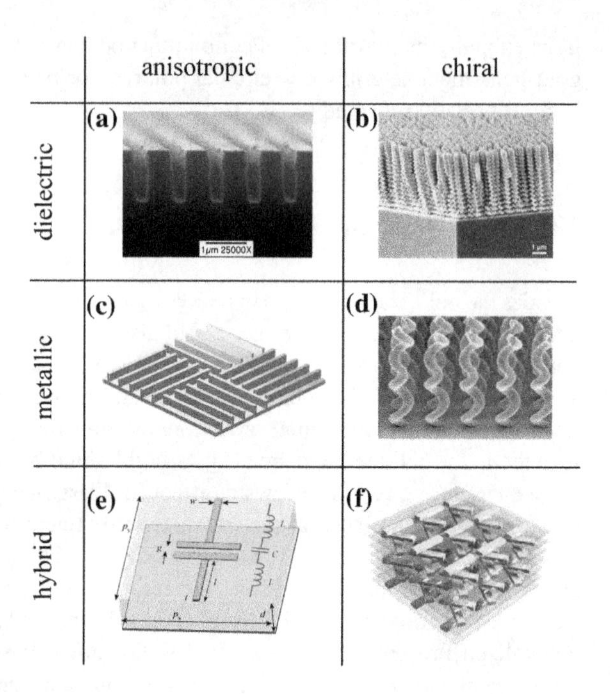

Fig. 12.1 Typical artificial anisotropic and chiral subwavelength structures. Reproduced from **a** [9], **b** [10], **c** [11], **d** [4], **e** [12], and **f** [13] with permission. **a** Copyright 1999, Optical Society of America. **b** Copyright 1996, Nature Publishing group. **c** Copyright 2018, WILEY-VCH Verlag GmbH & Co. KGaA, Weinheim. **d** Copyright 2015, WILEY-VCH Verlag GmbH & Co. KGaA, Weinheim. **e** Copyright 2013, American Institute of Physics. **f** Copyright 2012, Macmillan Publishers Limited

hertz, infrared, and visible with the advances of theories and fabrication technologies. Note that anisotropic structures can be utilized as building blocks for geometric phase structures [14, 15], as discussed in Chap. 2.

12.2.1.1 Transmissive Type

One kind of early used anisotropic materials is subwavelength gratings, which can be taken as homogeneous birefringent materials. For the case of one-dimensional gratings (the parameters are defined in Fig. 12.2), the following equations predict the birefringence versus linewidth-to-period ratio of the gratings [16]:

$$n_{\parallel} = \left[n_1^2 q + n_2^2 (1 - q)\right]^{1/2} \tag{12.2.1}$$

$$n_{\perp} = \left[\left(1/n_1^2\right) q + \left(1/n_2^2\right)(1 - q)\right]^{-1/2} \tag{12.2.2}$$

A potential application of such an artificial birefringent material is in QWP and HWP. These devices require that the grating thickness be $t = \lambda/4\Delta n$ for a QWP and $t = \lambda/2\Delta n$ for a HWP, where $\Delta n = |n_1 - n_2|$. A typical grating cross section is shown in Fig. 12.3a [9]. In the mid-infrared range, the etch depth is approximately 1.23 μm and the sidewalls are slightly sloped. The phase retardation and the transmission coefficients of the transverse electric (TE) and transverse magnetic (TM) modes were measured at normal incidence by using a Fourier transform infrared (FTIR) spectropolarimeter. As illustrated in Fig. 12.3b, the measured phase retardation of the QWP over the 3.5–5 μm wavelength range varies from 0.49π to 0.57π (89°–102°), and the rigorous coupled wave analysis (RCWA) simulation result agrees well with experimental measurement.

Another early anisotropic waveplate made of subwavelength structures is the multilayer meander-line [17], which possess inductive and capacitive impedances along two orthogonal directions (Fig. 12.4). When the electric field is polarized perpendicular to the lines, the structure acts as a capacitor and introduces a positive phase retardation. For the other polarization, the surface acts as an inductor and a negative phase shift is induced. The transmittance for both the two polarizations is guaranteed by the destructive interference in reflection [18].

Resonant antennas can also be utilized to realize anisotropic phase shift, which is attributed to their polarization-dependent responses. Ma et al. [19] reported a single-layer metasurface to convert the incident linearly polarized wave into circularly polarized one in a broadband, as shown in Fig. 12.5. The building block is composed of a 135°-oriented metallic strip located at the center of a metallic ring. Owing to the anisotropic impedance, anisotropic transmission coefficients can be obtained. By optimizing the geometries of the anisotropic structures to generate equal transmission amplitude and phase retardation of $\pi/4$, this metasurface can transform the incident linearly polarized light into circularly polarized one in the frequency range from 13.5 to 15.3 GHz. The 3-dB axial ratio (AR) bandwidth of this polarizer is 17%, and its thickness is only $\lambda/40$ (λ is the central working wavelength).

The above polarization manipulation principle can be extended to higher frequency band. One typical transmissive terahertz polarization converter is composed of anisotropic metallic cut-wire arrays, as shown in Fig. 12.6 [20]. Two-side metallic gratings are utilized to ensure only the orthogonal polarization component transmits (Fig. 12.6a). Extra super- and sub-cladding layers are utilized to generate multi-reflections. The consequent interferences of polarization couplings in the multi-reflections are utilized to enhance (reduce) the overall transmitted fields with cross-polarizations (co-polarizations). Figure 12.6b shows the theoretical results by multi-reflection model and measured reflection and transmission, from which one can see

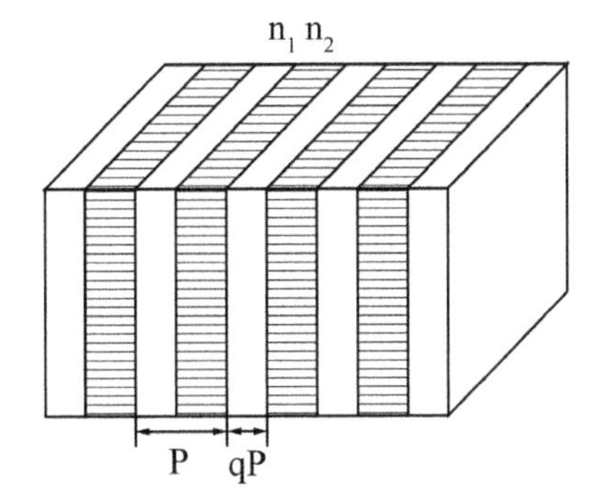

Fig. 12.2 Illustration of the geometry of a square profile grating on a substrate

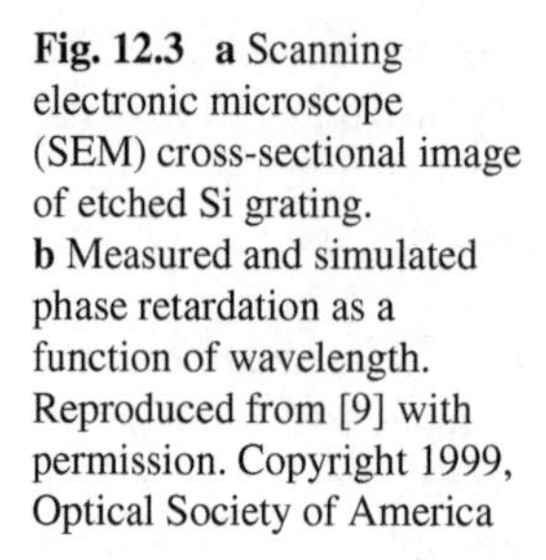

Fig. 12.3 **a** Scanning electronic microscope (SEM) cross-sectional image of etched Si grating. **b** Measured and simulated phase retardation as a function of wavelength. Reproduced from [9] with permission. Copyright 1999, Optical Society of America

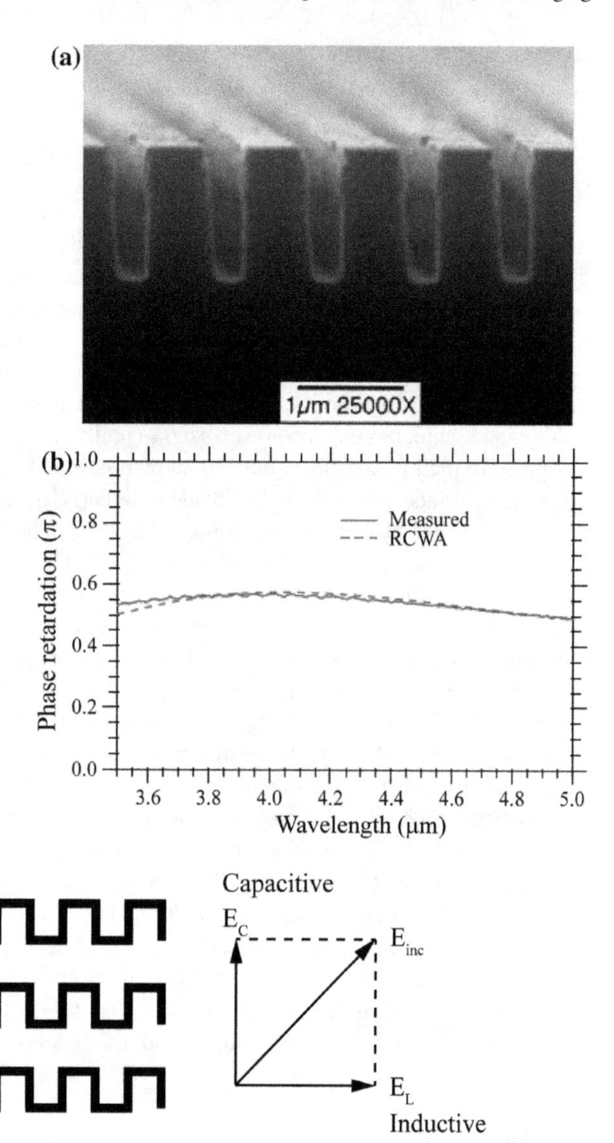

Capacitive

E_C

E_{inc}

E_L

Inductive

Fig. 12.4 A front view of the meander-line along with the incidence electric fields

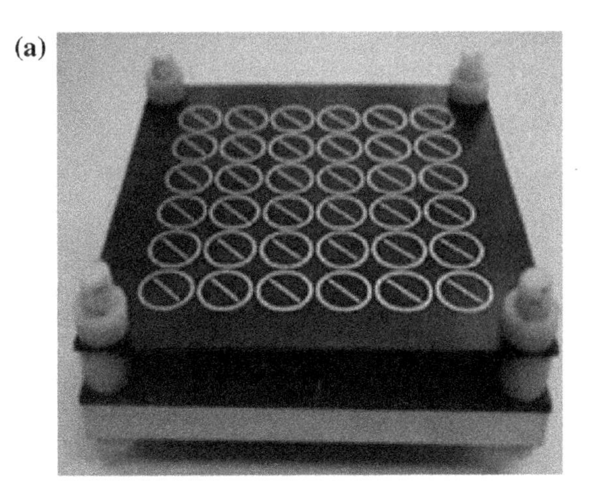

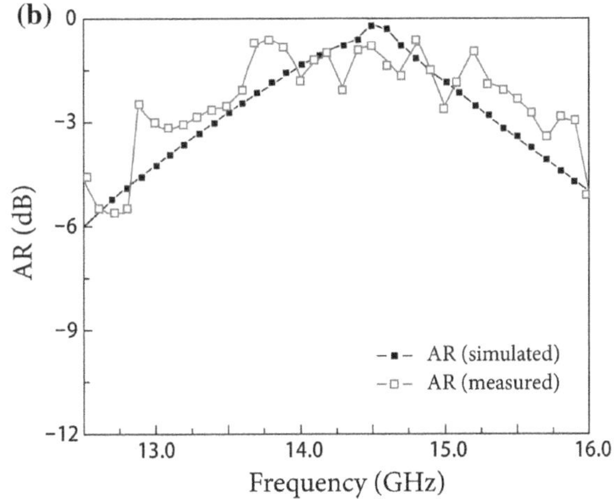

Fig. 12.5 **a** Photograph of the fabricated polarizer. **b** Measured and simulated axial ratio (AR) for the polarizer. Reproduced from [19] with permission. Copyright 2012, Wiley Periodicals, Inc.

that the device is able to rotate the linear polarization by 90°, with a conversion efficiency exceeding 50% from 0.52 to 1.82 THz. The highest efficiency of 80% appears at 1.04 THz. The device performance, limited by the dielectric loss and polarized reflection, can be further improved through optimizing the structural design and using lower-loss dielectric materials.

Along with the emergence of polarization-dependent phase-gradient metasurface [21, 22], it was found that one could control the polarization state almost arbitrarily by controlling different regions in the same metasurfaces. For instance, QWP operating at infrared band has been proposed by constructing two V-shaped antenna super-units with a horizontal offset d (Fig. 12.7). Each super-unit generated linear gradient phase for cross-polarization, which generate two co-propagating waves with equal amplitudes (ensured by symmetric super-units with respect to the incidence plane), orthogonal polarizations (due to the $\pi/4$ orientation difference between them), and a

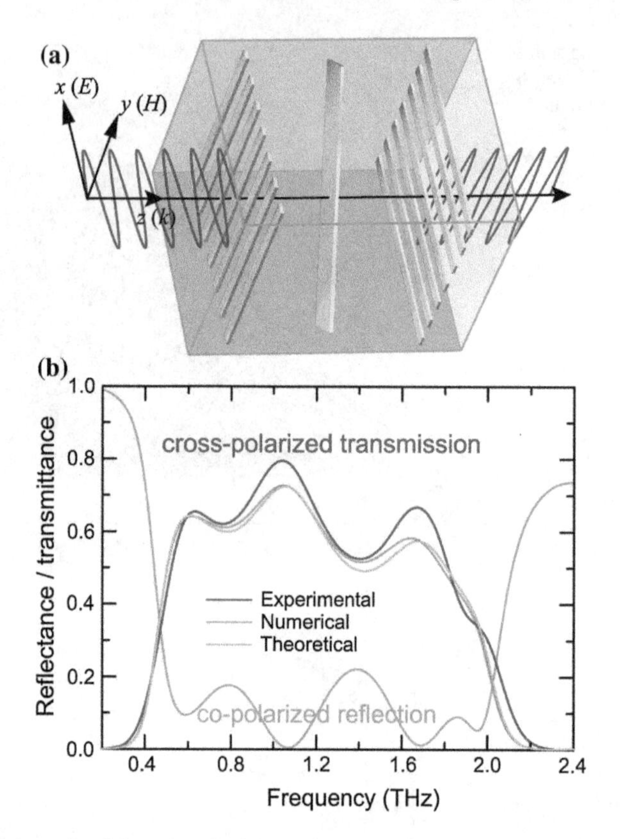

Fig. 12.6 **a** Schematic of the unit cell of the metamaterial linear polarization converter. **b** Cross-polarized transmittance obtained through experimental measurements, numerical simulations, and theoretical calculations. Also shown is the numerically simulated co-polarized reflectance. Reproduced from [20] with permission. Copyright 2013, American Association for the Advancement of Science

$\pi/2$ phase difference (stems from the horizontal offset $d = \Gamma/4$, where Γ is the period of super-unit cell). The waves coherently interfere, producing a circularly polarized extraordinary beam that bends away from the propagation direction of the ordinary beam, with a deflection angle of $\theta = \sin^{-1}(\lambda/\Gamma)$. Due to the spatial separation of the two beams, the extraordinary beam is almost background-free.

Besides polarization converters, polarization rotators were also demonstrated. In principle, a linearly polarized (LP) light with an angle of polarization φ can be written as a superposition of its circular components as follows [24]:

$$\mathbf{E} = E_0\left(\hat{\mathbf{x}}\cos\varphi + \hat{\mathbf{y}}\sin\varphi\right) = \frac{E_0}{\sqrt{2}}\left(\hat{\mathbf{r}}\mathrm{e}^{-i\varphi} + \hat{\mathbf{l}}\mathrm{e}^{-i\varphi}\right) \tag{12.2.3}$$

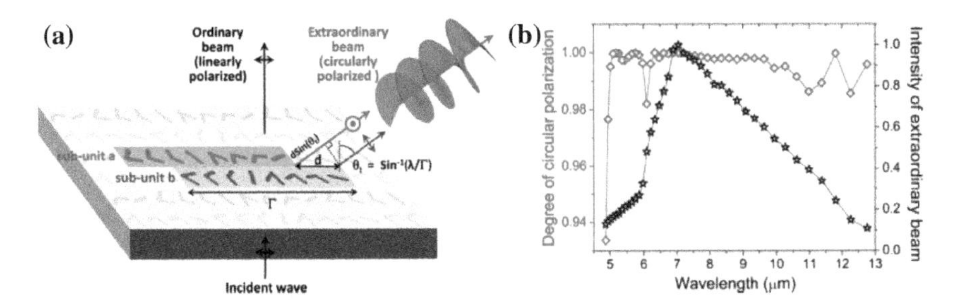

Fig. 12.7 **a** A background-free metasurface-based QWP consisting of two V-shaped antenna sub-units that generate two co-propagating waves with orthogonal linear polarizations, equal amplitudes, and a $\pi/2$ phase difference upon LP incidence. **b** Calculated degree of circular polarization and intensity of the extraordinary beam as a function of wavelength. Reproduced from [23] with permission. Copyright 2012, American Chemical Society

where $\hat{\mathbf{r}} = \left(\hat{\mathbf{x}} + i\hat{\mathbf{y}}\right)/\sqrt{2}$ and $\hat{\mathbf{r}} = \left(\hat{\mathbf{x}} - i\hat{\mathbf{y}}\right)/\sqrt{2}$ are the unit vectors of RCP and LCP light, respectively. Equation (12.2.3) indicates that a phase delay of 2φ introduced to the RCP component with respect to the LCP component will cause a rotation of polarization angle by a value of φ. If the RCP and the LCP coefficients are switched, the polarization angle will reverse from φ to $-\varphi$.

A metasurface constructed by two dumbbell-shaped antenna super-units with period p and horizontal offset d is proposed to realize polarization rotation [24]. The operation principle is illustrated in Fig. 12.8. For gradient metasurface composed of single super-units (Fig. 12.8a), it will transmit two beams upon excitation with circularly polarized light (CPL): normally transmitted co-polarized beam and cross-CPL deflected in an anomalous direction with $\theta = \sin^{-1}(\lambda/p)$. When the handedness of CPL is reversed, the cross-CPL will be deflected to an opposite direction (Fig. 12.8b). By superposition of the above two cases (Fig. 12.8c), the anomalous portion of the linearly polarized input beam is split into its circular components in two opposite diffraction directions. To retrieve linearly polarized output light in the anomalous direction, one can use two subarrays of anisotropic nanoantennas rotating in opposite directions as shown in Fig. 12.8d (two subarrays in blue and red). In this case, an offset distance $d = p/4$ is introduced between the two subunits to cause a $\pi/2$ phase shift between the RCP and LCP, leading to a 45° rotation of the output angle of polarization according to Eq. (12.2.3).

In another interesting experiment, it was shown that diffractive polarization rotators can be utilized to improve the efficiency of polarizers [25]. Conventional polarizers operate by rejecting undesired polarization, which limits their transmission efficiency to be much less than 50% when illuminated by unpolarized light. In order to realize a high-efficiency linear polarizer, it is desired a polarizer not only can transmit the polarization parallel to its principal axis but also can rotate light with polarization perpendicular to its principal axis by 90 degree. To perform the function above, an ultra-high-efficiency metamaterial polarizer was divided into 20 × 20

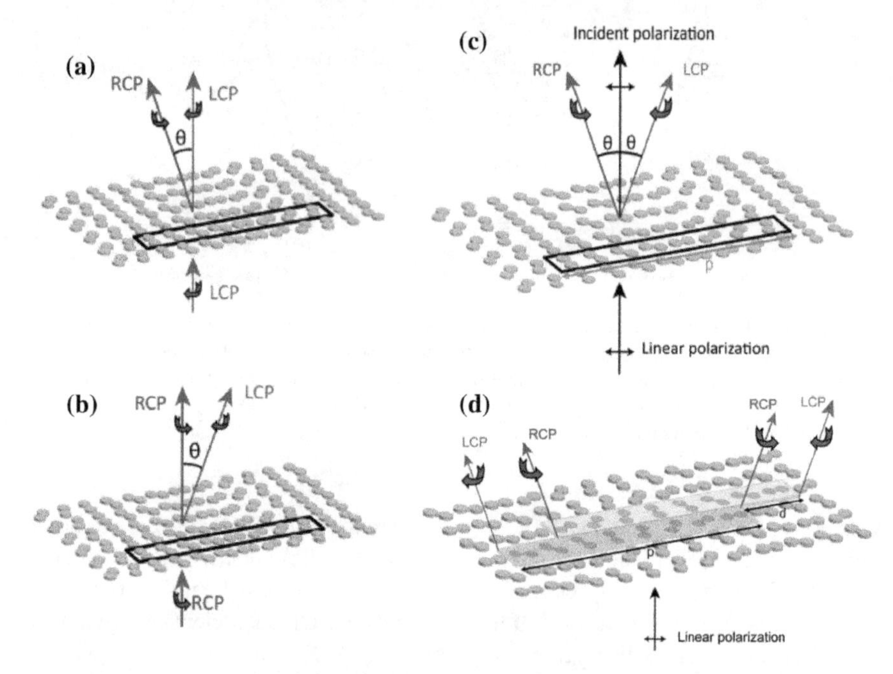

Fig. 12.8 **a**, **b** Effect of the metasurface on circularly polarized incident light and **c** applying superposition to obtain circular beam splitting effect for linearly polarized incident light; a part of the beam is transmitted normally with no change. **d** Metasurface structure that performs optical rotation. It consists of two subarrays (in blue and red) causing circular polarization splitting into two opposite diffraction directions. Reproduced from [24] with permission. Copyright 2014, American Chemical Society

pixels. Each pixel is composed of silicon with different etching depth covering an area of 200 nm × 200 nm pixels, as shown in Fig. 12.9 [25].

Optimization algorithm was utilized to maximize a figure of merit (FOM), which is defined as the transmission efficiency at the desired polarization (e.g., E_x), when the polarizer is illuminated by both polarizations (E_x and E_y) with equal amplitude. The optimized design is shown on the top left of Fig. 12.9a. The simulated performances of the fabricated device are shown in Fig. 12.9b–e. One can see that the optimized polarizer can effectively rotate the polarization component of E_y to the desired polarization E_x. Therefore, compared with the surrounding unpatterned silicon, the transmission of E_x is greatly enhanced.

12.2.1.2 Reflective Type

Compared with transmissive polarization converter, the reflective polarization converters possess higher efficiency. If we neglect the material loss, the polarization conversion performance only depends on the phase difference between them. By

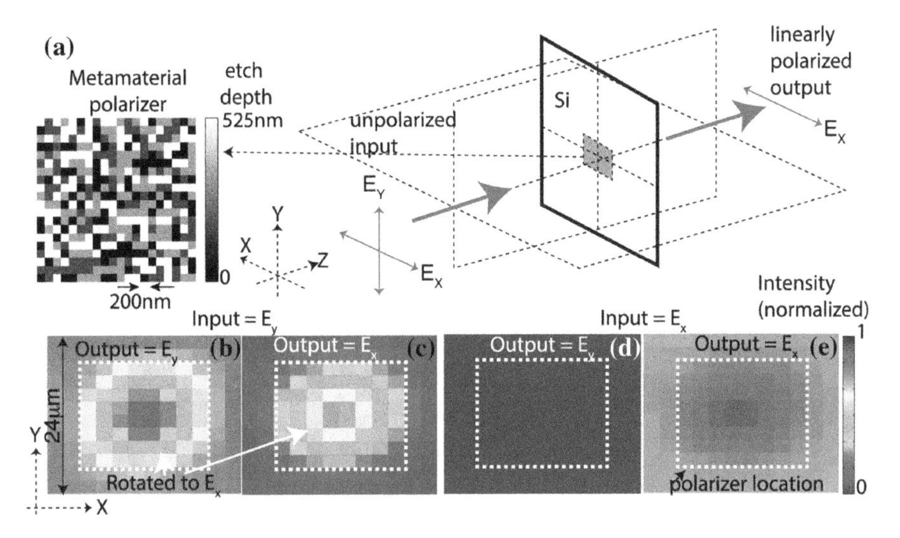

Fig. 12.9 **a** High-efficiency metamaterial polarizer. The design (left) is composed of etched square pixels in silicon. **b–e** Simulated light intensity distributions after transmission through the polarizer for **b** E_y and **c** E_x under E_y input and for **d** E_y and **e** E_x for E_x input. The white dashed lines in **b–e** indicate the boundaries of the finite device. Reproduced from [25] with permission. Copyright 2014, Optical Society of America

exploiting magnetic resonance formed in a reflective metasurface, thin polarization converter with nearly 100% conversion efficiency has been demonstrated [26]. Nevertheless, restricted by the magnetic resonance of high quality factor, the operation bandwidth is greatly limited. Subsequent investigations mostly resort to the electric resonance. It has been shown that by exploiting the dispersion engineering principle, the thickness–bandwidth limit can be approached and the operation bandwidth of reflective polarization converters can be greatly extended [27].

(1) One-dimensional dispersion engineering

One-dimensional dispersion engineering technology means that once the impedance along one axis is determined, the ideal impedance along the other axis can be theoretically deduced for dispersionless polarization conversion. Considering that all desired metamaterial response obeying the Kramers–Kronig relations can be approximated as a superposition of Lorentzian function and the Lorentzian function is a natural consequently of the LC resonance [28, 29], LC resonant unit cells are utilized as building block of dispersion engineering.

As an example to illustrate the implementation of dispersion engineering [30], one can assume single-serial LC resonance is excited in the metasurface when the electric field is along the x-direction. The frequency-dependent impedance $Z_x(\omega)=j\omega L + 1/(j\omega C)$ with $L = 2$ nH and $C = 100$ fF is plotted in red line in Fig. 12.10a. By submitting the impedance into the following equation:

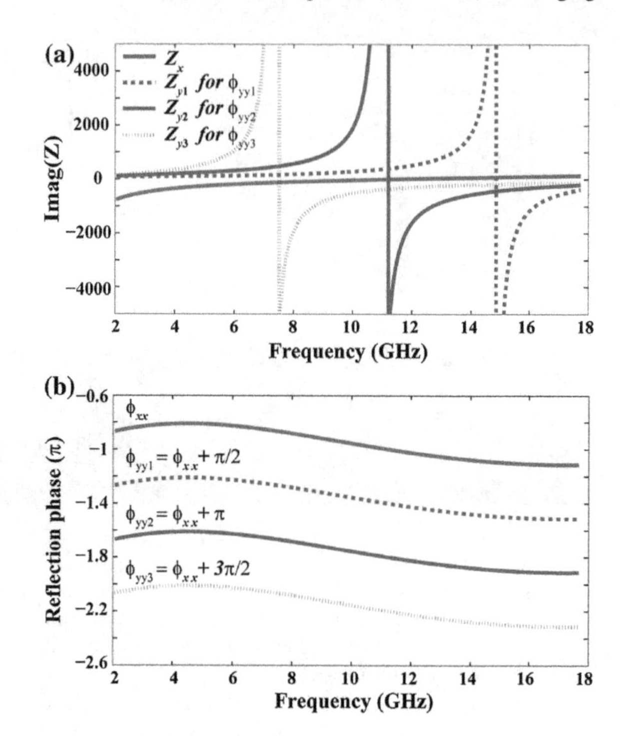

Fig. 12.10 **a** Given impedance Z_x (red solid line) and derived optimal impedance Z_y for 1/4 wave plate (blue dash line), 1/2 wave plate (blue solid line), and 3/4 wave plate (blue dot line). The real part is zero since no material loss is considered. The region above zero is inductive, and the region below zero is capacitive. **b** The given ϕ_{xx} and derived optimal ϕ_{yy} for achromatic 1/4, 1/2, and 3/4 wave plates. Reproduced from [30] with permission. Copyright 2015, The Authors

$$S_{ii} = \frac{-Z_0(1-\exp(2ikd)) - 2Z_i\exp(2ikd)}{Z_0(1-\exp(2ikd)) + 2Z_i}, \tag{12.2.4}$$

one can derive the dispersion $\phi_{xx}(\omega)$ and corresponding $\phi_{yy}(\omega)$ with different phase differences, e.g., $\pi/2$, π, and $3\pi/2$, which is shown in Fig. 12.10b. Subsequently, the optimal impedance of $Z_y(\omega)$ can be derived by submitting $S_{yy}(\omega) = \exp(-j\phi_{yy})$ into the following equation:

$$Z_i(\omega) = \frac{Z_0}{\frac{1-S_{ii}}{1+S_{ii}} - \frac{\exp(-ikd)+\exp(ikd)}{\exp(-ikd)-\exp(ikd)}}. \tag{12.2.5}$$

According to the variation trend of $Z_y(\omega)$ with frequency shown in Fig. 12.10a, a parallel *LC* circuit should be constructed along the *y*-direction. For the cases of more complex dispersion in the *x*-direction, multiple serial and parallel LC circuits would be needed.

For a typical anisotropic metasurface constructed by homogenous I-shaped stripe array (Fig. 12.11a) [12], we have $Z_x = \infty$ since it does not respond to the electric component along *x*-direction. In this case, the reflection phase along *x*-direction Φ_x is determined as $\pi + 2kd$, where *k* and *d*, respectively, are the wave number in the dielectric spacer and the thickness of spacer. Then the ideal impedance Z_y that can generate constant phase retardation $\Delta\Phi = \Phi_y - \Phi_x$ can be determined, as displayed

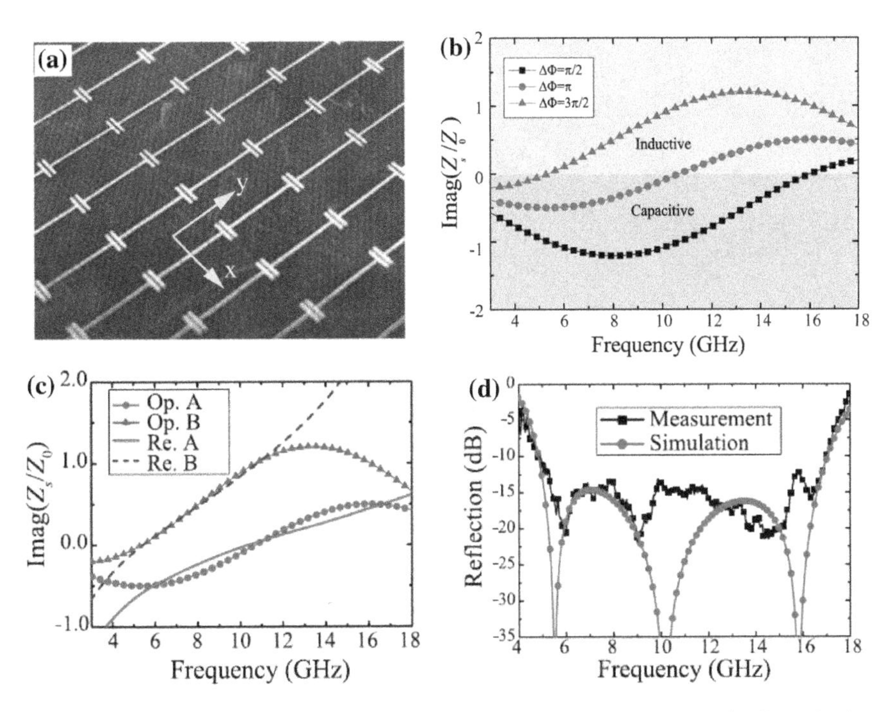

Fig. 12.11 **a** Photograph of the fabricated sample. **b** Retrieved ideal impedance for dispersionless phase difference. **c** Sheet impedances retrieved from S-parameters and the corresponding optimal impedances (Samples A and B, respectively, for QWP and HWP). **d** Measured and simulated co-polarization reflection coefficient. Reproduced from [12] with permission. Copyright 2013, American Institute of Physics

in Fig. 12.11b. By optimizing the geometries of I-shaped stripe array to approach the ideal impedance (Fig. 12.11c), broadband polarization conversion can be realized. For instance, the co-polarization reflection coefficient is suppressed below -15 dB in the frequency range from 5.5 to 16.5 GHz, as depicted in Fig. 12.11d.

After the experimental verification in microwave band, the one-dimensional dispersion engineering technology is extended to near-infrared band. A large-area (2 cm $\times$ 2 cm) linear polarization converter has been fabricated by virtue of orthogonal interference lithography [31], as shown in Fig. 12.12a, b. This fabrication strategy breaks the bottleneck of the fabrication of large-area metasurface in the optical regime, promising an unprecedented progress for optical communication and integrated optics. The major and minor axes of the ellipse-shaped plasmonic planar resonator were controlled by adjusting the exposure dosages and developing time. The period of metallic array is determined by the incident angle of laser beam. The measured polarization conversion ratio (PCR) is larger than 91.1% in a broadband from 730 to 1870 nm (Fig. 12.12c, d).

(2) Two-dimensional dispersion engineering

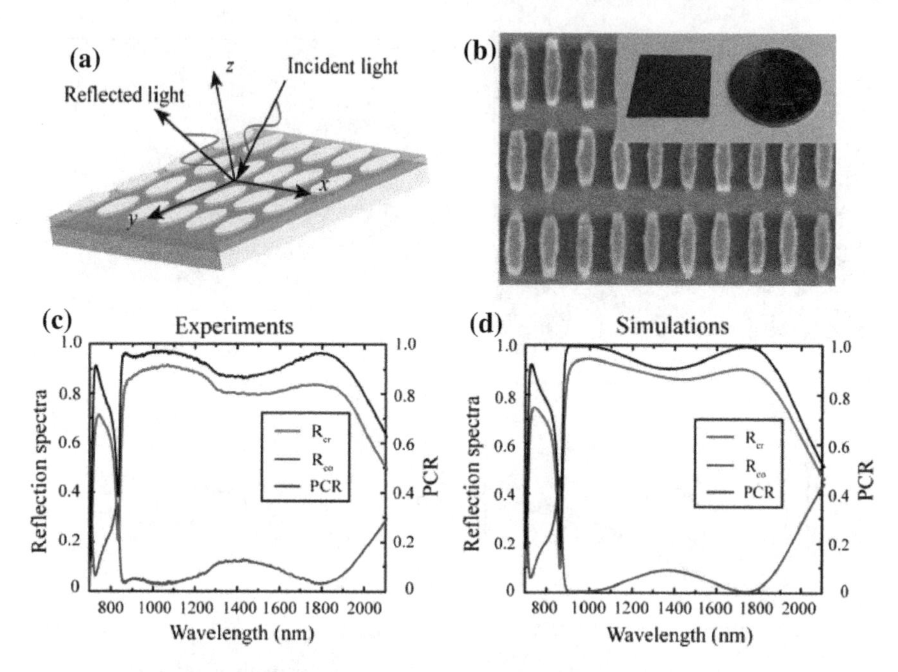

Fig. 12.12 **a, b** Schematic and SEM image of the linear polarization converter sample. Scale bar: 500 nm. The inset shows the photograph of the sample of the linear polarization converter. **c, d** Measured and simulated reflection spectrum and PCR with an incidence angle of 20°. Reproduced from [31] with permission. Copyright 2015, American Institute of Physics

It should be noted that dispersion engineering mentioned above is only implemented in 1D of the metasurfaces, which results in an operation bandwidth typically no more than two octaves. Therefore, it requires fully released dispersion engineered metasurface in both two dimensions to obtain broader bandwidth. Although it cannot directly deduce the ideal impedance pairs as there is only one constrained condition about phase difference, two-dimensional dispersion engineering may be implemented by the following method [32]. First, the dispersion of metasurface can be formulated with initial geometry according to Eq. (12.2.4). Subsequently, the requirement of impedance for target phase difference is derived according to Eq. (12.2.5). Finally, the geometry of metasurface is adjusted to approach the designated impedance. Since the adjustment of the geometry influences the dispersion in both dimension, sometimes we need to repeat the steps above to obtain the most optimized anisotropic dispersions. During the above processes, the relations between the circuit elements and geometry of meta-atoms are used in optimizing the final structure. For example, the inductor L is related to the current distribution in the metallic wires and increases with longer and thinner metallic structures. The capacitor C results from the electric field distribution in the gaps between metallic elements and tends to be larger for smaller distance between them.

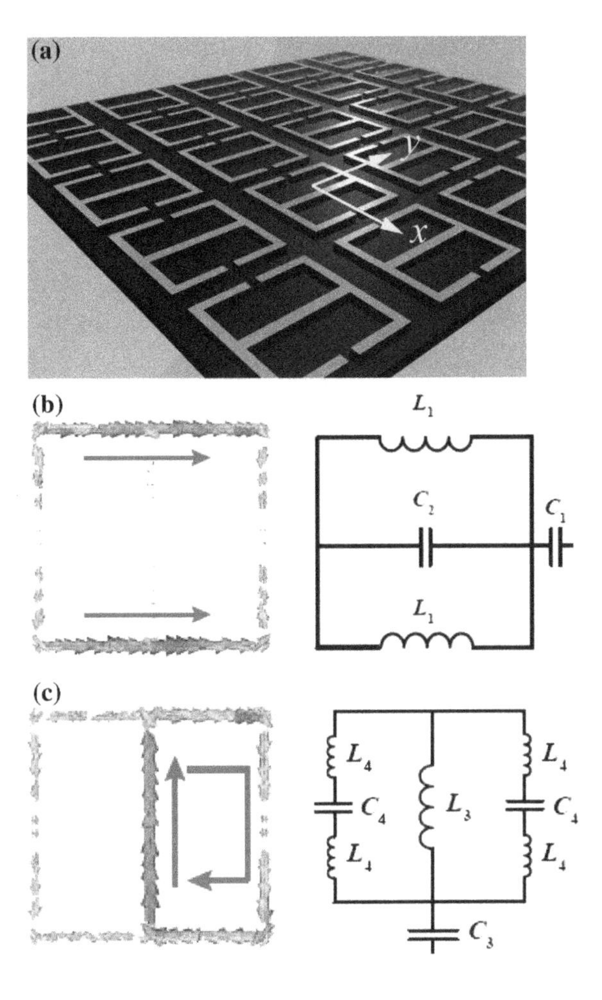

Fig. 12.13 **a** Perspective view of the broadband polarization converter through two-dimensional dispersion engineering. **b** Surface electric current under x-polarization illumination and corresponding equivalent circuit model. **c** Surface electric current under y-polarization illumination and corresponding equivalent circuit model. Reproduced from [30] with permission. Copyright 2015, The Authors

In order to construct the superimpositions of Lorentz dispersions in both dimensions, split-ring resonators (SRRs) sitting back to back are utilized [30], as shown in Fig. 12.13a. The surface current flow and equivalent circuit model under the x- and y-polarizations are presented in Fig. 12.13b and c. The frequency-dependent impedance can be expressed as:

$$Z_x(\omega) = \frac{1}{\frac{2}{j\omega L_1} + j\omega C_2} + \frac{1}{j\omega C_1}, \tag{12.2.6}$$

and

$$Z_y(\omega) = \frac{1}{\frac{1}{j\omega L_3} + \frac{1}{j\omega L_4 + \frac{1}{2j\omega C_4}}} + \frac{1}{j\omega C_3}. \tag{12.2.7}$$

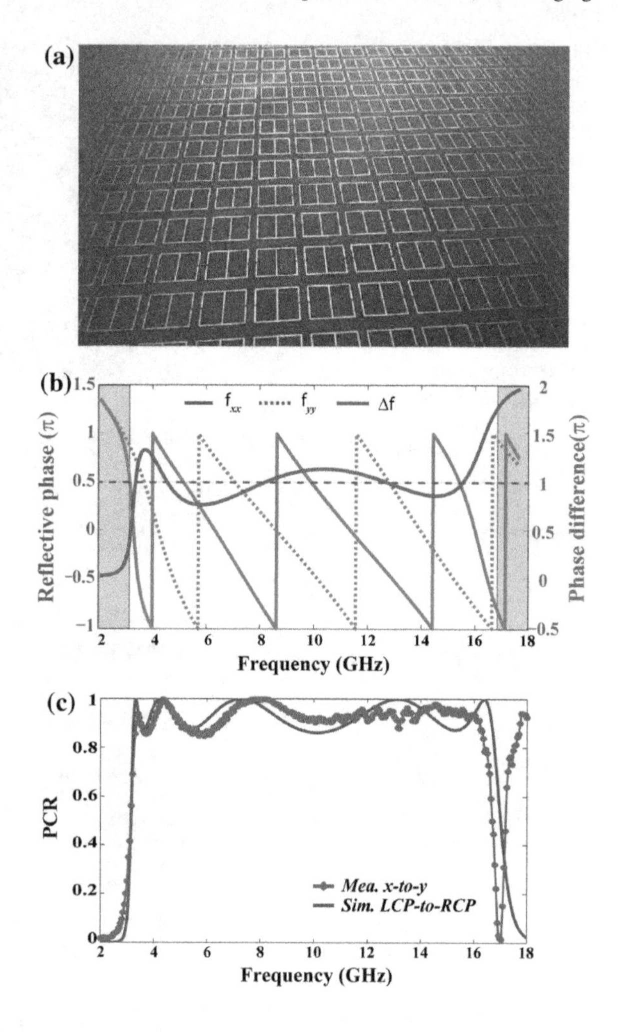

Fig. 12.14 **a** Photograph of the fabricated metasurface. **b** Simulated reflection phase of x- (blue solid line) and y-polarization (blue dash line). **c** Measured and simulated polarization conversion efficiency. Reproduced from [30] with permission. Copyright 2015, The Authors

By engineering, the impedances Z_x and Z_y, the reflection phase difference can be maintained around π with a fluctuation within $\pm 0.25\pi$, as indicated in Fig. 12.14b. Experimental results match well with the simulation results (Fig. 12.14 c), and we can see five peaks with 100% conversion ratio appear at frequencies of 3.3, 4.2, 7.3, 13.2 and 16.3 GHz, respectively. The PCR is higher than 88% from 3.17 to 16.9 GHz (beyond 2-octave bandwidth). This converter is also superior to the previous devices in the frequency-band selectivity because the operation band approximates an ideal rectangle. The rectangular coefficient, defined as the bandwidth ratio between high (>80%) and low (<20%) conversion efficiency here, is high up to 0.94.

Figure 12.15 shows another configuration to implement two-dimensional dispersion engineering. Different from the aforementioned SRRs structures, dual-layer metasurface is utilized. Since the dual metasurface is composed of orthogonal metallic stripes operates at different frequencies, the reflective phases of orthogonally electric components can be independently adjusted. From Fig. 12.16a, one can see that most LCP is transformed to the RCP after reflection by the meta-mirror. The conversion efficiency is higher than 90% from 5.4 to 32.7 GHz (bandwidth ratio is larger than 5:1). Especially, six conversion peaks with near 100% conversion efficiency appear at the frequencies of 6.3, 10.4, 19.8, 25.5, 29.6, and 31.5 GHz. Such converter also exhibits harp roll-off factor at the boundary of the operation band. By comparing the results in Figs. 12.16a and b, one can find that the permittivity of dielectric spacer has a significant influence on the conversion performance.

Figure 12.17a illustrates a schematic concept of broadband polarization-transforming electromagnetic mirror constructed by alternately stacked metallic cut-wires with orthogonal orientation, and each polarized set is arranged in a log-periodic configuration [34]. The scaling factor that governs the relation between consecutive antenna dimension and location is defined as τ. According to the transmission line circuit theory, the phase of the reflection coefficient of each set of arrays varies essentially linearly with the logarithm of frequency as follows:

$$\phi_x = \phi_0 - (2\pi/\log(\tau))\log(f/f_x), \tag{12.2.8}$$

$$\phi_y = \phi_0 - (2\pi/\log(\tau))\log(f/f_y), \tag{12.2.9}$$

where ϕ_0 is constant, f is the incidence frequency, and f_x and f_y are the resonant frequencies of x- and y-polarized array. The phase difference between them is expressed as:

$$\Delta\phi = 2\pi/\log(\tau)\log(f_x/f_y). \tag{12.2.10}$$

From Eq. (12.2.10), it can be found that the phase difference is frequency-independent, which is only a function of the scale factor ρ between adjacent orthogonally polarized arrays. By adjusting the scale factor, arbitrary polarization state can be achieved. Especially, if $\rho = \tau^{1/2}$ ($\tau^{1/4}$), $\Delta\phi = \pi$ ($\pi/2$) achromatic half- (quarter-) wave plate is realized. In order to obtain a balance between the design complex and operation bandwidth, a reflective dual-layer metasurface is adopted, as illustrated in Fig. 12.17b.

The simulated circular polarization reflectance is presented in Fig. 12.17c. It can be found that the cross-polarized reflection carries more than 80% of the incident power in the range of 0.5–3.1 THz (beyond 2-octave bandwidth), while the co-polarized component is mostly below 20%. The PCR is also calculated and plotted in Fig. 12.17d. The PCR is larger than 0.8 in the range of 0.5–1.8 THz, and larger than 0.9 in the range of 1.8–3.1 THz. Especially, the LCP is totally converter to the RCP at six frequency points of 0.54, 1.1, 1.9, 2.2, 2.5, and 3 THz.

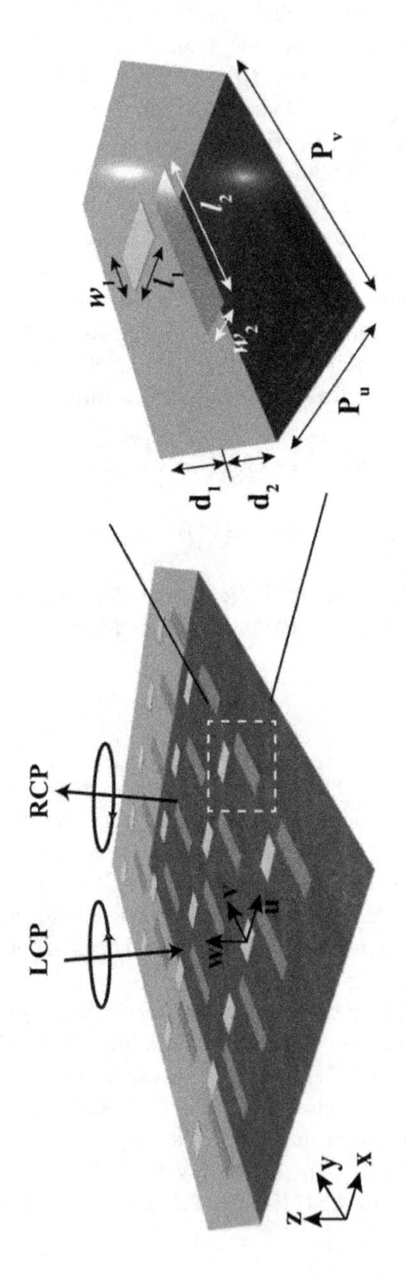

Fig. 12.15 Schematic of an ultra-broadband polarization converter. Reproduced from [33] with permission Copyright 2015, Optical Society of America

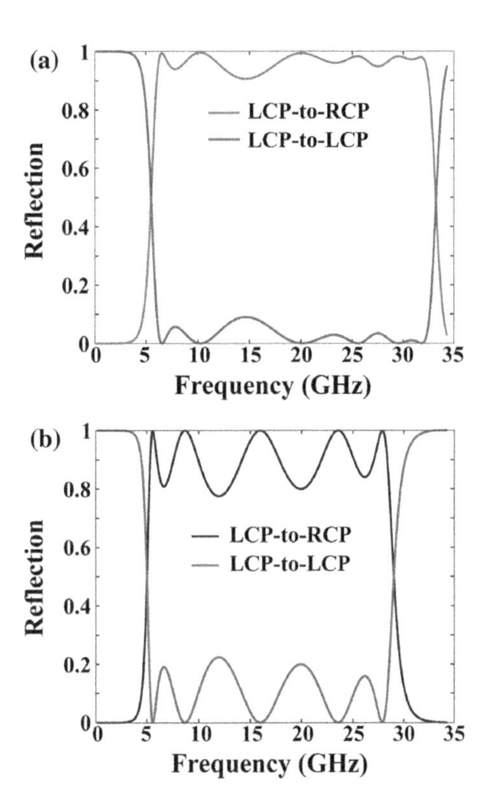

Fig. 12.16 Simulated reflection spectrum of meta-mirror with different dielectric material $\epsilon = 2.2$ (**a**) and $\epsilon = 4.6$ (**b**). Reproduced from [33] with permission. Copyright 2015, Optical Society of America

12.2.1.3 Coherent Polarization Conversion

Similar to coherent perfect absorption [35–37], the bandwidth of polarization conversion can be extended by coherent illuminations. Figure 12.18a shows the schematic of anisotropic coherent polarization converter, where the subscripts 1 and 2 denote the co-polarization and cross-polarization, respectively. Considering the conservation of energy and Maxwell continuous boundary condition, we can deduce the intensity of scattering satisfying:

$$I_{co} = A = B = 0,\ I_{cross} = C = D = 1 \tag{12.2.11}$$

for symmetrical input ($s = 1$), and:

$$I_{co} = A = B = 1,\ I_{cross} = C = D = 0 \tag{12.2.12}$$

for anti-symmetrical input ($s = -1$), where $A = |a|^2$,$B = |b|^2$, $C = |c|^2$, $D = |d|^2$, and I_{co} and I_{cross} are the intensity of co-polarized and cross-polarized output intensities. According to Eqs. (12.2.11) and (12.2.12), all the incident energy can be transformed into its orthogonal polarization for symmetrical input, while the

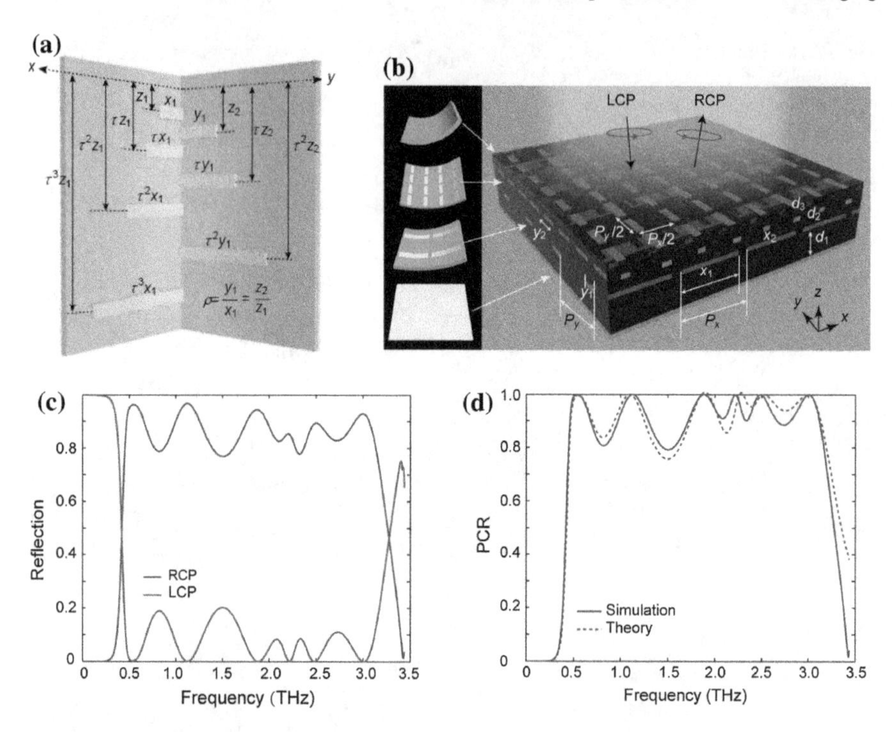

Fig. 12.17 **a** Schematic concept of wideband polarization-transforming electromagnetic mirror based on log-periodic antennas. **b** Schematic of the proposed anisotropic meta-mirrors for achromatic polarization manipulation. The period of the metasurface is ($P_x/2$, $P_y/2$) and (P_x, P_y) for the upper one and bottom one, respectively. **c** Simulated reflectance of the co- and opposite circular polarizations. **d** Simulated and theoretically calculated PCR. Reproduced from [34] with permission. Copyright 2015, Institute of Optics and Electronics, Chinese Academy of Sciences

polarization states are maintained for anti-symmetric input. When the phase of s is continuously changed, the polarization states of the outputs can be dynamically tuned accordingly. If the coherent condition is satisfied, the key of the design is to make sure the scattering coefficients along two directions have the same amplitudes, i.e., $|r| = |t| = 0.5$. To this end, metasurface composed of periodic aligned gold wire girds was utilized, where the period, width, and conductivity are, respectively 5, 2 μm, and 1e7 S/m. The intensities of co-polarized and cross-polarized outputs are calculated and shown in Fig. 12.19b. Obviously, almost all the energy are transformed to the cross-polarization in the whole frequency band. It should be noted that although the results are suitable for any long wavelength, the dimension of the whole structure may become unpractical since the dimension should be much larger than the working wavelength.

A coherent polarization converter was experimentally demonstrated in infrared band, as shown in Fig. 12.19. In order to decrease the complexity of coherent illumination, single-side illumination is adopted with combination of a metallic reflector.

An equivalent model is shown in Fig. 12.19a, and thus, the reflection electric field can be written as:

$$\vec{E}_{\text{ref}} = -\vec{E}_{\text{inc}} + \vec{E}_{\text{rad}}\left(-e^{ikd} + e^{-ikd}\right) \quad (12.2.13)$$

Considering the reflection coefficient is -1 for PEC, the above equation means the reflection field can be taken as coherent superimposition of two light experience different paths. Therefore, by controlling the product of kd, equivalent symmetric illumination can be realized. Nevertheless, the coherent illumination condition is only satisfied within a special operation band. As a consequence, the operation bandwidth is shrunk compared with the two-side illumination manner. The measured phase difference is $-90°$ for the x-polarized incidence, and it becomes $90°$ for the y-polarized incidence, as shown in Fig. 12.19b. From Fig. 12.19c, one can see the measured bandwidth of the half-wave plate reaches about 67% of the central wavelength.

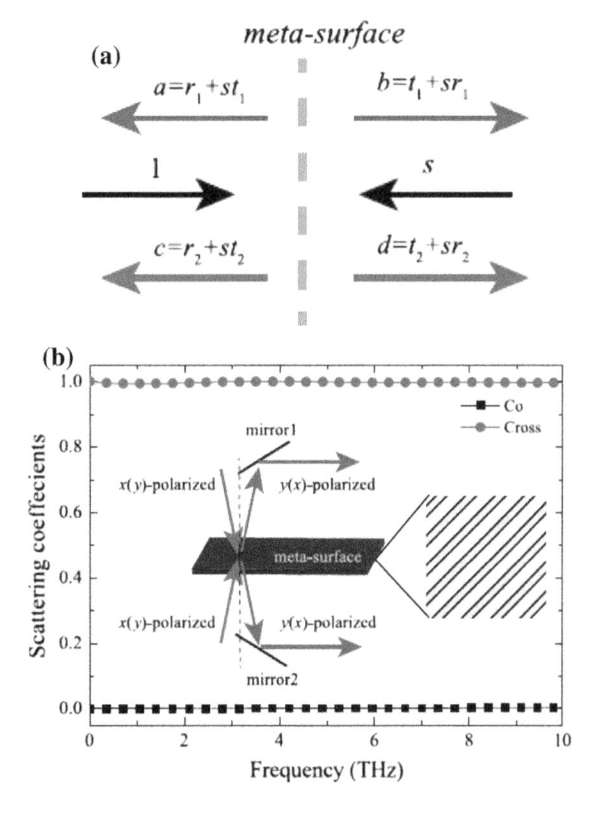

Fig. 12.18 **a** Schematic of anisotropic coherent polarization converter. **b** The output intensity of the co-polarized and cross-polarized waves for symmetrical input. Inset is the schematic of wire grids with equal scattering coefficients along two directions. Reproduced from [38] with permission. Copyright 2013, Elsevier B.V

12.2.1.4 Dynamic Polarization Conversion

Although the coherent illumination offers a way to realize dynamic polarization conversion, it generally requires strict control of the operation condition. Alternatively, dynamic and reconfigurable polarization conversion can be realized by active material or devices [40, 41].

Figure 12.20a shows the unit cell of an active polarization converter, where four identical PIN diodes are embedded into the metallic slot, and three different working states can be obtained by controlling the diodes. In the state 1, diodes 1 and 3 are switched on while diodes 2 and 4 are turned off. In state 2, the diodes 1 and 3 are shut off while diodes 2 and 4 are switched on. The equivalent physical model and circuit model are shown in Fig. 12.20c, from which one can see that two reversed states were realized. Consequently, a linear polarization can be dynamically converted into LCP or RCP. In this case, the material is both anisotropic and chiral, and the ellipticity spectra of the outgoing waves for state 1 and 2 are calculated by the formula:

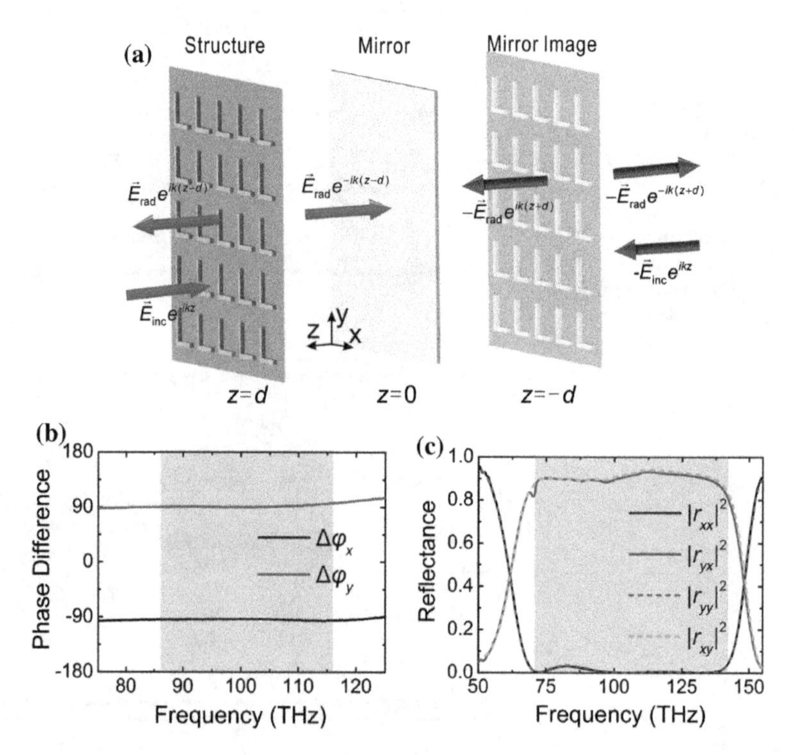

Fig. 12.19 **a** Schematic to show the interaction of light with each interface of the system made of an array of L patterns ($z = d$) in front of a perfect electric conductor ($z = 0$). **b** Experimentally measured phase difference under the illumination of x- and y-polarized light. **c** Experimentally measured reflectance. Reproduced from [39] with permission. Copyright 2014, American Physical Society

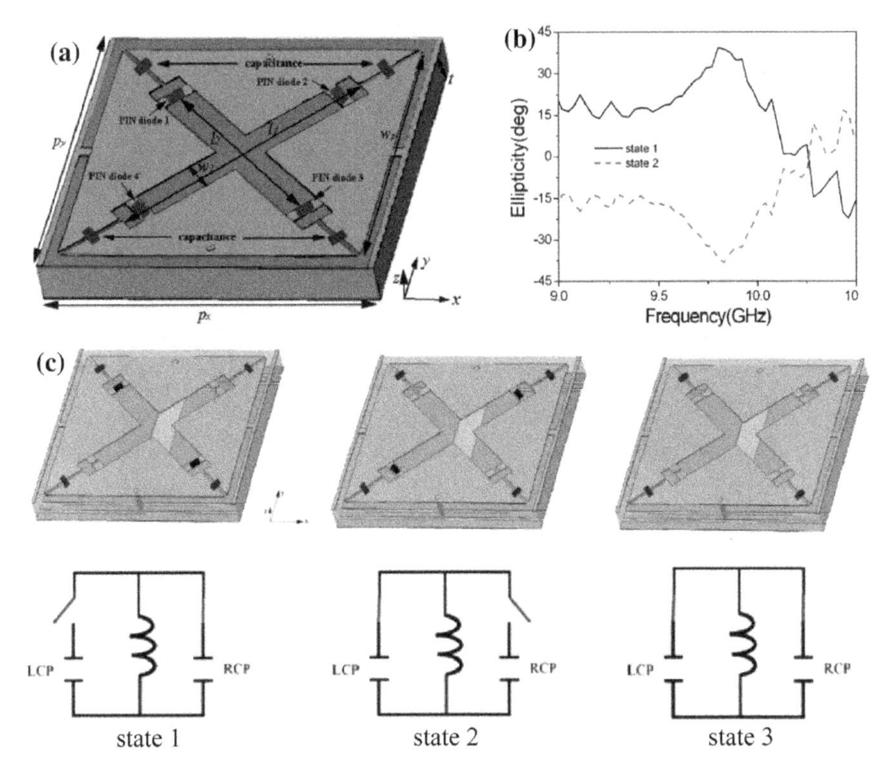

Fig. 12.20 **a** Schematic of the meta-molecule of the active metamaterial with side view of the top layer and dielectric substrate. **b** Measured ellipticity (solid line for state 1 and dashed line for state 2) of the active metamaterial for state 1 and state 2. **c** Schematic of equivalent physical model and circuit model. Reproduced from [42] with permission. Copyright 2014, WILEY-VCH Verlag GmbH & Co. KGaA, Weinheim

$$\eta = \frac{1}{2}\sin^{-1}\left(\frac{|T_+|^2 - |T_-|^2}{|T_+|^2 + |T_-|^2}\right), \tag{12.2.14}$$

where $T_\pm = T_{xx} \pm iT_{yx}$ stands for the RCP wave and LCP wave. From Fig. 12.20b, one can see that the ellipticity of state 1 is close to 40° at the resonance while is about −40° for state 2, which demonstrates that the outgoing fields are close to pure circular polarization, and the handedness is opposite between state 1 and 2. In state 3, all diodes are turned on and no polarization conversion occurs due to the isotropy of the equivalent model. Based on similar method, Cui et al. demonstrated that the linearly polarized incident wave can be dynamically converted to right-handed or left-handed circular polarization in the frequency range between 3.4 and 8.8 GHz [43].

Besides diodes, phase-change materials are also promising candidates to realize dynamic modulation. A switchable ultra-thin QWP is implemented in terahertz band by hybridizing a phase-change material, vanadium dioxide (VO_2), with a metasur-

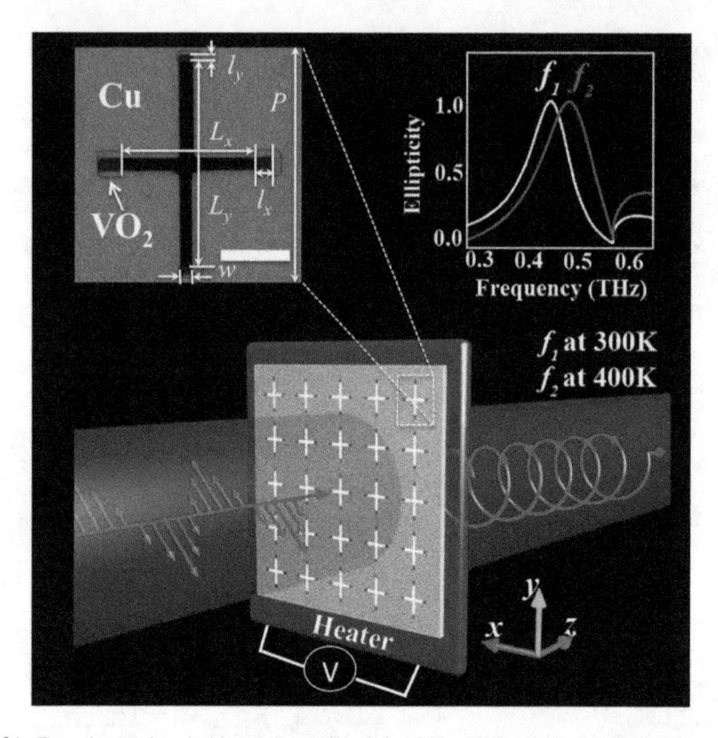

Fig. 12.21 Experimental switching schematic of the THz QWP. A linear normal incident THz wave polarized at $\theta = 45°$ to the two slots is converted into a CPL. The top left inset is a microscope image of one unit cell in the fabricated samples. The scale bar is 50 μm. The top right inset is the simulated ellipticities of the output THz waves, indicating that at both f_1 and f_2, the output THz waves are circularly polarized. Reproduced from [44] with permission. Copyright 2015, The Authors

face [44] (Fig. 12.21). Before the phase transition, VO_2 behaves as a semiconductor and the metasurface operates as a QWP at 0.468 THz. After the transition to metal phase via a resistive heater, the effective length of the resonators can be altered, and thus, the operating frequency changes to 0.502 THz. At the corresponding operating frequencies, the metasurface converts a linearly polarized light into a CPL. Similar to this approach, a dual-band switchable terahertz QWP was demonstrated by inserting the VO_2 into complementary electric ring resonators, which support multiple resonances at terahertz band [45]. Before the VO_2 phase transition, this phase-change metasurface achieves linear-to-circular polarization conversion at 0.45 and 1.10 THz with an ellipticity of 0.998 and −0.971, respectively. After the VO_2 phase transition, linear-to-circular polarization conversion is obtained at both 0.50 and 1.05 THz with an ellipticity of 0.999 and −0.999, respectively.

12.2.2 Circular Dichroism

Besides anisotropic artificial materials, there is another type of materials, namely chiral materials that can also be utilized to manipulate the polarization of electromagnetic waves. The structure of chiral materials is asymmetric and cannot be superposed with their mirror structures by rotation or displacement. Due to the lack of symmetry in chiral materials, the electric and magnetic fields would couple with each other and thus present unique electromagnetic properties. The constitutive relationship equations of chiral medium can be expressed as follows [46]:

$$\begin{pmatrix} D \\ B \end{pmatrix} = \begin{pmatrix} \varepsilon_0 \varepsilon & i\kappa\sqrt{\varepsilon_0 \mu_0} \\ i\kappa\sqrt{\varepsilon_0 \mu_0} & \mu_0 \mu \end{pmatrix} \begin{pmatrix} E \\ H \end{pmatrix}, \tag{12.2.15}$$

where ε_0 and μ_0 are the permittivity and permeability of vacuum, ε and μ the relative permittivity and permeability, and κ the strength of the coupling between the electric and magnetic fields. The eigenmodes of chiral medium are RCP and LCP, and the corresponding refraction indices are $n_\pm = \sqrt{\varepsilon\mu} \pm \kappa$, where the symbols "+" and "−", respectively, stand for the RCP and LCP. The difference in the imaginary parts results in the amplitude discrepancy of the transmitted circularly polarized waves, while the difference in the real parts leads to the phase difference. These two types of discrepancy, respectively, result in the circular dichroism (CD) and optical activity (OA). Unfortunately, in the natural chiral medium, the coupling between the electric and magnetic field is rather weak and the chirality is small, which leads to the extremely large thickness of chiral materials in real applications.

12.2.2.1 Three-Dimensional Chiral Structures

Chiral metamaterials exhibit chiro-optical effects with magnitudes much larger than those found in ordinary materials. Probably, the simplest chiral metamaterial may be an array of gold helix, which selectively transmits LCP or RCP according to the handedness of the helix [3]. Figure 12.22 shows a 3D-chiral metamaterial with each unit cell consists of a right-handed helix and left-handed helix coupled together. The measured circular polarization-conserving transmittances T_{LL} and T_{RR} as well as circular polarization conversions T_{LR} and T_{RL} are shown in Fig. 12.23. As expected, strong circular polarization conversions from RCP to LCP with the efficiency up to 75% can be observed for a bandwidth of more than one octave. Furthermore, another circular polarization conversion from LCP to RCP is well below 10% in the entire operation band.

Although 3D-chiral metamaterials have more appealing properties as circular polarizers, periodic arrays with subwavelength chiral features and precise alignment are challenging to realize at visible wavelengths. To overcome this problem, an alternative venue has been proposed [13], which is composed of stacked planar metasurfaces that can be realized with conventional lithographic techniques. The

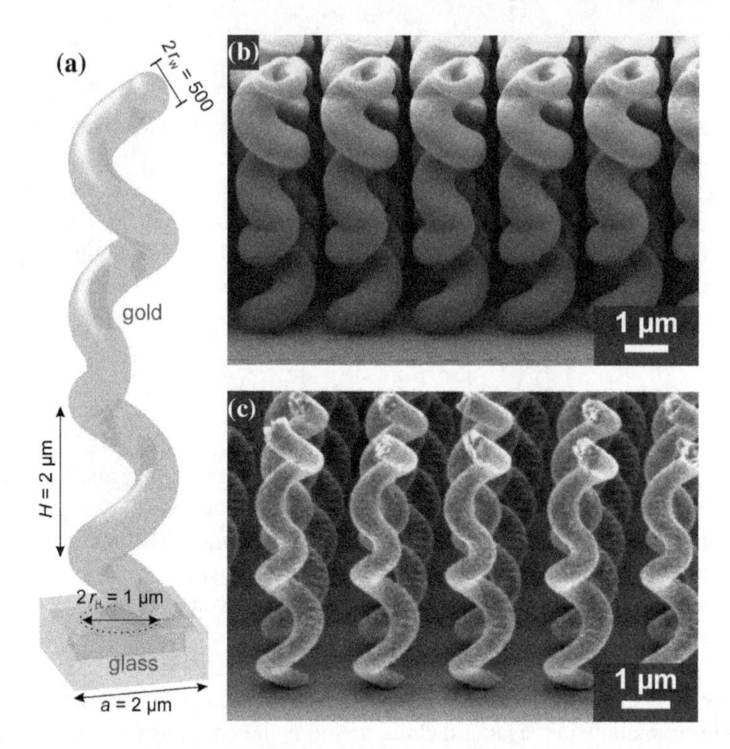

Fig. 12.22 **a** Illustration of one metamaterial unit cell. **b** SEM image of polymer templates. **c** SEM image of the final gold structures. Reproduced from [4] with permission. Copyright 2015, WILEY-VCH Verlag GmbH & Co. KGaA, Weinheim

key idea is by introducing a twist in the lattice orientation of metamaterial inclusions to achieve exotic effects analogous to 3D metamaterials, but with much simpler fabrication schemes.

The measured and simulated transmission curves with different metasurface layers are shown in Fig. 12.24. One can find that the single metasurface (Fig. 12.24a) exhibits identical transmission of LCP and RCP waves. But as soon as a second layer is cascaded with proper twist and distance (Fig. 12.24b), some forms of polarization selectivity were obtained. Furthermore, a larger number of layers (Fig. 12.24c and d) ensure better extinction ratios and increase the bandwidths.

12.2.2.2 Planar and Thin Chiral Structures

In recent years, some planar chiral metasurfaces without C4 symmetry have been proposed as a polarizer with simplified structures. Due to the lack of C4 symmetry and mirror symmetry, these chiral metasurfaces exhibit CD and OA at the same time. By properly designing the structural parameters, the electromagnetic coupling may contribute positively to the transmission of certain circular polarization, leading to

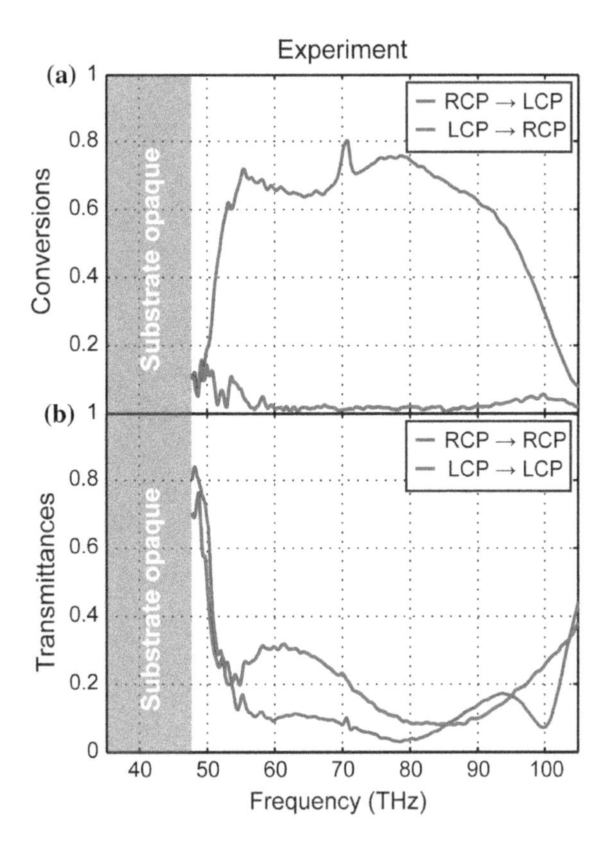

Fig. 12.23 Measured polarization conversions (**a**) and polarization-conserving transmittances (**b**). Adapted with permission from [4]. Copyright (2015) WILEY-VCH Verlag GmbH & Co. KGaA, Weinheim

the near-unit PCR with a linearly polarized incident wave. Figure 12.25 shows a triple-band circular polarizer, which is composed of three-layered metasurfaces with different twist angles separated by dielectric layers [47]. Each metasurface consists of an array of metallic arcs with a width of 0.85 mm and an open angle of $\theta = 80°$. The orientations among different metasurface layers are rotated about the z-axis.

The simulated results of the transformation responses for this triple-band polarizer are depicted in Fig. 12.26. There are three obviously resonant frequencies occurring at 13.33, 15.56, and 16.75 GHz. The transmitted waves are RCP at 13.33 and 16.75 GHz, while are LCP at 15.56 GHz. Figure 12.26b presents the measured transformation results of the fabricated sample. It is shown that RCP waves are produced at 12.75 GHz and 16.20 GHz, and LCP wave is transformed at 15.25 GHz.

The operating frequencies of the planar chiral metamaterial are directly related to the structural parameters. Figure 12.27 presents the transformation for LCP and RCP waves with different values of R. One can see that the positions of transformation frequencies for LCP and RCP waves all shift toward lower frequencies when R increases. Numerical analysis reveals that the resonant frequencies of the designed sample are closely related to the effective total length L_{eff} of the metallic arcs, which can be approximately determined by the following formula:

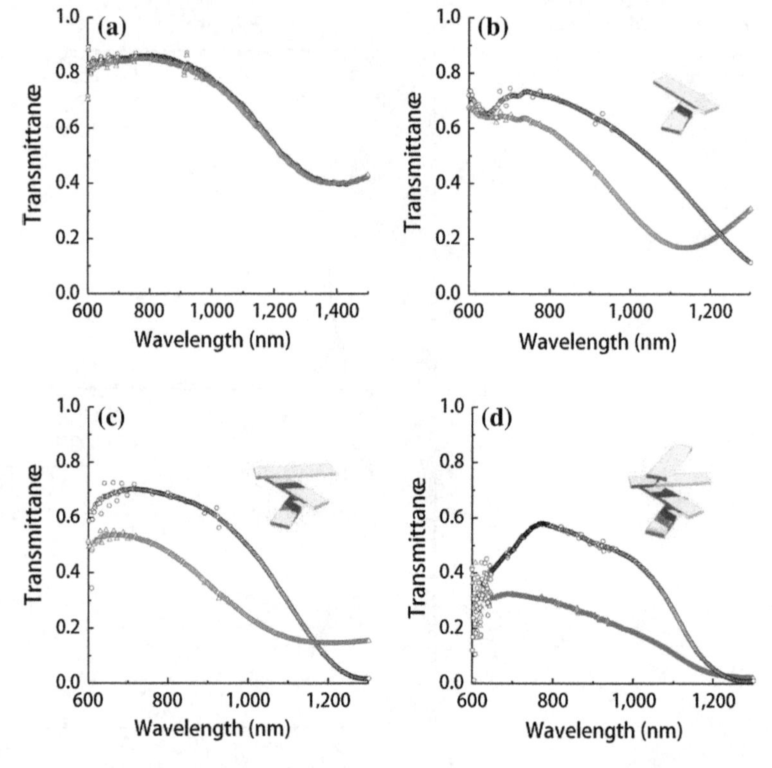

Fig. 12.24 **a–d** Experimental measurements of twisted metamaterial slabs with one-to-four layers. Reproduced from [13] with permission. Copyright 2012, Macmillan Publishers Limited

$$L_{\mathrm{eff}} \approx \frac{2\pi\theta}{120} \times \left(R - \frac{w}{2}\right) + 2h. \tag{12.2.16}$$

Since the effective wavelength is directly proportional to L_{eff}, the increase of effective length L_{eff} of the arcs would result in redshift of the resonant frequency when R increases.

Based on the planar arcs, a modified structure is proposed to realize three and four-band chiral transmissions. As shown in Fig. 12.28, the proposal is composed of two types of copper spiral structures with different dimensions, which are printed on each side of the dielectric lamina to respond different frequencies. Intrinsically, the chiral character stems from the electromagnetic coupling between these metallic arcs. Therefore, there exist two kinds of coupling behaviors: one is generated among different arcs in the same layer, while the other is generated among different layers.

When a linearly polarized wave is normally incident on the chiral metasurface, it can be decomposed into two circularly polarized waves with equal amplitude and phase. As the unit cell also lacks C4 symmetry, the two circular polarizations partially convert to their cross-polarizations at the resonance. The simulated results in

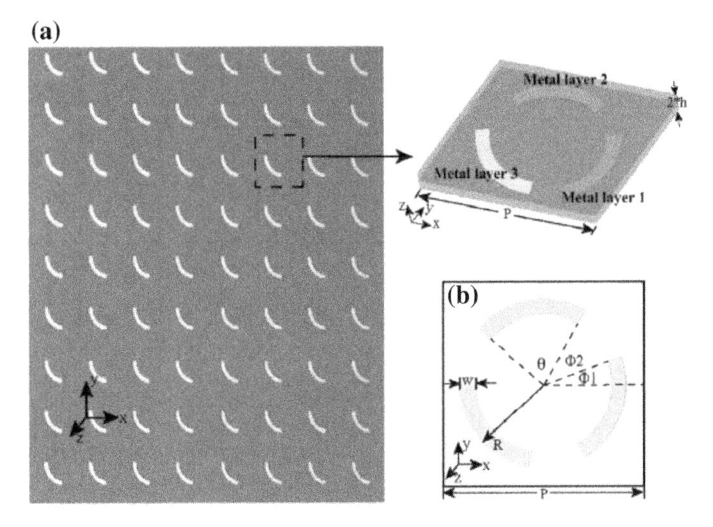

Fig. 12.25 Schematic configuration of the triple-band circular polarizer. **a** Photograph of the circular polarizer and the magnified unit cell. **b** Front view of the unit cell. Φ_1 represents the angle between the upper end of metal layer 1 and x-axis. Reproduced from [47] with permission. Copyright 2012, Optical Society of America

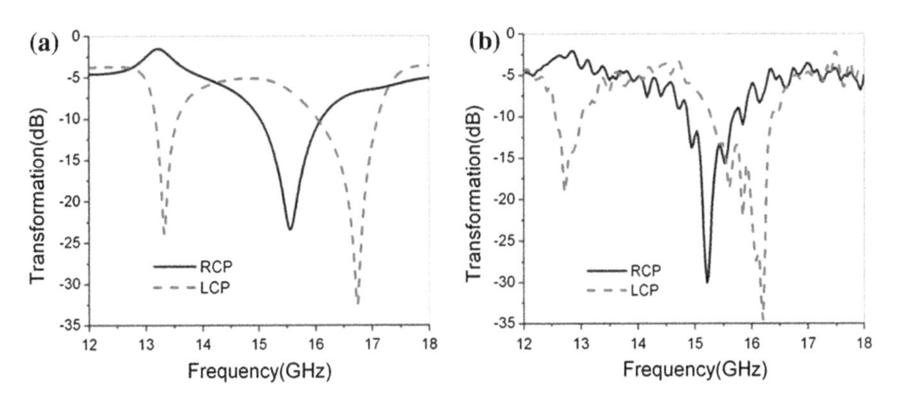

Fig. 12.26 Transformation coefficients for the fabricated circular polarizer. **a** The simulated results. **b** Measured results. Reproduced from [47] with permission. Copyright 2012, Optical Society of America

Fig. 12.29a reveal that the transmission of the RCP wave reaches its local minimum value of −17.52 and −23.4 dB at 12.25 and 15.57 GHz, where the corresponding transmission coefficients of the LCP wave are −2.4 and −1.9 dB, respectively. The tremendous difference between the transmissions of the RCP wave and the LCP wave means that the transmitted waves are the LCP waves at these two resonances.

At the other two resonant frequencies of 13.9 and 16.86 GHz, the outgoing circular polarization reverses. It can be seen that the transmission coefficients of RCP wave are −1.2 and −1.7 dB, while the transmission coefficients of the LCP wave are as low

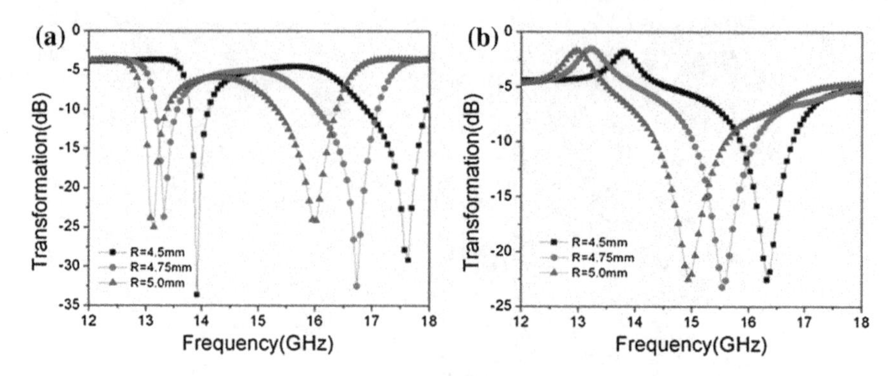

Fig. 12.27 Transformation spectra for different radius *R*. **a** Transformation of LCP. **b** Transformation of RCP wave. Reproduced from [47] with permission. Copyright 2012, Optical Society of America

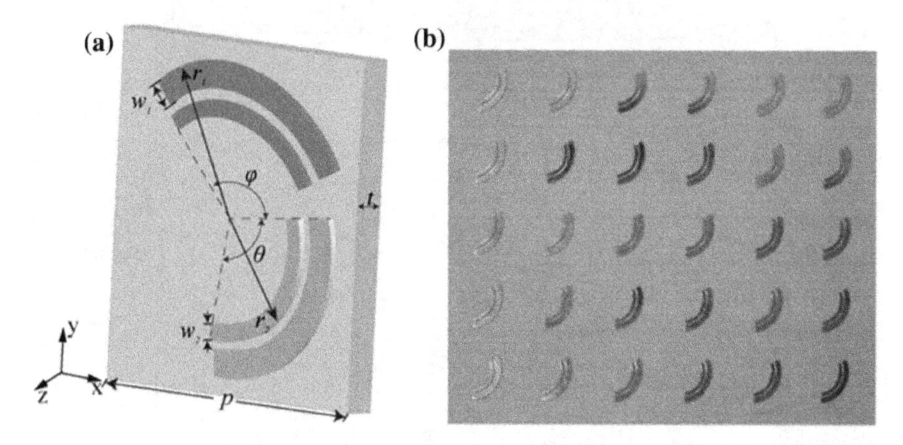

Fig. 12.28 Chiral metamaterials based on twisted arc structures. **a** Schematic of a unit cell geometry. **b** A photograph of the fabricated CMM structure. Reproduced from [48] with permission. Copyright 2012, American Institute of Physics

as −26.56 and −19.56 dB, respectively. The differences between the transmissions of the RCP wave and the LCP wave are 25.3 and 17.86 dB, implying the RCP wave is excited at 13.9 and 16.86 GHz. Intrinsically, the above phenomena are related to the constructive and deconstructive interferences between the converted and directed transmitted circular waves, which can be analyzed in the transmitted amplitude and phase difference shown in Fig. 12.29b. Owing to the simple yet superb performance, the twisted arc structures have also been demonstrated in near-infrared band [49].

Besides CD, chiral metasurfaces can also realize polarization rotation. As shown in Fig. 12.30a and b, the unit cell of the dual-band polarization rotator is composed of two pairs of the twisted SRRs, and the SRRs in diagonal direction are identical. The parameters of one pair SRR are scaled down by a factor of 0.9 with respect to another pair of the SRRs. Due to the lack of C4 symmetry, the response of this polarization

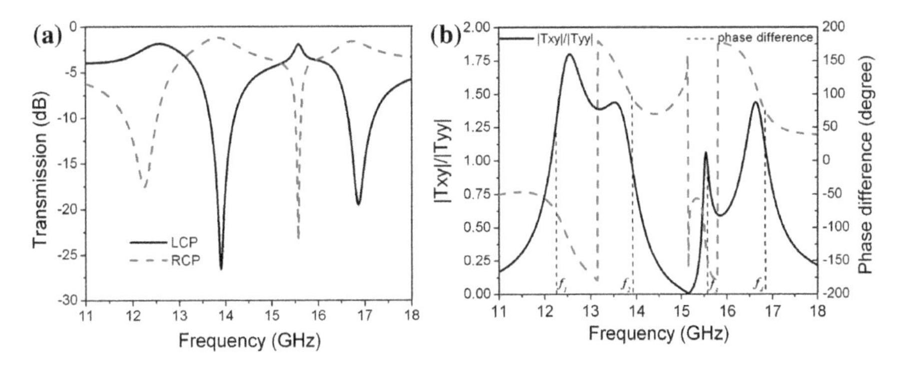

Fig. 12.29 **a** Transmission spectra of the LCP and RCP waves. **b** The ratio of $|T_{xy}|/|T_{yy}|$ and phase difference of $\Phi(T_{xy})-\Phi(T_{yy})$. Reproduced from [48] with permission. Copyright 2012, American Institute of Physics

rotator to the incident electromagnetic wave is not the same for different polarized electric fields and optical activity can only be exhibited for x-polarization [50].

The simulated and measured transmission coefficients of the co-polarized and cross-polarized emitted wave are depicted in Fig. 12.30c. There are two obvious transmission peaks at 13.54 and 15.48 GHz, where y-polarized emitted waves are produced. In order to clearly demonstrate the optical activity of the polarization rotator, the polarization azimuth rotation angle θ is calculated by the following formulas:

$$\theta = \frac{1}{2}\left[\arg(T_{+}) - \arg(T_{-})\right], \tag{12.2.17}$$

The polarization azimuth rotation angle θ represents the rotation angle between the polarization planes of the emitted and incident waves, while the ellipticity η denotes the polarization state of the emitted wave. When η equals to zero, the emitted wave is still linearly polarized, but the polarization plane has an angle rotation of θ with respect to the incident wave. It means that pure optical activity is produced. The emitted wave would be circularly polarized if η equals to $\pm 45°$. Figure 12.30d depicts the simulated azimuth rotation angle θ and the ellipticity η as functions of frequency for the dual-band polarization rotator. It can be seen that the values of ellipticity η are in the vicinity of 0° at 13.54 and 15.48 GHz, while the polarization rotation angles θ are close to 90° and −90° at these two resonances. It means that the polarization plane of the emitted wave has a 90° rotation angle with respect to the incident wave.

Lately, Ma et al. proposed a double-layer twisted Y-shape structure to achieve obvious circular dichroism and giant optical rotation at different frequencies [51]. The schematic geometry of the proposed metasurface is depicted in Fig. 12.31. The unit cell is composed of two Y-shaped metallic structures at a certain twisted angle printed on two sides of a dielectric lamina. The angle between the neighboring two branches of this Y-shaped structure is designed to be 120°.

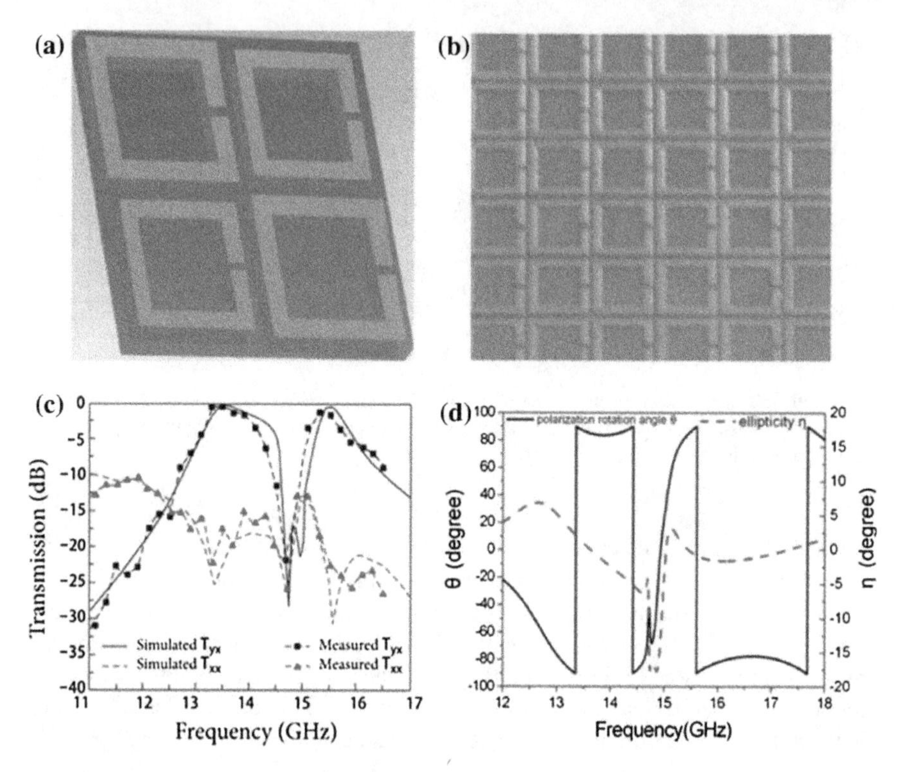

Fig. 12.30 **a** Unit cell and **b** photograph of the dual-band wideband polarization rotator. **c** The simulated and measured transmission of co-polarization (T_{xx}) and cross-polarization (T_{yx}). **d** The polarization rotation angle θ and the ellipticity η for the dual-band polarization rotator. Reproduced from [50] with permission. Copyright 2013, Elsevier B.V

The calculated η and θ are depicted in Fig. 12.32a. The value of η is very close to 45° at 12.28 GHz, which reveals that the circular polarized wave is produced and great circular dichroism is obtained at this frequency. Furthermore, it is found that η is almost 0° in the frequency band of 12.7–14.25 GHz, while the value of the polarization rotation angle θ is kept around −90°. Therefore, the radiated wave is still linearly polarized, but the polarization plane is rotated with angle of 90° with respect to the incident wave. Although 90° polarization rotation is realized in a broad frequency range, the peak of the transmission coefficient T_{xy} appears only at 12.7 GHz. Figure 12.32b shows the measured results of polarization rotation angle θ and the ellipticity η.

12.3 Polarization Manipulation in Integrated Waveguides

The focus of the research in metasurfaces has mainly been on the manipulation of the electromagnetic waves in free space during the past few years, which generally require thousands of subwavelength building blocks or even more. Recently, it is shown that the gradient metasurface can be integrated into silicon optical waveguide to control the polarization of the guided wave by strong, consecutive scattering of antenna array [52–54]. This collective effect of the gradient metasurface on guided waves enables us to substantially reduce the numerous building blocks and the footprint of photonic integrated devices.

12.3.1 Integrated Polarization Converter

One advantage of using metasurfaces to control waveguide modes is that the optical near-fields of nanoantennas contain both TE- and TM-polarized components. Therefore, nanoantennas are able to mediate a strong interaction between TE and TM waveguide eigenmodes, which otherwise cannot couple with each other in a bare waveguide [52].

Over a propagation distance of only a few times the wavelength, an effective wavevector many times larger than the phase gradient $d\Phi/dx$ can be imparted to the incident mode, resulting in a large overall wavevector change. If the wavevector change induced by gradient metasurface can effectively compensate the wavevector mismatch between different polarization modes, then polarization conversion

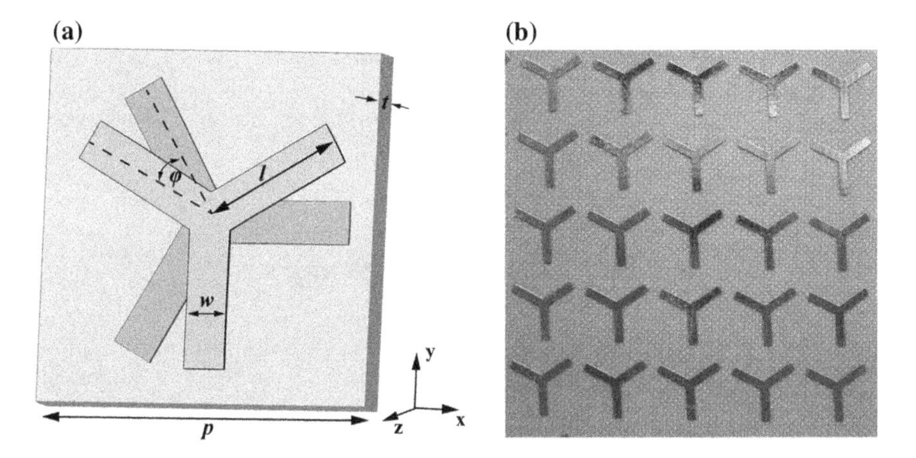

Fig. 12.31 **a** Schematic geometry of the unit cell. **b** The photograph of the fabricated twisted Y-shaped chiral metasurface. Reproduced from [51] with permission. Copyright 2013, The Japan Society of Applied Physics

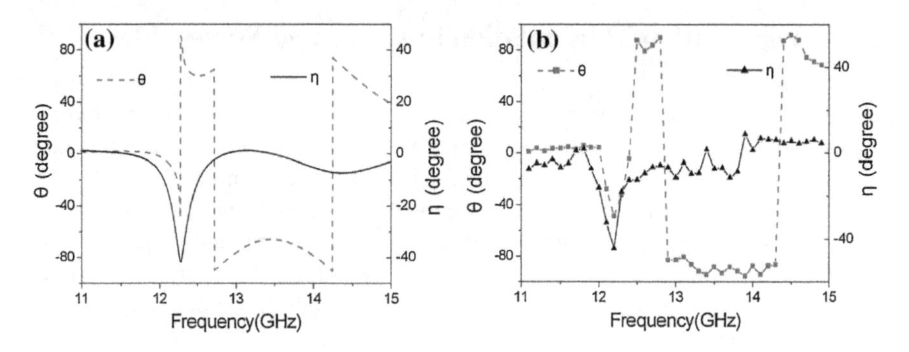

Fig. 12.32 **a** Schematic geometry of the unit cell. **b** The photograph of the fabricated twisted Y-shaped chiral metasurface. Reproduced from [51] with permission. Copyright 2013, The Japan Society of Applied Physics

between these modes will occur. Mathematically, it is expressed as, $d\Phi/dx = k_0|n_{in}-n_{out}|$, where n_{in} is the modal index of the input waveguide mode and n_{out} is the modal index of the converted waveguide mode. Following the above principle, a gradient phase metasurface composed of gold phased array antennas is patterned on silicon waveguide [52]. The schematic of the proposal and simulated electric field distribution shown in Fig. 12.33 demonstrates that high-purity modes are obtained at the output, indicating a high-efficiency polarization conversion.

The SEM image of TE_{00}-to-TM_{00} and TE_{00}-to-TM_{10} mode converters is shown in Fig. 12.34a and c. The output modes are detected by raster-scanning a single-pixel indium antimonide detector in front of the waveguide output facet. Figure 12.34b shows that the output from the TE_{00}-to-TM_{00} mode converter (polarization rotator) has only one far-field lobe with TM polarization, and that the TE-polarized far-field is comparatively weak, indicating that the incident TE_{00} mode has been converted into the TM_{00} mode with high efficiency. The far-field profile of the TE_{00}-to-TM_{10} converter has two lobes. The residual TE-polarized component in the far-field is negligible for the TE_{00}-to-TM_{10} mode converter (Fig. 12.34d), which indicates a complete mode conversion [52].

12.3.2 Integrated Polarization Splitters/Routers

Besides integrated polarization modes converter, integrated polarization beam splitter (PBS) for a design wavelength of 1.55 μm was also reported with a footprint of only 2.4 × 2.4 μm^2 [55]. As shown in Fig. 12.35a, the PBS is patterned on a CMOS-compatible silicon-on-insulator (SOI) substrate. The device is composed of 20 × 20 pixels, where each pixel is in the shape of a square of 120 × 120 nm^2. Each such pixel can occupy two states: silicon (black) or air (white). The PBS was designed using a nonlinear search algorithm, referred to as "direct-binary search." Specifically, a randomly chosen pixel is first perturbed so as to switch its state, then a FOM is calculated, which is defined as the average transmission efficiency for TE and TM polarization states. The pixel state is retained if the FOM is improved. If not, the perturbation is reversed and the algorithm proceeds to the next pixel. A single iteration comprises such inspection of all pixels. The iterations continue until the FOM does not improve further [55].

After the optimization, the electromagnetic fields within the device are simulated using a finite-difference time-domain method. As shown in Fig. 12.35b and c, the incident light generates resonant modes within the nanophotonic device that are polarization-dependent. The measured coupling efficiencies (Fig. 12.35d) are up to 71% and 80% at the design wavelength (1.55 μm), respectively. The measured TE and TM extinction ratios (Fig. 12.35e) at the design wavelengths are 11.8 and 11.1 dB, respectively. If the extinction ratio is allowed to fall 3 dB from the peak value, the devices can tolerate a variation in top silicon thickness of up to $\pm$ 20 nm, exhibiting a robust fabrication tolerance. Note that, more complex structures that can perform multiple functions in single devices can also be implemented based on the above optimization method [55].

Besides the integrated PBS above, an integrated polarization beam router has also been proposed [54], which is attributed to the spin–orbit interaction of the

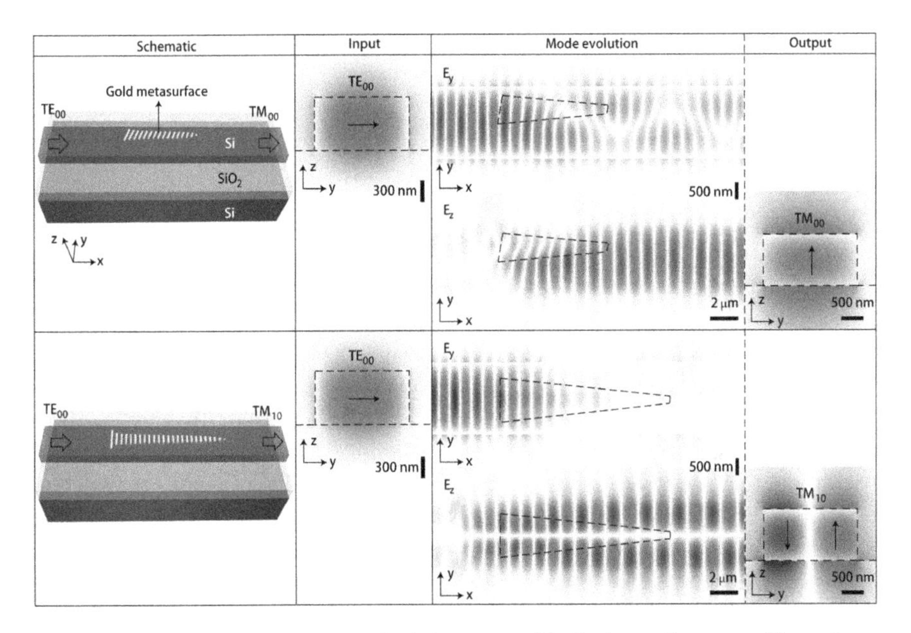

Fig. 12.33 Simulated polarization manipulating waveguide device performance. First column: device schematics. Second and fourth columns: waveguide modes at the input and output ports of the devices, respectively. The field distributions are plotted at $\lambda = 4.50\,\mu$m and 4.16 μm, respectively. Third column: mode evolutions as light propagates from left to right. Reproduced from [52] with permission. Copyright 2017, Macmillan Publishers Limited, part of Springer Nature

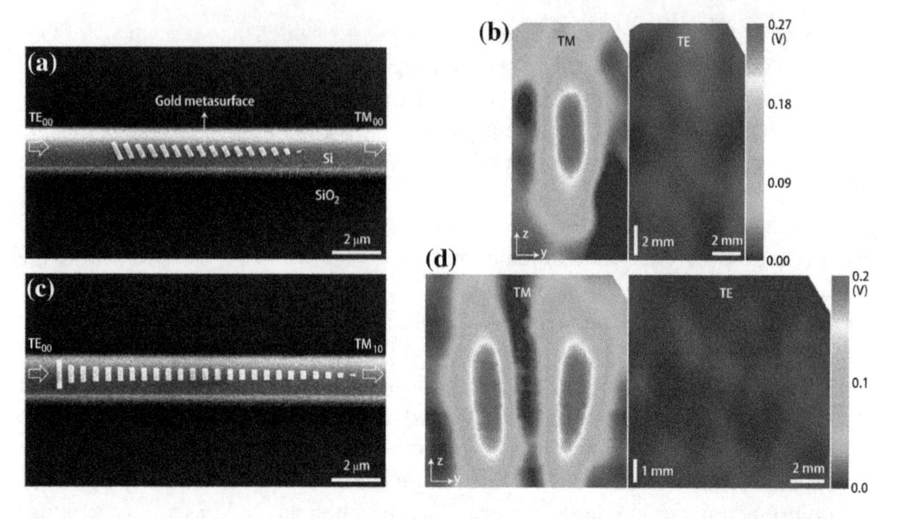

Fig. 12.34 **a** SEM image of the TE_{00}-to-TM_{00} mode converter. **b** Measured far-field emission patterns of the TE_{00}-to-TM_{00} mode converter (polarization rotator), showing that the TM-polarized component has a single lobe and that the TE-polarized component is very weak. **c** SEM image of the TE_{00}-to-TM_{10} mode converter. **d** Measured far-field emission patterns of the TE_{00}-to-TM_{10} mode converter, showing two lobes with TM polarization and weak far-field with TE polarization. Reproduced from [52] with permission. Copyright 2017, Macmillan Publishers Limited, part of Springer Nature

chip-integrated geometric phase metasurface. As indicated in Fig. 12.36, a metallic geometric metasurface is constructed by gold rods with a constant gradient of orientation angle θ along the x-direction to generate a linearly gradient phase. The whole metallic geometric metasurface is deposited on top of the center of a standard SOI slab waveguide into which light is launched. The SOI wafer has a buried oxide layer with a thickness of 1 μm and a silicon top layer with a thickness of 220 nm.

When a CPL is normally incident onto the anisotropic metallic rod from a focused lens or lensed fiber tip, linearly gradient wavefront will be generated along the waveguide due to the geometric phase shift stemming from spin–orbit interaction in metasurface. This effect is equivalent to introduce a unidirectional effective wavevector and thus leads to directional coupling. Thanks to the spin-dependent gradient wavefront, opposite light flow is excited in the integrated waveguide by switching the handedness of incidence, which may provide a potential pathway to integrated polarization sorters and switchers [54].

Since only the component with opposite handedness can be coupled into the SOI waveguide and the component with the same handedness without phase modulation are leaked into the substrate or directly reflected, the polarization conversion efficiency of the geometric metasurface is a significant factor that determines the coupling efficiency of the designed device. The building block of the metallic metasurface, as illustrated in Fig. 12.37a, is simulated and optimized using frequency-domain solver in CST MICROWAVE STUDIO with unit cell boundary conditions.

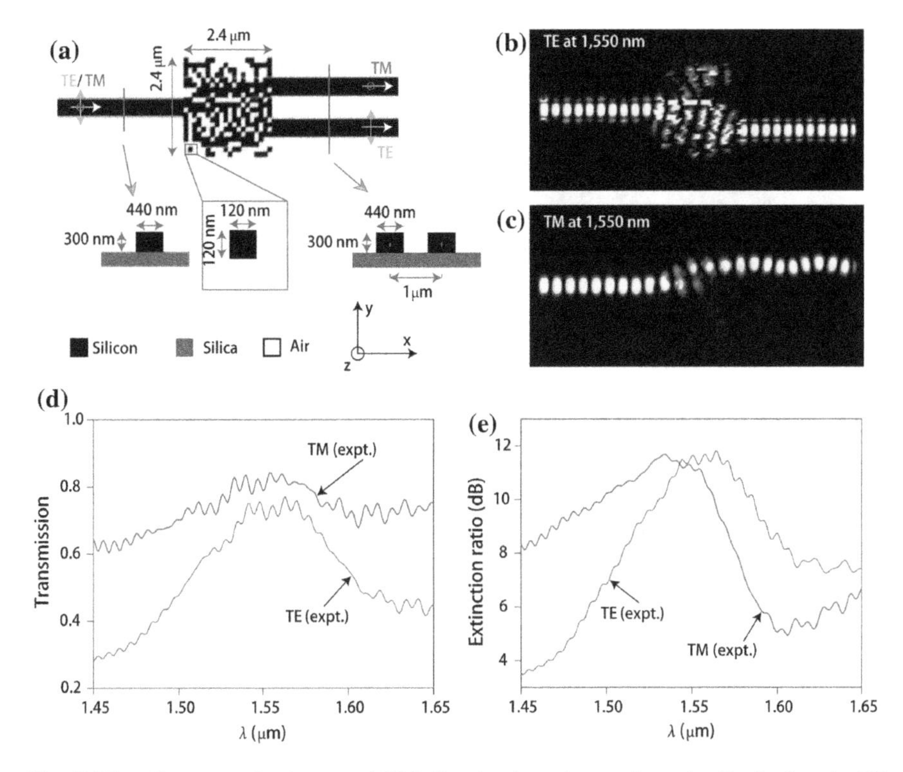

Fig. 12.35 **a** Geometry of an integrated PBS. Simulated steady-state intensity distributions for TE (**b**) and TM (**c**) polarized light at the design wavelength of 1.55 μm. **d**, **e** Measured transmission and extinction ratios of the PBS for both TE and TM polarization. Reproduced from [55] with permission. Copyright 2015, Macmillan Publishers Limited

The lattice constant is $P = 430$ nm, which is much smaller than the incident wavelength of 1550 nm. The dimensions of the metallic rod antennas are optimized as $l = 300$ nm, $w = 110$ nm, and $h = 110$ nm. For the normal illumination of LCP incidence from 1500 nm to 1600 nm, the amplitude and phase shift of opposite handedness (i.e., RCP) through different oriented nanoantennas (with a step of 30°) are shown in Fig. 12.37b and c, respectively. It can be found that the amplitude of transmitted RCP is nearly constant in the whole simulated wavelength range. Furthermore, the amplitude and phase response at 1550 nm as a function of rotation angle is plotted in Fig. 12.37d, which shows that the phase shift is proportional to the rotation angle and consistent with the theoretical prediction $\Delta\Phi = 2\sigma\Delta\varphi$.

Full-wave simulations were carried to verify the above theory. Figure 12.38a and b display the electric field distributions in the xy-cross section through the waveguide for RCP and LCP incidence at the wavelength of 1550 nm, respectively, which are retrieved from a pre-established field monitor. One can find that light flow is mainly coupled into the $-x$-direction for RCP incidence and $+x$-direction for LCP incidence.

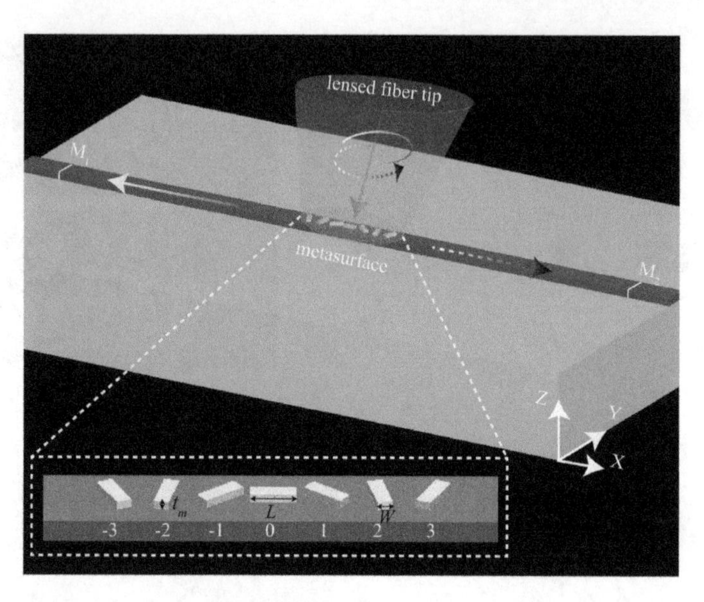

Fig. 12.36 Artistic depiction of the designed chip-integrated metallic geometric metasurface as a compact platform for direction coupler and polarization sorter. Insets show the seven rotating metallic antennas are located on single-mode SOI waveguide with a thickness of 220 nm and width of 570 nm. Reproduced from [54] with permission Copyright 2018, IEEE

Obviously, by reversing the handedness of incidence, one can flexibly switch the routing direction of light and realize circular polarization sorting [54].

The power that couples into the $\pm x$-directions of the waveguide as a function of incident wavelength is plotted in Fig. 12.39. Obviously, most of power scattered from the metallic metasurface are coupled to the left side of the waveguide, while the power coupled to the right side can be neglected around the wavelength of 1550 nm. The contrast ratio between them defined as $10 \cdot |\log_{10}(T_1/T_2)|$ (black curve) reaches its maximum value of 22 dB at around 1550 nm, which is larger than the previous reports based on coherent interferences [56]. Owing to spin–orbit interaction and dispersionless geometric phase, contrast ratio of proposed directional coupler can still maintain above 10 dB in a bandwidth of 50 nm.

Compared with the metasurfaces composed of tens to thousands of meta-atoms and even more for electromagnetic manipulation in free space, the number of the meta-atom of chip-integrated metasurface can be greatly reduced to seven, which is benefited from the fact that integrated optical waveguide offers a well-confined platform for wave manipulation. Recently, we have demonstrated even single optical catenary antenna integrated into the on-chip waveguide can also realize high-contrast polarization router [57]. Therefore, this proposal may release the large-scale fabrication requirement of traditional metasurfaces. For example, polarization controlled unidirectional propagation has been demonstrated by only a line array of catenary apertures [58]. Furthermore, by replacing the chip-integrated gold metasurface by

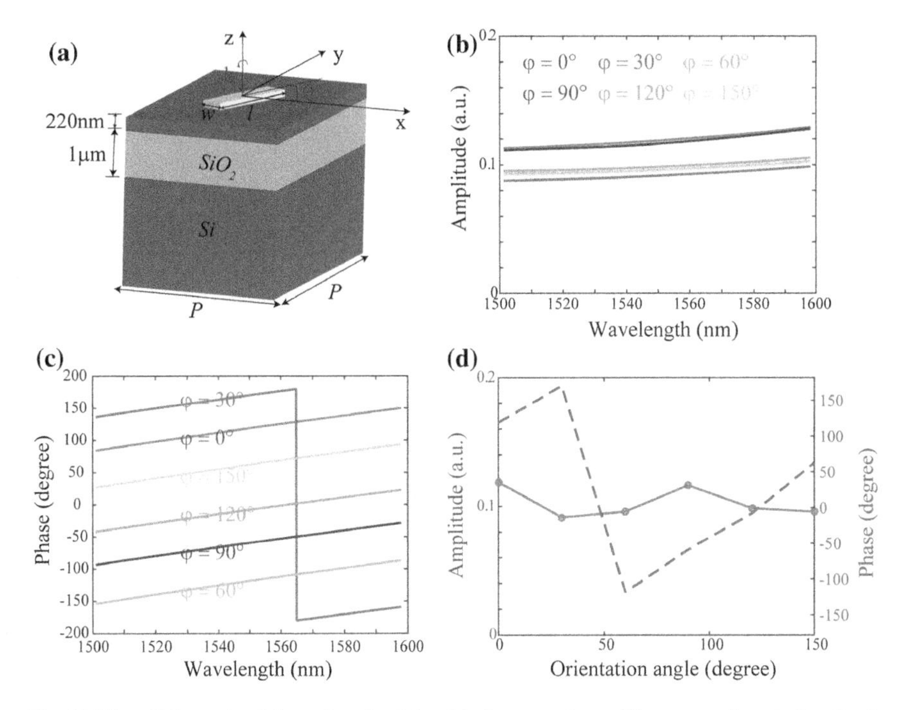

Fig. 12.37 **a** Schematic of the unit cell of the chip-integrated metallic metasurface. **b** Amplitude and **c** phase shift of CPL with opposite handedness through nanoantenna with variable orientation for normal LCP incidence from 1500 to 1600 nm. **d** Amplitude and phase shift as a function of orientation angle at the wavelength of 1550 nm. Reproduced from [54] with permission. Copyright 2018, IEEE

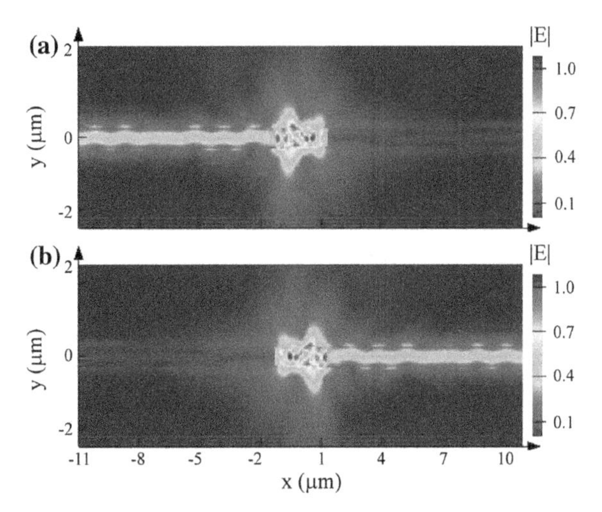

Fig. 12.38 Electric field intensity in the *xy*-plane through the waveguide at the wavelength of 1550 nm for **a** RCP incidence and **b** LCP incidence. Reproduced from [54] with permission. Copyright 2018, IEEE

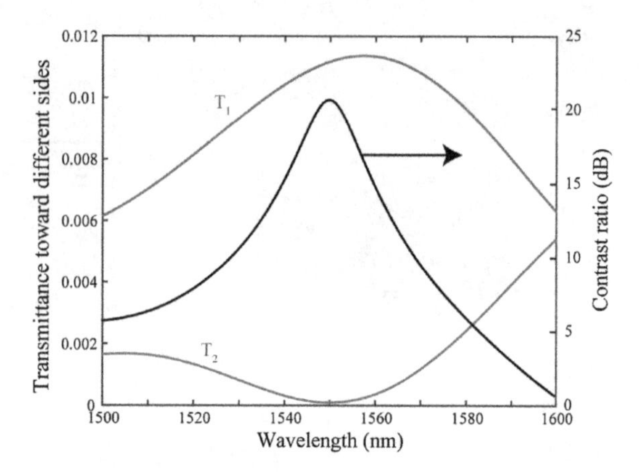

Fig. 12.39 Retrieved transmittance of coupled light toward either side of the waveguide (red and blue curves represent that along $-x$-direction and $+x$-direction, respectively) and calculated contrast ratio as a function of wavelength (black curve). Reproduced from [54] with permission. Copyright 2018, IEEE

chip-integrated silicon metasurface, CMOS-compatible polarization beam router can be realized.

12.4 Polarization Measurement

In this section, we discuss methods of measuring (or creating) the Stokes vector $\boldsymbol{S}$, which consists of four elements S_0, S_1, S_2, and S_3. Among them, S_0 represents the intensity of an optical field (i.e., $S_0 = I_x + I_y$); S_1 and S_2 denote the intensity toward 0° and 45° linear polarization, respectively (i.e., $S_1 = I_x - I_y, S_2 = I_a - I_b$); S_3 expresses the difference between right and left circular polarizations (i.e., $S_3 = I_r - I_l$).

The measurement process can be represented as [59]:

$$\mathbf{I} = \mathbf{A} \cdot \boldsymbol{S} \tag{12.4.1}$$

where $\mathbf{I}$ is the vector of flux measurements as made by the detector, $\mathbf{A}$ is a matrix whose dimensions depend on the number of measurements and whose elements depend on the optical system, and $\boldsymbol{S}$ is the incident Stokes vector. Then, the full state of polarization (SoP) of any input polarization could be retrieved as $\mathbf{S} = \mathbf{A}^{-1} \cdot \mathbf{P}$.

12.4.1 Polarization Measurement in Free Space

Figure 12.40a shows a proposal of spectropolarimeters, consisting of three gap-plasmon phase-gradient metasurfaces occupying 120° circular sector each (denoted as MS1, MS2, and MS3), which could selectively steer the different input SoP and spectral components into six separate spatial domains [60]. Each sector-shaped metasurface functions as an efficient polarization splitter for one of the three different polarization bases ($|x>$, $|y>$), ($|a>$, $|b>$), and ($|r>$, $|l>$). The basis ($|a>$, $|b>$) corresponds to a rotation of the Cartesian coordinate system ($|x>$, $|y>$) by 45° with respect to the x-axis, while ($|r>$, $|l>$) is the basis for CPL. The building block of the three gap metasurfaces is shown in Fig. 12.40b. In order to deflect the x- and y-polarizations to opposite directions (MS1), the lengths and widths of the 10 nanobricks are simultaneously optimized to generate opposite phase gradient, as indicated in Fig. 12.40c. The building blocks of MS2 are similar to that of MS1 just with a rotated orientation angle. In order to deflect the LCP and RCP incidence to opposite directions, geometric phase metasurface is adopted, which consists of spatially rotating nanobricks that can behave as a local waveplate.

Figure 12.41a shows the relevant far-field images of the diffraction spots for six different incident SoPs, each resulting in a unique intensity distribution. From the CCD images, one can clearly see that the bright spots from the ± 1 diffraction order change in intensity in response to altering the incident SOP. In order to quantify the far-field diffraction pattern, a diffraction contrast is defined by

$$D = \frac{I_{+1} - I_{+1}}{I_{+1} + I_{+1}} = \frac{A_u^2 - A_v^2}{A_u^2 + A_v^2} \tag{12.4.2}$$

where u, v represent the direction of ± 1 diffraction. Based on these experimental images, the determined diffraction contrasts are shown in Fig. 12.41b. Therefore, one can determine the SoPs of incidence from the measured D_1, D_2, and D_3 of the single metasurface, which will definitely release the complexity of measurement.

12.4.2 Polarization Measurement in Integrated Optical Waveguide

Recent theoretical investigations indicate that a SOI-based scatterer-waveguide system allows mapping the polarization of a light beam into different amplitudes of guided waves propagating along different optical paths or modes [59]. This system is not limited to the separation of circular polarizations, but also enables sorting linearly polarized photons as long as the incident radiation induces a spinning component in it. The above effect enables the recovery of the SoP of the incoming wave, as illustrated in Fig. 12.42a.

Considering the simplest form, a scatterer is coupled to a single waveguide supporting two propagation directions with (at least) two guided modes each. Breaking the mirror symmetry by introducing a T-block shaped scatterer enables SOI to perform as a nanopolarimeter (Fig. 12.42b). To ensure the full Stokes measurement, the geometries of the T-block shaped scatters were optimized via an optimization process that maximizes the volume of the tetrahedron inscribed within the Poincare sphere. For a serial of experiments performed at 1558 nm, the retrieved polarizations are

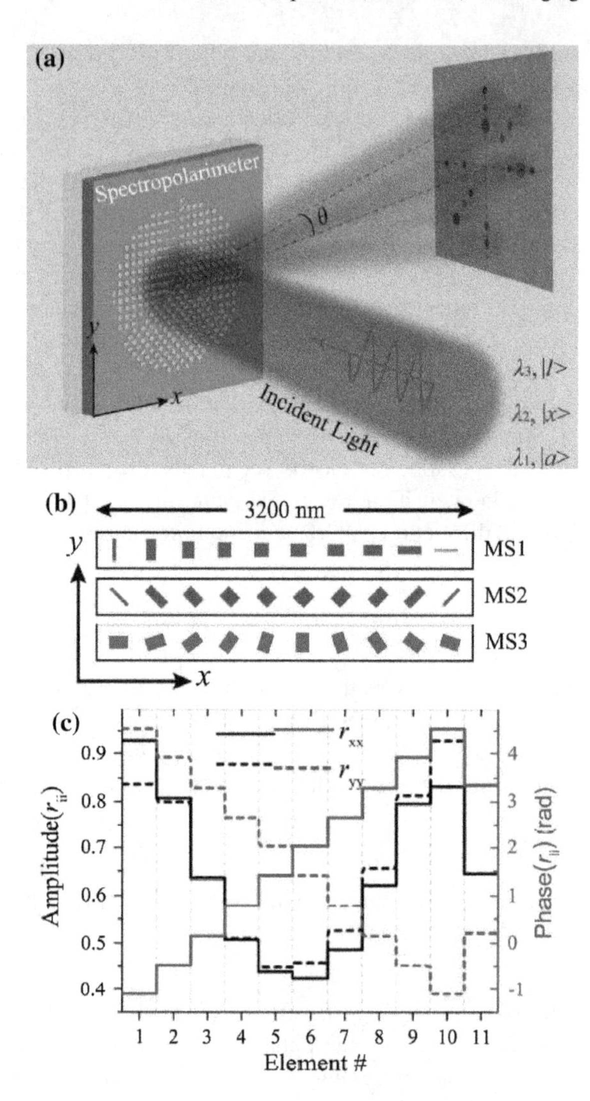

Fig. 12.40 **a** Artistic rendering of the working principle: Different polarization and spectral components are selectively diffracted into predesigned spatial domains with distinct spot contrasts. **b** Schematic of the three metasurface super-cells for the polarization bases (|*x*>, |*y*>), (|*a*>, |*b*>), and (|*r*>, |*l*>). **c** The corresponding reflection amplitudes and phases of the associated 11 nanobricks at λ = 800 nm. Reproduced from [60] with permission. Copyright 2017, American Chemical Society

shown in Fig. 12.46c. The agreement between the generated polarization (measured externally) and the polarization retrieved after measurement of the output power for each mode is remarkable, even for elliptical polarizations.

In the above configuration, owing to both the TE-like and TM-like guided modes are scattered in each output, separation of them is required to determine the Stokes parameters, which may increase the complex of the system. In order to avoid conversion or filtering processes, a metallic scatterer asymmetrically deposited on top of a waveguide crossing (see Fig. 12.43a) could be employed. In this case, by measuring the total optical power (TM + TE modes) at each of the four output ports (see the

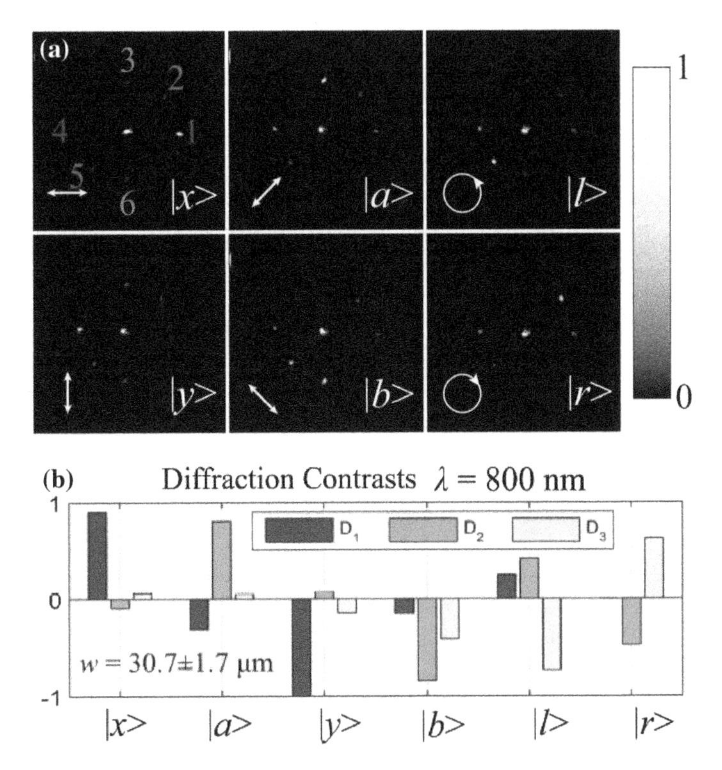

Fig. 12.41 **a** Normalized far-field images for the six incident SoPs at $\lambda = 800$ nm. The brightness and contrast have been adjusted for visualization. The six output channels of diffraction spots are marked by numbers. **b** Measured diffraction contrasts for the six polarizations that represent the extreme value of the three Stokes parameters. Reproduced from [60] with permission. Copyright 2017, American Chemical Society

SEM image of a fabricated sample in Fig. 12.43b), it is possible to retrieve the Stokes parameters of the incoming signal (Fig. 12.43c).

12.5 Polarimetric Imaging

Polarization-resolved imaging and spectroscopy have versatile applications ranging from remote and environmental sensing to biological studies. For example, resolving chirality is essential for extracting structural and functional properties of biochemical species at both molecular and bulk levels. Among different devices for imaging polarimetry, a division of focal plane polarization cameras (DoFP-PCs) is often adopted. In general, DoFP-PCs are based on an array of polarization filters in the focal plane (Fig. 12.44a), which are not only expensive, bulky but also suffer from low image quality and spatial resolution. Recently, it was shown that an optical

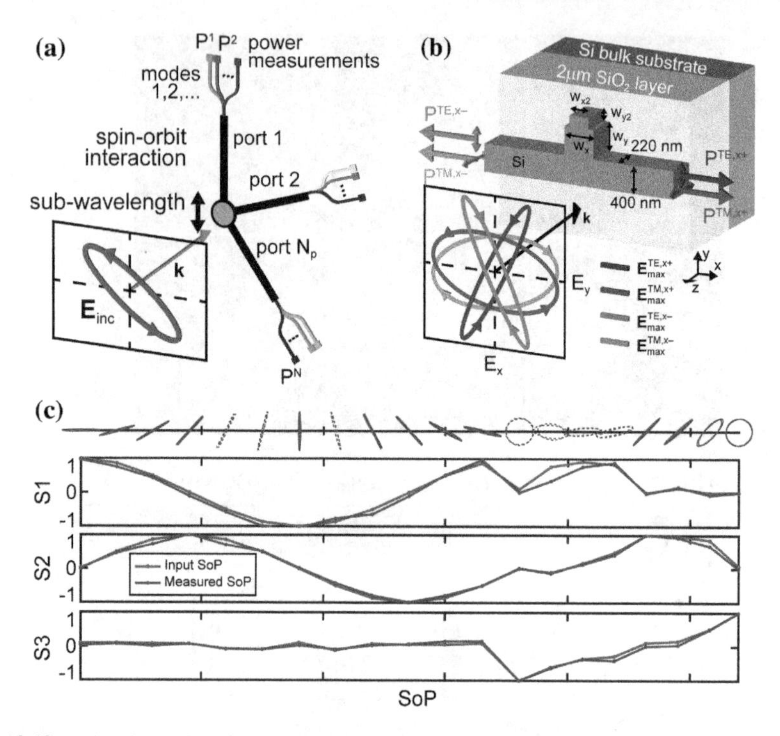

Fig. 12.42 **a** A subwavelength scatterer illuminated by a transverse wave will scatter light into different modes of a set of waveguides. **b** A SOI Stokes nanopolarimeter on a bimodal silicon waveguide (supporting TE-like and TM-like guided modes) laterally perturbed by a scatterer, which breaks the mirror symmetry and enables SOI. **c** Retrieved polarization for a set of experiments performed at 1558 nm. Reproduced from [59] with permission. Copyright 2017, American Chemical Society

metasurface with the ability to fully control phase and polarization of light can perform the same task over a much smaller volume and without changing any optical components. The metasurface can split any two orthogonal states of polarization and simultaneously focus them to different points with high efficiency and on a micron-scale (Fig. 12.44b). Figure 12.44c shows a possible configuration where the three metasurface PBSs are multiplexed to make a super-pixel, comprising of six pixels on the image sensor. Since the anisotropic α-Si nanopost can realize independent phase modulation along its two orthogonal axes, such metasurface PBS can be easily implemented. Each image sensor pixel can then be used to measure the power in a single polarization state.

The experimental results of polarimetric imaging using the DoFP metasurface mask are shown in Fig. 12.45. The mask converts x-polarized input light to an output polarization state characterized by the polarization ellipses and the Stokes parameters shown in Fig. 12.45a and b, respectively. Each Stokes parameter is +1 or −1 in an area of the image corresponding to the specific polarization. For instance, S_3 is +1 in the right half circle, −1 in the left half circle, and 0 elsewhere. The results measured by

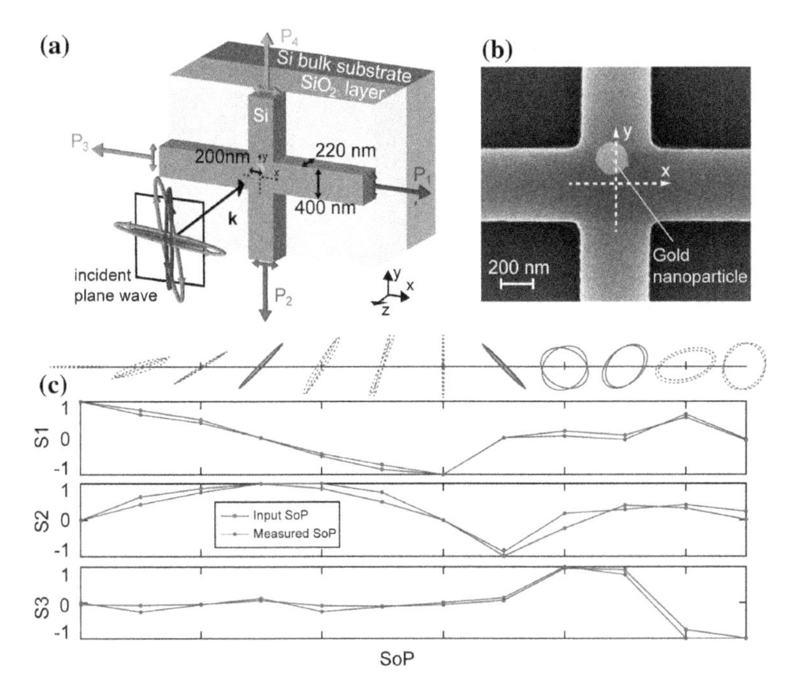

Fig. 12.43 **a** Scheme of the nanopolarimeter consisting of a gold microdisk asymmetrically placed on top of a silicon waveguide crossing. **b** SEM image of a fabricated device. **c** Experimental SoP retrieval at $\lambda = 1550$ nm: input (red) and retrieved (blue) SoP. Reproduced from [59] with permission. Copyright 2017, American Chemical Society

designed metasurface polarimetric camera are shown in Fig. 12.45d and are in good agreement with the results of regular polarimetric imaging (Fig. 12.45c). The lower quality of the metasurface polarimetric camera image is mainly due to the limited number of super-pixels that fit inside the single field of view of the microscope.

Figure 12.46a shows another implementation, so-called generalized Hartmann–Shack beam profiler, where each pixel is composed of 2×3 sub-metalens-array. Each metalens consisting of elliptical silicon pillars focus one of the basis polarizations (e.g., x-, y-, $\pm 45°$, LCP, and RCP) into the following camera with a position shift (d_x, d_y) from the pixel center. The optical micrograph of a fabricated metalens array is shown in Fig. 12.46b–e.

This device not only allows to fully determine the Stokes parameters in each pixel of the array but also could distinguish two common beam profiles with non-constant polarization, e.g., a radially polarized beam and with an azimuthally polarized beam. For local linear polarization of light, the Stokes parameters can be characterized by the polarization angle θ_p:

$$\theta_p = \frac{1}{2}\tan^{-1}\frac{I_a - I_b}{I_x - I_y} \tag{12.5.1}$$

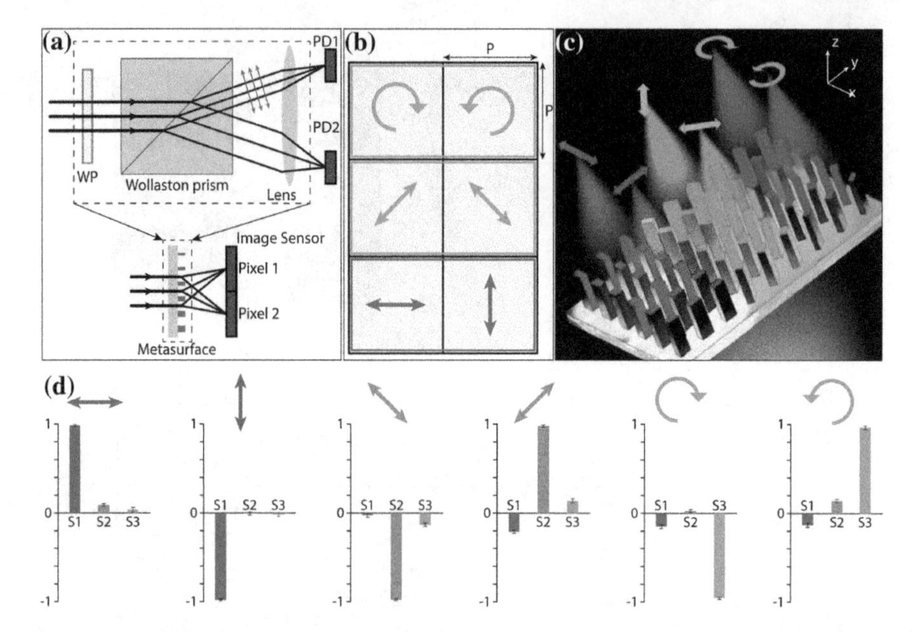

Fig. 12.44 **a** Schematics of a conventional and metasurface setup used for polarimetry. **b** A possible arrangement for a super-pixel of the polarization camera, comprising six image sensor pixels. Three independent polarization bases (H/V,±45°, and RHCP/LHCP) are chosen to measure the Stokes parameters at each super-pixel. **c** Three-dimensional illustration of a super-pixel focusing different polarizations to different spots. **d** Measured average Stokes parameters for different input polairzations (shown with colored arrows). Reproduced from [61] with permission Copyright 2018, American Chemical Society

Figure 12.47a, b show the beam intensity profiles for these incident two beams impinging directly onto the camera. After transmitting through the metalens array, the raw arrays of focal spots recorded in camera are shown in Fig. 12.47c, d, respectively. From these raw data, one can derive the angle profile $\theta_p(x,y)$ according to Eq. (12.5.1). The results are depicted by the black arrows in Fig. 12.47e, f, respectively. The red arrows shown in these figures correspond to the theoretical values.

Finally, as in ordinary Hartmann–Shack wavefront sensor, the local phase gradients along the x- and y-directions can be obtained via the expressions:

$$\frac{\partial \phi}{\partial x} = \frac{2\pi}{\lambda} \cdot \frac{\mathrm{d}_x}{\sqrt{f^2 + \mathrm{d}_x^2}} \tag{12.5.2}$$

$$\frac{\partial \phi}{\partial y} = \frac{2\pi}{\lambda} \cdot \frac{\mathrm{d}_y}{\sqrt{f^2 + \mathrm{d}_y^2}} \tag{12.5.3}$$

Following the above principle, a more complex wavefront, namely for a vortex beam with a twisted wavefront can also be imaged. Figure 12.48a shows the beam

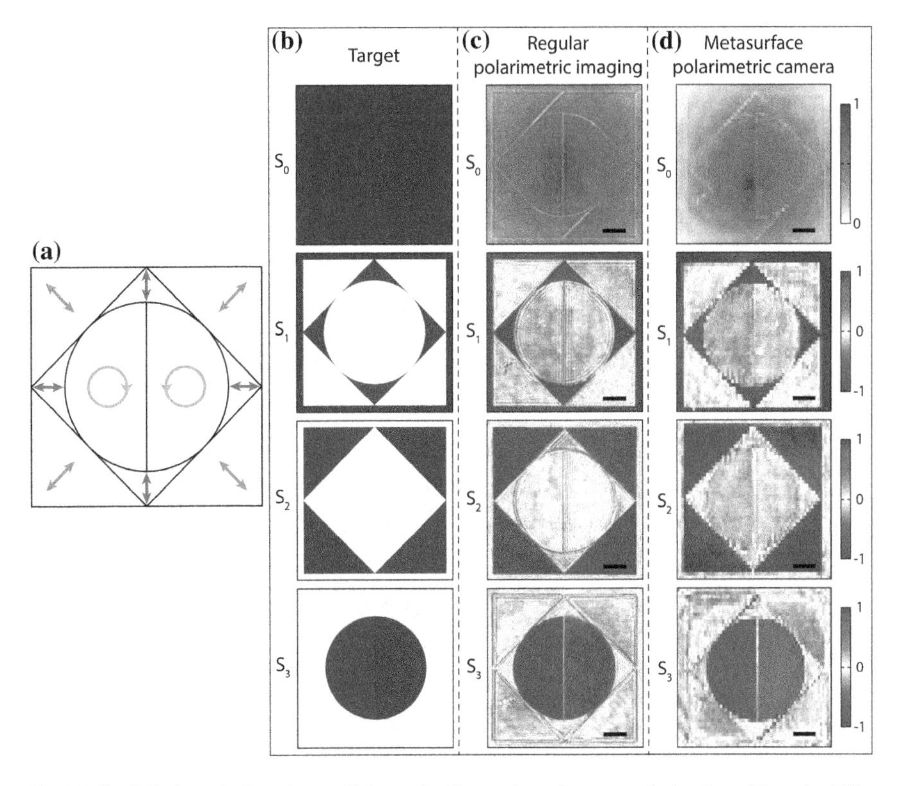

Fig. 12.45 Polarimetric imaging. **a** Schematic illustration of target polarization ellipse in different parts of the polarization sample. Stokes parameters of the polarization sample: **b** the targeted polarization mask, **c** the fabricated mask imaged using conventional polarimetry, and **d** the same mask imaged using the metasurface polarimetric camera. The scale bars denote 100 μm in the metasurface polarization camera mask plane. Reproduced from [61] with permission. Copyright 2018, American Chemical Society

intensity profiles for the incident beam impinging directly onto the camera without passing through the metalens array. Figure 12.48b shows the raw data of metalens focal spots and Fig. 12.48c the phase-gradient profile (see black arrows) derived from Eqs. (12.5.2) and (12.5.3). By integration, the phase profile is obtained (see false-color scale). The calculated topological charge of the vortex beam is 3.25, which deviates by only about 8% from the input value of 3.

Recently, a multispectral chiral lens is proposed, which can image light of different chirality into different spatial positions (Fig. 12.49a) [63]. To this end, metasurfaces composed of two interlaced arrays of blue and green nanofins on a glass substrate are utilized to response to the different CPL, as indicated in Fig. 12.49b. The required phase for focusing is imparted based on the geometric phase. Since the geometric phase is dispersionless and chiral-sensitive, a chiral beetel *Chrysinagloriosa*, which is known to exhibit high reflectivity of left CPL, can be probed across the visible

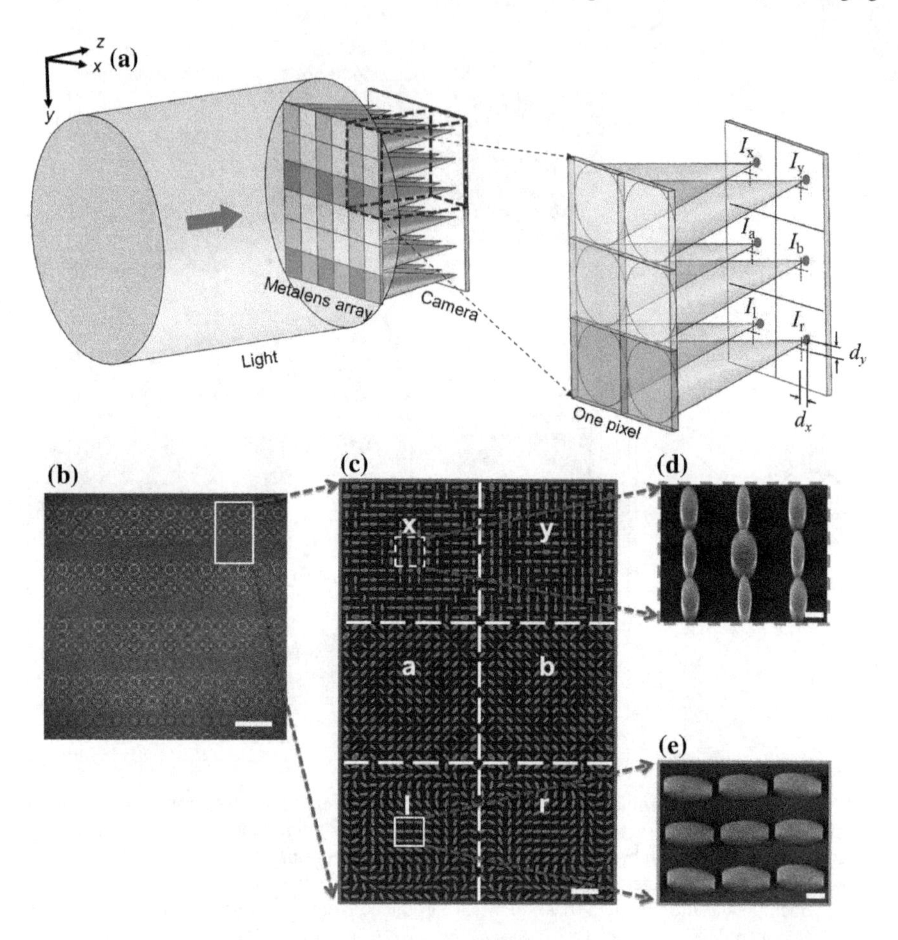

Fig. 12.46 **a** Schematic of the generalized Hartmann–Shack beam profiler. Each pixel of the array consists of six different polarization-sensitive metalenses. A camera records a magnification of the common plane of the metalens foci. **b** Optical micrograph of a fabricated metalens array. Scale bar: 50 μm. **c** Corresponding magnified electron micrograph. Scale bar: 5 μm. **d**, **e** Oblique-view of selected parts of the metalenses. Scale bar: 500 nm Adapted from [62] with permission. Copyright 2018, The Authors

spectrum using only the lens and a camera without the additional of polarizers or dispersive optical devices (Fig. 12.49c–e).

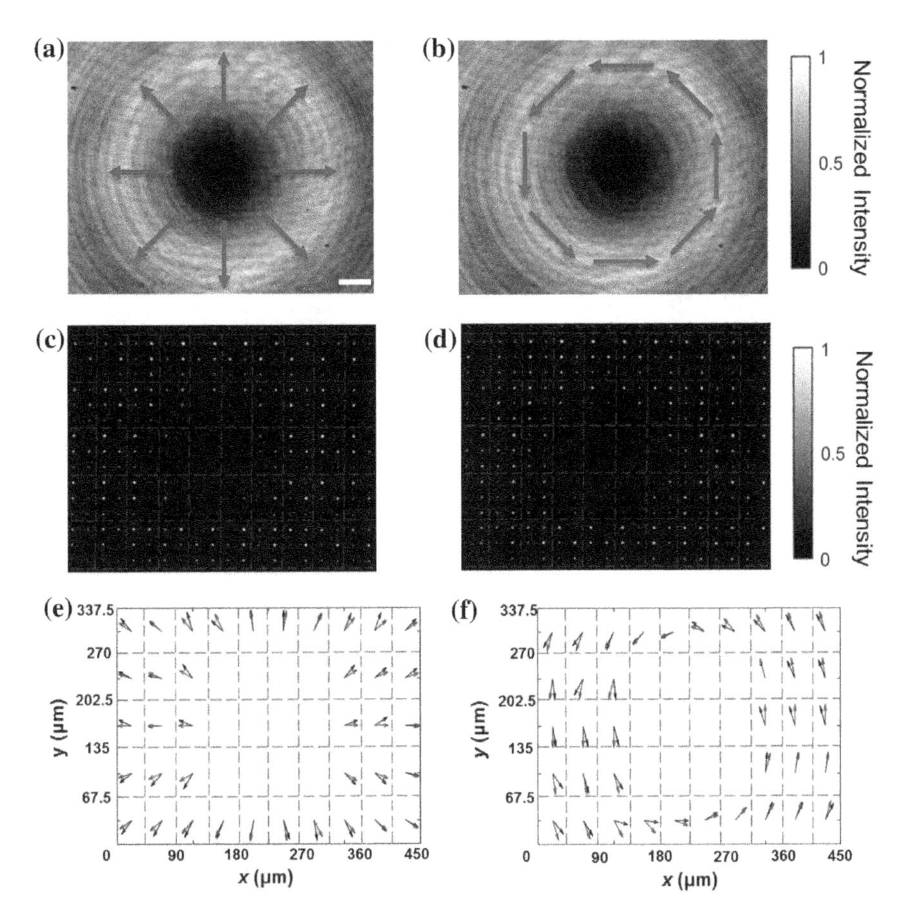

Fig. 12.47 Intensity distributions of focal spots for **a** radially polarized incident beam and **b** an azimuthally polarized beam, respectively. The blue arrows qualitatively indicate the local polarization states. Scale bar: 50 μm. Images of focal spots from the metalens array for **c** the radially polarized and **d** the azimuthally polarized beam, respectively. **e** and **f** Polarization profiles. The black arrows correspond to the measured local polarization vectors and the red arrows to the calculated ones. The dashed gray lines highlight the individual pixels of the metalens array Adapted from [62] with permission. Copyright 2018, The Authors

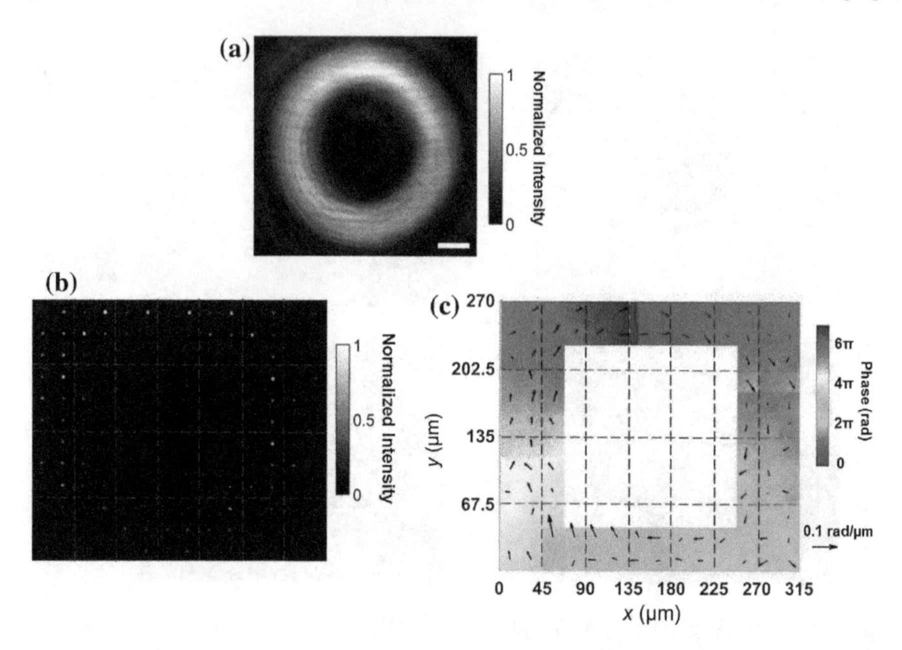

Fig. 12.48 **a** Intensity profile for a vortex beam without the metalens array. Scale bar: 50 μm. **b** Raw data of measured focal spots of the metalens array. **c** Phase gradients (arrows) and wavefront (false-color scale) reconstructed on this basis Adapted with permission from [62]. Copyright 2018, The Authors

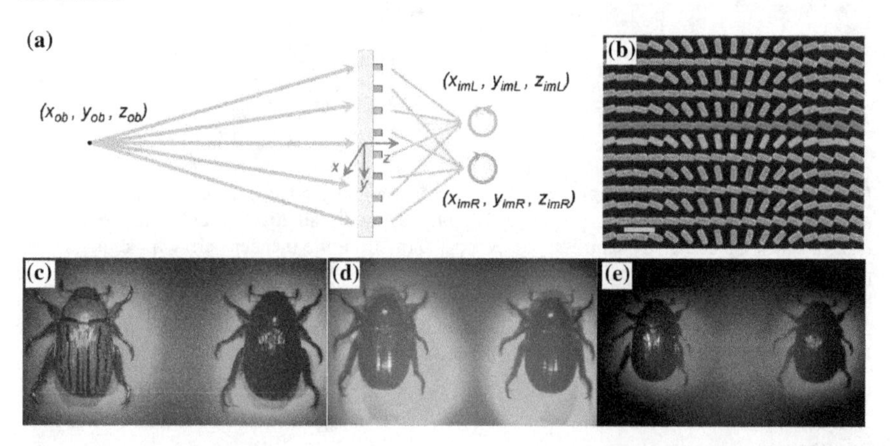

Fig. 12.49 **a** Schematic diagram illustrating the chiral imaging principle. **b** Top view of the building block. The blue and green nanofins impart the required phase profile required to focus RCP light and LCP light, respectively. **c**–**e** Chiral imaging of the beetle, *Chrysinagloriosa*, on the color camera at **c** 532 nm, **d** 488 nm, and **e** 620 nm. The left (right) image is formed by focusing LCP (RCP) light reflected from the beetle. Reproduced from [63] with permission. Copyright 2016, American Chemical Society

References

1. B.E.A. Saleh, M.C. Teich, *Fundamentals of Photonics*, 2nd edn. (Wiley, 2007)
2. G. Elert, The physics hypertextbook (1998–2019)
3. J.K. Gansel, M. Thiel, M.S. Rill, M. Decker, K. Bade, V. Saile, G. von Freymann, S. Linden, M. Wegener, Gold helix photonic metamaterial as broadband circular polarizer. Science **325**, 1513–1515 (2009)
4. J. Kaschke, L. Blume, L. Wu, M. Thiel, K. Bade, Z. Yang, M. Wegener, A helical metamaterial for broadband circular polarization conversion. Adv. Opt. Mater. **3**, 1411–1417 (2015)
5. X. Ma, M. Pu, X. Li, Y. Guo, X. Luo, All-metallic wide-angle metasurfaces for multifunctional polarization manipulation. Opto-Electron. Adv. **2**, 180023 (2019)
6. G. Ohman, The pseudo-Brewster angle. IEEE Trans. Antennas Propag. **25**, 903–904 (1977)
7. X. Luo, Principles of electromagnetic waves in metasurfaces. Sci. China-Phys. Mech. Astron. **58**, 594201 (2015)
8. D. Markovich, A. Andryieuski, M. Zalkovskij, R. Malureanu, and A. Lavrinenko, Metamaterial polarization converter analysis: limits of performance. Appl. Phys. B 1–10 (2013)
9. G. Nordin, P. Deguzman, Broadband form birefringent quarter-wave plate for the mid-infrared wavelength region. Opt. Express **5**, 163–168 (1999)
10. K. Robbie, M.J. Brett, A. Lakhtakia, Chiral sculptured thin films. Nature **384**, 616 (1996)
11. X. Xie, X. Li, M. Pu, X. Ma, K. Liu, Y. Guo, X. Luo, Plasmonic metasurfaces for simultaneous thermal infrared invisibility and holographic illusion. Adv. Funct. Mater. **28**, 1706673 (2018)
12. M. Pu, P. Chen, Y. Wang, Z. Zhao, C. Huang, C. Wang, X. Ma, X. Luo, Anisotropic meta-mirror for achromatic electromagnetic polarization manipulation. Appl. Phys. Lett. **102**, 131906 (2013)
13. Y. Zhao, M.A. Belkin, A. Alù, Twisted optical metamaterials for planarized ultrathin broadband circular polarizers. Nat. Commun. **3**, 870 (2012)
14. M. Pu, Z. Zhao, Y. Wang, X. Li, X. Ma, C. Hu, C. Wang, C. Huang, X. Luo, Spatially and spectrally engineered spin-orbit interaction for achromatic virtual shaping. Sci. Rep. **5**, 9822 (2015)
15. Y. Guo, J. Yan, M. Pu, X. Li, X. Ma, Z. Zhao, X. Luo, Ultra-wideband manipulation of electromagnetic waves by bilayer scattering engineered gradient metasurface. RSC Adv. **8**, 13061–13066 (2018)
16. D.C. Flanders, Submicrometer periodicity gratings as artificial anisotropic dielectrics. Appl. Phys. Lett. **42**, 492–494 (1983)
17. L. Young, L.A. Robinson, C. Hacking, Meander-line polarizer. IEEE Trans. Antennas Propag. **21**, 376–378 (1973)
18. R.-S. Chu, K.-M. Lee, Analytical method of a multilayered meander-line polarizer plate with normal and oblique plane-wave incidence. IEEE Trans. Antennas Propag. **35**, 652–661 (1987)
19. X. Ma, C. Huang, M. Pu, C. Hu, Q. Feng, X. Luo, Single-layer circular polarizer using metamaterial and its application in antenna. Microw. Opt. Technol. Lett. **54**, 1770–1774 (2012)
20. N.K. Grady, J.E. Heyes, D.R. Chowdhury, Y. Zeng, M.T. Reiten, A.K. Azad, A.J. Taylor, D.A.R. Dalvit, H.-T. Chen, Terahertz metamaterials for linear polarization conversion and anomalous refraction. Science **340**, 1304–1307 (2013)
21. N. Yu, P. Genevet, M.A. Kats, F. Aieta, J.-P. Tetienne, F. Capasso, Z. Gaburro, Light propagation with phase discontinuities: generalized laws of reflection and refraction. Science **334**, 333–337 (2011)
22. M. Pu, X. Li, X. Ma, Y. Wang, Z. Zhao, C. Wang, C. Hu, P. Gao, C. Huang, H. Ren, X. Li, F. Qin, J. Yang, M. Gu, M. Hong, X. Luo, Catenary optics for achromatic generation of perfect optical angular momentum. Sci. Adv. **1**, e1500396 (2015)
23. N. Yu, F. Aieta, P. Genevet, M.A. Kats, Z. Gaburro, F. Capasso, A broadband, background-free quarter-wave plate based on plasmonic metasurfaces. Nano Lett. **12**, 6328–6333 (2012)
24. A. Shaltout, J. Liu, V.M. Shalaev, A.V. Kildishev, Optically active metasurface with non-chiral plasmonic nanoantennas. Nano Lett. **14**, 4426–4431 (2014)

25. B. Shen, P. Wang, R. Polson, R. Menon, Ultra-high-efficiency metamaterial polarizer. Optica **1**, 356–360 (2014)
26. J. Hao, Y. Yuan, L. Ran, T. Jiang, J.A. Kong, C.T. Chan, L. Zhou, Manipulating electromagnetic wave polarizations by anisotropic metamaterials. Phys. Rev. Lett. **99**, 063908 (2007)
27. Y. Guo, M. Pu, X. Ma, X. Li, X. Luo, Advances of dispersion-engineered metamaterials. Opto-Electron. Eng. **44**, 3–22 (2017)
28. Q. Feng, M. Pu, C. Hu, X. Luo, Engineering the dispersion of metamaterial surface for broadband infrared absorption. Opt. Lett. **37**, 2133–2135 (2012)
29. C.A. Dirdal, J. Skaar, Superpositions of Lorentzians as the class of causal functions. Phys. Rev. A **88**, 033834 (2013)
30. Y. Guo, Y. Wang, M. Pu, Z. Zhao, X. Wu, X. Ma, C. Wang, L. Yan, X. Luo, Dispersion management of anisotropic metamirror for super-octave bandwidth polarization conversion. Sci. Rep. **5**, 8434 (2015)
31. Z. Zhang, J. Luo, M. Song, H. Yu, Large-area, broadband and high-efficiency near-infrared linear polarization manipulating metasurface fabricated by orthogonal interference lithography. Appl. Phys. Lett. **107**, 241904 (2015)
32. Y. Guo, M. Pu, X. Li, X. Ma, P. Gao, Y. Wang, X. Luo, Functional metasurfaces based on metallic and dielectric subwavelength slits and stripes array. J. Phys. Condens. Matter **30**, 14003 (2018)
33. Y. Guo, L. Yan, W. Pan, B. Luo, Achromatic polarization manipulation by dispersion management of anisotropic meta-mirror with dual-metasurface. Opt. Express **23**, 27566–27575 (2015)
34. P. Chen, Ultra-broadband terahertz polarization transformers using dispersion-engineered anisotropic metamaterials. Opto-Electron. Eng. **44**, 82–86 (2017)
35. M. Pu, Q. Feng, M. Wang, C. Hu, C. Huang, X. Ma, Z. Zhao, C. Wang, X. Luo, Ultrathin broadband nearly perfect absorber with symmetrical coherent illumination. Opt. Express **20**, 2246–2254 (2012)
36. C. Yan, M. Pu, J. Luo, Y. Huang, X. Li, X. Ma, X. Luo, Coherent perfect absorption of electromagnetic wave in subwavelength structures. Opt. Laser Technol. **101**, 499–506 (2018)
37. Y. Wang, X. Ma, X. Li, M. Pu, X. Luo, Perfect electromagnetic and sound absorption via subwavelength holes array. Opto-Electron. Adv. **1**, 180013 (2018)
38. Y. Wang, M. Pu, C. Hu, Z. Zhao, C. Wang, X. Luo, Dynamic manipulation of polarization states using anisotropic meta-surface. Opt. Commun. **319**, 14–16 (2014)
39. S.-C. Jiang, X. Xiong, Y.-S. Hu, Y.-H. Hu, G.-B. Ma, R.-W. Peng, C. Sun, M. Wang, Controlling the polarization state of light with a dispersion-free metastructure. Phys. Rev. X **4**, 021026 (2014)
40. M. Zhang, M. Pu, F. Zhang, Y. Guo, Q. He, X. Ma, Y. Huang, X. Li, H. Yu, X. Luo, Plasmonic metasurfaces for switchable photonic spin-orbit interaction based on phase change materials. Adv. Sci. **5**, 1800835 (2018)
41. A. Nemati, Q. Wang, M. Hong, J. Teng, Tunable and reconfigurable metasurfaces and metadevices. Opto-Electron. Adv. **1**, 180009 (2018)
42. X. Ma, W. Pan, C. Huang, M. Pu, Y. Wang, B. Zhao, J. Cui, C. Wang, X. Luo, An active metamaterial for polarization manipulating. Adv. Opt. Mater. **2**, 945–949 (2014)
43. J. Cui, C. Huang, W. Pan, M. Pu, Y. Guo, X. Luo, Dynamical manipulation of electromagnetic polarization using anisotropic meta-mirror. Sci. Rep. **6**, 30771 (2016)
44. D. Wang, L. Zhang, Y. Gu, M.Q. Mehmood, Y. Gong, A. Srivastava, L. Jian, T. Venkatesan, C.-W. Qiu, M. Hong, Switchable ultrathin quarter-wave plate in terahertz using active phase-change metasurface. Sci. Rep. **5**, 15020 (2015)
45. D. Wang, L. Zhang, Y. Gong, L. Jian, T. Venkatesan, C. Qiu, M. Hong, Multiband switchable terahertz quarter-wave plates via phase-change metasurfaces. IEEE Photonics J. **8**, 1–8 (2016)
46. X. Luo, M. Pu, X. Ma, X. Li, Taming the electromagnetic boundaries via metasurfaces: from theory and fabrication to functional devices. Int. J. Antennas Propag. **2015**, 204127 (2015)
47. X. Ma, C. Huang, M. Pu, C. Hu, Q. Feng, X. Luo, Multi-band circular polarizer using planar spiral metamaterial structure. Opt. Express **20**, 16050–16058 (2012)

48. X. Ma, C. Huang, M. Pu, Y. Wang, Z. Zhao, C. Wang, X. Luo, Dual-band asymmetry chiral metamaterial based on planar spiral structure. Appl. Phys. Lett. **101**, 161901 (2012)
49. Y. Cui, L. Kang, S. Lan, S. Rodrigues, W. Cai, Giant chiral optical response from a twisted-arc metamaterial. Nano Lett. **14**, 1021–1025 (2014)
50. C. Huang, X. Ma, M. Pu, G. Yi, Y. Wang, X. Luo, Dual-band 90° polarization rotator using twisted split ring resonators array. Opt. Commun. **291**, 345–348 (2013)
51. X. Ma, C. Huang, M. Pu, W. Pan, Y. Wang, X. Luo, Circular dichroism and optical rotation in twisted Y-shaped chiral metamaterial. Appl. Phys. Express **6**, 022001 (2013)
52. Z. Li, M.-H. Kim, C. Wang, Z. Han, S. Shrestha, A.C. Overvig, M. Lu, A. Stein, A.M. Agarwal, M. Lončar, N. Yu, Controlling propagation and coupling of waveguide modes using phase-gradient metasurfaces. Nat. Nanotechnol. **12**, 675–683 (2017)
53. C. Wang, Z. Li, M.-H. Kim, X. Xiong, X.-F. Ren, G.-C. Guo, N. Yu, M. Lončar, Metasurface-assisted phase-matching-free second harmonic generation in lithium niobate waveguides. Nat. Commun. **8**, 2098 (2017)
54. Y. Guo, M. Pu, X. Li, X. Ma, S. Song, Z. Zhao, X. Luo, Chip-integrated geometric metasurface as a novel platform for directional coupling and polarization sorting by spin-orbit interaction. IEEE J. Sel. Top. Quantum Electron. **24**, 4700107 (2018)
55. B. Shen, P. Wang, R. Polson, R. Menon, An integrated-nanophotonics polarization beamsplitter with $2.4 \times 2.4\ \mu m^2$ footprint. Nat. Photonics **9**, 378–382 (2015)
56. T.P.H. Sidiropoulos, M.P. Nielsen, T.R. Roschuk, A.V. Zayats, S.A. Maier, R.F. Oulton, Compact optical antenna coupler for silicon photonics characterized by third-harmonic generation. ACS Photonics **1**, 912–916 (2014)
57. Y. Guo, M. Pu, X. Li, X. Ma, X. Luo, Ultra-broadband spin-controlled directional router based on single optical catenary integrated on silicon waveguide. Appl. Phys. Express **11**, 092202 (2018)
58. J. Jin, X. Li, Y. Guo, M. Pu, P. Gao, X. Ma, X. Luo, Polarization-controlled unidirectional excitation of surface plasmon polaritons utilizing catenary apertures. Nanoscale (2019)
59. A. Espinosa-Soria, F.J. Rodríguez-Fortuño, A. Griol, A. Martínez, On-chip optimal stokes nanopolarimetry based on spin-orbit interaction of light. Nano Lett. **17**, 3139–3144 (2017)
60. F. Ding, A. Pors, Y. Chen, V.A. Zenin, S.I. Bozhevolnyi, Beam-size-invariant spectropolarimeters using gap-plasmon metasurfaces. ACS Photonics **4**, 943–949 (2017)
61. E. Arbabi, S.M. Kamali, A. Arbabi, A. Faraon, Full stokes imaging polarimetry using dielectric metasurfaces. ACS Photonics **5**, 3132–3140 (2018)
62. Z. Yang, Z. Wang, Y. Wang, X. Feng, M. Zhao, Z. Wan, L. Zhu, J. Liu, Y. Huang, J. Xia, M. Wegener, Generalized Hartmann-Shack array of dielectric metalens sub-arrays for polarimetric beam profiling. Nat. Commun. **9**, 4607 (2018)
63. M. Khorasaninejad, W.T. Chen, A.Y. Zhu, J. Oh, R.C. Devlin, D. Rousso, F. Capasso, Multispectral chiral imaging with a metalens. Nano Lett. **16**, 4595–4600 (2016)

Chapter 13
Perfect Absorption of Light

Abstract The absorption of light refers to the conversion of photons and electromagnetic waves to other kinds of energy such as heat and photo-generated carriers. In classic optics, absorbers are characterized by how black they are; thus, an ideal absorber should be black as much as possible. In EO 2.0, the elaborately designed subwavelength structures not only provide a mean to realize ultra-black absorbers, but also enable the precise control of absorption spectrum in the entire electromagnetic spectrum ranging from microwave to ultraviolet band. In this chapter, we give a discussion of various narrow and broadband optical absorbers with special attentions paid on wide-angle, transparent, and refractory absorbers, which show great advantages over their traditional counterparts. Their applications in bolometers, solar cells, and sensors are presented.

Keywords Metamaterial absorber · Broadband absorber · Wide-angle absorber · Transparent absorber · Refractory absorber

13.1 Introduction

Absorption of light is a fundamental process in the light–matter interaction. Perhaps, the most famous optical absorber is the blackbody proposed by Gustav Kirchhoff in 1860 [1]:

> ...the supposition that bodies can be imagined which, for infinitely small thicknesses, completely absorb all incident rays, and neither reflect nor transmit any. I shall call such bodies *perfectly black*, or, more briefly, *black bodies*.

The blackbody is a hypothetical concept. For light wave illuminating on ideal flat natural materials, there is always reflection and transmission. If the material is lossy and thick enough, all transmitted lights will be absorbed. Theoretically, these coefficients can be calculated using Fresnel's equations and Beer's law. As shown in Fig. 13.1, an approximate blackbody could be realized by an insulated enclosure with a small hole to let light in. After many times of reflection, all incidents could be completely absorbed.

X. Luo, *Engineering Optics 2.0*, https://doi.org/10.1007/978-981-13-5755-8_13

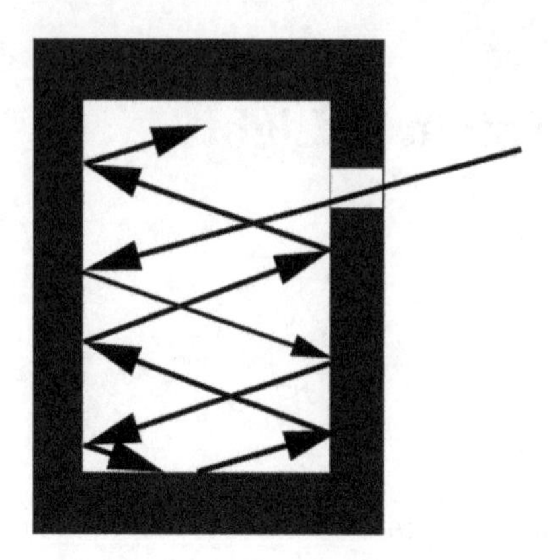

Fig. 13.1 Schematic of an approximate blackbody realized by an insulated enclosure with a small hole

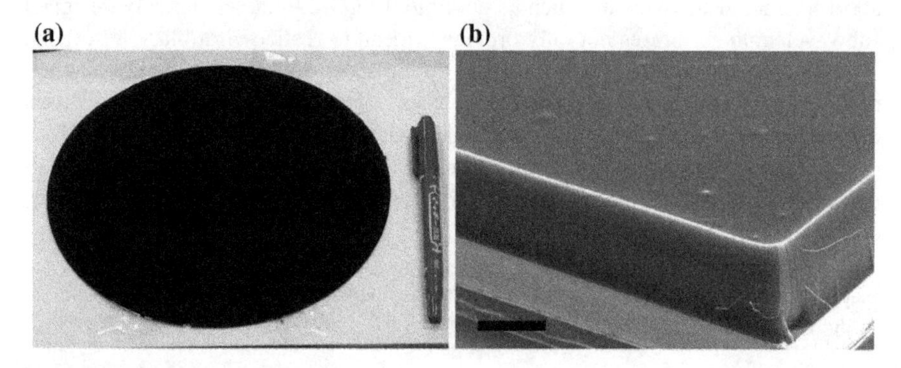

Fig. 13.2 **a** SWNT forest grown on an 8-in. silicon wafer. **b** SEM image of SWNT forest vertically standing on a silicon substrate. Scale bar, 0.5 mm. Reproduced from [2] with permission. Copyright 2009, The Authors

In practical applications, it is often desired that Kirchhoff's original concept could be realized; i.e., all incident lights could be absorbed in an ultra-thin material. However, this seems to be counterintuitive and not practical. On the one hand, according to Fresnel's equations, perfect absorption favors a small refractive index and attenuation coefficient. On the other hand, according to Beer's law, a small attenuation coefficient would lead to a very large thickness. For instance, vertically aligned carbon nanotube can be made to be very lightweight with a refractive index close to unity. Therefore, the interfacial reflection is greatly reduced. When the thickness approaches hundreds of micrometers, a near-perfect absorption can be realized in the entire ultraviolet (UV), visible, and infrared (IR) bands. Figure 13.2 shows a photograph and scanning electron microscope (SEM) image of single-walled carbon nanotube (SWNT) forest (height: 460 μm, density: 0.07 g/cm^3) [2]. The measured reflection spectrum in the UV, visible, and IR bands is shown in Fig. 13.3, revealing absorption larger than 99% in the entire range.

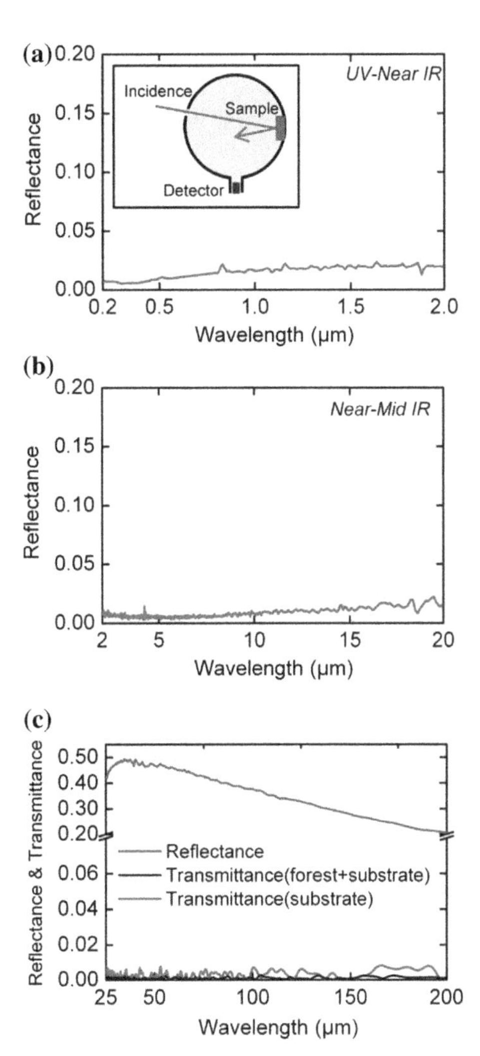

Fig. 13.3 **a** Reflectance in the UV-to-near IR region. The inset illustrates the configuration of the reflectance measurements. **b** Reflectance in the near-to-mid-IR region. **c** Reflectance in the mid-to-far IR region (red) and transmittance of the substrate (black) and forest + substrate (blue). Reproduced from [2] with permission. Copyright 2009, The Authors

Since optical absorbers are typically much thicker than the wavelength, Max Planck concluded that Kirchhoff's infinitely thin blackbody cannot be realized [3]. It is therefore a key problem to design perfect absorbing material with minimal thickness. Considering the fact that natural materials are not so efficient, it is natural to utilize structured material to trap light and enhance the absorption efficiency.

The investigation of subwavelength EM absorbers can date back to 1902 when Wood observed the anomalous dips in the reflection spectra of metallic gratings under illumination of a white light source [4]. Subsequent investigations show that the absorption of light is mainly attributed to the excitation of surface plasmon polaritons (SPPs) [5], which can tightly trap the incident light so that they can be fully absorbed.

In the microwave regime, electromagnetic absorbers can be realized by placing resistive sheets (a lossy thin metallic film with thickness smaller than the skin depth) in the structure in order to match the free-space impedance ($Z_0 = 377\ \Omega$) to that of reflective metallic ground plane ($Z = 0\ \Omega$). A Salisbury screen is one of the simplest electromagnetic absorbers that have a resistive sheet ($Z_s = 377\ \Omega$) placed at $\lambda/4$ (λ is the wavelength of the center operating frequency) over the ground plane [6]. Although it is very efficient at the resonance frequency, this kind of absorber is narrowband and dependent on the incidence angle. Another classical electromagnetic absorber is the Jaumann absorber, which utilizes a multilayer structure to increase the bandwidth [7]. However, to obtain a broader bandwidth, the structure becomes very thick and bulky, making them not practical for many applications.

In 2008, a novel concept of metamaterial perfect absorber (MPA) was proposed [8]. MPA has several superiorities over traditional absorbers. First, the thickness of the absorbers is much thinner than traditional Salisbury absorbers [6], which generally require a thickness of quarter wavelength. For example, the thickness of the first MPA is only about $\lambda/40$. Secondly, the absorption spectrum can be tuned on demand by changing the resonant frequency of metallic structures with specially designed geometries. After the first MPA was reported in microwave band, it was immediately extended to terahertz band [9–11] and even infrared [12, 13] and visible bands [14] by geometry scaling. Third, it has been demonstrated that by engineering the dispersion of surface structures and the metasurface wave (M-waves) therein, one can surpass the thickness–bandwidth limit under coherent illumination [15–19]. Fourth, the absorption frequency and efficiency can be dynamically tuned [20, 21].

Generally, the basic MPA is composed of two metallic structures separated by a deep subwavelength dielectric spacer. Among them, the top metallic film is designed as a special shape to generate necessary electric resonance, and the bottom film acts as a mirror to excite magnetic resonance while meanwhile suppresses the transmission (i.e., $T = 0$) to the structure. Therefore, the absorption (A) of MPA can be determined from the reflection (R) as $A = 1 - R$. During the past decade, numerous strategies have been proposed to realize low reflection, e.g., wave-impedance match and free-bound wave exchange.

13.2 Narrowband and Multi-band Metamaterial Absorbers

13.2.1 Wave-Impedance Match

Figure 13.4 shows the MPA proposed by Landy et al. [8], where metallic split-ring resonator (SRR) is utilized to interact with the incident electric field, while a wide metallic wire is coupled with the SRR to confine the magnetic field. Since the period of the unit cell is in subwavelength scale, the MPA was taken as a homogenous material with effective permittivity $\varepsilon_{\mathrm{eff}}$ and permeability μ_{eff}. In order to make $R = 0$, the equivalent homogenous material should satisfy the impedance match condition:

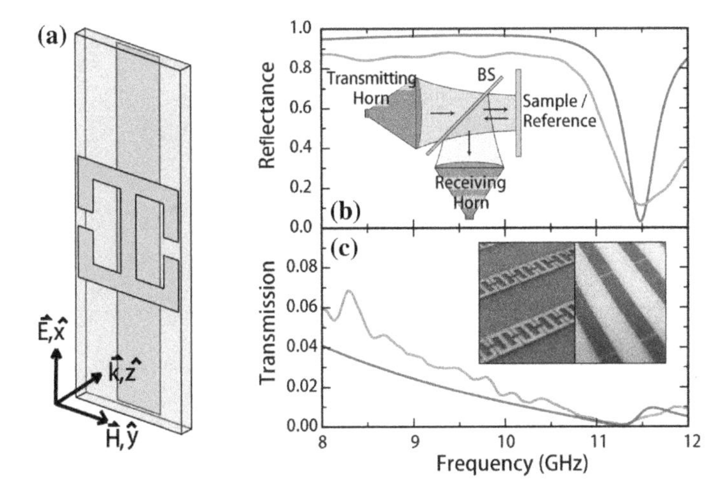

Fig. 13.4 **a** Schematic of a unit cell of MPA. **b**, **c** Simulated and measured reflectance and transmittance of the absorber. Inset is the schematic of the experiment (**b**) and photographs of both sides of the MPA (**c**). Reproduced from [8] with permission. Copyright 2008, American Physical Society

$$Z_{\text{eff}} = \sqrt{\varepsilon_{\text{eff}}/\mu_{\text{eff}}} = 1. \tag{13.2.1}$$

By simultaneously adjusting the electric and magnetic resonances, the impedance match condition can be satisfied at special frequency (for instance, 11.5 GHz), where reflection vanishes and perfect absorption is realized. By scaling the geometric parameters, the above design methodology was also extended to terahertz band [22, 23].

It should be noted that many subwavelength structured absorbers actually cannot be homogenized as metamaterials since they are not periodic in the propagation direction of light. Instead, each layer should be considered separately using equivalent impedance sheet. By engineering the dispersion of the impedance sheet, perfect absorption with broadband [12] and angle-insensitive [24] performances can be realized, which will be discussed in Sects. 13.3.3 and 13.5.1.

13.2.2 Free-Bound Wave Exchange

Inspired by Wood's anomalous observed in metallic gratings, free-bound two-wave exchange mechanism was proposed for perfect absorption at visible band [25, 26]. A simple method to realize the free-bound wave exchange is shown in Fig. 13.5a, where periodic square holes were drilled in the top metallic film to excite gap plasmons or the so-called catenary plasmons. Besides, the strong coupling between the top pattern gold film and the bottom gold reflective film supports the propagation of antisymmetric surface plasmons along the transverse plane.

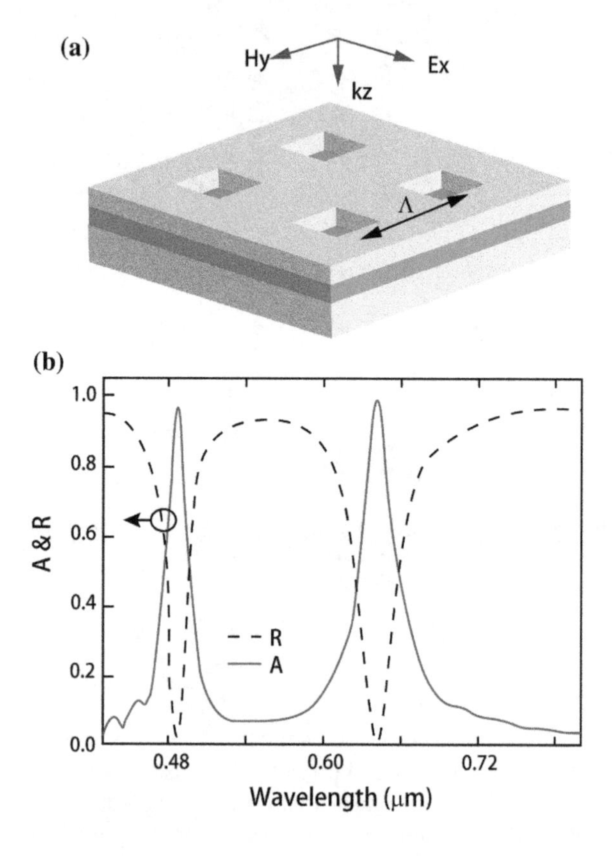

Fig. 13.5 **a** Schematic of MPA. **b** Simulated absorption and reflection. Adapted from [25] with permission. Copyright 2009, Optical Society of America

The period Λ of the holes was selected to satisfy the momentum match condition:

$$\beta = k\ \sin\theta + m\frac{2\pi}{\Lambda} \tag{13.2.2}$$

where $k = 2\pi/\lambda$ is the wave number of incidence, β is the wave number of SPP confined in metal–insulator–metal (MIM) waveguide, and θ is the angle of incidence. Moreover, the depth of the metallic holes is optimized to make the radiative decay rate of the MIM cavity equal to the rate associated with dissipation, so that they can absorb light completely. According to the general principles of the scattering theory developed in quantum mechanics, the reflectivity at an angular frequency, ω, reduces to [27]:

$$R = \frac{(\omega - \omega_0)^2}{(\omega - \omega_0)^2 + \Gamma^2/4} \tag{13.2.3}$$

where Γ is the resonance width. As anticipated, the reflectivity vanishes at the resonance angular frequency ω_0. As indicated in Fig. 13.5b, two near-perfect absorption peaks appear at 642.7 nm, corresponding to the first-order and second-order diffractions.

13.2.3 Super-Unit Cell Design

Owing to the resonant property, the original MPA is often narrowband. Many methods have been proposed to extend the bandwidth. Besides absorption in a continuous wideband, multi-band absorption is desired in many applications. In super-unit cell design, multiple unit cells with different resonance frequencies are combined into a super-cell [28–32]. There are two combination methods including horizontally cascading (Fig. 13.6) [32, 33] and vertically cascading [11, 34, 35]. The former is at an expense of expanded dimension of super-unit cell, usually being utilized at lower frequency since it needs to satisfy the subwavelength condition to suppress the high-order diffractions. The latter extends the bandwidth at a cost of increased thickness and thus is preferred at high frequency since the absolute thickness is not too thick. Nevertheless, the bandwidth extension is very limited in the performance, typically no more than 20% relative bandwidth, because only finite number of resonators can be integrated into a super-unit cell.

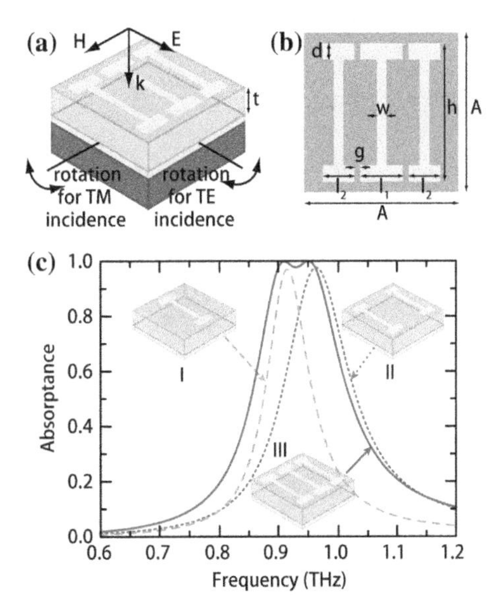

Fig. 13.6 **a** Schematic of the whole unit cell and **b** top view of the MPA. **c** Numerical simulation results of absorption spectra at normal incidence for three different configurations of the I-shaped resonators that are indicated in the inset. Reproduced from [32] with permission. Copyright 2012, Optical Society of America

13.3 Broadband Absorbers

In order to obtain an octave beyond absorption band, different band-expansion strategies have been proposed, which will be discussed in this section.

13.3.1 Broadband Absorption Based on Mode Hybridization

Figure 13.7 shows a vertically cascading design for mode hybridization, which consists of M super-unit cells with different lateral sizes w_i ($i = 1, 2, \ldots, M$ is the order of super-unit cell) [11]. Each super-unit cell is constructed by N sub-unit cells that are metal–dielectric pairs with the same lateral size but different refractive indices n_j ($j = 1, 2, \ldots, N$ is the order of sub-unit cell), represented by different colors. The thicknesses of the metal and dielectric films are t and h, respectively. The period P of the proposed structure is 25 μm, which is smaller than the wavelengths of terahertz frequencies (30–3000 μm). The conductivity and the thickness of the metal are assumed as $\sigma = 4.0 \times 10^7$ S/m and 0.2 μm, respectively. In this design, sub-unit cells have the same lateral dimension and super-unit cells have enough dimension variation among them to reduce the demand for fabrication precision, which is quite different with the sawteeth structure [35] that has a slow dimension variation. Here, the stacked metal–dielectric pairs could be regarded as multiple metal–dielectric–metal

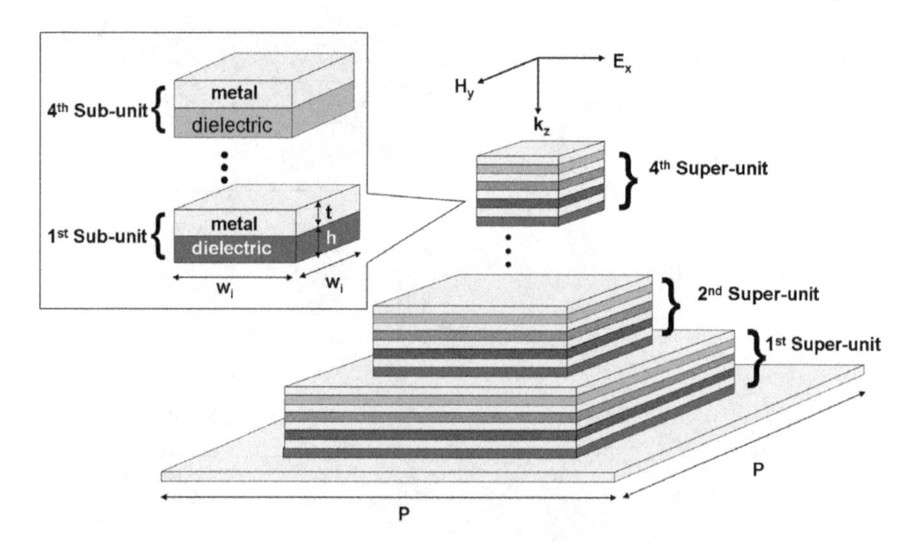

Fig. 13.7 Schematic unit cell of broadband terahertz absorber based on $M \times N$ cascading metal-dielectric pairs. Reproduced from [11] with permission. Copyright 2014, Springer Science+Business Media New York

electromagnetic absorption cells. The coupling of them will cause the mode hybridization.

To illustrate the formation process of broadband absorption, three cascading structures (e.g., 3 × 1, 3 × 2, and 3 × 3 displayed in insets of Fig. 13.8) are simulated. The absorption responses are shown with relevant geometric parameters listed in Table 13.1. For the case of 3 × 1, three distinct absorption peaks appear, which are attributed to three super-unit cells with different lateral sizes w_i. As the second sub-unit with different indices is inserted in the super-unit cells, more absorption peaks appear and the absorptivity of the structure increases. The variation of the absorptivity reflects the degree of impedance matching between the structure and the free pace. When the third sub-unit is introduced, original absorption peaks further broadened and merge together eventually due to their neighboring resonant frequencies.

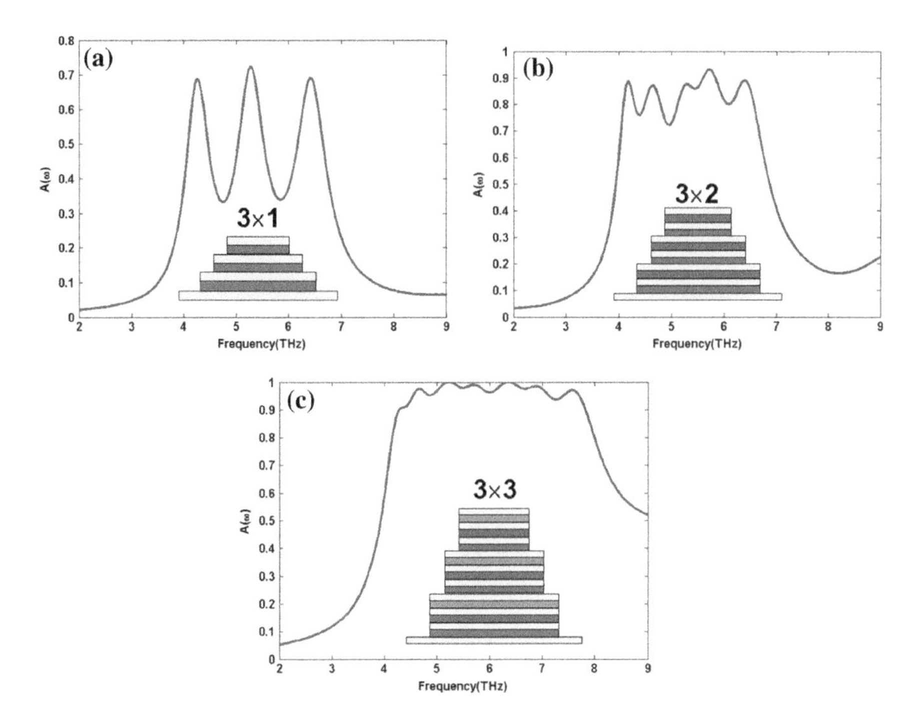

Fig. 13.8 Absorption spectra of different cascading structures. **a** 3 × 1, **b** 3 × 2, and **c** 3 × 3. Reproduced from [11] with permission. Copyright 2014, Springer Science+Business Media New York

Table 13.1 Geometric parameters (μm) for different cascading structures [11]

$M \times N$	h	w_1	w_2	w_3	n_1	n_2	n_3
3 × 1	0.7	16	13	10.5	2.2	–	–
3 × 2	0.7	16	13	10.5	2.2	2.0	–
3 × 3	0.7	16	13	10.5	2.2	2.0	1.8

Consequently, broadband absorption with absorptivity larger than 92% is obtained from 4.4 to 7.8 THz.

To extend the absorption band further, 4 × 4 cascading structure shown in Fig. 13.9 is investigated and the theoretical absorption bandwidth limit is also discussed. With optimized geometric parameters listed in Table 13.2, high absorption is obtained from 3.2 to 11.8 THz with absorptivity larger than 92%; i.e., the absorption band is about 115% of the center frequency. Analysis indicates that the maximum absorption bandwidth which can be achieved by this method is limited by the period. In particular, the minimum absorption frequency (3.2 THz) corresponding to the unit cell with maximum lateral dimension (20 μm) *cannot exceed the period* of the metamaterial structure. Meanwhile, the maximum absorption frequency (11.8 THz) corresponding to the minimum resonant wavelength (25.4 μm) *must be larger than the period* of the metamaterial structure. These conditions make sure the constituents of the metamaterial structure are in subwavelength scale to suppress the high-order diffraction effect [23].

In order to clearly show the resonant characteristics, magnetic amplitude at some frequencies of 4.1, 5.4, 6.9, and 10.6 THz is plotted in Fig. 13.10. We can see lower frequency is confined at the bottom side of the proposed MMA while higher frequency is at the top side, and detailed frequency distributions can be found in Table 13.3. Compared with Refs. [33, 35] where the resonant electromagnetic field diffuses into several metal–dielectric pairs nearby, most of the field energy is confined to single tri-layer resonance cell at these frequencies. This may be due to the refractive index difference between neighbor sub-unit cells that blocks the field diffusion. Such strong

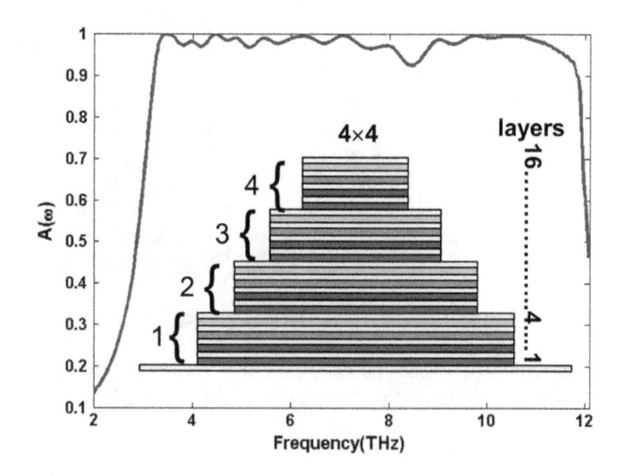

Fig. 13.9 Absorption responses of the 4 × 4 cascading structure. Reproduced from [11] with permission. Copyright 2014, Springer Science+Business Media New York

Table 13.2 Geometric parameters (μm) for the 4 × 4 cascading structure [11]

$M \times N$	h	w_1	w_2	w_3	w_4	n_1	n_2	n_3	n_4
4 × 4	0.7	20	15	11.7	7.8	2.2	2.0	1.8	1.6

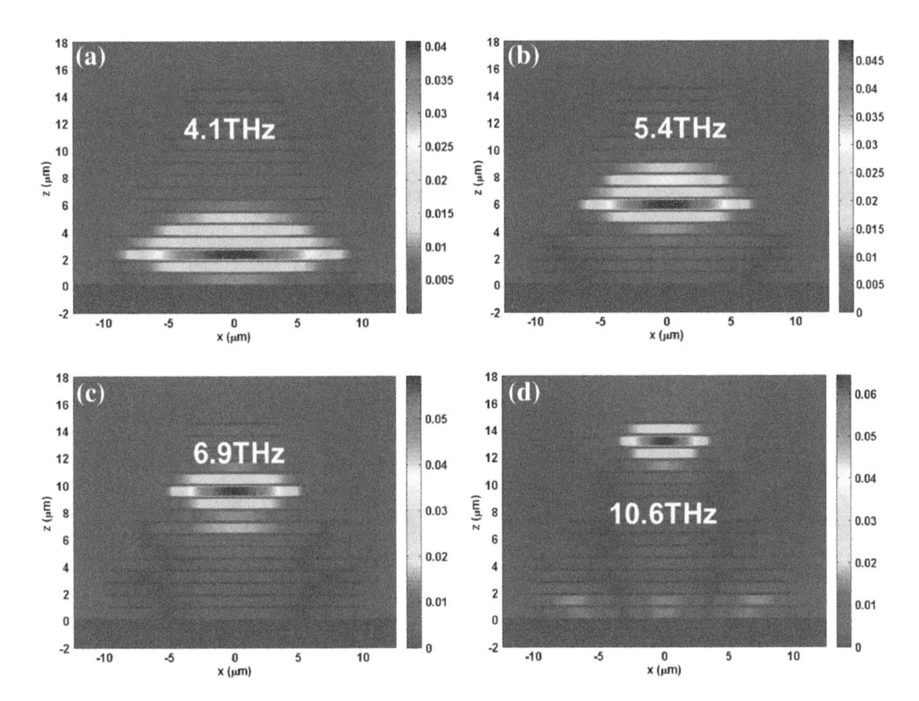

Fig. 13.10 Simulated magnetic amplitude distributions at the cross section (i.e., $y = 0$ plane) for different frequencies **a** 4.1 THz, **b** 5.4 THz, **c** 6.9 THz, and **d** 10.6 THz. Reproduced from [11] with permission. Copyright 2014, Springer Science+Business Media New York

Table 13.3 Resonant frequencies f (THz) confined by metal–dielectric–metal absorption cells [11]

Layer	1	2	3	4	5	6	7	8	9	10	11	12	13	14	15	16
f	3.3	3.7	4.1	4.3	4.6	4.9	5.4	5.8	6.0	6.4	6.9	7.7	9.0	9.7	10.6	11.9

confinements effectively trap the light energy and provide sufficient time to dissipate it. The fluctuations in the absorption spectra are mainly due to that the broadband absorption spectrum is formed by the superposition of multiple discrete absorption peaks. On the other hand, some fluctuations may be caused by the interferences between the fundamental mode of the upper resonant cells and the high-order mode of the bottom ones.

The absorber presented above has considerable potentials in many applications. On the one hand, compact and low-cost terahertz imaging system can be obtained if the absorber is integrated with suitable terahertz sources. On the other hand, if bringing such absorber into close contact with thermal microsensors or microelectromechanical systems, one can roughly detect and estimate the frequency of the incidence according to the correspondence in Table 13.3.

13.3.2 Broadband Absorption Based on Destructive Interferences

In principle, destructive interference is a basic mechanism in anti-reflection and absorption applications. Similar to traditional Dallenbach absorbers, metamaterials can provide new opportunities to control the interference effects. For instance, metamaterials composed of multilayered SRRs were used to realize a desirable refractive index dispersion spectrum, which can induce a successive anti-reflection and absorption in a wide frequency range [36]. The whole anti-reflection system is illustrated in Fig. 13.11a, where the total reflection is the superposition of the reflection at two interfaces.

$$r = \frac{r_{12}\exp(i\phi_{12}) + t_{12}r_{23}t_{21}(i\varsigma)}{1 - r_{21}r_{23}\exp[i(\phi_{12} + \phi_{23} + 2\beta)]}, \tag{13.3.1}$$

$$t = \frac{t_{12}t_{23}\exp[i(\Psi_{12} + \Psi_{23} + \varsigma)]}{1 - r_{21}r_{23}\exp[i(\phi_{12} + \phi_{23} + 2\beta)]} \tag{13.3.2}$$

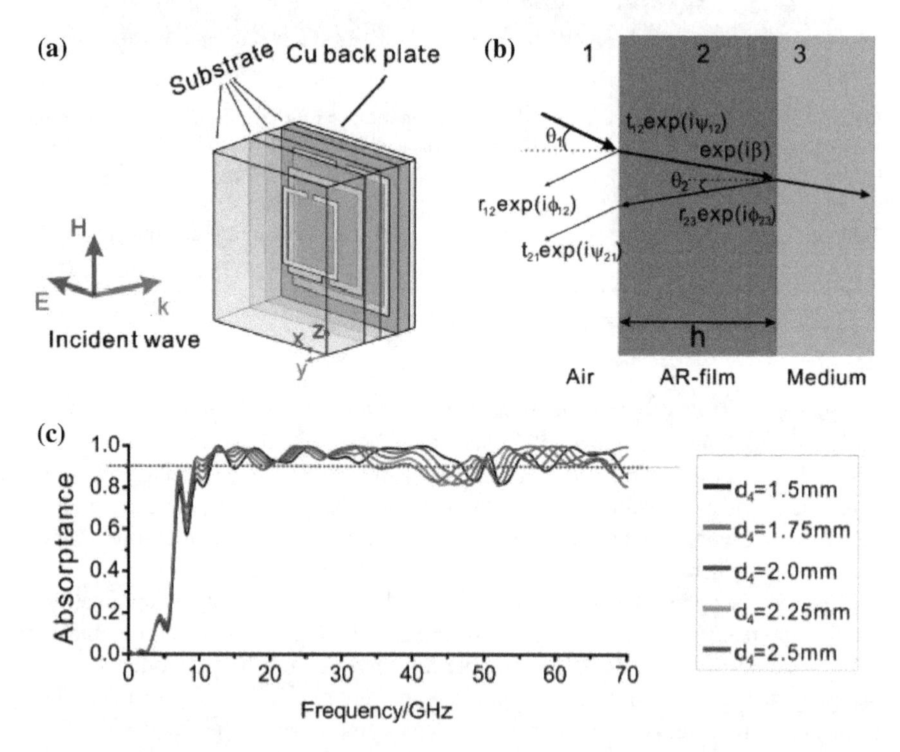

Fig. 13.11 **a** Schematic of broadband MPA composed of multilayered SRRs. **b** Illustration of multiple transmissions and reflections. **c** Simulated absorptance at the frequency range from 0 to 70 GHz. Reproduced from [36] with permission. Copyright 2011, Optical Society of America

where r_{12}, t_{12}, r_{21}, t_{21}, and r_{23} are Fresnel coefficients at two interfaces, as defined in Fig. 13.11b. Here, $\varsigma = \psi_{12} + \phi_{23} + \psi_{21} + 2\beta$ and $\beta = -nk_0h/\cos(\theta_2)$, ϕ and ψ represent the transmission and reflection phase, and the subscripts denote different interfaces. Then, the reflectance can be written as:

$$R = |r|^2 = \frac{r_{12}^2 + (t_{12}r_{23}t_{21})^2 + 2r_{12}t_{12}r_{23}t_{21}\cos(\phi_{12} - \varsigma)}{1 + (r_{21}r_{23})^2 - 2\cos(\phi_{12} + \phi_{23} + 2\beta)r_{21}r_{23}}, \tag{13.3.3}$$

where $\phi_{12} - \varsigma$ is the phase difference of the reflected waves from the two interfaces. For a large refractive index mismatch, the reflection coefficient at the second interface is almost one ($r_{23} \approx 1$). Assuming

$$r_{12} \approx t_{12}t_{21} \tag{13.3.4}$$

$$\cos(\phi_{12} - \varsigma) = -1 \tag{13.3.5}$$

Eq. (13.3.3) can be rewritten as:

$$R = |r|^2 = \frac{r_{12}^2 + (t_{12}r_{23}t_{21})^2 - 2r_{12}t_{12}t_{21}}{1 + (r_{21}r_{23})^2 - 2\cos(\phi_{12} + \phi_{23} + 2\beta)r_{21}r_{23}}, \tag{13.3.6}$$

Therefore, one can deduce the anti-reflection condition from $R = 0$ and approach the ideal anti-reflection condition by geometry optimization. Figure 13.11c shows the simulated results of S_{11} and absorptance of the structure with different covering layer thicknesses. This frequency range contains three anti-reflection periods divided by the intrinsic anti-reflection peaks of the substrate around 12, 32, and 60 GHz. The interval frequency region between the three intrinsic peaks is occupied by other peaks created by SRR resonance, forming an absorptive bandwidth of almost 60 GHz (absorptance > 90%).

It should be noted that transfer-matrix method (TMM) is equivalent to multiple interference analyses [37], but is often much easier to implement, so that in the following discussion, we will utilize TMM as a basic tool.

13.3.3 Broadband Absorption Based on Dispersion Engineering

13.3.3.1 Broadband Absorption Based on Multiple Lorentzian Resonances

Although the multi-reflection model above is intuitive to explain the broadband absorption, it poses great challenges to obtain the optimal design due to the complex interference process. In order to solve this problem, a simpler physical model and design principle were proposed [24], where the metasurface is taken as a uniform thin impedance sheet.

As shown in Fig. 13.12, a plane wave is normally projected on the meta-mirror with the amplitude of E-field denoted as A. Due to reflections occurring at the metasurface and background plane, both the forward- and the backward-going waves exist in the dielectric spacer and surrounding space. In order to utilize TMM, we assume the amplitude of E-field of forward (backward)-going wave at the reflection plane is 1, while the reflectivity of the metasurface is represented as S_{ii} ($i = x, y$ represent the electric field polarized along x- and y-directions, respectively). According to the boundary conditions of Maxwell's equations, electromagnetic fields at the upper side (E^{U} and H^{U}) and lower side (E^{L} and H^{L}) of metasurface should satisfy:

$$\begin{aligned} &A + AS_{ii} = E^{\mathrm{U}} = E^{\mathrm{L}} = \exp(-ikd) + R_{\mathrm{m}} \exp(ikd), \\ &Y_0(A - AS_{ii}) = H^{\mathrm{U}} = H^{\mathrm{L}} = Y_1(\exp(-ikd) - R_{\mathrm{m}} \exp(ikd)) + J, \\ &J = Y_{\mathrm{i}} E^{\mathrm{U(L)}} = Y_{\mathrm{i}}(A + AS_{ii}), \end{aligned} \tag{13.3.7}$$

where Y_0, Y_{i}, $Y_1 = \sqrt{\varepsilon_{\mathrm{d}}} Y_0$, and $Y_{\mathrm{m}} = \sqrt{\varepsilon_{\mathrm{m}}} Y_0$ are the intrinsic admittance of vacuum, metasurface, dielectric spacer, and metal, ε_{d} and ε_{m} are permittivity of dielectric and metal, d is the thickness of dielectric layer, k_0 and $k = \sqrt{\varepsilon_{\mathrm{d}}} k_0$ are the wave numbers in the vacuum and dielectric spacer, $R_{\mathrm{m}} = (Y_1 - Y_{\mathrm{m}})/(Y_1 + Y_{\mathrm{m}})$ is the reflection coefficient at the ground plane, and J is the surface current flowing in the metasurface.

By eliminating A in above equations, one can obtain the impedance of metasurface and reflection coefficient. For simplicity, assuming the dielectric layer is free space with permittivity $\varepsilon_{\mathrm{d}} = 1$ and the metal is perfect electric conductor (PEC) with $\varepsilon_{\mathrm{m}} = \infty$, we can derive $R_{\mathrm{m}} = -1$. Consequently, the reflection coefficient and impedance can be expressed as [24, 38]:

$$S_{ii} = \frac{-Z_0(1 - \exp(2ikd)) - 2Z_i \exp(2ikd)}{Z_0(1 - \exp(2ikd)) + 2Z_i}, \tag{13.3.8}$$

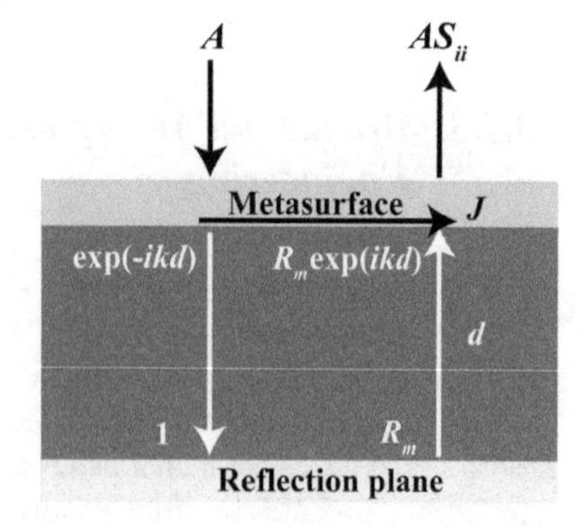

Fig. 13.12 An equivalent physical model of MPA. Reproduced from [24] with permission. Copyright 2015, The Authors

and

$$Z_i = \frac{Z_0}{\frac{1-S_{ii}}{1+S_{ii}} - \frac{\exp(-ikd)+\exp(ikd)}{\exp(-ikd)-\exp(ikd)}}, \tag{13.3.9}$$

Equations (13.3.8) and (13.3.9) provide valuable guidance for the design of broadband metasurfaces. On the one hand, the spectral response of the absorber can be deduced by inserting the equivalent impedance of metasurface in Eq. (13.3.8). On the other hand, the ideal impedance (dispersion) for desired response can be derived by making the reflection coefficient S_{ii} satisfy proper condition. For example, the optimal dispersion of a perfect impedance-matched sheet (PIMS) can be obtained by setting the reflection coefficient at all frequencies to be zero ($S_{ii} = 0$) in Eq. (13.3.9):

$$Z_{\text{ideal}} = 1/Y_{\text{ideal}} = 1/(Y_0 - Y_1 \frac{\exp(-ikd) - \exp(ikd)}{\exp(-ikd) + \exp(ikd)}) \tag{13.3.10}$$

According to Eq. (13.3.10), the ideal impedance ($Z_{\text{ideal}} = R - iX$, where R and X are resistance and reactance) for PIMS at normal incidence is depicted in Fig. 13.13, from which one can see that the impedance should be pure resistive ($R = 377\ \Omega, X = 0$) corresponding to the traditional Salisbury screen when d is near $\lambda/4n$. In Zone 1, the required resistance should be very small and the corresponding reactance should be capacitive ($R > 0$, $X < 0$). In contrast, the required reactance is inductive ($R > 0$, $X > 0$) for Zone 2. As the thickness of the absorber is much smaller than the working wavelength in Zone 1 as shown in Fig. 13.13, Zone 1 is typically the case for most practical applications. To mimic the impedance of PIMS, lossy metal should be deposited in appropriate geometric patterns.

As a proof of concept, a broadband absorber was demonstrated in infrared band based on dispersion-engineered cross-shaped nichrome patterns, as shown in Fig. 13.14a [12]. Polarization-insensitive absorption larger than 97% was numerically demonstrated over an octave band ranging from 21 to 44 THz (Fig. 13.14b).

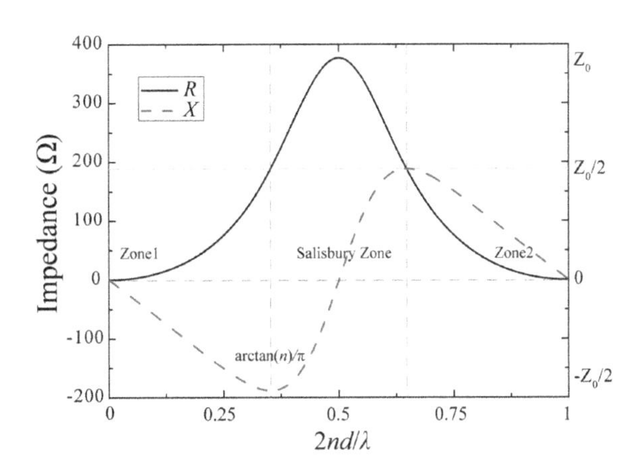

Fig. 13.13 Ideal impedances of PIMS versus the effective thickness of dielectric layer ($2nd/\lambda$). Adapted with permission from [24] with permission. Copyright 2011, Optical Society of America

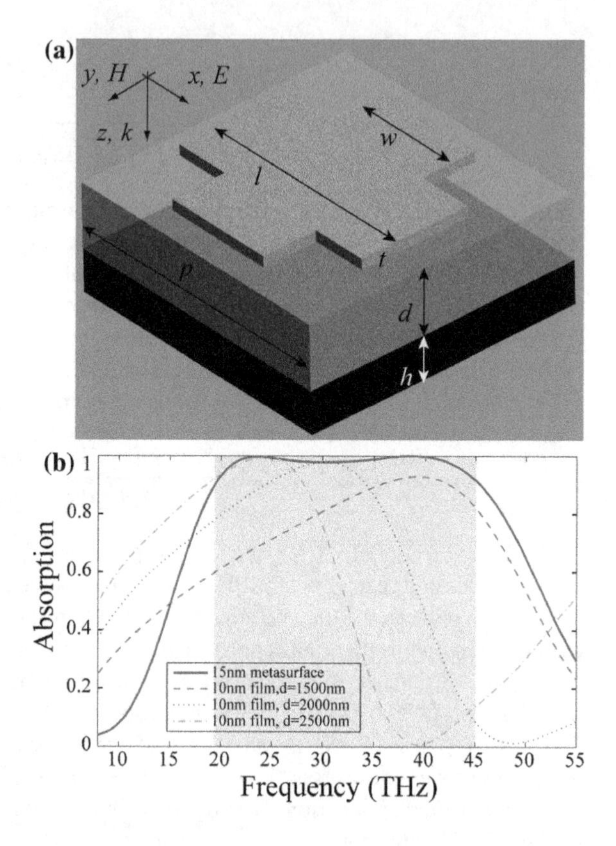

Fig. 13.14 **a** Unit cell of the broadband absorber. **b** Absorption at normal incidence as a function of frequency. The absorption curves for Salisbury-screen-type absorbers are shown to illustrate the bandwidth enhancement effect. Reproduced from [12] with permission. Copyright 2012, Optical Society of America

Compared with the Salisbury-screen-type absorbers (homogenous layer of nichrome without pattern) of different thickness, the dispersion-engineered structure exhibits larger bandwidth. The main reason for the broadband absorption can be illustrated from the viewpoint of equivalent surface impedance.

In order to explain the bandwidth enhancement effect of dispersion-engineered MPA, the effective impedance of the cross is retrieved by S-parameters. As illustrated in Fig. 13.15a, both the real and the imaginary parts of effective impedance approach the ideal impedance in a wide frequency range; thus, a broadband absorber is realized. From the viewpoint of equivalent circuit theory (ECT), the cross-shaped metallic pattern behaves as a lossy plasmonic element, which can be described by an inductor and resistor, while the electric coupling between the adjacent resonators forms a dipole, which can be approximated as an equivalent capacitor. These circuit elements are connected in a series; thus, the effective sheet impedance is written as: $Z_{\mathrm{eff}}(\omega) = R - j\omega L + j/(\omega C)$. By fitting the retrieved impedances, the circuit elements R, L, and C can be determined. In the circuit theory, these values can be approximately calculated using electrostatic calculations. In particular, it is found that the catenary of equal strength can describe the dispersion of metallic gratings with high accuracy [39]. Combining with the fact that the fields in the metallic gaps feature a hyperbolic

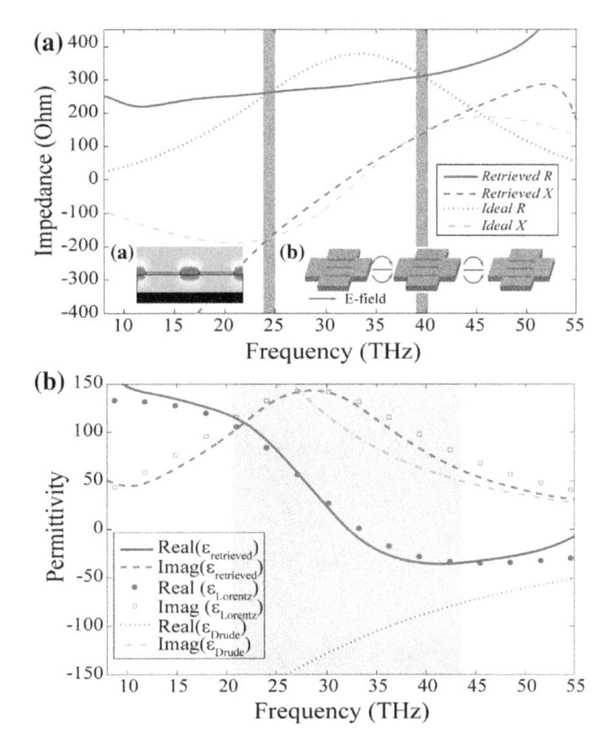

Fig. 13.15 **a** Retrieved sheet impedance and the corresponding impedance of an ideal absorbing sheet. The retrieved impedances overlap with the ideal one at 24.5 and 39.5 THz. Left inset is the side view of the *E*-field distribution at normal incidence, and right inset is the schematic of the *E*-field distribution in the metasurface. **b** Effective permittivity of the metasurface retrieved from the reflection coefficient and fitted by the Lorentz model. The permittivity described by the Drude model is also shown. Reproduced from [12] with permission. Copyright 2012, Optical Society of America

cosine function [40], a new research direction called catenary optics is forming. By stacking multilayered metasurface structures, the absorption bandwidth could be greatly increased to cover almost the entire microwave and terahertz bands [41, 42]. Owing to the particularity, we would like to address the catenary electromagnetics elsewhere.

Interestingly, the principle of material dispersion engineering appears when we change the viewpoint from surface impedance to effective permittivity via the following relationship:

$$\varepsilon_{\text{eff}} = 1 + \frac{i\sigma_{\text{eff}}}{\varepsilon_0 \omega} = 1 + \frac{i}{\varepsilon_0 \omega t Z_{\text{eff}}} \tag{13.3.11}$$

where $\sigma_{\text{eff}} = 1/(tZ_{\text{eff}})$ is the effective complex conductivity and t is the thickness of the metal film. In other works, the effective permittivity of nichrome layer is evolved from the Drude model to Lorentz form after dispersion engineering:

$$\varepsilon_{\text{eff}}(\omega) = 1 + \frac{\omega_1^2}{\omega_0^2 - \omega^2 - i\omega\gamma} \tag{13.3.12}$$

where $\omega_0 = 2\pi f_0 = (LC)^{-1/2}$ denotes the oscillation frequency of the bound electron in the subwavelength patterns, while $\omega_1 = 2\pi f_1 = (\varepsilon_0 L t)^{-1/2}$ and $\gamma = R/L$ are related to the density and damping of the electrons. As demonstrated in Fig. 13.15b, the effective permittivity retrieved from the reflection coefficients is consistent well with that fitted by the Lorentz model. Intuitively, the enhanced absorption can be understood by an analogue of interband transition in noble metals [43], considering the electrons are bounded in the metallic cross and form a bound state.

Besides engineering the effective permittivity ε, the effective permeability μ can also be engineered to realize broadband absorption. Different from the natural magnetic resonance induced by electronic spin, here magnetic response is realized by exciting antiparallel electric currents. It is well known that metamaterials exhibit Lorentzian dispersions, whose real and imaginary parts of the constitutive parameter $\chi(\omega)$ of either $\varepsilon(\omega)$ or $\mu(\omega)$ obey the Kramers–Kronig (K-K) relations [44]:

$$\mathrm{Re}(\chi(\omega)) = 1 + 2/\pi \int_{0}^{+\infty} \mathrm{d}\omega' \mathrm{Im}\left(\chi(\omega')\right)\omega' / \left(\omega'^2 - \omega^2\right) \tag{13.3.13}$$

$$\mathrm{Im}(\chi(\omega)) = -2\omega/\pi \int_{0}^{+\infty} \mathrm{d}\omega' \mathrm{Re}\left[\left(\chi(\omega') - 1\right)\right] / \left(\omega'^2 - \omega^2\right) \tag{13.3.14}$$

The K-K relations imply that the real part of a constitutive parameter can be tuned by controlling the imaginary part (or vice versa). Specifically, a metamaterial consisting of engineered subwavelength-scale cells makes it possible to modify dispersion by deliberately introducing multiple resonances with controlled dissipation.

To validate this approach, a MPA was proposed [44], as indicated in Fig. 13.16. The basic subwavelength unit cell is composed of metallic SRR and rod to generate multiple electric and magnetic resonances. To introduce independently tunable loss for each SRR, the outside corner of each SRR is cut and lumped resistors are soldered (Fig. 13.16a and b). The values of resistors were adjusted to tune the dispersion of $\varepsilon(\omega)$ or $\mu(\omega)$, as indicated in Fig. 13.16c. The final optimized impedances for resistors 1–4, respectively, satisfying the perfect absorption condition ($\mathrm{Re}[\varepsilon_r] = \mathrm{Re}[\mu_r]$, $\mathrm{Im}[\varepsilon_r] = \mathrm{Im}[\mu_r] > 2$), are 203, 122, 39, and 63 Ω. With these values, the dispersions of the complex permittivity and permeability are nearly equal in the frequency band from 1.2 to 1.8 GHz (panel IV of Fig. 13.16c). In this band, the real parts of permittivity and permeability approach zero, while the imaginary parts are all larger than 5.

The measured absorption under normal incidence and reflectance under oblique incidences are shown in Fig. 13.17. One can see that in the cases when incident angles are smaller than 30°, the power absorption below −20 dB still occurs over a wide bandwidth of 0.59 GHz (from 1.22 to 1.81 GHz), corresponding to a relative bandwidth of 39%.

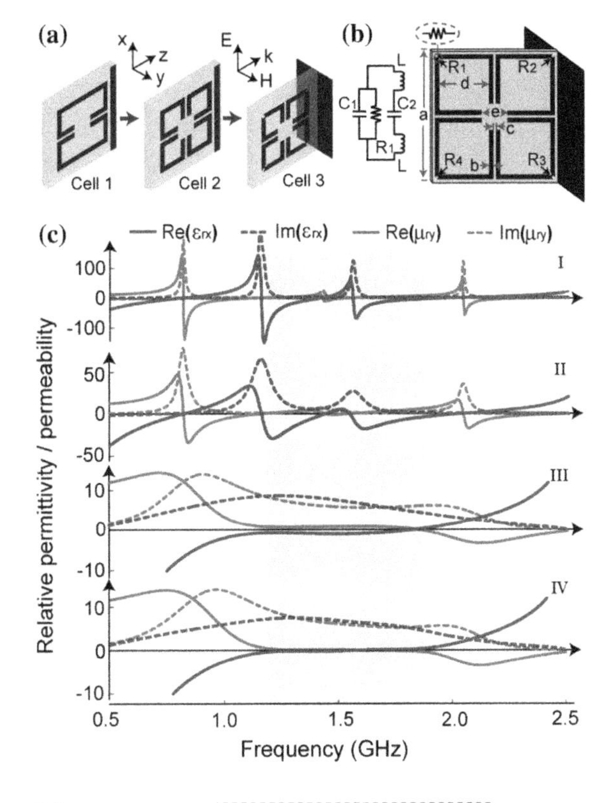

Fig. 13.16 Design of the unit cell of an ultra-wideband MPA. **a** Introducing multiple resonances in an SRR and rod cell. **b** Topology of the new cell and related equivalent circuit model. **c** Retrieved constitutive parameters for different resistances. Reproduced from [44] with permission. Copyright 2013, American Physical Society

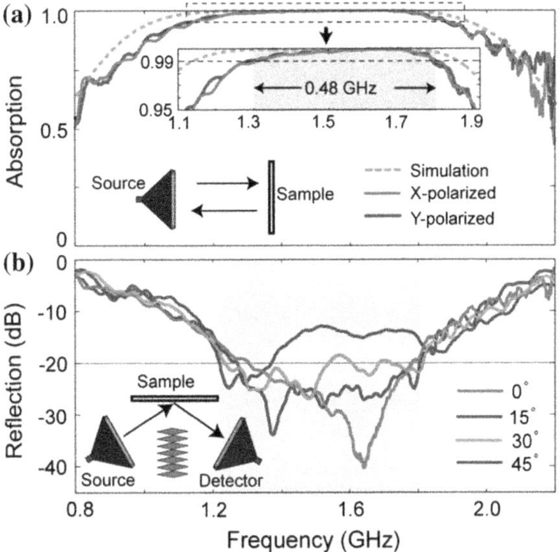

Fig. 13.17 Experimental absorption and reflection curves of the fabricated MPA sample. **a** Normal incidence. **b** Oblique incidences. Reproduced from [44] with permission. Copyright 2013, American Physical Society

13.3.3.2 Broadband Absorption Based on Dispersion-Engineered Doped Silicon Grating

Doped silicon is another important platform to realize broadband absorption. When the working frequency is larger than the plasmon frequency, doped silicon behaves as a highly lossy dielectric material. Yet, a doped silicon slab by itself is not a good absorber due to the impedance mismatch between the silicon slab and the free space. Therefore, anti-reflection techniques should be used. Generally, optical anti-reflection was realized by using single- or multiple-layer dielectric film coatings with certain required refractive indices and appropriate thickness profiles. For high refraction index substrates, index-matched transparent materials that are suitable for coating become rare and the anti-reflection bandwidth is narrow. Alternatively, graded surface relief structures have been investigated in various wavelength ranges. However, surface relief structures involve complicated fabrication processes and impose challenges for device integration. Therefore, it seems quite difficult to realize broadband absorption in doped silicon materials. Recently, by simultaneously exploiting the zero-order and high-order diffraction in silicon grating, broadband absorption has been demonstrated [45], as indicated in Fig. 13.18.

When the wavelength is larger than the period of grating, it can be taken as a homogenous layer with the effective medium theory (EMT):

$$\varepsilon_{\text{eff}} = \frac{(1+\eta)\varepsilon_{\text{si}}}{1+\eta\varepsilon_{\text{si}}} \tag{13.3.15}$$

By changing the fill ratio ($\eta = w/(p-w)$) of grating, the subwavelength structure can be treated as an equivalent medium with quarter-wavelength thickness to realize anti-reflection so that all the transmissions (zero-order diffraction) into the lossy substrate can be totally absorbed.

When the wavelength is comparable to the grating period, the EMT is not valid anymore. Instead, the grating structure may be viewed as a waveguide array, as indicated by the electric field distribution at 2.25 THz as shown in Fig. 13.19a. The electromagnetic fields concentrated in the grating region can be treated as a waveguide with an effective impedance of about $Z_{\text{eff}} = (w/p)Z_0 = 0.43Z_0$ and refractive index near 1. Meanwhile, the impedance of first-order diffraction at 2.25 THz is $Z_s = Z_0\cos\theta/n = 0.22Z_0$. Since the zero-order transmission T_0 is zero at 2.25 THz, the anti-reflection condition, defined as $Z_{\text{eff}} = \sqrt{Z_0 Z_s} = 0.47Z_0$, can be fulfilled (Fig. 13.19b).

The above theory model and numerical analysis have been verified by different groups. Some experimental investigations have been carried [46, 47]. As indicated in Fig. 13.20, broadband absorption beyond 1-octave band was realized based on doped grating patterns with the absorptance higher than 97%.

In principle, higher-order diffraction can also be utilized to extend the absorption bandwidth. As demonstrated in Fig. 13.21, the absorption is larger than 90% for frequencies between 1 and 4 THz and three absorption peaks located at 1.2, 2.2, and 3.7 THz are in coincidence with the peak of 0, ± 1, and ± 2 order diffractions.

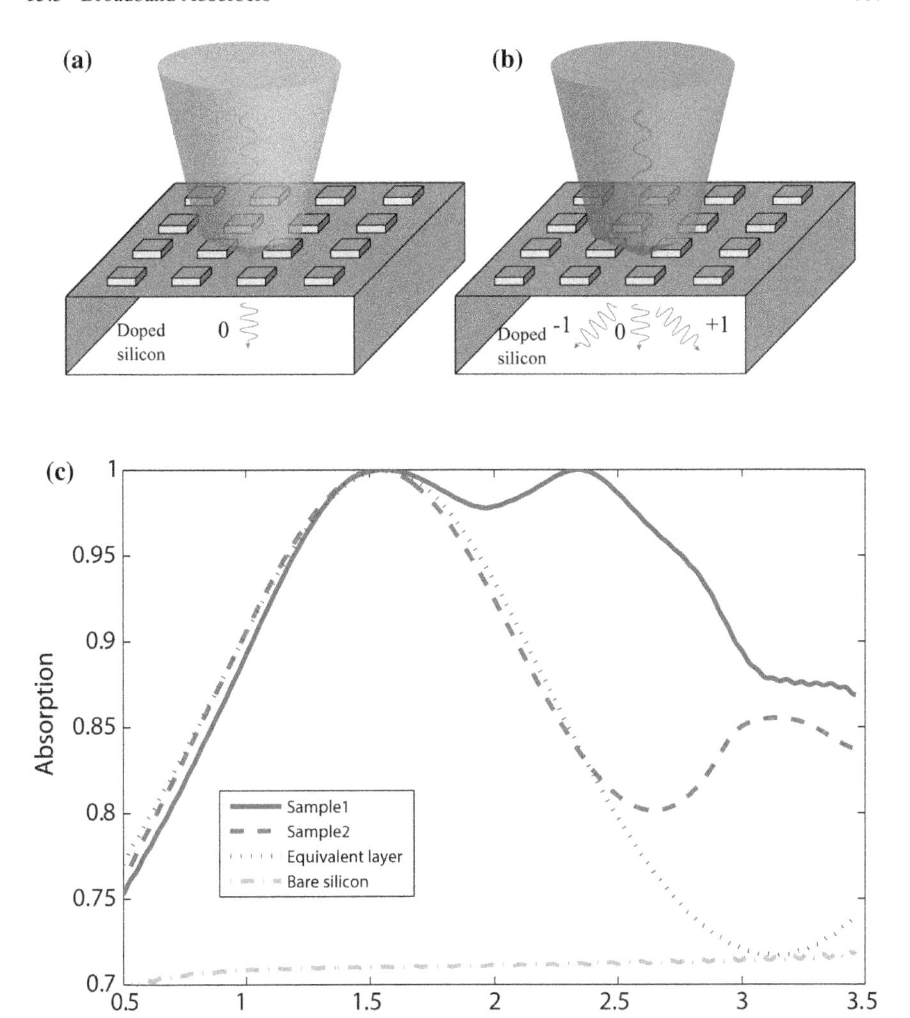

Fig. 13.18 Schematic of diffraction when illuminated at two different frequencies. **a** Only zero-order diffraction occurs in the sample 1. **b** Both zero- and first-order diffractions occur in the sample 2. **c** Absorption spectra of samples with different periods. The cases for a bare doped silicon slab and an absorber based on quarter-wavelength anti-reflection layer are also shown. Reproduced from [45] with permission. Copyright 2012, Optical Society of America

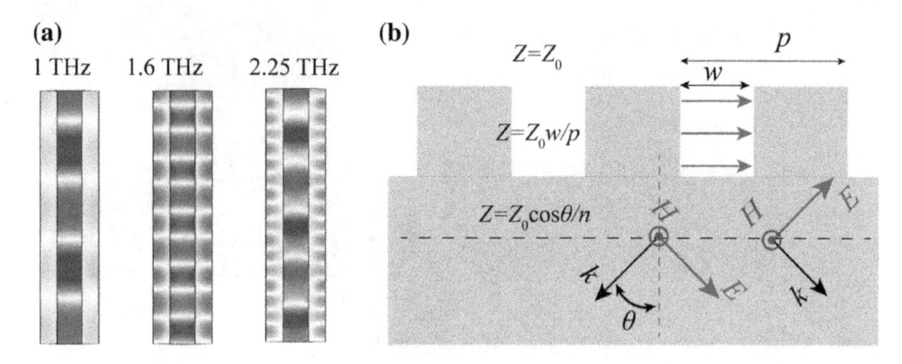

Fig. 13.19 **a** Horizontal electric field distributions at different frequencies for an infinite thick grating. **b** Schematic of the impedance matching for first-order diffraction. Reproduced from [45] with permission. Copyright 2012, Optical Society of America

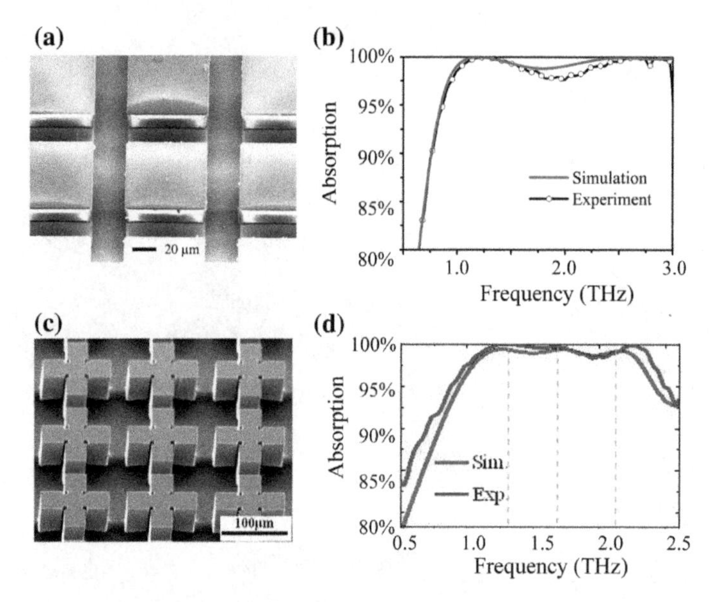

Fig. 13.20 **a**, **c** SEM of fabricated broadband absorbers based on doped silicon patterns. **b**, **d** Simulated and measured absorption spectrum. **a**, **b** Reproduced from [46] with permission. Copyright 2015, American Institute of Physics. **c**, **d** Reproduced from [47] with permission. Copyright 2015, The Authors

Nevertheless, the fabrication process may become more complex and the thickness will become larger.

By combining subwavelength grating and anti-reflection film, the absorption could also be extended. As shown in Fig. 13.22, absorption larger than 90% can be obtained in 0.44–10 THz, demonstrating a great improvement over traditional absorber [48].

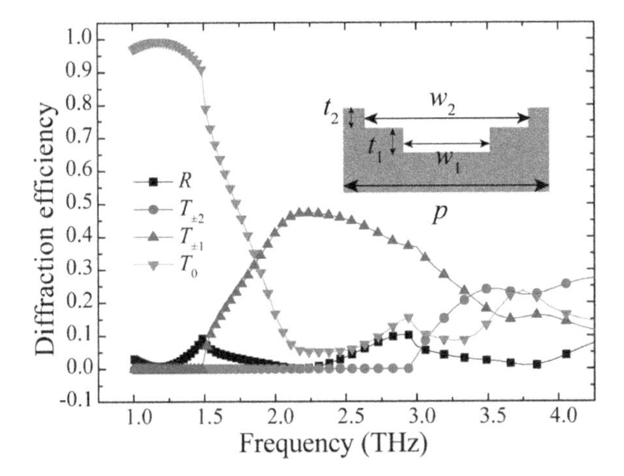

Fig. 13.21 Diffraction efficiencies of a two-layer doped silicon grating as shown in the inset. Reproduced from [45] with permission. Copyright 2012, Optical Society of America

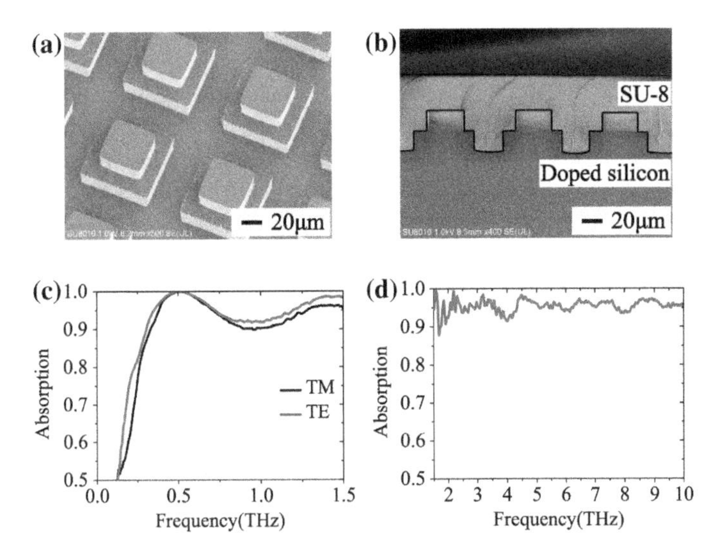

Fig. 13.22 SEM images and the measured absorption spectrums of the samples. **a** A stereogram SEM image of the sample without SU-8 colloid. **b** A side-view SEM image of the sample with SU-8 colloid. **c** The absorption spectrum from 0.1 to 1.5 THz whose incident angle is 15°. **d** The absorption spectrum from 1.5 to 10 THz whose incident angle is 11°

13.3.4 Broadband Absorption Based on Optimization Algorithm

In order to realize broadband absorption, multiple resonances should be excited within a single metasurface. To this end, genetic algorithm (GA) can be utilized to optimize the geometries of the absorber [49]. This optimization approach affords great flexibility to access complex and non-intuitive nanostructure features that fall well outside the bounds of conventional resonant elements such as crosses and rings. The structure of MPA is shown in Fig. 13.23. The bottom three layers of the structure create the resonant electromagnetic cavity of the absorber, which is composed of a doubly periodic array of metal nanostructures patterned on a thin dielectric layer that is backed by a solid metal ground plane. A second dielectric layer is included on top of the structure to provide an impedance match between the metallic nanostructures and the free space. This three-layer resonant electromagnetic cavity can be optimized to produce multiple overlapping resonances over a broad bandwidth in the near-to-mid-IR range. In addition, the metal ground plane prevents light transmission through the MPA, thereby causing the incident light to be either reflected or absorbed.

The MPA was optimized by encoding one unit cell of the nanostructure array as a 15×15 binary grid, where "1"("0") represents a pixel with (without) a metal feature [49]. The unit cell has eightfold mirrored symmetry, which ensures polarization-independent performances and greatly reduced computation time. In each optimization cycle, the optical properties of all candidate MPA unit cells in the population

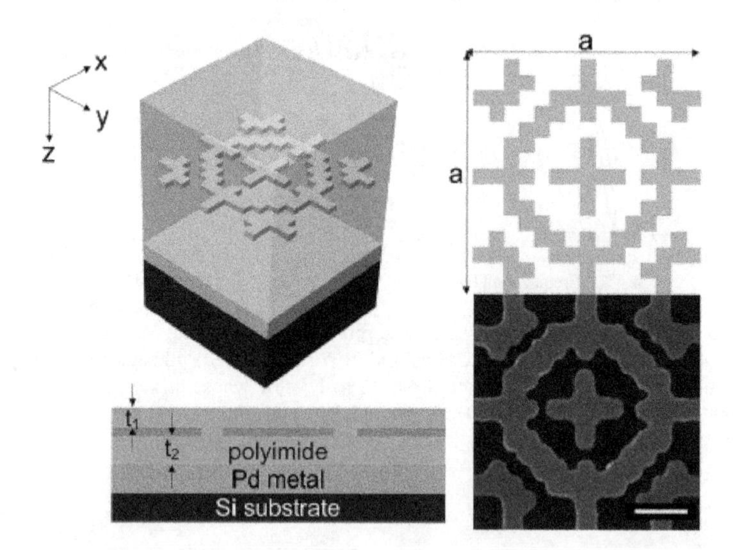

Fig. 13.23 Top left: schematic of the MPA structure optimized to have broadband absorption from 2 to 5 μm in the mid-IR. Bottom left: cross-sectional view of the four-layer EBG-based broadband absorber. Right: schematic and SEM image of one unit cell. Scale bar is 200 nm. Reproduced from [49] with permission. Copyright 2014, American Chemical Society

were evaluated with a full-wave electromagnetic solver and compared to the target properties. Sixteen test wavelengths are specified in approximately equally spaced frequency intervals over the range from 2 to 5 μm. The test incidence angles were chosen to be $(\theta, \varphi)_j = \{(0°, 0°), (40°, 0°), (40°, 45°)\}$, covering a field of view (FOV) up to ±40°. During the evolutionary process, structures that failed to meet predefined nanofabrication constraints were eliminated from subsequent candidate populations. Using a population of 24 chromosomes, the GA converged to an optimized structure with minimized cost in 86 generations.

The simulated and measured absorption spectra under normal incidence and oblique incidence are depicted in Fig. 13.24, which demonstrates a broadband,

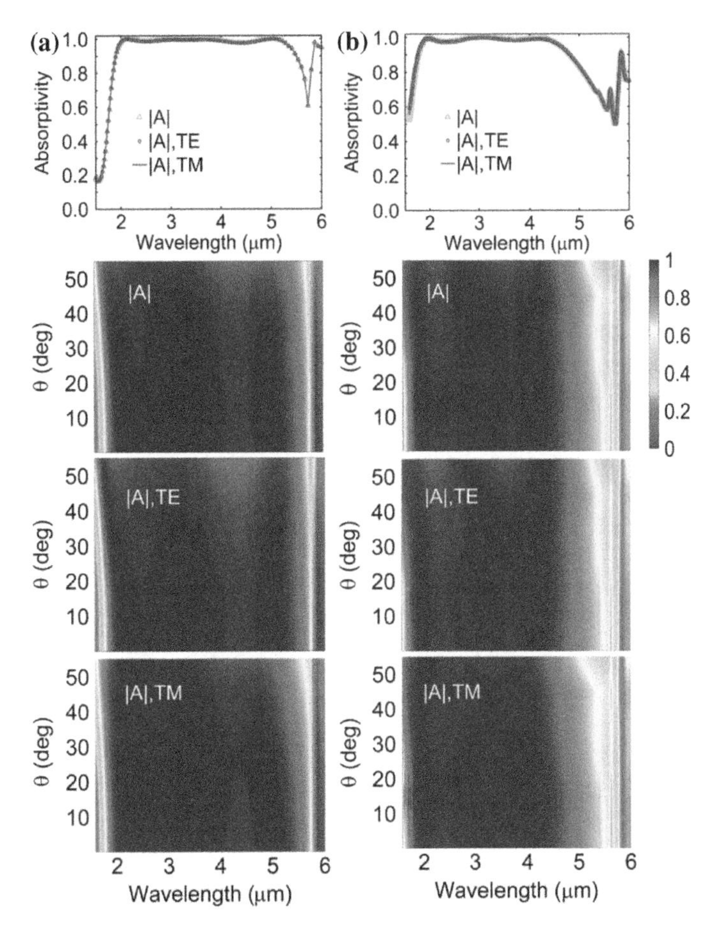

Fig. 13.24 **a** Simulated and **b** measured absorptivity. Top row: normal incidence absorptivity for unpolarized, TE, and TM illumination. Contour plot of absorptivity as a function of wavelength and incident angle from normal up to 55° under unpolarized (second row), TE (third row), and TM (bottom row) illumination. Reproduced from [49] with permission. Copyright 2014, American Chemical Society

polarization-insensitive absorption with measured average absorptivity greater than 98% over a wide ±45° FOV for mid-infrared (mid-IR) wavelengths between 1.77 and 4.81 μm.

13.4 Coherent Perfect Absorbers

In the above discussion, the absorbers are always illuminated from only one side. Recently, coherent perfect absorption of light was proposed and demonstrated in a planar intrinsic silicon slab when illuminated on both sides by two beams with equal intensities and correct relative phase [50]. Such a device is termed a "coherent perfect absorber" (CPA) and a "time-reversed laser." Like the lasing phenomena, classic CPA is narrowband and cannot be utilized in photovoltaic and stealth applications. However, as theoretically and experimentally demonstrated later [15, 16, 19], the bandwidth of thin-film CPA can be extremely broadband. As a result, this effect actually provides a mean to break down the traditional limit of absorbers set by Planck [3] and Rozanov [51]. Interestingly, the concept of CPA has also been extended to the acoustic domain to realize broadband absorption at the low frequencies [52].

Figure 13.25 shows a typical planar CPA structure, where two coherent incidences (A_1 and A_2) from two opposite sides of a planar structure are partially transmitted and reflected, generating two output beams (B_1 and B_2). By solving Maxwell's equations [53], one can obtain the relationship between input and output:

$$\begin{bmatrix} B_1 \\ B_2 \end{bmatrix} = S\begin{bmatrix} A_1 \\ A_2 \end{bmatrix} = \begin{bmatrix} r_{11} & t_{12} \\ t_{21} & r_{22} \end{bmatrix}\begin{bmatrix} A_1 \\ A_2 \end{bmatrix}, \tag{13.4.1}$$

where r_{ij} and t_{ij} are the reflection and transmission coefficients. For a symmetric optical structure, there are $r_{12} = r_{21} = r$ and $t_{12} = t_{21} = t$. Therefore, Eq. (13.4.1) can be rewritten as:

$$\begin{bmatrix} B_1 \\ B_2 \end{bmatrix} = S\begin{bmatrix} A_1 \\ A_2 \end{bmatrix} = \begin{bmatrix} r & t \\ t & r \end{bmatrix}\begin{bmatrix} A_1 \\ A_2 \end{bmatrix}, \tag{13.4.2}$$

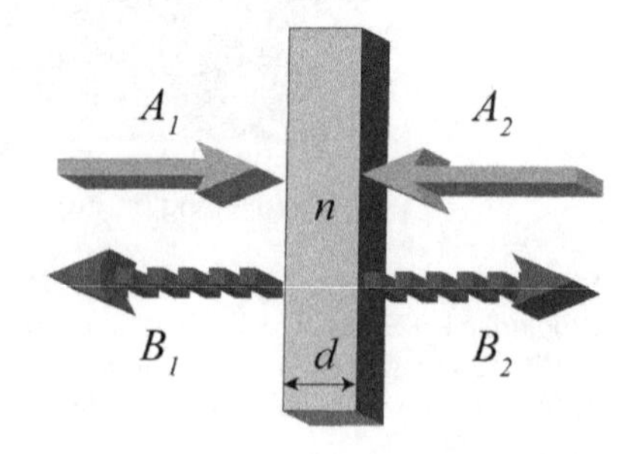

Fig. 13.25 Concept of coherent perfect absorption in thin film. Adapted from [15] with permission. Copyright 2012, Optical Society of America

Coherent perfect absorption means that both the output wave components vanish ($B_1 = B_2 = 0$), which can only be achieved for symmetrical inputs ($A_1 = A_2$, $r + t = 0$) or anti-symmetrical inputs ($A_1 = -A_2$, $r - t = 0$).

Based on Fresnel's equations, the reflection and transmission coefficients can be described by [15]:

$$r = \frac{(n^2 - 1)(-1 + e^{i2nkd})}{(n+1)^2 - (n-1)^2 e^{i2nkd}}, \tag{13.4.3}$$

$$t = \frac{4ne^{inkd}}{(n+1)^2 - (n-1)^2 e^{i2nkd}}, \tag{13.4.4}$$

where $k = \omega/c$, d is the thickness of the slab, and $n = n' + in''$ is the complex refractive index. Using Eqs. (13.4.3) and (13.4.4), the CPA condition for normal incidence can be obtained as:

$$\exp(inkd) = \pm\frac{n-1}{n+1}. \tag{13.4.5}$$

The $\pm$ sign corresponds to the symmetrical or anti-symmetrical inputs. The reflection and transmission coefficients for a single input beam can be written as:

$$r_s = -\frac{1}{2}\left(\frac{n^2-1}{n^2+1}\right), \tag{13.4.6}$$

$$t_s = \pm\frac{1}{2}\left(\frac{n^2-1}{n^2+1}\right). \tag{13.4.7}$$

Apparently, a phase shift of π is added to the reflection wave for both symmetrical and anti-symmetrical CPAs. However, the phase shifts introduced by transmission for these two conditions are distinctive, i.e., either 0 or π. Thus, for two coherent input beams meeting the CPA condition, the transmitted wave of one beam and the reflected wave of the other beam will interfere destructively, resulting in total absorption of incident energy.

In previous discussion [50], an infinite number of discrete solutions to Eq. (13.4.5) have been be found for $kd >> 1$. The bandwidth is defined as the frequency width between the maximum absorption and the adjacent minimum absorption and characterized by $\Delta f \approx c/(2nd)$. In this case, the CPA is rather narrowband and referred to as a time-reversed process of laser. Although such narrowband absorbers are potentially useful transducers, modulators, or optical switches, they are not applicable to broadband purpose.

For a thin-film CPA, i.e., stratifying $d \leq \lambda$, $|nkd| \leq 1$, the left and right sides of Eq. (13.4.5) can be approximated as $1 + inkd$ and $\pm(1 - 2/n)$. Obviously, only plus sign term in the right side should be selected to satisfy the CPA condition, which leads to:

$$n' \approx \text{n}'' \approx \frac{1}{\sqrt{kd}} = \sqrt{\frac{c}{\omega d}}. \tag{13.4.8}$$

where c is the speed of light in vacuum. Obviously, proper complex refractive index must be chosen to obtain the coherent perfect absorption at a specific frequency. Due to the extremely low quality factor of the thin film, such absorption may be wideband. However, material with specific dispersion characteristics should be used to obtain a broadband CPA since the required complex refractive index is frequency dependent.

For metallic or doped semiconductor, the complex dielectric function can be described by Drude model:

$$n^2 = \varepsilon_1 + i\varepsilon_2 = \varepsilon_\infty - \frac{\omega_\text{p}^2}{\omega(\omega + i\Gamma)}, \tag{13.4.9}$$

$$\varepsilon_1 = \varepsilon_\infty - \frac{\omega_\text{p}^2\tau^2}{1+\omega^2\tau^2},$$
$$\varepsilon_2 = \frac{\omega_\text{p}^2\tau^2}{\omega(1+\omega^2\tau^2)} \tag{13.4.10}$$

where ε_∞ is the dielectric constant, $\Gamma=1/\tau$ is collision frequency, and ω_p is the plasma frequency. In the very low-frequency range where $\omega << \tau^{-1}$, we have $\varepsilon_1 >> \varepsilon_2$, and the real and imaginary parts of the refractive index are of comparable magnitude:

$$n' \approx n'' \approx \sqrt{\frac{\varepsilon_2}{2}} = \sqrt{\frac{\tau\omega_p^2}{2\omega}}. \tag{13.4.11}$$

Inserting Eq. (13.4.11) into Eq. (13.4.8), the thickness for CPA at this frequency range can be arrived:

$$d_\text{w} \approx \frac{2c}{\omega_\text{p}^2\tau}. \tag{13.4.12}$$

This characteristic length is the so-called Woltersdorff thickness [54], which quantifies the thickness of a metallic film with maximum absorption of 0.5 for incoherent input in the low-frequency range. For $d < d_\text{w}$, most of the energy is transmitted; for $d > d_\text{w}$, most of it is reflected. Since the Woltersdorff thickness is independent with frequency, the absorption is very broadband, which is quite different from CPAs based on standard high-Q cavities. When the working frequency is close to the plasmon frequency ($\varepsilon_1 = 0$), the refractive index in Eq. (13.4.9) can be written as:

$$n' = n'' = \sqrt{\frac{\varepsilon_2}{2}} = \sqrt{\frac{\varepsilon_\infty}{2\omega\tau}} \tag{13.4.13}$$

and the second characteristic thickness of CPA, termed as Plasmon thickness, is obtained:

$$d_{\mathrm{p}} \approx \frac{2c\tau}{\varepsilon_{\infty}} \tag{13.4.14}$$

For doped semiconductors, the plasmon frequency can be tuned to satisfy $\omega_{\mathrm{p}}^2\tau^2 = \varepsilon_{\infty}$ so that the plasmon thickness is equal to the Woltersdorff thickness and the two absorptions couple together.

Figure 13.26 shows the coherent absorption properties of doped silicon film of different thicknesses at symmetrical inputs (same phase). The silicon is heavily doped with boron, and the doping concentration is chosen as 4×10^{19} cm^{-3}. The corresponding plasmon frequency ($\omega_{\mathrm{p}}^2 = N_{\mathrm{c}}e^2/(\varepsilon_0 m^*)$, here N_{c} is the carrier density, e is the electronic charge, ε_0 is the permittivity of vacuum, and m^* is the effective carrier mass taken as $0.26m_0$, m_0 is the free electron mass) and scattering time are 7.0×10^{14} rad/s and 8.1 fs. Obviously, there are many frequency-dependent narrow absorption peaks for $kd >> 1$, which is consistent with the previous literatures [55]. However, in the region where $kd << 1$, the absorption is nearly frequency independent. The calculated Woltersdorff thickness and plasmon thickness are 150 nm and 450 nm, in good agreement with the theoretical values (151 nm and 416 nm). Since the plasmon frequency can be tuned by doping concentration, the coherent absorption peak may vary with the doping concentration. Taking 40 THz as an example, the absorption can be tuned from near-zero to unit (Fig. 13.27).

For infrared or optical frequencies, doped silicon cannot be described by Drude model any more due to the phonon-assisted absorption. Alternatively, various metals can be chosen as the material of CPA. For a 17-nm-thick freestanding tungsten layer with AC, conductivity for tungsten is 1.79×10^7 S/m, and the corresponding Woltersdorff thickness is only 0.3 nm, far smaller than the practical thickness. Therefore, the absorption here is coherent plasmon absorption. Compared with the intrinsic silicon CPA, the tungsten CPA here has much larger operation ranges (Fig. 13.28), which is

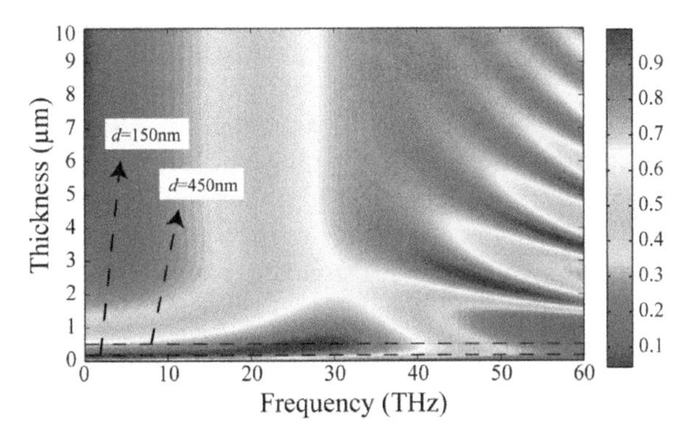

Fig. 13.26 Absorption as a function of the frequency and the thickness of the doped silicon film. The dashed lines illustrate the absorption at two different characteristic thicknesses. Reproduced from [15] with permission. Copyright 2012, Optical Society of America

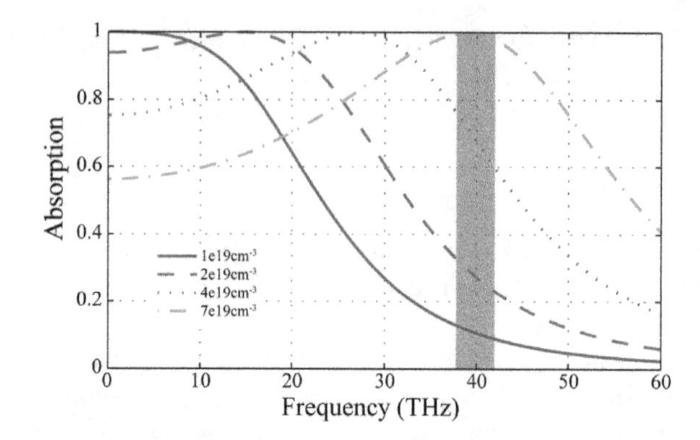

Fig. 13.27 Coherent absorption of a 450-nm-thick doped silicon film for different doping concentrations. The red solid curve shows the absorption when the plasmon thickness is equal to the Woltersdorff thickness and the two absorptions couple together. As the doping concentration increases, the absorption peak shifts to higher frequency. The shaded area around 40 THz illustrates the tunability of absorption provided by varying doping concentration. Reproduced with permission from [15] with permission. Copyright 2012, Optical Society of America

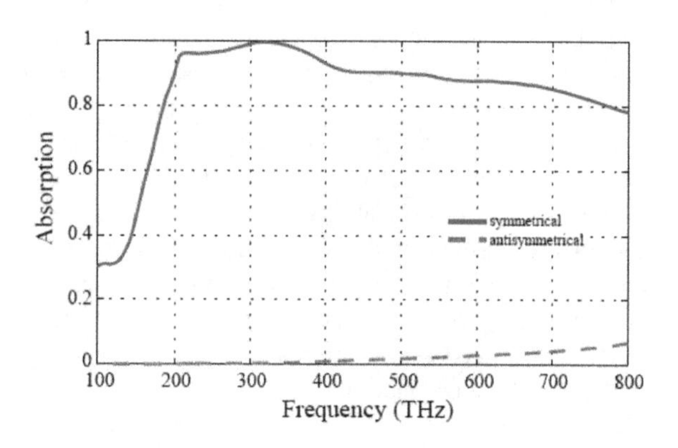

Fig. 13.28 Coherent absorption of 17-nm-thick tungsten for symmetrical and anti-symmetrical inputs. Reproduced from [15] with permission. Copyright 2012, Optical Society of America

partly due to the interband transition [56]. The anti-symmetrical absorption is also drawn as a lower bound of the coherent absorption.

Although the above thin-film CPA can work in a wide frequency range from microwaves up to near-infrared frequencies, the performances rely heavily on the optical property of the constitutive material, and the working frequency cannot be designed artificially. The combination of coherence with metamaterial may provide more freedom of absorption control. As a proof of concept, a schematic of CPA based on symmetric MIM structures is shown in Fig. 13.29 [16]. Each metallic patch is indeed one kind of plasmonic resonator and can be characterized by an electric dipole. The combination of the two resonators makes the single resonant mode split into two separate modes, namely the symmetrical ($|\omega_+>$) and anti-symmetrical ($|\omega_->$) modes with parallel and antiparallel surface currents.

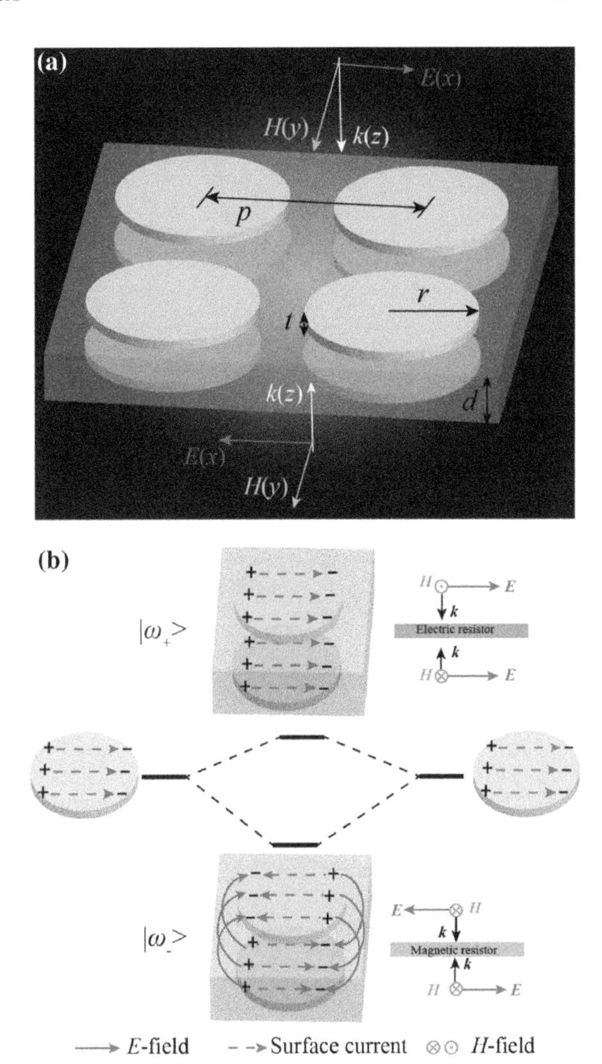

Fig. 13.29 **a** Schematic of MIM structure. **b** The plasmon hybridization model and the effective medium description. Reproduced from [16] with permission. Copyright 2012, Springer Science+Business Media New York

Intuitively, the overall structure can be approximated by electric conductor/resistor and artificial magnetic conductor/resistor for symmetrical and anti-symmetrical modes, respectively. Specifically, for the electric conductor/resistor, when the radiative decay rate is accurately equal to the rate of dissipation, the incident energy can be totally consumed by the damping of electrons through collision. In the case of magnetic conductor/resistor, the magnetic fields are parallel and able to make the equivalent "magneton" oscillate in the same way as the electron does in the case of electric conductor/resistor. Thus, all the incident energies can be absorbed by the collision of "magneton."

Figure 13.30 shows the absorption of two hybrid modes under different illuminations. In the simulation range, the optical property of gold is described by a standard

Fig. 13.30 Coherent absorption of infrared absorber for different phase shifts. Reproduced from [16] with permission. Copyright 2012, Springer Science+Business Media New York

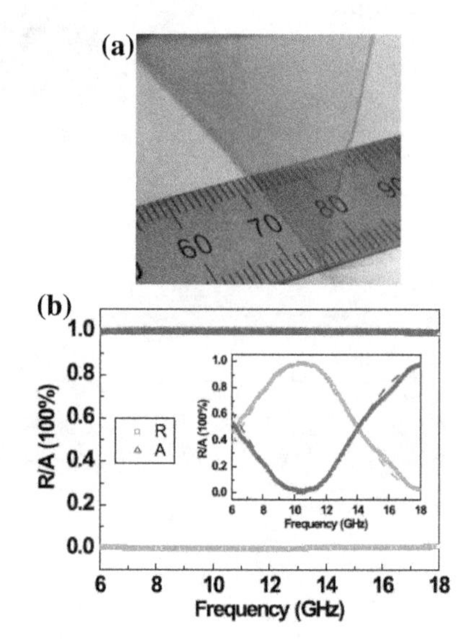

Fig. 13.31 **a** A photograph of the sample, where 2.6-μm-thick conducting layer and a 0.2-mm-thick polymer substrate are employed. **b** Measured (open symbols) and calculated (dashed lines) results in free space for the cases of $L = 0$ and $L = 14.5$ mm (inset). L denotes the difference in the distance of the film to the two horn antennas. The sample has the sheet resistance 180 Ω. Adapted with permission from [18] with permission. Copyright 2015, American Physical Society

Drude model ($\varepsilon_\infty = 5, \omega_\mathrm{p} = 1.23 \times 10^{16}$ rad/s, $\gamma = 1.33 \times 10^{14}$ Hz) and the dielectric material is set as Al_2O_3, which is lossless in the operating frequency. One can see that CPA, respectively, occurs at 200 and 300 THz for symmetric and antisymmetric modes.

Li et al. [18] experimentally show that broadband, perfect absorption of microwaves can be realized in a single layer of ultra-thin conductive film (~λ/10, 000) when illuminated coherently by two oppositely directed incident beams (Fig. 13.31). This is especially helpful in the low frequency, where thin and lightweight radar absorbers are not accessible [19]. For thin conductive film characterized by surface impedance Z_s, the reflection in the coherent illumination is expressed as:

$$R = \left| \frac{1 - \frac{Z_0}{2Z_s}}{1 + \frac{Z_0}{2Z_s}} \right|^2 \quad (13.4.15)$$

Obviously, $R = 0$ only when $Z_s = Z_0/2$, where $Z_0 = 377\ \Omega$ is the impedance of air. According to the relationship between impedance and conductivity ($Z_s = 1/\sigma h$), perfect absorption can be realized by adjusting the thickness of the thin film. Since the conductivity of metallic film is dispersionless in microwave band, frequency-independent absorption behavior is obtained in the whole spectrum. The absorption bandwidth can be adjusted by changing the difference in the distance of the film to the two horn antennas.

In many practical applications, perfect absorption under single-beam illumination has a greater benefit than the illumination of two opposite beams which have to satisfy the coherence condition (Fig. 13.32). In this case, a perfect magnetic conductor (PMC) surface may be utilized as a mirror boundary. By covering the PMC surface with an ultra-thin conductive film of sheet resistance 377 Ω, the perfect microwave absorption is achieved when the film is illuminated by a single beam from one side. Nevertheless, since the PMC condition can only be realized at special frequencies, the CPA is limited to single frequency or narrow operation band.

13.5 Perfect Absorbers Based on Special Materials

13.5.1 Wide-Angle and Omnidirectional Absorbers

As shown in Sect. 13.2.2, light absorption based on delocalized surface excitations such as SPPs is highly sensitive to the angle of incidence. This directionality prevents their application to photovoltaic cells and microscale lighting, where wide collection and emission angles are generally required. The intuitive idea to create an omnidirectional plasmonic absorber can be found in the old concept of the blackbody. When light is transmitted into a lossy cavity through a small hole, almost all energies would be converted into heat by multiple reflections and absorptions. Inspired by the classic blackbody model, nanostructured metasurface composed of close-packed voids was proposed for omnidirectional absorption [27]. Such nanostructured metasurface can sustain localized optical excitations over wide ranges of incidence angles, thus yielding a total absorption effect that is not only omnidirectional but also polarization independent, as indicated in Fig. 13.33.

Although omnidirectional absorption was realized in the above design through exploiting the localized surface plasmon, it only operates within a relatively narrow spectral range. In order to extend the operation bandwidth of omnidirectional absorption, a modified design based on truncated tungsten spherical voids was proposed [56]. Figure 13.34 shows the absorbance as a function of frequency and incidence angle under TE and TM polarizations. Obviously, such structure can realize broadband omnidirectional absorption regardless of TE or TM polarization. Compared with TE polarization, the electric field of TM polarization existing along perpendicular direction of voids can induce an additional electric field distribution in the voids and electric charge collection at the rim of the structure, resulting in an obvious improvement of the wide-angle absorbance.

It should be noted that the absorption mechanism is different for different incidence frequencies. For low frequency below 450 THz, the whole nanostructure can be taken as gradient material composed of a serial of homogenous thin layer. The effective permittivity of each layer can be calculated by Maxwell-Garnett theory:

$$\varepsilon_{\mathrm{i}} = \varepsilon_{\mathrm{tungsten}} \frac{\varepsilon_{\mathrm{air}}(1+2f) + 2\varepsilon_{\mathrm{tungsten}}(1-f)}{\varepsilon_{\mathrm{air}}(1-f) + \varepsilon_{\mathrm{tungsten}}(2+f)} \tag{13.5.1}$$

where f is the filling fraction of air. The absorptance of the whole structure is calculated from the TMM and shown in Fig. 13.35a, which is consistent well with the simulated results, illustrating such equivalent model is appropriate for long wavelength. However, for higher frequency, such physical model is not strictly valid, and the absorption may be explained from the multiple reflections in the spherical cavity,

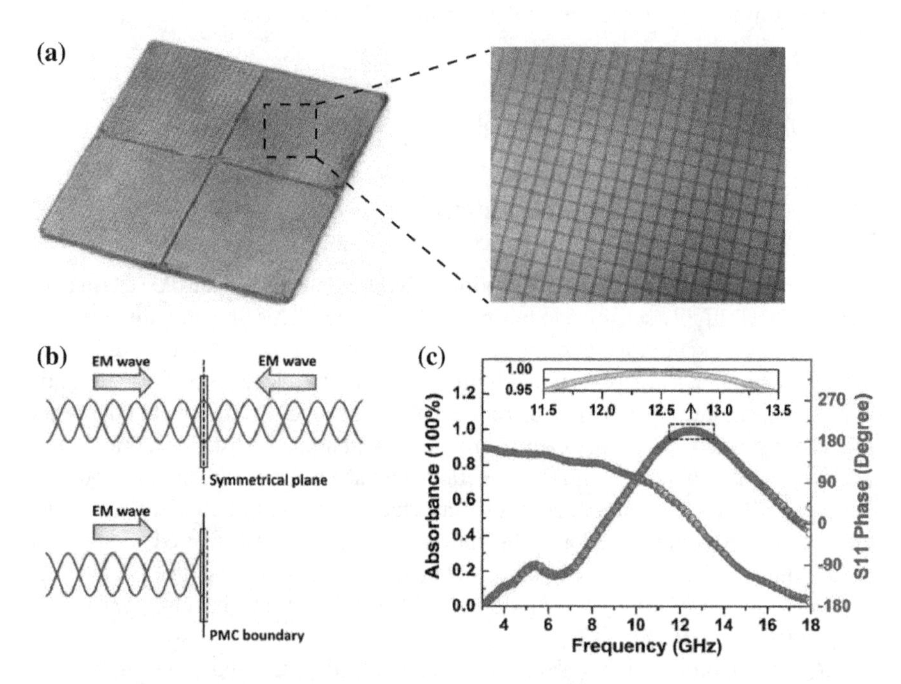

Fig. 13.32 **a** Photograph of the sample composed of four different sizes of mushroom patches. **b** Schematic of the equivalent realization of CPA under a single-beam illumination. **c** The measured absorption (magenta open symbols) of the composite structure with the conductive film of R_s = 383 Ω and the S_{11} phase (blue open symbols) of the bare high impedance surface. The inset shows the frequency band with absorbance greater than 95%. Reproduced from [17] with permission. Copyright 2014, The Authors

as indicated in Fig. 13.35b. Intuitively, optical resonances trap light energy for a period of time proportional to their quality factor, and the trapped light is absorbed as a result of dissipation in the metal when the dwell time is sufficiently large.

In some different applications, the cubic void absorbers shown above can be replaced by simpler configuration (e.g., MIM cavity). As shown in Fig. 13.36a, circular patches arrayed in hexagonal lattice were utilized due to its high azimuthal symmetry. Moreover, the dielectric spacer is very small to excite strong magnetic resonance in the MIM cavity. The angle dependences of the absorption of the dual-band absorber are shown in Fig. 13.36b and c. One can see that the absorption for TM (TE) polarization at 100 THz is larger than 90% at incidence angle of 70° (55°), while the absorption is still larger than 80% at incidence angle of 30° (45°) at 280 THz. The angle-independent absorption phenomenon can be explained using the effective medium theory. Considering the resonant characteristics, the effective refractive index is very large at the resonant frequency, which leads to the fact that the critical coupling conditions are independent of the incident angle [26].

To analyze the exciting resonant modes in this design, the field distributions and the surface electric current in the structure are explored at the resonant frequencies. Then, the equivalent circuit models of two resonant modes can be presented, as

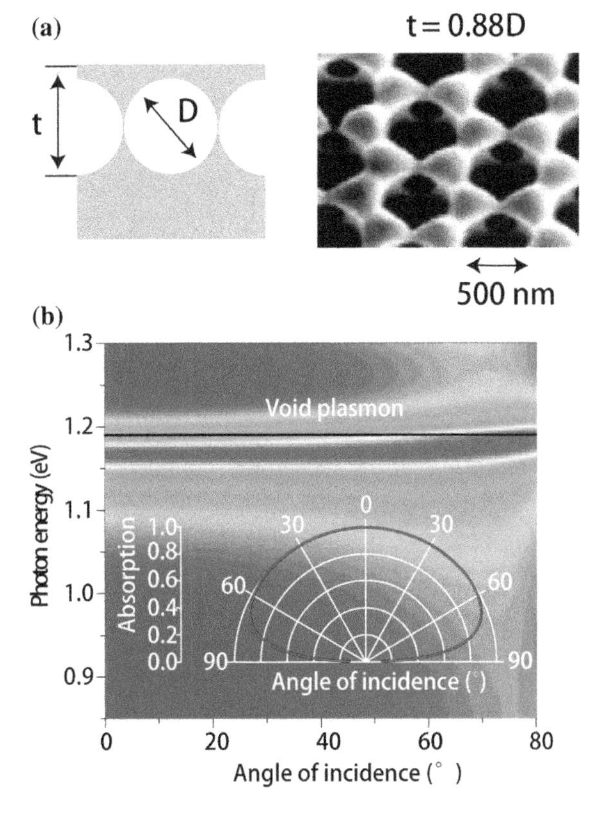

Fig. 13.33 **a** Sketch and SEM image of the mesoporous gold surfaces, consisting of a layer of close-packed voids. **b** Calculated incidence angle dependence of absorption. Reproduced from [27] with permission. Copyright 2008, Nature Publishing Group

indicated in Fig. 13.37. Obviously, at the lower frequency (100 THz), only the lowest resonance mode is responsible for the circuit model. However, the applicability to the single resonance model breaks down and higher-order modes must be taken into account at higher frequency (280 THz). Therefore, the effective impedance in the whole simulated frequency region can be written as:

$$Z_{\text{seff}}(\omega) = \frac{1}{1/(R + j\omega L) + j\omega C} + \frac{1}{j\omega C_0}, \quad f < f_c$$
$$Z_{\text{seff}}(\omega) = \frac{1}{1/(R_1 + j\omega L_1) + j\omega C_1} + \frac{1}{1/(R_2 + j\omega L_2) + j\omega C_2} + \frac{1}{j\omega C_3}, \quad f > f_c \tag{13.5.2}$$

where $f_c = 200$ THz is the frequency where mode begins to change.

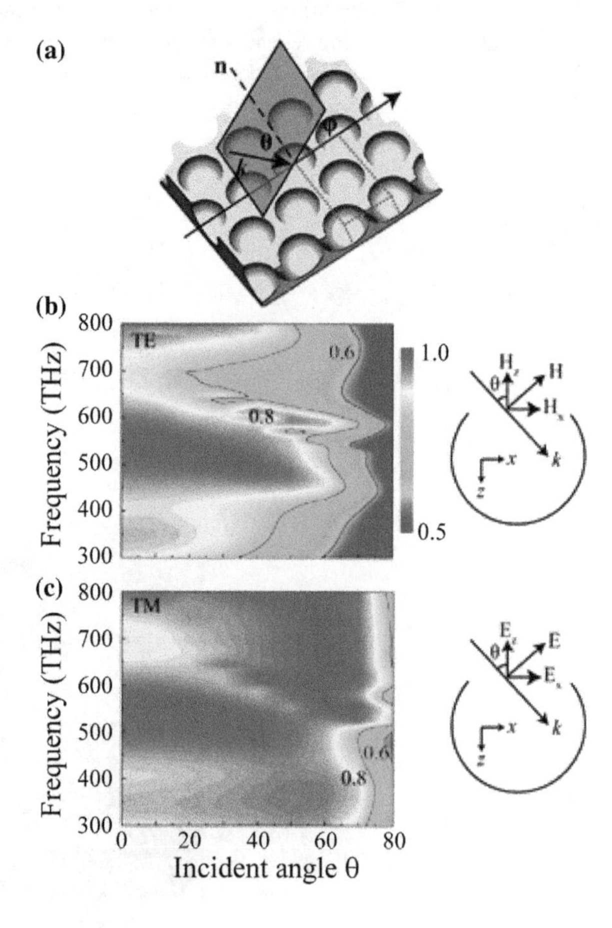

Fig. 13.34 **a** Schematic of broadband and omnidirectional absorber composed of truncated tungsten spherical voids arranged in hexagonal lattice. Simulated absorption as a function of incidence angle and frequency under **b** TE polarization and **c** TM polarizations. Reproduced from [56] with permission. Copyright 2011, Optical Society of America

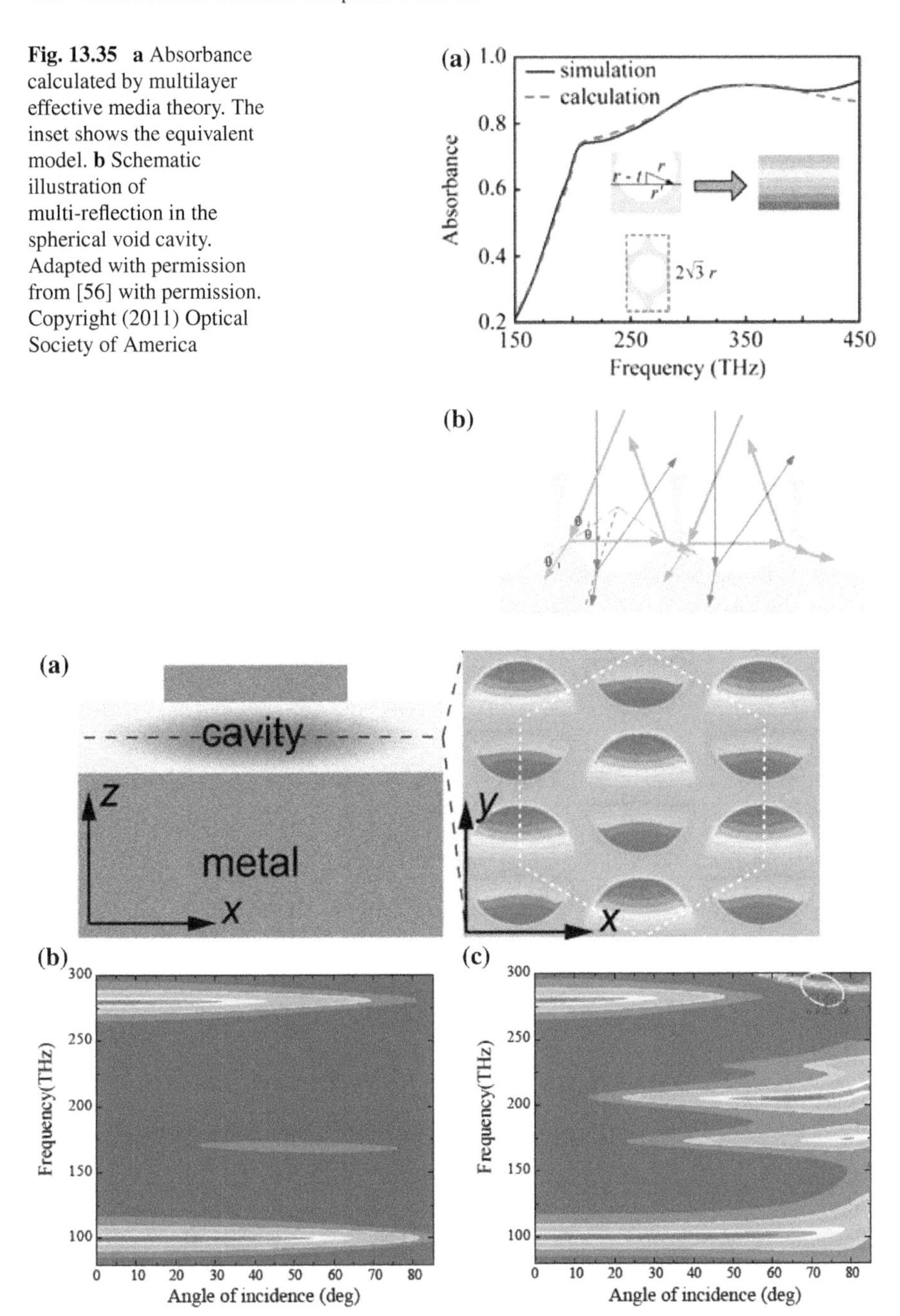

Fig. 13.35 **a** Absorbance calculated by multilayer effective media theory. The inset shows the equivalent model. **b** Schematic illustration of multi-reflection in the spherical void cavity. Adapted with permission from [56] with permission. Copyright (2011) Optical Society of America

Fig. 13.36 **a** Front view and side view of the absorber. Absorption of the dual-band absorber as a function of frequency and the angle of incidence for **b** TE and **c** TM polarizations. Adapted with permission from [24, 26] with permission. Copyright 2011, Optical Society of America. Copyright 2017, The Royal Society of Chemistry

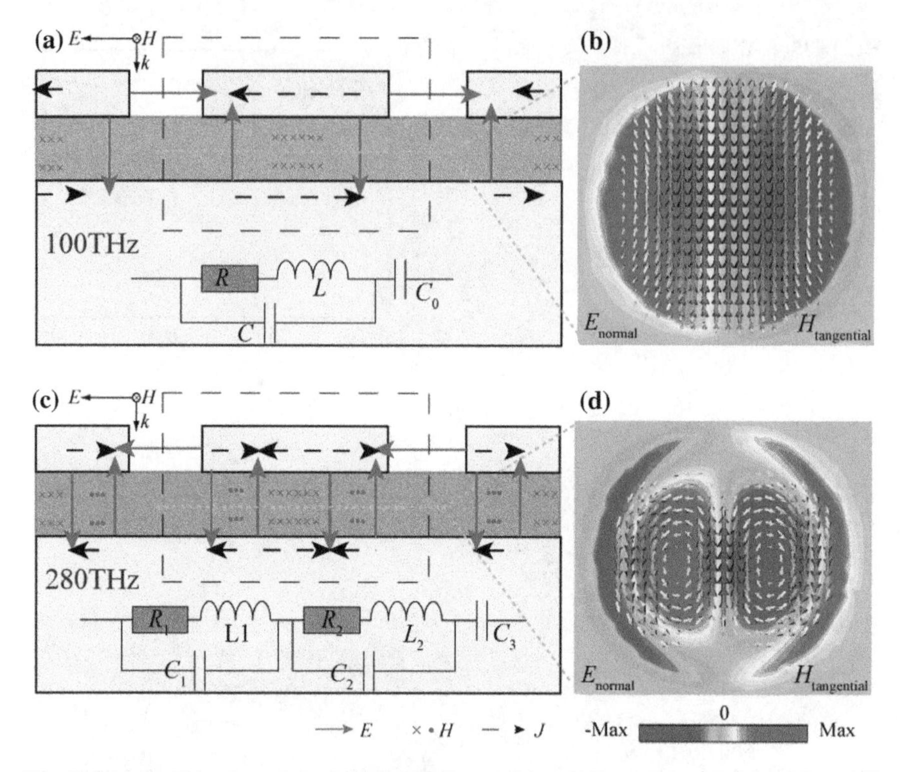

Fig. 13.37 **a**, **c** Side view of the field distribution and the corresponding circuit models at 100 and 280 THz. **b**, **d** Front view of the maximal normal electric field and tangential magnetic field at the center of dielectric spacer. There is 90° phase shift between *E*- and *H*-fields. Adapted with permission from [24] with permission. Copyright (2011) Optical Society of America

13.5.2 Transparent Absorbers

13.5.2.1 Transparent Infrared Absorbers

Transparent absorption is desired in some special cases such as microwave shielding window. To this end, optically transparent conducting materials are often required. For example, Huang et al. proposed a four-layer metamaterial with efficient light modulation in two distinctive spectra. To achieve efficient dual-band spectrum modulation, the structure of transparent infrared absorber depicted in Fig. 13.38a is composed of four alternating ITO and photoresist layers on top of a silica substrate. As all these materials are intrinsically transparent at visible frequencies, the structure is highly optically transparent. Note that, the fabrication process employed is simpler and more cost-effective because there is no need for lithography. An 80-cm-diameter sample is fabricated. A photograph of the sample is taken and shown in Fig. 13.38b, from which we can see that the pattern underneath is clearly visible in the photograph.

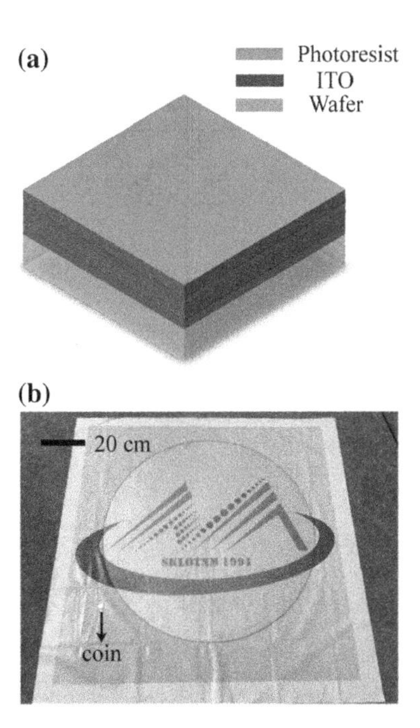

Fig. 13.38 **a** Schematic image of a transparent infrared absorber consisting of alternating indium tin oxide (ITO) and photoresist layers. The thicknesses of the layers from top to bottom are 1.5, 0.1, 1.8, and 0.3 μm. **b** A photograph of an 80-cm-diameter transparent absorber. The patterns covered by the sample are clearly visible. A coin is set beside the sample for comparison. Reproduced from [13] with permission. Copyright 2017, The Japan Society of Applied Physics

The wavelength dependence of transmittance between 400 and 800 nm is illustrated in Fig. 13.39a with the measured average transmittance above 80%. The ultra-broadband infrared absorption is achieved by the combination of the interband transition of charge carriers of the two materials and the structural effect induced by the alternating alignment of ITO and photoresist layers. The measured and simulated wavelength dependences of absorptivity in far infrared are depicted in Fig. 13.39b with more than 90%. The absorptivity of the bare silica substrate employed in the fabrication is also shown in Fig. 13.39b for comparison, which indicates that the absorption stems from both the alternatively stacked films and silica.

13.5.2.2 Flexible Transparent Microwave Absorber

Classic absorbers constructed from conventional materials are typically rigid and optically opaque. If the absorber can be made optically transparent and structurally flexible, it can provide higher design freedom for practical applications. For example, an optically transparent and flexible absorber can be applied to applications such as window glass and curved surfaces. Recently, a transparent and flexible polarization-independent microwave broadband absorber was proposed [57]. In order to balance the optically transparent, structurally flexible and microwave absorption, the proposed absorber is composed of top Al wire grid metallic patterned patches, polyethylene terephthalate (PET), polydimethylsiloxane (PDMS), and a metallic wire grid

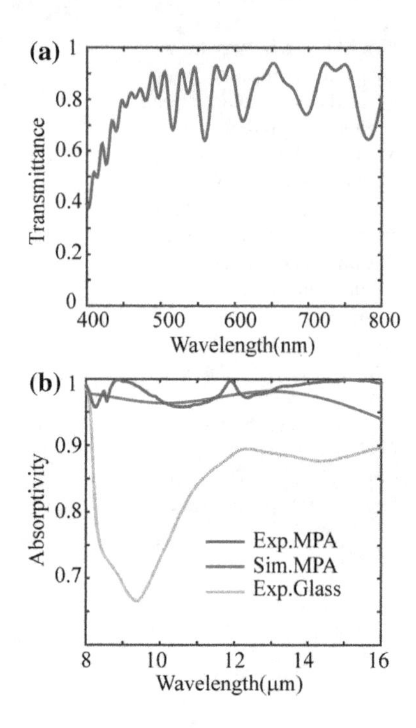

Fig. 13.39 **a** Transmittance of the metamaterial at visible wavelengths (400–800 nm). The structure has an average transmittance of more than 80% in this region. **b** Measured (red line) and simulated (blue line) absorptivities of the metamaterial at $\theta = 15°$. The yellow line is the measured absorptivity of the glass employed in the fabrication. Reproduced from [13] with permission. Copyright 2017, The Japan Society of Applied Physics

ground. To reduce the reflection from the absorber structure, good impedance matching to air is required. This can be achieved by varying the spacing between bow-tie structure and dielectric spacer layer thickness as well as using the optimized metal thickness. As a consequence, the optical transmittance of the total structure is more than 62%, as shown in Fig. 13.40a. Furthermore, bow-tie shape can offer a broader response range by exploiting not only its own resonance but also the coupling between the neighboring unit cells in a periodic array via the side of the bow ties. By merging the two resonances, an absorption above 90% in the frequency range 5.8–12.2 GHz was realized, and the bandwidth is 71.1% of the center frequency (Fig. 13.40b).

13.5.3 Refractory Absorber

It should be mentioned that most of the absorbers mentioned above are based on common metals, which suffer from the severe problem of low melting points. Nevertheless, many applications such as solar thermophotovoltaics (STPV) generally require the working temperature of at least 800 °C and more, approaching or even surpassing the melting points of bulk Au, Ag, and Al. In addition, scarcity, high price, and limitations on the material parameters of these metals also urge researchers to explore alternative materials for optical metamaterials with high performances.

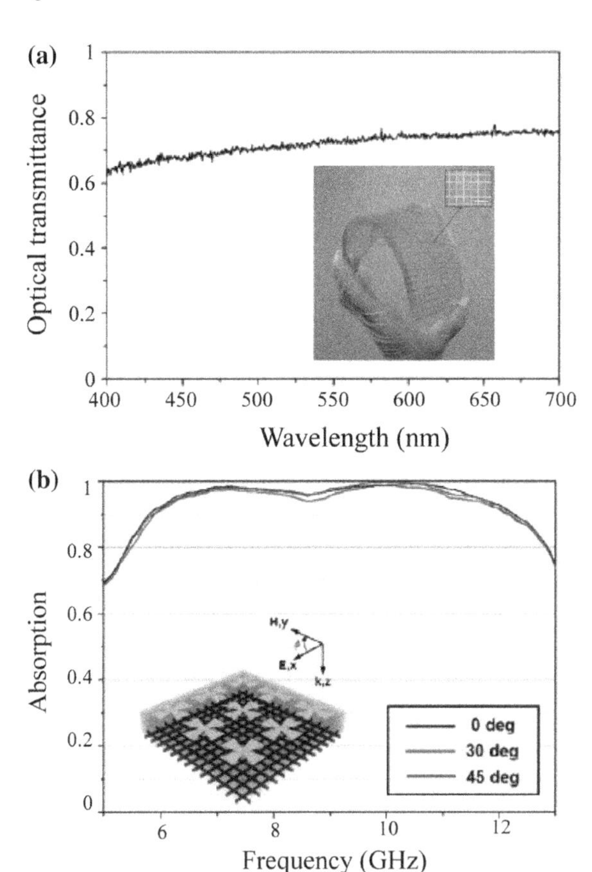

Fig. 13.40 **a** Optical transmittance. Inset is the photograph of the fabricated sample composed of metallic bow-tie array on top of a flexible and transparent PET layer. **b** Measured absorption at different polarization angles ϕ (0, 30, and 45°). Reproduced from [57] with permission. Copyright 2014, American Chemical Society

A schematic diagram of broadband refractory absorber and simulated absorption performances is shown in Fig. 13.41a [14]. The MPA consists of a periodic array of refractory tungsten (W) cylinders covered by a 40-nm-thick silicon carbide (SiC) layer and a 40-nm silicon dioxide (SiO_2) layer. In order to obtain broadband absorption, the diameters of the SiO_2, SiC, and W cylinders were optimized. The SEM image of fabricated sample and the measured absorption are shown in Fig. 13.41b. Apparently, the measurement results agree quite well with the simulation both in the structure profile and in the absorption performance.

To shed light on the physical origin, the absorption of different configurations is investigated and depicted in Fig. 13.42a. Obviously, a planar W film cannot absorb the incident wave efficiently. This is mainly because that the impedance between the W film and the environment is mismatched which will lead to high reflectance. This low absorbing performance is enhanced when subwavelength nanostructures are etched on the film. However, the absorption of W cylinder array is still far from perfect in a broadband spectrum as the resonances induced by the nanostructures are comparatively narrowbanded. Interestingly, after coated with two optically transpar-

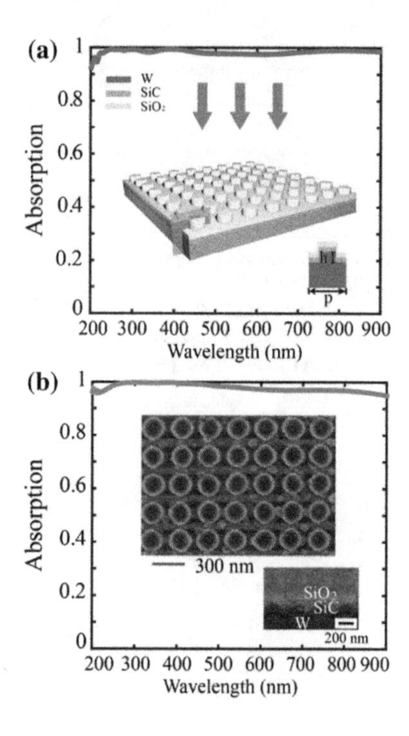

Fig. 13.41 **a** A schematic image of the proposed MPA and its simulated absorption in the designed region. The inset is the cross section of a unit cell with $h = 80$ nm and $p = 300$ nm. **b** The SEM of the fabricated MPA and the measured absorption. The inset in the right bottom shows the cross section of the sample that three different materials can be easily identified. Adapted with permission from [14] with permission. Copyright 2018, Royal Society of Chemistry

ent materials with nanometer thicknesses, the device becomes totally black with little reflectance. In fact, this ultra-broadband impedance matching behavior of this device attributes to two effects. The first is the diffraction effect induced by the resonances in the nanostructures at around 300 nm. As the period of the structure is larger than the wavelength in the substrate but still smaller than that in free space, the first-order diffraction will happen in the MPA. This diffraction effect would undoubtedly decrease the reflection of the structure and thus enhance the absorption. The second mechanism for the ultra-broadband absorption is the graded refractive index of the coating materials (SiO_2 and SiC) that forms a wideband anti-reflection layer. This layer will dramatically decrease the reflection at the structure surface and therefore increase the absorption. Such high absorption behavior can be well explained by the effective medium theory. In this case, the structure can be regarded as multilayered films and the effective permittivity of each layer ε_{i} can be calculated by the Maxwell-Garnett theory. The absorption of such effective medium was obtained by employing the transform matrix method. The comparison between the retrieved and the simulated absorption is plotted in Fig. 13.42b. The curves are well overlapped from 450 to 900 nm and separated at short wavelengths. The discrepancy can be addressed by the closer dimension of the period of the structure and the incident wavelength. As observed, such structure can be regarded as a five-layer film stack with incremental permittivity of each layer at long wavelengths.

To demonstrate that this proposal can stand the high temperature, the absorber is baked to different temperatures for an hour and annealed naturally for several hours.

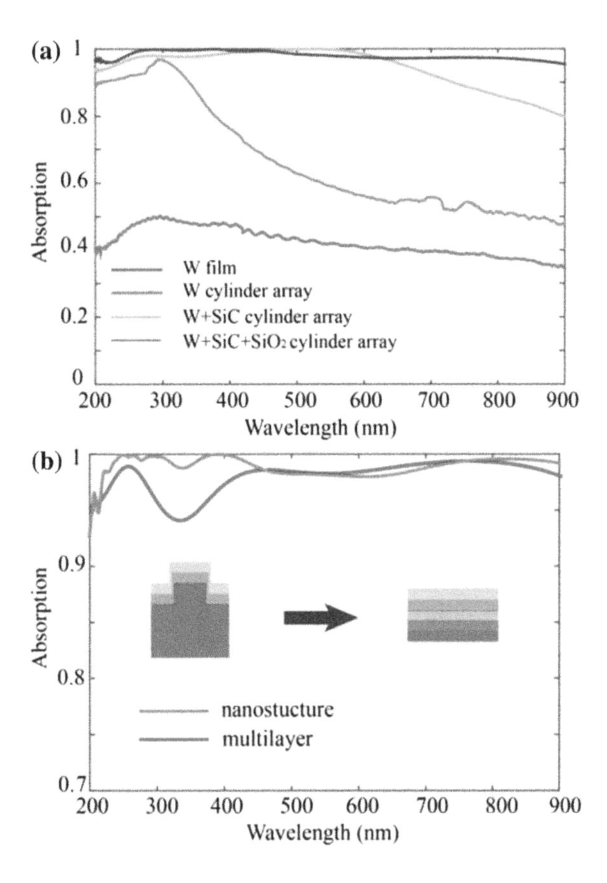

Fig. 13.42 **a** The absorption performance for various structures. The bandwidth and absorbing efficiency are remarkably improved compared with planar W film. **b** The absorption retrieved from the effective medium theory and numerical simulation. The nanostructure can be regarded as a multilayered film stack at long wavelengths. Reproduced from [14] with permission. Copyright 2018, Royal Society of Chemistry

The temperature was increased at a rate of 20 °C min^{-1}. The measured absorption and the SEM image after annealing at 600 °C are shown in Fig. 13.43 with their room temperature counterpart as comparison. It can be inferred that the absorption performance and the nanostructure configurations are nearly unchanged that manifest the stability of the device. In fact, the nanostructures are even stable at higher temperatures (800 °C), but the thermal expansion mismatch will lead to the shrinkage of the films. This high-temperature tolerance absorber is crucial for many applications such as thermophotovoltaic energy harvest and artificial thermal emitters.

Besides W and SiC, another refractory material titanium nitride (TiN) has also been utilized to construct refractory absorber [58]. As indicated in Fig. 13.44, the TiN sample retains its shape after annealing at 800 °C for 8 h and thus maintains its performance, while the rings in the Au metamaterial sample heated at 800 °C are melted into nanoparticles even only after 15 min of annealing, resulting in a degenerated performance. The absorption performance of TiN absorber can be explained by the fact that bulk TiN has a melting point of 2930 °C, much higher than that of Au (1063 °C). In practical solar applications, the heat source is sunlight instead of a uniform thermal field. The heat converted from sunlight will be distributed in the

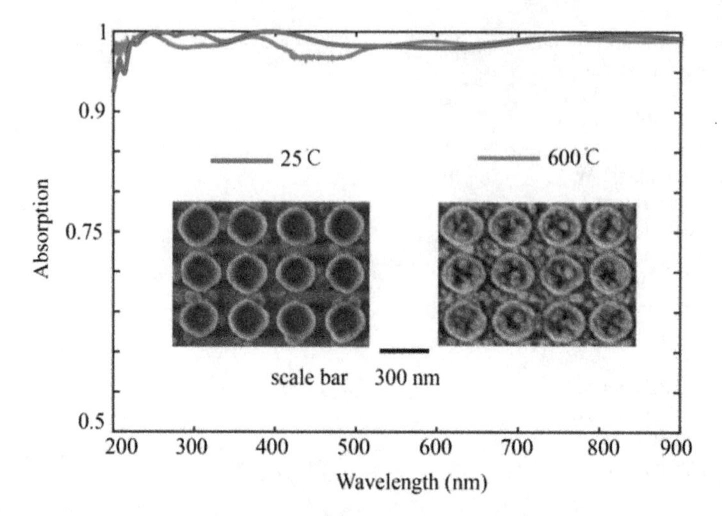

Fig. 13.43 Absorption and SEM image of the device at room temperature and after annealing at 600 °C. The absorption performance as well as the nanostructured configurations is nearly unchanged that demonstrate the high-temperature tolerance of the MPA. Reproduced [14] with permission. Copyright 2018, Royal Society of Chemistry

absorber inhomogeneously. As a consequence, the continuing generated local heat can still cause damage before the average temperature reaches the melting point of the nanostructures in the metamaterials. From Fig. 13.44e and f that shows the SEM images of absorber under pulsed laser excitation of 6.67 W/cm^2, one can see that the TiN absorber exhibits higher stability performance than gold absorber.

13.6 Solar Cells Based on Thin Absorber

Traditional solar cells based on crystalline silicon wafers generally suffer from large thicknesses (typically between 180 and 300 μm) [59]. As an alternative, thin-film solar cells may provide a viable pathway toward large-scale implementation of photovoltaic technology by offering low materials and processing cost. However, the reduced thickness causes an important challenge to maintain the high absorption efficiency. Fortunately, plasmonic structures offer several ways to reduce the physical thickness of the photovoltaic absorber layers while keeping their optical thickness constant [60]. As indicated in Fig. 13.45a, the first approach is based on the multiple scatterings, which could increase the optical path dramatically. Second, metallic nanoparticles can be used as subwavelength antennas in which the plasmonic near-field is coupled to the semiconductor, increasing its effective absorption cross section (Fig. 13.45b). Third, a corrugated metallic film on the back surface of a thin photovoltaic absorber layer can couple sunlight into SPP modes supported at the

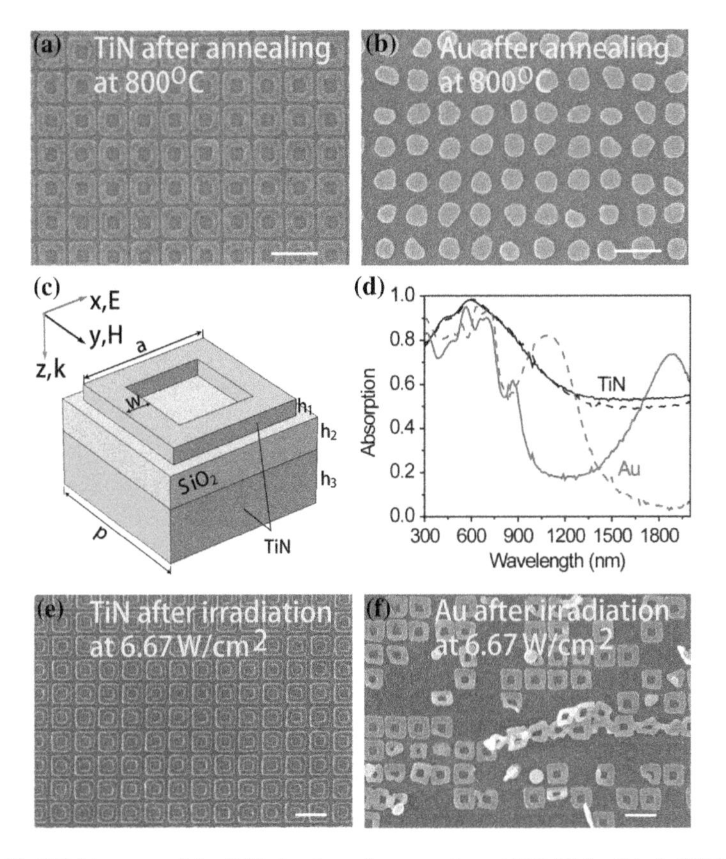

Fig. 13.44 SEM image **a** of the TiN absorber after annealing at 800 °C for 8 h. **b** SEM images for the Au absorber after annealing at 800 °C for 15 min. **c** Schematic representation of a unit cell of the three-layer TiN metamaterial absorber. **d** The measured absorption of TiN absorber before annealing. SEM of the TiN absorber (**e**) and Au absorber (**f**) shot by a laser with an intensity of 6.67 W/cm^2. Reproduced from [58] with permission. Copyright 2014, Wiley-VCH Verlag GmbH & Co. KGaA, Weinheim

metal/semiconductor interface as well as photonic modes that propagate in the plane of the semiconductor layer (Fig. 13.45c).

These light-trapping techniques may allow considerable shrinkage (possibly 10- to 100-fold) of the photovoltaic layer thickness, while keeping the optical absorption (and thus efficiency) as a constant. The simplest way to resonantly excite SPP is to use nanostructures, such as metallic nanoclusters and periodic nanowires. The field enhancement from nanoclusters is highly localized around the nanoclusters, and the possible exaction quenching can limit the utility of such nanoclusters in thin-film organic solar cell (OSC). Moreover, the metallic nanostructures used to excite the surface plasmon resonance can simultaneously act as transparent electrodes. Based on the above enhancement mechanism, an efficiency-enhanced OSC has been proposed [61], which exploits a coupled plasmon mode in a MIM sandwich structure

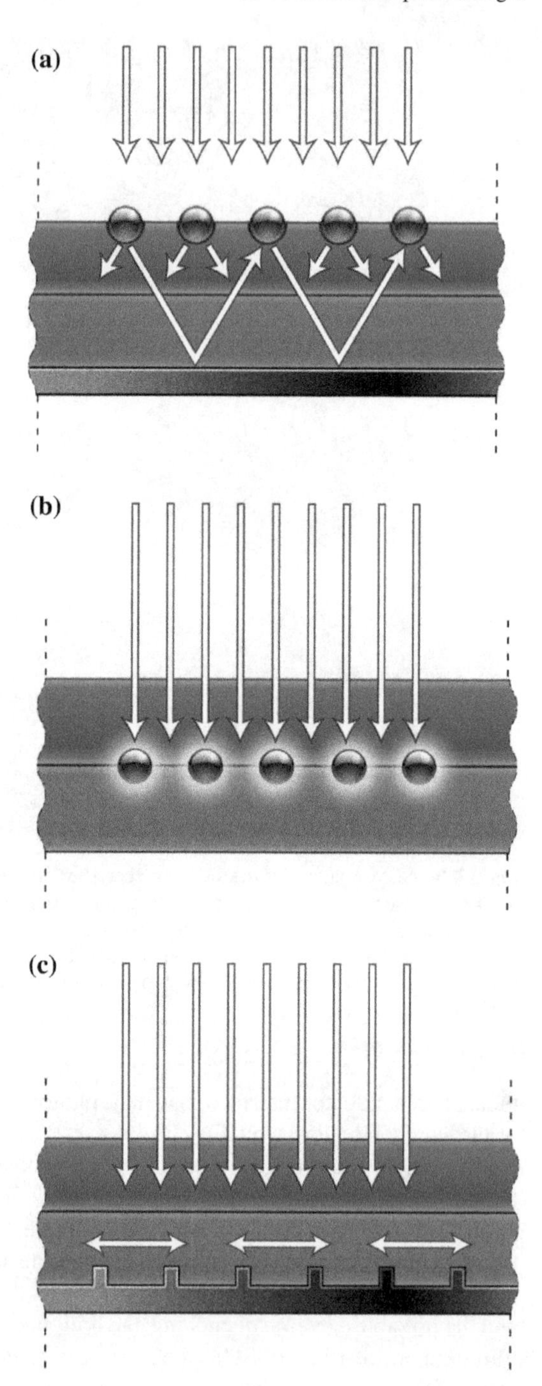

Fig. 13.45 **a–c** Several different light-trapping mechanisms in the solar cell. Reproduced from [60] with permission. Copyright 2010, Springer Nature

where the dielectric layer is composed of the organic semiconductors, and the metallic layers are chosen to be Ag due to its excellent plasmonic properties in the visible range. One metal layer is continuous and acts as cathode, and the other is a periodic nanowire structure acting as semitransparent anode, as shown in Fig. 13.46.

The obtained external quantum efficiency (EQE) spectra of two devices are shown in the inset of Fig. 13.47. It can be clearly seen that the EQE of nanowire device is higher than that of ITO device across the whole visible range, which can explain the higher short-circuit current (J_{sc}) observed in the nanowire device. The enhancement in EQE between the nanowire device and the ITO device as a function of the wavelength can be extracted from the ratio of the two EQE curves and is plotted in Fig. 13.47. A 2.5-fold EQE enhancement around the wavelength of 560 nm can be clearly observed. Since the EQE depends proportionally on the absorption efficiency

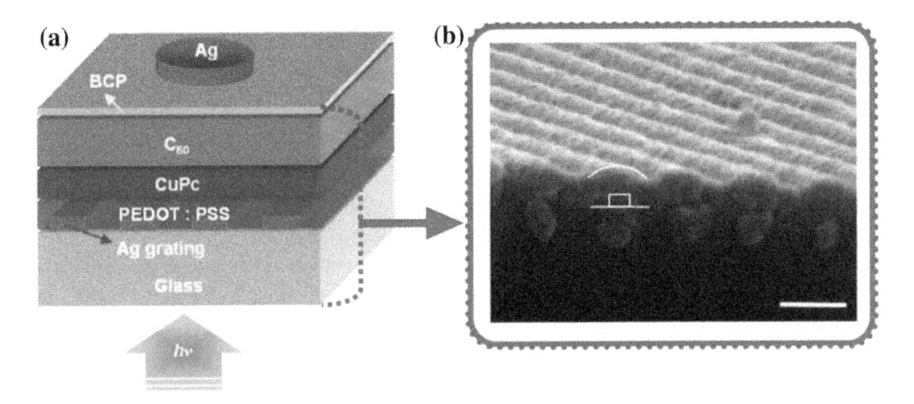

Fig. 13.46 **a** Schematic of the fabricated small molecular weight organic solar cell. **b** Cross-sectional view of one of the fabricated devices, but without 70-nm-thick Ag cathode. Reproduced from [61] with permission. Copyright 2010, Wiley-VCH Verlag GmbH & Co. KGaA, Weinheim

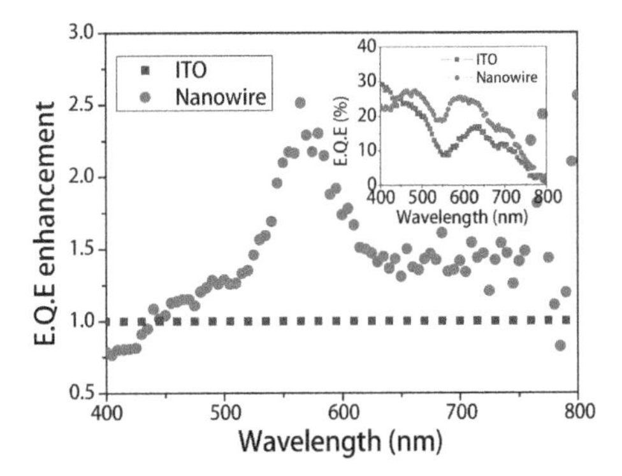

Fig. 13.47 EQE enhancement of nanowire device with reference to ITO device. The inset gives the measured EQE of the nanowire and ITO devices. Reproduced from [61] with permission. Copyright 2010, WILEY-VCH Verlag GmbH & Co. KGaA, Weinheim

of the active organic semiconductor layers and excludes the absorption effect by the Ag electrodes, the observed enhancement of J_{sc} can be attributed to the enhanced optical absorption caused by the surface plasmon resonance and waveguide effects in the Ag nanowire device.

Besides OSC, advancing perovskite solar cell technologies toward their theoretical EQE requires delicate control over the carrier dynamics throughout the entire device. By controlling the formation of the perovskite layer and careful choices of other materials, one can suppress carrier recombination in the absorber, facilitate carrier injection into the carrier transport layers, and maintain good carrier extraction at the electrodes [62].

In general, the standard theory of light trapping demonstrated that absorption enhancement in a medium cannot exceed a factor of $4n^2/\sin^2\theta$, where n is the refractive index of the active layer and θ is the angle of the emission cone in the medium surrounding the cell. Recent theory [63] reveals that the conventional limit can be substantially surpassed when optical modes exhibit deep subwavelength-scale field confinement, opening new avenues for highly efficient next-generation solar cells.

Figure 13.48a shows a thin absorbing film with a thickness of 5 nm, consisting of a material with $\varepsilon_L = 2.5$ and a wavelength-independent absorption length of 25 μm. The film is placed on a mirror that is approximated to be a PEC. In order to enhance the absorption in the active layer, a transparent cladding layer ($\varepsilon_H = 12.5$) is placed on top of the active layer. Such a cladding layer serves two purposes. First, it enhances density of state. Second, the index contrast between active and cladding layer provides nanoscale field confinement. In order to couple incident light into such nanoscale guided modes, a scattering layer with a periodicity much larger than the wavelength range of interest is introduced on top of the cladding layer. Each unit cell consists of a number of air grooves. These grooves are oriented along different directions to ensure that scattering strength does not strongly depend on the angles and polarizations of the incident light.

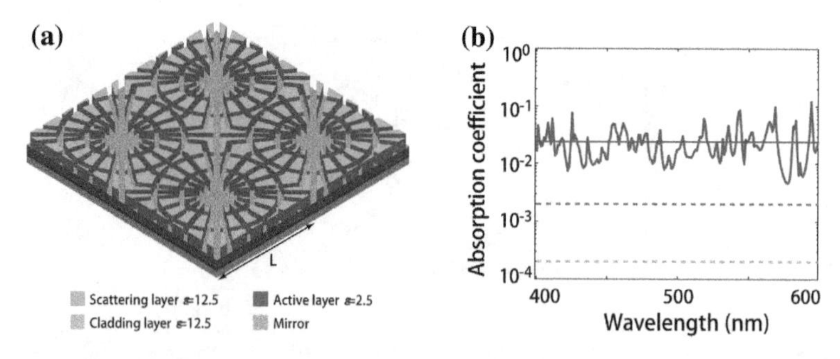

Fig. 13.48 **a** A nanophotonic light-trapping structure. The scattering layer consists of a square lattice of air groove patterns with periodicity $L = 1200$ nm. **b** Absorption spectrum with the light-trapping structures for normally incident light. The dark-gray dashed line represents the theoretical limit. Reproduced from [63] with permission. Copyright 2010, The Authors

The simulated absorption spectrum is shown in Fig. 13.48b. The spectrally averaged absorption (red solid line) is much higher than both the single-pass absorption (light-gray dashed line) and the absorption as predicted by the limit of $4n_L^2$ (dark-gray dashed line). The enhancement factor of about 119 is well above the conventional limit for both the active material ($4\varepsilon_L = 10$) and the cladding material ($4\varepsilon_H = 50$).

In the above implementation, one of the design cores is how to obtain an optimized scatter design by nanostructures. A general model of the nanostructured photovoltaic device in 2D is illustrated in Fig. 13.49a [64], where the geometries are generalized based on four nanostructured interfaces with numerous variables that define the device geometry. The absorber is P3HT:PCBM, and the cladding material is GaP, which has a higher refractive index than the absorber. These structures scatter light waves such that incident sunlight can be efficiently coupled into waveguide modes inside the low-index absorbing layer.

A generalized optimization method for deriving optimal geometries that allow for such enhancement is detailed in Fig. 13.50. The parameters are traversed in a random order within each iteration. The enhancement of short-circuit current J_{sc} is considered as the figure of merit (FOM) during optimization. Due to the complexity of the geometry, additional constraints are required to preserve realistic designs. After optimization, the short-circuit current density enhancement factor as a function of incident wavelength is computed. Proper termination conditions such as a minimum improvement in FOM are imposed to guarantee numerical convergence. Because of

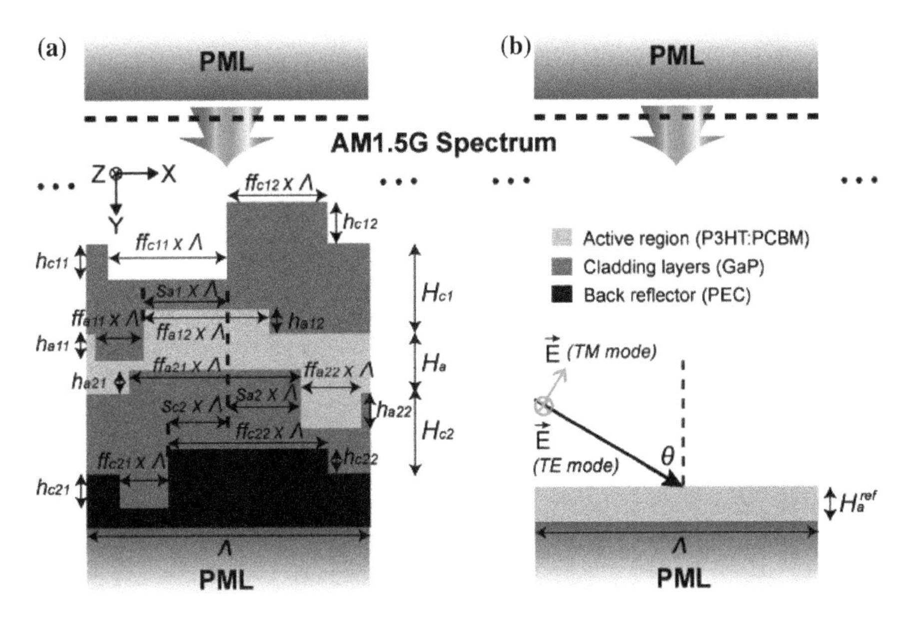

Fig. 13.49 **a** Photovoltaic device with light-trapping nanostructures defined by 22 geometric parameters. *Ha* is kept constant during optimization. **b** Reference device with the same volume of absorbing material as that of the structured design in **a**. Adapted with permission from [64] with permission. Copyright 2014, Optical Society of America

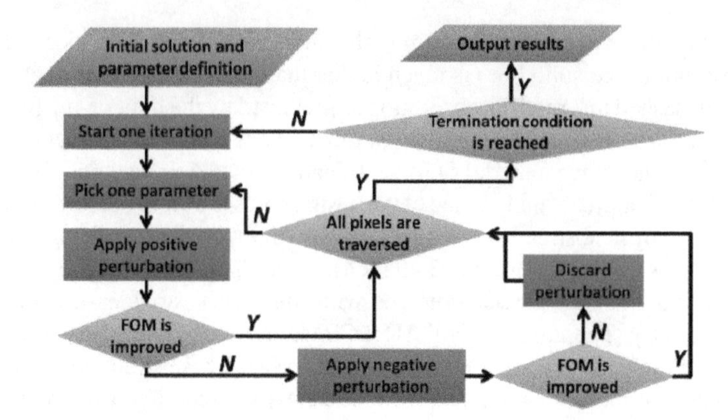

Fig. 13.50 Flowchart of the direct binary search-based algorithm adopted for optimizing nanophotonic light-trapping structures. Reproduced from [64] with permission. Copyright 2014, Optical Society of America

Fig. 13.51 Current density enhancement spectra of the optimized designs and those corresponding to the ergodic limit and the LDOS limit. Reproduced from [64] with permission. Copyright 2014, Optical Society of America

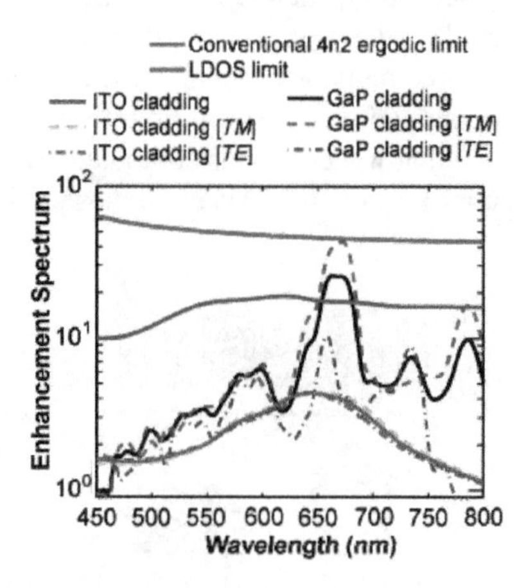

the algorithm's tendency of premature convergence to local maxima, the same optimization process was repeated with several randomly generated initial candidates, among which the best optimized solution was chosen.

Enhancement spectra are plotted in Fig. 13.51 in logarithmic scale, which shows that the optimized device with a high-index cladding (black solid line) can provide enhancement beyond the Ergodic limit ($4n_{\mathrm{L}}^2$, solid blue line) for wavelengths close to the bandgap of the absorber, where n_{L} is the real part of the refractive index of the active layer of P3HT:PCBM. Even more, the TM resonant mode at 672 nm (dark-green dashed line) allows for an enhancement factor that reaches the LDOS limit ($4n_{\mathrm{H}}^2$, red solid line), where n_{H} is the real part of the refractive index of the cladding material GaP.

13.7 Sensors Based on Narrowband Absorber

A potential application of narrowband absorption in the visible and near-infrared regime is plasmonic sensor, which is a promising platform for applications in biology and chemistry. In particular, the local electromagnetic field enhancement in metal nanostructures is highly correlated with the surrounding environment, providing a new way for the realization of high sensitivity biosensors [65].

Figure 13.52a illustrates the geometry of an absorber-based sensor structure [66]. It consists of two functional layers. The top layer is a two-dimensional gold disk array, and the bottom layer is a gold mirror. The two layers are separated by an MgF_2 dielectric spacer. The structure is designed to be polarization independent in x- and y-directions at normal incidence. Figure 13.52b shows the results of a proof-of-principle experiment, displaying the measured reflectance spectra with air ($n = 1$) and water ($n = 1.312$) on the sample surface. The experimental reflectance reaches a minimum of 1% at 185.6 THz (1.6 μm) in air, which corresponds to an experimental absorbance of 99%. When water (red curve) is applied to the sample surface, a clear increase of the reflectance intensity from 1 to 28.7% at 185.6 THz is visible, resulting from the refractive index change of the local dielectric environment. A FOM defined as $\max|[dI(\lambda)/dn(\lambda)]/(I(\lambda)|$ has a maximum value of about 87, as shown in Fig. 13.52c.

Generally, the sensitivity of plasmonic sensors is limited by broad spectral features due to large radiative loss of metallic nanostructures in visible region. Fano resonance provides a new way to increase the sensitivity [67]. In order to realize high-Q resonance, the radiation loss of the Fano resonance system can be greatly reduced or even completely inhibited, as a result of the interference between super-radiation and sub-radiation patterns. Such characteristic promises the Fano resonance a series of excellent electromagnetic properties such as narrow spectral line width, strong electromagnetic field enhancement, and high-refractive index sensitivity.

As shown in Fig. 13.53, a three-layer structure consisting of an ellipsoidal silver pair separated from the silver reflector by a layer of silica was proposed. Structural symmetry was broken by rotating one of the elliptical silver cylinders with respect to the other. Simulated results show that the distinct Fano-like line shape with sharp peak as narrow as 10.8 nm (FWHM) appears around a wavelength of 681 nm when the orientation angle $\alpha = 25°$. From Fig. 13.54, one can see that the Fano resonance in this structure is raised by the interference of dipole resonance aroused by the incident light and quadrupole mode aroused by the asymmetry of the ellipsoidal pair. The dipole mode and quadrupole mode represent the super-radiation and sub-radiation patterns, respectively.

The mid-IR spectrum is essential for sensing because of the presence of characteristic molecular absorption fingerprints originating from the intrinsic vibrational modes of chemical bonds. Although mid-IR spectroscopy allows nondestructive, label-free, and direct identifying biochemical building blocks, including proteins, lipids, and DNA, the sensitivity of mid-IR spectroscopy is limited because of the mismatch between mid-IR wavelengths and dimensions of nanometer-scale molecules [69].

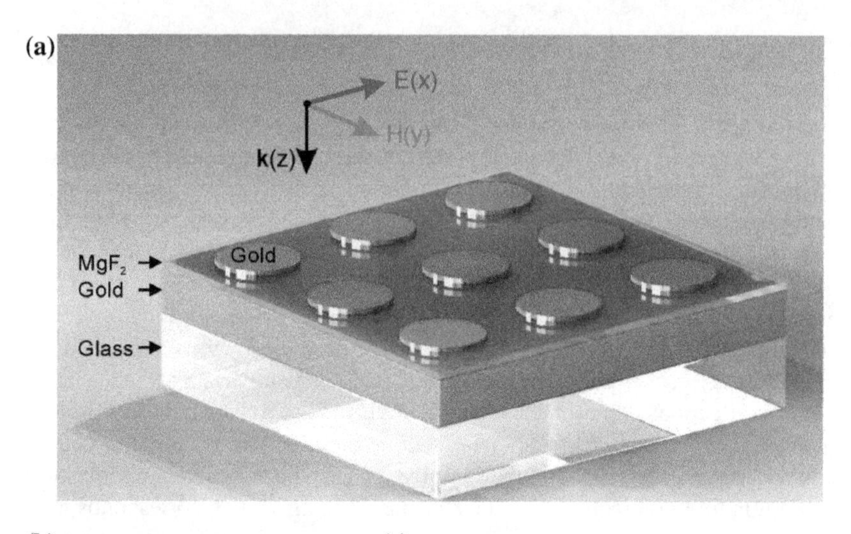

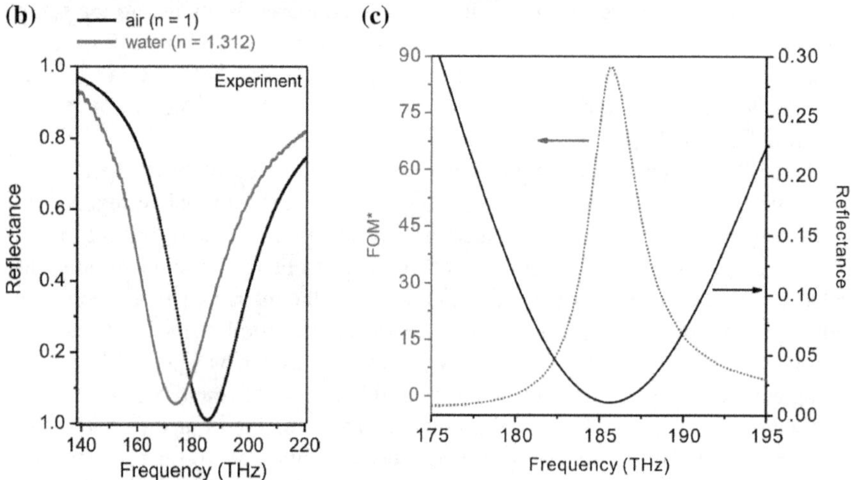

Fig. 13.52 **a** Schematic of the perfect absorber structure and the incident light polarization configuration. **b** Experimental tuning of the reflectance spectra by changing the dielectric environment which is adjacent to the gold disks from air to water is shown. **c** Experimental FOM as a function of frequency. Adapted with permission from [66] with permission. Copyright 2010, American Chemical Society

In order to further improve the detection sensitivity, low-loss and high-Q resonator arrays are highly desired [70, 71]. Recently, a two-dimensional pixelated dielectric metasurface was proposed [69]. Individual meta-pixels contain a zigzag array of anisotropic hydrogenated amorphous silicon (a-Si:H) resonators, which provide high-Q resonances when excited with linearly polarized light and allow for straightforward resonance tuning via scaling of the unit cell geometry. As a consequence, when a conformal protein layer covers the pixelated dielectric metasurface,

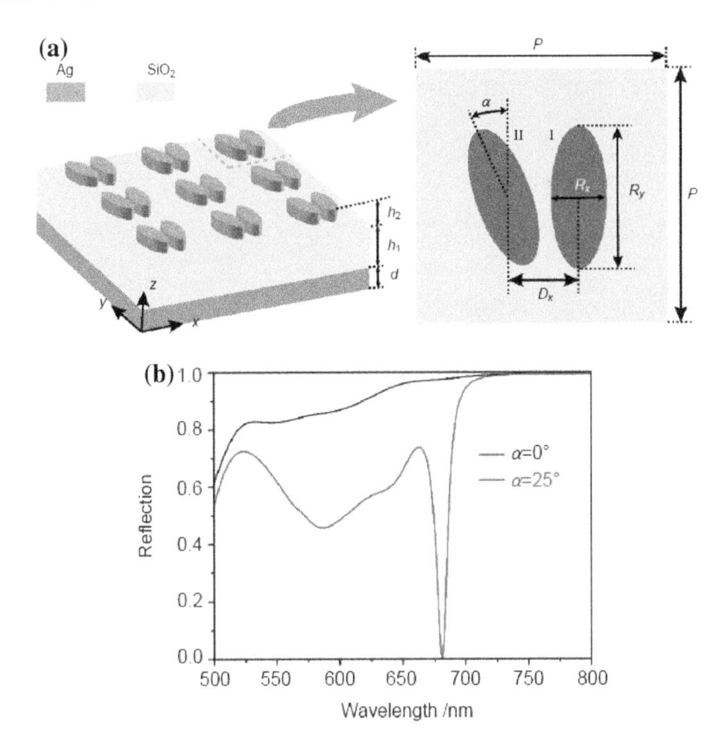

Fig. 13.53 **a** Arrays and unit cell of the asymmetric ellipsoidal pair. **b** Reflection spectra of the symmetric structure ($\alpha = 0°$) and the asymmetric structure ($\alpha = 25°$). Adapted with permission from [68] with permission. Copyright 2018, Institute of Optics and Electronics, Chinese Academy of Sciences

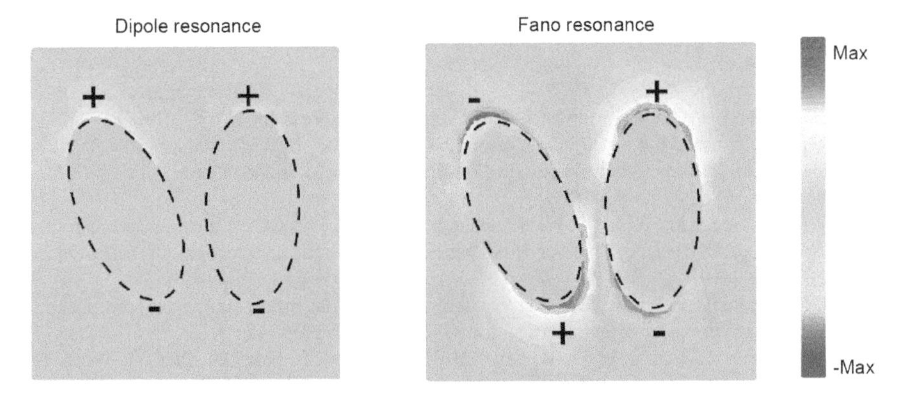

Fig. 13.54 E_z at the top of the asymmetric double elliptical cylinder structures in the case of $\alpha = 25°$ at the dipole resonance ($\lambda = 582$ nm) and the Fano resonance ($\lambda = 681$ nm). Adapted with permission from [68] with permission. Copyright 2018, Institute of Optics and Electronics, Chinese Academy of Sciences

the molecular absorption signatures can be read out at multiple spectral points, and the resulting information is then translated into a barcode-like spatial absorption map for imaging. The signatures of biological, polymer, and pesticide molecules can be detected with high sensitivity, covering applications such as biosensing and environmental monitoring.

References

1. Black body. https://en.wikipedia.org/wiki/Black_body
2. K. Mizuno, J. Ishii, H. Kishida, Y. Hayamizu, S. Yasuda, D.N. Futaba, M. Yumura, K. Hata, A black body absorber from vertically aligned single-walled carbon nanotubes. Proc. Natl. Acad. Sci. U. S. A. **106**, 6044–6047 (2009)
3. M. Planck, *The Theory of Heat Radiation* (P. Blakiston's Son & Co., 1914)
4. R.W. Wood, On a remarkable case of uneven distribution of light in a diffraction grating spectrum. Proc. R. Soc. Lond. **18**, 269 (1902)
5. R.H. Ritchie, E.T. Arakawa, J.J. Cowan, R.N. Hamm, Surface-plasmon resonance effect in grating diffraction. Phys. Rev. Lett. **21**, 1530–1533 (1968)
6. W.W. Salisbury, Absorbent body for electromagnetic waves, U.S. Patent 2599944, 1952
7. E.F. Knott, J.F. Shaeffer, M.T. Tuley, *Radar Cross Section*, 2nd edn. (SciTech Publishing, 2004)
8. N.I. Landy, S. Sajuyigbe, J.J. Mock, D.R. Smith, W.J. Padilla, Perfect metamaterial absorber. Phys. Rev. Lett. **100**, 207402 (2008)
9. C. Hu, X. Li, Q. Feng, X. Chen, X. Luo, Introducing dipole-like resonance into magnetic resonance to realize simultaneous drop in transmission and reflection at terahertz frequency. J. Appl. Phys. **108** (2010)
10. C. Hu, X. Li, Q. Feng, X. Chen, X. Luo, Investigation on the role of the dielectric loss in metamaterial absorber. Opt. Express **18**, 6598–6603 (2010)
11. Y. Guo, L. Yan, W. Pan, B. Luo, X. Luo, Ultra-broadband terahertz absorbers based on 4 $\times$ 4 cascaded metal-dielectric pairs. Plasmonics **9**, 951–957 (2014)
12. Q. Feng, M. Pu, C. Hu, X. Luo, Engineering the dispersion of metamaterial surface for broadband infrared absorption. Opt. Lett. **37**, 2133–2135 (2012)
13. Y. Huang, M. Pu, P. Gao, Z. Zhao, X. Li, X. Ma, X. Luo, Ultra-broadband large-scale infrared perfect absorber with optical transparency. Appl. Phys. Express **10**, 112601 (2017)
14. Y. Huang, L. Liu, M. Pu, X. Li, X. Ma, X. Luo, A refractory metamaterial absorber for ultra-broadband, omnidirectional and polarization-independent absorption in the UV-NIR spectrum. Nanoscale **10**, 8298–8303 (2018)
15. M. Pu, Q. Feng, M. Wang, C. Hu, C. Huang, X. Ma, Z. Zhao, C. Wang, X. Luo, Ultrathin broadband nearly perfect absorber with symmetrical coherent illumination. Opt. Express **20**, 2246–2254 (2012)
16. M. Pu, Q. Feng, C. Hu, X. Luo, Perfect absorption of light by coherently induced plasmon hybridization in ultrathin metamaterial film. Plasmonics **7**, 733–738 (2012)
17. S. Li, J. Luo, S. Anwar, S. Li, W. Lu, Z.H. Hang, Y. Lai, B. Hou, M. Shen, C. Wang, An equivalent realization of coherent perfect absorption under single beam illumination. Sci. Rep. **4**, 7369 (2014)
18. S. Li, J. Luo, S. Anwar, S. Li, W. Lu, Z.H. Hang, Y. Lai, B. Hou, M. Shen, C. Wang, Broadband perfect absorption of ultrathin conductive films with coherent illumination: superabsorption of microwave radiation. Phys. Rev. B **91**, 220301(R) (2015)
19. C. Yan, M. Pu, J. Luo, Y. Huang, X. Li, X. Ma, X. Luo, Coherent perfect absorption of electromagnetic wave in subwavelength structures. Opt. Laser Technol. **101**, 499–506 (2018)
20. M. Wang, C. Hu, M. Pu, C. Huang, X. Ma, X. Luo, Electrical tunable L-band absorbing material for two polarisations. Electron. Lett. **48**, 1002–1003 (2012)

21. X. Wu, C. Hu, Y. Wang, M. Pu, C. Huang, C. Wang, X. Luo, Active microwave absorber with the dual-ability of dividable modulation in absorbing intensity and frequency. AIP Adv. **3**, 022114 (2013)
22. N.I. Landy, C.M. Bingham, T. Tyler, N. Jokerst, D.R. Smith, W.J. Padilla, Design, theory, and measurement of a polarization-insensitive absorber for terahertz imaging. Phys. Rev. B **79** (2009)
23. H. Tao, N.I. Landy, C.M. Bingham, X. Zhang, R.D. Averitt, W.J. Padilla, A metamaterial absorber for the terahertz regime: design, fabrication and characterization. Opt. Express **16**, 7181–7188 (2008)
24. M. Pu, C. Hu, M. Wang, C. Huang, Z. Zhao, C. Wang, Q. Feng, X. Luo, Design principles for infrared wide-angle perfect absorber based on plasmonic structure. Opt. Express **19**, 17413–17420 (2011)
25. C. Hu, Z. Zhao, X. Chen, X. Luo, Realizing near-perfect absorption at visible frequencies. Opt. Express **17**, 11039–11044 (2009)
26. M. Pu, X. Ma, X. Li, Y. Guo, X. Luo, Merging plasmonics and metamaterials by two-dimensional subwavelength structures. J. Mater. Chem. C **5**, 4361–4378 (2017)
27. T.V. Teperik, F.J. García de Abajo, A.G. Borisov, M. Abdelsalam, P.N. Bartlett, Y. Sugawara, J.J. Baumberg, Omnidirectional absorption in nanostructured metal surfaces. Nat. Photonics **2**, 299–301 (2008)
28. C. Hu, L. Liu, Z. Zhao, X. Chen, X. Luo, Mixed plasmons coupling for expanding the bandwidth of near-perfect absorption at visible frequencies. Opt. Express **17**, 16745–16749 (2009)
29. Y.Q. Ye, Y. Jin, S. He, Omnidirectional, polarization-insensitive and broadband thin absorber in the terahertz regime. J. Opt. Soc. Am. B **27**, 498–504 (2010)
30. X. Shen, T.J. Cui, J. Zhao, H.F. Ma, W.X. Jiang, H. Li, Polarization-independent wide-angle triple-band metamaterial absorber. Opt. Express **19**, 9401–9407 (2011)
31. J. Grant, Y. Ma, S. Saha, A. Khalid, D.R.S. Cumming, Polarization insensitive, broadband terahertz metamaterial absorber. Opt. Lett. **36**, 3476–3478 (2011)
32. L. Huang, D.R. Chowdhury, S. Ramani, M.T. Reiten, S.-N. Luo, A.J. Taylor, H.-T. Chen, Experimental demonstration of terahertz metamaterial absorbers with a broad and flat high absorption band. Opt. Lett. **37**, 154–156 (2012)
33. Y. Cui, J. Xu, K.H. Fung, Y. Jin, A. Kumar, S. He, N.X. Fang, A thin film broadband absorber based on multi-sized nanoantennas. Appl. Phys. Lett. **99**, 253101 (2011)
34. F. Ding, Y. Cui, X. Ge, Y. Jin, S. He, Ultra-broadband microwave metamaterial absorber. Appl. Phys. Lett. **100**, 3506 (2012)
35. Y. Cui, K.H. Fung, J. Xu, H. Ma, Y. Jin, S. He, N.X. Fang, Ultrabroadband light absorption by a sawtooth anisotropic metamaterial slab. Nano Lett. **12**, 1443–1447 (2012)
36. J. Sun, L. Liu, G. Dong, J. Zhou, An extremely broad band metamaterial absorber based on destructive interference. Opt. Express **19**, 21155–21162 (2011)
37. M. Pu, P. Chen, Y. Wang, Z. Zhao, C. Wang, C. Huang, C. Hu, X. Luo, Strong enhancement of light absorption and highly directive thermal emission in graphene. Opt. Express **21**, 11618–11627 (2013)
38. Y. Guo, Y. Wang, M. Pu, Z. Zhao, X. Wu, X. Ma, C. Wang, L. Yan, X. Luo, Dispersion management of anisotropic metamirror for super-octave bandwidth polarization conversion. Sci. Rep. **5**, 8434 (2015)
39. M. Pu, X. Ma, Y. Guo, X. Li, X. Luo, Theory of microscopic meta-surface waves based on catenary optical fields and dispersion. Opt. Express **26**, 19555–19562 (2018)
40. M. Pu, Y. Guo, X. Li, X. Ma, X. Luo, Revisitation of extraordinary Young's interference: from catenary optical fields to spin-orbit interaction in metasurfaces. ACS Photonics **5**, 3198–3204 (2018)
41. M. Zhang, M. Pu, F. Zhang, Y. Guo, Q. He, X. Ma, Y. Huang, X. Li, H. Yu, X. Luo, Plasmonic metasurfaces for switchable photonic spin-orbit interactions based on phase change materials. Adv. Sci. **5**, 1800835 (2018)
42. Y. Huang, J. Luo, M. Pu, Y. Guo, Z. Zhao, X. Ma, X. Li, X. Luo, Catenary electromagnetics for ultrabroadband lightweight absorbers and large-scale flat antennas. Adv. Sci. 1801691 (2019)

43. S.A. Maier, *Plasmonics: Fundamentals and Applications* (Springer Science & Business Media, 2007)
44. D. Ye, Z. Wang, K. Xu, H. Li, J. Huangfu, Z. Wang, L. Ran, Ultrawideband dispersion control of a metamaterial surface for perfectly-matched-layer-like absorption. Phys. Rev. Lett. **111**, 187402 (2013)
45. M. Pu, M. Wang, C. Hu, C. Huang, Z. Zhao, Y. Wang, X. Luo, Engineering heavily doped silicon for broadband absorber in the terahertz regime. Opt. Express **20**, 25513–25519 (2012)
46. S. Yin, J. Zhu, W. Xu, W. Jiang, J. Yuan, G. Yin, L. Xie, Y. Ying, Y. Ma, High-performance terahertz wave absorbers made of silicon-based metamaterials. Appl. Phys. Lett. **107**, 073903 (2015)
47. X. Zang, C. Shi, L. Chen, B. Cai, Y. Zhu, S. Zhuang, Ultra-broadband terahertz absorption by exciting the orthogonal diffraction in dumbbell-shaped gratings. Sci. Rep. **5**, 8091 (2015)
48. J. Yuan, J. Luo, M. Zhang, M. Pu, X. Li, Z. Zhao, X. Luo, An ultra-broadband THz absorber based on structured doped silicon with antireflection techniques. IEEE Photonics J. 1–1 (2018)
49. J.A. Bossard, L. Lin, S. Yun, L. Liu, D.H. Werner, T.S. Mayer, Near-ideal optical metamaterial absorbers with super-octave bandwidth. ACS Nano **8**, 1517–1524 (2014)
50. W. Wan, Y. Chong, L. Ge, H. Noh, A.D. Stone, H. Cao, Time-reversed lasing and interferometric control of absorption. Science **331**, 889–892 (2011)
51. K.N. Rozanov, Ultimate thickness to bandwidth ratio of radar absorbers. IEEE Trans. Antennas Propag. **48**, 1230–1234 (2000)
52. Y. Wang, X. Ma, X. Li, M. Pu, X. Luo, Perfect electromagnetic and sound absorption via subwavelength holes array. Opto-Electron. Adv. **1**, 180013 (2018)
53. B.E.A. Saleh, M.C. Teich, *Fundamentals of Photonics*, 2nd edn. (Wiley, 2007)
54. W. Woltersdorff, Über die optischen Konstanten dünner Metallschichten im langwelligen Ultrarot. Z. Für Phys. Hadrons Nucl. **91**, 230–252 (1934)
55. Y.D. Chong, L. Ge, H. Cao, A.D. Stone, Coherent perfect absorbers: time-reversed lasers. Phys. Rev. Lett. **105**, 053901 (2010)
56. M. Wang, C. Hu, M. Pu, C. Huang, Z. Zhao, Q. Feng, X. Luo, Truncated spherical voids for nearly omnidirectional optical absorption. Opt. Express **19**, 20642–20649 (2011)
57. T. Jang, H. Youn, Y.J. Shin, L.J. Guo, Transparent and flexible polarization-independent microwave broadband absorber. ACS Photonics **1**, 279–284 (2014)
58. W. Li, U. Guler, N. Kinsey, G.V. Naik, A. Boltasseva, J. Guan, V.M. Shalaev, A.V. Kildishev, Refractory plasmonics with titanium nitride: broadband metamaterial absorber. Adv. Mater. **26**, 7959–7965 (2014)
59. J.A. Schuller, E.S. Barnard, W. Cai, Y.C. Jun, J.S. White, M.L. Brongersma, Plasmonics for extreme light concentration and manipulation. Nat. Mater. **9**, 193 (2010)
60. H.A. Atwater, A. Polman, Plasmonics for improved photovoltaic devices. Nat. Mater. **9**, 205 (2010)
61. M.-G. Kang, T. Xu, H.J. Park, X. Luo, L.J. Guo, Efficiency enhancement of organic solar cells using transparent plasmonic Ag nanowire electrodes. Adv. Mater. **22**, 4378 (2010)
62. H. Zhou, Q. Chen, G. Li, S. Luo, T. Song, H.-S. Duan, Z. Hong, J. You, Y. Liu, Y. Yang, Interface engineering of highly efficient perovskite solar cells. Science **345**, 542 (2014)
63. Z. Yu, A. Raman, S. Fan, Fundamental limit of nanophotonic light trapping in solar cells. Proc. Natl. Acad. Sci. **107**, 17491–17496 (2010)
64. P. Wang, R. Menon, Optimization of generalized dielectric nanostructures for enhanced light trapping in thin-film photovoltaics via boosting the local density of optical states. Opt. Express **22**, A99–A110 (2014)
65. T. Lai, Q. Hou, H. Yang, X. Luo, M. Xi, Clinical application of a novel sliver nanoparticles biosensor based on localized surface plasmon resonance for detecting the microalbuminuria. Acta Biochim. Biophys. Sin. **42**, 787–792 (2010)
66. N. Liu, M. Mesch, T. Weiss, M. Hentschel, H. Giessen, Infrared perfect absorber and its application as plasmonic sensor. Nano Lett. **10**, 2342–2348 (2010)
67. M. Pu, C. Hu, C. Huang, C. Wang, Z. Zhao, Y. Wang, X. Luo, Investigation of Fano resonance in planar metamaterial with perturbed periodicity. Opt. Express **21**, 992–1001 (2013)

68. J. Fang, M. Zhang, F. Zhang, H. Yu, Plasmonic sensor based on Fano resonance. Opto-Electron. Eng. **44**, 221–225 (2017)
69. A. Tittl, A. Leitis, M. Liu, F. Yesilkoy, D.-Y. Choi, D.N. Neshev, Y.S. Kivshar, H. Altug, Imaging-based molecular barcoding with pixelated dielectric metasurfaces. Science **360**, 1105 (2018)
70. M. Pu, M. Song, H. Yu, C. Hu, M. Wang, X. Wu, J. Luo, Z. Zhang, X. Luo, Fano resonance induced by mode coupling in all-dielectric nanorod array. Appl. Phys. Express **7**, 032002 (2014)
71. M. Song, H. Yu, C. Wang, N. Yao, M. Pu, J. Luo, Z. Zhang, X. Luo, Sharp Fano resonance induced by a single layer of nanorods with perturbed periodicity. Opt. Express **23**, 2895–2903 (2015)

Chapter 14
Radiation Engineering and Optical Phased Array

Abstract Radiation is a very important energy conversion process in engineering optics. Effective thermal radiation management can not only improve the efficiency of thermophotovoltaics but also realize passive radiative cooling, thermal cloak, and camouflage. Besides, light-emitting diode (LED) with small volume and high efficiency is poised to replace the traditional light bulb and liquid crystal display (LCD) in the next few decades. In addition, minimized micro-/nanolaser that can serve as coherent light sources in on-chip electro-photonic circuits can be widely applied in nanoscale applications. Finally, if the phase retardation of coherent emitters can be actively controlled in a compact manner, high-performance optical phased light detection and ranging will replace the traditional beam scanning technology based on mechanical steering. In this chapter, we give a detailed discussion about the radiation engineering technology and optical phased array.

Keywords Thermal radiations · Thermophotovoltaics · Light-emitting diodes · Micro- and nanolasers · Optical phased arrays

14.1 Introduction

Thermal radiation represents a ubiquitous aspect of nature. Conventional thermal radiation in macroscopic scale, such as that coming from the sun, is typically incoherent, broadband, unpolarized, near isotropic, as indicated in Fig. 14.1a [1]. Besides, there is a set of fundamental constraints in classic thermal radiation theory. For example, the spectral density of the thermal emission per unit emitter area is bounded by Planck's law of thermal radiation (Fig. 14.1c). In addition, thermal radiators are typically subject to Kirchhoff's law, which states that the angular spectral absorptivity and emissivity must be equal to each other at thermal equilibrium. These characteristics and constraints impose strong restrictions on the capabilities for controlling thermal radiation.

Recently, the advances of nanoscience and nanotechnology offer an opportunity to manipulate the thermal radiation in manners different from traditional methods. For example, nanophotonic structures could have coherent, narrowband, polarized, and

X. Luo, *Engineering Optics 2.0*, https://doi.org/10.1007/978-981-13-5755-8_14

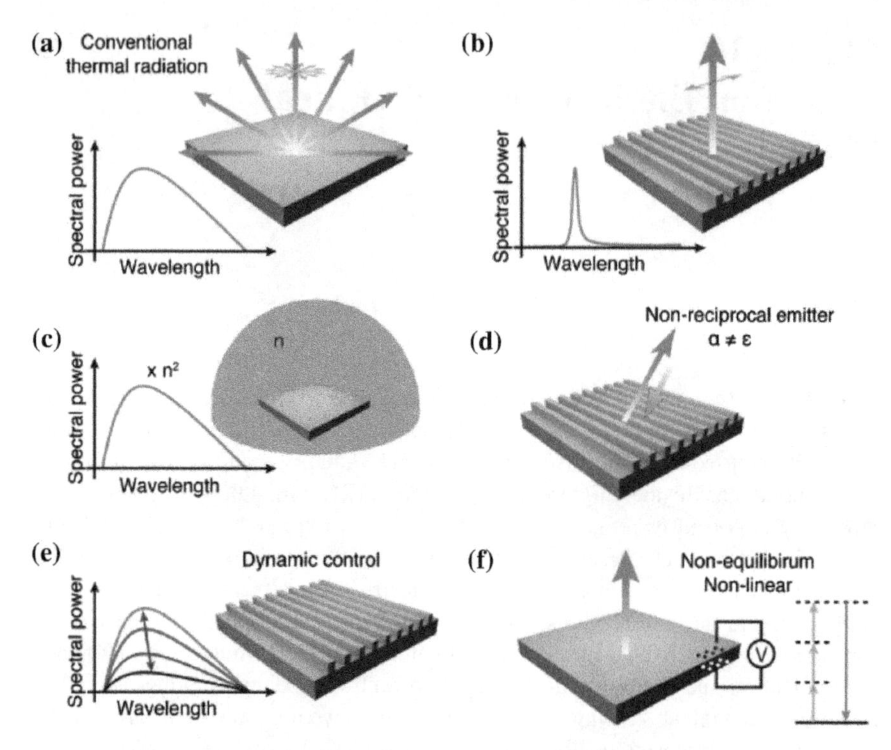

Fig. 14.1 **a** Conventional thermal radiation is incoherent, broadband, unpolarized and near-isotropic in its directionality. **b–f** Thermal radiations for EO 2.0 are drastically different from the conventional counterparts. **b** Nanophotonic structures could have control on coherence, bandwidth, polarization and directionality of thermal radiation. **c** Enhanced far-field thermal radiation by thermal extraction. **d** Violation of Kirchhoff's Law by breaking reciprocity. **e** Dynamic control of thermal radiation with nanophotonic structures. **f** Non-equilibrium and nonlinear thermal radiation. Reproduced from [1] with permission. Copyright 2018, The Authors

directional thermal radiation (Fig. 14.1b) [1]. With appropriate condition, Planck's law and Kirchhoff's law can be surpassed, leading to many unusual energy applications, including thermal cloak, daytime radiative cooling, and thermophotovoltaic systems.

Besides thermal radiation, spontaneous and stimulated emissions are two kinds of important energy conversion methods. Light-emitting diodes (LEDs) invented in the mid-twentieth century are one of the key technologies that are thriving in the past few decades. During the past few years, new materials including quantum dots and perovskites have been introduced in the high-efficiency LED design, with the highest external quantum efficiency (EQE) beyond 20%.

As coherent light source, the laser has played an important role in advancing the development of communication, imaging, and spectroscopy technologies due to its coherent, intense, and directional emission. However, conventional lasers suffer from cavity sizes much larger than their operating wavelength and thus are not suitable for on-chip integrations. Recently, there has been significant progress toward designing lasers with small sizes [2–4], as shown in Fig. 14.2. Miniaturized lasers are promising for diverse applications, including, optical communication, bioimaging, and sensing.

Beam steering is vital to realize various important functionalities, e.g., light detection and ranging (LIDAR), high-speed free-space point-to-point communications, and spatially resolved optical sensing. Traditionally, beam steering is realized by mechanical scanning, which is slow and bulky. Inspired by the microwave phased array, the concept of optical phased array (OPA) has been proposed and demonstrated, where the beam direction is tuned by changing the relative phase retardation of radiating elements. Therefore, OPAs allow very stable, rapid, and precise beam steering without mechanical motion, making them robust and insensitive to external constraints such as acceleration. Conventionally, OPA was implemented based on electro-optical crystal and liquid crystal. The former exhibits fast response time but requires high applied bias, while the later operates at a low applied bias but suffers from slow response time. Recently, silicon-based on-chip OPA (Fig. 14.3) has received a great deal of attention with the development of CMOS process, which may play important roles in auto-driving vehicles and android vision.

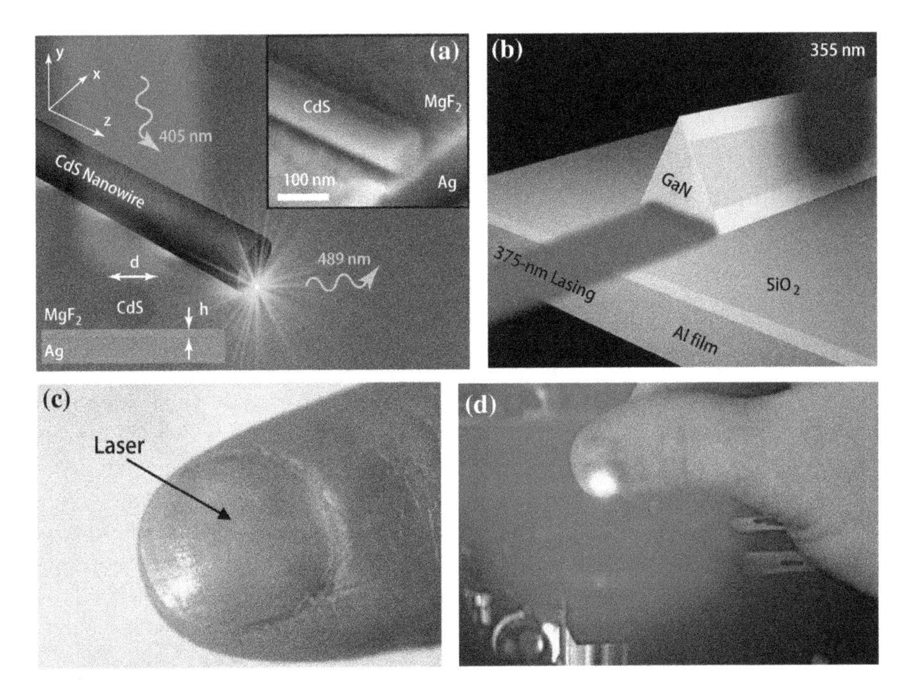

Fig. 14.2 Schematic of micro- and nanolasers. **a**, **b** Plasmonic lasers, and **c**, **d** membrane lasers. Reproduced from **a** [2], **b** [3], **c** and **d** [4] with permission. **a** Copyright 2009, Macmillan Publishers Limited. **b** Copyright 2014, Macmillan Publishers Limited. **c** and **d** Copyright 2018, The Authors

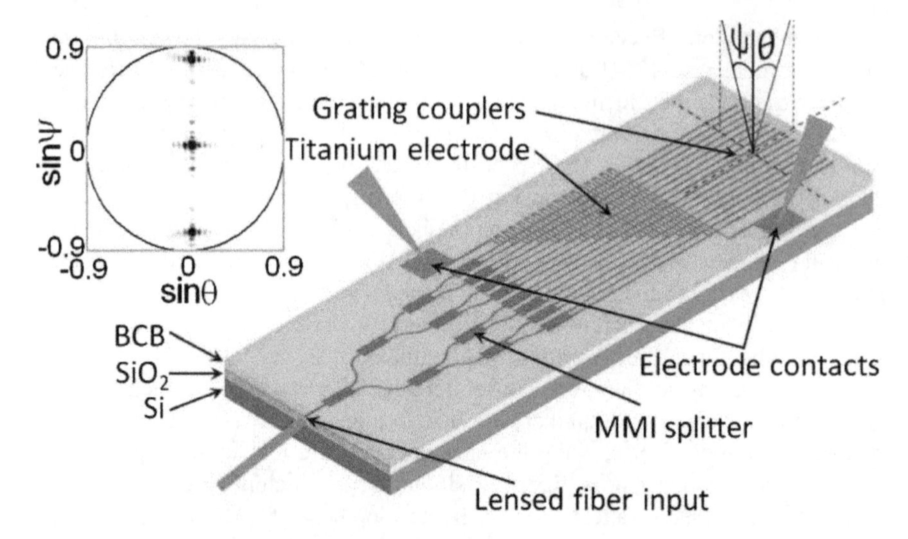

Fig. 14.3 Schematic of the on-chip OPA. The inset shows the far-field image. Reproduced from [5] with permission. Copyright 2009, Optical Society of America

14.2 Thermal Radiations

14.2.1 Spectrally Selective and Directional Thermal Radiations

According to Kirchhoff's law of thermal radiation, the emissivity of a material equals to its absorptivity at equilibrium [6, 7]. Therefore, photonic crystals (PhC) have been utilized to significantly suppress thermal radiation for a range of frequencies in the bandgap while enhance radiation outside the bandgap [8]. Alternatively, because of the resonant nature of metamaterials, the perfect absorber could yield sharp resonances with high absorption, enabling their use as high-Q emitters with high emissivity.

As shown in Fig. 14.4a, a cross-shaped metallic resonator separated from a ground plane by a dielectric layer can be utilized to realize single-band infrared absorption [9], which has an absorption peak at 5.8 μm with 97% absorption. Since the operation band of absorber could be feasibly engineered by the design, two-band absorption can be easily obtained by a super-cell design that incorporates different resonators, which exhibits absorptivity of 80% and 93.5%, respectively, at 6.18 and 8.32 μm (Fig. 14.4b) [9].

The measured emittances of the single- and dual-band metamaterial samples when heated at different temperatures are shown in Fig. 14.5a, b [9]. As a comparison, the emittance of a blackbody reference—black carbon at the same temperature is also presented. Obviously, the emitted energy of the blackbody is significantly greater than

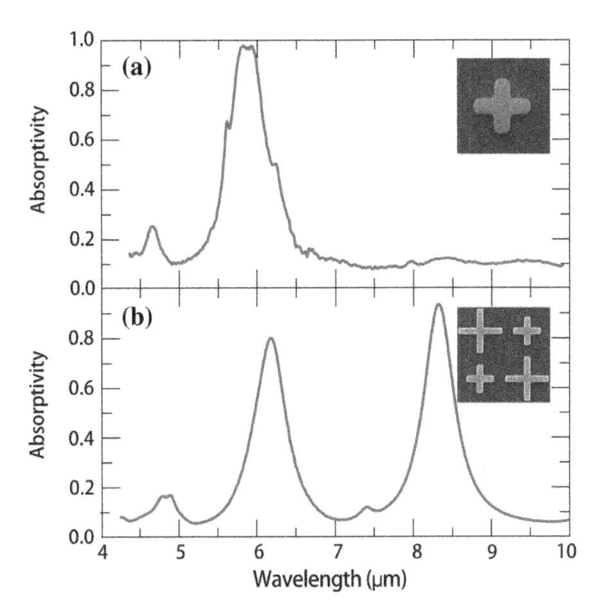

Fig. 14.4 **a** Experimental absorptivity of the single-band metamaterial absorber. **b** Experimental absorptivity of the dual-band metamaterial absorber. Insets display scanning electron microscope (SEM) images of one unit cell for the fabricated single- and dual-band absorbers. Reproduced from [9] with permission. Copyright 2011, American Physical Society

that of the metamaterial emitters at the same temperature, except at near 6 μm for the single band, and near 6 and 8 μm for the dual-band sample, where metamaterial emissivities are nearly equal to that of the blackbody [9]. Furthermore, the results shown in Fig. 14.5c, d demonstrate that the absorption curve and radiation curve match with each other well. Therefore, the emissive properties of metamaterials can be tuned by the absorption, i.e., by modifying the transmissive and reflective properties [9].

Just like the spectra control of the emissivity, one can also tailor the angular or directional properties of the emissivity using nanophotonic structures. Greffet et al. [10] demonstrated strong angular dependency of emissivity from a SiC grating structure, as indicated in Fig. 14.6. The SiC–air interface supports surface phonon-polaritons, and the use of grating enables a wavevector-selective resonant excitation of such surface waves. The relationship between the emission angle θ and the wavelength λ is simply given by the usual grating law:

$$\frac{2\pi}{\lambda}\sin\theta = k_{\parallel} + m\frac{2\pi}{\Lambda}, \tag{14.2.1}$$

where Λ is the grating period, m is an integer, and $k_{\parallel}$ is the wavevector of the surface wave. Thus, by modifying the characteristics of the surface profile, it is possible to modify the direction and the value of the emissivity of the surface at a given wavelength.

Similar with the optical beaming effect of bull's eye discovered in 2002 [11], thermal beaming effect has been demonstrated recently [12]. Tungsten (W) and molybdenum (Mo) that can withstand high temperatures were patterned to support

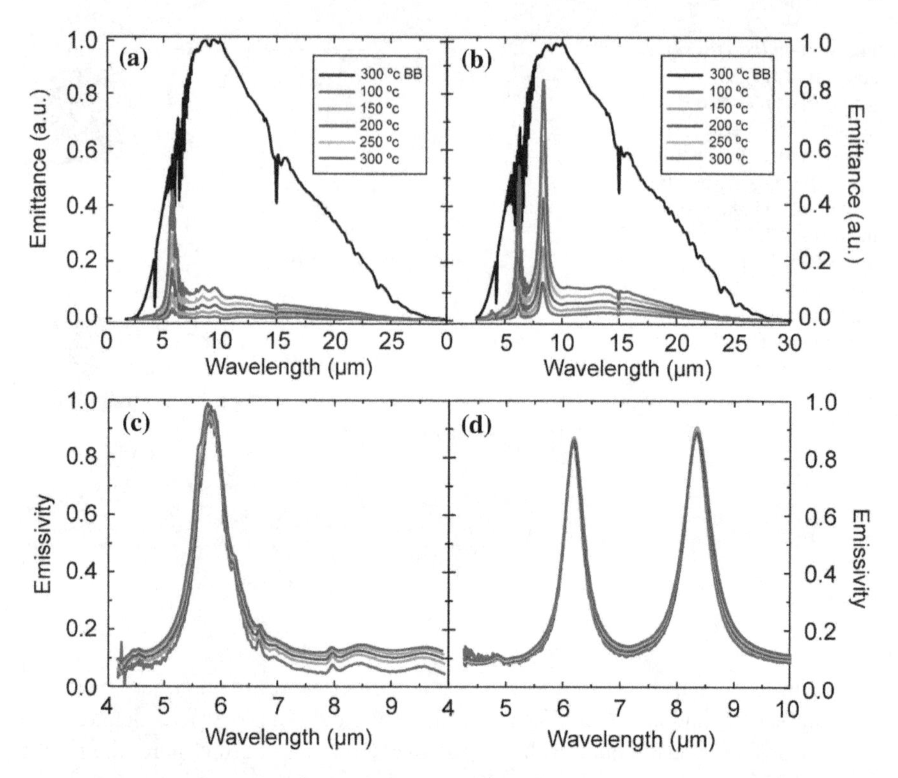

Fig. 14.5 Experimental emittance and normalized emissivity. **a** Emittance of the single-band metamaterial emitter at five different temperatures and emittance of the blackbody reference at 300 °C. **b** Emittance of the dual-band metamaterial emitter at five different temperatures and emittance of the blackbody reference at 300 °C. **c** Normalized emissivity of single-band metamaterial emitter. **d** Normalized emissivity of dual-band metamaterial emitter. Reproduced from [9] with permission. Copyright 2011, American Physical Society

surface plasmon polaritons (SPPs) at the target wavelength. The fabricated bull's eye contains 300 circular concentric grooves with a period of 3.52 μm and the diameter of the entire structure is ~2.1 mm, as shown in Fig. 14.7 [12].

Figure 14.8 shows the thermal emission spectra of the tungsten bull's eye at 900 °C. The bull's eye exhibited a sharp emission peak in the normal direction with a peak wavelength at 3.532 μm and full width at half maximum (FWHM) of 30 nm (Fig. 14.8a) [12]. The FWHM angular divergence ($\Delta\theta$) of the emission peak for the measured and calculated emission spectra were estimated as 1.4° and 1.2°, exhibiting a good directionality. Figure 14.8b shows the emission spectra of the bull's eye initially and heated at 900 °C for 12 h, from which we can see the emission spectrum maintains well after heated. In contrast, the emissivity of a flat tungsten is quite smaller than the bull's eye [12].

Besides the patterned structures, a novel directive thermal emission mechanism has been proposed based on a reflective homogenous graphene, as shown in Fig. 14.9

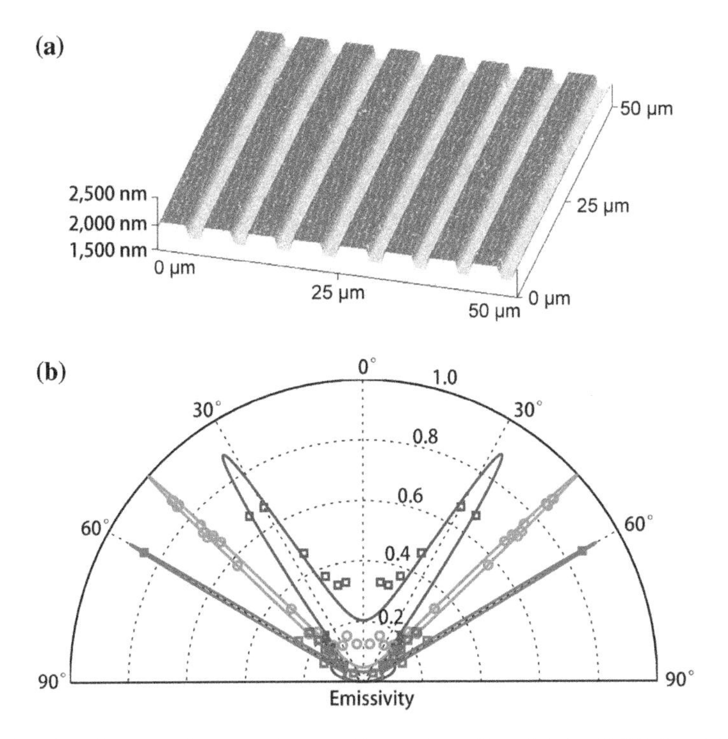

Fig. 14.6 **a** Atomic force microscope image of the grating with period $d = 0.55\lambda$ ($\lambda = 11.36\ \mu$m) and depth $h = \lambda/40$. **b** Emissivity of a SiC grating in p-polarization. Blue, $\lambda = 11.04\ \mu$m; green, $\lambda = 11.36\ \mu$m; red, $\lambda = 11.86\ \mu$m. The emissivity is deduced from measurements of the specular reflectivity using Kirchhoff's law. Reproduced from [10] with permission. Copyright 2002, Macmillan Magazines Ltd.

[13]. The surface (2D) conductivity of graphene is highly dependent on the working frequency and chemical potential (or Fermi energy). When there is no external magnetic field present, the local conductivity is isotropic; i.e., there is no Hall conductivity. In this case, the conductivity can be approximated for $k_B T < \mu_c, \hbar\omega$ as follows:

$$\sigma_{2\mathrm{D}}(\omega) \approx \frac{ie^2}{4\pi\hbar}\ln\left[\frac{2|\mu_\mathrm{c}| - (\omega + i2\Gamma)\hbar}{2|\mu_\mathrm{c}| + (\omega + i2\Gamma)\hbar}\right] + \frac{ie^2 k_\mathrm{B}T}{\pi\hbar^2(\omega + i2\Gamma)}\left[\frac{\mu_\mathrm{c}}{k_\mathrm{B}T} + 2\ln\left(e^{-\mu_\mathrm{c}/k_\mathrm{B}T} + 1\right)\right], \quad (14.2.2)$$

where $k_\mathrm{B}T$ is the thermal energy; μ_c is the chemical potential and Γ is the scattering rate; e, k_B, and $\hbar$ are electron charge, Boltzmann constant, and reduced Planck constant (Dirac constant), respectively [13]. The first term in Eq. (14.2.2) is due to the contribution of interband transition and the second term results from intraband transition. The frequency of transition between these two regimes is dependent on

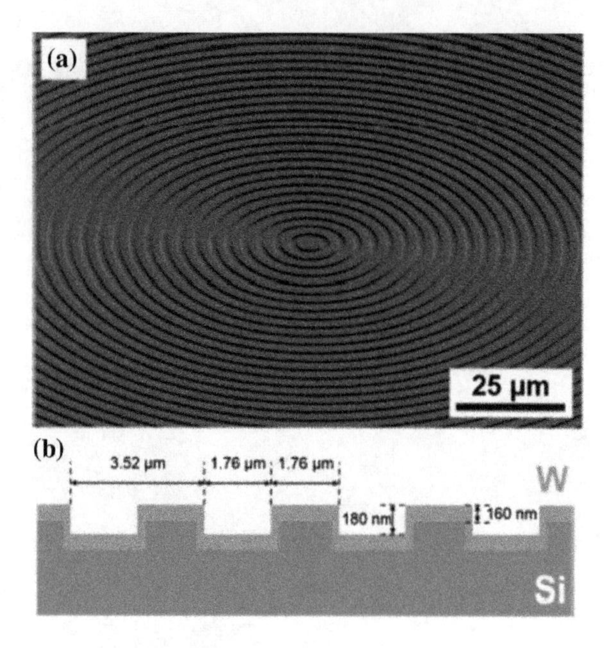

Fig. 14.7 **a** SEM of a tungsten bull's eye. **b** The structural parameters of the bull's eye. Reproduced from [12] with permission. Copyright 2016, American Chemical Society

the chemical potential. The scattering rate is assumed to be $\Gamma = 0.43$ meV. The sheet impedance of graphene can be calculated as [13]:

$$Z_s(\omega) = \frac{1}{\sigma_{2D}(\omega)}, \tag{14.2.3}$$

which can also be transformed into traditional form by writing the 3D conductivity as $\sigma_{3D}(\omega) = \sigma_{2D}(\omega)/t$, where $t \approx 0.5$ nm is the thickness of graphene.

Based on the transfer-matrix method (TMM) and boundary conditions, one can deduce that the following sheet impedance should be satisfied to realize perfect absorption [13]:

$$Y_s = \frac{Z_0}{Z_s} = \cos\theta - i\sqrt{\varepsilon - \sin^2\theta}\cot\left(\sqrt{\varepsilon - \sin^2\theta}kd\right) \tag{14.2.4}$$

for transverse electric (TE) polarization and

$$\frac{Z_0}{Z_s} = \frac{1}{\cos\theta} - i\frac{\varepsilon}{\sqrt{\varepsilon - \sin^2\theta}}\cot\left(\sqrt{\varepsilon - \sin^2\theta}kd\right) \tag{14.2.5}$$

for transverse magnetic (TM) polarization.

In order to realize large angle absorption and thus radiation, the chemical potential is chosen as 200 meV and the sheet impedance at 1 THz becomes 55 + 266*i*. As illustrated in Fig. 14.10 [13], periodically located absorption peaks are observed for TE polarization at $\theta \approx 83°$. For TM polarization, the absorption mainly located at

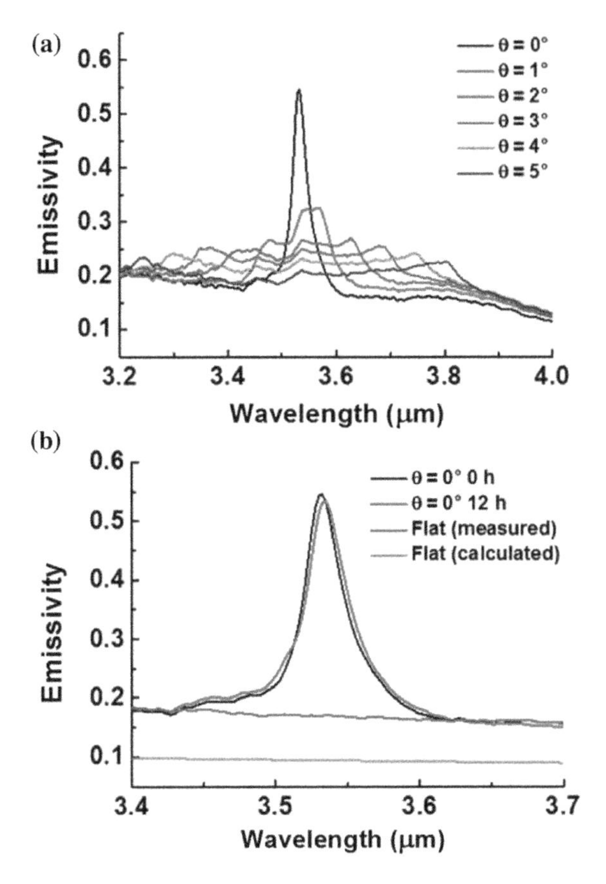

Fig. 14.8 Thermal emission spectra of a tungsten bull's eye at 900 °C. **a** Measured thermal emission spectra at various tilt angles (θ) from normal. The collection angle of the emission spectrum was $\pm 0.1°$ and a carbon pellet was used to calibrate the emissivity. **b** Comparison of the emission spectra initially and after 12 h at 900 °C. The measured and calculated emission spectra for a flat *W* film at 900 °C are also shown. Reproduced from [12] with permission. Copyright 2016, American Chemical Society

small angles of incidence. In general, when the sheet impedance is much larger than the vacuum impedance, the angle selectivity would be much better.

In fact, the radiation angle can be changed by altering the layer number of multilayer graphene. Since the optical conductivity is in proportional with the layer number, the conductivity of bilayer and tri-layer graphene at optical frequency would be twice and triple of $e^2/4\hbar$. In Fig. 14.11 [13], the radiation patterns at $f = 155$ THz ($d = 0.46$ μm) for 1, 5, and 10 layers of graphene are calculated by using TMM for different radiation angles. As the increase of the layer number, the radiation angle would decrease. Meanwhile, the radiation at small angle (such as normal direction) increases and the beam width would be larger. Thus, for practical applications, a trade-off between radiation angle and beam width is required.

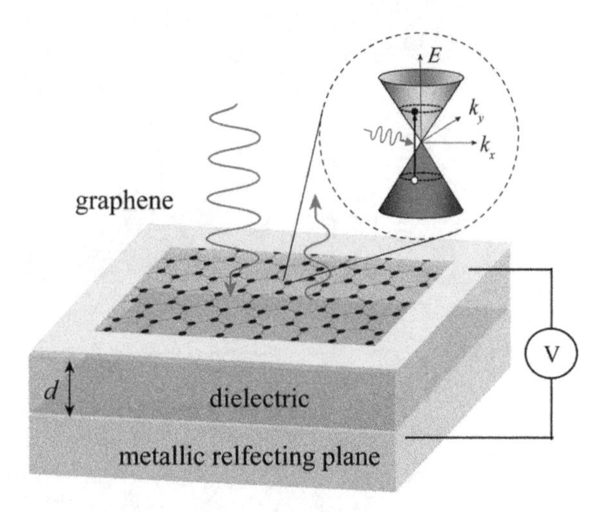

Fig. 14.9 Schematic of the biased graphene. The bottom gold layer is used as both the back-gate and reflecting plane to prevent transmission. Inset depicts the optical transitions between hole and electron bands in monolayer graphene. Reproduced from [13] with permission. Copyright 2013, Optical Society of America

14.2.2 Near-Field Thermal Radiations

In the previous section, we only consider the radiation distance far larger than the characteristic wavelength of thermal radiation (λ_T). As the distance decreases and becomes comparable with or even shorter than λ_T, near-field effects become important. Traditional Planck's law and Stefan-Boltzmann's law can no longer be used to obtain the energy transfer between two bodies due to the photon tunneling effects [14, 15].

The concept of photon tunneling and evanescent waves is illustrated in Fig. 14.12a where two media (1 and 2) are separated by a vacuum gap [14]. Note that the coupled evanescent waves would follow a catenary function, which is very important for the photon tunneling [16]. However, owing to its specialty, we would like to discuss this phenomenon elsewhere. Figure 14.12b shows the predicted heat transfer between two SiC plates. The plates are maintained at 300 and 0 K, respectively, and the net heat flux is calculated at different vacuum gaps. For comparison, the heat transfer between two blackbodies maintained at the same temperature as the SiC plates is also plotted [14]. It can be seen that the near-field heat transfer between the two SiC plates at 1 nm apart is around five orders of magnitude greater than that between the two blackbodies.

One limitation of the near-field thermal emission using PhC or surface electromagnetic excitations is that the energy transfer beyond the black body limit (super-Planckian thermal emission) only occurs in a narrow bandwidth. In order to overcome the bandwidth limitation, hyperbolic metamaterials (HMM) formed by stacks of SiC and SiO_2 layers were utilized [17] ($\varepsilon_{\|} = \varepsilon_{xx} + i\varepsilon''$, $\varepsilon_{\perp} = \varepsilon_{zz} + i\varepsilon''$). The emitted energy density of a HMM in equilibrium at temperature T, at a distance z restricted to the near-field and in the limit of low losses is given by

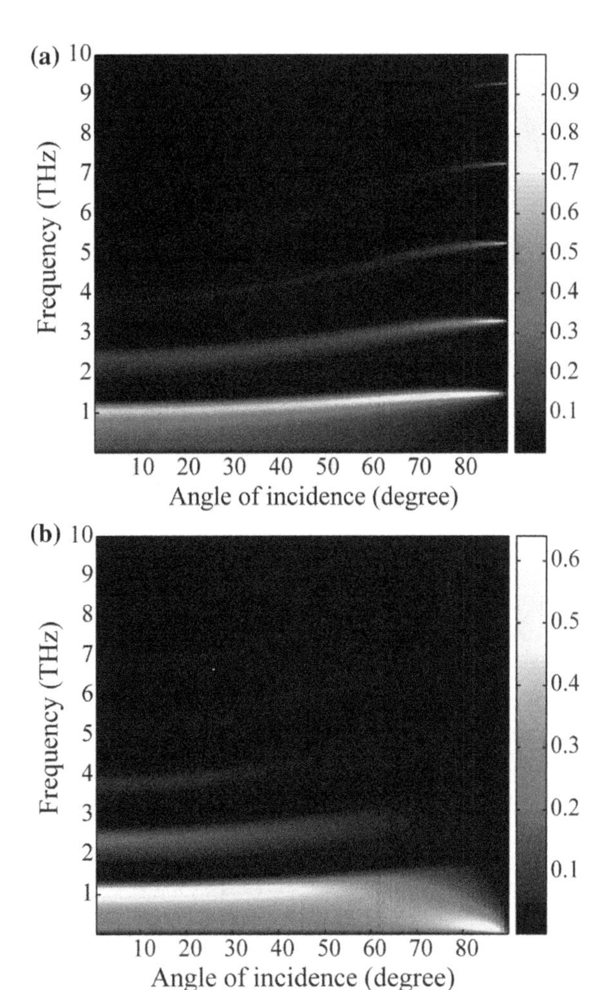

Fig. 14.10 **a** TE and **b** TM absorbance for different angles and frequencies. The thickness of dielectric layer is kept as 70 μm. Reproduced from [13] with permission. Copyright 2013, Optical Society of America

$$u(z,\omega,T)^{z\ll\lambda_0} \approx \frac{U_{\mathrm{BB}}(\omega,T)}{8}\left[\frac{2\sqrt{|\varepsilon_{xx}\varepsilon_{zz}|}}{(k_0z)^3(1+|\varepsilon_{xx}\varepsilon_{zz}|)} - \varepsilon''\frac{2(\varepsilon_{xx}+\varepsilon_{zz})}{(k_0z)^3(1+|\varepsilon_{xx}\varepsilon_{zz}|)^2}\right], \tag{14.2.6}$$

where $U_{\mathrm{BB}}(\omega, T)$ is the blackbody emission spectrum at the temperature T, and ε'' is the imaginary part of the permittivity, $\varepsilon_{xx}\varepsilon_{zz} < 0$. Figure 14.13 shows the analytical results and exact numerical solution of emitted energy density for metamaterials with fill fraction of 0.25 and 0.5. For the fill fraction of 0.25, two regions of broadband super-Planckian thermal emission are due to type I and type II HMM (Fig. 14.13a). When the fill fraction is 0.5, these two frequency ranges are merged into an ultra-broadband (Fig. 14.13b) [17].

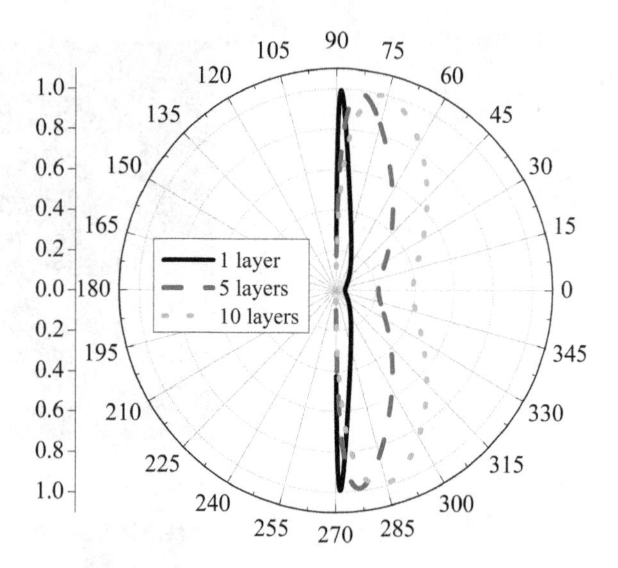

Fig. 14.11 Polar plot of the radiation pattern of graphene with 1, 5, and 10 layers. The sample is placed vertically (in the direction along 90°–270°). Reproduced from [13] with permission. Copyright 2013, Optical Society of America

Besides bulk HMMs, by patterning near-field coupled thin films into planar metasurfaces, the radiative heat flux can be further improved by more than one order of magnitude in certain range of thickness. Figure 14.14 gives the radiative heat flux for both configurations with varying volume filling ratios f, which is defined as W/P and W^2/P^2 for 1D and 2D metasurfaces, respectively [18]. Obviously, patterned thin metasurfaces can increase the heat flux over thin films and bulks. Especially, patterning the film into 1D metasurface can enhance thermal radiation for all practical volume filling ratios. While the 2D metasurface yields a radiative heat flux higher than that of thin films at moderate filling ratios, it does not support a heat flux as high as that of the 1D metasurface. Theoretical investigations show that the underlying mechanism for the enhancement can be attributed to the coupling of SPPs inside the thin film [18].

14.2.3 Thermophotovoltaics

Thermophotovoltaic is an important application of thermal radiation. Generally, solar thermophotovoltaics (STPVs) mainly includes several processes: Optically concentrated sunlight is converted into heat in the absorber, the absorber temperature rises and heat conducts to the emitter. Then the hot emitter thermally radiates toward the photovoltaic cell, where radiation is ultimately harnessed to excite charge carriers and generate power. According to the processes above, the overall efficiency (η_{stpv}) can be expressed as a product of the optical efficiency of concentrating sunlight (η_{o}), the thermal efficiency of converting and delivering sunlight as heat to the emitter (η_{t}),

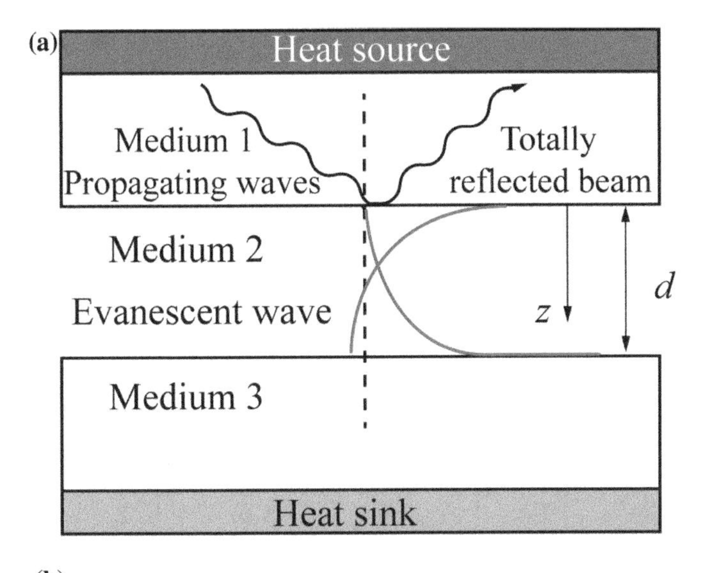

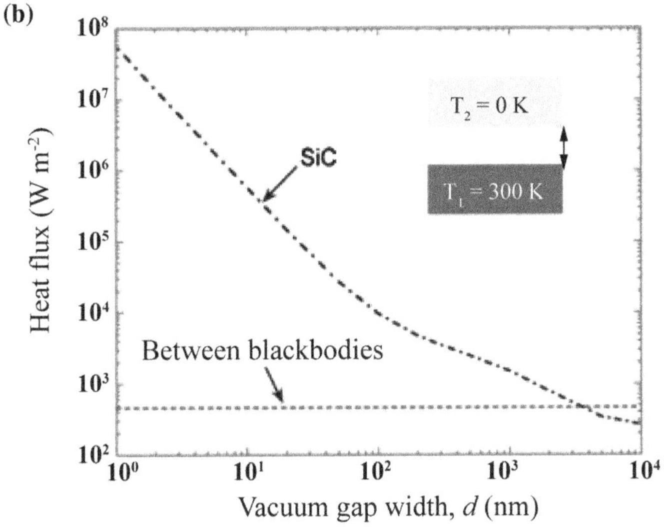

Fig. 14.12 **a** Schematic illustrating the concept of evanescent waves and photon tunneling. **b** Radiation heat transfer between two SiC plates maintained at 300 and 0 K. The energy transfer between two blackbodies has also been shown for reference. Reproduced from [14] with permission. Copyright 2009, John Wiley & Sons, Ltd.

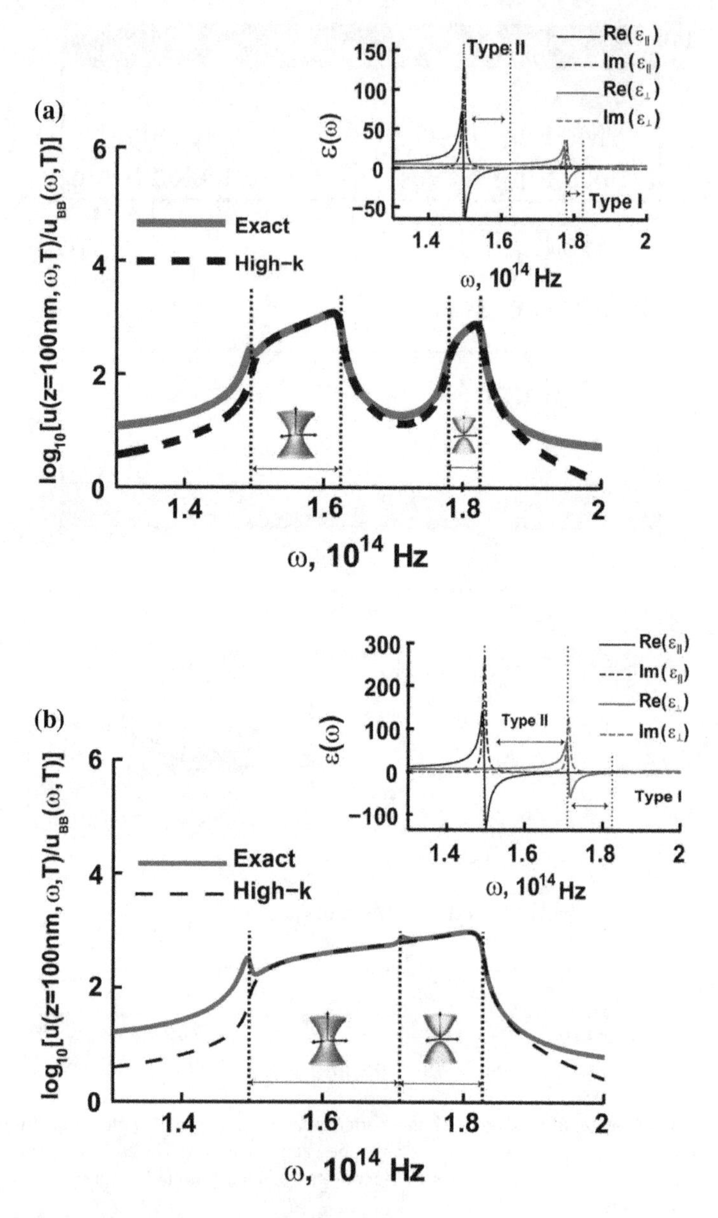

Fig. 14.13 Comparison of the analytical result in the near-field with an exact numerical solution for an effective medium HMM with metallic fill fractions of **a** 0.25 and **b** 0.5. Reproduced from [17] with permission. Copyright 2012, American Institute of Physics

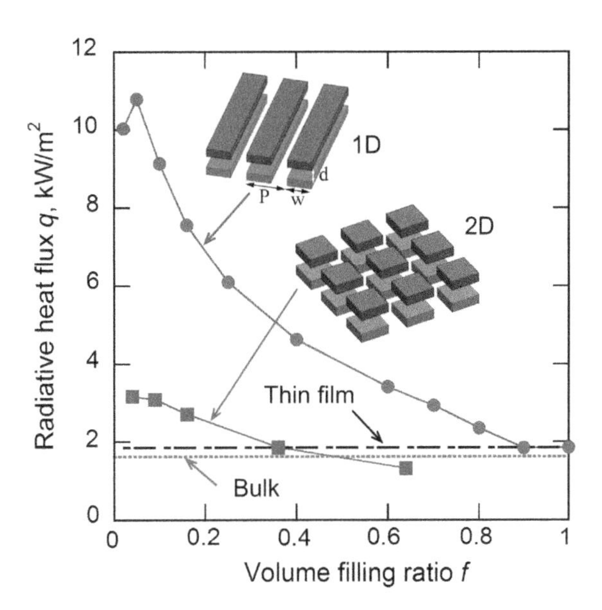

Fig. 14.14 Schematic of enhanced near-field radiative heat flux based on 1D and 2D metasurfaces. Reproduced from [18] with permission. Copyright 2015, American Chemical Society

and the efficiency of generating electrical power from the thermal emission (η_{tpv}) [19, 20]:

$$\eta_{\text{stpv}} = \eta_{\text{o}}\eta_{\text{t}}\eta_{\text{tpv}}. \tag{14.2.7}$$

In order to improve the whole conversion efficiency, Lenert et al. [19] proposed a device integrating a multiwalled carbon nanotube absorber and a 1D Si/SiO_2 photonic-crystal emitter on the same substrate, as shown in Fig. 14.15. The absorber, emitter, and the photovoltaic cells are integrated in a vacuum environment to minimize parasite heat loss due to convection and conduction.

The emitter was made of 1D Si/SiO_2 PhC, as shown in Fig. 14.16a. These materials were chosen for ease of fabrication and high-temperature compatibility with the silicon substrate. The spectrally selective emitter has high emittance for energies above the photovoltaic bandgap (E_g) and low emittance for energies below the bandgap, as indicated in Fig. 14.16b.

By using an improved InGaAsSb cell ($E_g = 0.55$ eV) and a sub-bandgap photon reflecting filter [21], the STPV efficiency could approach 20% at moderate optical concentrations (Fig. 14.17). In the solar PV process, high-energy electrons generate heat within the cell as they decay down to $E_{\text{conduction}}$. The STPV process generates an equivalent amount of free electrons, but they are localized to $E_{\text{conduction}}$, markedly reducing heat generation. A 1D PhC comprised of several Si/SiO_2 layers is used as the selective thermal emitter. Both constructive and destructive wave interferences provide a steep cutoff in the spectral emittance at the bandgap of the InGaAsSb PV cell. A rugate filter is utilized to suppress the high intrinsic emission of the underlying Si at lower energies [21].

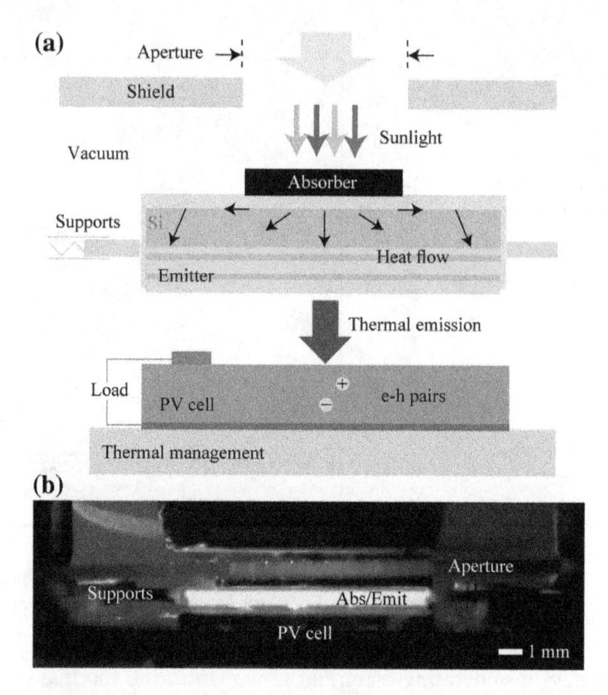

Fig. 14.15 Operating principle and components of the nanophotonic area ratio optimized STPV. Sunlight is converted into useful thermal emission and, ultimately, electrical power, via a hot absorber–emitter. Schematic (a) and optical image (b) of the vacuum-enclosed devices. Reproduced from [19] with permission. Copyright 2014, Macmillan Publishers Limited

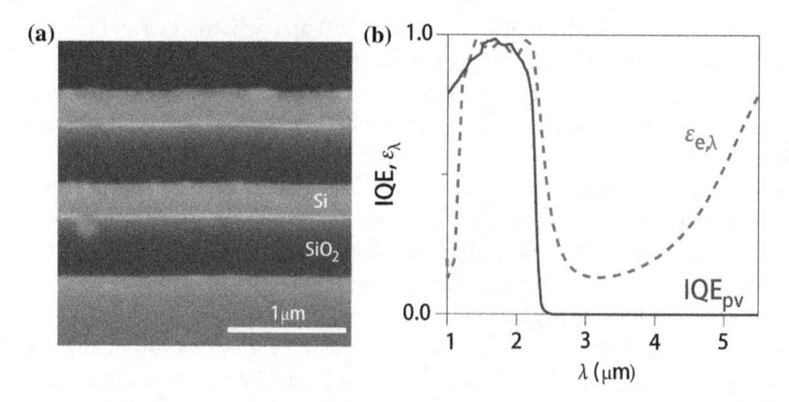

Fig. 14.16 **a** SEM cross section of the 1D PhC showing the alternating layers of silicon and SiO_2. **b** Internal quantum efficiency (IQE) spectrum of the InGaAsSb photovoltaic cell and the emissivity spectrum of the emitter. Reproduced from [19] with permission. Copyright 2014, Macmillan Publishers Limited

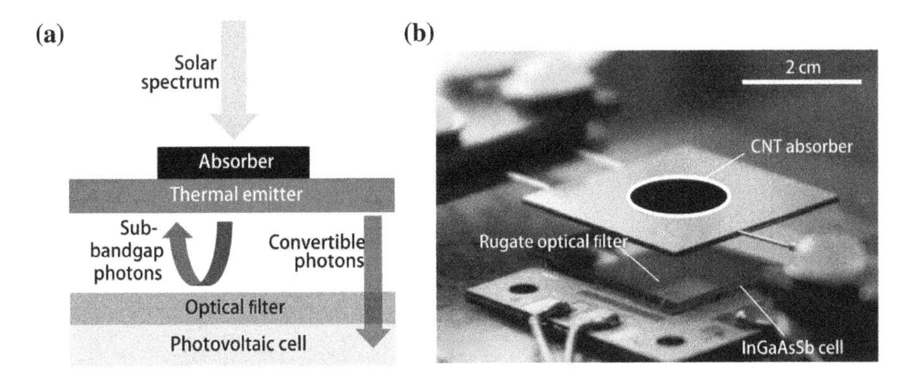

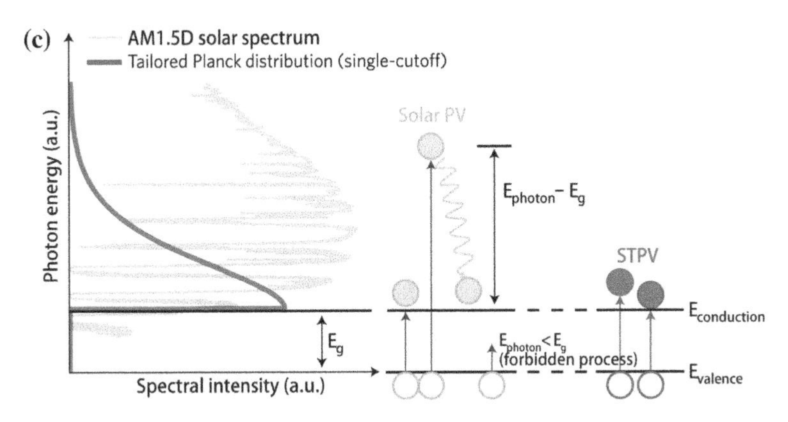

Fig. 14.17 **a** Schematic representation of a solar thermophotovoltaic device. **b** Optical image of the solar thermophotovoltaic device. **c** Energy conversion mechanisms in the cell comparing illumination by engineered thermal radiation (STPV) with direct solar (PV). The schematic on the right depicts electrons (filled circles) being excited by incident photons from the valence to the conduction band of the semiconductor diode. Reproduced with permission from [21] with permission. Copyright 2016, Macmillan Publishers Limited

Figure 14.18 shows the measured heat generation in the PV cell for the experiments. At 0.35 W cm^{-2} of electrical power density, the solar PV generated ~2× more heat in the cell than the STPV despite having the same conversion efficiency. Excessive heat loads must be dissipated with higher heat transfer coefficients to prevent the increase in cell temperature and the reduction in electrical performance [21].

14.2.4 Daytime Radiative Cooling

Radiative cooling is a process that lowers the temperature of the object via thermal radiation. Two necessary conditions have made it possible to realize radiative cooling

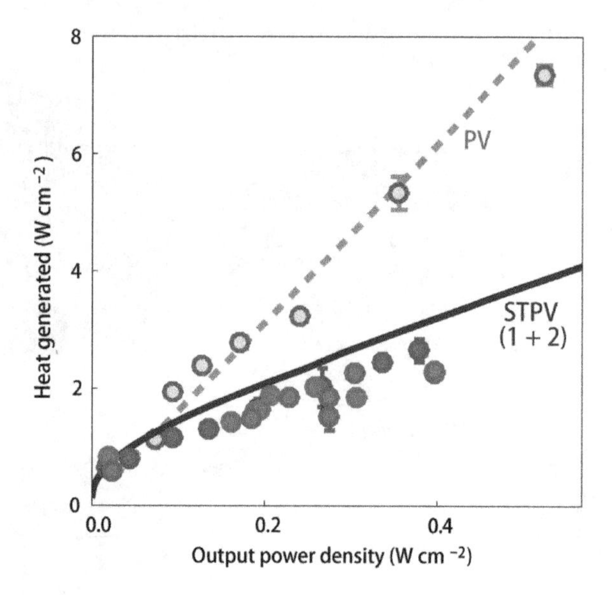

Fig. 14.18 Heat dissipated by the cooling loop in both the PV and STPV experiments to maintain the PV converter at an equilibrium temperature. Reducing the illumination of the PV cell with unusable photons improves efficiency and markedly reduces heat generation. Reproduced with permission from [21] with permission. Copyright 2016, Macmillan Publishers Limited

[22]. The first condition is that one can exploit the coldness of the universe at a temperature of 3 K as the ultimate heat sink both in terms of its temperature and its capacity. The second one is that there exists an atmosphere transparent window in the wavelength of 8–13 μm, where any object exposed to the sky can radiate heat out to the universe and thus lower its temperature.

Consider a radiative cooler at temperature T, whose spectral and angular emissivity is $\varepsilon(\lambda, \theta)$. When the radiative cooler is exposed to a daylight sky, it is subject to both solar irradiance and atmospheric thermal radiation (corresponding to ambient air temperature T_{amb}). The net cooling power P_{cool} of such a radiative cooler is given by [22]:

$$P_{cool}(T) = P_{rad}(T) - P_{atm}(T_{amb}) - P_{sun} - P_{cond+conv} \tag{14.2.8}$$

In order to realize a net positive power outflow, several constraints should be satisfied. First, it must reflect sunlight strongly to minimize P_{sun}. Therefore, it must be strongly reflecting over visible and near-infrared wavelength ranges. Second, it must strongly emit thermal radiation P_{rad} while minimizing incident atmospheric thermal radiation P_{atm}. Thus, the device must emit selectively and strongly only between 8 and 13 μm, where the atmosphere is transparent, and reflect all other wavelengths.

Raman et al. [22] have designed and fabricated a multilayer photonic structure for daytime radiation cooling, which consists of seven dielectric layers deposited on top of a silver mirror, as shown in Fig. 14.19. These layers are designed using a systematic optimization process taking into account fabrication constraints. The top three layers, with thicknesses on the order of hundreds of nanometers, are responsible

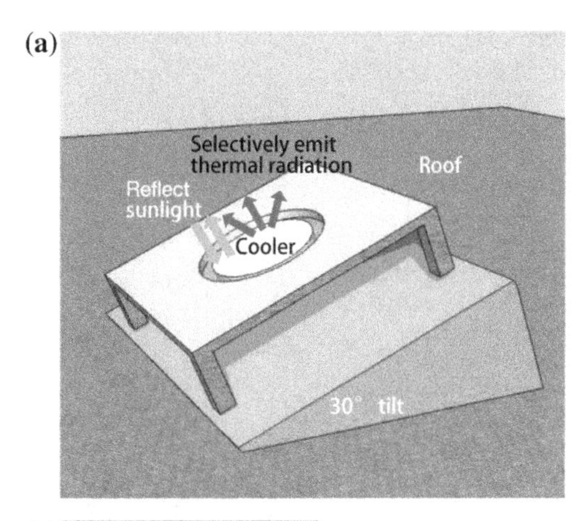

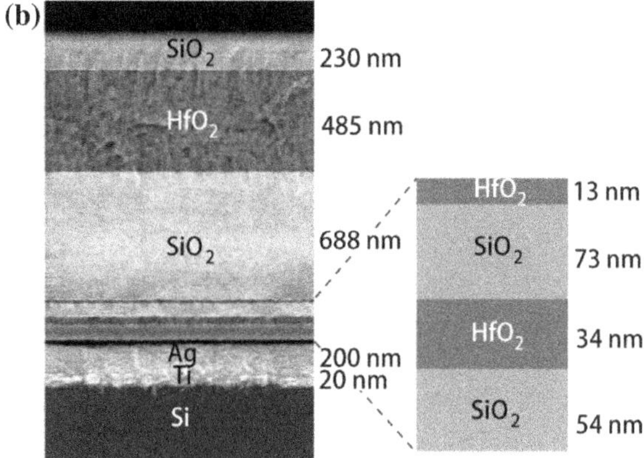

Fig. 14.19 **a** 3D schematic of the apparatus and radiative cooler. The apparatus is designed to minimize conductive and convective heat exchange to the cooler. **b** SEM image of the photonic radiative cooler. It consists of seven layers of HfO_2 and SiO_2. Reproduced from [22] with permission. Copyright 2014, Macmillan Publishers Limited

for generating strong thermal radiation. The bottom four layers are used to enhance the reflectivity of silver mirror especially in the ultraviolet wavelength range [22].

The experimentally measured absorptivity/emissivity spectrum of photonic radiative cooler is shown in Fig. 14.20 [22]. The cooler shows minimal absorption from 300 nm to 2.5 μm, where about 97% incident solar power at near-normal incidence is reflected. In addition, the cooler has strong and remarkably selective emissivity in the atmospheric window between 8 and 13 μm. The above spectral behavior promises a net positive power outflow [22].

The experimental results shown in Fig. 14.21 indicate that such design was able to reach a temperature 5 °C below the ambient air temperature, in spite of having about 900 W/m^2 of sunlight directly impinging upon it [22]. These results demonstrate that a tailored, photonic approach can fundamentally enable new technological possibilities for energy efficiency. Further, the cold darkness of the universe can be used as a renewable thermodynamic resource, even during the hottest hours of the day [22].

In order to meet the large area requirements of radiative cooling, scalable manufactured randomized glass polymer hybrid metamaterial has been proposed recently

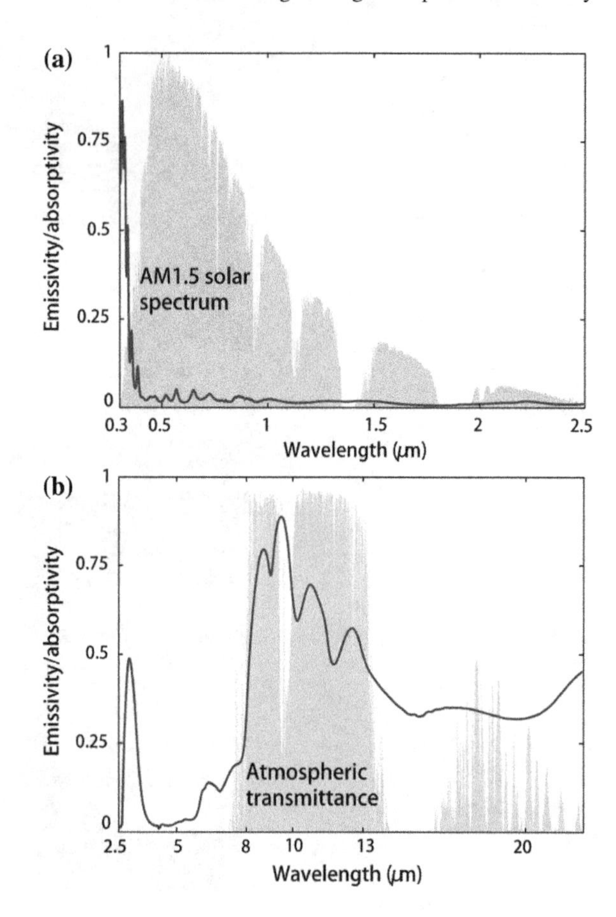

Fig. 14.20 **a** Measured emissivity/absorptivity of the photonic radiative cooler over optical and near-infrared wavelengths using an unpolarized light source, with the AM1.5 solar spectrum plotted for reference. The cooler reflects 97% of incident solar radiation. **b** Measured emissivity/absorptivity of incidence over mid-infrared wavelengths using an unpolarized light source, with a realistic atmospheric transmittance model. The photonic cooler achieves strong selective emission within the atmospheric window. Reproduced from [22] with permission. Copyright 2014, Macmillan Publishers Limited

[23], which is composed of a visibly transparent polymer encapsulating randomly distributed silicon dioxide (SiO_2) microspheres. The spectroscopic response can span two orders of magnitude in wavelength (0.3–25 μm).

14.2.5 Thermal Cloak and Camouflage

A compact planar device can serve as an "invisible" coat when its radiation is in the form of a step function, i.e., near zero in the atmosphere windows (3–5 and 8–12 μm) but near unity in the 5–8 μm. According to Kirchhoff's law, the above prerequisite means the device should possess an ideal band-pass absorption spectrum in the 5–8 μm. In order to implement the above functionality, a frequency filter located on a perfect absorber is adopted [24]. The frequency filter is composed of alternating layers of calcium fluoride (CaF_2) and germanium (Ge) of varying thickness, while the perfect absorber is constructed by a nichrome–Ge–nichrome triple layer structure,

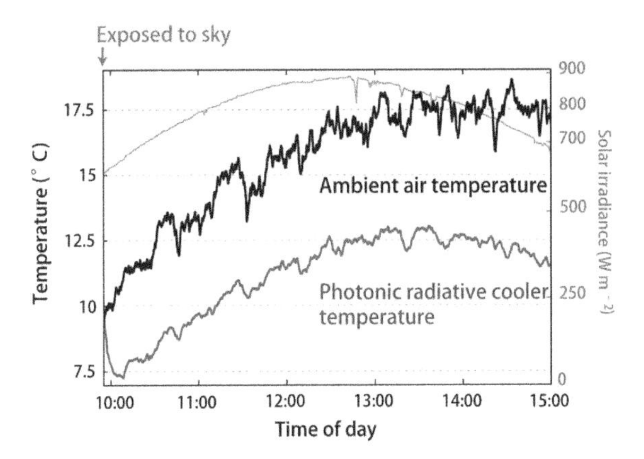

Fig. 14.21 Rooftop measurement of the photonic radiative cooler's performance (blue) against ambient air temperature (black). The photonic radiative cooler immediately drops below ambient once exposed to the sky and achieves a steady-state temperature of 4.9 ± 0.15 °C below ambient for over one hour where the solar irradiance incident on it (green) ranges from 800 to 870 W m^{-2}. Reproduced from [22] with permission. Copyright 2014, Macmillan Publishers Limited

as indicated in Fig. 14.22. Since there is no wave can transmit through the whole structure, the absorption (A) of the device can be expressed as $A = 1 - R$, where R is the reflectance [24].

In order to obtain an ideal band-pass absorption in 5–8 μm, robust genetic algorithm (GA) optimizer coupled with TMM is utilized to identify the thickness of each layer. The cost function is given by:

$$\text{optimize}(\lambda) = -\sum R(\lambda_1) + \sum R(\lambda_2) - \sum R(\lambda_3) \tag{14.2.12}$$

where R is the reflectance of the structure at a certain wavelength retrieved from the transfer matrix, and $\Sigma R(\lambda_i)$ ($i = 1, 2, 3$) is the sum of reflectance in three wavelength region: 3–5, 5–8 and 8–12 μm. For a 10 layers structure, the optimized thicknesses from top to bottom are 0.63 μm, 0.768 μm, 1.646 μm, 0.161 μm, 0.394 μm, 0.288 μm, 0.753 μm, 14 nm (metal film), 0.4 μm, 0.3 μm, respectively [24]. The simulated absorption or emissivity is shown in Fig. 14.22b, from which one can see that the emissivity is larger than 90% outside the atmosphere window (3–5 and 8–12 μm), while smaller than 5% in the atmosphere window. Therefore, this design can behave as a thermal "invisible" coat. As a comparison, the emissivity of isolate triple layer absorber is shown in Fig. 14.22c. Comparing Fig. 14.22b with Fig. 14.22c, one can find that with the help of the multilayer filtering structure, the emissivity property is more approaching an ideal step function [24].

For many applications, emission at large angle of emergent is needed for both TE and TM waves. Figure 14.23 shows the wavelength dependence of the emissivity for oblique emergence of TM and TE polarized waves with different color lines, respectively [24]. As the angle of emergence increases, the emissivity for both TM

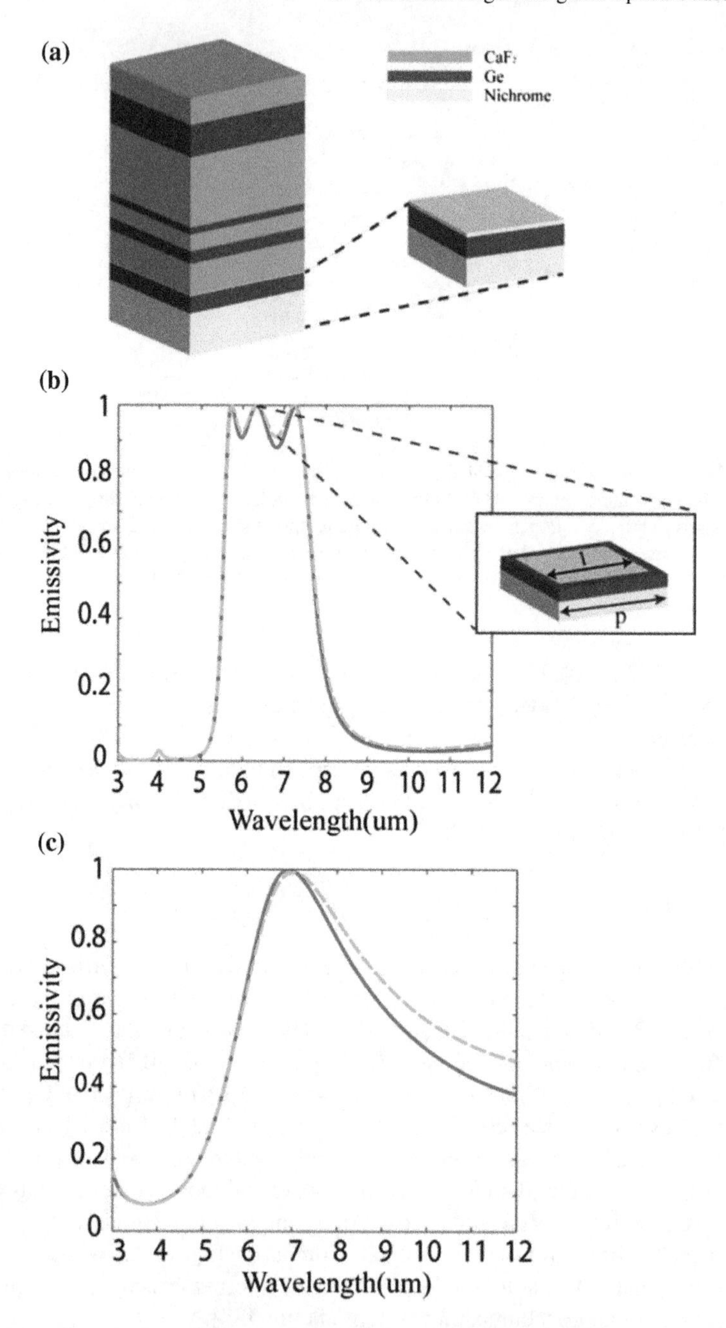

Fig. 14.22 **a** Schematic illustration of a planar thermal cloak, which can be taken as a combination of absorber (Nichrome–Ge–Nichrome triple layer) and band-pass frequency filter (alternating layers of CaF_2 and Ge). **b** Emissivity of the multilayer photonic structure. The red solid line is the emissivity of the structure depicted in (a) and the green dashed line is the emissivity when replacing the structure in (b) with $l = 0.7\ \mu m$, $p = 0.8\ \mu m$. **c** Emissivity of the isolate Nichrome-Ge-Nichrome triple layer. Reproduced with permission from [24] with permission. Copyright 2018, Published by Elsevier B.V.

and TE waves decrease gradually. The difference is that the high-emission band of TM waves move to shorter wavelengths with the bandwidth nearly unchanged, while the high-emission band of TE waves are narrowed but maintaining the high performance. It can be found that the emission band will not deviate from 5 to 8 μm if the emission angle is less than 40°, which indicates that this design has the capability to work at a wide range of emission angles [24].

Besides thermal cloak, adaptive thermal camouflage has aroused intense interests [25]. Figure 14.24a shows a schematic of the active thermal surface consisting of a multilayer-graphene electrode on a porous polyethylene (PE) membrane and a back gold-electrode. Figure 14.24b illustrates the working principle of the active thermal surface [25]. Under a voltage bias, the ionic liquid intercalates into the graphene layers and dopes them. As a result of doping, the charge density on graphene increases and Fermi-level shifts to higher energies, which suppress the IR absorption and thus the emissivity of the graphene electrode decreases. Figure 14.24c and d shows the thermal camera images of the fabricated device at 0 and 3 V, respectively. At 0 V, the temperature profile of the background can be seen through the device. However, at 3 V the emissivity of the device is significantly suppressed, which screens the background temperature profile [25].

In general, thermal camouflage requires low thermal emission, thus high infrared reflection, which may be harmful for other applications such as LIDAR invisibility. Based on subwavelength structures, virtual shaping technology has been proposed to overcome this problem [6].

14.3 Light-Emitting Diodes

After decades of research and development, organic light-emitting diodes (OLEDs) [26–28] and semiconductor quantum dots (QDs) [29, 30] have emerged as the leading technologies in the consumer flat-panel displays industry. Electroluminescent OLEDs offer superior contrast, energy savings, and wider viewing angles. However, the synthesis of OLED emitters is typically complex and requires the incorporation of heavy metal complexes (Ir, Pt), which coupled with the expensive fabrication process, contributes to high manufacturing costs. Quantum dots, on the other hand, possess narrow and tunable emission, which greatly improves color reproduction in both electroluminescent and photoluminescent devices [31]. Besides QLEDs, lead halide perovskites have been proposed as an alternative semiconductor for LED and aroused great interests.

14.3.1 Quantum Dot-Based LEDs

Figure 14.25a, b shows a schematic of QLED [29]. Thin films of colloidal ZnO nanocrystals are employed as electron-transport interlayers (ETLs) because of their

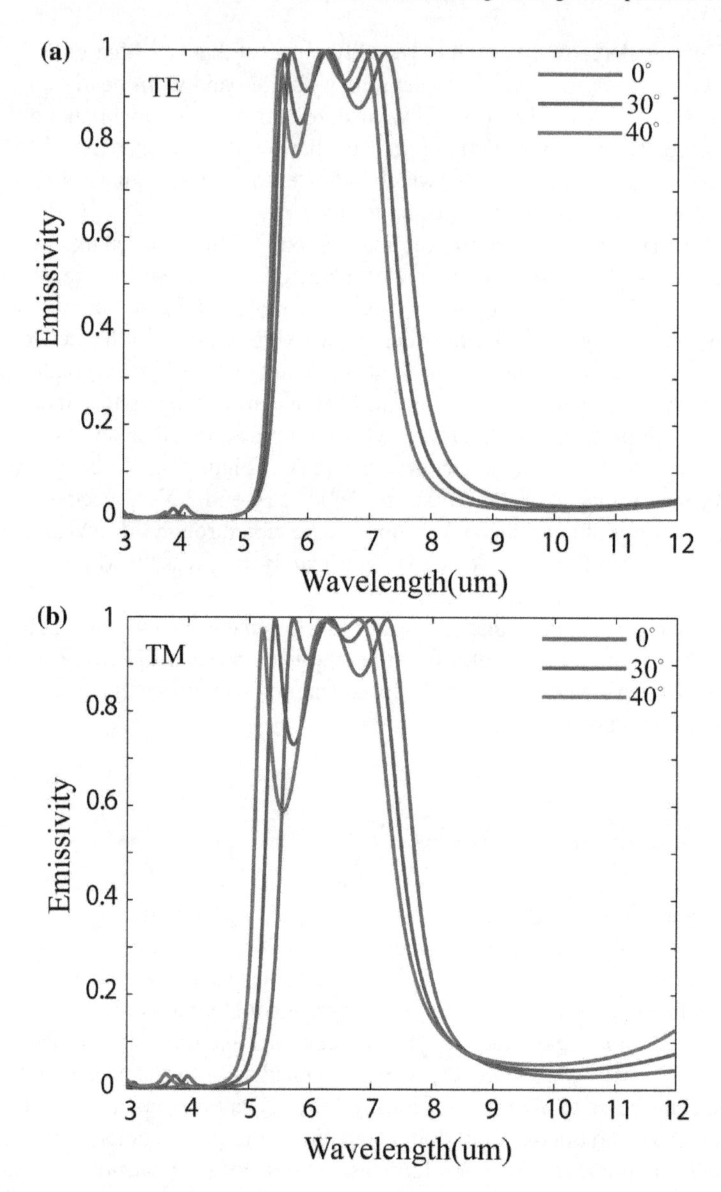

Fig. 14.23 Wavelength dependence of emissivity for oblique emergence of TE (a) and TM (b) polarized waves, respectively. Red, blue, and green solid lines refer to emergence with 0°, 30° and, 40°, respectively. Reproduced with permission from [24] with permission. Copyright 2018, Published by Elsevier B.V.

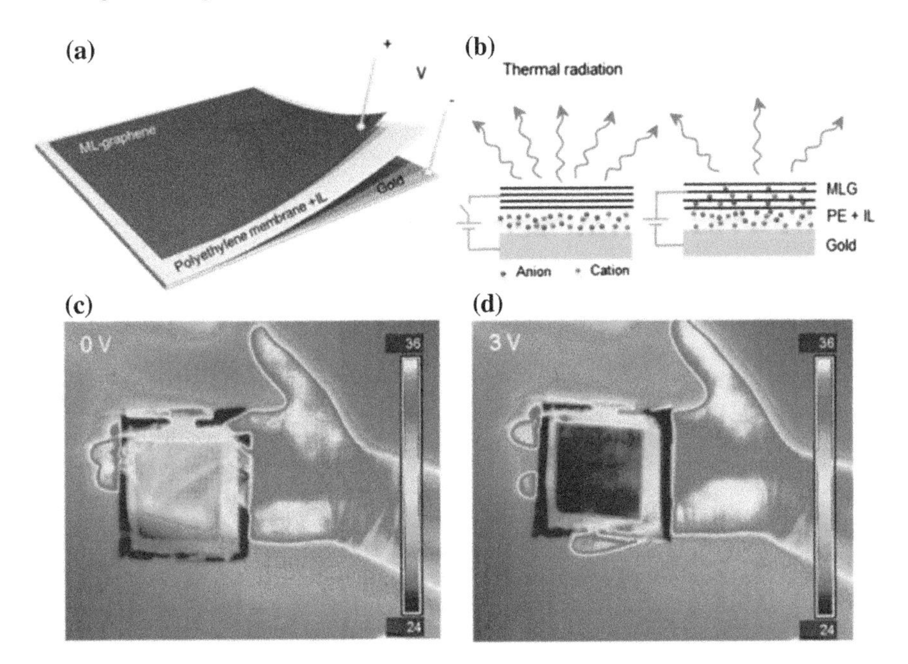

Fig. 14.24 **a** Schematic drawing of the active thermal surface consisting of a multilayer-graphene electrode, a porous polyethylene membrane, and a back gold-electrode coated on heat resistive nylon. **b** Schematic representation of the working principle of the active thermal surface. The emissivity of the surface is suppressed by intercalation of anions into the graphene layers. **c** and **d** Thermal camera images of the device placed on the author's hand under the voltage bias of 0 and 3 V, respectively. Reproduced from [25] with permission. Copyright 2018, American Chemical Society

unique combination of high electron mobility, ease of preparation and the previously identified benefit of efficient electron injection into the quantum dot layers. The key component of this device, a thin insulating poly(methylmethacrylate) (PMMA) layer, is inserted between the ZnO, ETL and the quantum dot emissive layer [29].

The normalized electroluminescence spectrum of the QLED is shown in Fig. 14.26a [29]. The symmetric emission peak at 640 nm with a narrow full width half maximum of 28 nm represents color-saturated deep-red emission, which is ideal for display applications (Fig. 14.26b) [29]. Figure 14.26c shows that the current density and luminance increase steeply once the voltage reaches ~1.7 V, yielding a maximum brightness of over 42,000 cd m^{-2} at 8 V. The peak EQE of 20.5% is achieved at a current density of ~7 mA cm^{-2} and a brightness of ~1200 cd m^{-2}. From Fig. 14.26d, one can find that high EQE (>18%) can be maintained in a wide range of current densities (1–42 mA cm^{-2}). When the current density reaches 100 mA cm^{-2}, an EQE of >15% is sustained. The established solution-processing protocol leads to devices with excellent reproducibility [29].

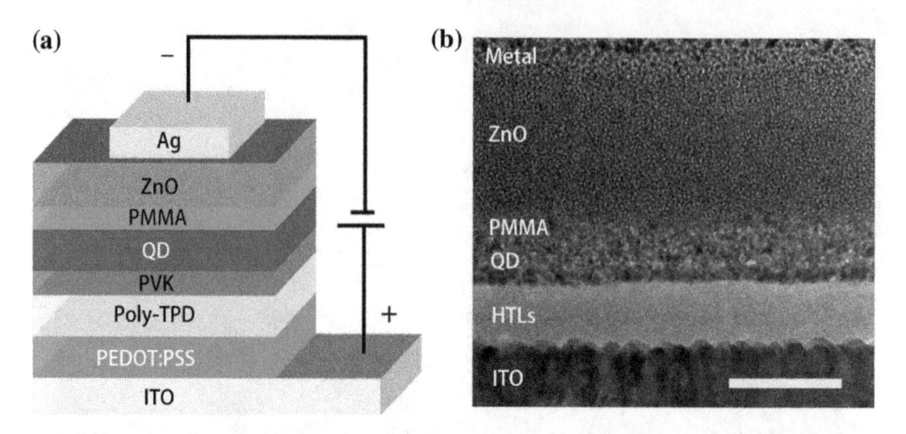

Fig. 14.25 Multilayer QLED device. **a** Device structure. **b** Cross-sectional transmission electron microscopy image showing the multiple layers of material with distinct contrast. Scale bar, 100 nm. Reproduced with permission from [29] with permission. Copyright 2014, Macmillan Publishers Limited

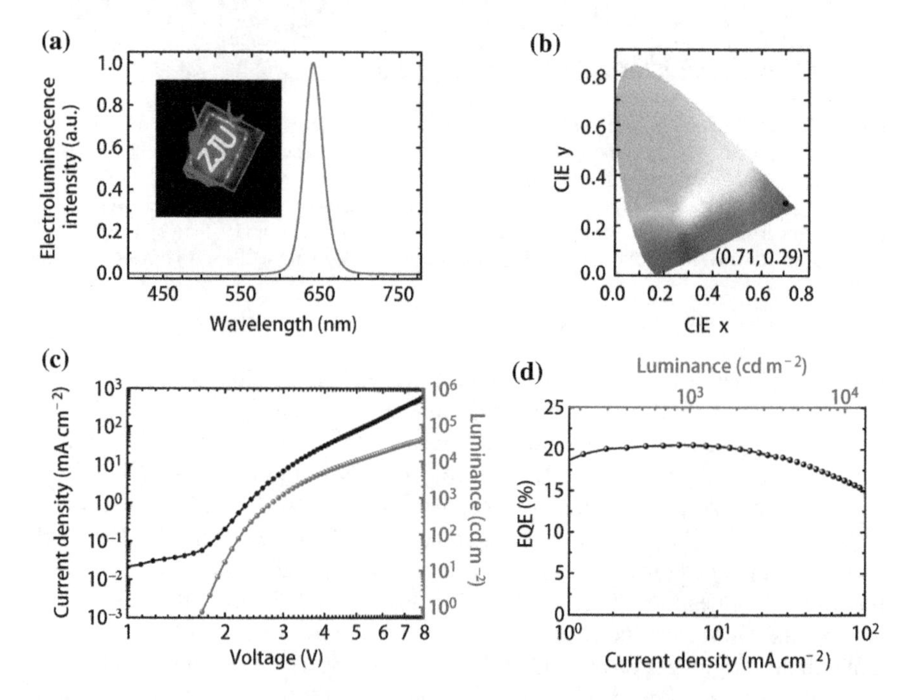

Fig. 14.26 **a** Electroluminescence spectrum at an applied voltage of 3 V and, inset, a photograph of a device with the Zhejiang University logo. a.u., arbitrary units. **b** The corresponding CIE coordinates. **c** Current density and luminance versus driving voltage characteristics for the device with the best efficiency. **d** EQE versus current density and luminance for the device with the best efficiency. Reproduced from [29] with permission. Copyright 2014, Macmillan Publishers Limited

14.3.2 Perovskite LEDs

Recently, lead halide perovskites have emerged as a highly promising class of semiconductor for LED [32–34]. The many advantages of perovskites include their low-cost, facile synthesis, solution processability, direct bandgap, ease of bandgap tuning, and high luminescence quantum yield. In particular, the narrow line-width emission from lead halide perovskites enables their devices to display rich and saturated colors across the entire visible spectrum [31]. Quite recently, Cao et al. [35] and Lin et al. [36] have independently developed perovskite LEDs (PLEDs) that break an important technological barrier: The EQE of the devices is greater than 20%.

Cao et al. used a perovskite mixed with an amino acid additive (aminovaleric acid) to control the size and orientation of the resultant perovskite crystals [35]. The cross-sectional scanning transmission electron microscope (STEM) image in Fig. 14.27a shows the formation of discrete sub-micrometer-structured perovskites in the emitting layer. SEM observations (Fig. 14.27b) further show that the perovskites are faceted platelets with roughly rectangular shapes [35]. The platelets are randomly tiled on the substrate, and the size of the platelets is between 100 and 500 nm. The computational modeling in Fig. 14.27c, d shows that this sub-micrometer structuring increases the outcoupling efficiency to 30%, compared with 22% for an equivalent flat perovskite device without sub-micrometer structuring [35].

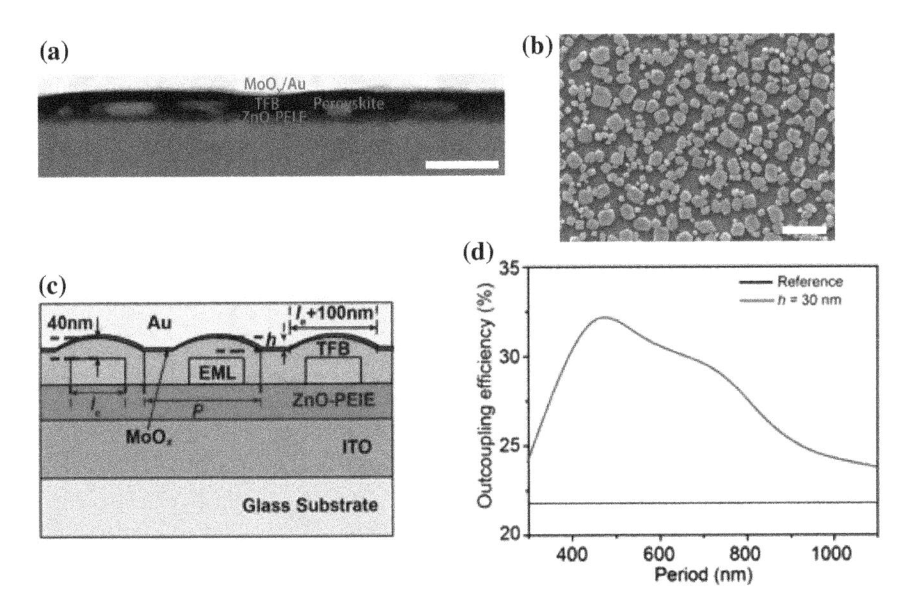

Fig. 14.27 Device fabrication and formation of sub-micrometer structure. **a** STEM image of the fabricated device. Scale bar, 200 nm. **b** SEM image of the perovskite. Scale bar, 1 μm. **c** An equivalent model in numerical simulation of the device. **d** Calculated outcoupling efficiency as a function of period P with convex height $h = 30$ nm. The reference is a device made from continuous perovskite film. Reproduced from [35] with permission. Copyright 2018, Springer Nature Limited

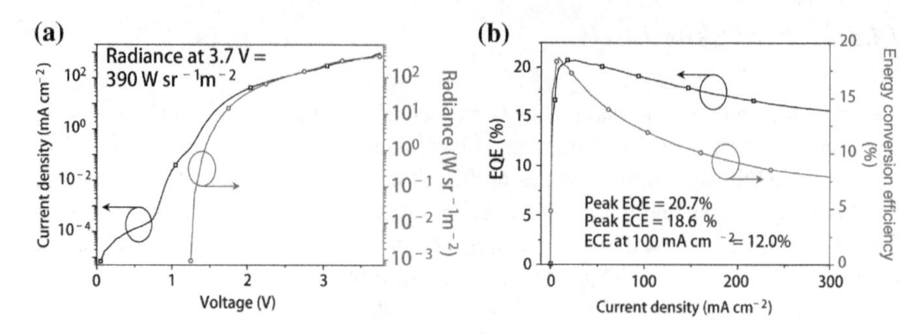

Fig. 14.28 **a** Dependence of current density and radiance on the voltage. **b** EQE and ECE plotted against current density. A peak EQE of 20.7% was achieved under a current density of 18 mA cm^{-2}. Reproduced with permission from [35]. Copyright 2018, Springer Nature Limited

The current-density/radiance/voltage characteristics of LEDs based on the structured perovskites are shown in Fig. 14.28a, from which one can see that the current density and radiance increase quickly once the device turns on at 1.25 V [35]. The peak EQE reaches 20.7% at a current density of 18 mA cm^{-2} (Fig. 14.28b) [35]. The brightness at the peak EQE is 18.4 W sr^{-1} m^{-2}. Statistical results for 100 devices show an average EQE of 19.2%, with a low relative standard deviation of 4%, indicating that the device performance is highly reproducible. The electroluminescence peak is located at 803 nm, consistent with the photoluminescence emission peak [35].

Lin et al. proposed a new approach combining a high photoluminescence quantum yield with balanced charge injection by constructing a compositionally graded perovskite based on a quasi-core/shell structure [36]. The cross-sectional TEM image (Fig. 14.29a) shows a well-defined layer-by-layer structure. The MABr shell in Fig. 14.29a passivates the non-radiative defects that would otherwise be presented in $CsPbBr_3$ crystals, boosting the photoluminescence quantum efficiency, while the MABr capping layer enables balanced charge injection [36]. By inserting an insulating layer of PMMA between the perovskite layer and the ETL to improve the device charge injection balance (Fig. 14.29b), a maximizing efficiency of 20.3% was realized (Fig. 14.29c) [36].

14.4 Micro- and NanoLasers

Shrinking coherent light sources to micro- and nanoscales can provide powerful tools for the control of light–matter interactions such as fluorescence, photocatalysis, quantum optics, and nonlinear optical processes. Although significant progress has been made to reduce the dimension size of lasers, they are still restricted, both in optical mode size and physical device dimension [2]. A way of addressing this issue is to make use of surface plasmons, which is capable of tightly localizing light, but ohmic

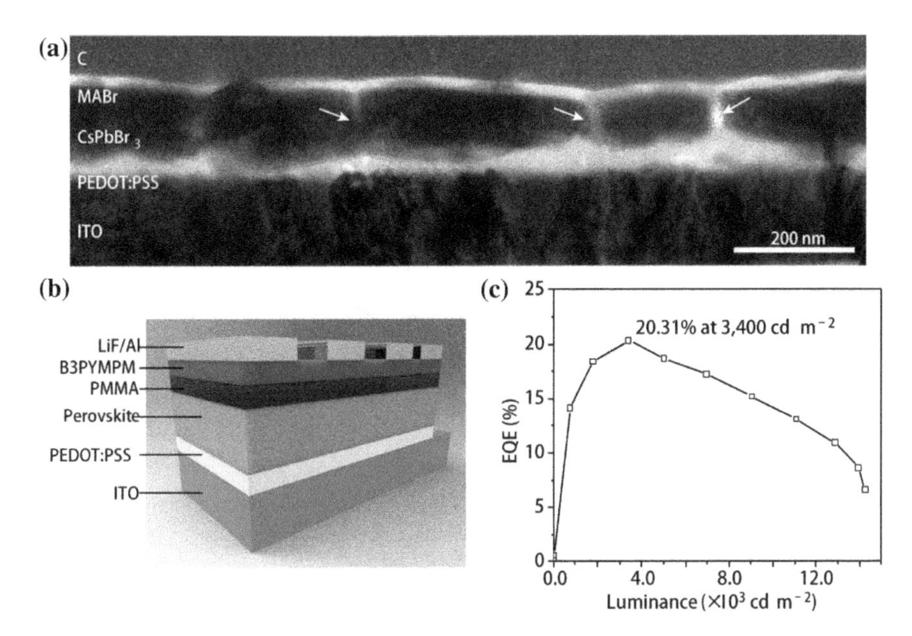

Fig. 14.29 **a** Cross-sectional TEM image of the PLED. **b** Configuration of a perovskite LED cell with a very thin PMMA layer inserted between the perovskite and the ETL. **c** EQE characteristics of the best-performing perovskite LEDs. Reproduced from [36] with permission. Copyright 2018, Springer Nature Limited

losses at optical frequencies have inhibited the realization of truly nanometer-scale lasers based on such approach. In order to decrease the ohmic losses and simultaneously maintain the ultra-small mode size, a hybrid plasmonic waveguide was utilized [2]. Figure 14.30a shows the schematic of a plasmon laser consisting of cadmium sulfide (CdS) nanowires on a silver film, where the gap layer is magnesium fluoride (MgF_2). The simulated electric field in Fig. 14.30b shows that the close proximity of the semiconductor and metal interfaces concentrates light into an extremely small area as much as a hundred times ($\sim\lambda^2/400$) smaller than a diffraction-limited spot [2]. The cross-sectional field plots (along the broken lines in the field map) illustrate the strong overall confinement in the gap region between the nanowire and metal surface with sufficient modal overlap in the semiconductor to facilitate gain [2].

When pumping these laser devices at a wavelength of 405 nm, amplified spontaneous emission peaks were obtained at moderate pump intensities (10–60 MW cm^{-2}), which correspond to the longitudinal cavity modes allowing plasmonic modes to resonate between the reflective nanowire end facets (Fig. 14.31) [2]. The four spectra for different peak pump intensities exemplify the transition from spontaneous emission (21.25 MW cm^{-2}) via amplified spontaneous emission (32.50 MW cm^{-2}) to full laser oscillation (76.25 and 131.25 MW cm^{-2}) [2].

In the above design, cryogenic operation is required to promote optical gain and prevent device damages, which considerably limits the practical application. Subse-

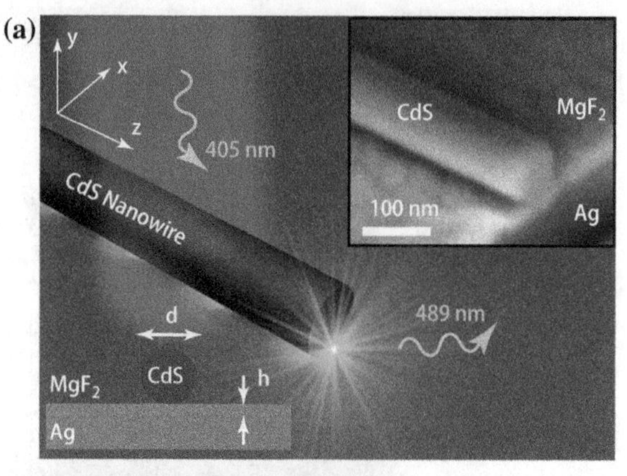

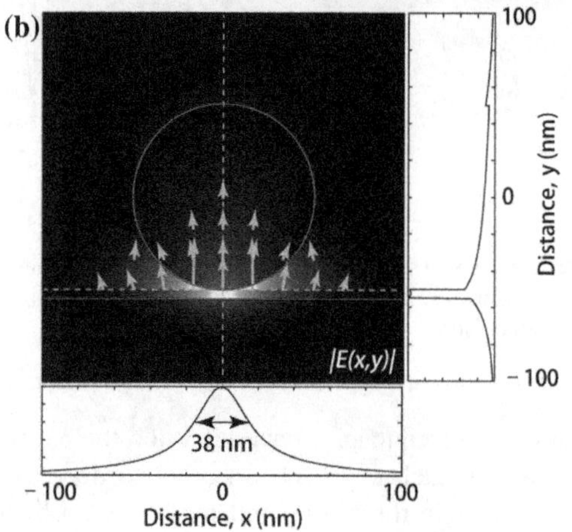

Fig. 14.30 **a** Plasmonic laser consists of a CdS semiconductor nanowire on top of a silver substrate, separated by a nanometer-scale MgF_2 layer. The inset shows a SEM image of the fabricated plasmonic laser. **b** The stimulated electric field distribution and direction $|E(x, y)|$ of a hybrid plasmonic mode at a wavelength of 489 nm, corresponding to the CdS I_2 exciton line. Reproduced from [2] with permission. Copyright 2009, Macmillan Publishers Limited

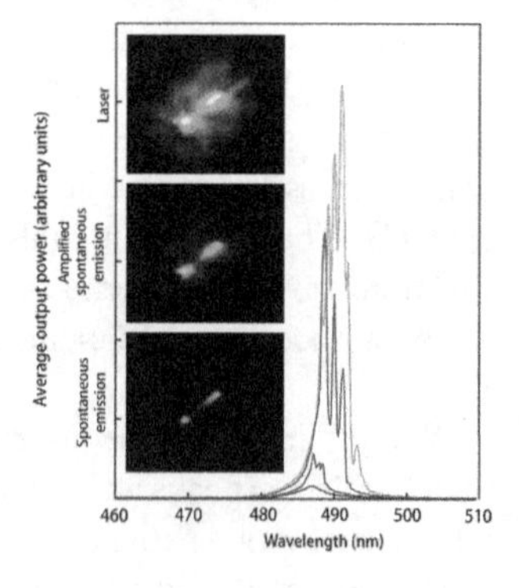

Fig. 14.31 Laser oscillation of a plasmonic laser, $d = 129$ nm, $h = 5$ nm (longitudinal modes). Reproduced from [2] with permission. Copyright 2009, Macmillan Publishers Limited

quently, an intense, single-mode, room temperature optically pumped ultraviolet SPP nanolaser was proposed [3]. The SPP lasing (~370 nm) threshold is ~3.5 MW cm^{-2}. It is found that the outstanding performance of the plasmonic laser device is attributed to high nanowire crystalline quality and a closed-contact planar semiconductor–insulator–metal interface, which can support efficient exciton–plasmon energy transfer, large regions of high Purcell factor and low scattering loss [3].

Figure 14.32a schematically shows the device's structure. Al is adopted as the plasmonic medium due to its smaller ohmic loss in the UV regime compared with silver. The large contact area and low scattering loss ensure to make full use of the optical gain from semiconductor [3]. The GaN nanowire edge length a is ~100 nm. The electromagnetic energy is concentrated in the thin 8 nm insulator gap region, exhibiting a deep-subwavelength modal volume of ~$\lambda^2/68$. The local Purcell factor distribution inside the GaN nanowires is displayed in Fig. 14.32c. One can see that most of the GaN nanowire regime exhibits a pronounced Purcell factor of ~7–10, supporting the effective coupling between the semiconductor waveguide and the SPP mode [3].

The fabricated plasmonic laser device shown in Fig. 14.33a is optically pumped by a 355 nm nanosecond-pulsed laser. The incidence excitation laser is circularly polarized. With $I_{ex} = 2.5 - 4.0$ MW cm^{-2}, the output power shows a super-linear increase, suggesting the occurrence of amplified spontaneous emission (Fig. 14.33b) [3]. As I_{ex} is further increased beyond 4 MW cm^{-2}, the spectrum is concentrated and dominant with the sharp peak. From the inset of Fig. 14.33b, one can find an

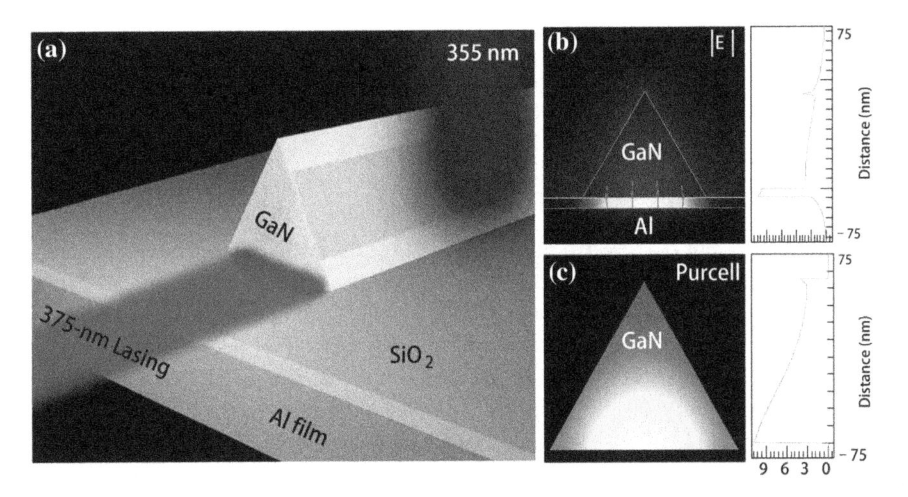

Fig. 14.32 **a** Schematic of semiconductor/dielectric/plasmonic devices. **b** Absolute electric field (|E|) distribution (left) around the plasmonic device with a wavelength of 370 nm. **c** Calculated local Purcell factor distribution around the GaN nanowire (left) and cross-sectional Purcell factor plots (right). Reproduced from [3] with permission. Copyright 2014, Macmillan Publishers Limited

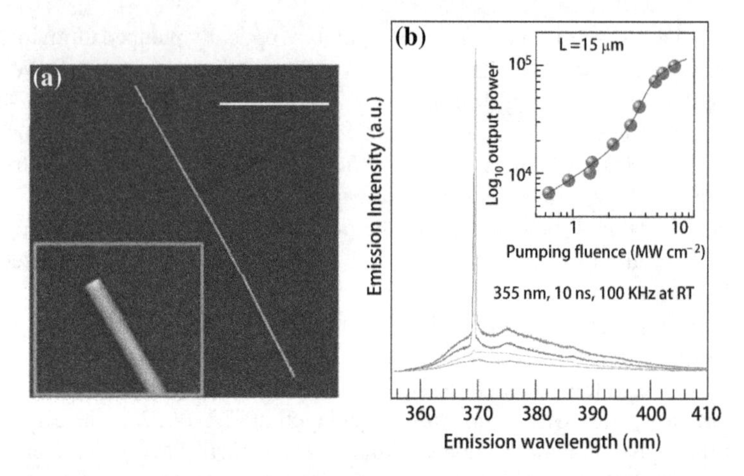

Fig. 14.33 **a** SEM image of a GaN nanowire sitting on SiO_2/Al film. Inset: magnified SEM image of one end of the GaN nanowire. The nanowire length and diameter is 15 μm and 100 nm, respectively. **b** Power-dependent emission spectra of the plasmonic devices. Inset: integrated emission versus pumping intensity. Reproduced from [3] with permission. Copyright 2014, Macmillan Publishers Limited

S-shaped relationship between the output power with the pumping fluence, which suggests the evolution from a spontaneous emission, amplified spontaneous emission to lasing process [3].

14.5 Optical Phased Arrays

The operating principle of OPA is the same as that of microwave phased arrays, where a beam direction is controlled by tuning the phase relationship between arrays of transmitters. For plane wave illumination, this could also be understood using the generalized Snell's law [37]. The concept of OPA was first demonstrated by Meyer [38], where a forty-six channel lithium tantalate phase modulator has been developed for 1D beam steering. So far, several different materials have been proposed for the active phase modulation, including electronic-optical crystals [38] (e.g., $LiNbO_3$, $LiTaNO_3$), liquid crystal [39, 40], and integrated optical waveguide (e.g., AlGaAs [41, 42] and SOI [43–57]). The OPAs based on electronic-optical crystals have short response times (~fs) and thus high modulation frequency. However, they suffer the disadvantages of high driving voltage high up to 10 kV [38], bulky volume, and small field of view. Although OPAs based on cascaded liquid crystal have a large field of view and low driving voltage, the main obstacle is the slow response velocity. Compared with the above approaches, OPAs-based integrated optical waveguide simultaneously possesses advantages of low driving voltage and high response velocity. Especially, those based on SOI waveguides are compact and compatible with the standard CMOS process and thus promise large-scale and low-cost implementation.

14.5.1 Silicon Chip-Based OPA and Solid-State LIDAR

A large-scale 2D OPA is shown in Fig. 14.34 [58]. A laser input is coupled into the main silicon bus waveguide through an optical fiber, and then evanescently coupled into 64-row waveguides. The optical power in each row waveguide is then similarly divided into 64 optical antenna units so that all 4096 optical nanoantennas (gratings) are uniformly excited. The emitted phase of each pixel is adjusted by the length of the optical delay line within the pixel. The far-field radiation field $E(\theta, \phi)$ of the phased array is calculated as the far-field of an individual nanoantenna $S(\theta, \phi)$ multiplied by the array factor $F_a(\theta, \phi)$ [58]:

$$E(\theta, \phi) = S(\theta, \phi) \times F_a(\theta, \phi) \tag{14.5.1}$$

where the array factor is a system factor related to the phase of optical emission from all the pixels. θ and ϕ are the far-field azimuth angle and polar angle, respectively. By assigning the optical phase ψ_{mn} of each pixel (where m and n are the pixel indices) in the phased array, the desired radiation pattern $E(\theta, \phi)$ can be achieved. The phase ψ_{mn} of each pixel can be determined by antenna synthesis through the Gerchberg–Saxton algorithm [59].

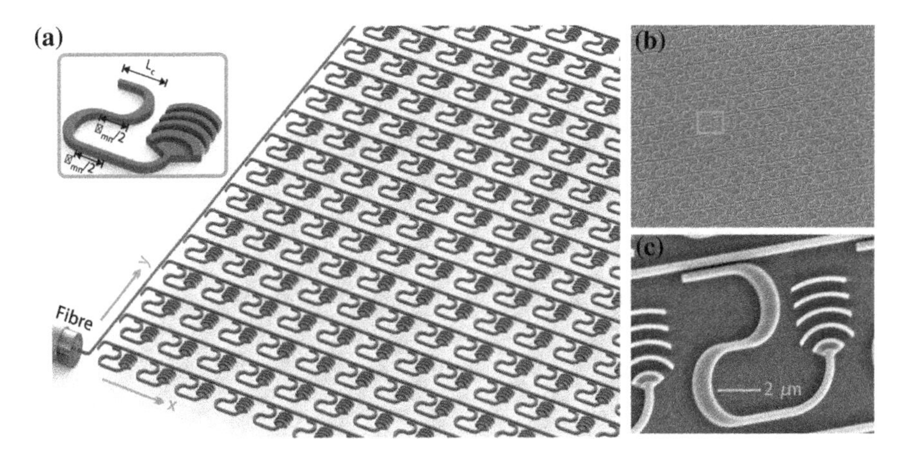

Fig. 14.34 **a** Schematic illustration of a 64 × 64 OPA system. **b** SEM of part of the phased array system fabricated at a CMOS foundry. **c** A close-up SEM of one pixel of the OPA system, indicated by the green rectangle in (b). Reproduced from [58] with permission. Copyright 2013, Macmillan Publishers Limited

Figure 14.35 shows the simulated and measured radiation pattern of the designed 64 × 64 nanophotonic phased array system to generate the MIT logo in the far-field. Limited by the optical diffraction limit in such waveguide, the dimension size of the unit cell is determined as 9 μm × 9 μm, several times of the operation wavelength 1.55 μm. As a consequence, higher-order interference patterns periodically present in the far-field [58].

Benefiting from the use of guided light in silicon instead of free-space light, active manipulation of the optical phase can be directly implemented to achieve dynamic far-field patterns with more flexibility and wider applications, for instance, by converting the pixel into a thermally phase-tunable pixel in a CMOS process (Fig. 14.36a, inset) [58]. Figure 14.36a illustrates an active 8 × 8 phased array in which each pixel has an independently tunable phase shifter, and the electrical controls are connected in rows and in columns to simplify the electrical circuitry. By applying different voltages on each pixel, different phase combinations can be achieved in the phased array to generate different radiation patterns dynamically in the far-field, as shown in Fig. 14.36b–f [58].

Solid-state LIDAR could be constructed based on OPA. In a recent demonstration [55], frequency-modulated continuous-wave (FMCW) LIDAR and triangular modulation were adopted to resolve the distance/velocity ambiguity caused by an induced Doppler shift from a moving target. For a moving target, two distinct beat frequencies, f_{IF1} and f_{IF2}, occur due to the Doppler shift on the received signal. As a consequence, simultaneous distance and velocity measurements can be performed according to the following equation [55]:

$$\begin{aligned} f_{\mathrm{IF1}} - f_{\mathrm{IF2}} &= 2f_{\mathrm{Doppler}} \approx 2vf_0/c \\ (f_{\mathrm{IF1}} + f_{\mathrm{IF2}})/2 &= f_{\mathrm{Dis}} \end{aligned} \tag{14.5.2}$$

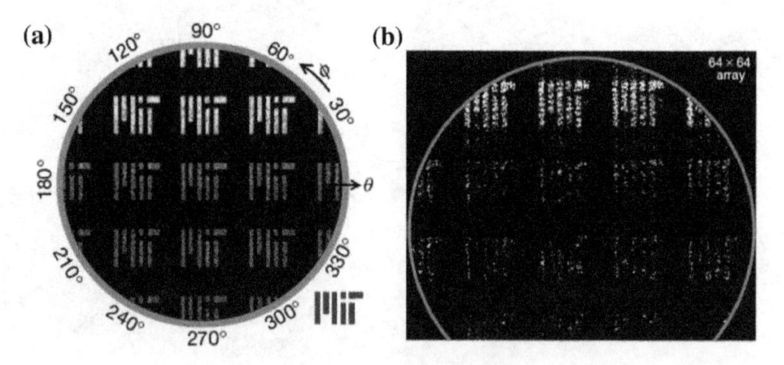

Fig. 14.35 **a** Simulated radiation pattern of the designed 64 × 64 nanophotonic phased array system to generate the MIT logo in the far-field. The inset on the lower right shows the targeted MIT logo pattern. **b** The measured far-field radiation pattern of the fabricated 64 × 64 OPA system. Reproduced from [58] with permission. Copyright 2013, Macmillan Publishers Limited

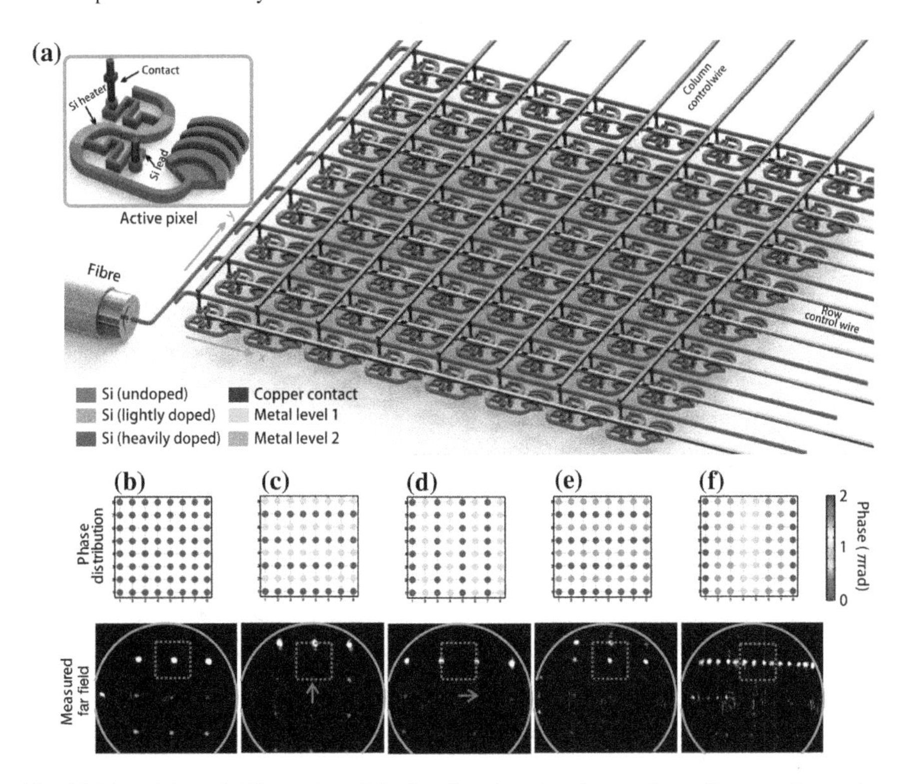

Fig. 14.36 **a** Schematic illustration of the 8 × 8 active phased array. Inset diagram of an active pixel with dimension size (9 μm × 9 μm). The optical phase of each pixel is continuously tuned by the thermo-optic effect through an integrated heater formed by doped silicon. **b–f** Experimental examples of the dynamic far-field patterns generated by the 8 × 8 active phased array by applying different voltage combinations to the pixels. Reproduced from [58] with permission. Copyright 2013, Macmillan Publishers Limited

Figure 14.37a shows a 3D rendering of the OPA designed for a solid-state LIDAR system, where a grouped cascaded phase shifter architecture is utilized to apply phase shifts to antenna elements. SEM images of the fabricated device are shown in Fig. 14.37b–d [55]. The phased array consists of 50 grating-based antennas placed at a 2 μm pitch. Figure 14.37e shows an optical micrograph of the chip. The system is controlled using nine copper electrical pads: three for the transmitter (TX) array, three for the receiver (RX) array, two for the signal and bias of the balanced photodetector, and one for the ground [55].

In order to test the steered LIDAR operation of the device, three targets were placed at different incident angles from the chip, as shown in Fig. 14.38a [55]. The targets were created with reflective tape. By steering the transmitting and receiving phased arrays simultaneously, each of the targets was measured separately. The distance to each target is shown in Fig. 14.38b [55].

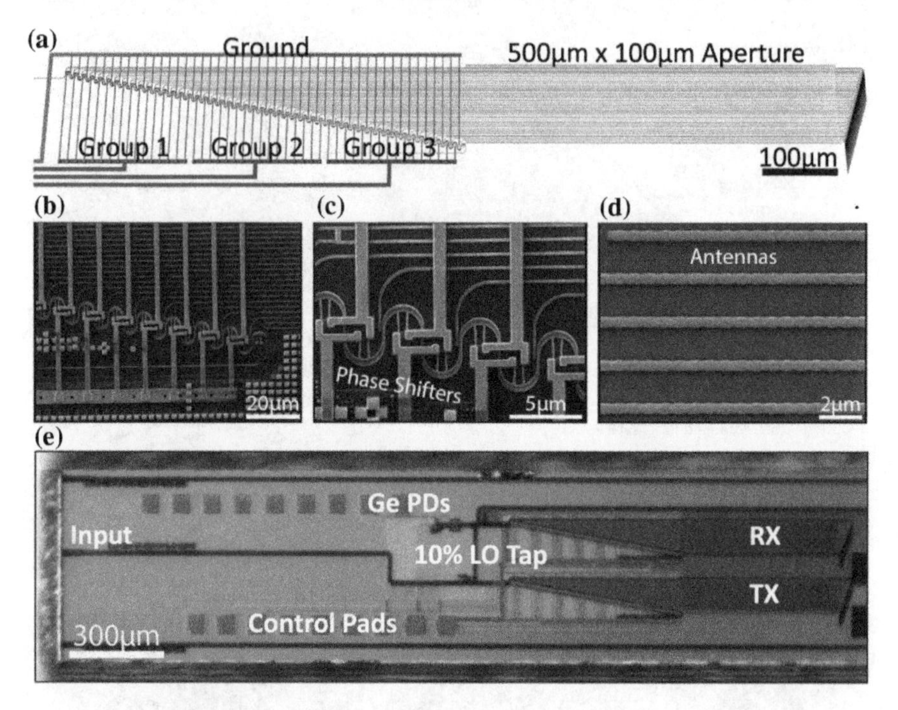

Fig. 14.37 **a** 3D rendering of the OPA with **b–d** SEM images. **e** Optical micrograph of the device. **f** Packaged system with epoxied fiber. Reproduced from [55] with permission. Copyright 2017, Optical Society of America

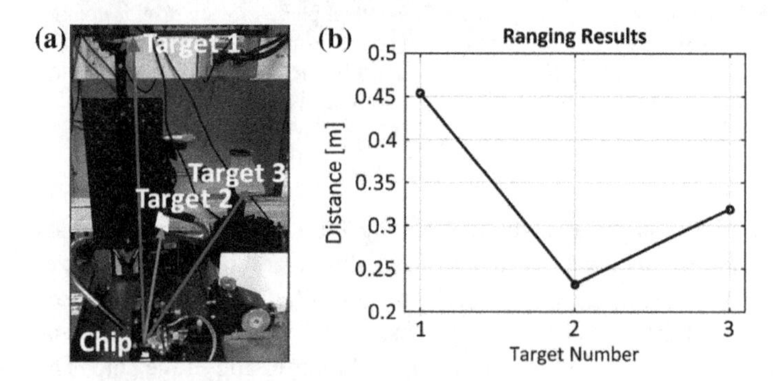

Fig. 14.38 **a** Far-field of the array at different thermal and wavelength tuning values. **b** Image of targets in the laboratory placed at different angles from the chip. Reproduced from [55] with permission. Copyright 2017, Optical Society of America

14.5.2 Nanophotonic Projection System Based on OPA

OPA technology can also be used to implement projection systems. Recently, an efficient integrated projection system based on a fast steering OPA was reported [60], which can project an image by vector or raster scan of the beam spot on the screen without use of a lens or any other optical components. The phase shifters of the OPA are *p–i–n* phase modulators with bandwidth of 200 MHz enabling ultra-fast beam steering. Using an integrated 4 × 4 OPA, image and video projections are demonstrated [60].

Figure 14.39 shows a 2D OPA, where $M \times N$ radiating elements with circular physical apertures of diameter d_s are spaced apart by d along the x- and y-axis. Assuming Gaussian beam profile for all elements and considering radiation along the z-axis at a wavelength of λ_0, the electric field for each element on the plane of the phased array ($z = 0$) can be written as [60]:

$$E_{mn}(x, y, 0) = E_{mn,0} e^{im\Delta\phi_m} e^{im\Delta\phi_n} e^{\frac{-4}{d_s^2}\left[(x-md)^2+(y-md)^2\right]} \tag{14.5.3}$$

where $E_{mn,0}$, m, n, $\Delta\phi_m$, and $\Delta\phi_n$ are the element field constant coefficient, element index along the x-axis and y-axis, and the constant phase difference between adjacent elements along the x- and y-axis, respectively. In this case, using the Fraunhofer

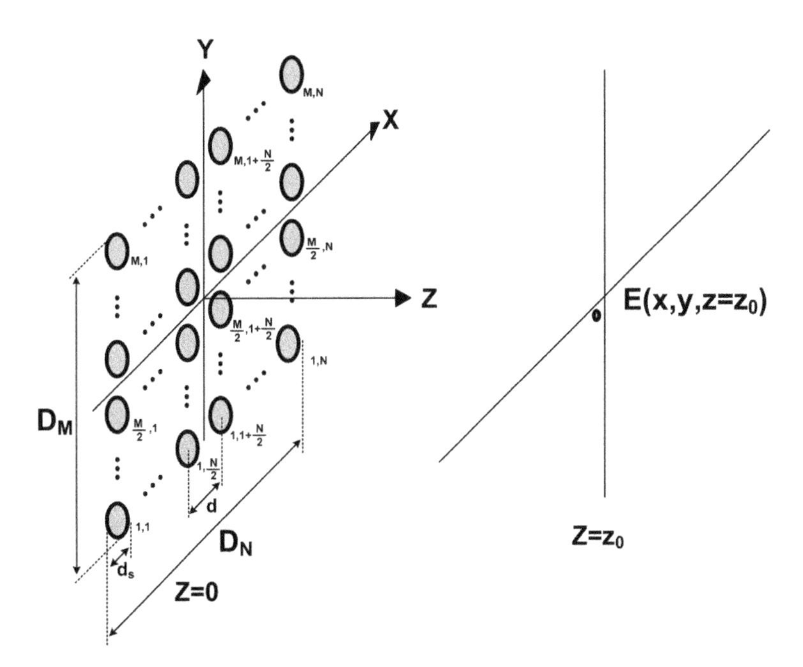

Fig. 14.39 $M \times N$ element OPA. Reproduced from [60] with permission. Copyright 2015, Optical Society of America

far-field approximation, and assuming all elements have the same filed constant coefficient of $E_{mn,0} = E_0$, the far-field intensity of the electric field at $z = z_0$ can be calculated as [60]:

$$I(x, y, z_0) = \frac{E_0^2 d_s^4 \pi^2}{16\lambda_0^2 z_0^2} e^{\frac{-\pi^2 d_s^2}{2\lambda_0^2 z_0^2}} \times |AF_x|^2 \times |AF_y|^2 \tag{14.5.4}$$

where $AF_x = \sum_{k=1}^{M} e^{i\left(\frac{2\pi d}{\lambda_0 z_0}x - \Delta\phi_m\right)k}$ and $AF_y = \sum_{k=1}^{N} e^{i\left(\frac{2\pi d}{\lambda_0 z_0}y - \Delta\phi_n\right)k}$ are, respectively, the normalized array factors along the x- and y-axis and can be calculated as:

$$|AF_x| = \frac{\sin\left[M\left(\frac{\pi d}{\lambda_0 z_0}x - \frac{\Delta\phi_m}{2}\right)\right]}{\sin\left[\frac{\pi d}{\lambda_0 z_0}x - \frac{\Delta\phi_m}{2}\right]} \tag{14.5.5}$$

$$|AF_y| = \frac{\sin\left[N\left(\frac{\pi d}{\lambda_0 z_0}y - \frac{\Delta\phi_n}{2}\right)\right]}{\sin\left[\frac{\pi d}{\lambda_0 z_0}y - \frac{\Delta\phi_n}{2}\right]} \tag{14.5.6}$$

Equations (14.5.5) and (14.5.6) indicates that by adjusting $\Delta\phi_m$ and $\Delta\phi_n$, the formed beam can be steered in the x- and y-directions, respectively [60].

Figure 14.40 shows the structure of the OPA [60]. The light is coupled into the input grating coupler through a single-mode optical fiber and then is uniformly splits and guides the coupled light into 16 *p–i–n* phase modulators. The phase modulated optical waves are then guided to the radiating antenna (grating coupler) array. The radiating elements are arranged in a 4 × 4 quadrate lattice with center-to-center spacing of 50 μm. The total area of the fabricated micro-projector chip is less than 1 mm^2 [60].

As a proof of concept, Fig. 14.41a and b shows the simulated and projected sad and smiley faces that are formed by fast vector scan of the beam spot, respectively. The entire image was captured in real time with a single snapshot of the IR camera. Figure 14.41c shows the projected images of letters “C”, “I”, and “T” [60].

14.5.3 Metasurface-Based OPA

The current OPAs, despite ever-increasing array size and element count, remain limited in beam quality and steering angle because of unwanted coupling between dielectric waveguides, which forces radiating emitters to be placed several microns apart. Emitter spacing greater than half a wavelength would draw optical power from the main beam, redirecting it into unwanted periodic grating lobes tens of degrees away. Intrinsically, the limitation of steering range is a consequence of diffraction limit of silicon optical waveguide, which impedes the realization of phase modulation

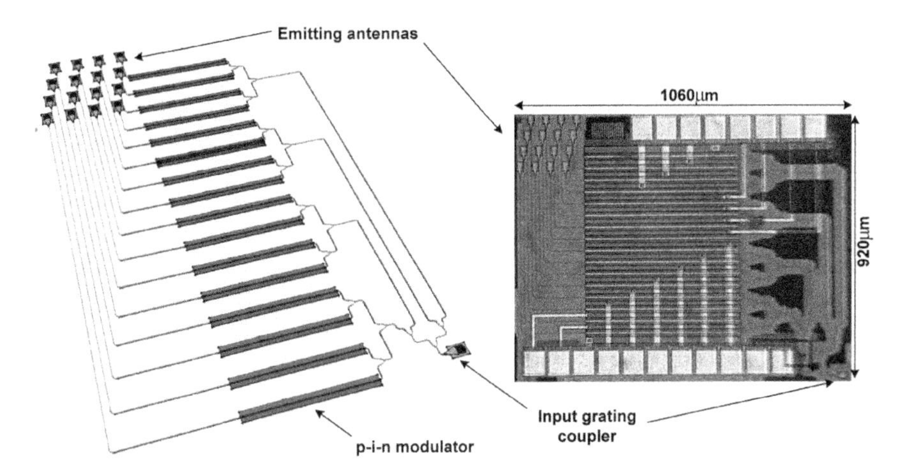

Fig. 14.40 Structure of the 4 × 4 integrated OPA with per-channel high-speed phase control and the chip micro-photograph of the fabricated OPA. Reproduced from [60] with permission. Copyright 2015, Optical Society of America

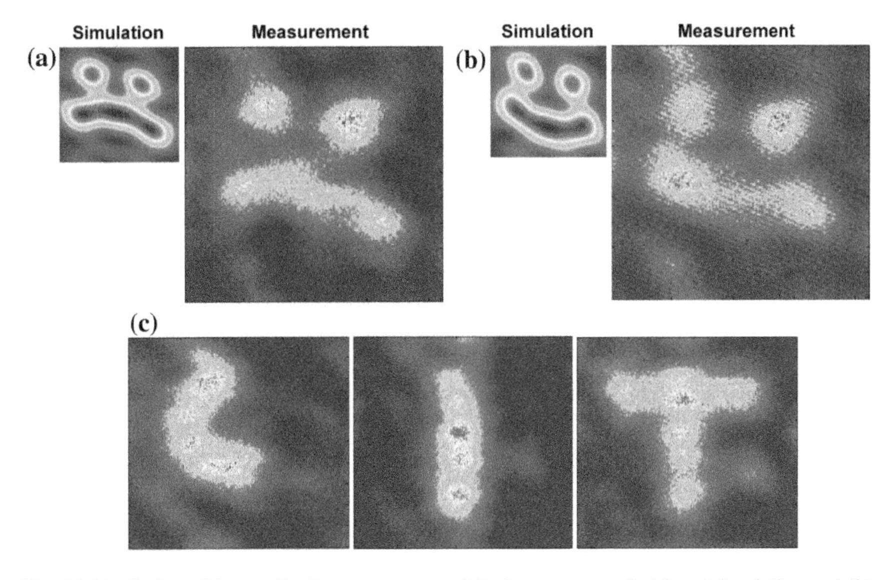

Fig. 14.41 Projected images by fast vector scan of the beam spot. **a** Sad face (simulation on left). **b** Smiley face (simulation on left). **c** Individual letters of CIT (California Institute of Technology). Reproduced from [60] with permission. Copyright 2015, Optical Society of America

in a size much smaller the wavelength. Moreover, thermo-optic control limites the modulation frequency of these phased arrays to less than 100 kHz, which is too slow for versatile beam steering in technologically important LIDAR applications. In addition, the thermal crosstalk between phase shifters and the photodetectors limited the detection range to 50° [61].

The emergence of metasurface offers an opportunity to realize local phase modulation in nanoscale, since the phase modulation does not depend on the propagation phase retardation anymore but rather stems from plasmonic retardation [62–65] or resonances [66, 67]. In other words, metasurfaces offer a different approach to the phased array architecture, in which the array is intrinsically 2D and the subwavelength antenna dimensions and antenna spacing can suppress side lobes [68]. Hence, it would be highly desirable to have a tunable metasurface platform for comprehensive and active control of scattered light. Different materials have been integrated in metasurfaces for active phase modulation, including doped semiconductors [69], phase-change materials [70], and graphene [71]. Among various physical mechanisms for active phase engineering, field-effect modulation possesses advantages of high-speed modulation and extremely low power dissipation [68].

Transparent conducting oxide (TCO) materials, such as indium tin oxide (ITO), have also been used as the active semiconductor layer [68]. By applying an electrical bias between metal and ITO, the sign of $\varepsilon_r^{\mathrm{ITO}}$ changes around the epsilon-near-zero (ENZ) region. Thus, a large electric field enhancement occurs in the accumulation layer for near-infrared wavelengths, providing an efficient way to electrically modulate the optical properties of nanophotonic devices with high modulation speed and low power consumption. The simulated and measured results show a phase change of $\sim$184° at an applied bias of 2.5 V.

A photograph of the gate-tunable device is shown in Fig. 14.42a. SEM images of a stripe antenna structure are depicted in Fig. 14.42b and c [68]. Adjacent stripe antennas are connected electrically in groups of three so that each group is connected to a different external gold-electrode. The electrodes are wire connected to a compact chip carrier and circuit board for electrical gating. When the phase difference between adjacent groups is equal to π (i.e., applied bias is equal to 2.5 V), the gate-tunable device behaves as a beam splitter and the split angle is determined by the grating period, as indicated in Fig. 14.42d. Modulation at frequencies exceeding 10 MHz is possible with this configuration. The device demonstrates an efficient beam steering with large range of angles (>40°) with 2–4 periodically gated nanoantennas, as indicated in Fig. 14.42e.

While the above proposal is conceptually promising as an approach to active metasurface design, in order to realize a comprehensively tunable metasurface, a phase shift from 0° approaching to 360° is required. Numerous efforts have been spent for this purpose. For example, a continuous phase shift from 0 to 230° by graphene gating under applied bias at a wavelength of 8.5 μm was recently reported [72]. More recently, dual-gated field-effect-tunable metasurface antenna arrays that enable phase shifts exceeding 300° at a wavelength of $\lambda = 1550$ nm have been proposed [73].

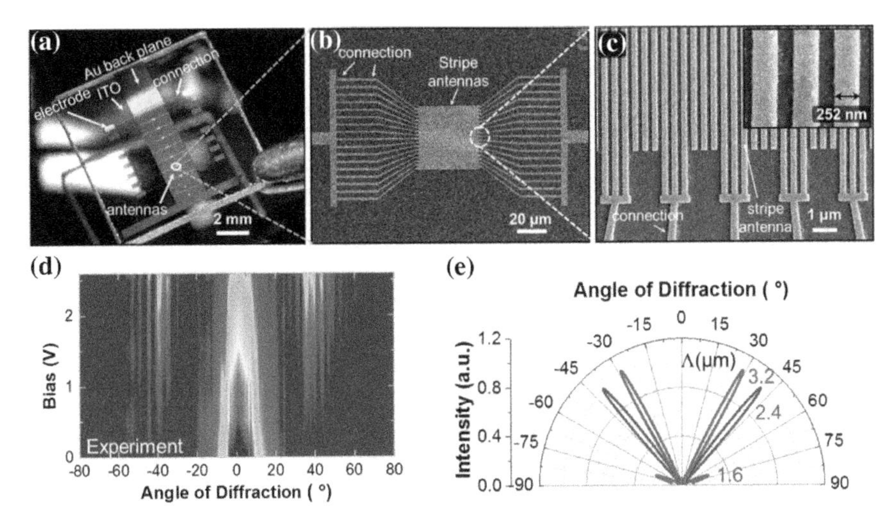

Fig. 14.42 **a** Photograph of the fabricated structure. For gate biasing, stripe antennas are connected to electrodes. **b** SEM image of the stripe antenna and connections. **c** SEM image of the close-up of stripe antenna. **d** Experimental measured far-field intensity profiles. **e** Simulated far-field intensity of the light beam reflected from the metasurface as a function of the diffraction angles for periodicities of 3.2, 2.4, 1.6 μm under a gate bias of 3.0 V. Reproduced from [68] with permission. Copyright 2016, American Chemical Society

Besides TCO, alternative materials such as phase-change material can be utilized for active phase engineering [74, 75]. Compared with the doped semiconductor, the phase-change material exhibits dramatic difference in optical and electrical characteristics between its crystalline and amorphous states. The phase transition can be induced by applying either heat, photon, or electric energy on the phase-change material. The reversibility between phase states can be realized by carefully modulating the input energy pulse. These properties can provide active modulation or switching when applied properly to metasurfaces.

Among these different phase-change materials (PCMs), $Ge_2Sb_2Te_5$ (GST) alloy is known to have advantages of low optical loss in the near-infrared (NIR) range and nonvolatility, high stability and quick response [75]. An active metasurface composed of addressable meta-molecules with constituent GST rods has been proposed [76], as shown in Fig. 14.43a. Near-field coupling between the GST rods within a meta-molecule can be tuned by modifying the phase state of these GST rods. A possible implementation configuration is shown in Fig. 14.43b, where the meta-atoms of the phase-change metasurface are made on the TiN electrodes connected with the conducting wire [76]. By applying electronic pulse on the specific meta-atoms, the phase state of the treated meta-atoms can be changed, resulting in the selective modification on the metasurface-based device. A reconfigurable gradient metasurface

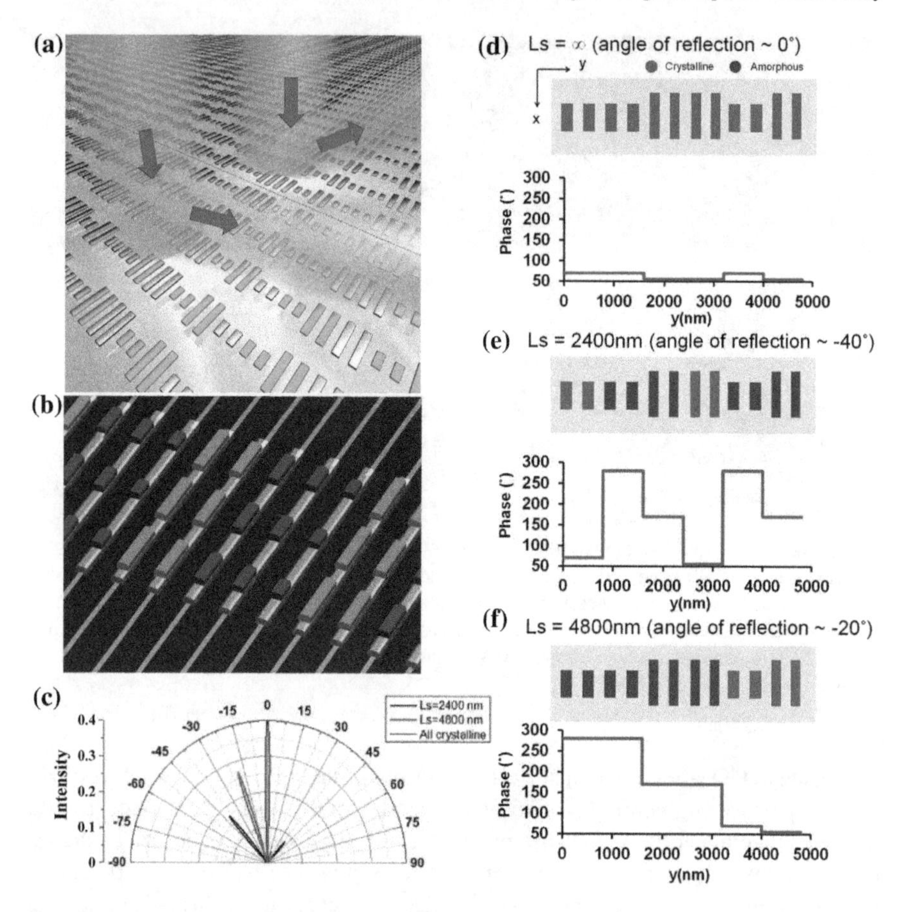

Fig. 14.43 **a** Schematic illustration of an active dielectric metasurface based on phase-change medium. **b** Schematic for realization of the selective modification of phase-change metasurface. **c** Scattered electric field intensity pattern. **d–f** Super-cell design of tunable gradient metasurface and their corresponding phase distribution under different states. Reproduced from [76] with permission. Copyright 2016, Wiley-VCH Verlag GmbH & Co. KGaA, Weinheim

capable of exhibiting the anomalous reflection for a normal incident light at 1550 nm is shown in Fig. 14.43c, which directs the normal incidence onto different angles of 0°, −40°, and −20° for three different combinations of crystalline and amorphous GST bars (Fig. 14.43d–f) [76].

References

1. W. Li, S. Fan, Nanophotonic control of thermal radiation for energy applications. Opt. Express **26**, 15995–16021 (2018)
2. R.F. Oulton, V.J. Sorger, T. Zentgraf, R.-M. Ma, C. Gladden, L. Dai, G. Bartal, X. Zhang, Plasmon lasers at deep subwavelength scale. Nature **461**, 629–632 (2009)
3. Q. Zhang, G. Li, X. Liu, F. Qian, Y. Li, T.C. Sum, C.M. Lieber, Q. Xiong, A room temperature low-threshold ultraviolet plasmonic nanolaser. Nat. Commun. **5**, 4953 (2014)
4. M. Karl, J.M.E. Glackin, M. Schubert, N.M. Kronenberg, G.A. Turnbull, I.D.W. Samuel, M.C. Gather, Flexible and ultra-lightweight polymer membrane lasers. Nat. Commun. **9**, 1525 (2018)
5. K. Van Acoleyen, W. Bogaerts, J. Jágerská, N. Le Thomas, R. Houdré, R. Baets, Off-chip beam steering with a one-dimensional optical phased array on silicon-on-insulator. Opt. Lett. **34**, 1477–1479 (2009)
6. X. Xie, X. Li, M. Pu, X. Ma, K. Liu, Y. Guo, X. Luo, Plasmonic metasurfaces for simultaneous thermal infrared invisibility and holographic illusion. Adv. Funct. Mater. **28**, 1706673 (2018)
7. X. Ma, M. Pu, X. Li, Y. Guo, X. Luo, All-metallic wide-angle metasurfaces for multifunctional polarization manipulation. Opto-Electron. Adv. **2**, 180023 (2019)
8. I. Celanovic, D. Perreault, J. Kassakian, Resonant-cavity enhanced thermal emission. Phys. Rev. B **72**, 075127 (2005)
9. X. Liu, T. Tyler, T. Starr, A.F. Starr, N.M. Jokerst, W.J. Padilla, Taming the blackbody with infrared metamaterials as selective thermal emitters. Phys. Rev. Lett. **107**, 045901 (2011)
10. J.-J. Greffet, R. Carminati, K. Joulain, J.-P. Mulet, S. Mainguy, Coherent emission of light by thermal sources. Nature **416**, 61–64 (2002)
11. H.J. Lezec, A. Degiron, E. Devaux, R.A. Linke, L. Martin-Moreno, F.J. Garcia-Vidal, T.W. Ebbesen, Beaming light from a subwavelength aperture. Science **297**, 820–822 (2002)
12. J.H. Park, S.E. Han, P. Nagpal, D.J. Norris, Observation of thermal beaming from tungsten and molybdenum bull's eyes. ACS Photonics **3**, 494–500 (2016)
13. M. Pu, P. Chen, Y. Wang, Z. Zhao, C. Wang, C. Huang, C. Hu, X. Luo, Strong enhancement of light absorption and highly directive thermal emission in graphene. Opt. Express **21**, 11618–11627 (2013)
14. S. Basu, Z.M. Zhang, C.J. Fu, Review of near-field thermal radiation and its application to energy conversion. Int. J. Energy Res. **33**, 1203–1232 (2009)
15. J.B. Pendry, Radiative exchange of heat between nanostructures. J. Phys. Condens. Matter **11**, 6621 (1999)
16. X. Luo, *Catenary Optics* (Springer Singapore, 2019)
17. Y. Guo, C.L. Cortes, S. Molesky, Z. Jacob, Broadband super-planckian thermal emission from hyperbolic metamaterials. Appl. Phys. Lett. **101**, 131106 (2012)
18. X. Liu, Z. Zhang, Near-field thermal radiation between metasurfaces. ACS Photonics **2**, 1320–1326 (2015)
19. A. Lenert, D.M. Bierman, Y. Nam, W.R. Chan, I. Celanovic, M. Soljacic, E.N. Wang, A nanophotonic solar thermophotovoltaic device. Nat. Nanotechnol. **9**, 126–130 (2014)
20. M. Song, H. Yu, C. Hu, M. Pu, Z. Zhang, J. Luo, X. Luo, Conversion of broadband energy to narrowband emission through double-sided metamaterials. Opt. Express **21**, 32207–32216 (2013)
21. D.M. Bierman, A. Lenert, W.R. Chan, B. Bhatia, I. Celanović, M. Soljačić, E.N. Wang, Enhanced photovoltaic energy conversion using thermally based spectral shaping. Nat. Energy **1**, 16068 (2016)
22. A.P. Raman, M.A. Anoma, L. Zhu, E. Rephaeli, S. Fan, Passive radiative cooling below ambient air temperature under direct sunlight. Nature **515**, 540–544 (2014)
23. Y. Zhai, Y. Ma, S.N. David, D. Zhao, R. Lou, G. Tan, R. Yang, X. Yin, Scalable-manufactured randomized glass-polymer hybrid metamaterial for daytime radiative cooling. Science **355**, 1062 (2017)
24. Y. Huang, M. Pu, Z. Zhao, X. Li, X. Ma, X. Luo, Broadband metamaterial as an "invisible" radiative cooling coat. Opt. Commun. **407**, 204–207 (2018)

25. O. Salihoglu, H.B. Uzlu, O. Yakar, S. Aas, O. Balci, N. Kakevov, S. Balci, S. Olcum, S. Süzer, C. Kocabas, Graphene based adaptive thermal camouflage. Nano Lett. **18**, 4541–4547 (2018)
26. H. Uoyama, K. Goushi, K. Shizu, H. Nomura, C. Adachi, Highly efficient organic light-emitting diodes from delayed fluorescence. Nature **492**, 234 (2012)
27. Q. Zhang, B. Li, S. Huang, H. Nomura, H. Tanaka, C. Adachi, Efficient blue organic light-emitting diodes employing thermally activated delayed fluorescence. Nat. Photonics **8**, 326 (2014)
28. K. Tuong Ly, R.-W. Chen-Cheng, H.-W. Lin, Y.-J. Shiau, S.-H. Liu, P.-T. Chou, C.-S. Tsao, Y.-C. Huang, Y. Chi, Near-infrared organic light-emitting diodes with very high external quantum efficiency and radiance. Nat. Photonics **11**, 63 (2016)
29. X. Dai, Z. Zhang, Y. Jin, Y. Niu, H. Cao, X. Liang, L. Chen, J. Wang, X. Peng, Solution-processed, high-performance light-emitting diodes based on quantum dots. Nature **515**, 96–99 (2014)
30. X. Gong, Z. Yang, G. Walters, R. Comin, Z. Ning, E. Beauregard, V. Adinolfi, O. Voznyy, E.H. Sargent, Highly efficient quantum dot near-infrared light-emitting diodes. Nat. Photonics **10**, 253 (2016)
31. X. Zhao, J.D.A. Ng, R.H. Friend, Z.-K. Tan, Opportunities and challenges in perovskite light-emitting devices. ACS Photonics **5**, 3866–3875 (2018)
32. Z.-K. Tan, R.S. Moghaddam, M.L. Lai, P. Docampo, R. Higler, F. Deschler, M. Price, A. Sadhanala, L.M. Pazos, D. Credgington, F. Hanusch, T. Bein, H.J. Snaith, R.H. Friend, Bright light-emitting diodes based on organometal halide perovskite. Nat. Nanotechnol. **9**, 687 (2014)
33. G. Li, F.W.R. Rivarola, N.J.L.K. Davis, S. Bai, T.C. Jellicoe, F. de la Peña, S. Hou, C. Ducati, F. Gao, R.H. Friend, N.C. Greenham, Z.-K. Tan, Highly efficient perovskite nanocrystal light-emitting diodes enabled by a universal crosslinking method. Adv. Mater. **28**, 3528–3534 (2016)
34. H. Cho, S.-H. Jeong, M.-H. Park, Y.-H. Kim, C. Wolf, C.-L. Lee, J.H. Heo, A. Sadhanala, N. Myoung, S. Yoo, S.H. Im, R.H. Friend, T.-W. Lee, Overcoming the electroluminescence efficiency limitations of perovskite light-emitting diodes. Science **350**, 1222 (2015)
35. Y. Cao, N. Wang, H. Tian, J. Guo, Y. Wei, H. Chen, Y. Miao, W. Zou, K. Pan, Y. He, H. Cao, Y. Ke, M. Xu, Y. Wang, M. Yang, K. Du, Z. Fu, D. Kong, D. Dai, Y. Jin, G. Li, H. Li, Q. Peng, J. Wang, W. Huang, Perovskite light-emitting diodes based on spontaneously formed submicrometre-scale structures. Nature **562**, 249–253 (2018)
36. K. Lin, J. Xing, L.N. Quan, F.P.G. de Arquer, X. Gong, J. Lu, L. Xie, W. Zhao, D. Zhang, C. Yan, W. Li, X. Liu, Y. Lu, J. Kirman, E.H. Sargent, Q. Xiong, Z. Wei, Perovskite light-emitting diodes with external quantum efficiency exceeding 20 per cent. Nature **562**, 245–248 (2018)
37. X. Luo, Principles of electromagnetic waves in metasurfaces. Sci. China Phys. Mech. Astron. **58**, 594201 (2015)
38. R.A. Meyer, Optical beam steering using a multichannel lithium tantalate crystal. Appl. Opt. **11**, 613–616 (1972)
39. P.F. McManamon, P.J. Bos, M.J. Escuti, J. Heikenfeld, S. Serati, H. Xie, E.A. Watson, A review of phased array steering for narrow-band electrooptical systems. Proc. IEEE **97**, 1078–1096 (2009)
40. X. Zhao, C. Liu, D. Zhang, Y. Luo, Direct investigation and accurate control of phase profile in liquid-crystal optical-phased array for beam steering. Appl. Opt. **52**, 7109–7116 (2013)
41. D.R. Wight, J.M. Heaton, B.T. Hughes, J.C.H. Birbeck, K.P. Hilton, D.J. Taylor, Novel phased array optical scanning device implemented using GaAs/AlGaAs technology. Appl. Phys. Lett. **59**, 899–901 (1991)
42. F. Vasey, F.K. Reinhart, R. Houdré, J.M. Stauffer, Spatial optical beam steering with an AlGaAs integrated phased array. Appl. Opt. **32**, 3220–3232 (1993)
43. K. Van Acoleyen, H. Rogier, R. Baets, Two-dimensional optical phased array antenna on silicon-on-insulator. Opt. Express **18**, 13655–13660 (2010)
44. J.K. Doylend, M.J.R. Heck, J.T. Bovington, J.D. Peters, L.A. Coldren, J.E. Bowers, Two-dimensional free-space beam steering with an optical phased array on silicon-on-insulator. Opt. Express **19**, 21595–21604 (2011)

45. K. Van Acoleyen, K. Komorowska, W. Bogaerts, R. Baets, One-dimensional off-chip beam steering and shaping using optical phased arrays on silicon-on-insulator. J. Light. Technol. **29**, 3500–3505 (2011)
46. C.T. DeRose, R.D. Kekatpure, D.C. Trotter, A. Starbuck, J.R. Wendt, A. Yaacobi, M.R. Watts, U. Chettiar, N. Engheta, P.S. Davids, Electronically controlled optical beam-steering by an active phased array of metallic nanoantennas. Opt. Express **21**, 5198–5208 (2013)
47. D. Kwong, A. Hosseini, J. Covey, X. Xu, Y. Zhang, S. Chakravarty, R.T. Chen, Corrugated waveguide-based optical phased array with crosstalk suppression. IEEE Photon. Technol. Lett. **26**, 991–994 (2014)
48. D. Kwong, A. Hosseini, J. Covey, Y. Zhang, X. Xu, H. Subbaraman, R.T. Chen, On-chip silicon optical phased array for two-dimensional beam steering. Opt. Lett. **39**, 941–944 (2014)
49. H. Abediasl, H. Hashemi, Monolithic optical phased-array transceiver in a standard SOI CMOS process. Opt. Express **23**, 6509–6519 (2015)
50. B.A. Nia, L. Yousefi, M. Shahabadi, Integrated optical-phased array nanoantenna system using a plasmonic rotman lens. J. Light. Technol. **34**, 2118–2126 (2016)
51. W.S. Rabinovich, P.G. Goetz, M.W. Pruessner, R. Mahon, M.S. Ferraro, D. Park, E.F. Fleet, M.J. DePrenger, Two-dimensional beam steering using a thermo-optic silicon photonic optical phased array. Opt. Eng. **55**, 111603 (2016)
52. D.N. Hutchison, J. Sun, J.K. Doylend, R. Kumar, J. Heck, W. Kim, C.T. Phare, A. Feshali, H. Rong, High-resolution aliasing-free optical beam steering. Optica **3**, 887–890 (2016)
53. J. Notaros, C.V. Poulton, M.J. Byrd, M. Raval, M.R. Watts, Integrated optical phased arrays for quasi-Bessel-beam generation. Opt. Lett. **42**, 3510–3513 (2017)
54. C.V. Poulton, M.J. Byrd, M. Raval, Z. Su, N. Li, E. Timurdogan, D. Coolbaugh, D. Vermeulen, M. Watts, Large-scale visible and infrared optical phased arrays in silicon nitride. Conference on lasers and electro-optics, OSA technical digest (Online) (Optical Society of America, 2017), p. STh1 M.1
55. C.V. Poulton, A. Yaacobi, D.B. Cole, M.J. Byrd, M. Raval, D. Vermeulen, M.R. Watts, Coherent solid-state LIDAR with silicon photonic optical phased arrays. Opt. Lett. **42**, 4091–4094 (2017)
56. C.V. Poulton, D. Vermeulen, E. Hosseini, E. Timurdogan, Z. Su, B. Moss, M.R. Watts, Lens-free chip-to-chip free-space laser communication link with a silicon photonics optical phased array. Frontiers in Optics 2017, OSA technical digest (Online) (Optical Society of America, 2017), p. FW5A.3
57. M. Raval, C.V. Poulton, M.R. Watts, Unidirectional waveguide grating antennas with uniform emission for optical phased arrays. Opt. Lett. **42**, 2563–2566 (2017)
58. J. Sun, E. Timurdogan, A. Yaacobi, E.S. Hosseini, M.R. Watts, Large-scale nanophotonic phased array. Nature **493**, 195–199 (2013)
59. R.W. Gerchberg, W.O. Saxton, A practical algorithm for the determination of phase from image and diffraction plane pictures. Optik **35**, 237–250 (1972)
60. F. Aflatouni, B. Abiri, A. Rekhi, A. Hajimiri, Nanophotonic projection system. Opt. Express **23**, 21012–21022 (2015)
61. A. Yaacobi, J. Sun, M. Moresco, G. Leake, D. Coolbaugh, M.R. Watts, Integrated phased array for wide-angle beam steering. Opt. Lett. **39**, 4575–4578 (2014)
62. H. Shi, C. Wang, C. Du, X. Luo, X. Dong, H. Gao, Beam manipulating by metallic nano-slits with variant widths. Opt. Express **13**, 6815–6820 (2005)
63. T. Xu, C. Wang, C. Du, X. Luo, Plasmonic beam deflector. Opt. Express **16**, 4753–4759 (2008)
64. Y. Guo, M. Pu, Z. Zhao, Y. Wang, J. Jin, P. Gao, X. Li, X. Ma, X. Luo, Merging geometric phase and plasmon retardation phase in continuously shaped metasurfaces for arbitrary orbital angular momentum generation. ACS Photon. **3**, 2022–2029 (2016)
65. M. Pu, X. Ma, X. Li, Y. Guo, X. Luo, Merging plasmonics and metamaterials by two-dimensional subwavelength structures. J. Mater. Chem. C **5**, 4361–4378 (2017)
66. N. Yu, P. Genevet, M.A. Kats, F. Aieta, J.-P. Tetienne, F. Capasso, Z. Gaburro, Light propagation with phase discontinuities: generalized laws of reflection and refraction. Science **334**, 333–337 (2011)

67. M. Pu, P. Chen, C. Wang, Y. Wang, Z. Zhao, C. Hu, X. Luo, Broadband anomalous reflection based on low-Q gradient meta-surface. AIP Adv. **3**, 052136 (2013)
68. Y.-W. Huang, H.W.H. Lee, R. Sokhoyan, R.A. Pala, K. Thyagarajan, S. Han, D.P. Tsai, H.A. Atwater, Gate-tunable conducting oxide metasurfaces. Nano Lett. **16**, 5319–5325 (2016)
69. P.P. Iyer, M. Pendharkar, J.A. Schuller, Electrically reconfigurable metasurfaces using heterojunction resonators. Adv. Opt. Mater. **4**, 1582–1588 (2016)
70. Q. Wang, E.T.F. Rogers, B. Gholipour, C.-M. Wang, G. Yuan, J. Teng, N.I. Zheludev, Optically reconfigurable metasurfaces and photonic devices based on phase change materials. Nat. Photonics **10**, 60–65 (2016)
71. C. Wang, W. Liu, Z. Li, H. Cheng, Z. Li, S. Chen, J. Tian, Dynamically tunable deep subwavelength high-order anomalous reflection using graphene metasurfaces. Adv. Opt. Mater. **6**, 1701047 (2018)
72. M.C. Sherrott, P.W.C. Hon, K.T. Fountaine, J.C. Garcia, S.M. Ponti, V.W. Brar, L.A. Sweatlock, H.A. Atwater, Experimental demonstration of >230° phase modulation in gate-tunable graphene-gold reconfigurable mid-infrared metasurfaces. Nano Lett. **17**, 3027–3034 (2017)
73. G. Kafaie Shirmanesh, R. Sokhoyan, R.A. Pala, H.A. Atwater, Dual-gated active metasurface at 1550 nm with wide (>300°) phase tunability. Nano Lett. **18**, 2957–2963 (2018)
74. K. Shportko, S. Kremers, M. Woda, D. Lencer, J. Robertson, M. Wuttig, Resonant bonding in crystalline phase-change materials. Nat. Mater. **7**, 653–658 (2008)
75. M. Zhang, M. Pu, F. Zhang, Y. Guo, Q. He, X. Ma, Y. Huang, X. Li, H. Yu, X. Luo, Plasmonic metasurfaces for switchable photonic spin-orbit interactions based on phase change materials. Adv. Sci. **5**, 1800835 (2018)
76. C.H. Chu, M.L. Tseng, J. Chen, P.C. Wu, Y.-H. Chen, H.-C. Wang, T.-Y. Chen, W.T. Hsieh, H.J. Wu, G. Sun, D.P. Tsai, Active dielectric metasurface based on phase-change medium. Laser Photon. Rev. **10**, 986–994 (2016)

CPSIA information can be obtained
at www.ICGtesting.com
Printed in the USA
LVHW081809070419
613269LV00001B/11/P